WELDING
Theory and Practice

Dr Jagadeesha T
Department of Mechanical Engineering
National Institute of Technology
NIT Calicut

TECHSAR™
NEW DELHI

Published and Printed by

TechSar Pvt. Ltd.
4435-36/7, Ansari Road, Daryaganj
New Delhi–110 002 (India)
E-mail: tech_info@techsarworld.com
Website: www.techsarworld.com

ISBN: 978-81-94867-61-6

Reprint 2025

This book is most affectionately dedicated to

Babuji Maharaj
Founder President, Shri Ramchandra Mission
Shahjahanpur, Uttar Pradesh, India

WELDING

Theory and Practice

Preface

Welding plays an important role in modern production methods, and has made possible the construction of plants, equipment and structures, unattainable by other known methods. It is almost impossible to make a product like automobile, aircraft, etc., without cast components. This book will serve as a bridges between the study of the processes and their applications in production industries.

Compared with the classical method of writing a book, here two similar fundamental processes, namely, theory and practice are blended and explained in such a way that the reader gets holistic approach to welding. First, basic processes are discussed, followed by the special processes and design, testing, weldability of systems are discussed at the end.

The approach to the various aspects of welding design that are dealt with here, has been kept as simple as practicable and wherever possible the fundamental features have been stressed. With these simple facts and fundamental principles firmly understood, the designer is well on road to success in mastering his calling. Every effort has been made to comply with the existing and accepted standards and in this respect the nomenclature adopted is consistent with standard specifications.

Industry data, foundry practices, and real-time industrial applications have been included.

A sound basis of fundamental theory is presented with up-to-date technical information with the help of simple line sketches.

This book has been designed as per the syllabus of B.E., B.Tech. M.Tech and other postgraduate core courses and electives of engineering streams in different Indian and foreign universities. This book will also serve as a useful reference for academics, manufacturing and material researchers, manufacturing, mechanical engineers, and professionals in manufacturing and related industries. Several GATE and IES questions are solved making the book suitable for the readers who take up competitive exams.

Dr Jagadeesha T
jagdishsg@nitc.ac.in

Acknowledgements

Several engineers and professors have contributed to my manufacturing engineering experience. I have received significant practical training from the machine tool design engineers, technologists and managers at TELCO, Machine Tool Division, Pune; IBM Pvt Ltd, Singapore; ASM Technologies, Singapore; APP Systems and Services, Singapore; Applied Materials, USA, and Singapore; and ST Microelectronics, Singapore. I am grateful to all of them.

The basic engineering education with manufacturing, thermal and design content from Karnataka Regional Engineering College, Surathkal, India, and the background of advanced manufacturing and materials engineering from National University of Singapore and SMA alliance have been most valuable in overall development of my manufacturing and design-research skills. I am grateful to all my teachers.

I am most influenced by the research and machine tool design and engineering style and philosophy of my teachers at KREC, PSG Tech and National University of Singapore (NUS) and Singapore MIT alliance (SMA). I express my sincere gratitude to all of them. I thank all my colleagues at National Institute of Technology NIT, Calicut for their support and cooperation.

I thank my sister Prabhavathi T, my wife Vasanthi and my son, Ram Kumar Jagadeesha for their support and motivation. My father Thimmaiah and mother Rangamma, who are hardworking, honest, and warm country folks living in the little town Chickmagalur (Karnataka, India) have not only been my role models but also a source of support for me. I am grateful to them.

Dr Jagadeesha T
jagdishsg@nitc.ac.in

Contents

SECTION E Advanced Welding Processes

SECTION F Allied Welding Processes

SECTION G Miscellaneous Topics in Welding

Section A

Basic Welding Process

1

CHAPTER

Introduction to Manufacturing Processes

1.0 INTRODUCTION

Making things has been an essential activity of human civilizations since before recorded history. Making things is known as Manufacturing. Certain technology is required to make things. Technology can be defined as the application of science to provide society and its members with those things that are needed or desired. Manufacturing is the critical factor that makes technology possible. Economically, manufacturing is an important means by which a nation creates material wealth. In India, the manufacturing industries account for more than 15% of gross domestic product.

The word manufacture derives from two Latin words, *manus* (hand) and *factus* (make); the combination means "made by hand". The English word manufacture is several centuries old, and "made by hand" accurately described the manual methods used when the word was first coined. Most modern manufacturing is accomplished by automated and computer-controlled machinery.

Manufacturing is the application of physical and chemical processes to alter the geometry, properties, and/or appearance of a starting material to make parts or products. Manufacturing also includes assembly of multiple parts to make products. The processes to accomplish manufacturing involve a combination of machinery, tools, power, and labour, as depicted in Figure 1.1.

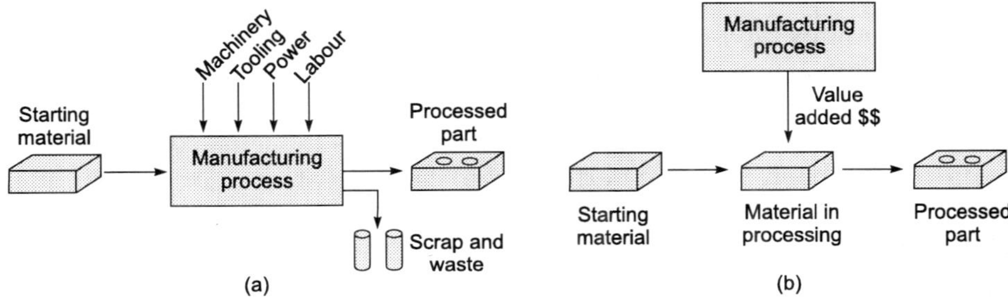

Figure 1.1 Two ways to define manufacturing: (a) As a technical process, and (b) as an economic process

Manufacturing is almost always carried out as a sequence of operations. Each operation brings the material closer to the final state. Economically, manufacturing is the transformation of materials into items of greater value by means of one or more processing and/or assembly operations, as depicted in Figure 1.1(b). The key point is that manufacturing adds value to the material by changing its shape or properties, or by combining it with other materials that have been similarly altered. The material has been made more valuable through the manufacturing operations performed on it. When iron ore is converted into steel, value is added. When sand is transformed into glass, value is added. When petroleum is refined into plastic, value is added. And when plastic is molded into the complex geometry of a patio chair, it is more valuable.

1.1 CLASSIFICATION OF MANUFACTURING PROCESSES

Manufacturing basically involves the sizing and shaping of the components. Manufacturing processes, based on the way by which the particular shape is given can be classified as

- Positive processes
- Negative processes
- Zero processes

Casting and forming involve mainly shifting of metal in a controlled way (using heat or pressure singly or in combination). Therefore, casting and forming are called zero processes. For example, upsetting of a bolt by forging, there is no material removal, rather there is adjustment of material from one zone to the other.

Machining is considered a negative process because unwanted material is removed from the stock in the form of small chips to get the desired size and shape.

Welding and joining are good examples of positive processes, in which simple components are joined to make assembly of parts. Manufacturing can be broadly classified into three categories.

Three categories of processing operations are distinguished:

1. Shaping operations
2. Property enhancing operations
3. Surface processing operations

Shaping operations alter the geometry of the starting work material by various methods. Common shaping processes include casting, forging, and machining. Property enhancing operations add value to the material by improving its physical properties without changing its shape. Heat treatment is the most common example. Surface processing operations are performed to clean, treat, coat, or deposit material onto the exterior surface of the work. Figure 1.2 shows the general classification of manufacturing processes.

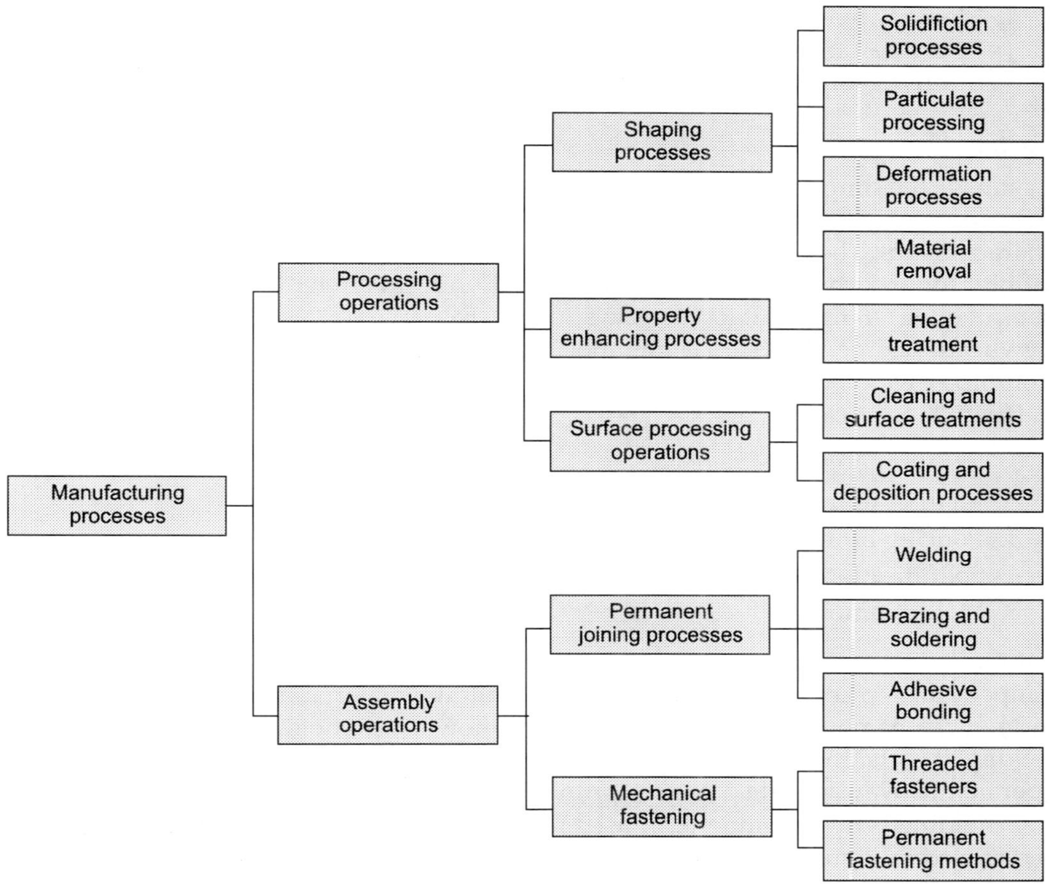

Figure 1.2 Classification of manufacturing processes

1.1.1 Shaping Processes

Most shape processing operations apply heat, mechanical force, or a combination of these to effect a change in the geometry of the work material. There are various ways to classify the shaping processes. The classification used in this book is based on the state of the starting material, by which we have four categories:

1. Solidification processes, in which the starting material is a heated liquid or semifluid that cools and solidifies to form the part geometry (**Casting processes**).
2. Particulate processing, in which the starting material is a powder, and the powders are formed and heated into the desired geometry (**Powder processes**).
3. Deformation processes, in which the starting material is a ductile solid (commonly metal) that is deformed to shape the part (**Metal forming processes**).

4. Material removal processes, in which the starting material is a solid (ductile or brittle), from which material is removed so that the resulting part has the desired geometry (**Machining processes**).

Casting Processes

Casting is the simplest way to get simple to complex shape, small to huge component. Castability of a material depends on fluidity, shrinkage, stress and segregation characteristics. The castability index is high if it has high fluidity, low shrinkage, low affinity for absorbing gases, low stresses and uniform strength. Castings have good compressive strength but low tensile strength. Examples of cast products are pulleys, flywheels, engine blocks, machine tool beds, large gear blanks, turbine blades, cast iron pipes, etc.

Basic steps in casting are:

1. Creation of mold cavity
2. Melting of material
3. Pouring of material
4. Solidification of material
5. Mould removal

Metals, ceramic glasses, and plastics can all be heated to sufficiently high temperatures to convert them into liquids. With the material in a liquid or semifluid form, it can be poured or otherwise forced to flow into a mold cavity and allowed to solidify, thus taking a solid shape that is the same as the cavity. Most processes that operate this way are called casting or molding. Casting is the name used for metals, and molding is the common term used for plastics. This category of shaping process is depicted in Figure 1.3.

Detailed classification of casting is shown in Figure 1.4.

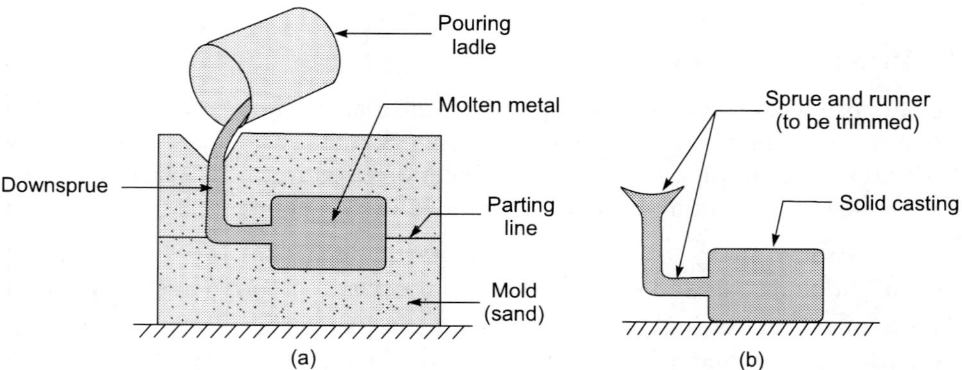

Figure 1.3 Casting process (a) Pouring the fluid into the mould cavity (b) Allowing the fluid to solidify, after which the solid part is removed from the mold

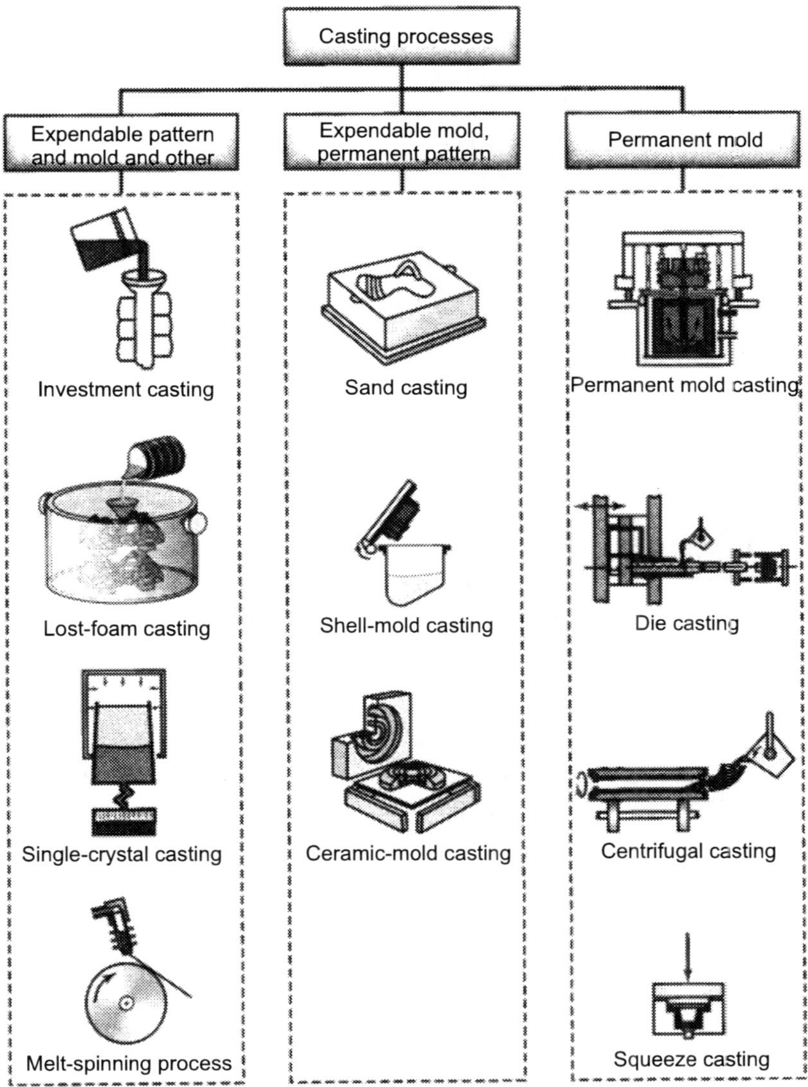

Figure 1.4 Classification of casting processes

Powder Process

In particulate processing, the starting materials are powders of metals or ceramics. Although these two materials are quite different, the processes to shape them in particulate processing are quite similar. The common technique involves pressing and sintering, as illustrated in Figure 1.5, in which the powders are first squeezed into a die cavity under high pressure and then, heated to bond the individual particles together.

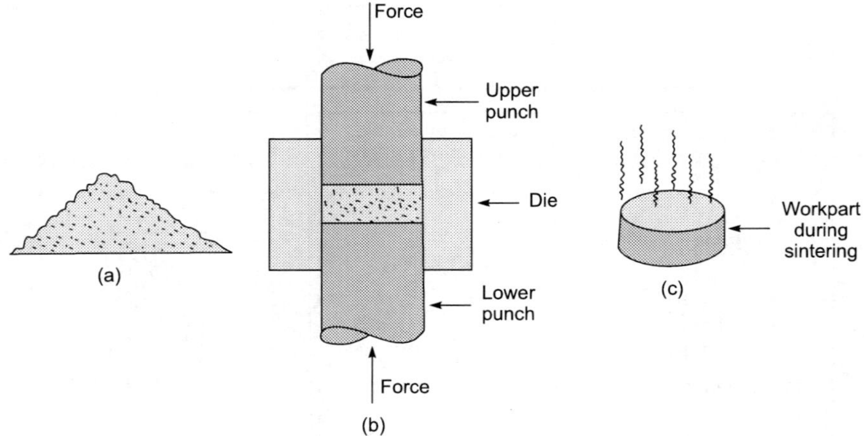

Figure 1.5 Powder processing (a) Starting material (powder) (b) Pressing (c) Sintering

Metal Forming Processes

Mechanical working of a metal below its recrystallization temperature is known as cold working and above the recrystallization, is called hot working. Materials, which are best suited to casting, have poor forming qualities.

The metal forming is based on plastic deformation of bulk material using compressive force. Formed components are generally stronger than those manufactured using other processes.

In deformation processes, the starting workpart is shaped by the application of forces that exceed the yield strength of the material. For the material to be formed in this way, it must be sufficiently ductile to avoid fracture during deformation. To increase ductility (and for other reasons), the work material is often heated before forming to a temperature below the melting point. Deformation processes are associated most closely with metal working and include operations such as forging and extrusion, as shown in Figure 1.6.

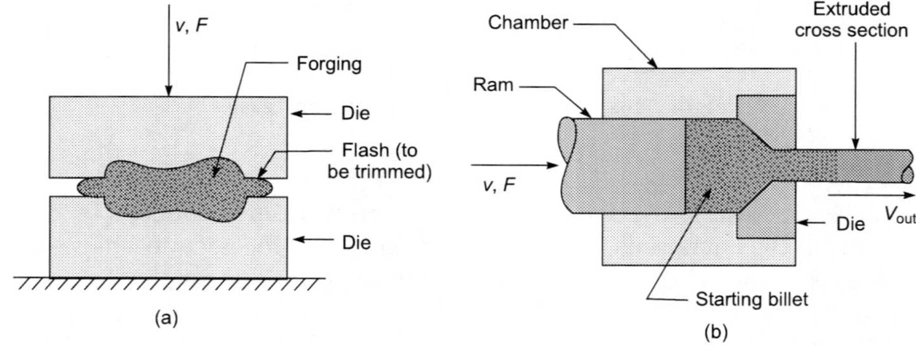

Figure 1.6 (a) Forging, in which two halves of a die squeeze the workpart (b) Extrusion, in which billet is forced to flow through a die orifice, thus taking the cross-sectional shape of the orifice

Detailed classification of forming process is shown in Figures 1.7 and 1.8.

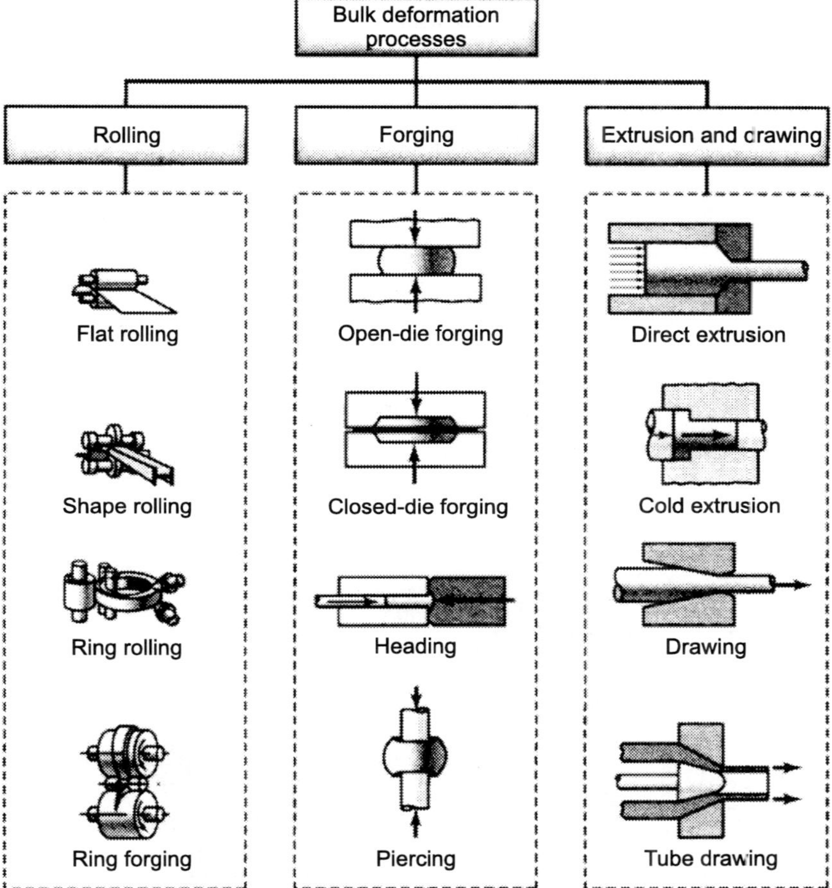

Figure 1.7 Classification of metal forming processes

Machining Processes

Machining is a subtractive process used to get desired size, shape and finish by removing surplus material in the form of chips. The most important processes in this category are machining operations, such as turning, drilling, and milling, shown in Figure 1.9. These cutting operations are most commonly applied to solid metals, performed using cutting tools that are harder and stronger than the work metal. Grinding is another common process in this category. Other material removal processes are known as non-traditional processes because they use lasers, electron beams, chemical erosion, electric discharges, and electrochemical energy to remove material rather than cutting or grinding tools.

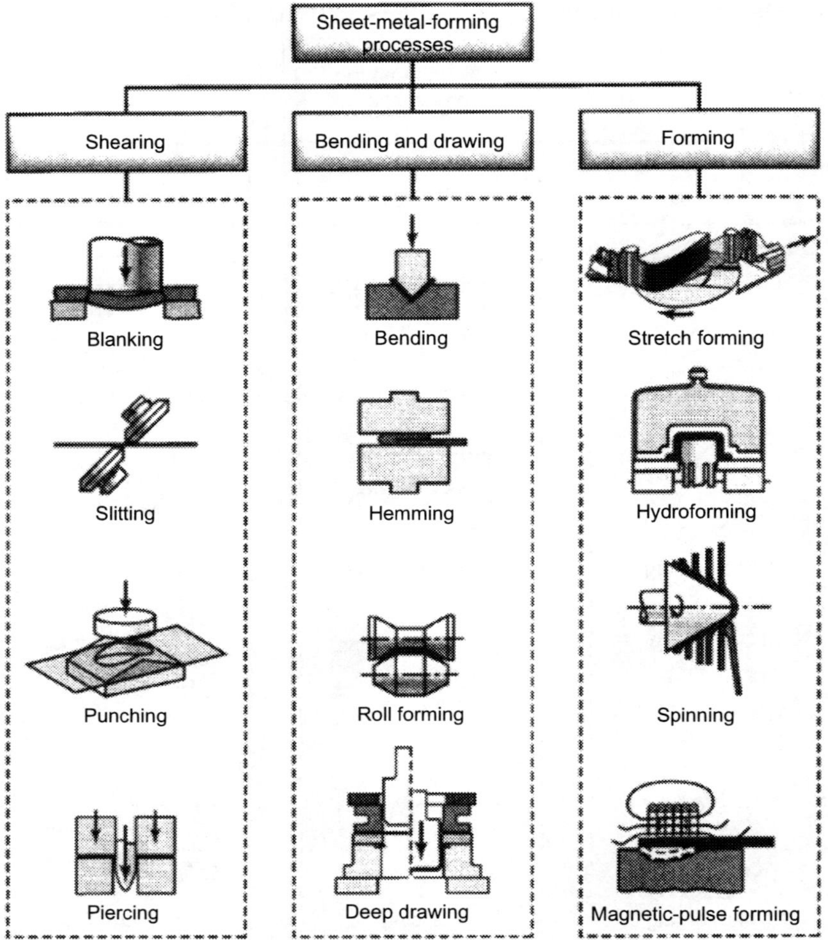

Figure 1.8 Classification of metal forming (sheet-metal) processes

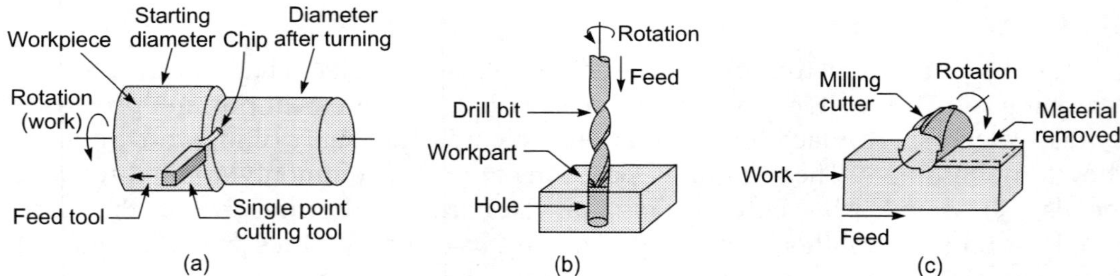

Figure 1.9 Common machining operations (a) Turning, in which a single-point cutting tool removes metal from a rotating workpiece to reduce its diameter (b) Drilling, in which a rotating drill is fed into the work to create a round hole (c) Milling, in which a workpart is fed fast a rotating cutter with multiple edges

Detailed classification of machining processes is given in Figure 1.10.

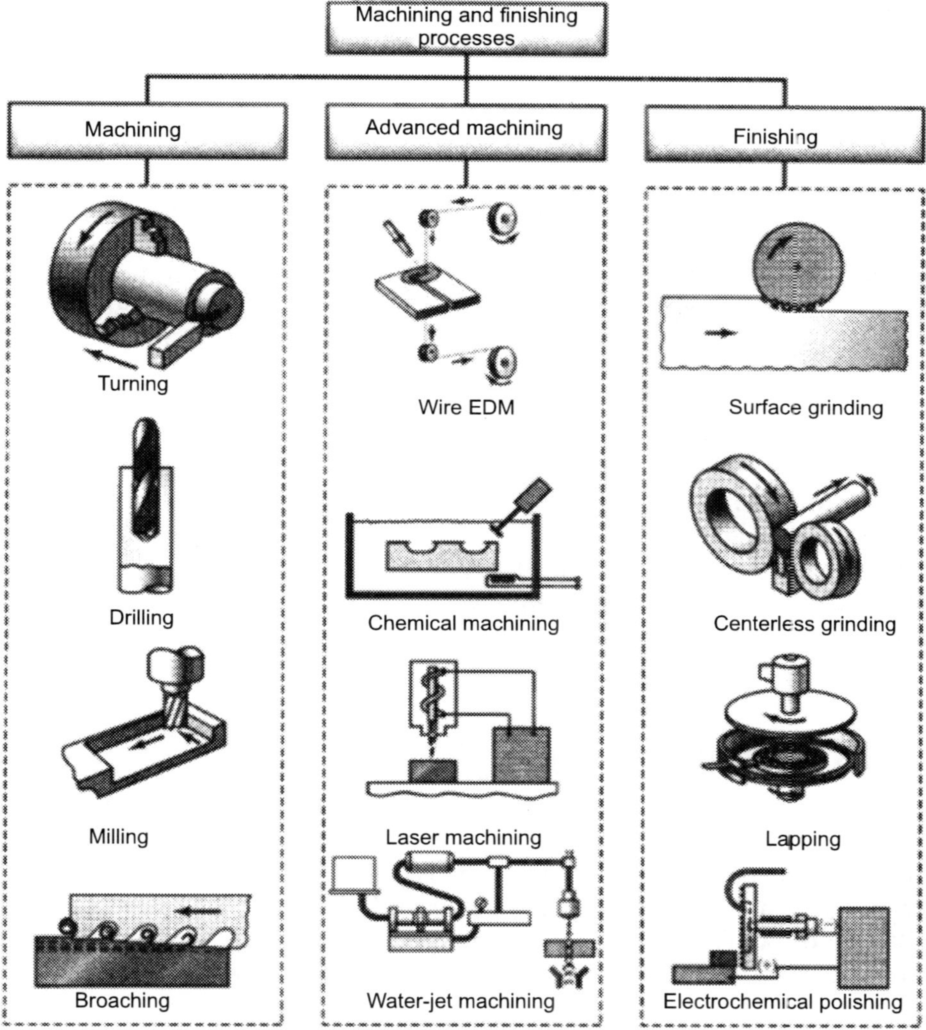

Figure 1.10 Classification of machining processes

1.1.2 Property Enhancing Processes

The second major type of part processing is performed to improve mechanical or physical properties of the work material. These processes do not alter the shape of the part, except unintentionally in some cases. The most important property enhancing processes involve heat treatments, which include various annealing and strengthening processes for metals and glasses. Sintering of powdered metals and ceramics is also a heat treatment that strengthens a pressed powder metal workpart.

1.1.3 Surface Processing

Surface processing operations include (1) cleaning, (2) surface treatments and (3) coating and thin film deposition processes. Cleaning includes both chemical and mechanical processes to remove dirt, oil and other contaminants from the surface. Surface treatments include mechanical working, such as shot peening and sand blasting, and physical processes, such as diffusion and ion implantation. Coating and thin film deposition processes apply a coating of material to the exterior surface of the workpart. Common coating processes include electroplating, anodizing of aluminum, organic coating (call it painting), and porcelain enameling. Thin film deposition processes include physical vapor deposition and chemical vapor deposition to form extremely thin coatings of various substances. Several surface processing operations have been adapted to fabricate semiconductor materials into integrated circuits for microelectronics. These processes include chemical vapor deposition, physical vapor deposition and oxidation. They are applied to very localized areas on the surface of a thin wafer of silicon (or other semiconductor material), to create the microscopic circuit.

1.2 ASSEMBLY OPERATIONS

The second basic type of manufacturing operation is assembly, in which two or more separate parts are joined to form a new entity. It is called positive process or addition process. Components of the new entity are connected either permanently or semi-permanently. Permanent joining processes include welding, brazing, soldering, and adhesive bonding. They form a joint between components that cannot be easily disconnected. Certain mechanical assembly methods are available to fasten two (or more) parts together in a joint that can be conveniently disassembled. The use of screws, bolts, and other threaded fasteners are important traditional methods in this category. Other mechanical assembly techniques form a more permanent connection; these include rivets, press fitting, and expansion fits. Special joining and fastening methods are used in the assembly of electronic products. Some of the methods are identical to or are adaptations of the preceding processes, for example, soldering. Electronics assembly is concerned primarily with the assembly of components, such as integrated circuit packages to printed circuit boards, to produce the complex circuits used in so many of products. Figure 1.11 gives a detailed classification of joining processes. Figure 1.12 gives the energy activation of joining mechanism.

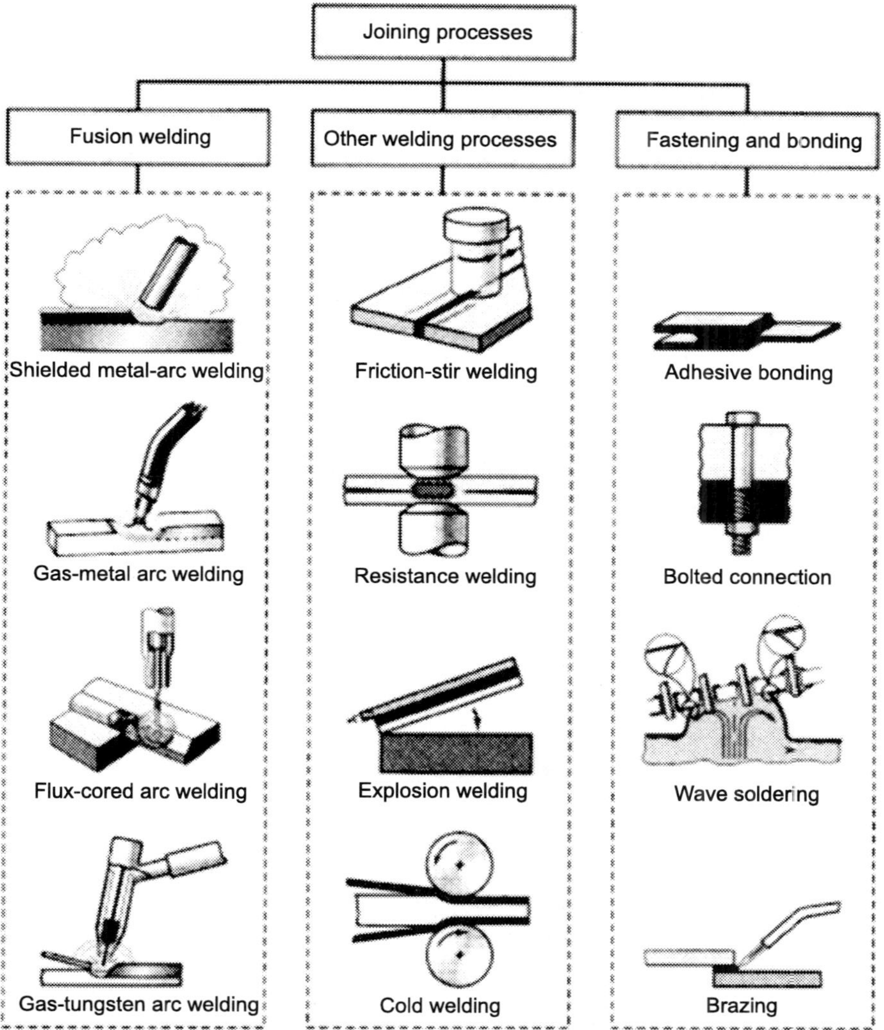

Figure 1.11 Classification of joining processes

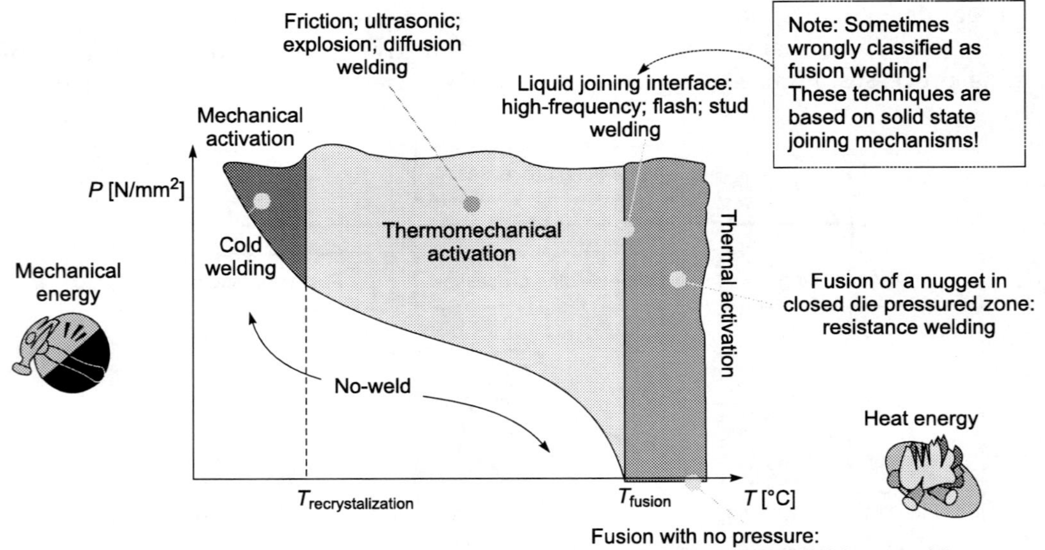

Figure 1.12 Energetic activation of joining mechanism

1.3 MACHINE TOOLS

Machine tool provides necessary relative motion between the workpiece and tool. Manufacturing operations are accomplished using machinery and tooling (and people). The extensive use of machinery in manufacturing began with the Industrial Revolution. It was at that time that metal-cutting machines started to be developed and widely used. These were called machine tools—power driven machines used to operate cutting tools, previously operated by hand. Modern machine tools are described by the same basic definition, except that the power is electrical rather than water or steam, and the level of precision and automation is much greater today. Machine tools are among the most versatile of all production machines. They are used to make not only parts for consumer products, but also components for other production machines. Both in a historic and a reproductive sense, the machine tool is the mother of all machinery.

Other production machines include presses for stamping operations, forge hammers for forging, rolling mills for rolling sheet metal, welding machines for welding and insertion machines for inserting electronic components into printed circuit boards. The name of the equipment usually follows from the name of the process.

Production equipment can be general purpose or special purpose. General-purpose equipment is more flexible and adaptable to a variety of jobs. It is commercially available for any manufacturing company to invest in. Special-purpose equipment is usually designed to produce a specific part or product in large quantities. The economics of mass production justifies large investments in special-purpose machinery, to achieve high efficiency and short cycle times. This is not the only reason for special-purpose equipment, but it is the dominant one. Another reason may be because the process is unique and commercial equipment is not available. Some companies with unique processing requirements develop their own special-purpose equipment.

Production machinery usually requires tooling that customizes the equipment for the particular part or product. In many cases, the tooling must be designed specifically for the part or product configuration. When used with general-purpose equipment, it is designed to be exchanged. For each workpart type, the tooling is fastened to the machine and the production run is made. When the run is complete, the tooling is changed for the next workpart type.

When used with special-purpose machines, the tooling is often designed as an integral part of the machine. Because the special-purpose machine is likely to be used for mass production, the tooling may never need changing, except for replacement of worn components or for repair of worn surfaces. The type of tooling depends on the type of manufacturing process. Table 1.1 lists examples of special tooling used in various operations.

TABLE 1.1 Production equipment and tooling used for various manufacturing processes

Process	Equipment	Special Tooling (Function)
Casting	Various casting setups and equipment	Mold (cavity for molten metal)
Molding	Molding machine	Mold (cavity for hot polymer)
Rolling	Rolling mill	Roll (reduce work thickness)
Forging	Forge hammer or press	Die (squeeze work to shape)
Extrusion	Press	Extrusion die (reduce cross section)
Stamping Machining	Press Machine tool	Die (shearing, forming sheet metal) Cutting tool (material removal) Fixture (hold workpart) Jig (hold part and guide tool)
Grinding	Grinding machine	Grinding wheel (material removal)
Welding	Welding machine	Electrode (fusion of work metal) Fixture (hold parts during welding)

1.4 SELECTION OF MANUFACTURING PROCESSES

The selection of a particular manufacturing process or, more often, sequence of processes, depends on the geometric features of the parts to be produced, including the dimensional tolerances and surface texture required, and on numerous factors pertaining to the particular workpiece material and its manufacturing properties. To emphasize the challenges involved, consider the following two cases:

- Brittle and hard materials cannot be shaped or formed without the risk of fracture, unless they are performed at elevated temperatures, whereas these materials can easily be cast, machined, or ground.
- Metals that have been preshaped at room temperature become less formable during subsequent processing, which, in practice, is often required to complete the part; this is because the metals have become stronger, harder, and less ductile than they were prior to processing them further.

Selection of manufacturing processes depends on:

- Complexity of the part
- Number of units to be produced
- Properties of the material
- Dimensional properties desired

Complex parts generally cannot be shaped easily and economically by such metal-working techniques as forging, whereas, depending on the part size and the level of complexity, the parts may be precision cast, fabricated from individual pieces, or produced by powder metallurgy techniques. Welding is not suitable for complex shapes. Flat parts and thin cross sections can be difficult to cast.

Production volume plays an important role in the selection of process. For example, if a few components are required then, we can use sand casting, if the production volume is high then, die casting can be employed.

Physical, chemical and mechanical properties of the material must be considered for selecting the manufacturing process. Melting point is important for the welding and casting processes as well. Chemical properties are important for the welding, forming processes and casting processes. Mechanical properties, like strength and the ductility, are important for machining and forming process.

Dimensional tolerances and surface finish in hot-working operations are not as fine as those obtained in operations performed at room temperature (called cold working), because of the dimensional changes, distortion, warping, and surface oxidation that occur at the elevated temperatures involved.

Process Substitution. It is a common practice in the industry that, for a variety of reasons and after a review of all appropriate and applicable processes, a particular production method (that may have been employed in the past) may have to be substituted with another. Consider, for example, the following products that can be produced by any of the set of processes indicated:

(a) cast versus forged crankshaft
(b) Stamped sheet metal versus forged or cast automobile wheels
(c) Cast versus stamped sheet metal frying pan

(d) Injection molded versus extruded or cast polymer bracket
(e) Welded versus riveted sheet metal safety hood for a machine

Many varieties of such products are widely available in the marketplace. However, a customer's preference will depend on his or her particular needs, which include factors such as the product's appeal to the customer, its price, whether maintenance is required, whether the product is for industrial or consumer use, the parameters to which the product will be subjected (such as temperatures and chemicals), and any environmental concerns that have to be addressed.

1.5 SELECTION OF MATERIAL

An increasingly wide variety of materials are now available, each type having its own (a) material properties and manufacturing characteristics, (b) advantages and limitations, (c) material and production costs, and (d) consumer and industrial applications.

The general types of materials used, either individually or in combination with other materials, are the following:

- Ferrous metals: Carbon, alloy, stainless, and tool and die steels.
- Non-ferrous metals: Aluminum, magnesium, copper, nickel, titanium, superalloys, refractory metals, beryllium, zirconium, low-melting-point alloys, and precious metals.
- Plastics (polymers): Thermoplastics, thermosets, and elastomers.
- Ceramics, glasses, glass ceramics, graphite, diamond, and diamond-like materials.
- Composite materials: Reinforced plastics and metal-matrix and ceramic-matrix composites.
- Nanomaterials.
- Shape-memory alloys (also called smart materials), amorphous alloys, semiconductors, and superconductors.

The selection of material depends on the following factors.

Properties of Materials: Mechanical properties of interest in manufacturing generally include strength, ductility, hardness, toughness, elasticity, fatigue, and creep resistance (Chapter 2). Physical properties are density, specific heat, thermal expansion and conductivity, melting point, and electrical and magnetic properties (Chapter 3). Optimum designs often require consideration of a combination of mechanical and physical properties. A typical example is the strength-to-weight and stiffness-to-weight ratios of materials for minimizing the weight of structural members. Weight minimization is particularly important for aerospace and automotive applications, in order to improve performance and fuel economy. Chemical properties include oxidation, corrosion, degradation, toxicity, and flammability. These properties play a significant role under both hostile (such as corrosive) and normal environments. Manufacturing properties indicate whether a particular material can be cast, formed, machined, joined, and heat treated with relative ease. As Table 1.2 illustrates, no one material has the same manufacturing characteristics. Another consideration is appearance, which includes such characteristics as color, surface texture, and feel, all of which can play a significant role in a product's acceptance by the public.

TABLE 1.2 General manufacturing characteristics of various materials

Alloy	Castability	Weldability	Machinability
Aluminum	E	F	E–G
Copper	G–F	F	G–F
Gray cast iron	E	D	G
White cast iron	G	VP	VP
Nickel	F	F	F
Steel	F	E	F
Zinc	E	D	E

Note: E, excellent; G, good; F, fair; D, difficult; VP, very poor. The ratings shown depend greatly on the particular material, its alloys, and its processing history.

Availability: The availability of materials is a major concern in manufacturing. Furthermore, if materials are not available in the desired quantities, shapes, dimensions, and surface texture, substitute materials or additional processing of a particular material may well be required, all of which can contribute significantly to the product's cost.

Reliability of Supply: It is important in order to meet production schedules. In automotive industries, for example, materials must arrive at a plant at appropriate time intervals. Reliability of supply is also important, considering the fact that most countries import numerous raw materials.

Service Life: Shortened service life of a product, which often can be traced to (a) improper selection of materials, (b) improper selection of production methods, (c) insufficient control of processing variables, (d) defective parts or manufacturing-induced defects, (e) poor maintenance, and (f) improper use of the product. Generally, a product is considered to have failed when it,

- stops functioning, due to the failure of one or more of its components, such as a broken shaft, gear, bolt, or turbine blade or a burned-out electric motor;
- does not function properly or perform within required specifications, for example, due to worn gears or bearings; and

- becomes unreliable or unsafe for further use, as in the erratic behavior of a switch, poor connections in a printed circuit board, or delamination of a composite material.

1.6 TYPES OF JOINTS AND THEIR SELECTION

There are three common types of joining techniques used to get desired size and shape. They are:

- Mechanical joints: bolts, nuts, clamps, etc.
- Adhesives: epoxy resin, fevicol, super glue, etc.
- Welding, brazing and soldering.

Each type of joint offers specific features with respect to load carrying capacity, reliability, compatibility in joining of similar or dissimilar materials, fitness for use in different environments and effort required to develop a joint.

Type of joint is selected on the following factors:

- Based on the requirement of a joint. If the parts are be joined temporarily, then rivets, nut-bolts and adhesive joints are used. On the other hand, welded joint, brazing and soldering give permanent joints.
- Based on compatibility between members to be joined. If similar material is to be joined, then welding is preferred. If dissimilar material is to be joined, then brazing, soldering, adhesive and mechanical joints are preferred.
- Based on the properties of material to be joined. Properties such as melting point, weld thermal cycle, size, thickness of the plates, thermal expansion coefficients affect the selection of joints.
- Based on the service condition for joint. Factors such as low/high temperature, corrosive environments, type of loading, decide the types of joint to be used.

Review Questions

1. Define manufacturing processes.
2. Explain the broad classification of manufacturing processes.
3. Distinguish between positive and negative manufacturing processes?
4. What are the main advantages of casting processes?
5. What are the main advantages of metal forming processes?
6. How do we select materials for manufacturing a product?
7. How do we select manufacturing processes?
8. What are the properties of material considered in manufacturing a product?
9. What is process substitution?
10. What are the three types of joints commonly used and how do we select them?

Introduction to Welding Technology

2.0 INTRODUCTION

Manufacturing processes involve shaping, sizing and imparting desired combination of the properties of material. The primary aim is to have a component or assembly or engineering systems being produced that can perform intended function for design life. For example, in case of four-wheeler, there are varieties of components which are made by the different manufacturing processes and each component is made up of different materials. Manufacturing processes can be termed positive, negative and zero processes. Casting and forming are zero processes. Casting and forming involve mainly shifting of metal in controlled manner, using heat and/or pressure.

Machining in which material is removed out is a negative process. Machining is considered a negative process because unwanted material is removed from the stock in the form of small chips to get the desired size and shape. Machine tool provides necessary relative motion between the workpiece and tool.

Joining is a positive or addition process used for assembling different members to get the desired configuration. There are three common types of joining techniques: mechanical joints using nuts, bolts and clamps; adhesive technique using epoxy, resin etc; and welding, brazing and soldering.

A wide range of manufacturing processes have been developed in order to produce the engineering components of different geometries, using materials of different physical, chemical and mechanical properties. Selection of a manufacturing process is dictated by complexity of geometry, number of parts to be manufactured and properties of the material.

Definition of Welding: Unlike the manufacturing processes employed to produce a single component, joining processes are used to assemble different members to yield the desired complex configuration. Such a complex geometry is either too difficult or impossible to obtain by using only manufacturing processes, such as casting or forming processes.

Welding is the most economical and efficient way to join metals permanently. It is the only way of joining two or more pieces of metal to make them act as a single piece. Welding ranks high among industrial processes and involves more sciences and variables than those involved in any other industrial process.

A weld occurs when pieces of metal are joined by causing the interface to melt and blend prior to solidifying as a uniform metal joint. This is often done by melting the workpieces and adding a filler material to form a pool of molten material (the weld puddle) that cools to become a strong joint, but sometimes pressure is used in conjunction with heat, or by itself, to produce the weld. This is in contrast with soldering and brazing, which involve melting a lower melting point material between the workpieces to form a bond between them, without melting the workpieces.

Welding was earlier confined to repair work but it has a vast field of application these days. Almost all major industries use welding as primary fabrication process. In welding, weld is stronger than the parent materials itself and this is the specific advantage of welding processes.

It is used in automobile industries, aircraft, machine frames, ship building, machine tools, boilers and pressure vessels, railways, structural members of bridges and buildings and many other fields.

A number of components like machinery beds, frames and brackets, which were made by casting, are now made by welding. Such construction has the advantages of being lighter and stronger than cast iron.

Welding is slowly replacing riveting and bolting for boilers and pressure vessels, tanks and structural members of bridges and buildings. In a riveted product, there will be reduction in the effective reaction due to rivet holes and this will be as high as 25%. This necessitates a heavier construction. In contrast, in welding the entire cross section of the metal is effective as there are no rivets holes. This results in the saving of materials.

While it is almost impossible to design a riveted joining with 100% effectiveness, it can be achieved in a welded joint. While riveting does not permit the use of high-strength steels due to difficulty in obtaining rivets and bolts of same strength, it can be done in welding.

Another important advantage is the saving in labour and welding is also a noiseless process compared to riveting. When certain structures are to be reinforced, welding is preferred against riveting because reinforcement can be welded without removing the rivets.

When the thickness of the plates is greater than those of that can be joined by riveting, welding is the only method of joining. This has greatly helped the development of certain fields which have been hampered due to lack of suitable fabrication method.

Welding can be defined as a joining process that produces coalescence of materials by heating them to the welding temperature, with or without the application of pressure or by application of pressure alone, and with or without the use of filler material. Coalescence means growing together or growth of one body of material to other. Weldments are an assembly of components joined by welding. Weldments can be made of many or fewer metal parts.

Welding can be explained as a process of joining together the pieces of metal by bringing them into intimate proximity and heating the places of contact to a state of fusion or plasticity.

Welding, in its broadest sense, can be defined as the process by which materials can be joined through the action of interatomic or intermolecular forces. Thus welding, brazing, soldering and adhesive bonding can be regarded as welding processes. Commonly used welding terminology is shown in Figure 2.1 along with weld bead.

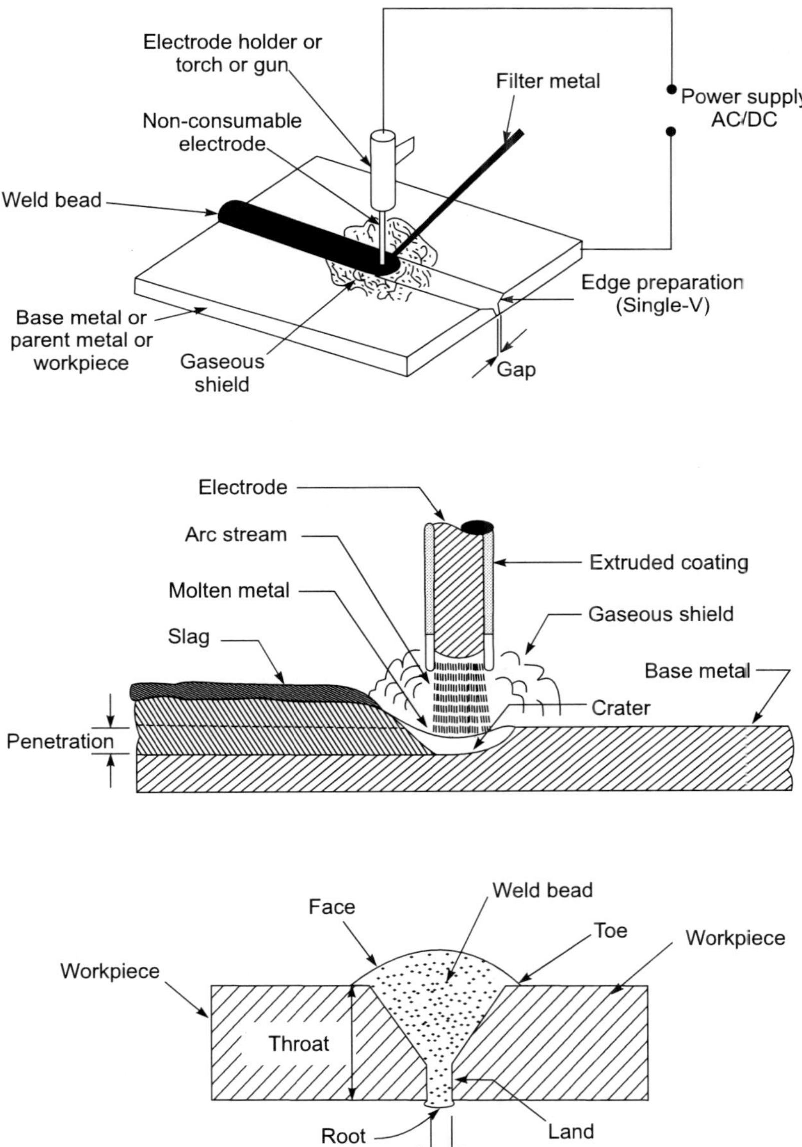

Figure 2.1 Welding terminology

2.1 HISTORY OF WELDING

To understand the evolution and development of welding, it is necessary to make a quick review of its history. The term 'weld' (the joining of materials by welding), appeared for the first time in the Old Testament several thousand years BC. Indeed, 4000 years ago, the Egyptians had already developed the art of welding (the uniting of two or more parts by heat or pressure, or both). The head of the sarcophagus of Tutankhamen (1361-1352 BC), whose tomb was discovered in 1922 in the Valley of the Kings, is a good example. Another notable example from the famous Colossus of Rhodes (built c. 292-280 BC), in its time one of the Seven Wonders of the World, which owed its height of 35 metres to a skeleton of welded iron. Also, in the Ukraine about 1200 years ago, craftsmen welded a magnificent steel sword which has survived to the present day. The first societies who knew how to work metal only had available, for assembly, the rudimentary methods of flow and forge welding. They also practised brazing. These processes remained almost unchanged up to the middle of the nineteenth century. At this time, methods of shaping were progressively perfected, mechanical methods of assembly were developed and gradually replaced the primitive and often, mediocre methods of welding. Then, in the second half of the nineteenth century, the forge welding process was perfected and for specialised applications, it proved highly productive.

A number of totally new welding principles emerged at the end of 19th century; sufficient electrical current could then be generated for resistance welding and arc welding. ***Arc welding*** was initially carried out using carbon electrodes, developed by Bemados, and was shortly followed by the use of steel rods. The Swede Oskar Kjellberg made an important advance when he developed and patented the coated electrode. The welding result was amazing and formed the foundation of the ESAB welding company.

Another early method of welding which was also developed at that time was ***gas welding***. The use of acetylene and oxygen made it possible to produce a comparatively high flame temperature, 3100°C, which is higher than that of other hydrocarbon-based gas.

The intensity of all these heat sources enables heat to be generated in, or applied to, the workpiece quicker than it is conducted away into the surrounding metal. Consequently, it is possible to generate a molten pool, which solidifies to form the unifying bond between the parts being joined.

Later, in the 1930s, new methods were developed. Up until then, all metal-arc welding had been carried out manually. Attempts were made to automate the process using a continuous wire. The most successful process was ***submerged arc welding*** (SAW), where the arc is "submerged" in a blanket of granular fusible flux.

During the Second World War, the aircraft industry required a new method for welding magnesium and aluminum. In 1940, experiments began in the USA with the shielding of the arc by inert gases. By using an electrode of tungsten, the arc could be struck without melting the electrode, which made it possible to weld with or without a filler material. The method is called ***TIG welding*** (Tungsten Inert Gas).

Some years later, the ***MIG welding process*** (Metal Inert Gas) was also developed using a continuously fed metal wire as the electrode. Initially, the shielding gases were inert, such as helium or argon. Zaruba and Potapevski tried to use CO_2 as this was much easier to obtain and by using the "dip transfer" method, they did manage to reduce some of the problems caused by the intense generation of spatter; however when using a relatively reactive gas such as CO_2 or mixed gases such as argon/CO_2, the process is generally called ***MAG welding*** (Metal Active Gas).

The power-beam processes, ***electron beam*** (EB) welding and laser welding, have the most intensive heat sources. The breakthrough of EB welding came in 1958. The aircraft and nuclear power industries were the first to utilise the method. The main characteristics of EB welding are its deep and narrow penetration. Its one limitation is the need for a vacuum chamber to contain the electron beam gun and the workpiece.

In some respects, ***laser welding*** (and cutting) have ideal characteristics. The laser beam is a concentrated heat source, which permits high speed and very low distortion of the workpiece, unfortunately, a high power laser is large and expensive. The beam must also be conducted to the joint in some way. The light from a CO_2 laser must be transmitted by mirrors, while that from a Nd:YAG-laser can be carried by a thin glass fibre, which makes it attractive for use with robotic welding.

In the future, it should be possible to utilise lightweight diode lasers with sufficient power for welding. The diode laser has a higher efficiency in converting electrical energy into the light beam. Although it has not yet been possible to produce diode lasers with the same power output and beam quality as present welding laser sources; these are already being used for welding metal up to about 1 mm thick. The low weight and size make them an interesting power source for use with ***robotic welding***.

The development of different welding processes is given in Figure 2.2.

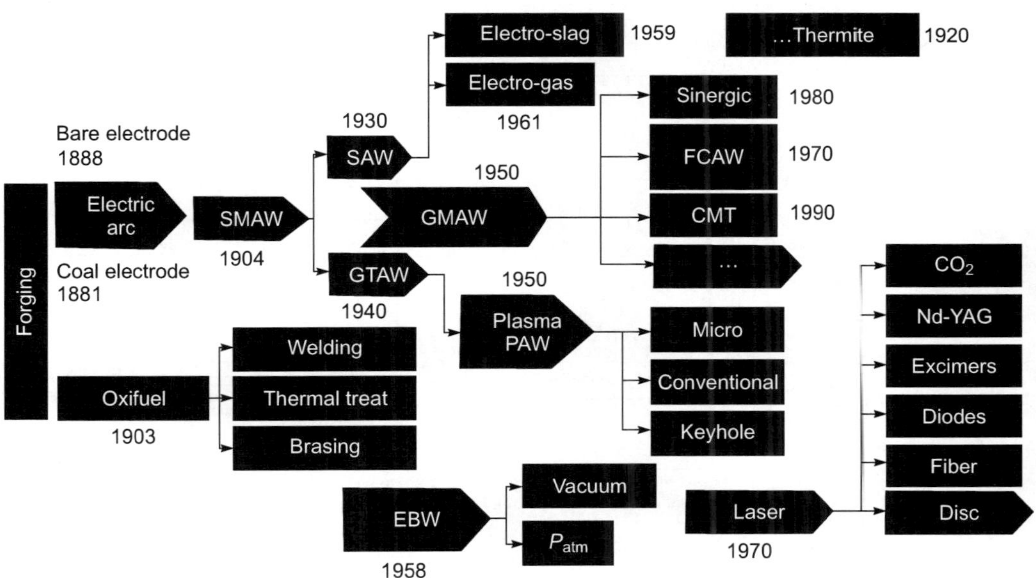

Figure 2.2 Development of welding processes

2.2 CLASSIFICATION OF WELDING AND ALLIED PROCESSES

The joining processes are so intimately related to the overall production system that these are also considered to form a class of manufacturing techniques.

Welding processes can be divided into:

Fusion welding – coalescence is obtained by melting the two parts to be joined.

Solid state welding – heat or pressure is applied to obtain gluing and no melting is involved.

Fusion welding can further be classified into arc welding, resistance welding, oxyfuel gas welding and other fusion welding processes. Both plasma arc welding and laser beam welding fall into the last category of other fusion welding processes, since they cannot be classified as any of the first three types. Each of these processes uses a unique technology to generate heat for melting, and typically, the applications are also unique in nature. Figure 2.3 shows the broad classification of welding processes. Figure 2.4 shows the master chart of welding and allied processes. On the basis of the composition of the joints, joining process can be of three types:

(i) **Homogeneous Process:** In this process, filler rod, also called electrode, is used and two same metals are joined without the application of pressure.

(ii) **Heterogeneous Process:** In this process, filler metal is soluble in both the parent metals, which themselves are soluble with each other.

(iii) **Autogeneous Process:** In this type of the process, no filler rod is used but pressure is extensively used. Resistance welding, forge welding, friction welding, diffusion welding are some examples of autogeneous welding processes.

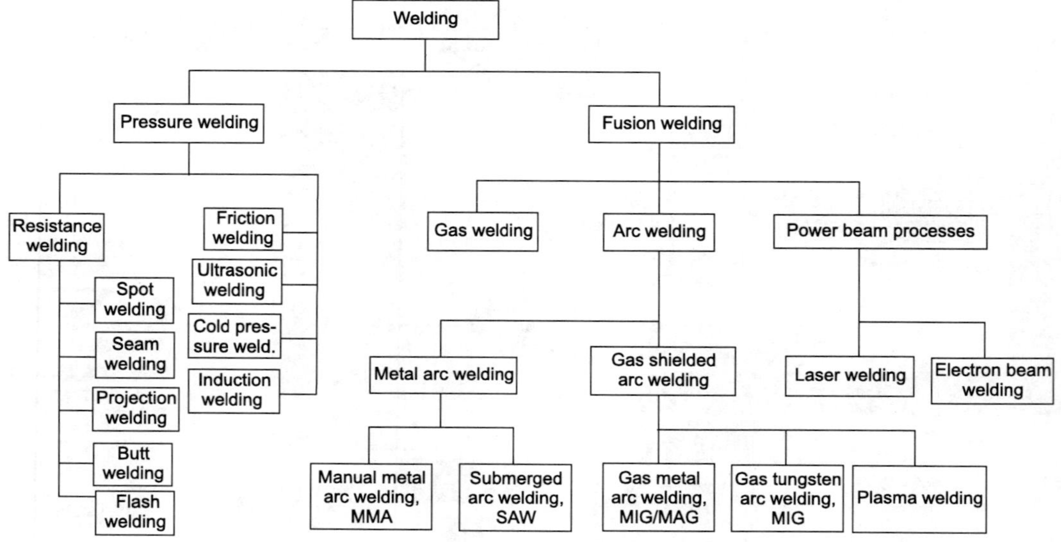

Figure 2.3 Broad classification of welding processes

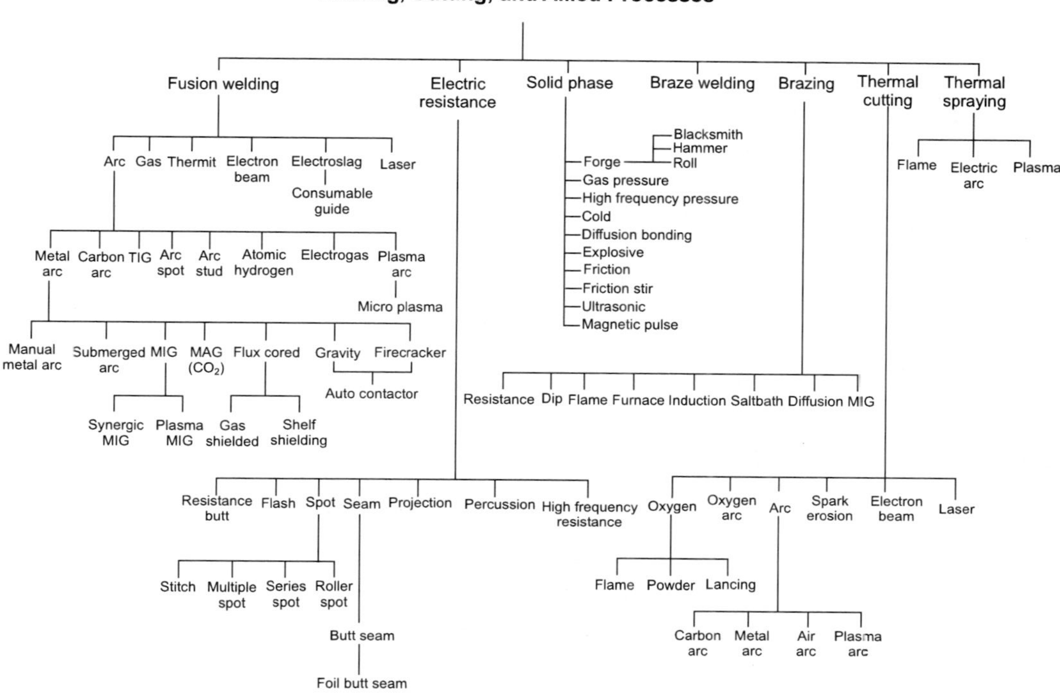

Figure 2.4 Complete classification of welding and allied processes

Welding processes can also be classified as follows:

1. **Gas Welding**
   Oxyacetylene
   Oxyhydrogen
2. **Arc Welding**
   Carbon arc
   Metal arc
   Submerged arc
   Inert gas welding: TIG and MIG
   Plasma arc
   Electroslag
3. **Resistance Welding**
   Spot welding
   Seam welding
   Projection welding
   Butt welding
   Induction welding

4. **Solid State Welding**
   Frictions stir welding
   Ultrasonic welding
   Explosive welding
   Forge and diffusion welding
5. **Thermochemical Welding**
   Thermit welding
   Atomic hydrogen welding
6. **Radiant Energy Welding**
   Electron beam welding
   Laser beam welding

2.3 OPERATING PARAMETERS IN WELDING

Essential variables to be controlled in various welding processes are:

- The welding process and its variation
- The method of applying the process
- The base metal type, specification or composition
- The base metal geometry, normally thickness
- The base metal need for preheat or postheat
- The welding position
- The filler metal and other materials consumed in making the weld
- The weld joint, that is, the joint type and the weld
- Electrical or operational parameters involved
- Welding technique

Some specifications also include non-essential variables and these are usually the following;

- The travel progression (uphill or downhill)
- The site of the electrode or filler wire
- Certain details of the weld-joint design
- The use and type of weld backing
- The polarity of the welding current

2.4 CONDITIONS FOR OBTAINING GOOD WELDS

To obtain satisfactory welds, it is desirable to have

(a) a source of energy to create union by fusion or pressure;
(b) a method of removing surface contaminants;

(c) a method for protecting metal from atmospheric contamination; and
(d) control of weld metallurgy.

(a) Energy Sources

Energy supplied is usually in the form of heat generated by arc, gas flame, the resistance to the electric current, radiant energy or by mechanical means such as friction, ultrasonic vibrations or by explosion. In fusion welding, the metal parts to be joined, melt and fuse together in the weld region. The word fusion is synonymous with melting but in welding, fusion implies union. The parts to be joined may melt but not fuse together and this fusion welding may not take place.

(b) Surface Contaminants

Surface contaminants may be organic films, absorbed gases and chemical compounds of the base metal, usually in the form of oxides. Heat, when used as a source of energy, effectively removes organic films and absorbed gases and only oxide films remain to be cleaned. Fluxes are used to clean the oxide film and other contaminants to form slag, which floats and solidifies above the weld bead, protecting the weld from further oxidation.

(c) Protecting Metal from Atmospheric Contamination

To protect the molten weld pool and filler metal from atmospheric containments, especially the oxygen and nitrogen present in the air, some shielding gases are used. These gases could be argon, helium or carbon dioxide supplied externally. Carbon dioxide could also be produced by the burning of the flux coating on the consumable electrode, which supplies the molten filler metal to the weld pool.

(d) Control of Weld Metallurgy

When the weld metal solidifies, the microstructure formed in the weld and the heat-affected zone (HAZ) region determines the mechanical properties of the joint produced. Preheating and postheating heat welding treatments can be used to control the cooling rates in the weld and HAZ regions and thus, control of microstructure and the properties of the weld produced. Deoxidants and alloying elements are added to control the weld metal properties.

2.5 SELECTION OF WELDING PROCESS

Welding is basically a joining process. Ideally, a weld should achieve complete continuity between parts being joined, such that the joint is indistinguishable from the metal in which the joint is made. Such an ideal situation is not possible practically but welds giving satisfactory service can be made in many ways.

The selection of a suitable welding process can best be done in four easy steps:

1. The joint to be welded is analysed in terms of its requirements.
2. The joint requirements are matched with the capabilities of available processes. One or more processes are selected for further examination.
3. A checklist of variables is used to determine the ability of the selected processes to meet the particular application.
4. Finally, the proposed process or processes deemed most efficient are reviewed with an informal representative of the equipment manufacturer for verification of sustainability.

Step 1: Analysis of Joint Requirements

The first thing is to look at whether the weld joint is large or small, whether the joint is out position or not, and whether the base metal is thick or thin. In welding, the needs of any joints are expressed in four terms:

1. ***Fast-Fill*** – It is required when the large amount of metal is to be deposited. A heavy bead can be laid down in minimum arc time with a high deposition rate.
2. ***Fast-Freeze*** – It implies that the joint is out of position, it could be overhead or vertical. In such cases, molten metal should be freezed very fast to avoid falling of material on the welder.
3. ***Fast-Follow*** – High arc speed and very small welds are the characteristics of fast-follow joints, such as thin sheet metal joints.
4. ***Penetration*** – The depths the weld penetrates the base metal, should be adequate to give sound weld joint. Burnout and cracking of base metal should not be allowed.

Step 2: Matching Joints with the Requirements of Process

The equipment manufacturer's literature usually provides the information about the ability of various processes to fulfil the needs of the joint. At this stage, the deposition rate and arc speed of each process are known and only the selection of process is left.

Step 3: Preparation of the Checklist

1. **Volume of Production:** If the rate of production is less, then the cost of joint/ weld will be expensive and uneconomical.
2. **Weld Specifications:** Material should be weldable. Properties of the material should be such that it should not create defects.
3. **Operator Skill:** Some welding requires special operator skills, such as laser welding, etc.
4. **Auxiliary Equipment:** Cost of auxiliary equipment must be considered in the selection of joints. If the process makes use of the existing equipment, then cost of weld is reduced.

5. **Accessory Equipment:** Cost of accessories, such as chipping hammers, deslagging tools, flux lay down and pickup equipment, exhaust systems, etc., must be considered.
6. **Base Metal Condition:** Condition of the supplied materials should be good and free from oil, dust, and grease, etc., otherwise cost of preparation increases.
7. **Arc Visibility Open:** Arc processes are used if the arc visibility is good. Submerged arc welding can be used if there is no problem in arc visibility and path of the weld bead is known or fixed by the traveling equipment of submerged arc welding.
8. **Fixturing Requirements:** Cost of fixturing is to be considered, when automation or semi-automation of the process is called for.
9. **Production Bottlenecks:** If the process reduces unit fabrication cost but creates production bottlenecks, its cost is high. Highly complicated equipment that requires frequent servicing will slow down the process and increase the cost of the weld.

Step 4: Review of Applications by Manufacturer's Representative

The talents of the experts should be used. "You know your application best and your welding expert knows equipment best". Together both should be able to confirm or modify the checklist.

2.6 WELD QUALITY AND PERFORMANCE

Welding is one of the principal activities in metal fabrication, aerospace, ship building, offshore and nuclear industries. The performance of these industries regarding product quality, delivery schedule and productivity depends upon the structural design, production planning, welding technology adopted and distortion control measures implemented during fabrication. The quality of the welding depends on the following parameters:

1. Skill of the welder
2. Welding parameters
3. Shielding medium
4. Working environment
5. Work layout
6. Edge preparation
7. Fit up and alignment
8. Protection of wild winds during on-site welding
9. Dimensional accuracy
10. Correct processes and procedures
11. Suitable distortion control procedures

2.7 IMPORTANCE OF WELDING

Welding is used as a fabrication process in every industry, large or small. It is a principal means of fabricating and repairing metal products. The process is efficient, economical and dependable as a means of joining metals. This is the only process which has been tried in the space.

Welding is a major need for most manufacturing industries. Almost any industry that works with metal cannot survive without welding. Welding, in a big way, is the backbone of all metal products. Every industry, from underwater construction to space explorations, relies on welding. Just look around right now how many things around you are made of metal? If it's made of metal, there is a good chance it was welded.

2.8 ADVANTAGES OF WELDING

1. Welded joints are stronger than base metal.
2. Finishing of welds is possible with grinding, shaping, etc.
3. The efficiency of the weld joint is high.
4. Appearance of welded joints is good.
5. Alteration is easy with weldments; another structure can be added easily.
6. Similar and dissimilar metals can be welded.
7. Weld joints are light in weight and reliable.
8. Welded joints are permanent joints.
9. Leak-proof joints are possible with welded joints.
10. Compared to casting, welding is a simple process.

2.9 DISADVANTAGES OF WELDING

1. Skilled operators are required to do welding.
2. Welding is not a safe processes. Both heat and light are by-products of welding.
3. Distortion of workpieces is possible if not clamped properly.
4. Thermal stresses and residual stress develop due to welding.
5. Welding produces metallurgical changes in material, such as HAZ, TMAZ, etc.
6. Edge preparation is required before welding, which is time consuming.
7. Special jigs and fixtures are sometimes needed.

2.10 COMPARISON OF WELDING WITH OTHER JOINING PROCESSES

1. Compared to casting and riveting, welded joints are light in weight.
2. Welding is faster and economical as compared to casting and riveting.

3. Welded joints are stronger as compared to casting and riveting.
4. There is no pattern, etc., in welding, so it is economical.
5. There is no need to drill the hole as in riveting, so weakening of structure is avoided.
6. Welded joints are permanent joints, whereas riveted joints are not.
7. Less use of jigs and fixtures for welding.

2.11 SIMILARITIES AND DIFFERENCES BETWEEN CASTING AND WELDED JOINTS

Differences Between Casting and Welding

1. Two dissimilar metals cannot be casted but two dissimilar metals can be joined.
2. Set-up cost is high for casting. It involves core making, pattern making, etc.
3. Due to welding, heat-affected zone (HAZ), grain refinement and transition zones are produced. Such problems are not there in casting.
4. Welding processes are more economical as compared to casting.
5. Welding processes are faster as compared to casting.
6. Most of the welding equipment are portable and welding can be done at site.

Similarities Between Casting and Welding

1. Both processes are melting processes. In casting, molten material is poured into the casting and in welding, parent metals are melted usually above melting temperature.
2. Both processes are not free from defects.
3. Secondary processes, like machining can be performed on welded joints as well as in casting.
4. Both casting and welding require skilled operator.
5. Both processes require extra safety precautions.
6. Both can be used for repair work.

2.12 APPLICATIONS OF WELDING

1. **Automobile and Transportation Industries**: Most of the automobile components are fabricated using welding. Examples include chassis, body of automobile, seat frames, etc.
2. **Aircraft Industries**: Wide varieties of materials are welded in aircraft industries. These include stainless steel, aluminum alloys, magnesium alloys, etc. Welding is used in fabrication of aircraft body, frames, ducts, fuel storage tank, etc. Welding is also used for material handling equipment of aircraft.

3. **Building Industries:** Welding is used widely in construction industries. RCC structures, frames, doors, windows, etc. are built using welding.
4. **Petrochemical Industries:** Chemical and petroleum industries make good use of welding for fabrication of plant and machinery. Stainless steel vessels and storage tanks are constructed using welding operation.
5. **Earthmoving Equipment Industries:** Earthmoving equipment manufacturing industries largely depend on welding for the fabrication of earthmoving machines, such as bulldozers, loaders, drilling units for oil exploration and water tube wells.
6. **Pressure Vessels and Tanks:** Pressure vessels and tanks are widely used in industries to carry fuel and other liquids, such as milk, chemical solutions, etc. These pressure tanks and vessels are made by welding processes.
7. **Rail Road Industries:** Welding processes are extensively used in fabrication of railway bogies, frames, trolleys, electric frame network, signal equipment, lightning towers, platform shades, storage tanks, etc.
8. **Material Handling Equipment Industries:** It is used in fabrication of overhead cranes, jib cranes, tower cranes, etc., along with various auxiliary equipment like trolley, lifting aids, etc.

   Welding finds its application in the manufacture of water turbines, fabrication of fission chamber of nuclear power plants, manufacture of household products like refrigerators, kitchen cabinets. It also finds its applications in the fabrication and repair of farm, mining, and oil machinery, machine tools , jigs and fixtures, boilers, furnaces, railway coaches and wagons, anchor cars, earthmoving machinery, ships, underwater construction and repair.
9. **Electrical Industries:** Most of the engineering systems used in electrical industry are fabricated using welding. ***Generation***: penstocks, water control gates, condensers, turbine blades and cooling fins. ***Distribution:*** electrical transmission towers and distribution system equipment transformer. ***Utilization of electrical energy:*** electrical motor starters.
10. **Electronic Industries:** Micro-joining techniques (ultrasonic, soldering) are used in electronic industry for joining. Soldering is used for joining electronic components to printed circuit boards (PCBs). Robot soldering is very common for joining parts to printed circuit boards of computers, television, communication equipment and other control devices.
11. **Nuclear Industries:** Welding is used for fabrication of spheres for nuclear reactors, pipeline bends, joining of two pipes carrying heavy water, etc.
12. **Defence Industries:** Welding is used for fabrication of tank bodies, joining of turret mounting to the main body of the tanks.
13. **Oil and Natural Gas Industries:** Welding is used for joining crude oil and gas pipes, construction of tankers for their storage and transportation, offshore structures, dockyards, loading and unloading cranes.

Review Questions

1. Define welding processes.
2. Explain the broad classification of welding processes.
3. What are the differences between arc and resistance welding processes?
4. What are the main advantages of welding processes?
5. What are the main applications of welding processes?
6. What are the similarities between welding and casting?
7. How do we select welding processes?
8. What are the properties of material considered in welding?
9. What are the differences between welding and casting?
10. What are the main conditions to get a good weld?

3 CHAPTER

Physics of Arc

3.0 INTRODUCTION

Arc welding is one of the several fusion processes for joining metals. Arc welding is a type of welding that uses a welding power supply to create an electric arc between an electrode and the base material to melt the metals at the welding point. They can use either direct (DC) or alternating (AC) current, and consumable or non-consumable electrodes. The welding region is usually protected by some type of shielding gas, vapor, or slag. Arc welding processes may be manual, semi-automatic, or fully automated (Figure 3.1).

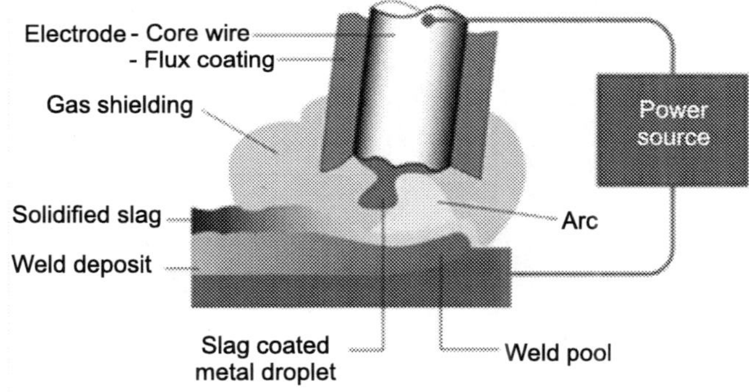

Figure 3.1 Basic arc-welding circuit

Definition: An arc is an electric discharge between two electrodes which takes place through an electrically conducting hot ionised gas, known as plasma. An electric arc used for welding is called the welding arc and is usually between a thin rod and a plate. A welding arc is a high current low voltage electric discharge, operating generally in the range of 10 to 2000 amperes and at 10 to 50 volts. In a welding circuit, the arc acts as a load resistor.

In arc welding processes, the heat generated by the welding arc is used for melting the faying surfaces of the base material. The arc can be established between the electrode and the base material. The electrode can be of the consumable or non-consumable type. Sometimes, this arc can also be established between the nozzle and electrode, as in the case of plasma arc welding.

- When an arc is established between the base metal and the non-consumable electrode like tungsten, in case of the plasma arc welding process and the gas tungsten arc welding process, it offers the better arc stability but, the production rate is low.
- When the welding arc is established between the base metal and the consumable electrode like steel or aluminum, in case of GMAW and the submerged arc welding processes, it offers the advantage of higher production rate. But, the stability of arc is low.

The heat generated in the arc welding processes is used for melting the faying surfaces of the base metal. That arc must be initially ignited and thereafter arc must be maintained, so that the efficient joint is produced. Success of the welding depends on how well the arc is created, maintained and how well electrodes are melted and deposited. It also depends on the rate of heat transfer from arc to electrode as well as the base metal (Figure 3.2).

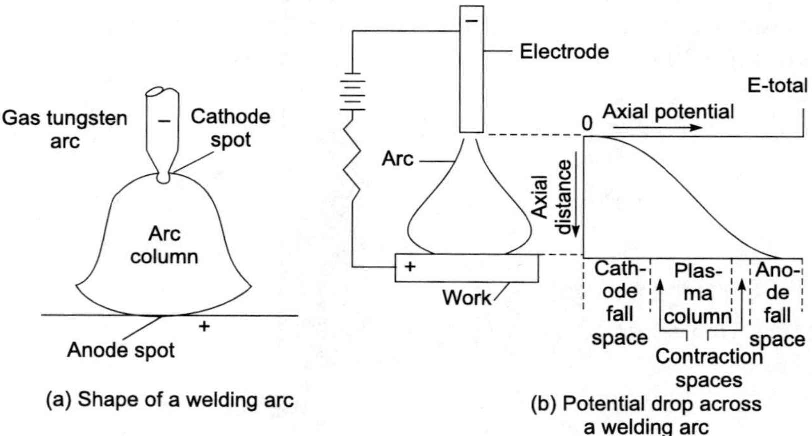

Figure 3.2 Arc characteristics

3.1 PHYSICS OF ARC

The success of welding depends upon many factors such as:

1. How effectively the heat is generated by welding arc, which affects arc efficiency.
2. Forces generated in arc region, which affects deposition efficiency.
3. Metal transfer, which dictates penetration and spattering.
4. Solidification of weld metal, which affects the heat-affected zone and strength of weld.

Therefore, it becomes important to understand the various aspects related with physics of the welding arc, which includes the mechanisms of the arc initiation and its maintenance, the various zones which are developed in the arc region, and their role in welding. The following topics are to be considered:

1. Arc initiation
2. Various zones in arc and their role
3. Arc maintenance
4. Arc forces and arc blow
5. Metal transfer
6. Melting rate
7. Shielding gases and their role

3.1.1 Arc Initiation

Welding arc is nothing but an electric discharge due to flow of current from cathode to anode. Flow of current through the gap between the electrode and workpiece requires column of charged particles. These charged particles are ions and electrons. The density of charged particles in the gap affects the electrical conductivity of gaseous column. Figure 3.3 shows the various charged particles in the gap.

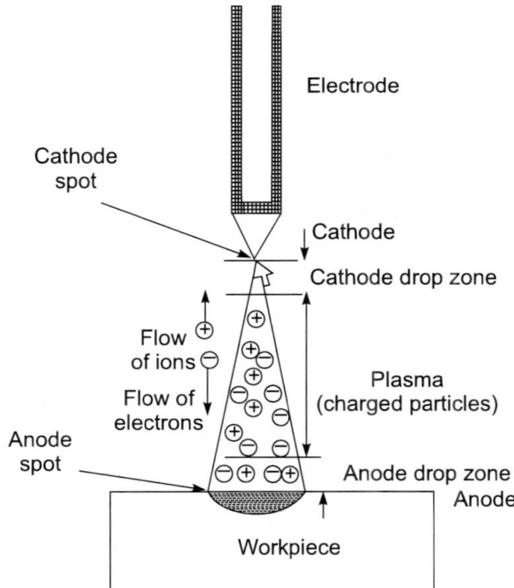

Figure 3.3 Various charged particles in gap

These charged particles are generated by various mechanisms, such as thermal emission, field emission and secondary emission. In an electric arc, electrons released from cathode (due to electric field or thermionic emission) are accelerated towards the anode because of potential difference applied between the electrode and workpiece. The high velocity electrons collide with gaseous molecules available in the gap and thereby decompose them into ions and electrons. These charged particles move as per polarity and form a part of welding current. Ion current becomes only about 1% of electron current as ions are heavier than electrons. Eventually, electrons merge into anode. Arc gap between electrode and workpiece acts as a resistive load. Heat generated in a welding arc depends on arc voltage and welding current.

Electron Emission Mechanisms

Electron emission from the cathode can be any one of the several mechanisms such as thermionic emission, autoelectronic or field emission, photoelectric emission and secondary emission.

- **Thermionic Emission**

  It involves the liberation of electrons from the heated electrodes. As the temperature of the electrode is raised, the kinetic energy of the free electrons increases to a point where they can escape from the surface of the cathode electrode. The emission of electrons from the carbon and tungsten cathodes is believed to be thermionic in character, but most other metals boil at temperatures well below that required for thermionic emission.
- **Autoelectronic Emission**

  This type of electron emission is produced by sufficiently strong electric field (10,000 volts), that is, when the voltage across the electrodes is so high that the air between them is ionized under its influence and the electric discharge ensues with the emission of the electrons from the cathode surface.
- **Photoelectric Emission**

  It occurs when energy in the form of a beam of light falls on the cathode surface and results in increased kinetic energy of the electrons and thus, results in their emission from the cathode into vacuum or another material. Such a mechanism of electron emission is utilized in generating X-rays.
- **Secondary Emission**

  It refers to the emission of electrons under the impact of rapidly moving ions. When the velocity of incident ions exceeds the orbital velocities of electrons in the atoms of the material of the cathode, it results in expulsion of electrons.

In the welding process, the electron emission is either of the thermionic type, for example, in gas tungsten arc welding, plasma arc welding and carbon arc welding, or it is of auto-emission type in conjunction with auxiliary means of ionizing the air gap between the electrode and the workpiece, such as for shielded metal arc welding, submerged arc welding and gas metal arc welding.

The emission of electrons from a cathode spot depends upon the excitation energy or the work function of a material, which is defined as the energy required in electron volts or joules, to get one electron released from the surface of the material to the surrounding space. Ionization potential, which is defined as the energy per unit charge in volts and required to remove an electron from atom to an infinite distance, also plays an important role in sustaining an electric discharge.

3.1.2 Various Zones in Arc Gap and their Role

Once the welding arc is initiated, there will be voltage across the arc gap. The voltage drop in arc voltage varies with the distance from electrode tip to the weld pool. There are five different zones observed in the arc gap, they are cathode spot, cathode drop zone, plasma zone, anode drop zone, and anode spot. Cathode spot and anode spot are the regions from where either electrons are released or the electrons are merged with. The voltage drop for other three regions is shown in Figure 3.4.

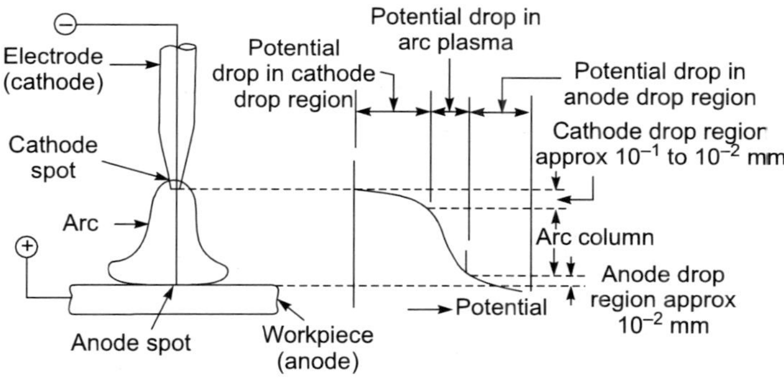

Figure 3.4 Voltage drop in arc gap

The sharp drop in the potential takes place in the cathode drop zone as well as in the anode drop zone. There is heat transfer from arc to cathode. This results in lower number of charged particles and high potential drop. Similar phenomenon will occur at anode also. In the plasma zone, the temperature gradient is constant, which results in uniform potential drop. The heat generated in the cathode drop zone is used for melting of the cathode side. The heat generated in the anode drop is used for melting the workpiece. Most of the heat generated in the plasma region is transferred towards the surrounding or to the shielding gasses. Table 3.1 gives electron work function and ionization potential of pure metals.

Cathode Drop Zone

Gaseous region is immediately adjacent to cathode in which a sharp voltage drop occurs. Combined size of cathode and anode drop zone is of order 10^{-3} mm, which is nearly equal to the electron mean free path. The voltage drop in the cathode drop zone for argon shielded tungsten electrode has been found to be about 8 volts at 100 amps and it increases as current decreases.

TABLE 3.1 Electron work function and ionization potential of pure metals

Metal	Work Function (eV)	Ionization Potential (V)
Aluminum	4.0	5.96
Barium	2.1	5.19
Calcium	2.2	6.09
Copper	4.3	7.68
Iron	4.5	7.83
Nickel	5.0	7.61
Potassium	2.2	4.30
Sodium	2.3	5.12
Tungsten	4.5	8.10

Arc Column

It is the bright visible portion of arc, which has high temperature and low potential gradient. The temperature of arc column depends upon the gases present in it and the amount of welding current flowing in the circuit. Usually, column temperature varies from 6000°C for iron vapors to 20000°C for argon shielded tungsten arc. At this high temperature, all molecular gases in the column get split into atomic form and atoms themselves further dissociate into ions and electrons. However, the number of electrons in any given volume of arc remains same, keeping the arc electrically neutral. As average ion is about 1000 times heavier than an electron, it is far more mobile & hence, carry most of the current across the arc column. The potential gradient in the column is lower than across the cathode drop zone or anode drop zone and it varies generally between 0.5 and 5 volt/mm for argon shielded tungsten arcs, whereas for shielded metal arc welding, it is normally around 1 volt/mm.

The welding arc is almost invariably between a rod or a wire electrode and a flat or wide workspace. This irrespective of electrode polarity, results in a bell or cone-shaped, arc with apex of cone at or near the tip of rod electrode. Due to this construction of arc near the rod electrode, it has highest energy density there but due to the cooling effect by the proximity of electrode, maximum temperature is at the core of the column. The region wherein the constricted column meets electrode is called arc root. Temperature distribution in the arc column for a 200 ampere argon shielded tungsten arc is shown in Figure 3.5.

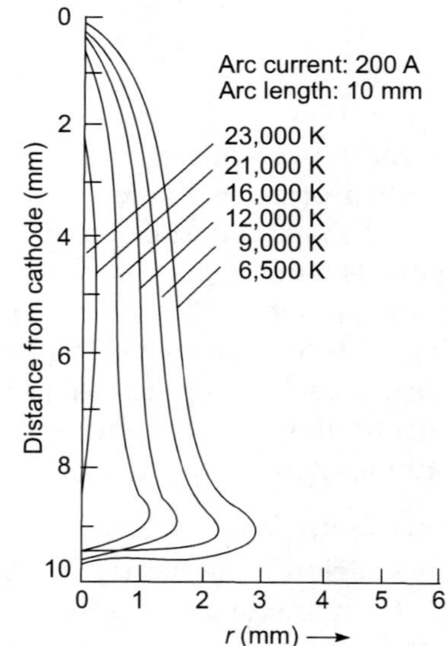

Figure 3.5 Temperature distributions in arc column

The flow of current in an arc column results in the development of electromagnetic forces. Since two parallel conductors, carrying current in same direction, attract each other and also the current is conducted by gaseous cylinder, it can be regarded as consisting of a large number of annular cylindrical conductors and there will be mutual attractions between the cylinders with all force acting inwards, due to high current acting at the core of the conductor. These constricting forces are balanced by a static pressure gradient established in gaseous conductor with zero pressure in outer periphery and maximum along the axis. However, in the present case, due to cone shape of the arc, the electromagnetic forces acting on it have two components, with the static pressure having two opposite components; one of which is along the arc axis and is the reason for formation of plasma jet that flows with a velocity of 10^4 cm/sec towards the workpiece. The axial plasma velocity decreases as the arc periphery approached as shown in Figure 3.6.

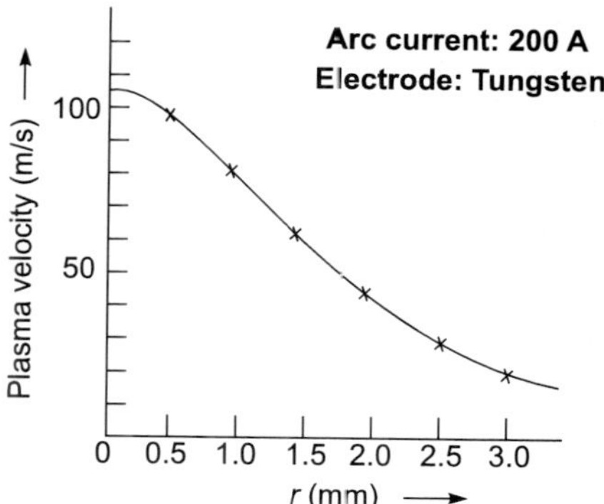

Figure 3.6 Plasma jet velocity at cross section of an arc

In a steady state, the plasma jet has stream line flow with flow velocity approximately proportional to the welding current. Figure 3.7 shows the pattern of gas flow lines and velocity lines in 200 A carbon arc. A considerable amount of heat energy is believed to be conveyed to workpiece through convective currents of plasma jet.

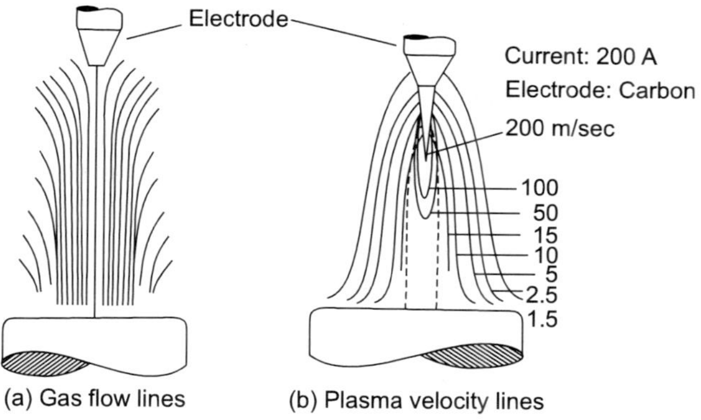

Figure 3.7 Gas flow lines and plasma velocity pattern in carbon arc welding

When the current flow in the arc is not symmetrical, it results in the setting up of magnetic forces, which deflect the arc column. If this occurs in a welding arc, it is known as *arc blow* and often results in unseemly and misplaced welds.

Anode and the Anode Drop Zone

On reaching the anode, the electrons lose their heat of condensation. However, unlike cathode spot, it is rare to define a well-defined anode spot and current density is also low, as shown in Figure 3.8 for a 200 A argon shielded tungsten cathode and copper plate anode. The current carrying area of the anode is slightly smaller than the widest spread of the arc at the anode end, and mean current density is also low. Voltage drop at the anode drop zone of this type of arc, appears to be between 1 and 3 volts. The depth of anode drop zone is of the order of 10^{-2} to 10^{-1} mm. When the rod electrode acts as anode, then it occupies lower hemisphere of molten droplet at the tip of electrode. However, for low pressure plasma jet, the anode appears to envelop molten droplet.

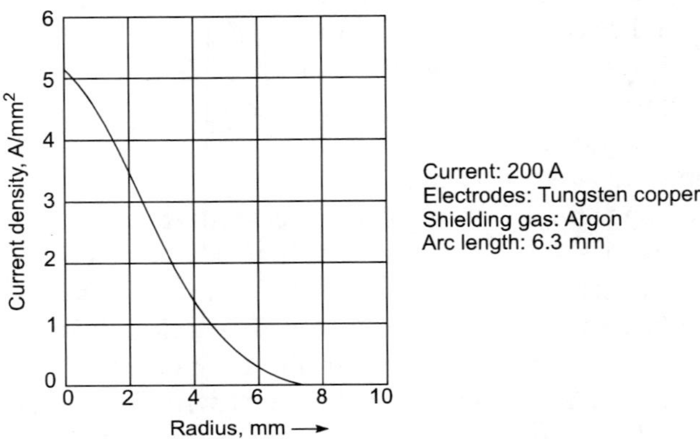

Figure 3.8 Current distributions at the anode of an arc

The total heat input at anode is due to condensation of electrons as well as conduction and convection by plasma jet. In DC arc, with non-consumable electrode like that of tungsten or carbon anode, heat is greater than heat liberated at cathode, as shown in Figure 3.9. With the increase in arc length, arc voltage increases and, due to this for current about 100 A, the heat input increases with increase in column, particularly for cathode spot mode as shown in Figure 3.10. However with increase in column length, column width also increases and results in still lower current density at anode and thus, anode becomes more diffused.

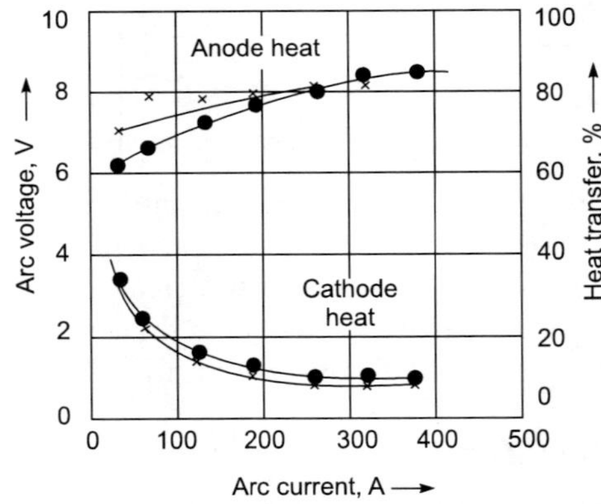

Figure 3.9 Effect of welding current on cathode and anode heats

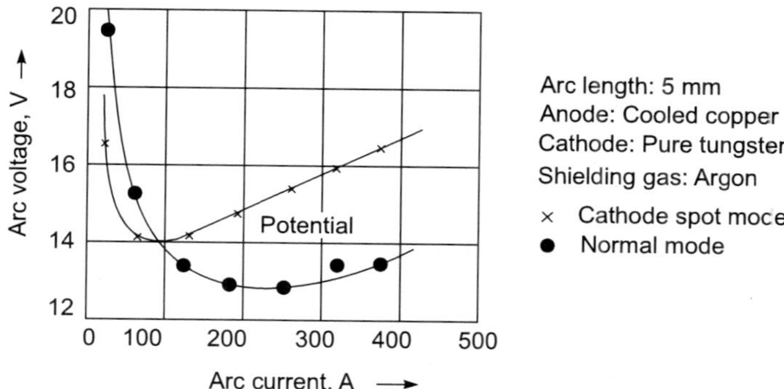

Figure 3.10 Volt ampere characteristics for argon shielded tungsten arc

3.1.2.1 Electrical Fundamentals of Welding Arc

For arc stability, arc voltage and current must have a definite relationship to each other. This relationship is known as the static volt-ampere.

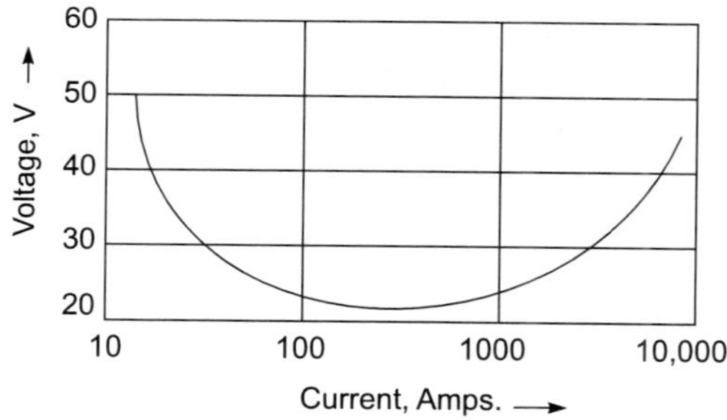

Figure 3.11 Static volt-ampere characteristic of an electric arc

Characteristic of the arc and its typical shape is shown in Figure 3.11. It shows that as the current increases up to about 100 A, the arc voltage decreases, which is explained on the basis that with the increase in current, in this range, the arc stream broadens, while the temperature rises with the result that the arc voltage decreases. This is known as drooping or negative volt-ampere characteristic. From 100 to 1000 amperes, the arc V-I characteristic is nearly flat or slightly rising. This is because in this range, the increase in arc stream cross section with increase in current remains nearly proportional and hence, the current density and the arc voltage remain constant. This part of the arc is said to have a plateau or flat volt-ampere characteristic. With the further increase in arc current, beyond 1000 A, there is hardly any expansion of arc stream and hence, the current density increases and so does the arc voltage. This part of the arc is said to have a positive or rising volt-ampere characteristic.

Welding arc acts as impedance for the flow of arc current. The impedance of arc is found inversely proportional to the density of charged particles and their mobility. Therefore, the radial and axial distribution of charged particles affect the total impedance of arc.

$$\text{Power of arc} = \text{Potential drop across the arc} \times \text{Welding current}$$

$$\text{Power of arc } (P) = V_c + V_p + V_a \tag{1}$$

where, V_c, V_p and V_a are voltage drops across cathode zone, plasma zone and anode zone, respectively. Arc power equation suggests that the distribution of heat in three zones (cathode, anode and plasma) can be varied by varying the voltage drop and welding current. Voltage drop, in turn, is influenced by arc length, shielding gas and presence of low-ionisation potential elements, like sodium and potassium, in electrode coating.

Heat generation at the anode and cathode is governed by the type of welding processes and polarity associated with the welding arc. Electrode can be connected to negative terminal or positive terminal of DC power supply. If the negative terminal is connected to electrode, it is called DCEN, on the other hand, if positive terminal of DC source is connected to electrode, it is called DCEP. Difference between DCEN and DCEP is given in Table 3.2.

TABLE 3.2 Difference between DCEN and DCEP

DCEN	DCEP
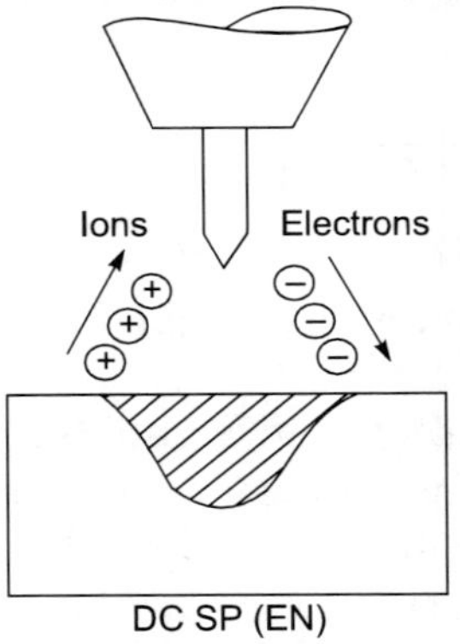 Ions Electrons DC SP (EN) No cleaning action 70% heat at work 30% heat at W Excellent electrode current capacity	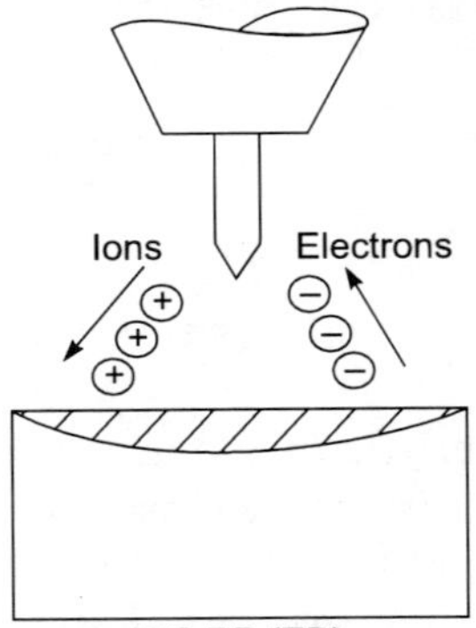 Ions Electrons DC RP (EP) Strong cleaning action 30% heat at work 70% heat at W Poor electrode current capacity
In direct current straight polarity, electrode is connected with the negative terminal of the power source and base metals are connected with the positive terminal.	In direct current reverse polarity, base metals are connected with the negative terminal of the power source and electrode is connected with the positive terminal.
Under sufficient potential difference, electrons liberate from the tip of the electrode and strike the base plate surface.	Here, electrons liberate from base plate surface and strike the electrode tip.

Contd.

Almost 2/3rd of the total arc heat is generated near the base plate and rest is generated at the electrode tip.	Here, 2/3rd of the total arc heat is generated at the electrode tip and rest is generated near the base plate.
Proper fusion of the base metal can be achieved easily. So, it eliminates lack of fusion and lack of penetration defects.	Due to less heat generation near-base plate, incomplete fusion plate may be encountered.
In case of consumable electrodes, filler metal deposition rate is quite low.	Filler metal deposition rate is quite high, as greater portion of heat is generated at the electrode tip.
Oxide cleaning action by arc is poor.	Oxide cleaning action by the arc is good.
Inclusion defects may arise if base plate surfaces are not cleaned properly prior to welding.	Due to good arc cleaning action, tendency of inclusion defects reduces.
DCEN may cause high distortion and broader HAZ in the welded component.	Distortion is less with DCEP and also HAZ is quite thin.
DCEN is not suitable for welding thin plates, as it may cause cutting instead of welding.	DCEP is suitable for welding thin plates.
Metals with high melting temperature (such as stainless steel, titanium) can be suitably joined by DCSP.	Metals with low melting temperature (such as copper, aluminum) can be suitably joined by DCSP.

3.1.2.2 Effect of Arc Length

Increase in spacing between the electrodes increases the potential of the arc because of increased loss of the charge carriers by radial migration to cool boundary of the plasma. Smooth and stable arc is possible if there is sufficient number of charged particles in the arc gap. More the number of the charged particles present in the arc gap, better will be the conductivity and easier will be the flow of the electrons. The decrease in the density of the charged particles in the arc gap mainly occurs due to the increased losses of these charged particles from the arc surface. These losses of electrons from the arc surface must be accommodated to stabilize the arc by increasing the applied voltage. Most of the heat generated in consumable arc welding processes goes to the weld pool, which, in turn, results in higher thermal efficiencies.

3.1.2.3 Arc Column Theory

The arc column is generated between an anode—the positive pole of a DC (direct current) power supply, and the cathode—the negative pole. Generation of the arc column has been explained with the help of two contradicting theories.

Ion Theory

The electrical theory of the arc column is that the ions pass from the positive pole to the negative pole because metal ions are positively charged and thus, they are attracted to the negative pole. The cathode and anode are touched together, and an air gap is established after they are drawn apart. Ions pass through this air gap because of their attraction to the negative cathode (Figure 3.12).

As they travel through this air gap, the ions collide with molecules in the atmosphere, which produces a thermal ionization layer. This ionized gas column acts as a high-resistance conductor that enables more ions to flow from the anode to the cathode. Heat is generated as ions strike the cathode.

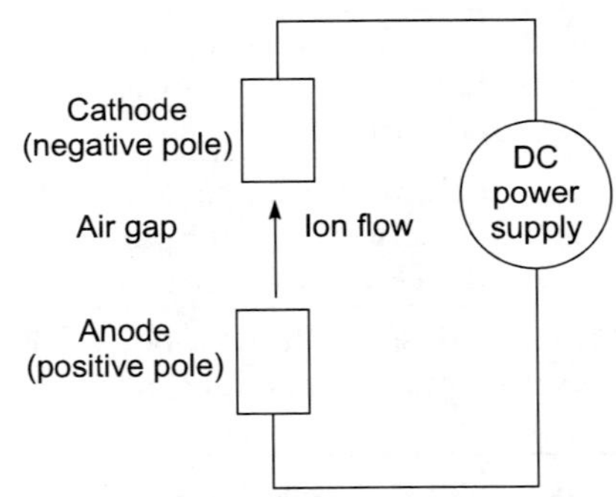

Figure 3.12 Ion theory of arc column

Drawbacks

This theory does not explain the tremendous amount of heat generated in the arc column.

Electron Theory

This theory is based on the principle of "Thermionic Emission" — metals release electrons readily when temperature of metal ions is increased. Thus in welding, an electron gets easily disassociated from the metal at the negative pole or the cathode. This small amount of negatively charged mass is accelerated away from the cathode to the positive pole or anode, striking it at a highly accelerated velocity, releasing considerable amount of heat energy (Figure 3.13).

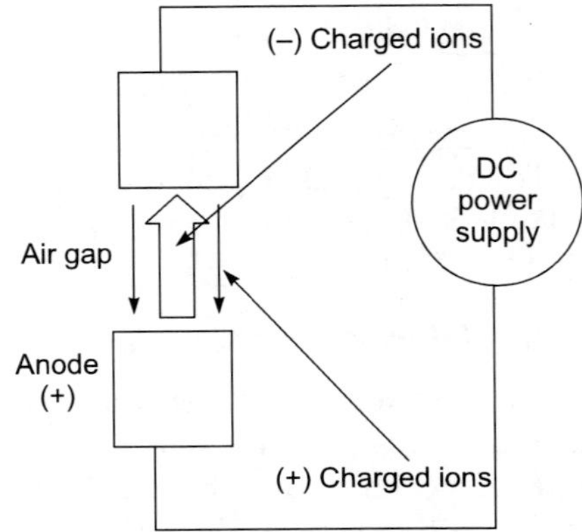

Figure 3.13 Electron theory of arc column

This flow of negatively charged mass is generally in the interior of the arc column, which is the hotter portion of the arc column.

At the same time, ions that are positively charged return from the anode to the cathode and hence, produce an ionized-gas layer, which also protects the electron and the electrostatic unit within the electron in the interior of the arc column. These ions strike the cathode, releasing medium amounts of heat energy.

Because of the characteristics of electrons and ions in the arc column, three areas of heat are liberated in the arc stream: *the cathode area, the plasma area* and *the anode area.* Of the three areas, the anode area is the high-heat area, where approximately 10,000 to 11,000 F of heat is liberated. The liberation of heat results from the combination of the impingement of the electrons upon the anode and the current carrying capacity of the electrons. The plasma area is heated mainly as a result of atomic collision of the few electrons and the many ions that pass through the ionized gas column. The cathode or the negative pole is subject mainly to ionic bombardment, which produces the state of medium heat in the arc column.

3.1.2.4 Arc Efficiency

Arc welding basically involves melting of faying surfaces of base metal using heat generated by the arc under a given set of welding conditions, i.e., welding current and arc voltage. However, only a part of heat generated by the arc is used for melting purpose to produce weld joint and remaining is lost in various ways, namely; through conduction to base metal, by convection and radiation to the surroundings.

Heat generation on the workpiece side depends on the polarity in case of DC welding while it is equally distributed in the workpiece and electrode side in case of AC welding. Further, it can be recalled that heat generated by arc is dictated by the power of the arc (VI), where V is arc voltage, i.e., mainly sum voltage drop in cathode drop (V_c), plasma (V_p) and anode drop regions (V_a), apart from work function related factor and I is welding current. Product of welding current (I) and voltage drop in particular region, governs the heat generated near anode, cathode and in plasma region. In case of DCEN polarity, high heat generation at workpiece facilitates melting of base metal to develop a weld joint of thick plates. Figure 3.14 shows that the basic reason for heat loss in the arc gap is due to the heat loss at cathode/anode which decreases the temperature and thereby, reducing the number of charged particles. Another reason is reduction in the number of charged particles in the arc zone with the increase of arc length. As arc length increases, the charged particles density decreases.

Power of arc = Potential drop across the arc × Welding current

$$\text{Power of arc, } P = V_c + V_p + V_a$$

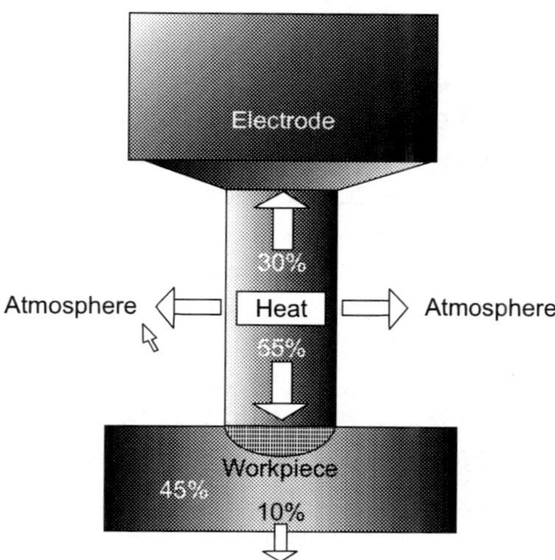

Figure 3.14 Heat distribution in arc gap

In the case of non-consumable electrode welding, only $V_a \times I$ is used to melt the workpiece. So arc efficiency, which is the ratio of the heat being used in the anode drop region divided by the heat generated in the welding arc. Non-consumable arc welding processes include GTAW, PAW, electron and laser beam welding.

$$\text{Arc efficiency} = \frac{V_a \times I}{VI} \tag{2}$$

In the case of consumable electrode welding, $V_a \times I$ and $V_c \times I$ are used to melt the workpiece and electrode (cathode). So arc efficiency, which is the ratio of the heat being used in the anode drop region and cathode drop region is divided by the heat generated in the welding arc. Consumable arc welding processes include SAW and GMAW (Figure 3.15).

$$\text{Arc efficiency} = \frac{(V_a \times I) + (V_c \times I)}{VI} \tag{3}$$

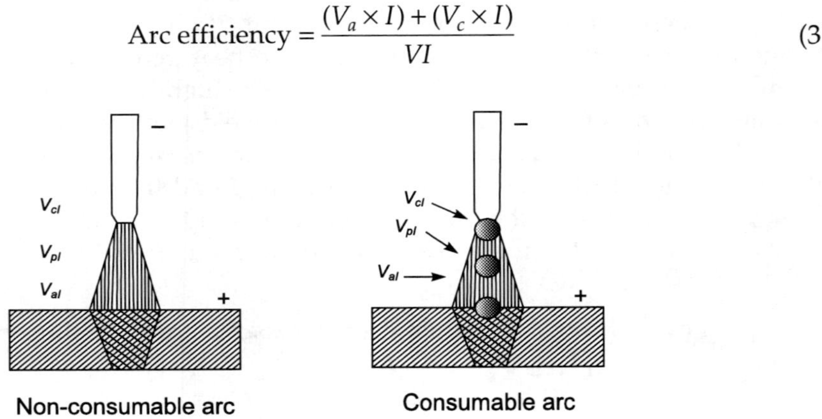

Figure 3.15 Heat distribution in non-consumable and consumable welding

Therefore, arc efficiency of consumable arc welding processes is higher than the non-consumable processes. The thermal efficiency of metal arc welding is found to be in the range of 70 to 80%, whereas that for non-consumable arc welding processes, is found to be in the range of 40 to 60%.

Additionally, in the case of consumable arc welding processes (SMAW, SAW), heat generated is more effectively used because of reduced heat losses to the surroundings as the weld pool is covered by molten flux and slag. Welding processes, in ascending order of arc efficiency, are GTA, GMA, SMA and SAW. GTA produces lower arc efficiency (21 to 48%) than SMA/GMA (66 to 85%) and SA welding (90 to 99%).

Heat generated at anode is found as the sum of heat generated due to electron emission and that from anode drop zone. $q_a = \phi + V_a$, where ϕ is work function of base metal at temperature $T = \phi_0 + 1.5KT$. ϕ_0 is work function of base metal at temperature 0 K. K is Boltzmann constant and T is temperature in kelvin. V_a is anode voltage drop and I is welding current.

Heat generated in plasma region = $q_p = V_pI$. Let fraction m% of the heat generated in plasma region goes to anode/workpiece for melting $= mV_pI$.

So arc efficiency = Total heat at anode/Total heat generated in arc

$$\eta = \frac{(q_a + m(V_p I))}{VI}$$

3.1.2.5 Arc Initiation Methods

There are two most commonly used methods to initiate an electric arc, namely, touch start and field start. Touch start is commonly used in all manual welding processes.

In touch start method, electrode is touched with the workpiece and separated quickly so as to initiate the arc. Touching of the electrode causes short circuiting, so flow of heavy current causes heating, partial melting and even slight evaporation of the electrode tip. This entire phenomenon occurs in a few seconds. Pulling of the electrode apart from the workpiece, flow of current starts through these charged particles and for a moment, arc is developed. Figure 3.16 shows the touch start method.

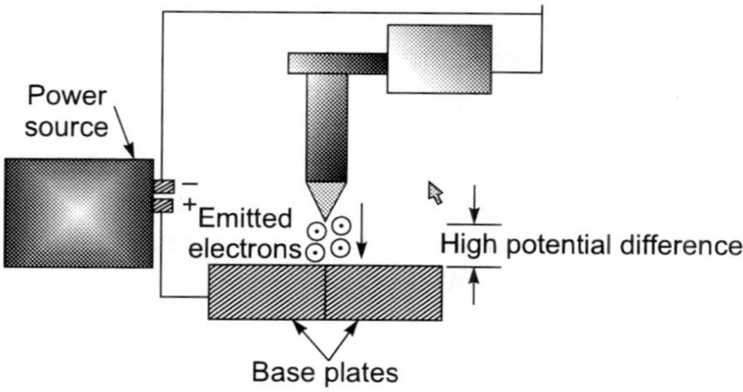

Figure 3.16 Touch start method

In field start method, high strength electric field in the order of 107 volts is applied to establish field and to initiate arc. Electrons are released from cathode by field emission. Once the free electrons are available in arc gap, normal potential difference between electrode and workpiece ensures flow of charged particles to initiate a welding arc. The field start is used for automatic welding processes and TIG welding. In TIG welding it avoids tungsten contamination. If we use touch start method for tungsten, tungsten melts and gets into work to form inclusions, which weaken the welded joint.

3.1.3 Arc Maintenance

After the initiation of the welding arc, the arc has to be maintained in the arc gap to use the heat generated for welding purpose. To maintain the arc, two conditions must be fulfilled, they are:

- Energy dissipation rate from the arc region should be equal to the energy generated, to maintain the temperature.
- Number of electrons produced should be equal to that of the electrons lost to the workpiece and the surroundings.

An electric arc primarily involves flow of current through the gap between the workpiece and electrode; hence there must be sufficient number of electrons. However, some of the electrons are lost from the arc surface to the weld pool and the surroundings and some reunite with ions. Loss of these electrons must be compensated by generating new free electrons. In case of direct current, magnitude and direction of current do not change with time, therefore, maintaining arc is not difficult. While in case of alternating current, both magnitude and direction of current changes with respect to time, therefore re-ignition of electric arc with AC is somewhat difficult. There are two methods commonly used for maintaining arc in AC welding:

- Use of low ionization potential elements in coating.
- Use of proper power factor.

3.1.3.1 Low Ionization Potential Elements

In this method, low ionization potential elements, such as potassium, calcium and sodium, are added in the flux covering of the electrode (coating). These elements facilitate easy release of free electrons needed for maintaining welding arc even with small potential difference between the electrode and workpiece. Figure 3.17 shows the electrode with and without low ionization potential element.

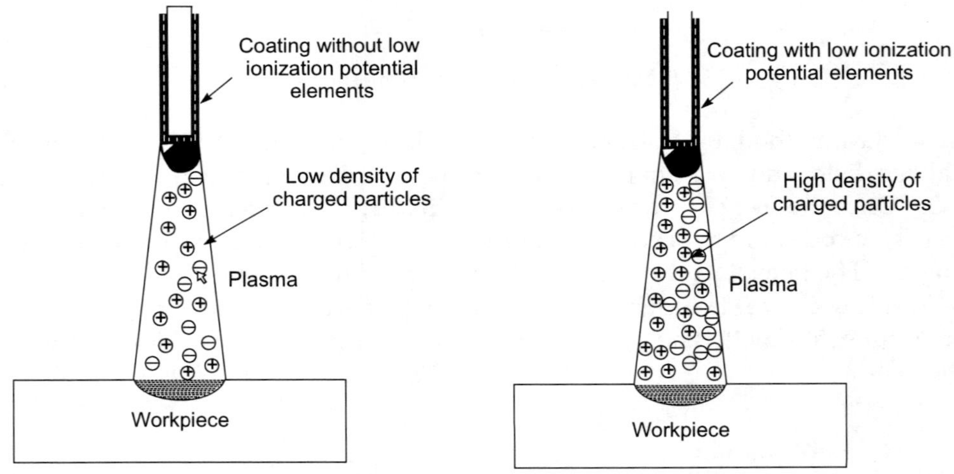

Figure 3.17 Use of low ionization potential elements

3.1.3.2 Use of Proper Power Factor

In this method, current and voltage are made out of phase by using proper power factor (0.3) so that when current is zero; full open circuit is available across the gap between electrode and workpiece. Full open circuit voltage across the electrode and work helps in release of free electrons to maintain the arc, which is a prerequisite for the maintenance of the arc. Power factor can be obtained from the difference of degree in phase of voltage and current as shown in Figure 3.18.

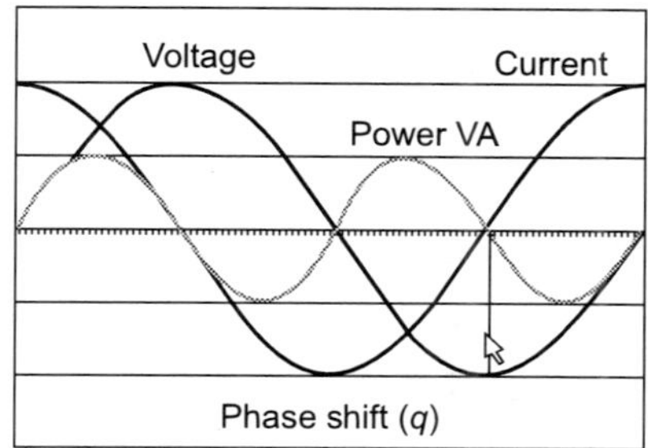

Figure 3.18 Power factor

3.1.3.3 Arc Characteristics

Welding arc characteristics show the variation in the arc voltage with welding current. There are three different regions in arc characteristics curve, namely, dropping, flat and rising zones, as shown in Figure 3.19.

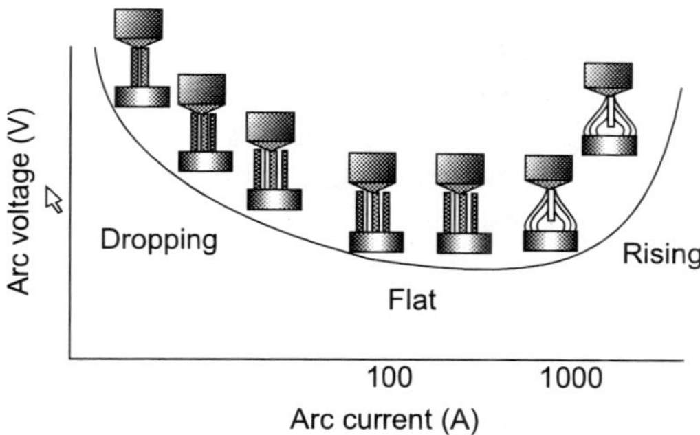

Figure 3.19 Arc characteristics

Dropping: When arc is thin, with an increase in welding current the temperature of arc zone increases, which in turn, increases the number of charged particles in plasma zone of the arc due to thermal ionization and thermally induced emission of electrons. As a result, the electrical conductivity of arc zone increases and hence, arc voltage decreases with initial increase in welding current. This trend continues up to certain level of current and beyond that increase in current increases the diameter of cylindrical arc, which in turn, increases the surface area of the arc and so, loss of heat from the arc surface. So arc tends to be stable in this region.

Flat Region: There is no significant rise in arc temperature with increased current in this region and hence, arc voltage is not affected appreciably over a range of current.

Rise Region: Further increase in current bulges the arc, which in turn, increases the resistance to the flow of current due to increased loss of charge carriers and heat from the arc. This leads to the rising part of curve.

Arc Characteristics with Arc Length: Increase in arc length increases arc voltage during welding. However, the extent of increase in arc voltage with increase in arc length varies with process, primarily due to variation in the presence of free electrons associated with particular process. The relationship between arc length and arc voltage is shown in Figure 3.20.

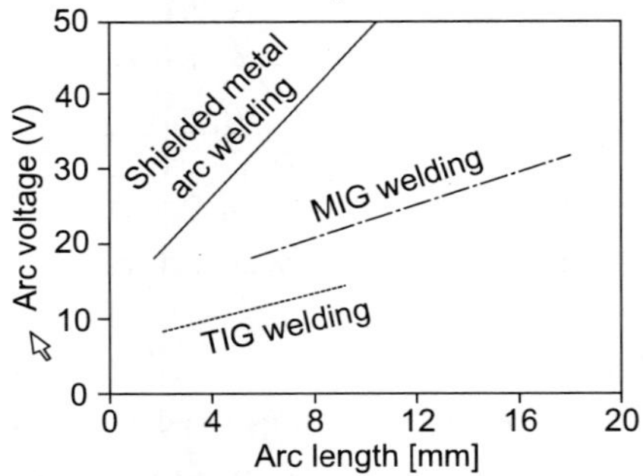

Figure 3.20 Relationship between arc length and arc voltage

In tungsten inert gas welding process, free electrons are released easily in the arc gap and increase in arc length does not affect the arc voltage appreciably. Whereas in the case of submerged shielding metal arc welding process, free electrons and charged particles in the arc gap are not easily released. Therefore, the increase in arc length leads to the deficiency of charged particles in the arc gap, which in turn, leads to the significant increase in arc voltage with the increase in arc length.

Temperature in Arc Zone: Temperature in the arc zone ensures melting and pace of melting of the faying surfaces during welding. It varies with location in arc region and is not uniform. This variation is primarily due to the presence of charged particles in specific zone. Figure 3.21 shows the temperature variation in the arc zone.

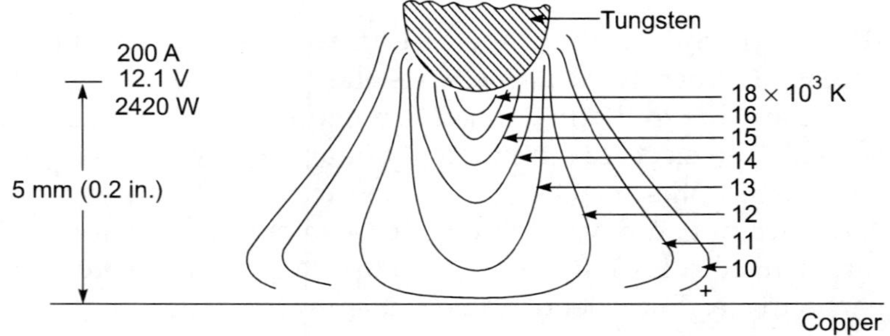

Figure 3.21 Temperature distribution in arc region

There are many factors which affect the temperature of arc zone, such as:

1. Arc voltage
2. Current
3. Thermal conductivities of the shielding gases present in arc zone, such as He, N, Ar. Thermal conductivity of most of the gases increases with rise in temperature. However, this increase is not continuous for some gases like He. Thermal conductivity governs temperature gradient in the arc region. Reduction in thermal conductivity increases the temperature gradient. Therefore, at low temperature of arc, there is rapid decrease in temperature with increase in distance from the centre of arc, primarily due to low *k* of shielding gases. Maximum temperature is observed at the centre, that is, along the axis of electrode in the arc and it decreases rapidly with distance away from the core. Temperatures in anode and cathode drop zones are generally lower than the plasma region due to cooling effect of electrode/workpiece. Temperature of arc can vary from 30,000 to 50,000 K, depending upon the current and plasma gas. For SMAW, temperature of arc is about 6000 K, while for TIG/MIG welding, it is around 20,000 to 30,000 K.

3.1.4 Arc Forces and their Significance

Forces acting in arc zone are known as arc forces. In respect of welding, influence of these forces on resisting or facilitating the detachment of molten metal drop at the electrode tip is important, which in turn, can affect the mode of metal transfer. Metal transfer is basically detachment and movement of molten metal drop from the tip of the electrode to the workpiece. This duration of molten metal drop in arc region affects the quality of weld and metal transfer efficiency. Figure 3.22 shows the various arc forces.

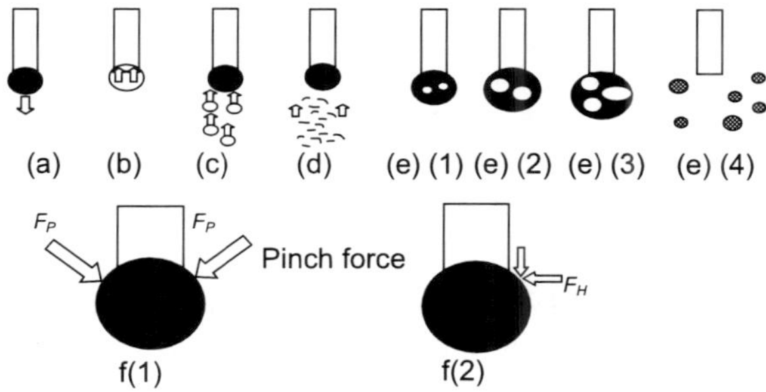

Figure 3.22 Schematic diagram showing different arc forces; (a) Gravitational force (b) Surface tension force (c) Force due to the impact of charged particles (d) Force due to metal vapors (e) (1) to (e) (5). Stages in force generation due to gas eruption. f(1) and f(2) Electromagnetic pinch force

Gravitational Force: This force is due to the gravitational force acting on molten metal drop hanging at the tip of the electrode. Gravitational force depends on the volume of the drop and density of metal. In case of downhand welding, gravitational force helps in detachment or transfer of molten metal drop from electrode tip.

$$\text{Gravitational force} = \rho \times V \times g$$

Surface Tension Force: It is due to the viscosity and surface tension forces acting on the liquid molten metal hanging at the tip of the electrode. This force is responsible for the detachment of the molten metal drop hanging at the tip of the electrode. Magnitude of surface tension force is influenced by the size of the droplet and surface tension coefficient. This force tends to resist the detachment of molten metal drop from the electrode tip and usually acts against the gravitational force. In case of vertical and overhead welding positions, high surface tension force helps in placing the molten weld metal at the required position more effectively by reducing fall out of molten melt. Accordingly, electrode composition for odd-position welding purpose must be designed to have viscous and high surface tension weld metal.

$$\text{Surface tension force} = \frac{2\pi\sigma R_e^2}{4R}$$

where, σ is the surface tension coefficient, R is drop radius and R_e is the radius of electrode tip. An increase in temperature of the molten metal reduces the surface tension coefficient. Hence, this reduces the hindering effect of the surface tension force on detachment of the drop.

Force Due to the Impact of Charge Carriers: As per polarity, charged particles (ions and electrons), move towards anode or cathode and eventually impact or collide with them. Force generated owing to this impact of charged particles tends to hinder the detachment of droplet. Force due to the impact of charged particles is given by

$$F_m = m\frac{dv}{dt}$$

where, m is the mass of the charged particles, v is the velocity and t is the time.

Force Due to Gas Eruption: Gases present in molten metal, such as oxygen, hydrogen, etc., may react with some of the elements, such as carbon, present in molten metal drop and form gaseous molecules (carbon dioxide). The growth of these gases in molten metal drop as a function of time ultimately leads to bursting of metal drops, which in turn, increases the spattering and reduces the control over handling of molten metal.

Force Due to Electromagnetic Field: Flow of current through the arc gap develops the electromagnetic field. Interaction of this electromagnetic field with charge carriers produces a force which tends to pinch the drop hanging at the tip of the electrode, also called pinch force. The pinch force reduces the cross section of molten metal drop near the tip of the electrode and thus, helps in detachment of the droplet from the electrode tip. A component of pinch force acting in downwards is generally responsible for the detachment of droplet, and is given by

$$\text{Pinch force, } F_p = \frac{\mu I^2}{8\pi}$$

μ is the magnetic permeability of metal and I is the welding current flowing through the arc gap.

Electrode Polarity in Welding: In the case of DC welding, polarity depends on the way electrode is connected to a power source. If electrode is connected to negative terminal of the power source, then it is called direct current electrode negative (DCEN) or straight polarity. If electrode is connected to positive terminal of the power source, then it is called direct current electrode positive (DCEP) or reverse polarity. Polarity, in case of AC welding, does not remain constant as it changes every half cycle of current. Selection of appropriate polarity is important for successful welding, as it affects distribution of heat generation at anode and cathode, stability of arc and cleanliness of weld. In general, more heat is generated at anode than the cathode. Out of the total DC welding arc heat, two-thirds is generated at the anode and one third at the cathode. The differential heat generation at the anode and cathode is due to the fact that the impact of high velocity electrons with anode generates more heat than that of ions with cathode, as electrons possess higher kinetic energy than the ions. Ions being heavier than electrons cannot accelerate and move at high velocity in the arc region. Therefore, DCEN is commonly used with non-consumable electrode welding processes so as to reduce the thermal degradation of the electrode due to low heat generation. Moreover, DCEP facilitates higher melting rate in the case of consumable electrode welding processes, such as SAW, MIG, etc. All those processes (SMAW, PAW, GTAW) in which electrode is expected to emit free electrons are required for easy arc initiation and their stability, selection of polarity affects the arc stability. Shielded metal arc welding, using covered electrode having low ionization potential elements, provides better stable arc with DCEN than DCEP. However, SMA welding with DCEP gives smooth metal transfer. Similarly, in the case of GTAW welding, tungsten electrode is expected to emit electrons for providing stable arc and therefore, DCEN is commonly used.

Arc Blow: It is basically a deflection of a welding arc from its intended path, i.e., axis of the electrode, because of magnetic forces. Deflection of arc during welding reduces the control over the handling of molten metal by making it difficult to apply the molten metal at the right place. A severe arc blow increases the spattering, which in turn, decreases the deposition efficiency of the welding process. Figure 3.23 shows the arc blow. It is usually found in DC welding and can be minimized by AC welding.

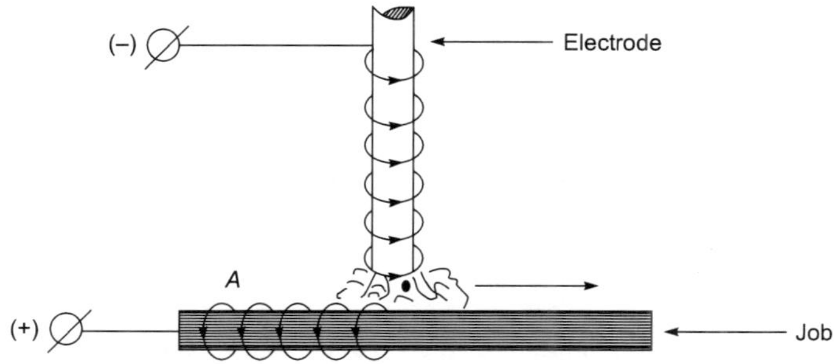

Figure 3.23 Magnetic field of the arc blow

Definition of Arc Blow: Normally, the welding arc is oriented along the axis of the electrode rather than along the shortest distance between the electrode and the workpiece. However, an electric arc is an extremely flexible conductor, which maybe deflected from its intended path by a variety of factors. When such a phenomenon occurs during welding, it's called arc blow.

Types of Arc Blow (Figures 3.24 and 3.25)

According to the direction of deflection of arc with respect to welding direction, an arc blow may be forward or backward arc blow. Deflection of arc ahead of the weld pool in the direction of the welding, is called forward arc blow and that in reverse direction, is called backward arc blow. Arc blow is mainly encountered during DC arc welding due to the interaction between different electromagnetic fields, in and around the welding arc. Incidences of interaction between electromagnetic fields mainly occur in areas where these fields are localized. There are two common situations of interaction between electromagnetic field that lead to arc blow; interaction between electromagnetic field due to flow of current (a) through the arc gap and (b) through plates being welded. Electromagnetic field is also generated around the arc in the arc gap. Any kind of interaction of this field with other electromagnetic fields deflects the arc from its intended path.

There will be interaction between electromagnetic field due to flow of current through the arc gap and localized concentration of lines of electromagnetic forces while welding near the edge of the plates. The lines of electromagnetic fields can flow more easily through the metal than the air and therefore, the distribution of lines of electromagnetic forces does not remain uniform around the arc. These lines are concentrated near the edge of the plate.

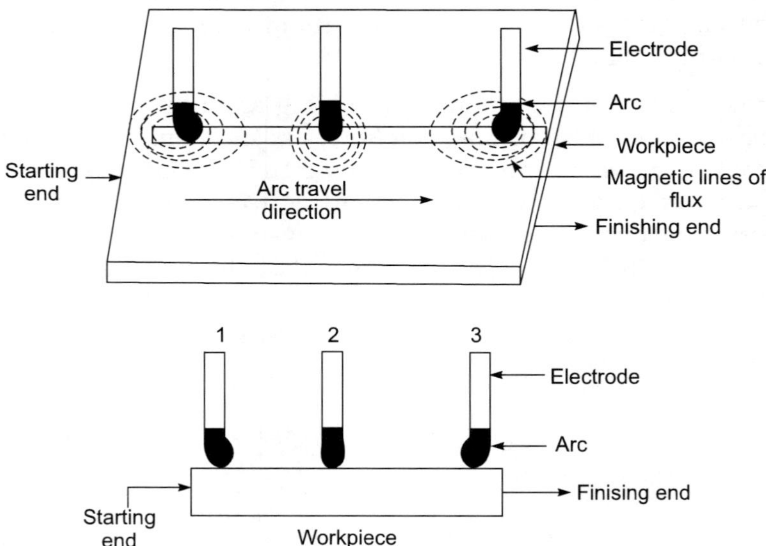

Figure 3.24 Mechanism of forward and backward arc blows

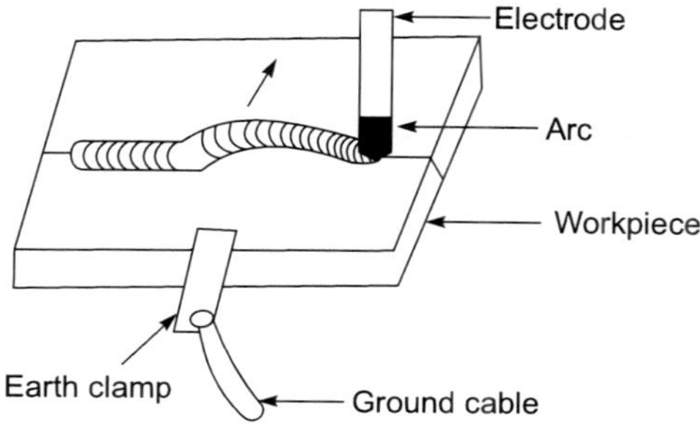

Figure 3.25 Sideward arc blow

Causes for Arc Blow

Arc blow may be caused due to the following factors:

1. Improper positioning of ground connections
2. The shape of the workpiece
3. Presence of gap at the joint
4. Large ferromagnetic mass close to the welding site
5. Wind or gas flow
6. Long arc length
7. Absence or inadequate electrode coating
8. High welding current
9. Use of DC supply

Mechanism of Arc Blow (Figure 3.26): Electromagnetic field is generated in a plane perpendicular to the direction of current flowing through a wire. Intensity of self-induced magnetic field,

$$H = \frac{i}{2\pi r}$$

due to flow of current depends upon the distance of point from the centre of wire (r) and magnitude of current (i). There can be two types of polarities, namely, like polarity and unlike polarity, as far as electromagnetic field due to current flow and interaction between them is concerned.

In the case of like polarity, the direction of flow of current is same in two conductors. Electromagnetic fields in case of like polarities repel each other while those of unlike polarities attract each other. Arc tends to deflect away from area where flux concentration exits. In practice, such kind of localization of electromagnetic fields and deflection of arc depend on the position of ground connection as they affect the direction of current flow and related field. Arc can blow towards or away from the earthling point, depending upon the orientation of electromagnetic field around the welding arc. Effect of ground connection on arc blow is called ground effect. Ground

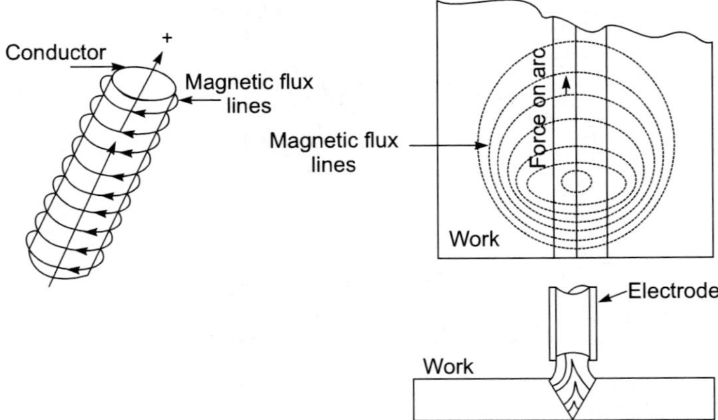

Figure 3.26 Mechanism of arc blow

effect may add or reduce the arc blow, depending upon the position of arc and ground connection. In general, ground effect causes the deflection of arc in the direction opposite to the ground connection. Arc blow occurring due to interactions between electromagnetic field around the arc and that of localized electromagnetic field near the edge of the plates, always tend to deflect the arc away from the edge of the plate. So, the ground connection in opposite side of the edge causing deflection helps to reduce the arc blow. Figure 3.27 shows backward arc blow.

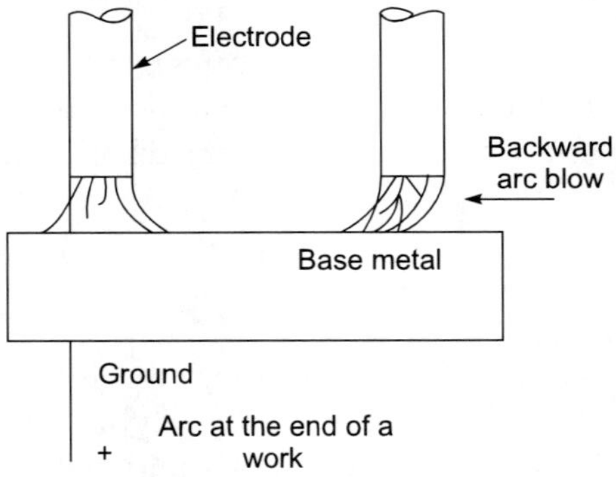

Figure 3.27 Backward arc blow

Reduction of the arc length reduces the extent of misplacement of molten metal. Adjust the ground connection as per the position of arc. Shift AC by neutralizing the arc blow occurring in each half and direct the tip in direction opposite to the arc blow.

Ways to Eliminate Arc Blow

To eliminate or at least to reduce the severity of an arc blow the following steps may be taken:

1. Preference should be given wherever possible to the use of AC current for welding.
2. Place ground connection as far away as possible from the joint.
3. Point the electrode tip in the direction of the arc blow.
4. Vary the angle between the electrode and the work to counteract the arc blow.
5. Use short arc length.
6. Use heavy coated electrode in shielded metal arc welding.
7. Do not weld very close to the heavy ferromagnetic mass.
8. Use low welding current consistent with good fusion.
9. Weld toward a heavy tack or a tab-out (Figure 3.28).

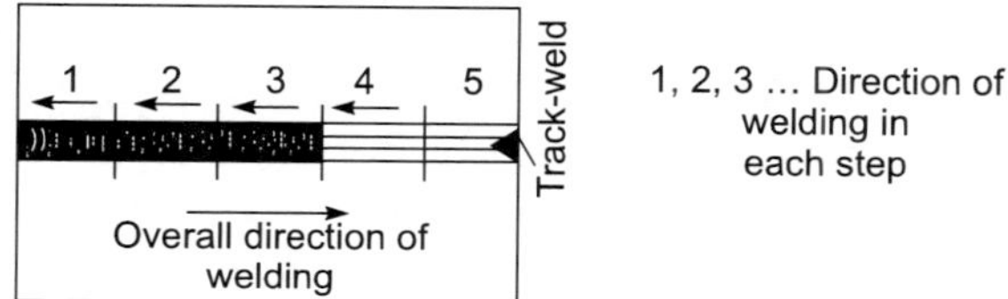

Figure 3.28 Back-step welding to minimize arc blow

10. Use the sequence of back-step welding.
11. Keep the ground cables around the work, so that current flows in a direction that will establish the magnetic field which will neutralize the magnetic field causing the arc blow.
12. Place the ground connections as far as possible from the weld.
13. Use appropriate ground connection for the vertical (Figure 3.29).

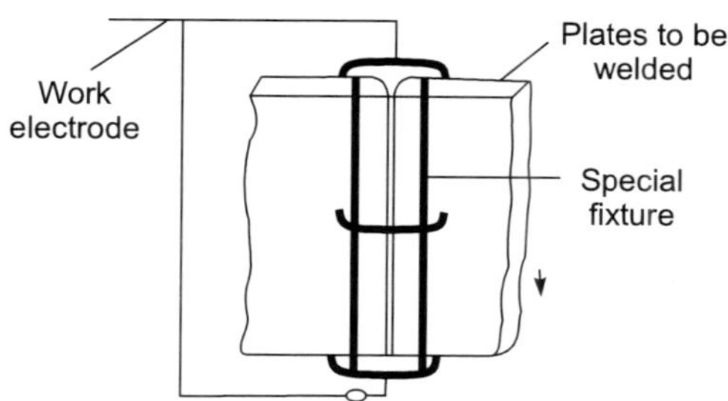

Figure 3.29 A special ground connection for vertical welding

3.1.5 Metal Transfer

Metal transfer refers to the transfer of molten metal from the tip of the electrode to the weld pool. It is of great importance for consumable electrode welding process, as it directly affects the control over handling of molten metal, slag and spattering. Metal transfer can be studied with motion pictures and by the analysis of short circuit oscillograms.

Four common modes of metal transfer are generally observed in consumable arc welding processes, they are:

1. Short-circuit metal transfer
2. Globular metal transfer
3. Spray metal transfer
4. Dip metal transfer
5. Pulse transfer

The arc welding related parameters affecting the mode of metal transfer for a given power setting (*V* and *I*) are shielding gas, composition of the electrode, diameter of the electrodes and extension of the electrodes.

Short-Circuit Transfer: This kind of metal transfer takes place, when the welding current is very low and arc gap is small. Under these welding conditions, molten metal droplet grows slowly at the tip of electrode and then as soon as drop touches the weld pool, short circuiting takes place. Due to narrow arc gap, molten drop does not attain a size big enough to fall down on its own due to gravitational force. On occurrence of short circuit, welding current flowing through the droplet to the weld pool increases abruptly, which in turn, results in excessive heat generation that makes the molten metal of droplet thinner (low surface tension). Touching of the molten metal drop to the weld pool leads to the transfer of molten metal into the pool by surface tension effect (Figure 3.30).

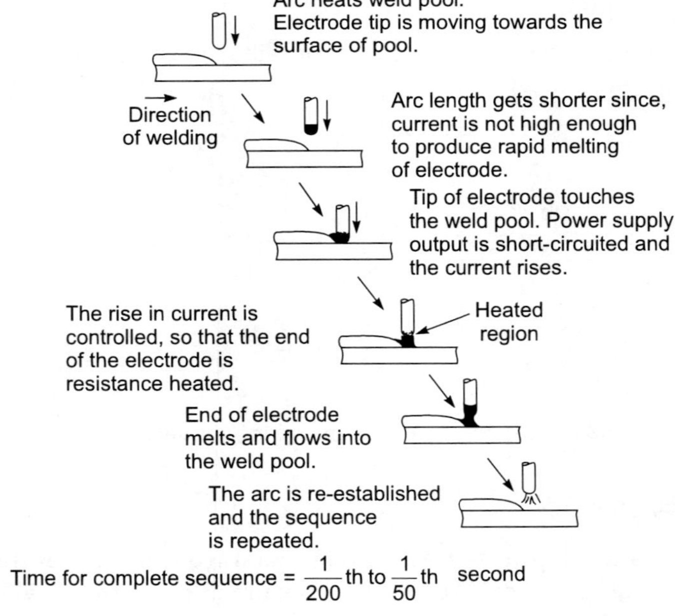

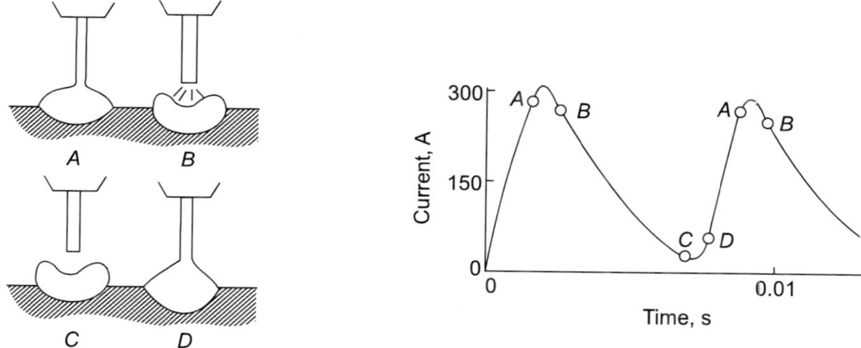

Figure 3.30 Short-circuit metal transfer

Once molten metal is transferred to the weld pool, an arc gap is established, which in turn, increases arc voltage abruptly. This increase in arc voltage (due to setting-up of the gap) re-ignites the arc and flow of current starts. This whole process is repeated during the welding (Figure 3.31).

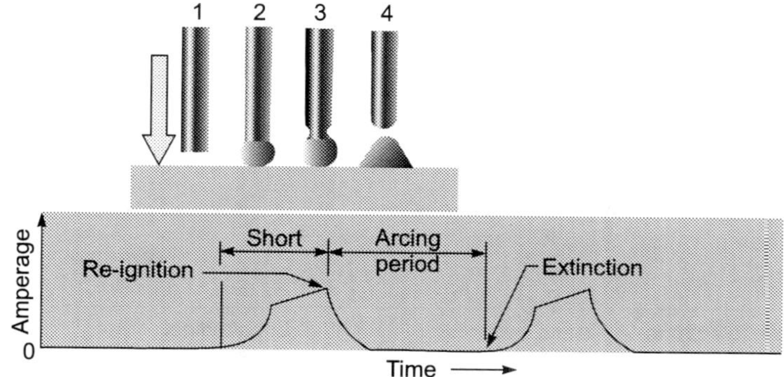

Figure 3.31 Current variation in short-circuiting metal transfer

Short-circuiting transfer of metal from the electrode tip to the molten weld pool has several advantages. Metal deposited in this way is less fluid and less penetrating than that formed with the spray transfer. It is easily handled by the welder in all positions, and it is particularly useful for joining thin materials. The spatter normally associated with short circuits is minimized by using electrical inductance to control the rate of current rise, when the wire and pool are in contact. As a result, the peak value of current at short circuit is relatively low. The average current is kept low by using small diameter electrodes.

With the proper adjustment of equipment, the rate of short circuiting is high (in the order of hundreds of shorts per second). There is less time available to melt the electrode, so the drops formed on the tip are very small. The drops are transferred to the weld by surface tension when the electrode tip and weld pool come in contact.

Globular Transfer: Globular metal transfer takes place when the welding current is low (but higher than that for short-circuit transfer) and arc gap is large enough. The molten metal droplet can grow slowly (at the tip of the electrode) with melting at the electrode tip. Drop continues to grow until gravitational force on the drop (due to weight of the drop) exceeds the surface tension force. As soon as the drop attains enough large size and gravitational force becomes more than the other force, such as surface tension force, drop detaches from the electrode tip and gets transferred to the weld pool. The transfer of molten metal drop normally occurs when it attains size larger than the electrode diameter. No short circuiting takes place in this mode of metal transfer (Figure 3.32).

Globular transfer takes place at lower welding currents than spray transfer. There is a transition current where the transfer changes to globular even when shielding gases using a high percentage of argon are used. When carbon dioxide (CO_2) is used as a shielding gas, the transfer is always globular. In globular transfer, a molten drop larger than the electrode diameter formed at the end of the electrode, moves to the outer edge of the electrode and falls into the molten puddle. Occasionally, a large drop will "short circuit" across the arc, causing the arc to extinguish momentarily, and then instantaneously reignite. As a result, the arc is somewhat erratic, spatter level is high, and penetration is shallow. Globular transfer is not suitable for out-of-positioning welding.

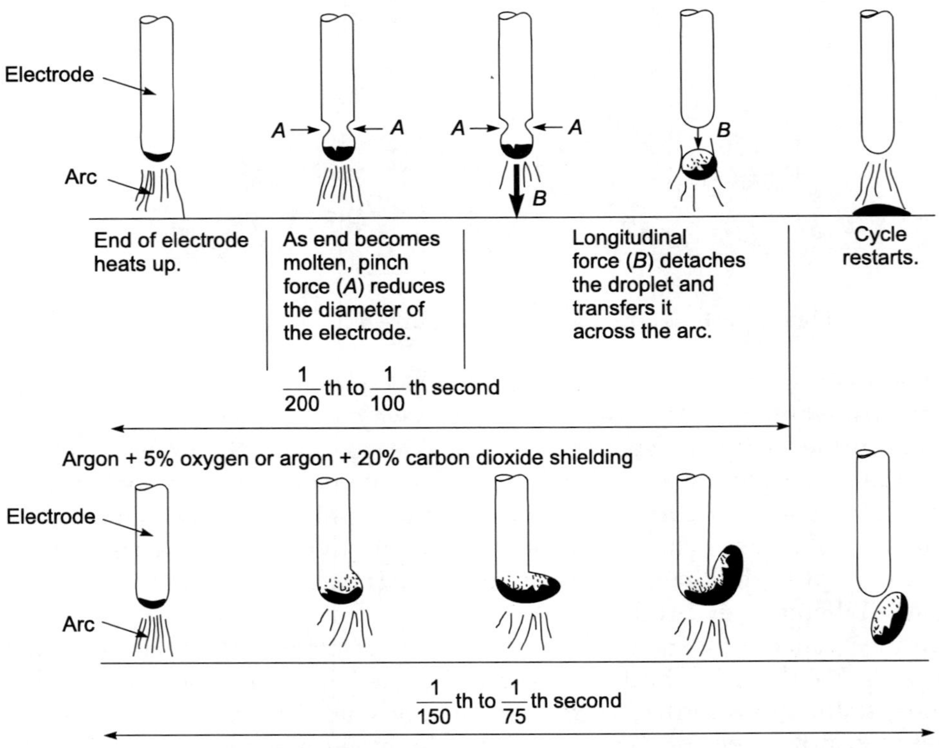

Figure 3.32 Metal spray

Spray Transfer: Another reason for the detachment of small droplets is that high welding current increases the temperature of arc zone, which in turn, lowers the surface tension force. Reduction in surface tension force decreases the resistance to detachment of drop from the electrode tip. Hence, transfer of molten metal from electrode tip appears similar to that of spray in line of axis of the electrode. This feature helps to direct the molten metal in proper place where it is required, like in odd position (Figure 3.33). The spray transfer is a high-current density process that rapidly deposits weld metal in droplets smaller than electrode diameter. They are propelled in a straight line from the center of the electrode. A shielding gas mixture of argon with 1% to 2% oxygen is used for welding mild and low alloy steel, and pure argon or argon-helium mixtures are used for welding aluminum, magnesium, copper, and nickel alloys. Welding current at which spray transfer takes place is relatively high and varies vary with the metal being welded, electrode diameter, and the shielding gas being used. Deposition rates are high and welding is usually limited to the flat or horizontal fillet position.

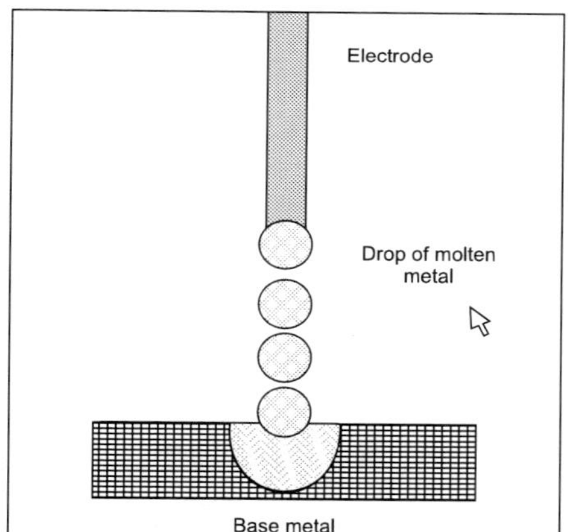

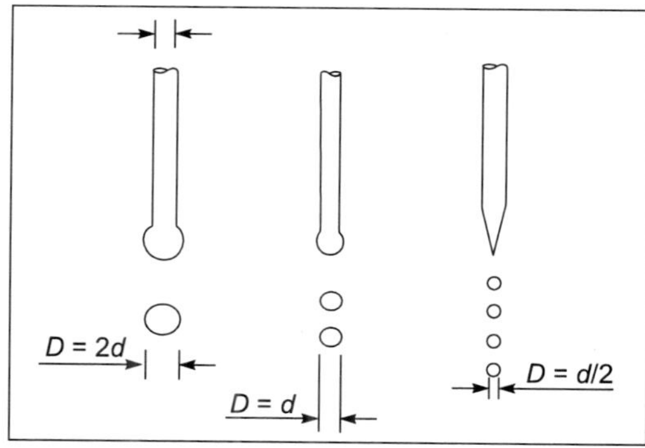

Figure 3.33 Spray mode

Effect of Polarity on Metal Transfer in Argon

Electrode Positive: At low welding currents in argon, liquid metal from the electrode is transferred in the form of drops having a diameter greater than that of the electrode. With electrode positive, the drop size is roughly inversely proportional to the current, and the drops are released at the rate of a few per second. With a sufficiently long arc to minimize short circuits, drop transfer is reasonably stable and is associated with relative absence of spatter.

Above a critical current level, however, the characteristics of this transfer change to the axial spray mode. In axial spray transfer, the tip of the electrode becomes pointed, and minute drops are transferred at the rate of hundreds per second. The current at which this occurs is called the transition current. Often, as in the case of steel, this change is abrupt.

The axial spray transfer is unique not only because of its good stability but also because of the absence of spatter. Furthermore, the drops are transferred in line with the electrode rather than along the shortest path between the electrode and workpiece. The metal, therefore, can be directed where needed, when making vertical or overhead welds.

The key to spray transfer is the pinch effect which automatically squeezes the drops off the electrode. This occurs as a result of the electromagnetic effects of the current, as illustrated in Figure 3.34.

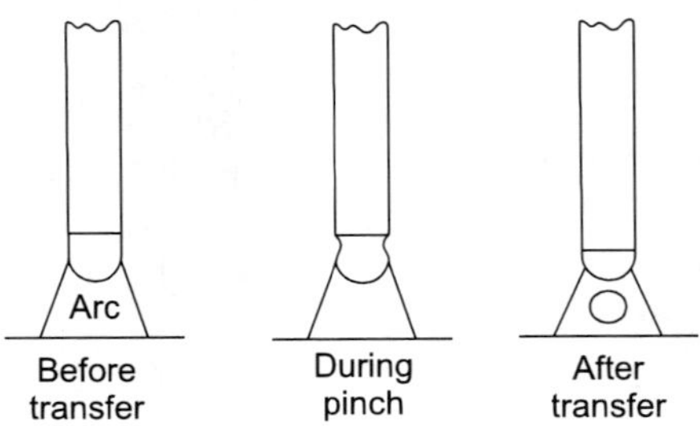

Figure 3.34 Individual drop formation sequence in spray transfer

The transition current depends on a number of variables, including the electrode composition, diameter, electrode extension, and the shielding gas composition. A great difference in transition current is found with various metal systems. Transition currents for various sizes of steel and aluminum electrodes are shown in Table 3.3.

TABLE 3.3 Approximate arc currents for transition from drop to spray metal transfer

Electrode Diameter		Transition Current*, A	
mm	In.	Steel Ar + 2% O_2	Aluminum Argon
0.75	0.030	155	90
0.90	0.035	170	95
1.15	0.045	220	120
1.6	0.062	275	170

* The transition current varies with electrode extension, alloy content, and shielding gas composition.

The transition current is approximately proportional to the diameter of the electrode, as shown in Figure 3.35. Transition current does not depend on current density, but is mildly dependent on the electrode extension. An increase in the extension allows a slight decrease in the current at which spray transfer develops.

In GMAW of steels, the spray arc mode is most often used with argon based shielding gas. Small additions of oxygen to the shielding gas lower the transition current slightly while CO_2 additions raise it. The transition current defines the lower limit of useful current for spray transfer.

At high welding current densities, a rotary arc mode takes place. With appropriate mixtures of shielding gases, wire feeding controls, and welding guns that perform well at high wire feed speeds, rotating mode GMAW can be used to deposit 7 kg/h or more steel weld metal. The useful upper current limit is the value where the rotational arc becomes unstable with loss of puddle control and high amounts of spatter.

The welding current at which axial spray disappears and rotational spray begins is proportional to the electrode diameter and varies inversely with electrode extension. The significance of these variables is illustrated in Figure 3.36. Spray transfer can also be achieved at average current levels below the transition current using pulsed welding current. One or more drops of filler metal are transferred at the frequency of the current pulses. This technique increases the useful operating range of a given electrode size.

Solid state power sources that simplify the set-up for pulsed power welding have increased the applications of pulsed-spray welding. The relatively low average current levels permit out-of-position welding of steel at relatively high deposition rates. It is also utilized for aluminum welding, with larger diameter electrodes and lower average currents and wire feed rates.

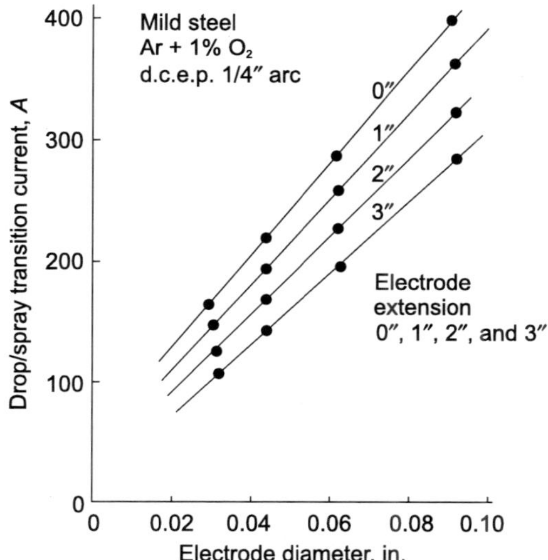

Figure 3.35 Influence of electrode diameter and extension on drop to spray transition current for mild steel

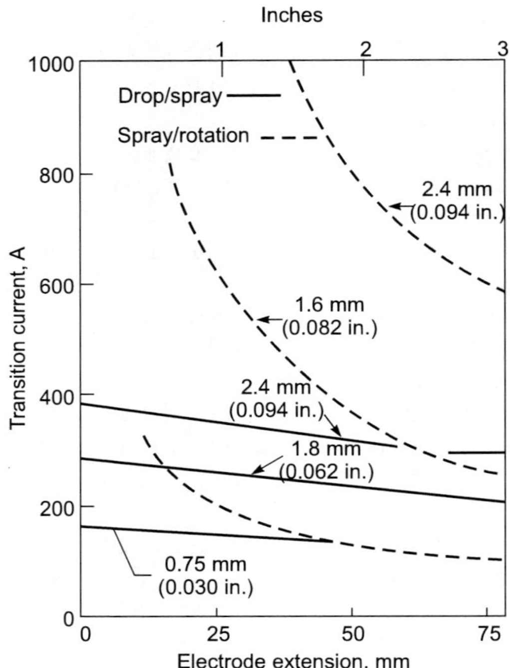

Figure 3.36 Effect of electrode extension and diameter on transition current of steel filler metals

Electrode Negative: The GMAW process is normally used with direct current electrode positive power. When the electrode is negative, the arc becomes unstable, and spatter is excessive. The drop size is large, and arc forces propel the drops away from the workpiece. This action appears to result from a low rate of electron emission from the negative electrode. If the thermionic properties of the electrode are enhanced by light coatings of alkali metal compounds, metal transfer is significantly improved. Although, the use of emissive coating allows spray transfer in GMAW with DCEN. Commercial filler metals are not generally available with such coatings.

Effect of Other Gases on Metal Transfer

Although helium is inert, it is unlike argon for shielding a welding arc because it does not usually produce an axial spray arc. Instead, the transfer is globular at all current levels and with both polarities. Helium-shielded arcs are useful, nevertheless, because they provide deep penetration. Spray transfer is produced in helium by mixing relatively small quantities of argon with it. Using dilute mixtures, the deep penetration is not adversely changed. Although 20 per cent argon in helium is sufficient to achieve these results, the normal commercial mixtures contain 25% argon as a safety factor. Argon-helium mixtures are used for welding non-ferrous materials, such as aluminum and copper. Generally, the thicker the material to be joined, higher is the percentage of helium in the shielding gas used.

Active gases, such as carbon dioxide and nitrogen, are much like helium in their effects on the arc. Spray transfer cannot be achieved without the treatment of wire surface. In addition, greater instabilities in the arc and chemical reactions between the gas and superheated metal drops cause considerable spatter. The difficulty with spatter can be minimized by welding with the buried arc technique. This technique is common when carbon dioxide is used to shield copper, and when nitrogen is mixed with argon to shield aluminum alloys.

To offset the harsh globular transfer and spatter associated with CO_2 shielding, argon may be added to stabilize the arc and improve metal transfer characteristics. Short-circuiting transfer is optimized using mixtures of 20 to 25% CO_2 in argon. Higher percentages of CO_2 are used for joining thick steel plates. Small amounts of oxygen (2 to 5%) or carbon dioxide (5 to 10%) are added to argon to stabilize the arc, alter the spray transition current, and improve wetting and the bead shape. These mixtures are commonly used for welding steel.

Dip Transfer: Dip type of metal transfer is observed when welding current is very low and feed rate is high. Under these welding conditions, electrode is high and it is short-circuited with weld pool, which leads to the melting of electrode and transfer of molten drop. Approach-wise dip transfer is similar to that of short-circuit metal transfer however, these two differ in respect of welding conditions that lead to these two types of metal transfers (Figure 3.37).

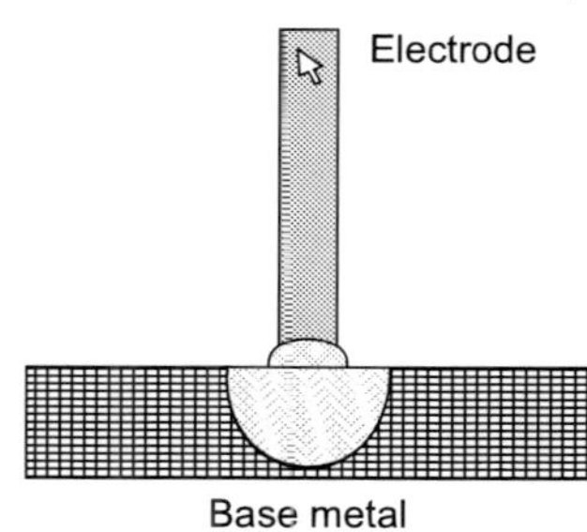

Figure 3.37 Dip transfer mode and sequence

Pulse Transfer: Pulse transfer is a mode of metal transfer, somewhat between spray and short-circuiting. The specific power source has two output levels: a steady background level, and a high output (peak) level. The later permits the transfer of metal across the arc. This peak output is controllable between high and low values up to several hundred cycles per second. The result of such a peak output produces a spray arc and a typical transition current (Figure 3.38).

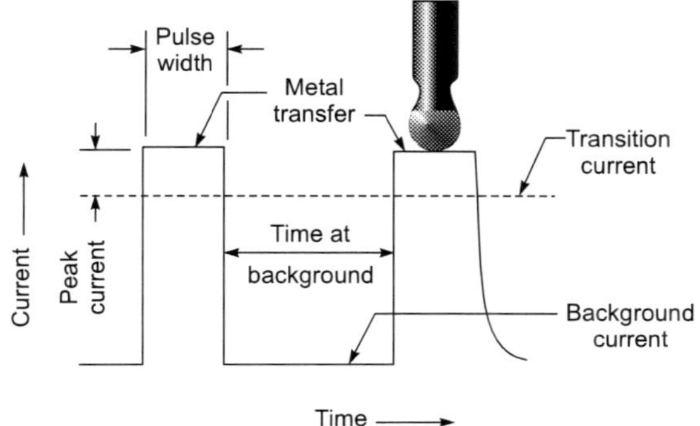

Figure 3.38 Pulse transfer

Pulsed current transfer is achieved by pulsing the welding current back and forth between the globular- and spray-transfer current ranges. To suppress globular transfer, the time period between consecutive pulses must be less than that required for transfer by the globular mode. The time period between pulses produced by positive half cycles from a 60 Hz power source is short enough to suppress globular transfer at all current levels. Conversely, the pulse duration is long enough to ensure that transfer by the spray mode will occur with an appropriate current in the spray transfer range. The pulsed current mode of transfer differs from normal spray transfer as metal transfer is interrupted between the current pulses.

Pulse shape and frequency may be varied over a wide range. Using microcomputer technology and solid state power sources, the pulse power can be coordinated with the electrode feed rate, such that a single drop of molten metal is transferred with each pulse. Both synergic and adaptive systems have been developed to employ pulsed power GMAW with relative ease on both ferrous and non-ferrous materials. A one-knob system can be used when proper relationship between the pulse rate and the feed rate is maintained.

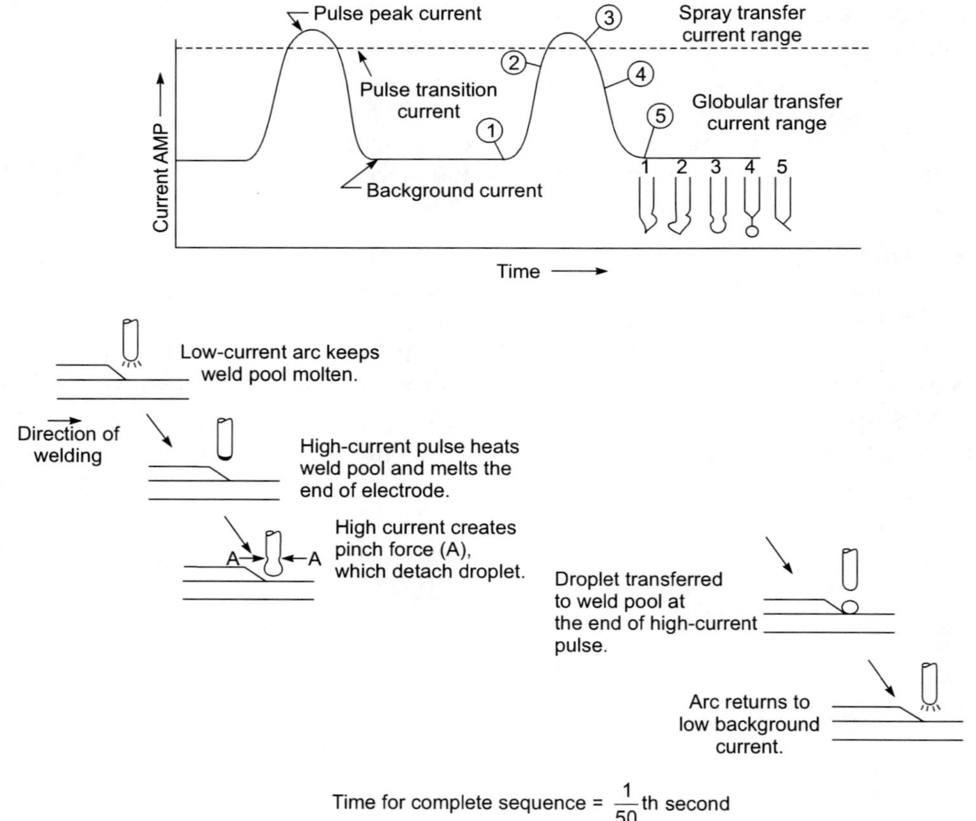

Figure 3.39 Output current wave form of a pulsed-current power source and metal transfer sequence

There are many ways of generating a modulated DC current for pulsed current transfer. The current is comprised of a background and a pulse current. Pulsating currents as illustrated in Figure 3.39, also shows the metal transfer sequence. With pulsed power, currents and deposition rates can be decreased to permit welding of sheet as thin as 1.0 mm (0.04 in.) or even thinner with mechanized welding.

3.1.6 Arc length

Arc length is the distance between the tip of the electrode and the workpiece surface upon which the molten globules are deposited as shown in Figure 3.40. Short arc deposits more weld metal with advantage such as high penetration, maximum strength, slight overlay, minimum porosity, maximum ductility and better behavior of alloy steel.

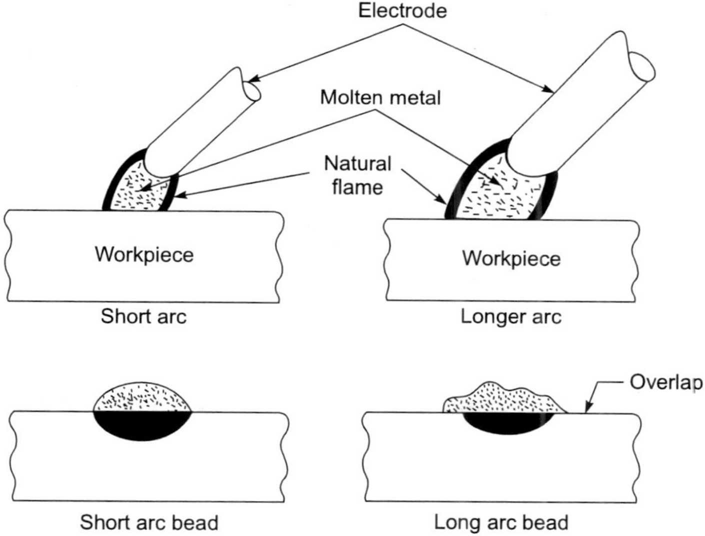

Figure 3.40 Configuration of weld bead

Important variable of any welding is the length of arc globules of molten metal, should not be allowed to come in contact with air for a longer time (period), as oxygen has adverse effect upon the weld properties. If the arc is longer than normal, arc flame gases can no longer protect the arc stream from atmosphere metal and this will produce oxide and nitride, resulting into a weak and brittle weld. A short arc gives little chance for air to come into contact with molten metal whereas in a long arc, time of contact of molten metal with air is longer. In such a way, longer arc gives a wide shallow bead and enough heat is wasted into the air and penetration is poor and more spattering of metal is produced, as shown in Figure 3.40. Although, electrode may be speedily consumed with long arc but effective weld production is low and inefficient. Similarly, high voltage gives a longer arc with spatters.

3.1.7 Melting Rate

Heat in an arc is generated by electrical reactions at the anode and cathode regions and within the plasma. Portions of this energy melt the electrode, which supports the arc unless it is adequately cooled. The greatest portion of the arc energy used for melting is obtained from cathode or anode reactions, but substantially more heat can be released from the cathode. Furthermore, when the electrode is the cathode (negative) terminal, good control of the energy release is possible. Little can be done to modify the release of energy at the anode (positive) terminal. It is related mostly to the current magnitude and less to composition and other factors.

Most commercial metals and their alloys form what is called a cold cathode; its area is rather small, but great quantities of energy are generated to release the electrons needed to support an arc. However, metals having very high melting points easily supply electrons to sustain the arc at high temperature. These metals are called thermionic, included in this category are molybdenum and tungsten.

In the cold cathode, low melting point metals can be made to supply electrons more easily by coating them with compounds that reduce the surface work function. They then become thermionic at lower temperatures. In some cases, the mechanism of electron release depends on the surface oxide emission. In others, it is due to the formation of an atomic film of an alkali metal on the surface. The degree of change is regulated by the selection of the oxides or metals and their quantities. In general, the change from cold cathode to thermionic emission is accompanied by lowering of the heating energy, and therefore, a reduction in the melting rate. Any improvement in the mode of metal transfer with DCEN or in arc stability with AC is associated with reduction in melting rate.

The arc plasma may or may not supply radiant heat to melt the electrode. When it does, the effect is relatively small in comparison with other sources. Therefore, the variables affecting the plasma, such as the shielding gas, flux, or the arc length, do not directly affect the melting rate. If a change in melting rate is demonstrated, it is more likely caused by another factor, such as a change in current or cathode heating.

In addition to the energy supplied by the welding arc, electrical resistance heating of the electrode by welding current affects the melting rate of the electrode. This effect is particularly significant in welding processes that use small diameter electrodes. This heating is caused by the resistance of the electrode to the flow of current. Electrical resistance is greater with small electrode diameters, long electrode extensions, and low-conductivity metals and alloys.

In consumable arc welding processes, weld metal deposition rate is governed by the rate at which electrode is melted during welding. Melting of the electrode needs sensible and latent heat, which is supplied by the electrical reactions, i.e., heat generated at anode (IV_a), cathode (IV_c), and plasma zone (IV_p)

Accordingly, melting rate of an electrode of consumable arc welding processes is given by the following equation.

$$\text{Melting rate} = aI + bLI^2$$

where, a and b are constants (independent of electrode extension L and welding current I). Value of constant "a" depends on ionization potential of electrode material (ability to emit the charge carriers), polarity, and compositions of electrode and anode/cathode voltage drops. While another constant b accounts for electrical resistance of electrode (which in turn, depends on electrode diameters and resistivity of electrode metal), L is the electrode extension and I is the welding current. Melting rate equation suggests that first factor (aI) accounts electrode melting due to heat generated by anode/cathode reactions and second factor (bLI^2) considers the melting rate owing to heat generated by electrical resistance heating. Difference in values of constants a and b and welding parameters leads to the variations in melting rate of electrode in different welding processes. To increase the melting rate, welding current for a specific welding process can be increased up to a limit. The upper limit of welding current is influenced by two factors: (a) extent of overheating of electrode, caused by electrical resistance heating and related thermal degradation of the electrode, and (b) required mode of metal transfer for smooth deposition of weld metal with minimum spatter. Lower level of welding current is the level at which short-circuit metal transfer starts and upper level of current is limited by the appearance of rotational spray transfer. For a given electrode material and diameter, upper limit of current in case of SMAW is dictated by thermal damage to tungsten electrode. Lower limit of current is limited by arc stability and other factors. Besides arc stability, lower level of current is generally determined by other requirements, such as penetration, proper placement of the weld metal, control over the weld pool, especially in vertical and overhead welding positions and poor accessibility. Typical constant values are given in Table 3.4.

TABLE 3.4 Relative magnitude of heating coefficients in the melting rate of 1.6 mm diameter wire diameter

Metal	a	b
	kg/h A	**kg/h, A^2. mm**
Aluminum (DCEP)	5.4×10^{-3}	4.4×10^{-6}
Mild steel (DCEP)	8.6×10^{-3}	2.5×10^{-5}
Mild steel (DCEN)	1.8×10^{-2}	2.5×10^{-5}

a = kg/hour, Amp. b = kg/hour, Amp.^{2} mm.

Rate of Heat Generation

The quantity of heat required to melt a given volume of metal is sum of the heat to raise temperature of the solid metal to its melting point, which depends on the metal's volumetric heat and heat to transform the metal from solid to liquid phase at the melting point, which depends on the metal's heat of fusion. To a reasonable approximation, this quantity of heat can be estimated by an empirical relationship given as

$$U_m = KT_m^2$$

where, U_m is the unit energy for melting, i.e., the quantity of heat required to melt a unit volume of the metal starting from room temperature (J/mm^3), T_m is the melting point of the metal on an absolute temperature scale k and K is constant, whose value is 3.33×10^{-6} when the kelvin scale is used.

Not all the input energy is used to melt the weld. There are two heat transfer mechanisms at work, both of which reduce the amount of heat available to the welding process.

1. The first mechanism is the transfer of heat between the heat source and the surface of work. This process has a certain heat transfer coefficient (f_1), which is defined as the ratio of actual heat received by the workpiece divided by the total heat generated at the source.
2. The second mechanism involves the conduction of heat away from the weld area to be dissipated through the work metal, so that a portion of heat transferred to the surface is available for melting. This melting efficiency (f_2) is the proportion of heat received at the work surface that can be used for melting. This combined effect of these two efficiencies is to reduce the heat energy available for welding as follows:

$$H_w = (f_1)(f_2)H$$

where, H_w = heat available for welding

f_1 = heat tranfer coefficent

f_2 = melting efficiency

Also,

$$H_w = U_m V$$

where, V is the volume to be melted in mm^3

Rate of heat generation is given by

$$H_{wr} = U_m(VWR)$$

VWR is the volume rate of metal welded in mm^3/s

$$H_{wr} = (f_1)(f_2)H = U_m A_w v$$

A_w = area of weld in mm^2

and v = welding speed in mm/s

Worked Examples

Example 1: Derive an expression for arc efficiency.

Solution: From the description of the characteristics of different parts of a welding arc, it is possible to determine arc efficiency, mathematical treatment of which follows:

We have,

$$\text{Arc voltage} = V_c + V_p + V_a$$

where,

V_c is cathode voltage drop
V_p is voltage drop in plasma column
V_a is anode voltage drop

Now,

$$V_c I = \phi I + q_e$$

where, ϕ is the work function at temperature T degrees kelvin

Also,
$$\phi = \phi_0 + \frac{3}{2}kT$$

where, ϕ_0 = work function at 0 degree kelvin
k = Boltzmann constant = 8.620×10^{-5} eV/K = 1.381×10^{-23} joule/K
T = temperature in degrees kelvin.

q_e is heat going to rod electrode.

Now, the total heat energy developed at the anode, q_a is given by the sum of energy received through the electrons and the energy gained by passing through the anode drop zone, i.e.,

$$q_a = \left(\phi_0 + \frac{3}{2}kT\right)I + V_a I = \left(\phi_0 + \frac{3}{2}kT + V_a\right)I$$

Heat developed in arc column, $q_p = V_p I$, a part of which is transferred to the anode. Let that be equal to mq_p.

Therefore, rate of heat input to the anode can be expressed as

$$h_a = q_a + mq_p$$

$$h_a = \left(\phi_0 + \frac{3}{2}kT + V_a\right)I + mq_p$$

$$\text{Total heat developed} = VI$$

Therefore,

$$h_a = VI - q_e - (1-m)q_p$$

$$h_a = VI - (q_e + (1-m)q_p)$$

$$h_a = VI - (q_e + (1-m)V_p I)$$

$$\text{Arc efficiency}, \eta = \frac{\text{Heat developed at anode}}{\text{Total heat developed}}$$

$$\eta = \frac{h_a}{VI} = \frac{VI - (q_e + (1-m)V_p I)}{VI} = 1 - \frac{(q_e + (1-m)V_p I}{VI}$$

Example 2: Find the arc efficiency for GTAW process if the welding current is 150 amperes and the arc voltage 20 volts. Assume a cathode drop of 8 volts and anode drop of 3 volts with 30% of the arc column energy being transferred to the anode. Take arc temperature as 15,000 K. Work function, ϕ_0 for tungsten = 4.5 eV.

Solution: $V_c I = \phi I + q_e$

where, ϕ is the work function at temperature T degrees kelvin

Also, $$\phi = \phi_0 + \frac{3}{2}kT$$

where, ϕ_0 = work function at 0 degree kelvin

k = Boltzmann constant = 8.620×10^{-5} eV/K = 1.381×10^{-23} joule/K

T = temperature in degrees kelvin.

q_e is the heat going to rod electrode.

$$V_c I = \phi I + q_e$$

$$8 \times 150 I = \phi \times 150 + q_e \tag{1}$$

We have,

$$\text{Arc voltage} = V_c + V_p + V_a$$

where,

V_c is cathode voltage drop

V_p is voltage drop in plasma column

V_a is anode voltage drop

$$\text{Arc voltage} = V_c + V_p + V_a$$

$$20 = 8 + V_p + 3$$

$$V_p = 9 \text{ volts}$$

$$q_a = \left(\phi_0 + \frac{3}{2}kT\right)I + V_a I = \left(\phi_0 + \frac{3}{2}kT + V_a\right)I$$

$$q_a = 8 \times 150 = \left(4.5 + \frac{3}{2} \times 15{,}000\right) \times 150 + q_e$$

Solving we get

$$q_e = 234.075 \text{ joules}$$

$$\eta = \frac{h_a}{VI} = \frac{VI - (q_e + (1-m)V_p I)}{VI} = 1 - \frac{234.075 + (1-0.3)9 \times 150}{3000} = 0.607 = 60.7\%$$

Example 3: A heat source transfers 3000 W to the surface of a metal part. The heat impinges on the surface in a circular area, with intensities varying inside the circle. The distribution is as follows: 70% of power is transferred within a circle of diameter = 5 mm and 90% is transferred within a concentric circle of diameter = 12 mm. What are the power densities in (a) the 5 mm diameter inner circle and (b) the 12 mm diameter ring that lies around the inner circle?

Solution:

(a) The inner circle has an area,

$$A = \frac{\pi \times 5^2}{4} = 19.63 \text{ mm}^2$$

The power inside this area $P = 0.70 \times 3000 = 2100$ W

$$\text{Thus power density } = \frac{2100}{19.63} = 107 \text{ W/mm}^2$$

(b) The area of the ring outside the inner circle,

$$A = \frac{\pi \times (12^2 - 5^2)}{4} = 93.4 \text{ mm}^2$$

The power inside this area $P = 0.90 \times 3000 - 2100 = 600$ W

$$\text{Thus power density } = \frac{600}{93.4} = 6.4 \text{ W/mm}^2$$

The power density seems high enough for melting in the inner circle, but probably not sufficient in the ring that lies outside this inner circle.

Example 4: The power source in a particular welding setup generates 3500 W that can be transferred to the work surface with a heat transfer factor = 0.7. The metal to be welded is low carbon steel, whose melting temperature is 1760°C. The melting factor in the operation is 0.5. A continuous fillet weld is to be made with cross-sectional area 20 mm^2. Determine the travel speed at which welding operation can be accomplished.

Solution: Let us find the unit energy required to melt the metal (u)

$$u = 3.33 \times 10^{-6} \times 1760^2 = 10.3 \text{ joules/mm}^3$$

$$\text{Travel velocity} = \frac{\text{Heat input} \times \text{heat transfer factor} \times \text{melting factor}}{\text{Unit energy} \times \text{area}}$$

$$\text{Travel velocity} = \frac{3000 \times 0.7 \times 0.5}{10.3 \times 20} = 5.95 \text{ mm/s}$$

Example 5: Make the calculations and plot on linearly scaled axes the relationship for unit melting energy as a function of temperature. Use temperatures as follows to construct the plot: 200°C, 400°C, 600°C, 800°C, 1000°C, 1200°C, 1400°C, 1600°C, 1800°C and 2000°C. On the plot, mark the positions of some of the welding metals.

Solution: $U_m = 3.33 \times 10^{-6}\, T_m^2$, the plot is based on the following calculated values:

For $T_m = 200°\text{C} = (200 + 273) = 473°\text{K}: U_m = 3.33 \times 10^{-6} \times 473^2 = 0.75 \text{ J/mm}^3$

For $T_m = 400°\text{C} = (400 + 273) = 673°\text{K}: U_m = 3.33 \times 10^{-6} \times 673^2 = 1.51 \text{ J/mm}^3$

For $T_m = 600°\text{C} = (600 + 273) = 873°\text{K}: U_m = 3.33 \times 10^{-6} \times 873^2 = 2.54 \text{ J/mm}^3$

For $T_m = 800°\text{C} = (800 + 273) = 1073°\text{K}: U_m = 3.33 \times 10^{-6} \times 1073^2 = 3.83 \text{ J/mm}^3$

For $T_m = 1000°\text{C} = (1000 + 273) = 1273°\text{K}: U_m = 3.33 \times 10^{-6} \times 1273^2 = 5.40 \text{ J/mm}^3$

For $T_m = 1200°\text{C} = (1200 + 273) = 1473°\text{K}: U_m = 3.33 \times 10^{-6} \times 1473^2 = 7.23 \text{ J/mm}^3$

For $T_m = 1400°\text{C} = (1400 + 273) = 1673°\text{K}: U_m = 3.33 \times 10^{-6} \times 1673^2 = 9.32 \text{ J/mm}^3$

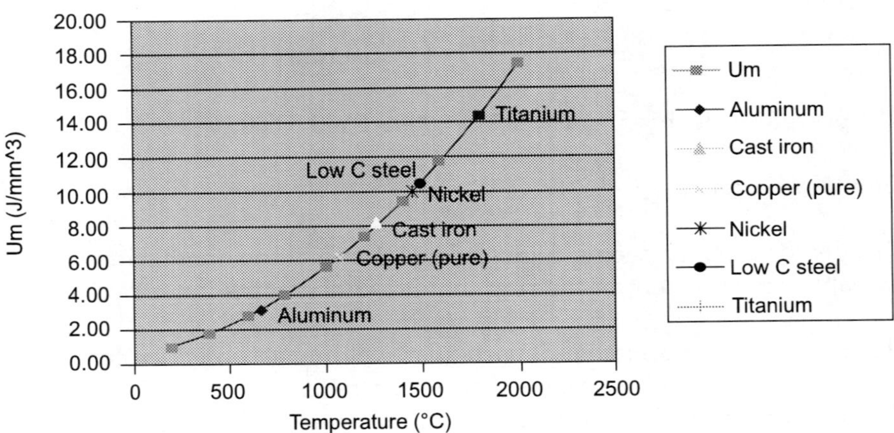

Figure 3.41 Melting energy vs melting temperature

Example 6: A fillet weld has a cross-sectional area of 25 mm^2 and is 300 mm long. (a) What quantity of heat (in joules) is required to accomplish the weld, if the metal to be welded is low carbon steel? (b) How much heat must be generated at the welding source, if the heat transfer factor is 0.75 and the melting factor = 0.63?

Solution: $U_m = 3.33 \times 10^{-6}\, T_m^2$

For low carbon steel, melting point is 1760 K

$$U_m = 3.33 \times 10^{-6}\, T_m^2 = 3.33 \times 10^{-6}\, 1760^2 = 10.32\ \text{J/mm}^3$$

Volume of metal melted,

$$V = 25 \times 300 = 7500\ \text{mm}^3$$

$$\text{Heat input} = 10.32 \times 7500 = 77{,}360 \text{ joules at weld}$$

Given heat transfer factor = 0.75 and melting factor = 0.63

$$\text{Heat input} = \frac{10.32 \times 7500}{0.75 \times 0.63} = 163{,}700 \text{ joules at source}$$

Example 7: A U-groove weld is used to butt weld 2 pieces of 7 mm thick titanium plate. The U-groove is prepared using a milling cutter so the radius of the groove is 3 mm. During welding, the penetration of the weld causes an additional 1.5 mm of material to be melted. The final cross-sectional area of the weld can be approximated by a semicircle with a radius of 4.5 mm. The length of the weld is 200 mm. The melting factor of the setup is 0.57 and the heat transfer factor is 0.86. (a) What is the quantity of heat (in joules) required to melt the volume of metal in this weld (filler metal plus base metal)? Assume the resulting top surface of the weld bead is flush with the top surface of the plates. (b) What is the required heat generated at the welding source?

Solution: Melting temperature for titanium is 2070 K

$$U_m = 3.33 \times 10^{-6}\, T_m^2$$

$$U_m = 3.33 \times 10^{-6}\, T_m^2 = 3.33 \times 10^{-6}\, 2070^2 = 14.29\ \text{J/mm}^3$$

$$A_{\text{weld}} = \pi \times \frac{r^2}{2} = \pi \times \frac{4.5^2}{2} = 31.8\ \text{mm}^2$$

Volume of metal melted,

$$V = 31.8 \times 200 = 6360\ \text{mm}^3$$

$$\text{Heat input} = 14.29 \times 6360 = 90{,}770 \text{ joules at weld}$$

Given heat transfer factor = 0.75 and melting factor = 0.63

$$\text{Heat input} = \frac{90770}{0.86 \times 0.57} = 182{,}200 \text{ joules at source}$$

Example 8: In a controlled experiment, it takes 3700 joules to melt the amount of metal that is in a weld bead with a cross-sectional area of 6.0 mm^2 that is 150 mm long. (a) Using Table 3.5, what is the most likely metal? (b) If the heat transfer factor is 0.85 and the melting factor is 0.55 for a welding process, how much heat must be generated at the welding source to accomplish the weld?

Solution: Volume of metal melted,

$$V = 6 \times 150 = 900 \text{ mm}^3$$

$$U_m = \frac{3700}{900} = 4.111 \text{ J/mm}^3$$

$$U_m = 3.33 \times 10^{-6} T_m^2$$

$$4.111 = 3.33 \times 10^{-6} T_m^2$$

Solving we get

$$T_m = 1111 \text{ K}$$

From the table, metal with the closest melting point to 1111 K is bronze.

Given heat transfer factor = 0.85 and melting factor = 0.55

$$\text{Heat input} = \frac{3700}{0.85 \times 0.55} = 7914 \text{ joules at source}$$

TABLE 3.5 Melting temperature of selected materials

Metal	Melting Temperature		Metal	Melting Temperature	
	°K[a]	°R[b]		°K[a]	°R[b]
Aluminum alloys	930	1680	Steels		
Cast iron	1530	2760	Low carbon	1760	3160
Copper and alloys			Medium carbon	1700	3060
Pure	1350	2440	High carbon	1650	2960
Brass, navy	1160	2090	Low alloy	1700	3060
Bronze (90 Cu–10 Sn)	1120	2010	Stainless steels		
Inconel	1660	3000	Austenitic	1670	3010
Magnesium	940	1700	Martensitic	1700	3060
Nickel	1720	3110	Titanium	2070	3730

[a]Kelvin scale = Centigrade (Celsius) temperature + 273.
[b]Rankine scale = Fahrenheit temperature + 460.

Example 9: A fillet weld is used to join 2 medium carbon steel plates, each having a thickness of 5 mm. The plates are joined at a 90° angle using an inside fillet corner joint. The velocity of the welding head is 6 mm/sec. Assume the cross section of the weld bead approximates a right isosceles triangle with a leg length of 4.5 mm, the heat transfer factor is 0.80, and the melting factor is 0.58. Determine the rate of heat generation required at the welding source to accomplish the weld.

Solution: $A_{\text{weld}} = \dfrac{bh}{2} = \dfrac{4.5^2}{2} = 10.125 \text{ mm}^2$

$$U_m = 3.33 \times 10^{-6}\, T_m^2$$

From Table 3.5, melting point for medium carbon steel is 1700 K

$$U_m = 3.33 \times 10^{-6}\, T_m^2 = 3.33 \times 10^{-6}\, 1700^2 = 9.62 \text{ J/mm}^3$$

Given heat transfer factor = 0.8 and melting factor = 0.58

$$\text{Rate of heat generation} = \frac{9.62 \times 10.125 \times 5}{0.8 \times 0.58} = 1260 \text{ joules/sec} = 1260 \text{ watts} = 1.26 \text{ kW}$$

Example 10: The welding power generated in a particular arc welding operation = 3000 W. This is transferred to the work surface with a heat transfer factor = 0.9. The metal to be welded is copper whose melting point is given in Table 3.5. Assume that the melting factor = 0.25. A continuous fillet weld is to be made with a cross-sectional area = 15.0 mm^2. Determine the travel speed at which the welding operation can be accomplished.

Solution: From Table 3.5, melting temperature for copper is 1350 K

$$U_m = 3.33 \times 10^{-6}\, T_m^2 = 3.33 \times 10^{-6}\, 1350^2 = 6.07 \text{ J/mm}^3$$

$$H_{wr} = (f_1)(f_2)H = U_m A_w v$$

$$0.9 \times 0.25 \times 3000 = 6.07 \times 15 \times v$$

Solving we get, $v = 7.4$ mm/sec

Review Questions

1. Define welding arc.
2. Explain the process of arc initiation.
3. Explain different electron emission mechanisms.
4. Explain various zones in the arc gap.
5. What are the main modes of metal transfer?
6. Explain globular and spray metal transfer.
7. What are the forces involved in metal transfer?
8. Define arc efficiency.
9. What are the advantages of pulse transfer?
10. Explain how melting rate is calculated.

Power Sources for Arc Welding

4.0 INTRODUCTION

The main role of power source is to deliver controllable current at the desired voltage according to demands of welding processes. Each welding process needs a specific kind of control over current, voltage to satisfy the transient operating condition experienced by welding arc.

Power is needed to supply the current that supports the arc column for fusion welding. These are available in a wide variety of types and sizes. Each type of power source has fundamental electrical differences that best suit particular processes. A welding power source:

- must deliver exact amount of electric current precisely at the right time to the welding arc; and
- must meet changing arc load and environmental conditions instantly.

4.1 REQUIREMENT FOR AN ARC WELDING POWER SOURCE

For striking a welding arc, it is essential to have the desired current in the welding circuit. Depending upon the type of current required (AC or DC), it may be obtained from the following sources:

- A welding transformer or a frequency changer (for AC welding).
- A DC generator or a rectifier (for DC welding).

The requirements which every welding power source is required to satisfy are:

1. The no load or open circuit (OCV) should be high enough to strike an arc, but not so high as to endanger the welding option. For example, for a 2000 A transformer, the OCV should not be over 90 volts, though for lower currents say upto 600 A, OCV may be upto 100 volts.

2. With direct current, the OCV must at least be 30 to 35 volts, while with alternating current, it should not be lower than 50 to 55 volts. An open arc welding can be sustained at 18 to 40 volts.
3. The minimum voltage for arc ignition is 25 volts, which must be attained within 0.05 seconds for maintaining a stable arc.
4. The power of welding source should be sufficient to provide the required welding current.
5. It should be possible to tap the desired current within the available limits, either in infinite steps (stepless regulation) or in a large number of discrete steps (stepped regulation).
6. The voltage of the power source should adjust itself rapidly with the charge in arc length.
7. The short-circuit current must be within the limits of safety for power source windings.
8. It should have output characteristics which match the arc characteristics.
9. It should incorporate a low voltage supply for the operation of auxiliary units.

4.2 FACTORS AFFECTING THE SELECTION OF WELDING POWER SOURCE

- Type of current to be used
- Welding process
- Welding consumable
- Requirement of open circuit voltage (OCV)
- The OCV normally ranges from 70 to 90 volts for welding transformers while in case of rectifiers, it is 50 to 80 volts.

4.3 CLASSIFICATION OF WELDING POWER SOURCE

Welding power sources can be classified based on the following parameters:

- Type of current: AC, DC, AC and DC
- Cooling medium: Air, water, oil cooled
- Cooling system: Forced or natural
- Static characteristics: Constant current/constant voltage

Detailed classification for power sources is given in Table 4.1.

TABLE 4.1 Classification system for power sources

Method of Classification	Features			
1. Type of basic design	Rotating type		Static type	
	Motor-driven generator	Engine-driven generator	Transformer	Rectifier/inverter
2. Method of current control	Field control, variable reactor, variable coupling, tap changer switch, saturable reactor, electronic control with feedback (thyristor control, transistor control, IGBT control, etc.)			
3. Type of output current	Direct current (DC) Alternating current (AC) Combination of both (AC and DC).			
4. Type of static characteristics (V-A curve)	Constant current type (CC) Constant voltage type (CV) Combination of both (CC and CV) type.			
5. Rated current and duty cycle	Capability of the power source to deliver current at specified duty cycle, say 60% duty cycle, 100% duty cycle, and range of welding current.			
6. Suitability for the process	Capability of the power source to be used for manual metal arc welding (MMAW), gas metal arc welding (MIG/CO_2), gas tungsten arc welding (GTAW or TIG), submerged arc welding (SAW), etc.			
7. Suitability for no. of operators	Capability of the power source to be used simultaneously by a single operator, by double operators or multioperators.			
8. Type of cooling method	Natural air-cooled (NA) Forced air-cooled (FA) Oil-cooled natural (ON) Oil-cooled forced (OF).			
9. Type of insulation	Class A, Class B, Class E, Class F, and Class H, depending on the max. allowable temperature rise over a specified ambient temperature (usually 40°C).			

4.4 CHARACTERISTICS OF WELDING POWER SOURCE

Basic characteristics of a welding power source are:

- Open circuit voltage
- Power factor
- Static characteristics

- Dynamic characteristics
- Current rating and duty cycle
- Class of insulation

Welding transformers, rectifiers and DC generators are used in job shops. Engine-coupled AC generators/DC generators are used at sites where the supply is not available. Normally, rectifiers and transformers are preferred because of less noise, higher efficiency and lower maintenance as compared to generators.

4.4.1 Open Circuit Voltage (OCV)

OCV is potential difference between the two terminals of power source when there is no load. Setting of correct OCV is important for the stability of welding arc, especially when AC is used. Too high OCV may cause electric shock. OCV of 50 to 100 V is used depending upon the type of metal, composition of electrode coating, type of welding current and shielding gas. Base metal of low ionization potential needs lower OCV than that of high ionization potential metal.

4.4.2 Power Factor (PF)

Power factor is the ratio of actual power (kW) used to produce the rated load (which gets registered on the power meter) and apparent power drawn from the supply line (KVA). It is always desired to have high PF. Lower power factor means unnecessary wastage of power and less utilization of power. Welding transformer with low power factor is used intentionally to increase the stability of AC welding arc. This is done by making current and voltage out of phase. High PF reduces the reactive power in a system, which in turn, decreases power consumption and thus, power costs. High PF gives improved voltage quality. Higher PF gives fewer voltage drops, smaller cable cross section and smaller transmission losses.

4.4.3 Static Characteristics of Power Supply (SCPS)

Shows the variation in voltage and current when power source is connected to pure resistive load. This variation is of three types, namely, constant current, constant voltage and rising voltage. In CCPS, variation in welding current with arc voltage (due to fluctuations in arc length) is very small and therefore, welding current remains more or less constant, in spite of fluctuations in arc voltage/length.

Constant Current Power Supply: This type of power sources are suitable for all welding processes, where large fluctuation arc length is observed like in manual metal arc welding (MMAW) processes. Relationship between arc voltage and welding current is shown in Figure 4.1.

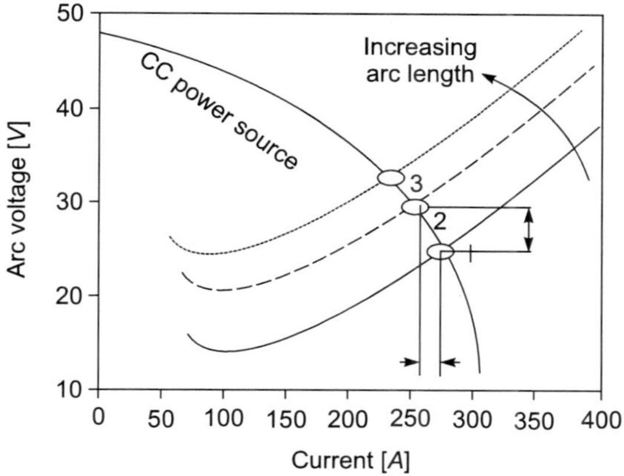

Figure 4.1 Relationship between arc voltage and welding current

The volt-ampere output curves for constant current power source are called dropper because of the negative slope of the curves. With a change in arc voltage, the change in current is small. Therefore, in a consumable electrode welding process, electrode melting rate would remain fairly constant with a change in arc length. Limited short-circuit current avoids damage to power source when consumable electrode gets stubbed to workpiece. Non-consumable tungsten electrode, when touches the workpiece for starting the arc may damage the electrode, if current is unlimited. The example is given in Figure in 4.2.

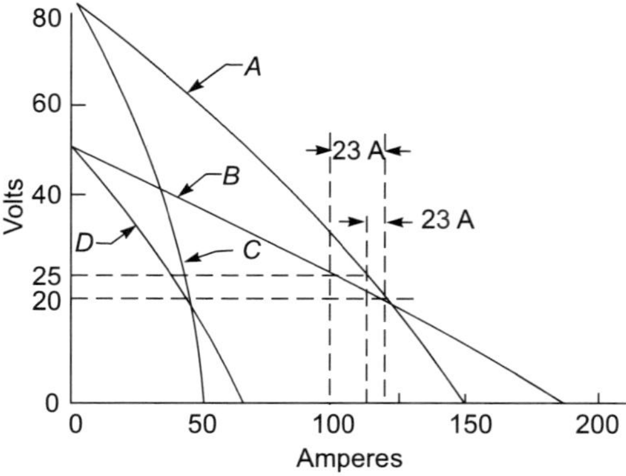

Figure 4.2 Constant current power supply, an example

With curve **A,** which has an 80 V open circuit, a steady increase in arc voltage from 20 to 25 V (25 percent) would result in a decrease in current from 123 to 115 A (6.5 percent). The change in current is relatively small. Therefore, in a consumable electrode welding process, electrode melting rate would remain fairly constant with a slight change in arc length. Setting the power source for 50 V open circuit and more shallow slope intercepting the same 20 V, 123 A position will give volt-ampere curve **B**. In this case, the same increase in arc voltage from 20 to 25 V would decrease the current from 123 to 100 A (19 percent), a significantly greater change. In manual welding, the flatter **V-A** curve would give a skilled welder the opportunity to substantially vary the current by changing the arc length. This could be useful for out-of-position welding because a welder can control the electrode melting rate and molten pool size. Current control is used to provide lower output. It would result in volt-ampere curves with a greater slope, as illustrated by curves **C** and **D**. They offer the advantage of more nearly constant-current output, allowing greater changes in voltage with minor changes in current.

Constant Voltage (CV) Power Supply: In CV power sources, a small variation in arc voltage (due to fluctuations in arc length) causes a significant change in the welding current. Since arc voltage remains almost constant during welding, despite fluctuations in arc length, therefore this type of power source is called constant voltage type. Moreover, the constant voltage power sources do not offer true constant voltage output as current voltage relationship curve shows slightly downward or negative slope. This negative slope is attributed to internal electrical resistance and inductance in the welding circuit that causes a minor drop in the output volt-ampere characteristics of the power source (Figure 4.3). The power source shall supply necessary current to melt the electrode at the rate required to maintain the preset voltage or arc length. The variation of arc voltage changes the average welding current. Use of such power source in conjunction with a constant speed electrode wire feed drive results in a self-regulating arc system. Hence, this type of power source is suitable for all those welding processes where small fluctuations in arc length can take place, like in semi-automatic welding processes like MIG, SAW and PAW. The example is given in Figure in 4.4.

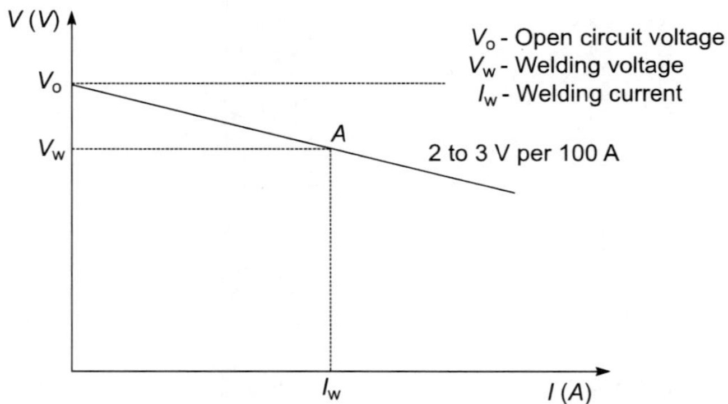

Figure 4.3 Constant voltage characteristics

It can be seen from Figure 4.4 that an increase or decrease in voltage to **A** or **C** (5 V or 25 percent) produces a large change in amperage (100 A or 50 percent). This V-A characteristic is suitable for constant feed electrode processes, such as gas metal arc, submerged arc, and flux cored arc welding, in order to maintain a constant arc length. A slight change in arc length (voltage) will cause a fairly large change in the welding current. This will automatically increase or decrease the electrode melting rate to regain the desired arc length (voltage). This effect is called *self-regulation.* Adjustments are sometimes provided with constant-voltage power sources to change or modify the slope or shape of the **V-A** curve. If done with inductive devices, the dynamic characteristics will also change.

The curve shown in Figure 4.4 can also be used to explain the difference between static and dynamic power supply characteristics. For example, during GMAW short circuiting transfer, the welding electrode tip touches the weld pool, causing a short circuit. At this point, the arc voltage approaches zero and only the circuit resistance or inductance limits the rapid increase of current. If the power supply responds instantly, very high current would immediately flow through the welding circuit, quickly melting the short-circuited electrode and freeing it with an explosive force, dispelling the weld metal as spatter. Dynamic characteristics designed in this power source compensate for this action by limiting the rate of current change, thereby decreasing the explosive force.

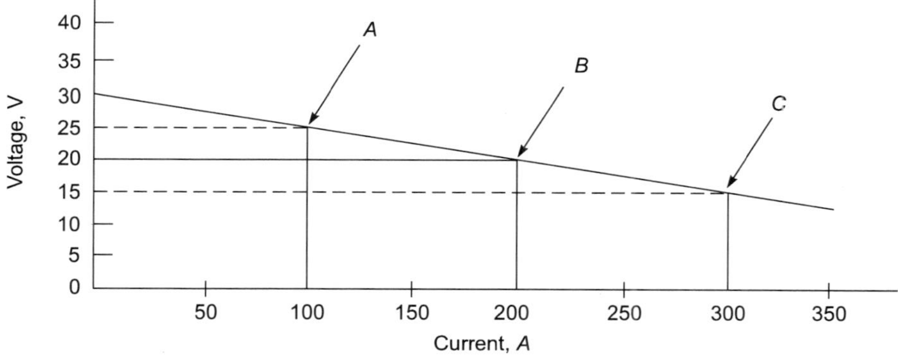

Figure 4.4 Combined constant current and constant voltage characteristics

Electronic controls can be designed to provide either a constant voltage or a constant current output from a single power source. Such a power supply can be used for a variety of welding processes. Electronically controlled outputs can also provide output curves that are a combination of constant current and constant voltage as shown in Figure 4.5. The top part of the curve is essentially constant current: below a certain trigger voltage, however, the curve switches to constant voltage. This type of curve is beneficial for SMAW to assist starting and to avoid electrode stubbing (sticking in the puddle) if welder use too short arc length.

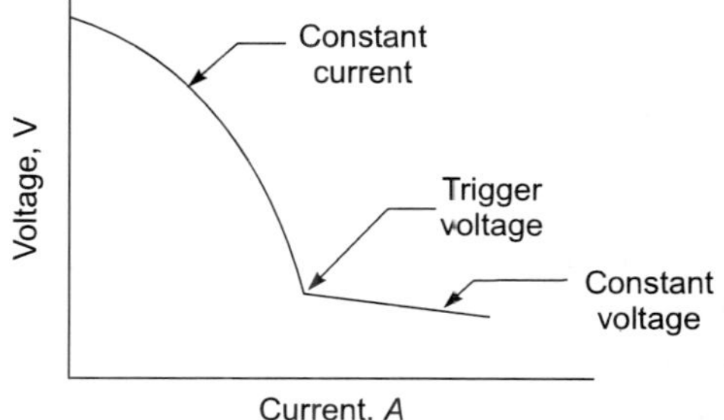

Figure 4.5 Combination volt-ampere curve

Rising Voltage Characteristics

Power sources with rising characteristics show increase in arc voltage with increase of welding current. In automatic welding processes where strict constant voltage is required, power sources with rising characteristics are used (Figure 4.6).

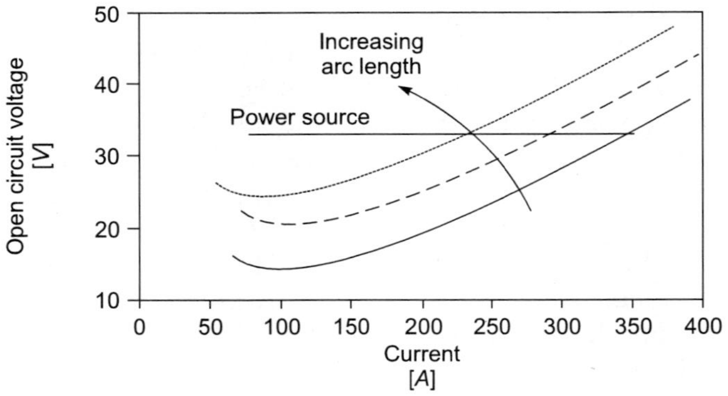

Figure 4.6 Rising voltage characteristics

TABLE 4.2 Welding processes and commonly used power sources

Welding Process	Type of Current	Type of Power Source
Manual metal arc welding	DC/AC	Constant current
Tungsten arc welding	DC/AC	Constant current
Plasma arc welding	DC	Constant current
Submerged arc welding	AC/DC	Constant current if electrode dia. >2.4 mm
	DC	Constant voltage if electrode dia. <2.4 mm
Gas metal arc welding	DC	Constant voltage

4.4.4 Dynamic Characteristics of Power Supply

A welding arc is never in steady state as it is subjected to severe and rapid fluctuations in arc voltage (due to changes in arc length) and welding current (within 0.1 micro sec), which in turn, causes transients in starting, extinction and reignition after each half cycle in AC welding. The power source must have good dynamic characteristics to satisfy these conditions so as to obtain stable and smooth arc. The dynamic characteristics of power supply show how fast the variation in arc voltage occurs with change in welding current over extremely short duration of time. A power source with good dynamic characteristics results in an immediate change in arc voltage and the welding current corresponding to new welding conditions, i.e., arc length, shielding gas, etc., in order to get smooth and stable arc (Figure 4.7).

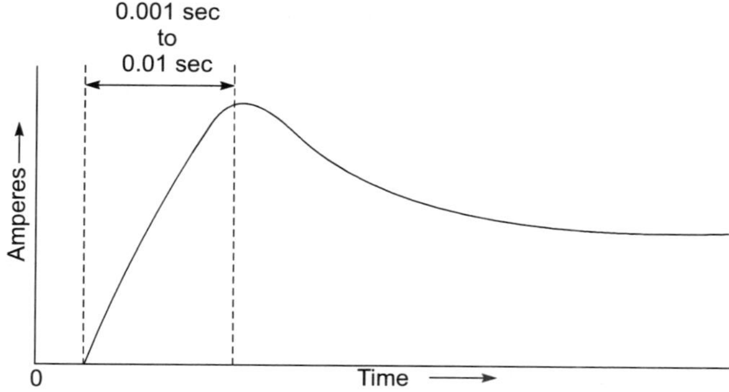

Figure 4.7 Transient nature of welding current

4.4.5 Duty Cycle (Figure 4.8)

Duty cycle refers to the percentage of welding time to total welding cycle, i.e., welding time plus rest time. Total welding time of 5 minutes is normally taken as a standard time. For example, welding for 3 minutes and rest of 2 minutes in total welding cycle of 5 minutes gives a duty cycle of 60%. Duty cycle and associated current are important as it ensures that power source remains safe and its windings are not damaged due to overheating. Duty cycle can also be defined as ratio of arcing time to the weld cycle time, multiplied by 100. If arcing time is continuously 5 minutes, then as per European standard it is 100 % duty cycle and 50% as per American standard. At 100% duty cycle, minimum current should be drawn. At low duty cycles, high current can be drawn. The welding current which can be drawn at a duty cycle can be obtained from the following equation:

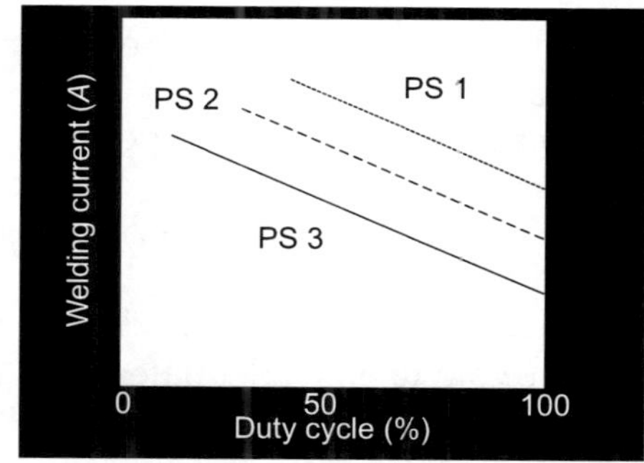

Figure 4.8 Duty cycle versus welding current

$$D_R \times I_R^2 = I_{100}^2 \times D_{100}$$

where, I_{100} is current at 100% duty cycle

D_{100} is 100% duty cycle

I_R is current at required duty cycle

D_R is required duty cycle

During the welding, heavy current is drawn from the power source. Flow of heavy current through the transformer coil and connecting cables causes electrical heating. Heating for long time may damage coils and cables. Therefore, welding operation should be stopped for some time depending upon the level of welding current. Hence, total welding cycle is taken as the sum of effective welding time and rest time. The maximum current which can be drawn from a power source depends on the size of winding wire, type of insulation and cooling system of the power source.

4.4.6 Class of Insulation

The duty cycle of a power source for a given current setting is largely governed by the maximum allowable temperature of various components (primary and secondary coils, cables, connectors, etc.), which in turn, depend on the quality and type of insulation and material of coils used for manufacturing power source. The insulation is classified as A, E, B, F and G in the increasing order of their maximum allowable temperature 60, 75, 80, 100 and 125°C.

4.5 AC WELDING POWER SOURCES

AC power sources are normally single-phase transformers that take AC supply from the mains and transform both voltage and current to the required values for arc welding.

4.5.1 Operating Principles of a Welding Transformer

In an AC welding arc, the current remains nearly sinusoidal while the voltage is distorted as shown in Figure 4.9.

Considering the transients, point *M* indicates the voltage required to strike an arc. The time during which the voltage rises from zero to the voltage sufficient for reignition of arc is termed the Arc Recovery Time. On the arc voltage transient it is denoted by '*t*'. If the arc is to be steady and quite, the time '*t*' should be as short as possible, because otherwise during the intervening interval, the cathode might become too cold to emit adequate number of electrons and ions to reignite and sustain the arc. One way to reduce '*t*' is to raise the open circuit-voltage of the welding power source, as is evident from Figure. 4.10. Voltage curve 2 has lower peak value than the voltage curve 1. In curve 1, the arc striking voltage is *E* and the arc recovery time is t_1. In the case of curve 2, with the same reigniting voltage *E*, the arc recovery time t_2 is considerably larger than t_1.

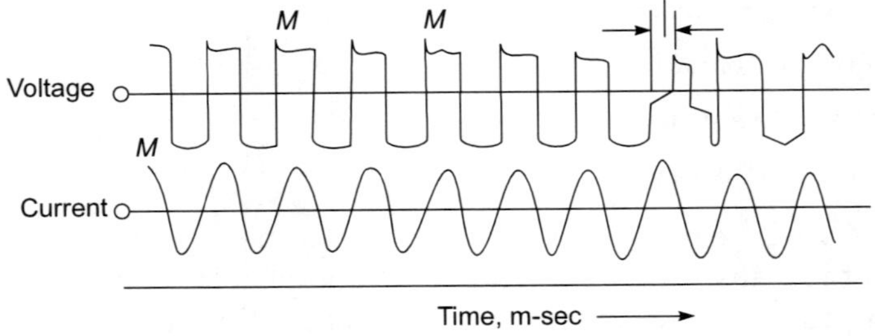

Figure 4.9 Arc voltage and current transients for AC welding

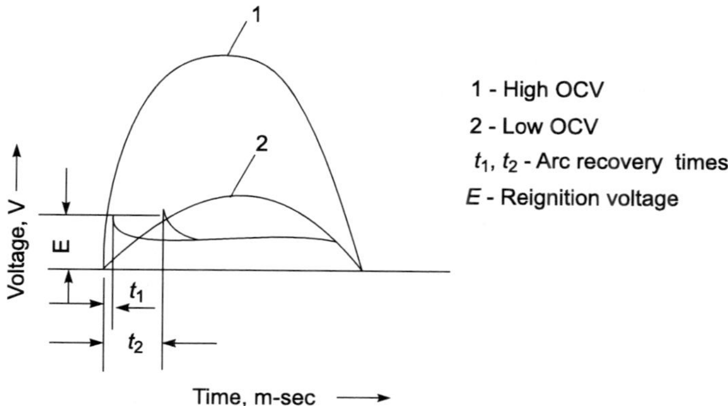

Figure. 4.10 Effect OCV on arc recovery time in AC welding

4.5.2 Requirements of a Welding Transformer

A welding transformer should satisfy the following requirements:

1. It should have a drooping static volt-ampere characteristic.
2. To avoid spatter, the surge of welding current during a short circuit should be limited to the least possible above the normal arc current.
3. The open circuit voltage should not normally exceed 80 volts and in no case, 100 volts.
4. The output current should be controllable continuously over the full available range.
5. The open circuit voltage should be just sufficiently high for ready initiation of an arc and not too high to impair the economics of welding.

4.5.3 Basic Types of Welding Transformers

The four basic types of welding transformers are:

1. The high reactance type
2. The external reactor type
3. The integral reactor type
4. The saturable reactor type

The High Reactance Type Welding Transformer

When a transformer supplies current, magnetic fluxes are produced around its windings. The lines of resultant magnetic flux, Φ, traverse the magnetic circuit and cut the primary (I) and secondary (II) windings, as shown in Figure. 4.11. However, not all the magnetic flux lines do so. Some of the lines of magnetic flux due to primary current do not cut secondary turns and vice-versa, since both have their paths in the air. In the diagram, these partial fluxes have been marked as Φ_{L1} and Φ_{L2}.

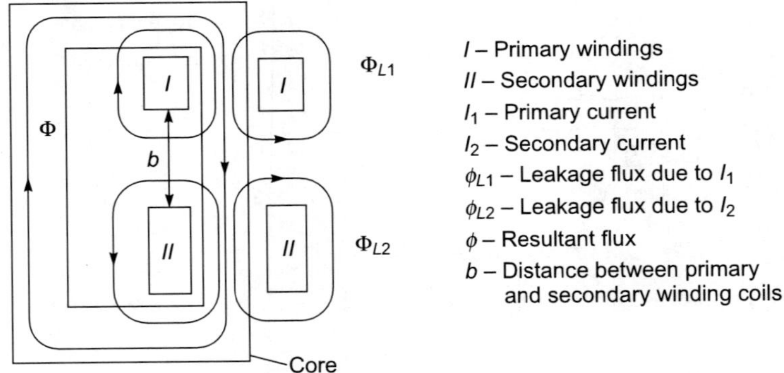

Figure 4.11 Schematic of a high reactance-type welding transformer windings

For a welding transformer to have a steeply drooping volt-ampere characteristic, both the primary and secondary windings should have a high reactance, i.e., they should have considerable leakage fluxes. This condition is satisfied by placing the primary and secondary windings either on separate limbs or on the same limb but spaced some distance apart, for example, distance '*b*' in Figure 4.11.

Control of current in high reactance welding transformers can be affected by three methods. One of them involves a moving primary coil as shown in Figure. 4.12. As the spacing between the windings is varied, so does the reactance and hence, the output welding current.

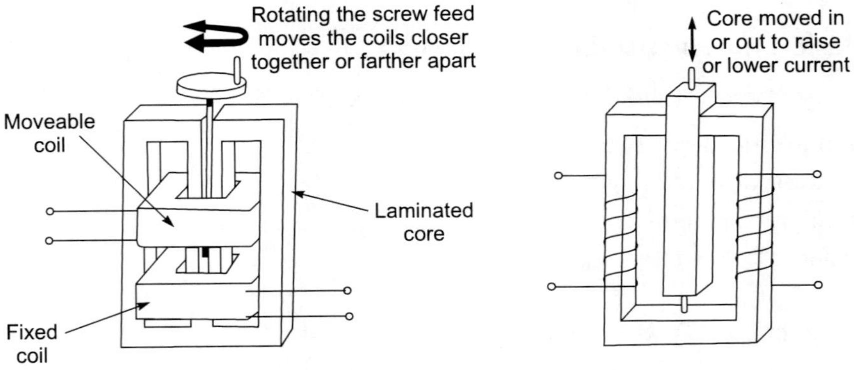

Moving coil and moving core type

Contd.

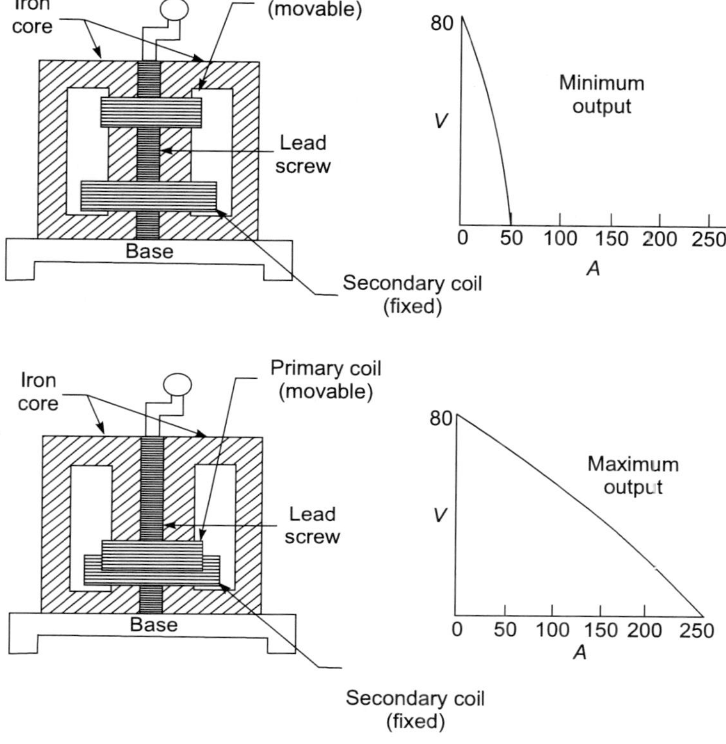

Figure 4.12 Moving coil high reactance welding transformer

The second method is based on the use of tapped windings either on the primary or the secondary side and the variation of transformation ratio can be done by bringing in or out of circuit the requisite number of turns, as shown in Figure 4.13.

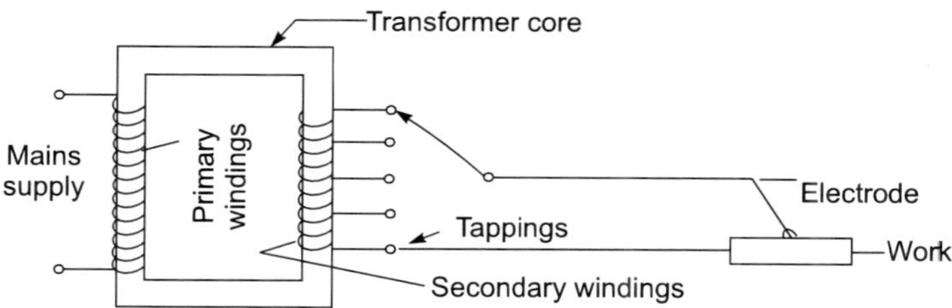

Figure 4.13 Tapped-type high reactance welding transformer

The third method utilizes movable magnetic shunt. The position of the shunt placed in the paths of the leakage fluxes, as shown in Figure 4.14, controls the output welding current through control of reactance.

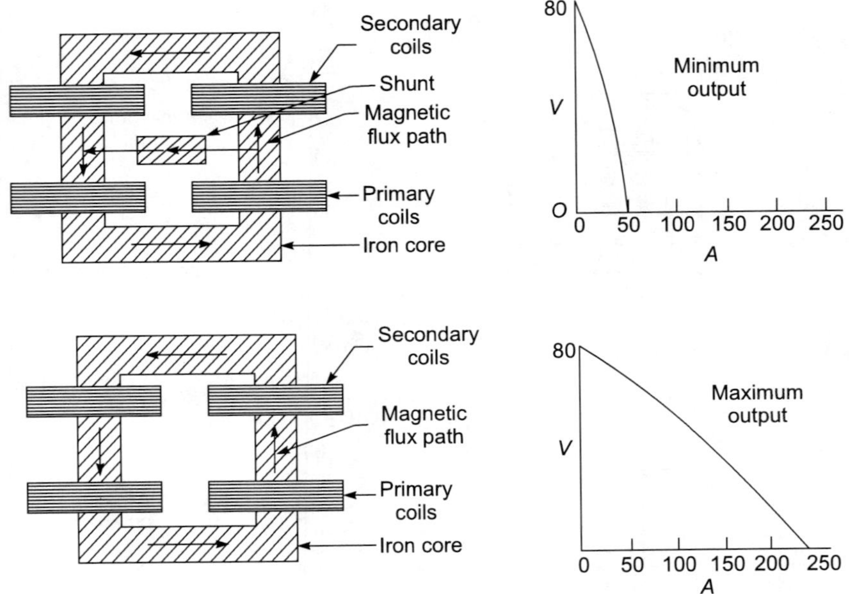

Figure 4.14 High reactance welding transformer with movable magnetic shunt

External Reactor Type Welding Transformer

This type of welding transformer consists of a normal reactance, single phase, step-down transformer and a separate reactor or choke.

The inductive reactances and resistances of the windings in such a welding transformer are low, so that its secondary voltage varies but a little with the welding current. The required drooping or negative volt-ampere characteristic is ensured by the reactor placed in the secondary of the welding circuit. The reactor consists of a steel core and a winding wound with a wire, designed to carry the maximum allowable current.

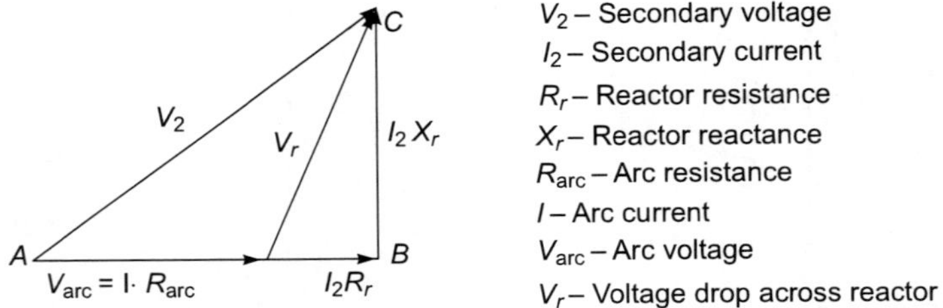

Figure 4.15 Vector diagram relating different voltages in an external reactor welding transformer

If the secondary voltage of the welding transformer is $\overline{v}_2$, the arc voltage is $\overline{v}_{arc}$, and the total resistive-cum-reactive drop across the reactor is $\overline{v}_r$, then the three quantities can be diagrammatically shown as in Figure 4.15 and are related mathematically as follows.

$$\overline{v}_2 = \overline{v}_{\text{arc}} + \overline{v}_r$$
$$\overline{v}_{\text{arc}} = \overline{v}_2 - \overline{v}_r$$

Thus, the arc voltage decreases with increase in current or with increase in voltage drop across the reactor. This gives a negative or drooping volt-ampere characteristic.

Control of welding current can be achieved by two methods, viz., by varying the reluctance of the reactor (the moving core reactor) or by varying the number of turns of the winding in the circuit (the tapped reactor).

The core of moving core reactor consists of a fixed portion carrying the winding, and a moving limb, which can be shifted towards or away from the fixed core by a suitable arrangement, thus varying the air gap between them. An increase in air gap adds to reluctance of the magnetic circuit of the reactor, while its self-induction and inductive reactance drop, so that the welding current increases.

When the air gap is reduced, the reluctance of the magnetic circuit is also reduced, the magnetic flux increases, as does the inductive reactance of the coil, and the welding current drop. In this way, the welding current can be adjusted accurately and continuously.

In the tapped reactor, the core is made solid but the coil is divided into a number of sections, each section having a tap brought out to the regulator point, as shown in Figure 4.16. Moving a contact arm across the taps will vary the number of turns brought in circuit and with that the magnitude of welding current. Thus, the current is controlled in steps.

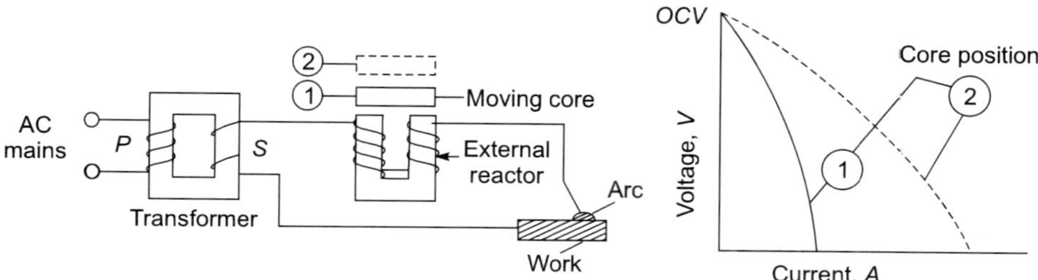

Figure 4.16 Circuit diagram of a tapped external reactor welding transformer

Integral Reactor Type Welding Transformer

The welding transformer of the integral reactor type, shown in Fig. 4.18, has a primary winding I, a secondary winding II, and a reactor winding III. Apart from main limbs, the core has additional limbs carrying the reactor winding. The current is adjusted by the means of moving core C placed between the additional limbs.

The part that carries winding I and II is thus, the transformer proper and part carrying winding III is the reactor.

The reactor can be connected with the secondary either in series aiding, or in series opposition.

When the reactor is connected in series aiding, Figure 4.17(a), the open circuit voltage of the transformer will be

$$E_t = E_2 + E_r$$

where, E_2 is the secondary voltage of the transformer and E_r is the reactor voltage.

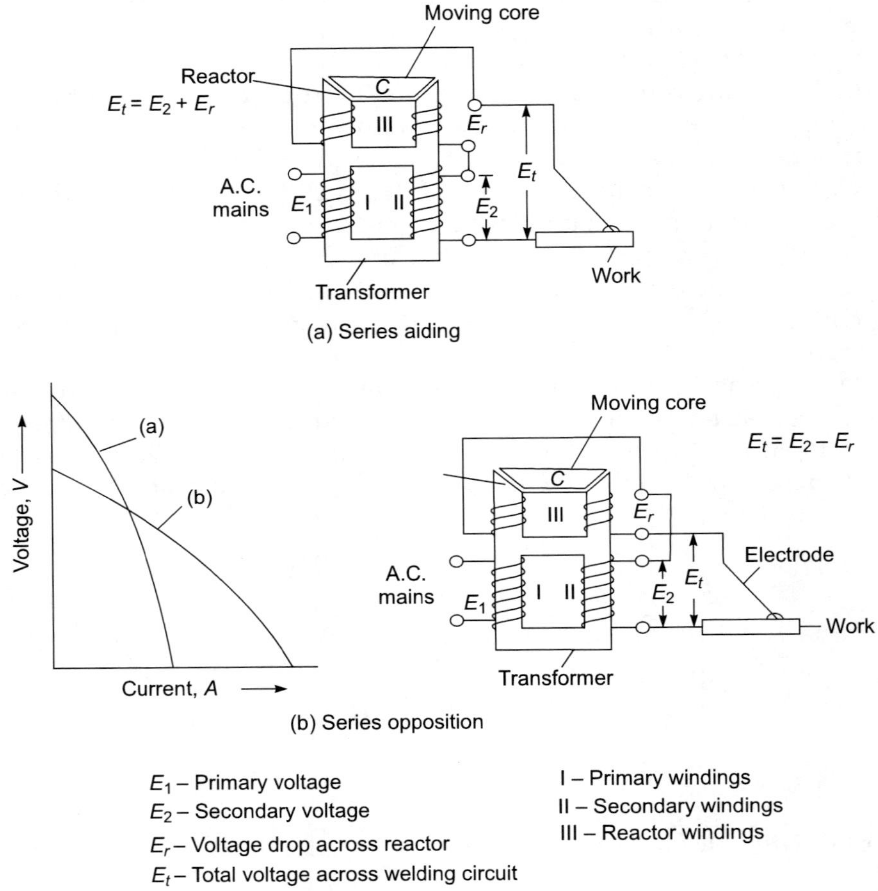

Figure 4.17 Circuit diagrams of integral reactor-type welding transformers

Series aiding connection produces a stable arc at low currents and is employed for welding thin plates.

When reactor is connected in series opposition, as shown in Figure 4.17(b), its voltage is subtracted from the open circuit voltage of the transformer, that is,

$$E_t = E_2 - E_r$$

Series opposition connection is used for welding thick plates with heavy currents.

Saturable Reactor Type Welding Transformer

In this welding transformer, an isolated low voltage, low amperage DC is employed to change the effective magnetic characteristics of the magnetic core. Thus, a large amount of AC is controlled by using a relatively small amount of DC, hence making it possible to adjust the output volt-ampere characteristic curve from minimum to maximum. For example, there is no DC flowing in the reactor coil, it has its minimum impedance and thus, maximum output of the welding transformer. As the magnitude of DC is increased with the help of a rheostat in the DC circuit, there are more continuous magnetic lines of force and thus, the impedance of reactor is increased and the output current of welding transformer is decreased. This method has advantage of removing movable parts and flexing conductors and is often used for gas tungsten arc welding power supplies.

Figure 4.18 shows the basics of a circuit for a simple saturable reactor power source. To achieve the desired aim of low voltage and high current, the reactor coils are connected in opposition to the DC control coil.

With AC, the wave form for gas tungsten arc welding is quite important. Saturable reactor tends to cause severe distortion of the sine wave supplied from the transformer. Placing an air gap, as shown in Figure 4.18, in the reactor core is one method of reducing this distortion. Alternatively, a large choke can be inserted in the DC control circuit. Either method, or a combination of the two, will produce the desired result.

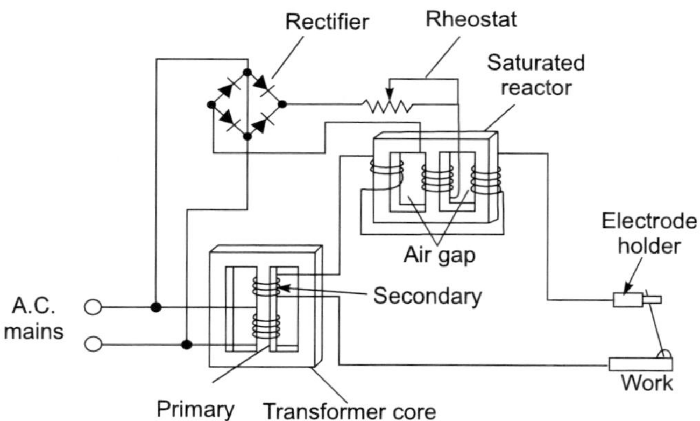

Figure 4.18 Circuit diagram of a saturable reactor welding transformer

Direct current supplied to this winding affects the offered impedance to alternating current flowing in the main coil in such a way that welding current may be continuously regulated by regulating direct current in the control winding. Saturable reactors are very costly and can be easily controlled by remote. In DC rectifier, if the arc current is controlled by resistance instead of the traditional reactance system:

1. No time delay in current response after a change in arc voltage.
2. Absence of this time delay and comparatively higher maximum current value achieved results in more pinch force.

3. It tends to prevent cold starting between the electrode and workpiece.

A stabilizer is used in DC circuit to minimize excessive current surges in load current. Constant current type rectifiers are employed for TIG, SAW, SMAW, carbon arc gauging, stud welding and PAW.

Parallel Operation of Welding Transformers

In welding operation, sometimes there is a need for current exceeding the maximum welding current obtainable from one transformer. In such a case, the desired welding current can be obtained by parallel operation of two or more welding transformers. The precaution needed for such a parallel operation is that no-load or open circuit voltages of the transformers should be the same. This is particularly essential in the case of high reactance-type welding transformers where the open circuit voltage and the transformation ratio vary to some extent according to the adjustment conditions and the control step.

When two transformers are connected for parallel operation, as shown in Figure 4.19, the like terminals of primary windings are to be connected to the identical line wires *A*, *B*, *C* of the supply mains, thus ensuring the coincidence of e.m.f phases in the secondary windings.

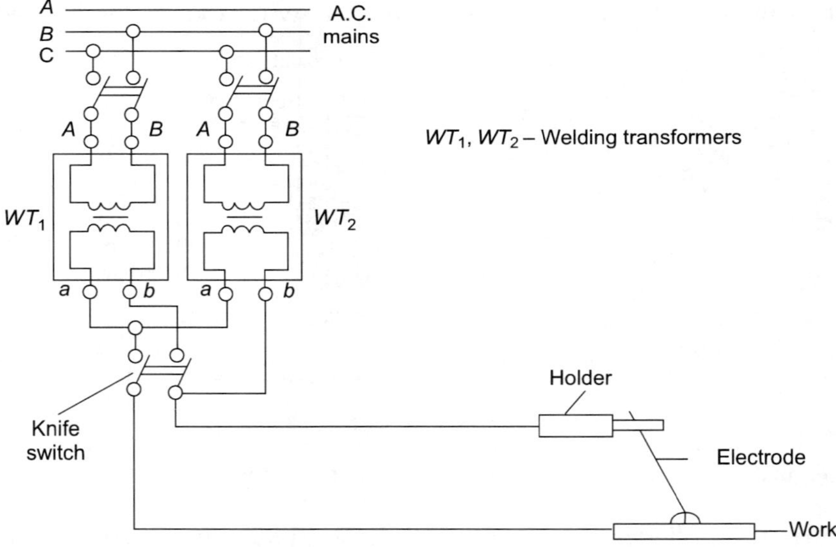

Figure 4.19 Circuit diagram for parallel operation of welding transformers

Multi-Operator Welding Transformers

A multi-arc or multi-operator welding transformer system utilizes a high current constant voltage power source for providing a number of welding circuits at the same time. Such a system is used when there is a large concentration of welding points in a relatively small operating area, for example, in ship-building construction sites for power stations, refineries, and chemical plants.

A multi-operator welding transformer with a flat volt-ampere characteristic may be of single-phase or 3-phase variety. A disadvantage of a single phase multi-operator welding transformer is that it puts an unbalanced load on the 3-phase supply mains. If a multi-operator welding transformer is to have a voltage which will not vary with the load (the maximum variation should not exceed 5%), it should have low magnetic leakage, that is, a low inductive reactance.

The number of arcs or welding circuits which can be connected to a welding transformer may be found by the relationship,

$$n = \frac{I_t}{I_{a \cdot K}}$$

where, n = number of arcs or welding circuits
I_t = rated output current of the welding transformer
I_a = average arc current in each welding circuit
K = diversity factor

The diversity factor (K) takes into account the fact that all the welders operating from one and the same power source do not work simultaneously. The diversity factor is related to average duty cycle and the laws of probability but is reduced as the number of welders operating from the same transformer increases. Usually, K is assumed to be anywhere between 0.6 and 0.8.

Each welding station is connected through a separate variable choke (current regulator), which provides a steeply drooping static volt-ampere characteristic curve for each welding circuit. The welding circuits are connected in parallel, because with this arrangement the source is better utilized when welding with low currents, of the order of 70 to 100 amperes.

Worked Examples

Example 1: The arc length-voltage characteristic of a DC arc is given by the equation, $V = 24 + 4l$, where, V is the arc voltage and l is the arc length in mm. The static volt-ampere characteristic of the power source is approximated by a straight line with no load voltage of 80 volts and the short circuit current of 600 amperes. Determine the optimum arc length for maximum power.

Solution: For welding arc, $V = 24 + 4l$ (i)

For power source, $V = 80 - \frac{80I}{600}$ (ii)

For stable arc (i) = (ii)

$$24 + 4l = 80 - \frac{80I}{600}$$

Which gives

$$I = 420 - 30l$$

$$\text{Power, } P = VI = (24 + 4l) \times (420 - 30l)$$

$$P = 120(84 + 8l - l^2)$$

For optimum arc length at maximum power,

$$\frac{dP}{dl} = \frac{d}{dl}(120(84 + 8l - l^2)) = 0$$

$$120 \times 8 = 120 \times 2l$$

Arc length, l = 4 mm

Example 2: The static volt-ampere characteristic of a welding power source is given by the parabolic equation $I^2 = -500(V - 80)$ and the arc characteristic is represented by the straight line equation $I = 23(V - 18)$. Determine, (a) the power of a stable arc, (b) if the arc length (l) and the arc voltage (V) are related by the expression $V = 20 + 4.5l$ determine the optimum arc length for power, and (c) if the convective and radiative losses for the arc in (b) are 15% the arc power, then determine if it will be advantageous to have an arc length of 4 mm, wherein these losses are only 20% of those for the arc in (b) are 15% of arc length. Comment briefly on the two cases.

Solution:

Case 1: Arc characteristics are given by

$$I_1 = 23(V - 18)$$

For power source

$$I_2^2 = -500(V - 80)$$

For a stable arc $I_1 = I_2$

$$[23(V - 18)]^2 = -500(V - 80)$$

$$519(V^2 + 18^2 - 2 \times V \times 18) = -500V + 500 \times 80$$

$$(519 \times V^2 + 519 \times 18^2 - 2 \times V \times 18 \times 519) = -500V + 40000$$

$$(V^2 - 35V + 2484) = 0$$

Solving we get,

$$V = 25.1 \text{ volts} \quad \text{or} \quad 9.9 \text{ volts.}$$

As 25.1 volt is the only feasible solution

Arc voltage = 25.1 volts

$$I = 23(V - 18)$$
$$I = 23(25.1 - 18) = 163.3 \text{ A}$$
$$\text{Arc power} = I \times V = 163.3 \times 25.1 \text{ VA} = 4.01 \text{ kVA.}$$

Case 2:
$$V = 20 + 4.5l \quad \text{(i)}$$
$$I^2 = -500(V - 80) \quad \text{(ii)}$$

Putting the value of V from (i) in (ii), we get,

$$I^2 = -500(20 + 4.5l - 80)$$
$$I^2 = 2250(13.33 - l)$$
$$I = 47.43\sqrt{(13.33 - l)}$$
$$\text{Power, } P = V \cdot I$$
$$\text{Power, } P = (20 + 4.5l) \times 47.43\sqrt{(13.33 - l)}$$

For maximum power,

$$\frac{dP}{dl} = \frac{d}{dl}(20 + 4.5l) \times 47.43\sqrt{(13.33 - l)} = 0$$

Solving this we get,

$$\text{Arc length} = 7.4 \text{ mm}$$

Case 3: Arc length, l = 4 mm

$$V = 20 + 4.5l = 20 + 4.5 \times 4 = 38 \text{ V}$$
$$I = 47.43\sqrt{(13.33 - l)} = 144.875 \text{ A}$$
$$\text{Power} = V \times I = 38 \times 144.875 = 5505.25 \text{ watts} = 5.5 \text{ kW}$$

For l = 7.4 mm

$$\text{Arc length, } l = 7.4 \text{ mm}$$
$$V = 20 + 4.5l = 20 + 4.5 \times 7.4 = 53.3 \text{ V}$$
$$I = 47.43\sqrt{(13.33 - l)} = 115.5 \text{ A}$$
$$\text{Power} = V \times I = 53.3 \times 115.5 = 6156.15 \text{ watts} = 6.156 \text{ kW}$$
$$\text{Net power} = \text{Gross power} - \text{losses}$$
$$\text{Net power} = 0.85 \times 6156.15 \text{ (losses} = 15\%) = 5232.73 \text{ watts}$$
$$\text{Losses} = 6156.15 - 5232.73 = 923.42 \text{ watts}$$
$$\text{Losses for } l = 4 \text{ mm}$$
$$= \frac{20}{100} \times 923.42 = 184.68 \text{ watts}$$

$$\text{Net power} = 5505.25 - 184.68 = 5320.57 \text{ watts}$$

The net power for an arc length of 4 mm will be higher than with arc length of 7.4 mm. Hence l = 4 mm should be preferred.

Example 3: Determine the change in the welding current if the arc length changes from 4 mm to 5 mm for the power sources with the following static volt-ampere characteristics.

$$I^2 = -400(V - 100)$$
$$I^2 = -8000(V - 80)$$
$$V = 48 - \frac{I^{1.05}}{50}$$
$$V = 30 + \frac{I^{1.05}}{50}$$

Assume that arc length (l) and arc voltage (V) are related by the expression $V = 20 + 4l$.

Solution:

Case 1: For arc, $V = 20 + 4l$ (1)

For power source, $I^2 = -400(V - 100)$ (2)

When $l = 4$ mm

Putting value of $l = 4$ mm in (1): we get

$$V = 20 + 4 \times 4 = 36 \text{ volts} \quad (3)$$

From (2) and (3),

$$I^2 = -400(V - 100)P = -400(36 - 100)$$
$$I = 160 \text{ A} \quad (4)$$

When $l = 5$ mm

Putting the value of l in (1), we get

$$V = 20 + 4 \times 5 = 40 \text{ volts} \quad (5)$$

From (2) and (5), we get

$$I^2 = -400(40 - 100)$$

Solving we get,

$$I = 154.9 \text{ A} \quad (6)$$

Change of current, I = (4)-(6),

$$\text{I} = 160 - 154.99 = 5.1 \text{ A}$$

Case 2: For arc, $V = 20 + 4l$

For power source, $I^2 = -8000(V - 100)$ (7)

When $l = 4$ mm

Putting value of $l = 4$ mm in (1): we get

$$V = 20 + 4 \times 4 = 36 \text{ volts} \quad (3)$$

From (3) and (7),

$$I^2 = -8000\,(V - 100) = 8000 \times 44$$

$$I = 593.3 \text{ A} \quad (8)$$

When $l = 5$ mm

Putting the value of l in (1), we get

$$V = 20 + 4 \times 5 = 40 \text{ volts} \quad (5)$$

From (2) and (5), we get

$$I^2 = -8000(40 - 100)$$

Solving we get

$$I = 565.7 \text{ A} \quad (9)$$

Change of current, I = (8)-(9)

$$I = 593.3 - 565.7 = 27.6 \text{ A}$$

Case 3: For arc, $V = 20 + 4l$ (1)

For power source,

$$V = 48 - \frac{I^{1.05}}{50} \quad (10)$$

When $l = 4$ mm

Putting value of $l = 4$ mm in (1): we get

$$V = 20 + 4 \times 4 = 36 \text{ volts} \quad (3)$$

From (3) and (10),

$$36 = 48 - \frac{I^{1.05}}{50}$$

$$I = 442.4 \text{ A} \quad (11)$$

When $l = 5$ mm

Putting the value of l in (1), we get

$$V = 20 + 4 \times 5 = 40 \text{ volts} \quad (5)$$

$$40 = 48 - \frac{I^{1.05}}{50}$$

$$I = 300.7 \text{ A} \tag{12}$$

Change of current, I = (11)-(12),

$$I = 442.4 - 300.7 = 141.7 \text{ A}$$

Case 4: For arc, $V = 20 + 4l$ (1)

For power source, $V = 30 + \dfrac{I^{1.05}}{50}$ (13)

When $l = 4$ mm

Putting the value of $l = 4$ mm in (1): we get

$$V = 20 + 4 \times 4 = 36 \text{ volts} \tag{3}$$

From (3) and (13),

$$36 = 30 + \frac{I^{1.05}}{50}$$

$$I = 283.5 \text{ A} \tag{14}$$

When $l = 5$ mm

Putting the value of l in (1), we get

$$V = 20 + 4 \times 5 = 40 \text{ volts} \tag{5}$$

$$40 = 30 + \frac{I^{1.05}}{50}$$

$$I = 470.1 \text{ A} \tag{15}$$

Change of current, I = (15)-(14),

$$I = 470.1 - 283.5 = 186.6 \text{ A}$$

Example 4: Current rating for a welding power source is 400 A at 60% duty cycle. Determine the welding current for automatic continuous welding, i.e., 100% duty cycle.

Solution: Rated current, $I_R = 400$ A

Rated duty cycle, $D_R = 60\%$

Desired duty cycle = 100%

$$D_R \times I_R^2 = I_{100}^2 \times D_{100}$$

where, I_{100} is current at 100% duty cycle

D_{100} is 100% duty cycle

I_R is current at required duty cycle

D_R is required duty cycle

$60 \times 400^2 = I_{100}^2 \times 100$

Solving we get

$$I_{100}^2 = 96000$$

$$I_{100} = 309.84 \text{ A}$$

Example 5: At what duty cycle needs a 300 A welding power source rated at 60% cycle to be operated to get an output of 400 A.

Solution: Rated current = I_R = 300 A

Rated duty cycle = D_R = 60%

Desired duty cycle to be calcuted

$$\text{Desired current} = 400 \text{ A}$$

$$D_R \times I_R^2 = I_D^2 \times D_D$$

$$60 \times 300^2 = 400^2 \times D_D$$

Solving we get

$$D_D = 0.3375$$

Desired duty cycle = 33.75%

Example 6: In an arc welding process, the voltage and current are 25 V and 300 A, respectively. The arc heat transfer efficiency is 0.85 and welding speed is 8 mm/sec. What is the net heat input (in J/mm). **(GATE 2006)**

Solution: Heat input in J/mm is given by

$$\text{Heat input} = \frac{VI\,(\text{in J/sec} = \text{watts})}{\text{Speed of welding}\left(\dfrac{\text{mm}}{\text{sec}}\right)} \times \eta$$

$$\text{Heat input} = \frac{25 \times 300}{8} \times 0.85 = 797 \text{ J/mm}$$

Example 7: A DC welding machine with a linear power source characteristic provides open circuit voltage of 80 V and short circuit current of 800 A. During welding with the machine, the measured arc length is 5 mm and the measured arc current is 460 A, corresponding to arc length of 7 mm. What is the relation between the linear voltage (L) and arc length (L) characteristics, given that E is measured in volts and L in mm. **(GATE 2007)**

Solution: $V = a + bL$

$$V_{\text{drop}} = V_0 - \frac{V_0}{I_s} \times I_m$$

where, V_0 = open circuit voltage
I_s = short circuit current
I_m = measured current
V = linear voltage
V_{drop} = V under steady state condition

$$a + bL = V_0 - \frac{V_0}{I_s} \times I_m$$

At arc length of 5 mm

$$a + b \times 5 = 80 - \frac{V_0}{800} \times 500 \quad (1)$$

At arc length of 7 mm

$$a + b \times 7 = 80 - \frac{V_0}{800} \times 460 \quad (2)$$

Solving (1) and (2), we get $a = 20$ and $b = 2$
Using

$$V = a + bL = 20 + 2L$$

Example 8: The DC power source for the welding has the characteristics $240 = 3V + I$, where, v is the voltage and I is the current in amperes. For the maximum arc power at the electrode, what should be voltage setting. **(GATE 2007)**

Solution: $240 = 3V + I$ (1)

$$240 - 3V = I$$

Power, $VI = V(240 - 3V) = 240 - 3V^2$
For optimum power

$$\frac{dP}{dV} = \frac{d}{dV}(240 - 3V^2) = 0$$

Solving we get, $V = 40$ volts

Example 9: In a DC arc welding process, the length characteristics were obtained as V arc $-20 + 5L$, where the arc length L was varied between 5 mm and 7 mm and V arc denotes the arc voltage in volts. The arc current was varied from 400 A to 500 A. Assuming linear power source characteristics, what are the open circuit voltage and short circuit current. **(GATE 2012)**

Solution: $V = a + bL = 20 + 5L$

$$V = V_p \text{ (under steady state condition)}$$

where, V_0 = open circuit voltage
I_s = short circuit current
I_m = measured current
V = linear voltage

$$a + bL = V_0 - \frac{V_0}{I_s} \times I_m$$

At arc length of 5 mm

$$20 + 5L = V_0 - \frac{V_0}{I_s} \times 500 \quad (1)$$

At arc length of 7 mm

$$20 + 5L = 80 - \frac{V_0}{I_s} \times 400 \quad (2)$$

Solving (1) and (2), we get

$$I_s = 950 \text{ A and } V = 95 \text{ volts}$$

Example 10: A DC welding power source has a linear voltage current characteristics with open circuit voltage of 80 volts and short circuit current of 300 A. For maximum arc power, what should be current value setting. **(GATE 2015)**

Solution: For maximum arc power

Current setting is half of short circuit current = 150 A

Example 11: During a TIG welding process, the arc current and arc voltage were 50 A and 60 V, respectively, when the welding speed was 150 mm/min. In another process, the TIG welding is carried out at a welding speed of 120 mm/min at the same arc welding voltage and heat input to the material so that weld quality remains the same. What should be the welding current in amperes for this process.

(GATE 2015)

Solution: Voltage = 60 volts and current = 50 A

Speed = 150 mm/min during first process

Speed = 120 mm/min during second process

Arc voltage is same, so we can write

$$V_1 = V_2$$

Heat input during the first process = Heat input during the second process

$$\left[\frac{VI}{\text{speed}}\right]_1 = \left[\frac{VI}{\text{speed}}\right]_2$$

$$\left[\frac{60 \times I_1}{150}\right]_1 = \left[\frac{60 \times I_2}{120}\right]_2$$

$$\left[\frac{60 \times 50}{150}\right]_1 = \left[\frac{60 \times I_2}{120}\right]_2$$

Solving we get,

$$I_2 = 40 \text{ A}$$

Example 12: The voltage length characteristics of direct current arc welding process are given by 100 + 40*L*, where, *L* is the arc length in mm and *v* is the arc voltage in volts. During the welding operation, the arc length varies between 1 mm to 2 mm and welding current is in the range of 200 to 250 A. Assuming a linear power source, what is short circuit current? **(GATE 2016)**

Solution: $V = a + bL = 100 + 40L$

$$V_{\text{drop}} = V_0 - \frac{V_0}{I_s} \times I_m$$

where, V_0 = open circuit voltage
I_s = short circuit current
I_m = measured current
V = linear voltage
$V_{\text{drop}} = V$ under steady state condition

$$a + bL = V_0 - \frac{V_0}{I_s} \times I_m$$

At arc length of 2 mm

$$100 + 40 \times 2 = 180 = V_0 - \frac{V_0}{I_s} \times 200 \quad (1)$$

At arc length of 1 mm

$$100 + 40 \times 1 = 140 = V_0 - \frac{V_0}{I_s} \times 250 \quad (2)$$

Solving (1) and (2) we get,

$$I_s = 425 \text{ A}$$

Example 13: An arc welding DC power source has a linear power source characteristic with open circuit voltage of 80 volts and short circuit current of 1000 A. The voltage length characteristics of arc are given by 20 + 4*L*, where, *L* is the length of the arc in mm. Calculate the optimum length of the arc for obtaining maximum arc power at welding. What voltage and current setting should be done on the power source for maximum arc power. **(GATE 1999)**

Solution: For welding arc, $V = 20 + 4l$ (i)

For power source, $V = 80 - \dfrac{80I}{1000}$ (ii)

For stable arc (i) = (ii)

$$24 + 4l = 80 - \frac{80I}{600}$$

Which gives

$$I = 750 - 50l$$

$$\text{Power, } P = VI = (20 + 4l) \times (750 - 50l)$$

$$P = 12.5(1200 + 160l - 16l^2)$$

For optimum arc length at maximum power,

$$\frac{dP}{dl} = \frac{d}{dl} 12.5(1200 + 160l - 16l^2) = 0$$

$$160 = 32l$$

Solving

Arc length, $l = 5$ mm

$$V = 20 + 4l = 20 + 4 \times 5 = 40 \text{ V}$$

$$I = 750 - 50l = 750 - 50 \times 5 = 500 \text{ A}$$

$$P = 12.5(1200 + 160l - 16l^2) = 20{,}000 \text{ watts} = 20 \text{ kW}$$

$$P = VI = 40 \times 500 = 20{,}000 \text{ watts} = 20 \text{ kW}$$

Example 14: The arc voltage characteristics of a DC power source has a linear power source characteristics of $V = 20 + 40L$, where, *V* is the arc voltage in volts and *L* is the arc length in cm. The static volt ampere characteristics of power source are approximated by a straight line with an open circuit of 80 V and short circuit current is 1000 A. Determine the arc length for maximum power. **(GATE 1991)**

Solution: Similar to the above Example 13. (Answer: Length of arc = 5 mm) and left as exercise for students.

Example 15: A low-carbon steel is to be welded by the manual metal arc welding process using a linear *VI* characteristics of DC power source. The following data is available.

Open circuit voltage = 62 V, short circuit current = 130 A, arc length = 4 mm,. Traverse speed of the welding = 15 cm/min, efficiency of heat input = 85%, the relationship between arc length L in mm and arc voltage is $V = 20 + 1.5L$. Calculate the heat input into the workpiece. **(GATE 1992)**

Solution: $V = a + bL = 20 + 1.5L$

$$V_{\text{drop}} = V_0 - \frac{V_0}{I_s} \times I_m$$

where, V_0 = open circuit voltage

I_s = short circuit current

I_m = measured current

V = linear voltage

$V_{\text{drop}} = V$ under steady state condition

$$a + bL = V_0 - \frac{V_0}{I_s} \times I_m$$

At arc length of 4 mm

$$20 + 1.5 \times 4 = 180 = 62 - \frac{62}{130} \times I_m$$

$$I_m = 75.5 \text{ A}$$

$$P = VI = 26 \times 75.5 = 1963 \text{ watts} = 1.963 \text{ kW}$$

Actual heat input = 0.85 × 1963 = 1668.55 watts

Example 16: The voltage arc length characteristic of a DC arc is given by $V = 20 + 4L$, where, L is the arc length, during the welding operation, arc length is expected to vary from 4 to 6 mm with the welding current limited between 450 and 550 A. If the power source has linear characteristics, what is the open circuit voltage? What is the arc power at the arc length of 5 mm. **(GATE 1997)**

Solution: $V = a + bL = 20 + 4L$

$$V_{\text{drop}} = V_0 - \frac{V_0}{I_s} \times I_m$$

where, V_0 = open circuit voltage

I_s = short circuit current

I_m = measured current

V = linear voltage

$V_{\text{drop}} = V$ under steady state condition

$$20 + 4L = V_0 - \frac{V_0}{I_s} \times I_m$$

Solution: At arc length of 4 mm

$$20 + 40 \times 4 = V_0 - \frac{V_0}{I_s} \times 550 \tag{1}$$

At arc length of 6 mm

$$100 + 40 \times 6 = 140 = V_0 - \frac{V_0}{I_s} \times 450 \tag{2}$$

Solving (1) and (2), we get

$$I_s = 1000 \text{ Amperes and } V_0 = 80 \text{ volts}$$

$$V = a + bL = 20 + 4 \times 5 = 40 \text{ volts}$$

$$V_{\text{drop}} = V_0 - \frac{V_0}{I_s} \times I_m = 80 - \frac{80}{100} \times I_m = 40$$

Solving we get

$$I_m = 500 \text{ A}$$

$$\text{Power} = VI = 40 \times 500 = 20{,}000 \text{ watts} = 20 \text{ kW}$$

Example 17: The arc voltage characteristics of a DC power source has a linear power source characteristic of $V = 20 + 40L$, where, V is the arc voltage in volts and L is the arc length in cm. What is the open circuit voltage and short circuit current for arc lengths ranging from 3 to 5 mm and current ranging from 400 to 500 A during the operation are. **(GATE 1993)**

Solution: Similar to Example 16 and left as an exercise. (Answer: Voltage = 72 volts and current is 900 A).

Example 18: The arc voltage characteristics of a DC power source has a linear power source characteristic of $V = 20 + 40L$, where, V is the arc voltage in volts and L is the arc length in cm. The static volt ampere characteristics of power source are approximated by a straight line with a open circuit of 80 V and short circuit current is 600 A. Determine the arc length for maximum power. **(GATE 2003)**

Solution: Welding arc, $V = 20 + 40l$ (i)

For power source, $V = 80 - \dfrac{80I}{600}$ (ii)

For stable arc (i) = (ii)

$$20 + 40l = 80 - \frac{80I}{600}$$

Which gives

$$I = 450 - 300l$$

$$\text{Power, } P = VI = (20 + 40l) \times 450 - 300l$$

$$P = 7.5(9000 + 12000l - 12000l^2)$$

For optimum arc length at maximum power,

$$\frac{dP}{dl} = \frac{d}{dl}7.5(9000 + 12000l - 12000l^2) = 0$$

$$160 = 32l$$

Solving

Arc length, l = 0.5 mm.

$$V = 20 + 4l = 20 + 40 \times .5 = 40 \text{ V}$$

$$I = I = 450 - 300l = 450 - 300 \times 0.5 = 300 \text{ A}$$

$$P = 7.5(9000 + 12{,}000l - 12{,}000l^2) = 12{,}000 \text{ watts} = 12 \text{ kW}$$

$$P = VI = 40 \times 30 = 12{,}000 \text{ watts} = 12 \text{ kW}$$

Example 19: A DC welding power source has a linear voltage current VI characteristics with open circuit voltage of 80 volts and short circuit current of 300 A. For the maximum arc power, the current in amperes should be set as ____________.

(GATE 2015-S2)

Solution: V_0 = 80 V and I_S = 300 A

$$\frac{V}{V_0} + \frac{I}{I_0} = 1 \quad \text{(power source characteristics)}$$

$$\frac{V}{80} + \frac{I}{300} = 1$$

$$7.5\,V + 2I = 600$$

$$V = \frac{1}{7.5}(600 - 2I)$$

$$\text{Power } (P) = VI = \frac{1}{7.5}(600 - 2I)I$$

For maximum power,

$$\frac{dP}{dt} = 0$$

$$\frac{dP}{dt} = \frac{1}{7.5}(600 - 4I) = 0$$

Solving we get current = 150 amperes.

Example 20: A welding operation is being performed with voltage of 30 V and current of 100 A. The cross-sectional area of weld bead is 20 mm^2. The workpiece and filler of titanium for which specific energy of melting is 14 J/mm^3.

Assuming a thermal efficiency of the welding process as 70%. The welding speed (in mm/s) is __________ (Correct to two decimal places). **(GATE 2018 – S2)**

Solution: $V = 30$ volts

$I = 100$ amperes

Cross-sectional area, $A = 20 \text{ mm}^2$

Specific energy of melting = 14 J/mm^3

Arc efficiency = 0.7 or 70%

We know that

Power = efficiency × voltage × current

Power = 0.7 × 30 × 100 = 2100 J/s

$$\text{Specific energy} = \frac{\text{Power}}{\text{Area} \times \text{Welding speed}}$$

$$\text{Welding speed} = \frac{\text{Power}}{\text{Area} \times \text{Specific energy}}$$

$$= \frac{2100}{20 \times 14} = 7.5 \text{ mm/sec}$$

Review Questions

1. What is the requirement for an arc welding power source?
2. What are the factors affecting the selection of a welding power source?
3. Explain different welding power sources.
4. Explain the difference between constant voltage and constant current welding power source.
5. Explain dynamic characteristics of power supply.
6. What are the requirements of a welding transformer?
7. What are the various types of welding transformer?
8. Define power factor.
9. Explain saturable reactor type welding transformer.
10. Explain the advantages of multi-operator welding transformers.

5 CHAPTER

Arc Welding Consumables

5.0 INTRODUCTION

All arc welding processes need some consumables to achieve the desired weld joint. The processes like carbon arc welding and gas tungsten arc welding consume electrodes at relatively low rate, hence it is not common to include such electrodes as consumables. However, the filler wire if used with these so called consumable electrodes is obviously a consumable item. The major welding consumables include four items, viz.:

- Coated electrodes
- Fluxes
- Filler wires
- Shielding gases

5.1 COATED ELECTRODES

Almost all general-purpose welding is done by shielded metal arc welding using coated electrodes. The coated electrodes consist of a core wire with a covering of coated material. Electrodes for manual and automatic arc welding of low and medium carbon steels as well as low alloy steels are mostly made from 0 to 12% carbon steel.

The core wire diameter for electrodes usually varies between 3.50 and 12.50 mm with the intermediate standard diameters being 4.00, 5.00, 6.30, 8.00 and 10.00 mm. The length of these coated electrodes varies between 350 and 450 mm with bare portion (without coating) ranging between 20 and 30 mm from where it is held in an electrode holder. The chemical composition of core wires for low and medium carbon steel is given in Table 5.1 and for mild steel electrode core wires in Table 5.2.

TABLE 5.1 Core wire composition for welding electrodes (IS: 2680-1964)

Element	Low Carbon Steels (%)	Medium Carbon Steels (%)
Carbon (max.)	0.15	0.25-0.30
Silicon	0.3-0.7	0.30-0.50
Manganese	0.9-1.5	0.30-1.60
Sulphur (max.)	0.03	0.03
Phosphorus (max.)	0.03	0.03

TABLE 5.2 Rimming and non-rimming MS for metal arc welding electrode core wire

Constituent	Rimming Steel(%)	Non-Rimming Steel (%)
Carbon	0.10 (Max.)	0.10 (Max.)
Manganese	0.38-0.62	0.38-0.62
Silicon	0.03 (Max.)	0.03 (Max.)
Sulphur	0.03 (Max.)	0.025 (Max.)
Phosphorus	0.03 (Max.)	0.02 (Max.)
Copper	0.03 (Max.)	0.15 (Max.)

5.1.1 Electrode Coatings

The coatings of welding electrodes serve four main purposes:

- To facilitate the initiation and maintenance of welding arc.
- To protect the molten metal from the deleterious effects of oxygen and nitrogen by producing a shield of gas around the arc and weld pool.
- To provide a slag blanket on the weld to reduce its rate of cooling.
- To provide a means of introducing alloy elements not contained in the core wire.

Depending upon the thickness of coatings, electrodes are classified into two types:

- Light-coated electrodes
- Heavy-coated electrodes

Light-Coated Electrodes

Coatings applied in a thin layer to a metal rod serve only the purpose of stabilizing the arc and that is why they are called stabilizing coatings. No ingredient is included in these coatings and hardly any slag is formed on the weld, nor are the mechanical properties of the weld metal improved. Therefore, these electrodes may only be used for non-essential jobs.

Of all the stabilizing coatings, widest use is made of the one prepared by dissolving 80 to 85 parts by weight of ground and screened chalk (calcium carbonate) in 20 to 15 parts (by weight) of sodium silicate (water glass), which acts as a binder.

Heavy-Coated Electrodes

Heavy-coated electrodes are sometimes referred to as shielded arc electrodes. These are used to obtain a weld metal of high quality, comparable with and even superior to the parent metal in terms of mechanical properties.

Coatings for heavy-coated low carbon arc electrodes can be classified into 5 main types:

- High-cellulose electrodes
- Titanium based coatings
- Acid coatings
- Basic coatings
- Oxide coatings

High-Cellulose Coatings

These are based on volatile substances (wood or cotton cellulose) plus natural silicates (like kaolin, mica, talc, feldspar) and ferroalloys (like ferromanganese, ferrosilicon, and ferrotitanium) as reducing agents. These electrodes produce less volume of slag and the reducing reactions take place in an atmosphere of hydrogen surrounding the weld pool. These reactions are of two types:

On iron oxide $FeO + H_2 \rightarrow Fe + H_2O$

On iron nitride $2Fe_4N + 3H_2 \rightarrow 8Fe + 2NH_3$

The metal deposited by electrodes with these coatings is fine grained and contains hardly any oxygen (less than 0.02%) but it does contain a high proportion of hydrogen (15-25 ml /100 g of weld metal). The weld produced is of deep penetration with rough appearance and high spatter. These are used particularly for position welding where a good degree of penetration is required. These coatings provide good mechanical properties particularly after aging.

Features of cellulosic electrodes are:

1. Flux rich in cellulose.
2. Burns to produce hydrogen and carbon monoxide, provides shielding to the arc.
3. Suitable with DC power and electrode positive polarity.
4. Presence of these gases in the arc with high ionization potentials results in a high arc voltage and therefore, a high arc energy. This results in a deeply penetrating arc and a rapid burn-off rate for high weld speed.
5. Suitable for all position welding.
6. Gives reasonably good mechanical properties.
7. High level of hydrogen generated – causes hydrogen embrittlement in HAZ.

Titanium Based Coatings or Rutile Coatings

These type of coatings, rutile (natural titanium dioxide, TiO_2, 95% pure) or ilmenite (iron titanate, $FeTiO_3$), also include natural silicates and ferroalloys as refining agents. The slags formed tend to have an acidic action as they tend to dissolve basic oxides.

Electrodes with medium base coatings give medium penetration welds of good appearance with high mechanical properties. They produce a quiet arc with a low level of spatter. They also have further advantages of high arc stability and positional welding properties.

Features of rutile coating electrodes are:

1. Contain high proportion of titanium oxide (rutile) in its coating. Titanium oxide promotes easy arc ignition, smooth arc operation and low spatter.
2. Good mechanical properties.
3. Because of the rutile and the ionizers in the coating, these electrodes can be used with either polarity and in all positions. They are most suitable for fillet welding in horizontal and vertical positions.
4. Moderate weld metal mechanical properties.
5. Good bead shape produced due to viscous slag.
6. Easy slag removal.
7. Positional welding possible with a fluid slag.

Acid Coatings

These coatings are based on oxides and natural silicates but they include a high proportion of deoxidizers and denitriders in the form of ferroalloys. The slags produced have acid reaction and thus, dissolve basic oxides like MnO as a result of which, a large amount of manganese is transferred to the slag. The manganese built up in the slag lowers its viscosity; this tends to improve the appearance of the weld and makes it possible to deposit the weld metal in all positions.

Oxide Coatings

These coatings consist of a mixture of iron oxides, silica, natural silicates (kaolin, talc, mica, feldspar) with little or no deoxidizers. The electrodes with these coatings are said to be of oxidizing-type because the molten metal absorbs a large amount of oxygen or ferrous oxide, FeO and nitrogen in the form of nitrides, Fe_4N. The welds produced are of medium penetration with low spatter. These are used on the most common types of electrodes and they have low mechanical properties but give a pleasing weld appearance in fillets.

Basic Coatings

These coatings consist of mixtures containing fluorites, calcium carbonates and magnesium carbonates, which have high heats of formation. They also contain a flux along with reducing and nitrogen removing agents in the form of ferroalloys. In these coatings, no use is made of cellulose, clay, asbestos, and other minerals that contain moisture. This ensures lowest possible content of hydrogen in the weld metal. That is why they are known as low-hydrogen electrodes.

The slags produced by these electrodes are highly basic in reaction which is highly stable. It is also known that a part of the molten iron can combine to form calcium ferrite, which has high heat of formation. Alumina used must be in the combined state, since at high temperatures its basic properties predominate over its acidic properties. These electrodes have superior weld properties with highest ductility of any of the weld deposits.

Features of basic coating electrodes are:

1. High proportion of calcium carbonate and calcium fluoride.
2. Referred to as low-hydrogen electrodes.
3. Make the slag more fluid than that of rutile coatings.
4. Slag is of fast-freezing type.
5. Suitable for vertical and overhead positions.
6. The coating composition is produced with low-moisture content. Hydrogen content of the deposited metal is less than other types of electrodes.
7. Weld deposits have low-hydrogen content.
8. Make them suitable for welding low-alloy steels susceptible to HAZ cracking.
9. Weld deposit has a high resistance to hot cracking.
10. Suitable for welding thicker steels and steels with higher-carbon content, weld metal has excellent mechanical properties, especially impact strength.
11. These electrodes are used for high quality applications, which call for a low-hydrogen content in weld deposit, the moisture content of the electrode coating should be kept to a minimum.
12. To prevent the electrode coating from moisture absorption, they should be carefully stored and dried.
13. These electrodes give poor bead profile and slag removal is difficult.

Iron Powder Coatings

Iron powder is added to electrode coatings to increase their deposition efficiency up to maximum of about 210%. This also helps in higher welding speed by virtue of higher current carrying capacity of such coatings. These coatings are coded depending on the extent of iron powders in them. Iron powders are added to increase deposition rate and to improve arc behavior.

In a conventional electrode, current is carried wholly by the core wire, whereas with iron powder addition in the flux, the coating becomes conductive near the arc, thus it provides an additional path to the current. Arc tends to spread out and metal deposition takes place over a wider area. It reduces the current density at the tip, thus reduces penetrating force of the arc. These electrodes give less penetration.

Iron powder also provides additional conducting area in the arc and limits the current surge when a short circuit takes place between electrode wire and the job.

The occurrence of the spatter is less. It provides more stable and smooth arc. It also improves sidewall fusion. It gives less undercut and spatter.

High deposition rate is actually achieved not because of additional metal in the flux but because of the ability to carry more current for the same core diameter. With increase in welding current, spatter increases, but in iron-powder electrodes, the rise in spatter with rise in current is much less.

Features of iron powder coating electrodes are:

1. Increased deposition rate.
2. More stable arc.
3. Flatter weld deposit.
4. Less penetration.
5. Less chances of undercut.
6. Less spatter.
7. Improved side wall fusion.
8. Most widely used welding process for short welds in production, maintenance and repair work.

5.2 WELDING RODS AND WIRES

Bare welding wires and rods are used in short lengths of about 1 metre or in coiled form in spools. Short lengths are used in processes like GTAW and plasma arc welding, wherein they are not of the welding circuit. Long wires are employed for processes like GMAW and SAW, where a part of the wire conducts current. When a welding wire forms a part of the electrical circuit, it is called a welding electrode otherwise, it is referred to as a welding rod.

Most wires used for welding structural steel usually contain 0.1% carbon and 0.35-0.60% manganese. Others have increased manganese content.

Excess silicon in welding wire results in heavy spatter, gassing in the weld pool and non-metallic materials in the weld pool. Therefore, maximum silicon content permitted is 0.95%. The content of harmful impurities, like sulphur and phosphorous, allowed is 0.3% each.

The welding wires are usually coated with copper to prevent rusting and improve current pickup from the contact tube, it also helps in drawing wires through dies. To avoid harmful effects of copper coating, it is usually kept thin.

5.3 WELDING FLUXES

Flux is an essential part of arc welding processes and it serves the following purposes:

- The portion of the flux which meets floats as a liquid blanket over the molten metal, protects it from the deleterious effect of the surrounding atmosphere, thereby reducing the pickup of oxygen and nitrogen. In welding with bare electrodes, the nitrogen content of the weld metal is as high as 0.18%. Heavy-coated electrodes keep the figure down to 0.026% while in SAW, there is a maximum of 0.005% nitrogen in the weld deposit, which gives an idea of the effectiveness of flux.

- It acts as a good insulator and concentrates heat within a relatively small welding zone. Thus, it improves the fusion of the molten metal from the welding electrode and the parent material.
- It acts as a cleanser for the weld metal, absorbs impurities, and adds alloying elements, such as manganese and silicon.
- Owing to flux the metal is not only clean, but also denser and has excellent physical properties.
- The blanket of flux improves the process efficiency by reducing spatter and burning losses.

5.4 SHIELDING GASES

The major shielding gases used for gas tungsten arc welding, gas metal arc welding and plasma arc welding processes are argon, helium, and carbon dioxide. Apart from these, nitrogen, oxygen, hydrogen and their mixtures with the first three gases are also used for achieving the desired bead configuration and weld metal properties. These shielding gases may be classed in two groups, namely,

- Inert gases, like argon and helium
- Gases which dissolve in and react with the metal, for example, CO_2, O_2, H_2 and N_2.

Argon

It is a non-combustible, and non-explosive gas obtained from air by deep refrigeration and fractionation, wherin it is present to the extent of 9.3×10^{-1} %. It is about 23% heavier than air.

Generally, argon is marketed in three grads, say, A, B and C, containing 0.01, 0.04, 0.1% of impurities, respectively. Commercial argon contains 16.7% impurities. For high class welding work, the purity of argon required is about 99.995%.

Argon is non-toxic but can cause asphyxiation in confined spaces by replacing the air.

Argon is stored and shipped in gaseous state in standard steel cylinders. Cylinders for pure argon are painted black at the bottom and white at the top, on which the words 'Pure Argon' are stenciled. In a cylinder, the gas is kept under a pressure of 150 atmosphere (15 N/mm^2 approx.), when fully filled. A standard 40 litre cylinder holds 6 cubic metres (6000 litres of argon. From a cylinder, argon is fed to the point of welding operation through a pressure regulator attached to the valve in the neck of the cylinder. The pressure regulator brings the pressure of the gas down to the figure required for welding (which is usually below 0.5 atmosphere) and also maintains the working pressure constant, irrespective of the pressure in the cylinder. The pressure regulators for argon cylinders are painted black.

The flow rate of argon is measured with a *flowmeter* called *rotameter*, which is attached to the regulator.

The specific uses of different grades of argon for welding are listed below:

Grade A (99.99% pure or more) argon is used for welding active and rare metals and also for welding of components made from other materials, at the final stage of fabrication.

Grade B (99.96% pure) argon is used for welding aluminum and magnesium base alloys.

Grade C (99.9% pure) argon is used for welding stainless and other high-alloy steels.

Helium

Helium is a rare gas. It is present in atmosphere to the extent of only $0.52 \times 10^{-3}\%$. In addition, helium is present upto 10% in natural gas. It also results from the decay of certain radioactive elements and is found in some uranium minerals.

On account of its high price, helium is comparatively less used inert gas. It is a light gas weighing only 1/7th that of air. This complicates the weld pool protection and results in increased gas consumption.

Helium is sold in two grades. Grade I is 99.6 to 99.7% pure and Grade II is 98.5 to 99.5% pure. It is stored and shipped in gaseous state in standard cylinders under a pressure of 15 MPa (150 atoms). Commercial Helium I is stored in cylinders which are painted brown and bear no inscription. Commercial Helium II is kept in cylinders which are painted brown and bear the word 'Helium' stenciled with white paint.

Helium has the highest ionization potential of any of the shielding gases and therefore, a welding arc can be operated at much higher potential than argon. Thus, arc with helium shielding produces a larger amount of heat. Because of its light weight, helium tends to float away from the arc zone and hence, produces an inefficient shield unless higher flow rates are maintained. However, its light weight is helpful for overhead welding. Because of the greater flow rate required for helium, higher welding speeds are attainable. It is possible to weld approximately 35 to 40% faster with helium than when using argon as the shielding gas. It is used often in gas tungsten arc welding, gas metal arc welding and automatic MIG welding processes.

Carbon Dioxide

It is a colorless gas with a slightly perceptible odour. When dissolved in water, it gives an acidic taste. It is about 1.5 times heavier than air.

Industrially, CO_2 is prepared by calcination of coke or anthracite in specially designed boiler fireboxes, and by extracting from natural sources. It is also obtained as a by-product of manufacture of ammonia and fermentation of alcohol.

Under pressure CO_2 becomes a liquid, and thus by sufficient cooling, solidifies to a snow-like substance (called dry ice). It vaporizes at –57°C.

The dry ice and gaseous CO_2 used commercially are obtained from liquid CO_2, which is a colorless liquid. When allowed to vaporize at 0°C and normal pressure (760 mm of Hg), one kg of CO_2 produces 509 litres of gaseous CO_2.

Liquid CO_2 is shipped in steel cylinders, in which it occupies 60 to 80% of the total space. A standard 40-litre cylinder holds 25 kg of liquid, which produces about 15 cu.m. of gas on evaporating. The pressure of the gas in the cylinder depends on the temperature which goes down as more gas is tapped from the cylinder.

CO_2 used for welding purposes may be of two grades. Grade I must contain at least 99.5% (by volume) of pure CO_2 and not over 0.178 g/m^3 of moisture. The corresponding figures for Grade II are 99% and 0.515 g/m^3.

The penetration characteristics of CO_2 are similar to the penetration characteristics of helium because of similarities of weights of the gases. CO_2 used for welding must be free of all moisture because moisture releases hydrogen, which produces porosity in the weld metal. Because CO_2 has greater electrical resistance, the current setting must be 20 to 30% higher than those used with argon and helium.

CO_2 is considered inert at the normal temperature and pressure. At elevated temperatures, however, it dissociates to the extent of 20 to 30% into CO and O. CO is toxic and has a safe concentration of only 175 ppm compared with 5000 ppm of CO_2. This calls for an effective exhaust system to safeguard against the ill effects of CO. The negative characteristic of the formation of oxygen is that it can reduce the normal strength of the metal. Another major disadvantage of the use of CO_2 is its extreme resistance to current flow. Because of this resistance, the arc length is sensitive. When the arc length is too long, it will extinguish more readily than when an inert gas, like argon or helium, is used.

Defective welds are most often produced when using CO_2 from the top or bottom portion of a cylinder. This is because the gas at the top carries most of the impurities (nitrogen, oxygen and moisture). After the liquid CO_2 has been fully used up, the gas coming out of the cylinder contains excessive moisture.

To avoid defects due to impurities in CO_2, it will be a good plan to allow the freshly delivered CO_2 to settle for 15 to 20 minutes and to release the top portion of the contents into atmosphere. It is also a good practice to turn a cylinder upside down and allow it to rest in this position for about 15 minutes. After that period, open the valve carefully, all the water in the cylinder will flow out.

When CO_2 is tapped at a rate of over 1000 litres/hour, it will be well advised for an operator to use at least two cylinders connected in parallel.

Other Gases

Generally Ar, He and CO_2 are used singly or in mixtures as shielding gases for welding. However, quite often other gases like O_2, H_2 and N_2 are added to these gases, to achieve certain desired shapes and characteristics of weld deposits.

Oxygen is a colorless, odorless and tasteless active gas which combines with many elements to form oxides. In steel, it may combine with carbon to form CO which may get entrapped in the solidifying weld metal and result in pores or voids. This defect is usually overcome by adding deoxidizers, like Mn and Si.

Hydrogen is the lightest gas in atmosphere to the extent of 0.01%. Hydrogen, however, may also be present in the arc atmosphere from moisture or hydrocarbons present in the base metal or the filler wire. It dissolves in molten steel but its solubility in steel at room temperature is very low. Thus, the hydrogen that escapes, goes to the grain boundaries and may result in cracks. It also causes underbead cracking in HAZ.

Nitrogen is in plenty in the atmosphere. It is colorless, odorless, non-toxic and almost an inert gas. It is soluble in molten steel but its solubility in steel at room temperature is very low. Thus, it can also cause pores and voids. In very small amounts, nitrites if formed can increase the strength and hardness of steel but reduce its ductility which can lead to cracking. Nitrogen is used sometimes for welding copper because it provides a high heat arc. Because of its low cost compared to argon, it is often used for purging stainless steel pipes and tubing systems.

Gas Mixtures (Figure 5.1)

Gas mixtures commonly employed in arc welding processes include: Ar-He, Ar-CO_2, Ar-O_2, Ar-H_2, Ar-CO_2-O_2 and the like.

Ratio of argon in Ar-He mixtures may vary from 25 to 95% Ar. However for welding aluminum, a combination that is often used is a mixture of Ar-75% He or Ar-80% He. Argon, as a shielding gas, helps to remove oxides and it also exhibits a certain amount of porosity control over the weld deposit. Helium helps to give a good bead configuration. Most ferrous and non-ferrous metals can be welded with either helium or argon, or their mixtures. Helium is particularly used for welding heavier sections of aluminum, magnesium and copper as well as for overhead welding.

A mixture of CO_2-75% Ar or CO_2-80% Ar is quite popular for welding structural and low alloy steels. Argon improves the metal transfer characteristics and CO_2 helps in improving the bead shape and economics of the process. These mixtures are however, used on thinner sections of steels when bead appearance is important. These are also found useful for out-of-position welding on extremely thin sheet metal.

Oxygen is sometimes added to argon with a view to improving head shape in welding low carbon steels. Small amount of oxygen added to argon produces significant changes. For example, it broadens the deep penetration finger at the centre of the bead. It also improves bead contour and eliminates the undercut at the edge of the weld.

Oxygen is normally added in amounts of 1%, 2% or 5%. Maximum amount of oxygen employed for Ar-O_2 mixtures is 5%. Higher amounts, if added, may lead to porosity in the weld metal. The main aim of adding oxygen to argon is to form thermionic iron oxide on the surface of steel electrode, which improves its emissivity and makes the cathode spot wider and stable, if electrode is made negative.

Hydrogen is sometimes added to argon but its amount is limited to a maximum of 5%. Normally, the mixtures used are Ar-2% H_2, or Ar-4% H_2. Addition of hydrogen causes increase in arc voltage, which leads to high heat in the arc. The Ar-H_2 mixture should not be used for low carbon or low alloy steels, as it may lead to hydrogen cracking often referred to as hydrogen embrittlement. It is mainly used for welding nickel or Ni-alloys. Heavy sections of stainless steels are also welded with this mixture.

Mixture of CO_2-O_2 or Ar-CO_2-O_2 is also sometimes used for welding mild steels. This improves the mode of metal transfer and bead shape. Addition of oxygen results in hotter running of the arc and thus, avoids lack of fusion.

Use of chorine in small amounts as shielding gas for aluminum, improves arc stability. Also nitric oxide as a very small addition (0.03%) to the shielding gas, for welding aluminum helps in reducing ozone content in the weld zone.

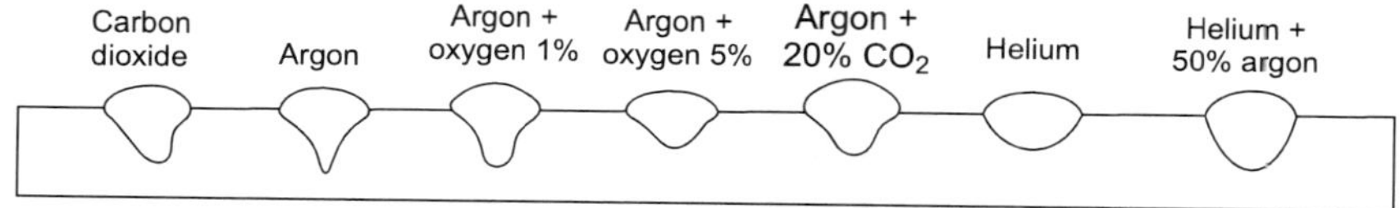

Figure 5.1 Effect of shielding gas on weld bead shape

In general, the composition of the shielding gas for gas shielded arc welding of different metals and their alloys can be based on the guidelines provided in Table 5.3.

TABLE 5.3 Gases for gas shielding arc welding processes

Metal	Shielding Gas	Gas Effects
Mild steel	CO_2	Low spatter
	CO_2 + 1 - 2% O_2	Avoids lack of fusion
	Ar + 1 - 2% O_2	Better bead shape
	Ar + 20 - 25% CO_2	No spatter
Low alloy steels	Ar + 2% O_2	Improved emissivity, hotter arc
	Ar + 5% O_2	Eliminates undercutting, improves bead shape
	Ar + 20% CO_2	Lower cost
Stainless steel	Ar + 1 - 5% O_2	Stable arc, reduced undercutting
	Ar + 2 - 4% H_2	Hotter running arc
	He + 7.5% Ar + 2.5% CO_2	Improved bead shape

Contd.

Aluminum	Argon He + 25% Ar	Removes surface oxides Oxide removal and porosity control
Magnesium	Argon Argon + 50% He	Removes surface oxides, hotter arc, reduced porosity
Nickel	Argon Ar + 50 – 75% He Ar + — 5% H_2	Improved control on base metal fluidity, good wettability, hotter running arc
Copper	Argon He + 25% Ar Nitrogen	Reduces sensitivity to surface cracks Improved bead shape. Counteracts high thermal conductivity by high heat input arc
Titanium	Argon	Good metal transfer

Review Questions

1. Explain the advantages of coated electrodes.
2. Describe the differences between light-coated and heavy-coated electrodes?
3. Describe the features of cellulosic electrodes.
4. Describe the features of rutile-coating electrodes.
5. Describe the features of basic-coating electrodes.
6. Compare the mechanical properties of basic and rutile electrodes.
7. Describe the advantages of iron powder coatings.
8. Describe the important functions of flux.
9. Why shielding gas is used in welding?
10. Compare different shielding gases.

6
CHAPTER

Shielded Metal Arc Welding

6.0 INTRODUCTION

Arc welding is a welding process wherein coalescence is produced by heating with an electric arc. Mostly arc welding is done without pressure and with or without the use of filler metal, depending on the plate's thickness. Electric arc is formed when an electric current passes between two electrodes separated by a short distance from each other. In arc welding, one electrode is welding rod or wire, while other is the metal to be welded.

Shielded Metal Arc Welding (SMAW) is an electric arc welding process where the heat for welding is generated by an electric arc between a covered metal electrode and the work. An electric arc is produced when two current carrying conductors are brought together forming an electric circuit and then separated from a distance, such that current continues to flow through the gaseous medium between solid conductors. The electric energy is thus converted into heat which attains a temperature of 3600 to 6000°C.

When the amount of heat is concentrated at the point of welding, it melts a small pool of metal in the work. Additional filler metal is deposited from the electrode and the electrode covering provides the shielding. The shielded metal arc welding process is one of the most simple and versatile arc welding processes. This process can be used to weld both ferrous and non-ferrous metals and it can weld thicknesses above approximately 18 gages in all positions. The arc is under the control of the welder and is visible. The welding process leaves a slag on the surface of the weld bead, which must be removed. The most popular use for this process is for the welding of mild and low alloy steels. The equipment is extremely rugged and simple, and the process is flexible in that the welder only needs to take the electrode holder and work lead to the point of welding.

6.1 PRINCIPLES OF OPERATION (FIGURE 6.1)

The shielded metal arc welding process uses the heat of the electric arc to melt the consumable electrode and the work being welded. An arc occurs when the electrode comes in contact with the workpiece and completes the circuit .The welding circuit includes a power source, welding cables, an electrode holder, a work clamp and a welding electrode. One of the welding cables connects the power source to the electrode holder and the other cable connects to the workpiece.

The welding begins when the arc is initiated by momentarily touching the electrode to the base metal, which completes the electrical circuit. The welder guides the electrode manually, controlling both the travel speed and the direction of travel. The welder maintains the arc by controlling the distance between the work material and the tip of the electrode (length of the arc). Some types of electrodes can be dragged along the surface of the work, so that the coating thickness controls the arc length, which controls the voltage.

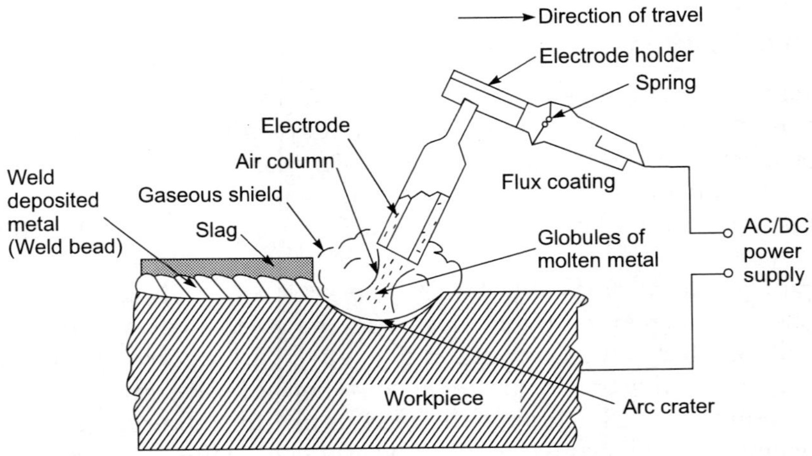

Figure 6.1 Nomenclature used in shielded metal arc welding

The heat of the arc melts the surface of the base metal and forms a molten weld puddle. The melted electrode metal is transferred across the arc and becomes the deposited weld metal. The deposit is covered by a slag produced by components in the electrode coating. The arc is enveloped in a gas shield provided by the disintegration of some of the ingredients of the electrode coating. The shielding gas protects the molten puddle from the atmosphere while stabilizing the arc. Most of the electrode core wire is transferred across the arc, but small particles escape from the weld area as spatter, and a very small portion leaves the welding area as smoke.

Functions of Slag in SMAW

- Slag is a combination of the flux coating and impurities from the base metal that float to the surface of the weld.
- Slag quickly solidifies to form a solid coating.
- The slag also slows the cooling rate of the weld.
- The slag can be chipped away and cleaned with a wire brush when hard.

6.2 ARC SYSTEMS

The constant current type of power source is preferred for manual shielded metal arc welding. The constant current welding machines provide a nearly constant welding current for the arc. The constant current output is obtained with a drooping volt-ampere characteristic, which means that the voltage is reduced as the current increases. The changing arc length causes the arc voltage to increase or decrease slightly, which in turn, changes the welding current. Within the welding range, the steeper the slope of the volt-ampere curve, the smaller is the current change for a given change in the arc voltage. Under certain conditions, there is a need for variations in the volt-ampere slope.

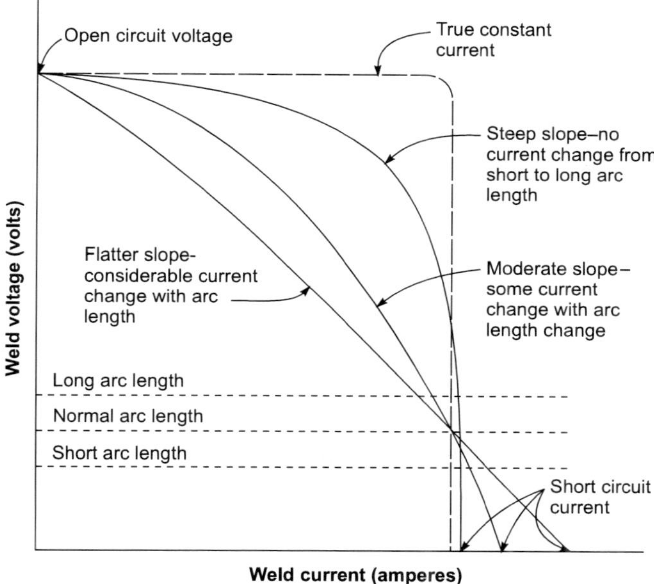

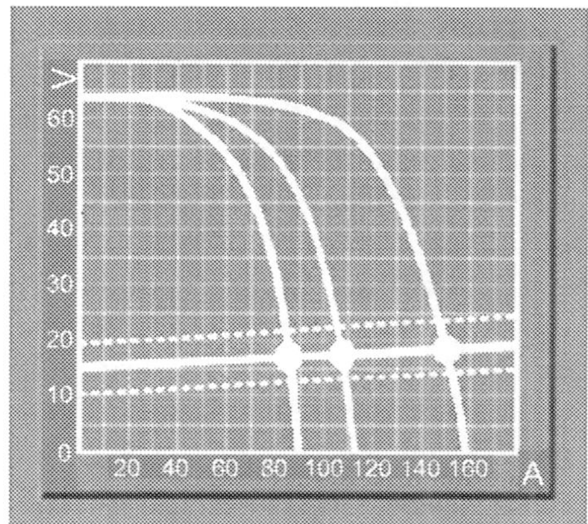

Figure 6.2 Arc systems for SMAW processes

A steep volt-ampere characteristic is desirable welding speed on some welding jobs. The steeper slope gives less current variation with changing arc length and it gives a softer arc. The types of machines that have this kind of curve are especially useful on sheet metal. Machines with this characteristic are typically used with large diameter electrodes and high amperages. On some applications, a less steep volt-ampere characteristic is more desirable, such as welding over rust, or all position pipe welding where better arc control with high penetration capability is desired. Machines with the less steep volt-ampere curve are also easier to use for depositing the root passes on joints that have varying fit up. This type of power source characteristic allows the welder to control the welding current in a specific range, by changing the arc length and producing a more driving arc. The variations in the power sources are caused by the differences in the basic power source designs. Figure 6.2 shows volt-ampere curves for different performance characteristics. This shows several slopes, all of which can provide the same normal voltage and current. The flatter slopes give a greater current variation for a given voltage change or arc length change. More positive starting is given by machines that have a higher short circuit current.

6.3 EQUIPMENT

Shielded metal arc welding may utilize either alternating current (AC) or direct current (DC), but in either case, the power source selected must be of the constant current (CAV) type. This type of power source delivers a relatively constant amperage or welding current regardless of arc length variations by the operator. The amperage determines the amount of heat at the arc and since it remains relatively constant, the weld beads produced will be uniform in size and shape. Whether to use an AC, DC or AC/DC power source, depends on the type of welding to be done and the electrodes used. The following factors should be considered:

1. Electrode selection—Using a DC power source allows the use of a greater power range of electrode types. While most of the electrodes are designed to be used on AC or DC, some will work properly only on DC.
2. Metal thickness—DC power sources may be used for welding both, heavy sections and light gage work. Sheet metal is more easily welded with DC because it is easier to strike and maintain the DC arc at low currents.
3. Distance from work—If the distance from work to the power source is great, AC is the best choice since the voltage drop through the cable is lower than with DC. Even though welding cables are made of copper and aluminum (both good conductors), the resistance in the cables becomes greater as the cable length increases. In other words, a voltage reading taken between the electrode and the work will be somewhat lower than a reading taken at the output terminal of the power source. This is known as voltage drop.
4. Welding position—Because DC may be operated at lower welding currents, it is more suitable for overhead and vertical welding than AC. AC can successfully be used for out-of-position work if proper electrodes are selected.

5. Arc blow—When welding with DC, magnetic fields are set up throughout the weldment. In weldments that have varying thickness and protrusions, this magnetic field can affect the arc by making it stray or fluctuate in one direction. This condition is especially troublesome when welding in corners. AC seldom causes this problem because of the rapidly reversing magnetic field produced.

   Combination power sources that produce both AC and DC are available and provide the versatility necessary to select the proper welding current for the application. When using a DC power source, the question of whether to use electrode negative or positive polarity arises. Some electrodes operate on both DC straight and reverse polarity and others on DC negative or DC positive polarity only. Direct current flows in one direction in an electric circuit and the direction of the current flow and the composition of the electrode coating will have a definite effect on the welding arc and weld bead.

Direct current flows in one direction continuously through the welding circuit. The advantages it has over alternating current are:

- Direct current is better at low currents and with small diameter electrodes.
- All classes of covered electrodes can be used with satisfactory results.
- Arc starting is generally easier with direct current.
- Maintaining a short arc is easier.
- Direct current is easier to use for out-of-position.

Alternating current is a combination of both polarities (positive and negative) in regular cycles (say 50 Hz). In each cycle the current starts at zero, builds up to a maximum value in one direction, decays back to zero, builds up to a maximum value in other direction, and again decays to zero. The direction of current changes 100 times with the 50 Hertz current that is used in India. Depth of penetration and deposition rates for alternating current are generally intermediate between those for DC electrode positive and DC electrode negative.

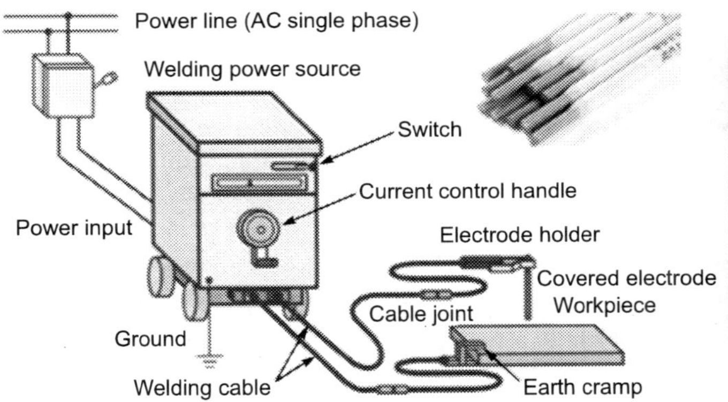

Figure 6.3 Equipment

Some of the advantages of alternating current are:

- Arc blow is rarely a problem with alternating current.
- Alternating current is well suited for the welding of thick sections using large diameter electrodes.

Comparison between AC and DC arc welding is given in Table 6.1.

TABLE 6.1 Comparison of AC arc welding and DC arc welding

S. No.	AC Arc Welding	DC Arc Welding
1.	Striking of arc with electrode is relatively difficult and maintenance of a short arc is also difficult except with iron powder electrode.	Striking an arc is easier and maintenance of short arc is also easy.
2.	No problem of arc blow in AC arc welding. Workpieces do not get magnetized as in DC.	Arc blow, a severe problem, is minimized with the use of proper corrective measures. Workpiece may get magnetized due to current flow in one direction.
3.	Arc is never stable.	Arc is more stable.
4.	No polarity change is possible and hence, not suitable for welding all metals. It is used for welding ferrous materials.	Polarity can be changed (DCEN/DCEP) and hence, suitable for both ferrous and non-ferrous metals.
5.	Most suitable for higher current values. It is less suitable for use at low current with small diameter of electrode.	It is most suitable for lower current values. Low current with smaller electrode can be used.
6.	Bare electrodes cannot be used. Only flux coated electrodes with arc stabilizing agent in flux can be used	Bare and coated electrodes can be used.
7.	Not suitable for thin sheets due to difficulty in striking arc.	It is suitable for welding of sheet metal as striking arc is easier and arc remains steady.
8.	Distribution of heat in arc is equal at electrode and job.	Most of the heat (up to 66.67%) is liberated when DCEP is used.
9.	Voltage drop in welding is less and hence, welding is suitable for longer distance from welding plant using long welding cable.	Voltage drop is relatively higher and hence, short cables are used to weld only close to the welding plant.
10.	Welding transformers have no moving part and working is quiet.	A DC generator set has several moving parts and therefore operation is noisy.
11.	AC transformer welding set is not costly. Simpler in operation, maintenance cost is also low.	A DC generator set is costly, difficult to operate and very high maintenance cost.
12.	Low initial cost.	High initial cost.
13.	Small floor area is required.	Large floor area is required.

6.3.1 Equipment Accessories (Figure 6.4)

Electrode Holder

The electrode holder connects to the welding cable and conducts the welding current to the electrode. The insulated handle is used to guide the electrode over the weld joint and feed the electrode into the weld puddle as it is consumed. Electrode holders are available in different sizes and are rated on their current carrying capacity.

Ground Clamp

The ground clamp is used to connect the ground cable to the workpiece. It may be connected directly to the work or to the table or fixture upon which the work is positioned. Being part of the welding circuit, the ground clamp must be capable of carrying the welding current without overheating due to electrical resistance.

Welding Cables

The electrode cable and the ground cable are important parts of the welding circuit. They must be very flexible and have a tough heat-resistant insulation. Connections at the electrode holder, the ground clamp, and at the power source lugs must be soldered or well-crimped to assure low electrical resistance. The cross-sectional area of the cable must be of sufficient size to carry the welding current with a minimum of voltage drop. Increasing the cable length necessitates increasing the cable diameter to lessen resistance and voltage drop.

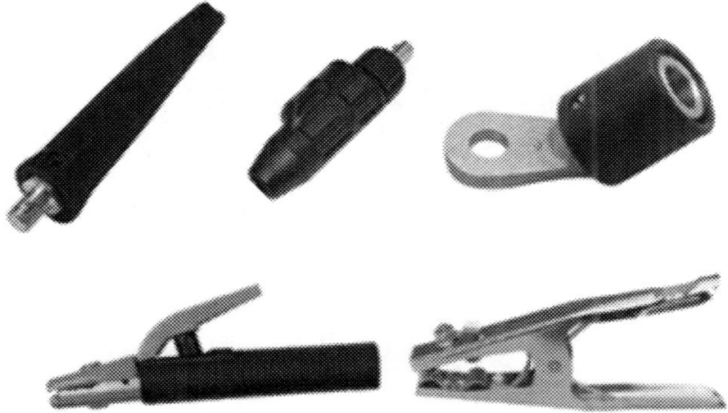

Figure 6.4 Equipment accessories for SMAW

Coated Electrodes

The SMAW coated electrode consists of a straight piece of solid metallic wire, called core wire, having a concentric covering (also called coating) of flux. As much as 25 mm to 40 mm of the wire is left bare at one end to serve as a grip in the electrode holder during welding. At the other end, called the striking end, only the core face is left exposed to facilitate arc striking. The electrode size refers to the diameter of its core wire. The electrode length is determined by the convenience of the welder to strike the arc and manipulate the electrode, and by the capacity of the electrode to carry a reasonably high welding current without getting overheated until it is consumed. The flux covering thickness varies with each class and brand of electrode and is usually expressed as coating factor, which is the ratio of coating diameter to the core wire diameter.

6.3.2 Operator Accessories (Figure 6.5)

The essential operator accessories include chipping hammer, wire brush, and a welding shield to protect the face. The chipping hammer is chisel-shaped and is pointed on the other end to help in the removal of slag. The wire brush is used to remove the tenacious slag sticking usually at the weld bead edges. It is generally made of hardened steel wire pieces embedded in a wooden block.

Welding shield is an essential accessory for successful and continuous welding. It not only protects the eyes but also shields the face from the harmful effects of infrared and ultraviolet rays, which are emitted by the welding arc. The shield is designed to cover the entire face and the throat.

Figure 6.5 Operator accessories for SMAW

6.3.3 Electrodes

The rod filler metal should be physically and metallurgically compatible with the base material. Selection of electrode depends on:

- Workpiece thickness and groove shape
- Welding current, type, value and polarity
- Welding position

Electrodes are often referred to as light-coated, medium-coated and heavy-coated, depending on their coating factor (outer diameter of electrode/diameter of core wire), as shown in Table 6.2.

TABLE 6.2 Coating factors

Light-coated	1.2 to 1.35
Medium-coated	1.4 to 1.70
Heavy-coated	1.8 to 2.2

Every electrode has a certain current range within which it operates satisfactorily. On a light job where overheating must be avoided, current on the lower side should be used. For heavy work where maximum heat for adequate fusion is necessary, current at the top of the range is used. Current in the middle of the range is used for normal work. Welding current increases with electrode size. For a given electrode size, the current increases with coating thickness, because as the coating thickness increases, a higher current is required to melt it. Too low a current gives an unstable arc and results in lack of fusion in the welded joint. Too high a current overheats the electrode, increases spatter and results in weld metal porosity. Current amount recommended by the electrode manufacturer must be used.

Purpose of Coating

The composition of the coating on the electrode determines the usability of the electrode, the composition of the deposited weld metal, and the specification of the electrode. An electrode coating is designed to provide the following characteristics:

1. Specific composition of the deposited weld metal.
2. Specific mechanical properties of the deposited weld metal.
3. Elimination of weld metal porosity.
4. Elimination of weld metal cracking.
5. Desirable weld deposit contour.
6. Desirable weld metal surface finish, i.e., smooth, with even edges.
7. Elimination of undercut adjacent to the weld.
8. Minimum spatter adjacent to the weld.
9. Ease of manipulation to control slag in all positions.
10. Providing a stable welding arc.
11. Providing penetration control, i.e., deep or shallow.
12. Providing for initial immediate arc striking and restriking capabilities.
13. Providing a high rate of metal deposition.
14. Eliminate noxious odors and fumes.
15. Reducing the tendency of the coating to pick up moisture when in storage.
16. Reducing electrode overheating during use.
17. Providing a strong tough durable coating.
18. Providing easy slag removal.
19. Providing a coating that will ship well and store indefinitely.

According to a recent estimate, almost 80% of the electrodes used in welding are either mild-steel electrodes or low-alloy steel electrodes. The remaining 20% of electrode usage is divided among specific metal or special-purpose electrodes. Therefore, only mild and low-alloy steel electrodes have been discussed here in detail.

Functions of flux used for coating of electrodes

1. Coating provides an inert atmosphere, such as carbon dioxide under the arc protects the weld pool from atmospheric contamination and reaction of the weld pool with oxygen, hydrogen and nitrogen.
2. Coating provides flux to the molten metal pool, which is mixed with oxides and other impurities present in the metal pool to form slag. Slag being light floats on the top of puddle and protects it against the surrounding air during weld bead solidification. Slag also provides annealing effect to the weld bead.
3. Some alloying element is added in the coating, which is used for stabilizing the arc. Different flux coating is used for AC welding machine and DC welding machine.
4. Some special alloying elements are added in flux coating, which improves the properties of the weld bead.
5. As the electrode coating is consumed at a slow rate compared to the electrode, coating would be extended beyond the electrode. It helps in concentrating the arc and directing the filler metal in the joint. Even thermal losses to the atmosphere from the electrode tip can be reduced. In this way, flux coating also improves metal deposition rate.
6. Flux coating also helps in the flow of electrons in a narrow groove and other different locations without any short circuiting problem because flux coatings are insulators of electricity.
7. Flux coatings reduce the spatter.
8. Flux increases weld deposition rate (by addition of metallic powder in coating).

The metallurgical, electrical and mechanical functions of flux coating are as follows:

Metallurgical Functions	Electrical Functions	Mechanical Functions
• Provides additional alloying component (and if possible, some additional filler metal) to the weld pool. • Produces shielding gas atmosphere to the weld pool and filler metal drops, avoiding oxidation of metallic materials involved in the joint.	• Non-conductor avoids parasite arcs (for example, avoids lateral arcs in narrow gap welds). • Confines the arc due to consumable delay. promoting the electric arc stability. • Produces low ionization potential atmosphere, enabling a more stable electric arc.	• Gas metal reactions promote the metallic transfer of the core rod. • Forms slag which helps to support the weld pool. • Reduces undercut susceptibility. • Decreases cooling rate of the weld bead.

Coating Constituents

The coating of electrodes for welding mild and low-alloy steels may have from 6 to 12 ingredients, such as:

- Cellulose, to provide a gaseous shield with a reducing agent. The gas shield surrounding the arc is produced by the disintegration of cellulose.

- Metal carbonates, to adjust the pH of the slag and to provide a reducing atmosphere.
- Titanium dioxide, to help form a highly fluid but quick-freezing slag. It will also provide ionization for the arc.
- Ferromanganese and ferrosilicon, to help deoxidize the molten weld metal and to supplement the manganese content and silicon content of the deposited weld metal.
- Clays and gums, to provide elasticity for extruding the plastic coating material and to help provide strength to the coating.
- Calcium fluoride, to provide shielding gas to protect the arc, adjust the pH of the slag, and provide fluidity and solubility of the metal oxides.
- Mineral silicates, to provide slag and give strength to the electrode covering.
- Alloying metals, including nickel, molybdenum, chromium and so on, to provide alloy content to the deposited weld metal.
- Iron or manganese oxide, to adjust the fluidity and properties of the slag. In small amounts, iron oxide helps to stabilize the arc.
- Iron powder, to increase the productivity by providing additional metal to be deposited in the weld.

By using combinations and different amounts of these constituents, it is possible to provide an infinite variety of electrode coatings. The binder used for most electrode coatings is sodium silicate, which is chemically combined and hardens to provide a tough, strong coating.

Classification of Coated Electrodes (Table 6.3)

The AWS classification system classifies mild and low-alloy steel electrodes. Each coated electrode is assigned a letter 'E' prefix and a 4 or 5 digit number. The 'E' prefix indicates "covered electrode" and the number represents each of following classificatory factors:

Example: E 60 1 0
(1) (2) (3) (4)

1. E – stands for electrode.
2. Second digit indicates tensile strength × 1000PSI.
3. Third digit refers to the welding position. EXX1X; all positions (flat, horizontal, vertical, overhead). EXX2X; horizontal and flat only. EXX3X; flat position only. EXX4X; flat, overhead, horizontal, vertical down.
4. Last digit indicates unstability of the electrode, i.e., type of current of the electrode, i.e., type of current and type of covering. In some cases both third and fourth digits are significant.

The above discussion clearly gives the significance of the last digit in determining the type of power supply, type of slag, type of arc, amount of penetration, presence of iron powder in the coating, and so on. But all these characteristics of the electrodes

depend on the type of flux coating used, as the coating constituents control the parameters such as type of power supply, type of slag, type of arc action, amount of penetration, welding position and so on . So it can be concluded here that the last digit clearly indicates the type of coating used on the electrode, which in turn, defines the other characteristics of that electrode.
ISO designation is given in the Table 6.4.

TABLE 6.3 Classification of coated electrodes

Classification	Current	Arc	Penet-ration	Covering & Slag	Iron Power
EXX10	DCEP	Digging	Deep	Celluloe-sodium	0-10%
EXXX1	AC or DCEP	Digging	Deep	Celluloe-potassium	0%
EXXX2	AC or DCEN	Medium	Medium	Titanium-sodium	0-10%
EXXX3	AC or DCEN or DCEP	Soft	Light	Titanium-potassium	0-10%
EXXX4	AC or DCEN or DCEP	Soft	Light	Titanium-iron power	25-40%
EXXX5	DCEP	Medium	Medium	Low hyd-sodium	0%
EXXX6	AC or DCEP	Medium	Medium	Low hyd-potassium	0%
EXXX8	AC or DCEP	Medium	Medium	Low hyd-iron power	25-40%
EXX20	AC or DCEP	Medium	Medium	Low hyd-iron power	0%
EXX22	AC or DCEN or DCEP	Medium	Medium	Iron oxide-sodium	0%
EXX24	AC or DCEN or DCEP	Soft	Light	Titanium-iron power	50%
EXX27	AC or DCEN or DCEP	Medium	Medium	Iron oxide-iron power	50%
EXX28	AC or DCEP	Medium	Medium	Low hyd-iron power	50%
EXX48	AC or DCEP	Medium	Medium	Iron oxide-sodium	25-40%

Note: Iron powder percentage is based on the weight of the covering.
DCEN – Direct Current Electrode Negative
DCEP – Direct Current Electrode Positive

TABLE 6.4 ISO designation of electrode

Sample: EN ISO 2560-A – E 46 3 1Ni B 5 4 H5

E – Flux-coated electrode for MMA
46 – Yield stress and elongation
3 – Impact energy (47 J @ temperature)
1Ni – Chemical composition of filler metal
B – Type of flux coating
5 – Efficiency and type of current
4 – Welding position
H5 – Hydrogen content in weld metal

Contd.

EN ISO 2560-A – E <u>46</u> 3 1Ni B 5 4 H5

Symbol	Minimum Yield[1] Strength N/mm²	Tensile Strength N/mm²	Minimum Elongation%[2]
35	355	440–570	22
38	380	470–600	20
42	420	500–640	20
46	460	530–680	20
50	500	560–720	18

1. For yield strength the lower yield ($R_{el.}$) shall be used when yielding occurs, otherwise the 0, 2% proof strength (R_{Po2}) shall be used.
2. Gage length is equal to five times the specimen diameter.

EN ISO 2560-A – E 46 <u>3</u> 1Ni B 5 4 H5

Symbol	Temperature for Minimum Average Impact Energy of 47°C
Z	No requirement
A	+20
0	0
2	–20
3	–30
4	–40
5	–50
6	–60

EN ISO 2560-A – E 46 3 <u>*1Ni*</u> B 5 4 H5

Alloy Symbol	Chemical Composition[1,2,3]		
	Mn	Mo	Ni
No symbol	2, 0	–	–
Mo	1, 4	0, 3 – 0, 6	–
MnMo	> 1, 4 – 2, 0	0, 3 – 0, 6	–
1Ni	1, 4	–	0, 6 – 1, 2
2Ni	1, 4	–	1, 8 – 2, 6
3Ni	1, 4	–	> 2, 6 – 3, 8
Mn1Ni	> 1, 4 – 2, 0	–	0, 6 – 1, 2
1NiMo	1, 4	0, 3 – 0, 6	0, 6 – 1, 2
Z	Any other agreed composition		

1. If not specified Mo < 0, 2, Ni < 0, 3, Cr < 0, 2.
   V < 0, 05, Nb < 0, 05, Cu < 0, 3.
2. Single values shown in the table are mean maximum values.
3. The results shall be rounded off to the same number of significant figures as in the specified value, using the rules according to ISO 31–0, annex B Rule A.

EN ISO 2560-A – E 46 3 1Ni *B* 5 4 H5

A = acid covering
C = cellulosic covering

R = rutile covering
RR = rutile thick covering
RC = rutile-cellulosic covering
RA = rutile-acid covering
RB = rutile-basic covering

B = basic covering

EN ISO 2560-A – E 46 3 1Ni B *5* 4 H5

Symbol	Nominal Electrode Efficiency, %	Type of Current[a, b]
1	⩽ 105	A.C. + D.C.
2	⩽ 105	D.C.
3	> 105 but ⩽ 125	A.C. + D.C.
4	> 105 but ⩽ 125	D.C.
5	> 125 but ⩽ 160	A.C. + D.C.
6	> 125 but ⩽ 160	D.C.
7	> 160	A.C. + D.C.
8	> 160	D.C.

[a] In order to demonstrate operability on A.C., tests shall be carried out with no load voltage higher than 65 V.
[b] A.C. = alternating current D.C. = direct current.

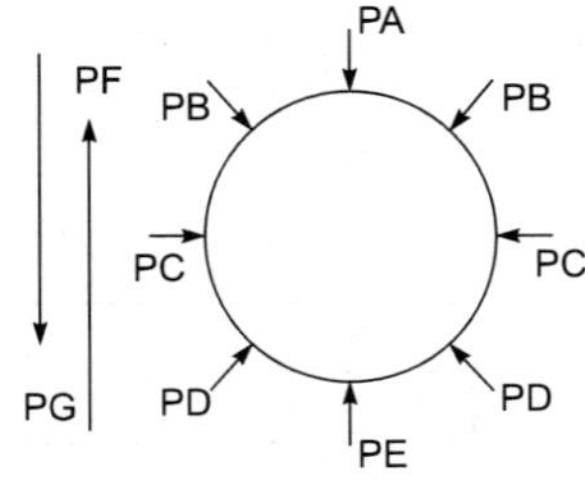

Symbol	Positions
1	PA, PB, PC, PD, PE, PF, PG
2	PA, PB, PC, PD, PE, PF
3	PA, PB
4	PA
5	PA, PB, PG

Symbol	Hydrogen content ml/100 g all-weld metal max.
H5	5
H10	10
H15	15

Types of Electrode Coating

The principal types of electrode coating for mild and low-alloy steels are described below:

1. **Cellulosic:** The cellulosic type of electrodes contain a large quantity of combustible organic substance having cellulose and some titanium oxide. The major constituent of this covering is cellulose, usually exceeding 30% by weight. Other organic materials such as wood flour, charcoal cotton, yarn, paper wrapping, starches and gums have also been used to partially replace cellulose. Other materials generally used are titanium dioxide metallic deoxidizer such as ferromanganese, various types of magnesium or aluminum silicate and liquid sodium silicates as binder.

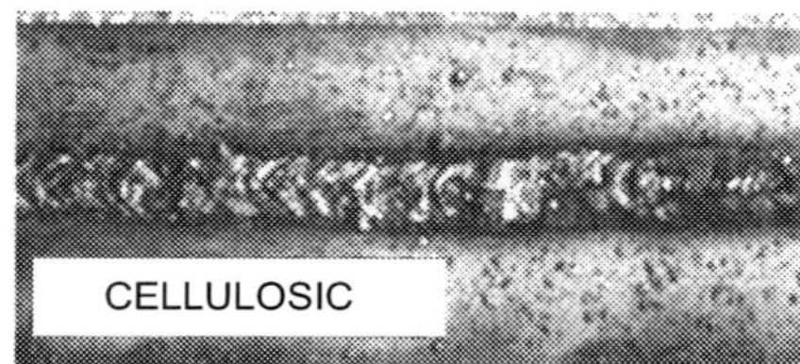

Figure 6.6 (a) Flux coating based on cellulose (organic materials), silicates and some reducing oxide elements. Irregular thin slag easy to remove. Bad visual appearance of weld seam. Very significant spatter. Good amount of shielding gases. Intermediate mechanical properties. Very high penetration/width ratio. Very sensitive to hydrogen

2. **Acidic:** The covering of acidic type electrodes contains oxides and carbonates of iron or manganese, along with fairly high percentage of ferro-manganese and other deoxidizers. The major constituent of this covering is iron oxide. Other constituents are manganese compounds and silica, along with sufficient deoxidizers. Organic materials are absent. Potassium silicate is used as a binder.

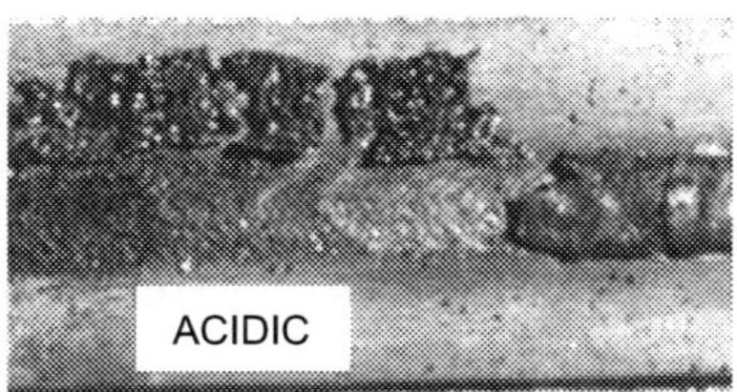

Figure 6.6 (b) Flux coating based on ferrous oxides, silicates and deoxidant elements. Porous slag. Good visual appearance of weld seam. Some spatter. Good amount of shielding gases needed. Intermediate mechanical properties

3. **Rutile:** These electrodes contain a large quantity of rutile component, derived from titanium oxide, usually exceeding 50% by weight. Rutile is naturally

occurring sand and contains around 95% titanium oxide. It has a natural dark black, shining appearance, unlike pure titanium dioxide, which is snow white and is therefore, also known as titanium white. The covering also contains slag forming siliceous materials like feldspar and china clay and calcium and potassium compounds to ensure arc stability in arc welding. Cellulose and ferro-magnetic in varying amount are also included. The binder is sodium silicate or potassium silicate.

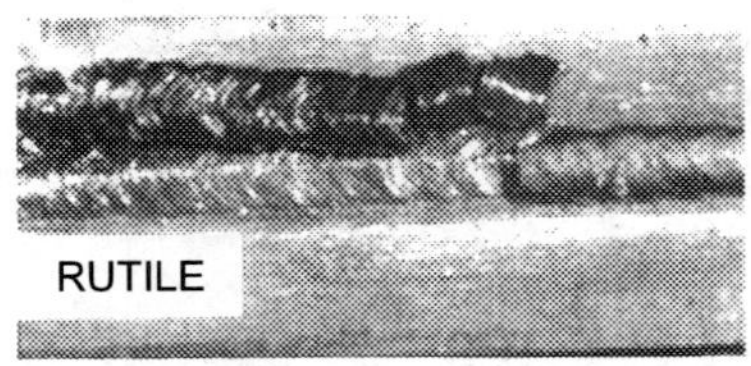

Figure 6.6 (c) Flux coating based on titanium oxide silicates and deoxidized elements. Porous slag is easy to remove. Good visual appearance of a slightly convex weld seam. Some spatter. Good amount of shielding gases needed. Good mechanical properties

4. **Oxidizing:** These types of electrodes have a thick covering, composed mainly of iron oxide with or without manganese oxide and silicate. The covering produces an oxidizing slag, so that the weld metal has a low carbon and manganese content, which is referred to as 'dead soft'. The resultant weld metal is therefore dead soft and has low mechanical properties. In these types of electrodes, the core wire melts up inside the coating forming a cup that can be used for 'touch welding'.

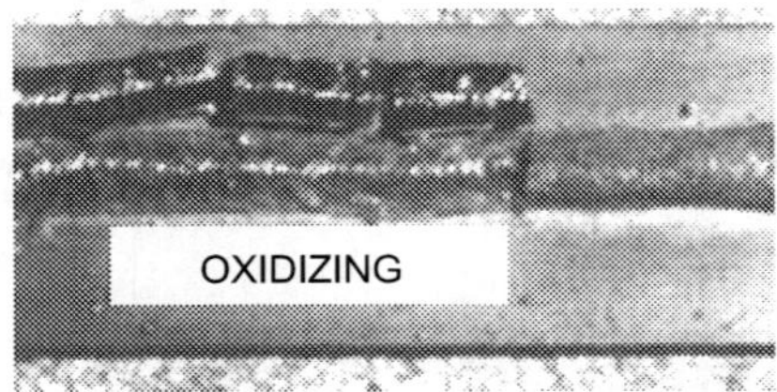

Figure 6.6 (d) Flux coating based on ferrous oxides. Compact slag. Good visual appearance of weld seam. Some spatter. Good amount of shielding gases. Bad mechanical properties

5. **Basic:** Electrodes of the basic type usually have thick covering of considerable quantity of calcium or other basic carbonate of fluorspar. The so called hydrogen controlled mild steel electrode are generally made with this type of coating, and this coating is also preferred in low-alloy high tensile, in order to minimize the risk of hydrogen, including cracking or cold cracking in base metal.

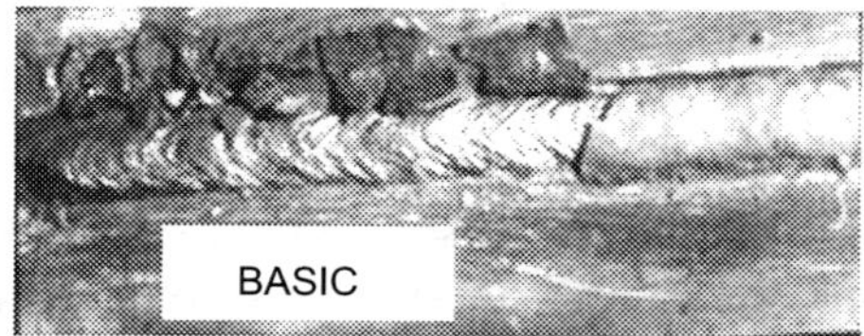

Figure 6.6 (e) Flux coating based on calcium and magnesium carbonates and deoxidized elements. Compact slag hard to remove. Good visual appearance of a slightly convex weld seam. Significant spatter. Good amount of shielding gases. The best mechanical properties. Very sensitive to hydrogen

6.4 OPERATING PARAMETERS

Shielded metal arc welding is a versatile welding process, and is used to weld a variety of ferrous and non-ferrous metals. The selection of the type of power source, current and voltage setting, polarity, and the type of electrode depends on the metal to be welded, joint thickness and geometry, welding position and intended purpose of the joint. Nowadays, the only selection to be made by the welder in SMAW process is the type of electrode required to weld, rest of the operating parameters are dictated on the electrode packaging, and are well defined by the electrode manufacturer.

6.4.1 Selection of Coated Electrodes

Many factors must be considered when selecting the proper electrode for a given application. Some items to be considered are:

1. **Type of Base Metal:** Welding mild steels or low-carbon steels (carbon content below 0.30%) with mild steel-coated electrodes presents no problems as far as tensile strength is concerned, since the tensile strength of the weld metal usually exceeds the tensile strength of the base metal. However, chemistry of the base metal is important. For example, welds made on free machining steels that have relatively high sulphur content, will be porous unless welded with a low hydrogen-type electrode, such as E7018. Therefore, matching the mechanical properties of the base metal and chemical composition as closely as possible is necessary.
2. **Position of the Weld:** Weld position will determine whether an all-position electrode or a flat and horizontal type electrode should be used. Higher welding currents and therefore, higher deposition rates are possible when welding flat or horizontally. Whenever possible, the work should be positioned both for ease of welding and to attain the highest welding speed.

3. **Available Equipment:** Electrode choice will depend on whether AC or DC welding machines are available. If both currents are available, consider these general facts:
   (a) For deepest penetration, use DC reverse polarity (electrode positive).
   (b) For lower penetration and higher deposition rate, use DC straight polarity (electrode negative).
   (c) For freedom of arc blow, use AC.
4. **Service Conditions:** Observe the service conditions and select electrode accordingly – high temperature, low temperature, corrosive atmosphere, impact loading, etc. These conditions are best met by using low-hydrogen electrodes.
5. **Plate Thickness:** When welding sheet metal, low penetration electrodes should be chosen. Heavier plate may demand an electrode with deep penetration. Very heavy plate may require a deep penetrating electrode for the initial or root pass, and a higher deposition type for succeeding passes.
6. **Fit Up:** Some electrodes are more suitable than others for bridging gaps between the members to be welded. This is termed "poor fit-up" and some electrode manufacturers make electrodes that are especially formulated for this purpose.
7. **Welding Costs:** The major factors that affect welding costs are labor and overheads, deposition rate, efficiency of the electrode being used and the cost of the electrodes. The cost of electrical power is also a factor to a lesser degree. By far, the largest factor is labor and overheads.
8. **Welder Appeal:** Welder appeal is definitely important, although this factor must not be allowed to subordinate other more significant criteria.

6.4.2 Polarity and Penetration (Figure 6.7)

The polarity indicates the direction of the current in the DC circuit. Since some welding operations demand that the flow of current be changed, a polarity switch or indicator is there on a DC welding machine. When the electrode cable is fastened to the negative pole on the generator and the piece to be welded is the positive pole, the polarity is negative; or, more commonly, is referred to as straight polarity (DCSP) or direct current electrode negative (DCEN), electron flow goes from the electrode into the work. When the electrode is positive and the work is negative, the electron flow is from the work to the anode, which is called reverse polarity or direct current electrode positive (DCEP).

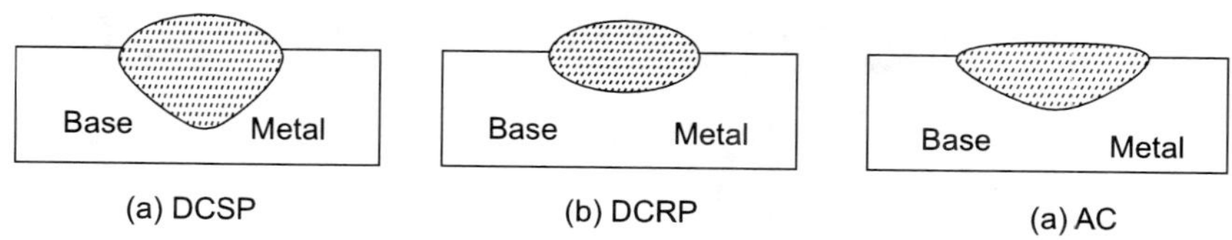

Figure 6.7 Polarity and penetration

Nearly two-thirds of the heat is developed near the positive pole while the remaining one-third is developed near the negative pole. As a result, an electrode that is connected to the positive pole will burn away approximately 50% faster than one that is connected to the negative pole. This is helpful in obtaining the desired penetration of the base metal. If straight polarity (DCSP) is used, the penetration will be greater because of the amount of heat energy supplied by the electrode force to the work. At the same time, the electrode will burn away slowly. If the poles are reversed and reverse polarity (deep) is applied, two-thirds of the heat will remain in the tip of the electrode. For this reason, the penetration of the heat zone in the base metal will be shallow when compared to the penetration depth of the DCSP arc column.

Alternating current combines the characteristics of both, the deep penetration and the shallow penetration, yielding a penetration depth that is approximately half way between that achieved by the positive ground and negative ground. The electron flow switches ground every time the AC cycle changes, yielding a bead penetration depth approximately between the two DC types.

In both the AC and DC power sources, the arc serves the same purpose. It produces heat to melt the metal. If two pieces of metal that are to be joined are placed so that they touch or almost touch one another, and the arc form the electrode is directed at this junction, the heat generated by the arc causes a small section of the edges of both pieces of the metal to melt. These molten portions along with the electrode flow together. As the arc column is moved, the molten puddle solidifies, joining the two pieces of metal with the combination of electrode metal and base metal.

6.5 WELD JOINT DESIGN FOR SMAW

The weld joint design used for shielded metal arc welding is determined by the design of the workpiece, metallurgical considerations, and by codes or specifications. Joints are designed for accessibility and economy during construction. Good accessibility during construction also helps reduce the cost and generally raises the quality of the weld joint. A weld joint consists of a type of weld made in a type of joint. There are five basic types of joints, but these can be used in combinations. A joint is defined as being the junction of members which are to be joined or have been joined. Figure 6.8 shows the five basic joints classification. Each of the different types of joints can be joined by many different types of welds. Figure 6.8 shows the most common types of welds. The type of weld is governed by the joint configuration. Figure 6.9 shows the weld nomenclature used for groove and fillet welds.

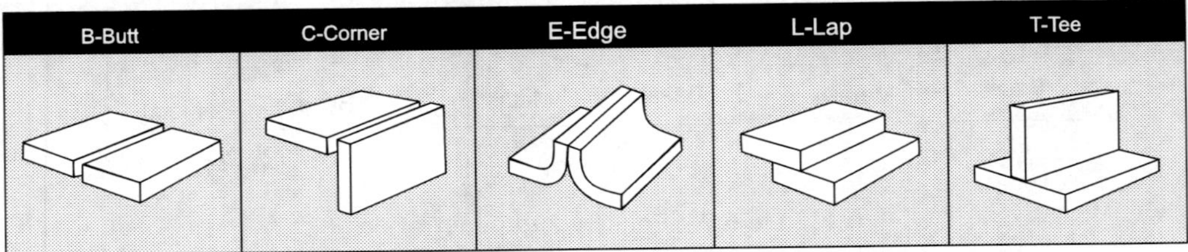

There are only five basic joints. They can, however, be used in combinations.

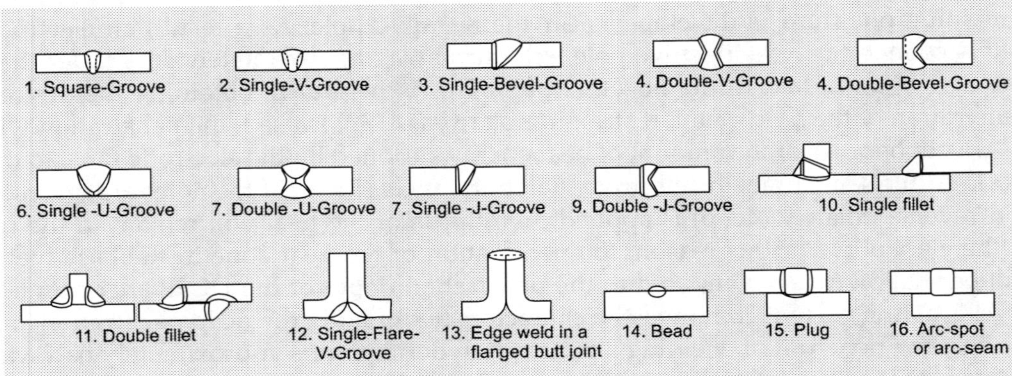

Figure 6.8 Weld joint designs for SMAW

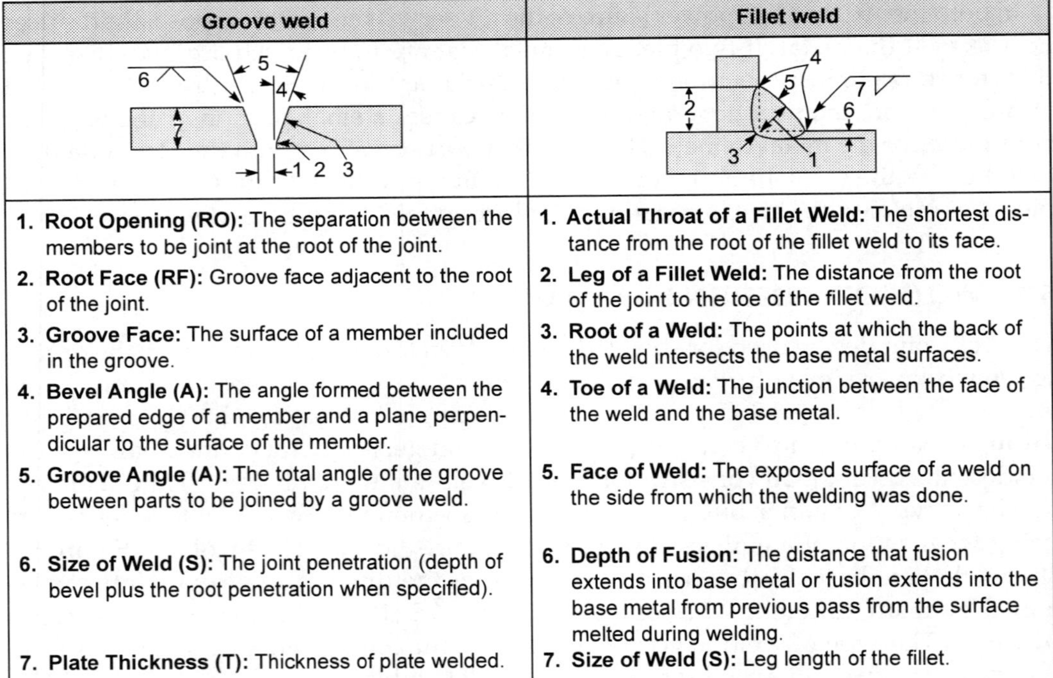

Groove weld	Fillet weld
1. **Root Opening (RO):** The separation between the members to be joint at the root of the joint.	1. **Actual Throat of a Fillet Weld:** The shortest distance from the root of the fillet weld to its face.
2. **Root Face (RF):** Groove face adjacent to the root of the joint.	2. **Leg of a Fillet Weld:** The distance from the root of the joint to the toe of the fillet weld.
3. **Groove Face:** The surface of a member included in the groove.	3. **Root of a Weld:** The points at which the back of the weld intersects the base metal surfaces.
4. **Bevel Angle (A):** The angle formed between the prepared edge of a member and a plane perpendicular to the surface of the member.	4. **Toe of a Weld:** The junction between the face of the weld and the base metal.
5. **Groove Angle (A):** The total angle of the groove between parts to be joined by a groove weld.	5. **Face of Weld:** The exposed surface of a weld on the side from which the welding was done.
6. **Size of Weld (S):** The joint penetration (depth of bevel plus the root penetration when specified).	6. **Depth of Fusion:** The distance that fusion extends into base metal or fusion extends into the base metal from previous pass from the surface melted during welding.
7. **Plate Thickness (T):** Thickness of plate welded.	7. **Size of Weld (S):** Leg length of the fillet.

Figure 6.9 Weld joint designs for SMAW

6.5.1 Butt Welds (Figure 6.10)

When the material thickness does not exceed 5 mm, it is often welded by simply placing the sheared and straightened edges close to each other with a gap of 1 to 2 mm. This type of joint is referred to as square butt joint. When the material thickness exceeds 8 mm, the edges are then prepared by machining to 'Vee' configuration. When the material thickness is more than 20 mm, it is better to have double **'Vee'** edge preparation.

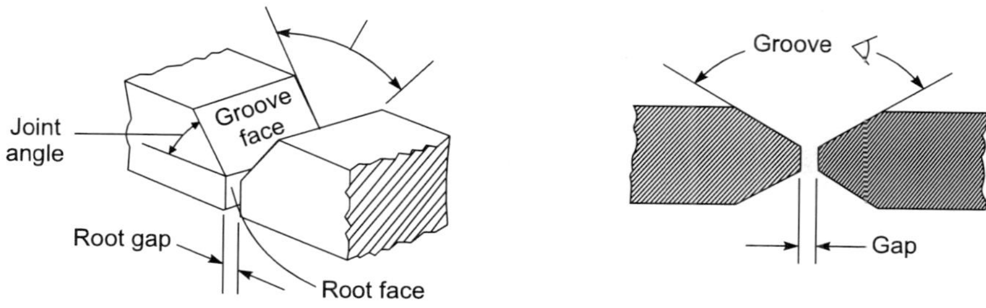

Plate Thickness (mm)	Electrode Size (mm)	Joint Preparation
Up to 5-8	3-4	2-4
4-26	4-5	60°, 2, 2-5
12-60	4-6	60°, 3-5, 3-4

Figure 6.10 Butt welds

6.5.2 Lap Welds (Figure 6.11)

A lap joint is one in which the plates overlap each other. The width of the lap is usually between 3 and 5 times the thickness, *t*. Proper cleaning should be done for a lap joint. Lap joints are not recommended for plate thickness above 10 mm.

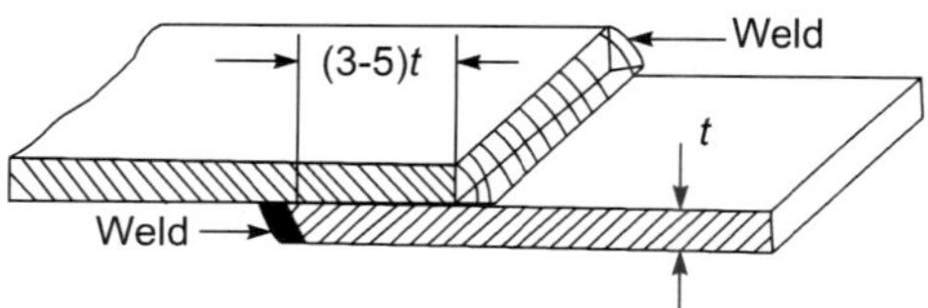

Figure 6.11 Lap welds

6.5.3 Fillet Welds (Figure 6.12)

Depending upon the thickness of the joining plates, the vertical member may have square or bevel or *J* edge preparation. These joints may be obtained with single-run or multi-run welds and that depends upon the thickness of the plat and the edge-preparation.

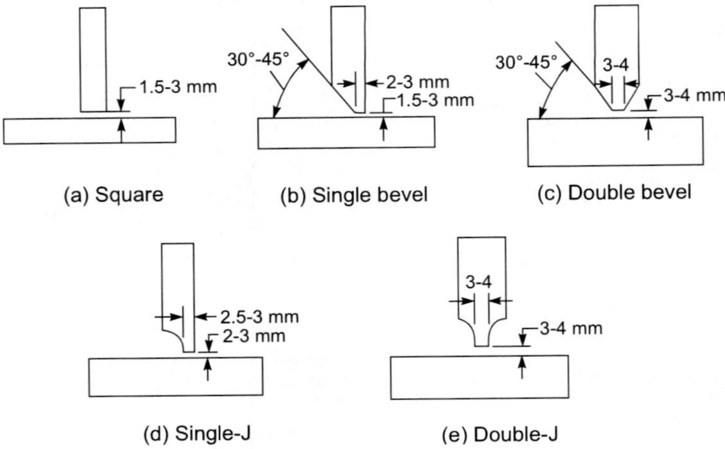

Figure 6.12 Fillet welds

6.6 WELDING CIRCUIT (FIGURE 6.13)

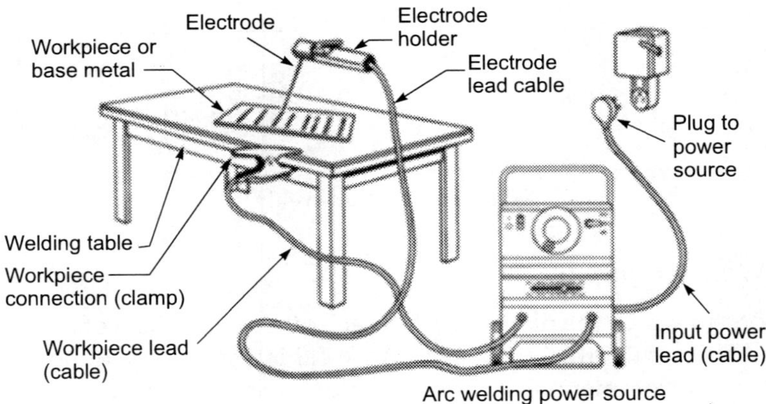

Figure 6.13 Equipment

SMAW Operation (Figure 6.14)

Once the electrical connections are made and a coated electrode is held in an electrode holder, the system is ready for the welding operation. There are usually two settings for open circuit voltage (OCV), 80 V and 100 V. The OCV is selected depending upon the type of coating of the electrode used. The current settings are usually available in steps of 5 to 10 A and need to be adjusted depending upon the core wire diameter of the electrode. After setting the OCV and the welding current, the only thing left off is to make provision for the protection of the eyes before initiating the arc to commence welding operation.

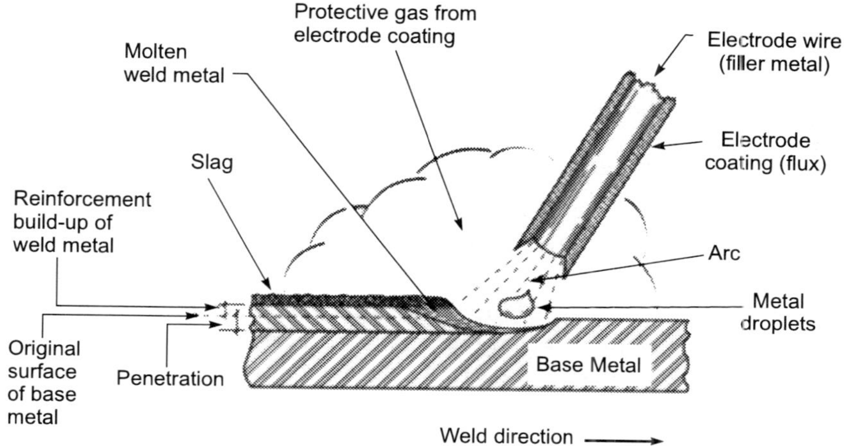

Figure 6.14 SMAW operations

6.6.1 Striking the Arc (Figure 6.15)

To initiate an arc, it is essential to ionize a small space around the point where welding is to commence. This is particularly so when the workpiece and the electrode are both cold.

There are two methods known as 'touch-and-draw' method and 'drag' method. In touch-and-draw method, the electrode is made to touch the workpiece at the desired spot and is quickly drawn back through a distance of 3 to 4 mm. In 'drag' method of arc initiation, the electrode is made to rub against the job at the required spot and that helps in ionizing a small volume of space around the spot and also some metal vapors are released. These conditions help in striking an arc normally in two or three attempts.

Having initiated a stable arc, the next step to achieve a good weld by SMAW is to control the arc length. For SMAW, an arc length of 2 to 4 mm is found to give the best results. Alternatively, it may be determined by the relationship

$$L = (0.5 - 1.1)\,d$$

where, L is the arc length and d is the electrode diameter.

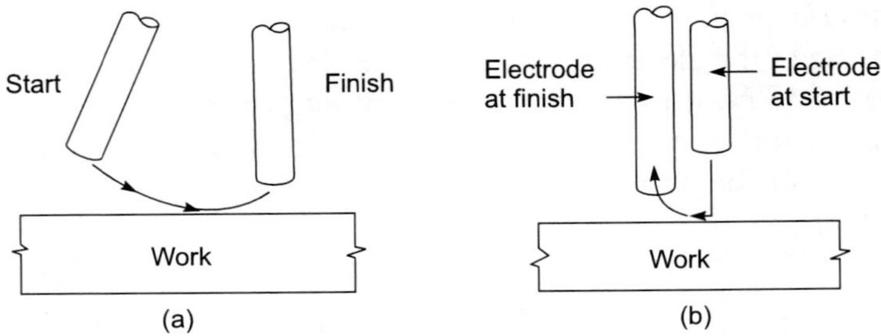

Figure 6.15 (a) Scratch method (b) Tapping method of striking the arc

6.6.2 Restriking the Arc (Figure 6.16)

Restriking an arc with the semi-consumed hot electrode on the hot workpiece is comparatively easy as the electron emission is facilitated by the high temperature of the electrode. The restriking of an arc should always be attempted at a spot 15 to 25 mm ahead of the crater away from the weld bead. Once a stable arc is established, it is moved towards the weld crater where it is held momentarily to fill the crater and then quickly moved on its intended path to achieve the weld. Restriking the arc ahead of the crater is done to avoid entrapment of slag in the weld metal.

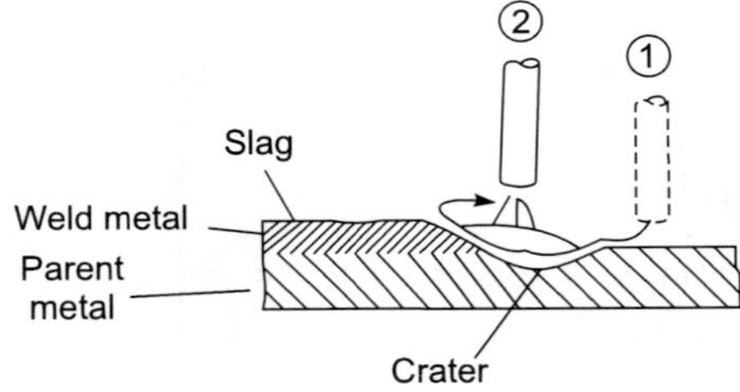

Figure 6.16 Restriking an interrupted arc

6.6.3 Metal Fusion and Weld Penetration (Figure 6.17)

To obtain a good weld, the depth of penetration must not be less than 1.5 to 2 mm. In SMAW, depending upon the welding current, the penetration usually varies between 1.5 and 5 mm. The depth of penetration depends upon the heat input into the workpiece per unit time, and thus depends upon the welding current. A cross section of a number of weld beads deposited on a plate with varying currents can depict the influence of welding current on the depth of penetration.

For controlling penetration, welding current is chosen according to the electrode grade and diameter. However for downhand welding of butt joints in low-carbon steel, the welding current may be determined as rough guide from the following relationships,

$$I = (40 - 20d)$$

$$I = (20 + 60d)$$

where, I is the welding current in amperes and d is the electrode diameter in mm.

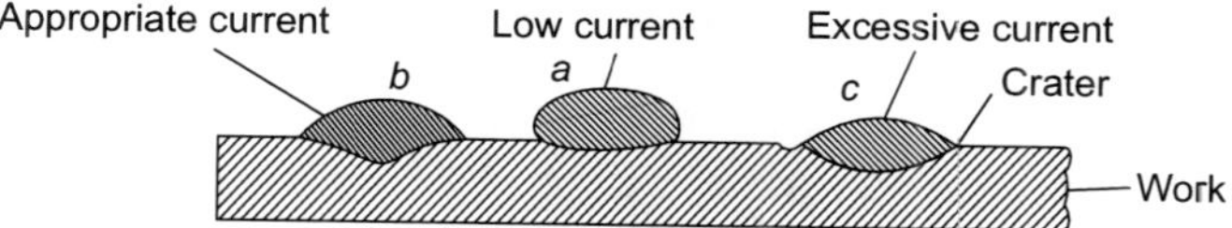

Figure 6.17 Effect of welding current on bead penetration

Welding current required for a thinly coated electrode is lower than that for heavily coated electrode. The optimum current for a given electrode and workpiece may be found by trial and error, by depositing a number of beads, using the equations above. Larger currents are required to be set for both heavier section and electrode size to achieve the desired penetration, because a heavy section acts as an efficient heat sink.

In a multi-run weld, the first run should be made with an electrode of not more than 2 to 3.15 mm diameter. For overhead and vertical welding, the electrode should have a maximum diameter of 4 mm. Electrodes of 5 mm diameter can be used to expedite welding in downhand welding position, particularly the filler and finishing runs.

6.6.4 Electrode Motions (Figure 6.18)

The width of the weld bead formed under normal welding conditions in SMAW is between 1.5 and 2.5 times the diameter of the electrode; with well penetrated and smooth passage of the deposited metal to the workpiece surface. To achieve this, arc length is kept as short as possible without the electrode sticking to the workpiece and by giving the electrode three types of motions simultaneously. One such motion is the continuous downhand feeding of the electrode towards the weld pool, the second motion is the advancing motion of the arc along the joint and the third motion is the sidewise and lateral oscillating motion or waving motion across the arc.

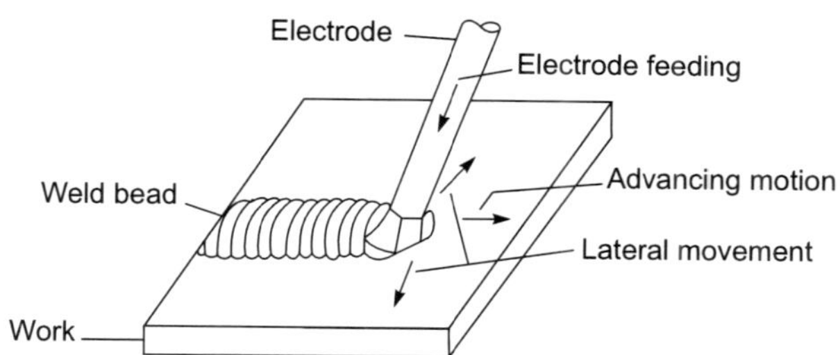

Figure 6.18 Three motions of an electrode during welding

When the arc is advanced without any waving motion, the width of the bead is usually 1 to 2 mm more than the electrode diameter. The bead so obtained is called a 'stinger bead' (Figure 6.18).

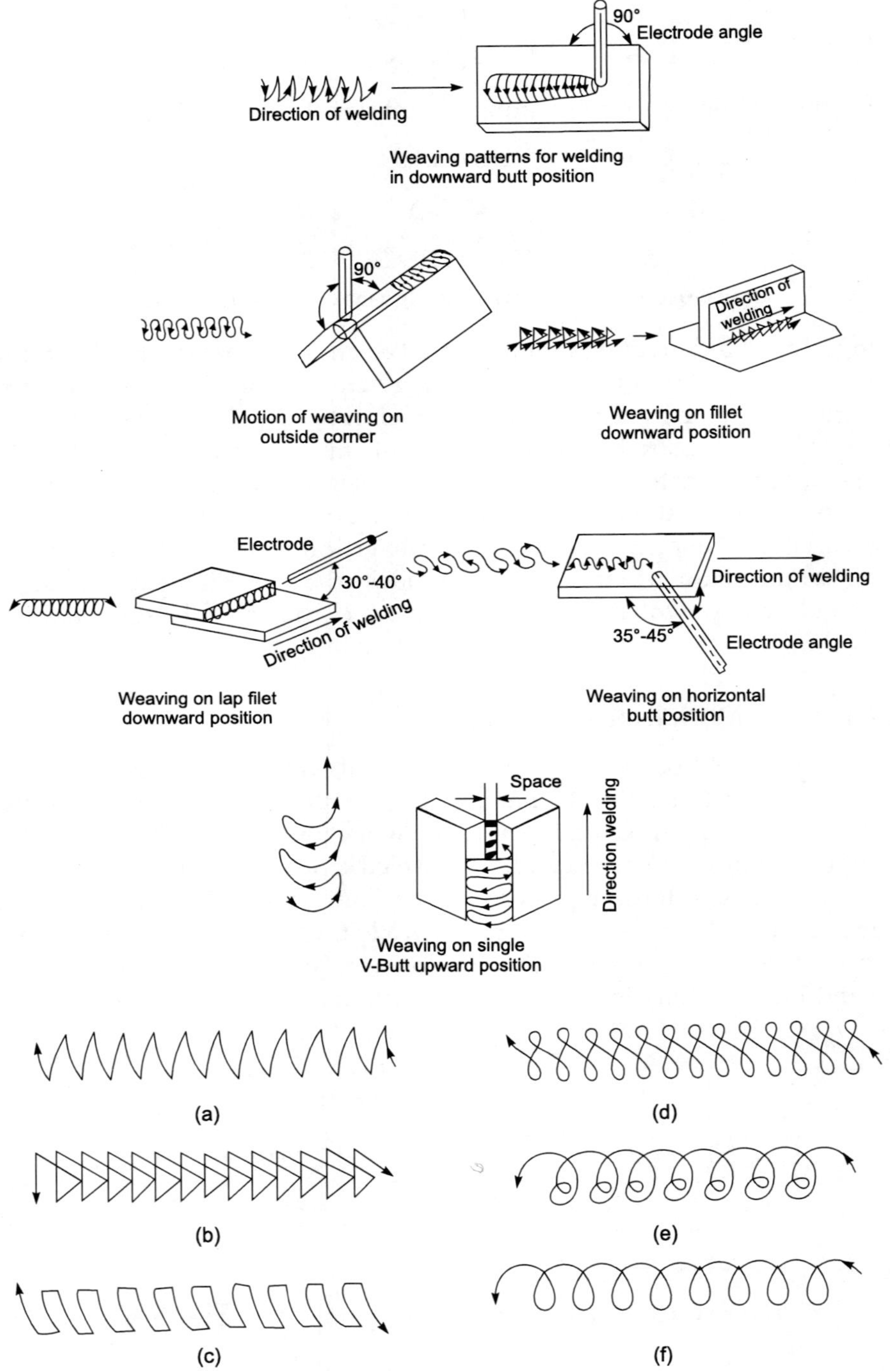

Figure 6.19 Diagram of weaving motion of electrodes

Weaving motion during welding is used when a spread bead or weave bead is required. Weave beads are commonly employed for making butt and fillet welds.

Weaving can be accomplished in a variety of patterns, depending upon the type of weld, joint preparation, and the skill of the operator. For consistency in bead width, it is essential that the swing of the waving motion is kept constant. A correct, well-penetrated and sound weld of high quality can be obtained only if the operator's movements are well controlled in all the three directions.

6.6.5 Positions in Welding (Figure 6.20)

Depending upon the position during welding, all welds are classified into four basic groups, viz., downhand or flat, horizontal, vertical and overhead welds.

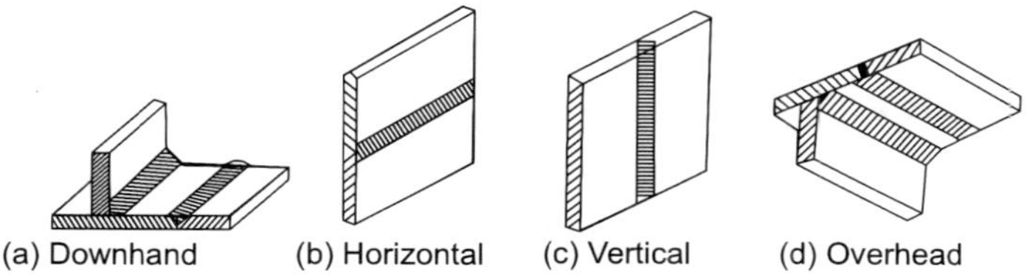

Figure 6.20 Different positions in welding

6.6.6 Downhand or Flat Welding (Figure 6.21)

This type of welding is the most common one. Welding done not in flat place is called 'out of space' welding. This requires least skill for a sound weld. There is no risk of molten metal running out of pool. It is most convenient to watch welding in this position. Elaborate fixtures, called positioners, are employed to rotate the work to bring it into downhand position.

The electrode should be held at 90° to the workpiece, with electrode tilted through 10° to 20° in the direction of welding. This selection of angles depends on the power source, the voltage and current settings, and the thickness of the workpiece. The stringer bead motion or the hesitation type movements are the common types of movement in flat position. Downhand welding is mainly used for butt welds, fillet welds and padding welds.

Downhand Welding of Butt Welds (Figure 6.22)

Up to 5 mm, square butt welding is employed and the gap between plates is kept between 2 and 4 mm. A slightly spreading bead with well melted fusion faces is deposited along the joint and the reinforcement height is limited up to 2 mm. If a sealing run is required to be laid, then the job is turned over, excess metal is chipped off and the joint is cleaned with a steel wire brush before welding.

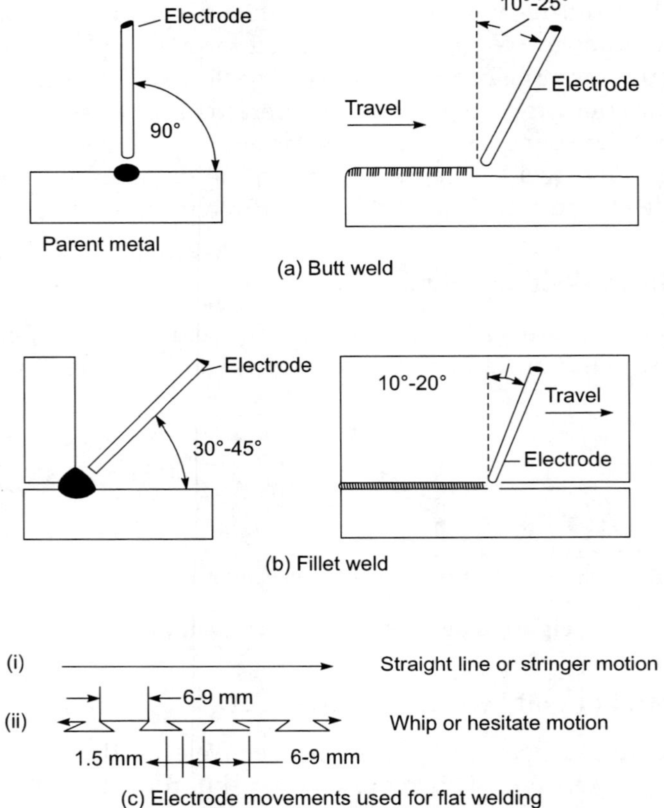

Figure 6.21 Downhand SMAW techniques

For 6 to 8 mm thick V edge preparation, a single-run weld is deposited. Thorough fusion of groove spaces is needed for full penetration. The arc should be started near the edge of the bevel and then moved to the groove at slow pace so as to obtain a good penetration at the root. To avoid any burn through, pace should quicken when going from one groove face to another. More than one weld run is required for plate thickness of more than 8 mm.

After cleaning the root run, the next run is made with an electrode of 4 or 5 mm diameter. The cross-sectional area of the weld run being deposited is usually correlated with the diameter of the electrode used. Cross-sectional area of the root run is given by equation $Fr = (6 \text{ to } 8d)$ and for the subsequent runs, $Fs = (8 \text{ to } 12d)$, where d is the diameter in mm.

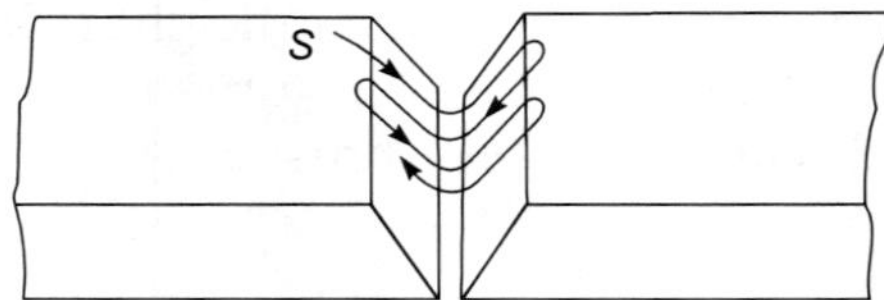

Figure 6.22 Electrode movements for a single V butt weld

In multi-run welds, slags and spatter should be cleaned before the subsequent runs and a final or cosmetic run should be made with appropriate reinforcement for the sake of appearance. To make the sealing run, the job is turned over and the metal is chipped off manually or with a pneumatic chisel. The double V edge preparation is the same as the single V edge but the job must be turned over a number of times if the reverse side is to be welded in the downhand position (Figure 6.23).

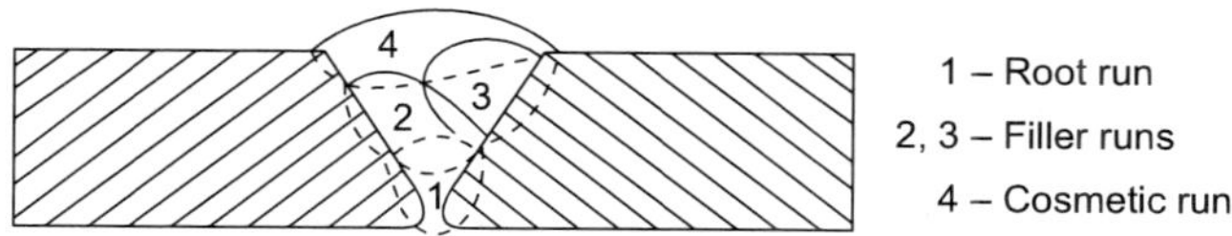

Figure 6.23 Multi-run butt weld

Downhand Fillet Welds (Figures 6.24 & 6.25)

The fillet welds are made in the downhand or horizontal position. One member is placed horizontally and the other, perpendicular to it. The weld is put at the intersection of two members, either on one or both sides. This type often suffers from poor penetration at the root and poor fusion at one surface of the weld. This angle may, however, be varied to get more heat. Fillet welds with a leg length up to 8 mm are made in a single pass. To initiate fillet welding, the arc is stuck on the horizontal surface at a distance equal to the fillet leg plus 3 or 4 mm. The weld should not be started on the vertical member or corners, this may lead to unmelted parent metal.

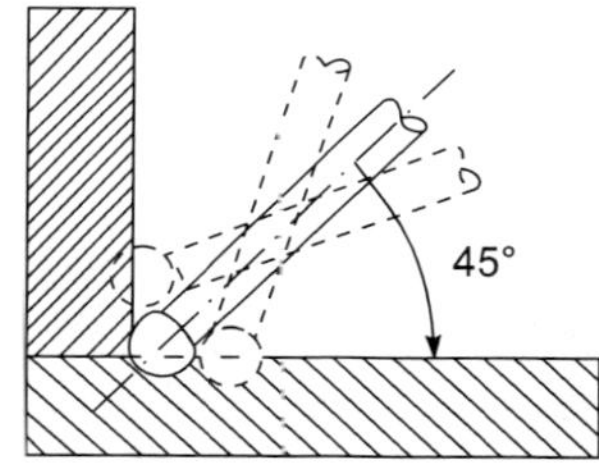

Figure 6.24 Electrode angles in fillet welds

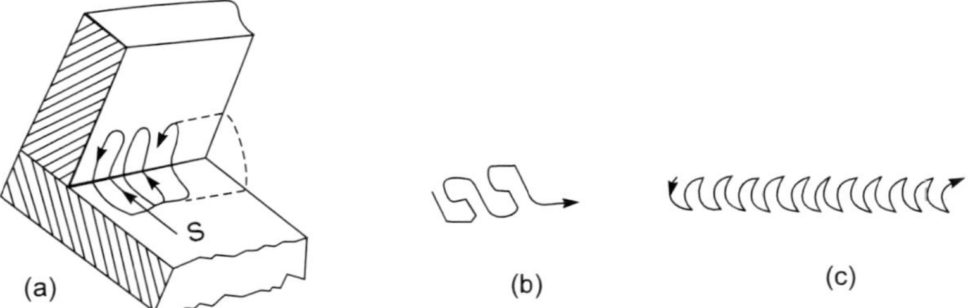

Figure 6.25 Electrode motions in fillet welds; (a) Electrode movement for a single-run weld, (b, c) Electrode movement for filler runs in a multi-run fillet weld

Padding Weld (Figure 6.26)

A padding weld consists of successive layers of overlapping weld beads. It is used for building up of broken or worn parts, filling in large cavities when heavy metals are welded, etc. The surface is thoroughly cleaned before depositing the first run, using narrow or slightly spreading bead. This is followed by subsequent runs carefully. If the two adjacent padding beads are separated by a depression, the padding will be discontinuous. Before the next laying, the beads already laid should be thoroughly deslagged. In multi-layer padding, care should be given to cleaning the beads laid with heavy coated electrodes, as they produce more slag which causes depression or undercuts.

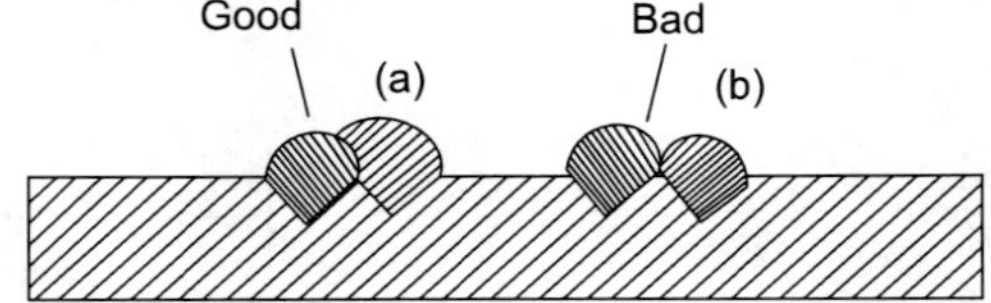

Figure 6.26 Bead placements in padding welds

6.6.7 Horizontal Welding (Figure 6.27)

Most commonly used in welding vessels and reservoirs, horizontal welding has a deposited rate next to that in downhand welding, making them popular. The edge preparation adopted is usually single bevel. The lower plate is not beveled to prevent the metal from running down. The arc initiation is done on the horizontal edge on the lower plate and then moved to beveled face while tilting the electrode backwards. In plates thicker than 8 mm, multi-layers are used. The preferred weld movements are *C*, *J*, *O* and whip or hesitation types. Angles are between 5° and 25° with the electrode

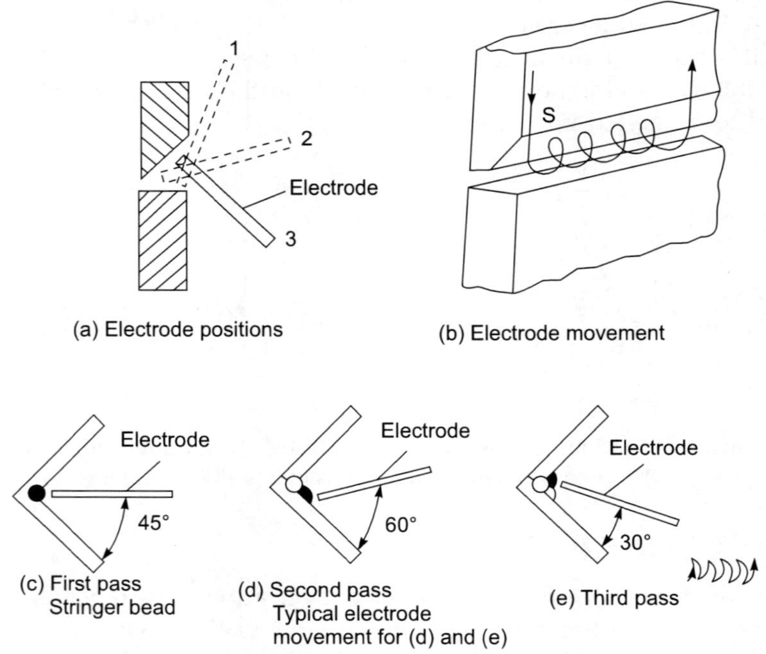

Figure 6.27 Electrode inclinations in horizontal welding

tip pointed upwards to reduce the effect of gravity and the inclination in the direction of welding is 10° to 25°. The sagging of molten puddle may be prevented by shorter arc length and quicker electrode movement. Improper horizontal welding, meanwhile leads to undercuts and overlaps.

6.6.8 Vertical Welding (Figure 6.28)

There are two types of vertical welding; vertical-up welding, which is most commonly used because it allows heat to penetrate deeper and thus providing stronger welds and vertical-down welding, which is used for sealing operations and for using sheet metal. Vertical welds with double V and single V as well as vertical fillet welds are made the same way as downhand welding but smaller diameter electrodes are used to avoid spilling of molten metals downwards. The electrode is inclined 10° to 20° to assess the progress of vertical-up welding.

Vertical welding is the laying of one weld puddle over the other, normally by short circuit mode of metal transfer, hence keeping shorter arc length. Normal movements are oval, C with hesitation, or whip movement. The major problems to avoid are breaking the arc, losing the arc column and restarting without cleaning. The main downfall of the vertical down welding is that the slag often runs ahead of the molten metal entrapping in it, resulting in poor penetration.

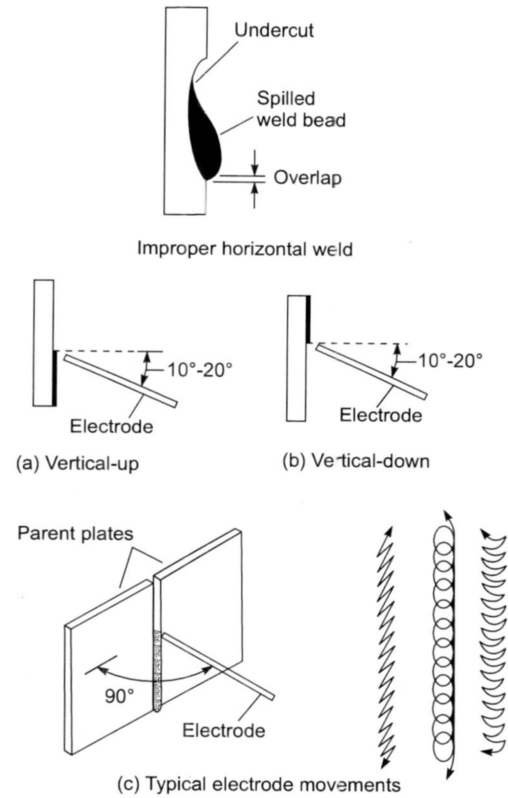

Figure 6.28 Electrode inclinations in vertical welding

6.6.9 Overhead Welding (Figure 6.29)

This is difficult to carry out because (a) dripping of molten metal downwards, and (b) flying sparks and spatter. Therefore, it is essential to use a very short arc length obtained by short circuiting mode of metal transfer. To cause fast solidification of molten metal, the electrode should be tilted 10° to 25° in the direction of welding with rapid electrode manipulation. Oval, whip and zig-zag are the common movements used.

It is a good practice to use coated metal, as this type of coating melts at a lower rate and thus, provides a protective barrel for the metal to be projected to the weld pool. This also helps lesser spatter. The current used in this is around 20% to 25% lower than in downhand welding. To reduce the fatigue of arm and hand, welder should drape the electrode over his shoulder to avoid the downward pull due to the weight of the cable.

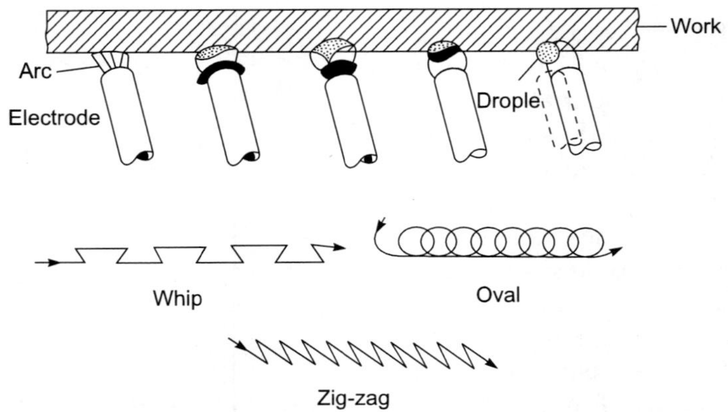

Figure 6.29 Typical electrode movements for overhead welding

6.7 WELDING PROCEDURE VARIABLES

The welding procedure variables are those that control the welding process and the quality of the welds. Three major types of welding variables used for welding are the fixed or preselected, primary adjustable, and the secondary adjustable variables. The fixed or preselected welding variables are those that are set before the actual welding starts. These are items such as electrode type, electrode size, and type of current. These variables cannot be changed after welding starts.

The primary adjustable variables are the major variables used to control the welding process, once the fixed variables have been selected. The primary variables control the formation of the weld bead by affecting the bead width, bead height, penetration, arc stability, and weld soundness. The primary welding variables are welding current, arc voltage, and travel speed. These can be easily adjusted and measured and so they can be used effectively to control the welding process.

The secondary adjustable variables are the minor adjustable variables that are used to control the welding process. These variables are usually more difficult to measure. Secondary adjustable variables are the work angle and the travel angle of the electrode.

The penetration of the weld is defined as the greatest depth below the surface of the base metal that the weld metal reaches. The bead height or reinforcement is the height of the weld metal above the surface of the base metal. The deposition rate is the weight of the metal that is deposited per unit of time. The definitions of bead height, bead width, and penetration are shown in Figure 6.30. The welding variables are discussed with particular attention to the three major characteristics of penetration, deposition rate, and bead shape. Figure 6.31 shows the effects of welding variables on the three major characteristics.

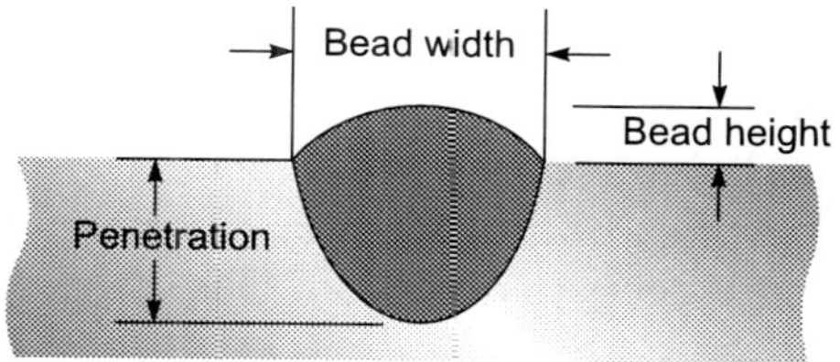

Figure 6.30 Bead height, width and penetration

Desired Characteristic		Welding Variable/Change Required				
		Arc Length	Welding Current	Travel Speed	Travel Angle	Electrode Diameter
Deeper penetration		–	Increase	–	Drag	Smaller
Shallower penetration		–	Decrease	–	Push	Larger
Bead height and width	Larger bead	–	Increase	Decrease	–	–
	Smaller bead	–	Decrease	Increase	–	–
	Higher, narrower	Decrease	–	Increase	Drag	–
	Flatter, wider	Increase	–	Decrease	Push	–
Faster deposition rate		–	Increase	–	–	Smaller
Slower deposition rate		–	Decrease	–	–	Larger

Note: These are the effects of changing only one variable at a time

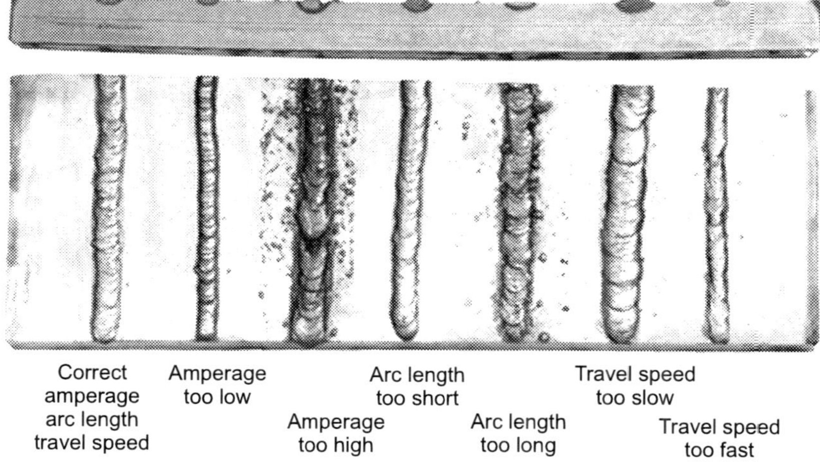

Figure 6.31 Effect of welding variables on the penetration, bead size and deposition rate

6.7.1 Fixed Variables

Electrode: The type of electrode used has an effect on the penetration, deposition rate, and shape of the weld bead. The type of electrode used usually depends upon the specific application. Some types of electrodes have a digging arc and produce welds with deep penetration. The deep penetrating effect of these electrodes is because of the high amount of cellulose in their coatings. Some types of electrodes produce moderate penetration while others produce light penetration. The deposition rate of the electrode is also greatly influenced by the type of electrode used. Iron powder is the main influence on deposition rate. Electrodes with high amounts of iron powder in their coatings can operate at higher welding currents than other types of electrodes. The iron powder goes into the weld metal and helps give the electrodes that have iron powder in their coatings, the highest deposition rates of any of the mild steel electrodes. The electrodes with the thinnest electrode coatings generally have the lowest deposition rates. The bead size and shape is also influenced by the type of electrode used. The iron powder electrodes generally produce wider and flatter beads.

Electrode Size: Larger diameter electrodes use higher welding currents, so the deposition rates and penetration depths increase as the electrode wire diameter is increased. Lower deposition rates and penetration depths are obtained with smaller electrode wire diameters because lower welding currents are used. Figure 6.32 shows typical deposition rates for different sizes of E6010, E7018, and E7024 electrodes.

Core Wire Diameter		E6010			E7018			E7024		
Inches	(mm)	Lb/hr	(kg/hr)	Amps	Lb/hr	(kg/hr)	Amps	Lb/hr	(kg/hr)	Amp
3/32	(2.4)	1.3	(0.6)	60	1.7	(0.8)	90	–	–	–
1/8	(3.2)	2.1	(0.9)	100	3.1	(1.4)	150	5.3	(2.4)	150
5/32	(4.0)	3.1	(1.4)	130	4.7	(2.1)	230	6.9	(3.1)	225
3/16	(4.8)	4.0	(1.8)	160	6.5	(3.0)	300	9.5	(4.3)	280
7/32	(5.4)	–	–	–	9.5	(4.3)	350	9.8	(4.5)	320
1/4	(6.4)	–	–	–	9.6	(4.4)	400	12.2	(5.5)	360

Figure 6.32 Typical deposition rates of different sizes of E6011, E7016 and E7024 electrodes

Current Type: The melting rate in the arc zone is directly related to the electrical energy in the welding arc. Part of the energy is used to melt the electrode and coating, some is lost to the atmosphere, and the rest goes to melt the base metal. The energy balance is determined by the electrical polarity and the constituents of the electrode covering. The different types of current affect the deposition rate and the depth of penetration. Direct current electrode positive gives the most penetration at a given welding current setting, followed by alternating current, with direct current electrode negative giving the least penetration. However, for a given welding current setting, direct current electrode negative gives the highest deposition rates for covered electrodes, followed by alternating current, with direct current electrode positive giving the lowest deposition rates.

6.7.2 Primary Variables

Welding Current

The welding current is the most important factor in determining the characteristics of the weld. The welding current is controlled by a knob or handle on the welding machine. The amount of current is determined by the type of electrode, size of electrode, position and joint design. The welding current is the best variable to be used for controlling the depth of penetration and the volume of the weld metal. As the welding current is increased and the other factors remain constant, the penetration and the deposition rate increase. The size of the weld bead also increases. Excessive weld current can produce large amounts of spatter, undercutting, and an irregular weld deposit. Lowering the welding current, while the other factors remain constant, reduce the deposition rate, the amount of penetration, and the size of the weld bead. An extremely low weld current can produce excessive piling up of weld metal and lack of fusion with the base metal. The manufacturers recommend the current ranges that should be used for each type and size of electrode. The effects of the welding current are shown in Figure 6.33.

Travel Speed

The travel speed is another important factor to consider for controlling the weld characteristics. The travel speed is determined by the welder, who manually controls the rate that the arc travels along the work. Increasing the travel speed while the other variables remain constant, reduces the width of the weld bead and increases the penetration. There is an optimum travel speed where the penetration is maximum. Increasing the travel speed beyond this point will decrease the penetration. Excessive travel speed will produce a weld bead that is too small with an irregular contour. This can produce welds that have too small a cross section. A very slow travel speed can result in excessive piling up of weld metal and lack of fusion at the edges of the weld. The effects of travel speed are also shown in Figure 6.33.

Welding Voltage (Arc Length)

The welding voltage is another important variable for shielded metal arc welding. The arc voltage is determined by the arc length between the end of the electrode and the base metal. The arc voltage is controlled manually by the welder who can move the tip of the electrode close to, or away from, the surface of the base metal. As the arc length is increased, the arc voltage increases, whereas decreasing the arc length decreases the arc voltage. The welding voltage primarily affects the shape of weld bead cross section and the general appearance of the weld. Increasing the welding voltage produces a wider and flatter weld bead and increases the susceptibility to arc blow. When the arc length is too long, which makes the welding voltage too high, the weld bead can look irregular with poor penetration and spatter. Also, the weld metal may not be properly shielded by the gas from decomposition of the electrode coating, and much of the heat may be lost to the atmosphere. Decreasing the arc length produces a stiffer and more easily controlled arc. A very short arc length can cause electrode sticking to the base metal. The effects of too long an arc length are also shown in Figure 6.33.

GOOD Proper current voltage and speed	BAD Welding current too low	BAD Welding current too high	BAD Arc too long (voltage too high)	BAD Travel speed too fast	BAD Travel speed too slow
Cross section weld bead	*Cross section weld bead*	*Cross section weld bead*	*Cross section weld bead*	*Cross section weld bead*	*Cross section weld bead*
Face weld bead	*Face weld bead*	*Face weld bead*	*Face weld bead*	*Face weld bead*	*Face weld bead*
A smooth, regular, well formed bead.	Excessive pilling up of weld metal.	Excessive spatter to be cleaned off.	Bead is very irregular with poor penetration.	Bead too small, with contour irregular.	Excessive pilling up of weld metal.
No undercutting, overlapping or pilling up.	Overlapping bead has poor penetration.	Undercutting along edges weakens joint.	Weld metal not properly shielded.	Not enough weld metal in cross section.	Overlapping without penetration at edges.
Uniform in cross section.	Slow up progress.	Irregular deposit.	An inefficient weld.	Weld not strong enough.	Too much time consumed.
Excellent weld at minimal material and labor cost.	Wasted electrodes and productive time.	Wasted electrodes and productive time.	Wasted electrodes and productive time.	Wasted electrodes and productive time.	Wasted electrodes and productive time.

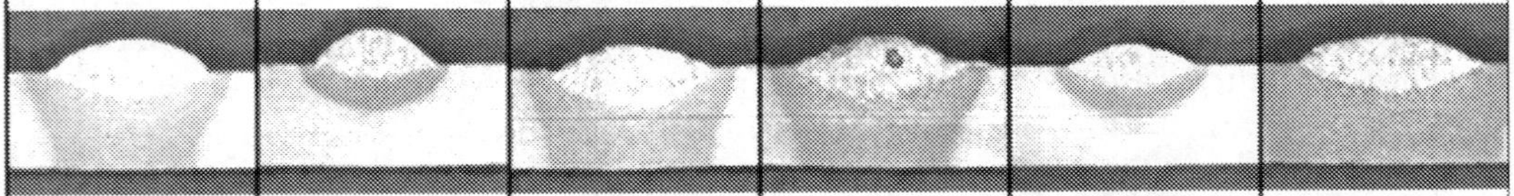

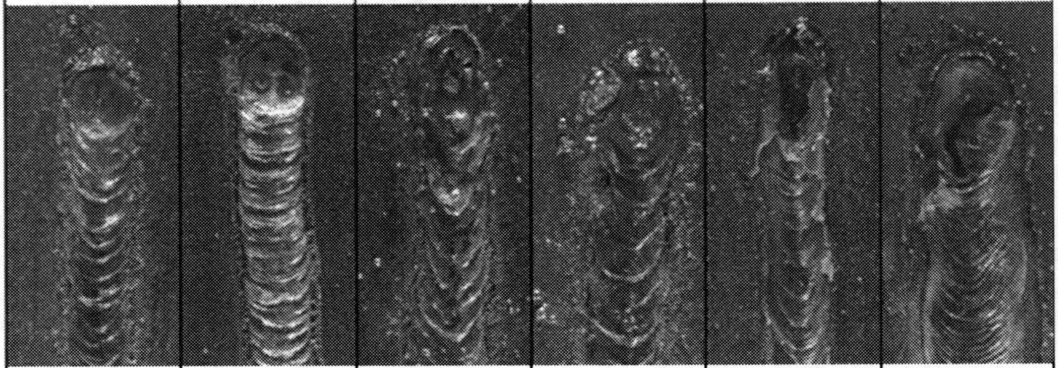

Figure 6.33 Effects of the welding variables on the weld bead

6.7.3 Secondary Variables

Electrode Angles: The angular position of the electrode in relation to the work, may have an effect on the quality of the weld deposit. The position of the electrode determines the ease at which the filler metal is deposited, freedom from undercutting and slag inclusions, and the uniformity of the bead.

The electrode angles are called the travel angle and the work angle. The travel angle of the electrode is the angle between the joint and the electrode in the longitudinal plane. The work angle is the angle between the electrode and the perpendicular plane to the direction of travel. These are shown in Figure 6.34. Increasing the travel angle in the direction of welding generally builds up the bead height. A work angle that is too large may result in undercutting. Especially with the low-hydrogen types of electrodes, the electrode angles are important in maintaining weld quality.

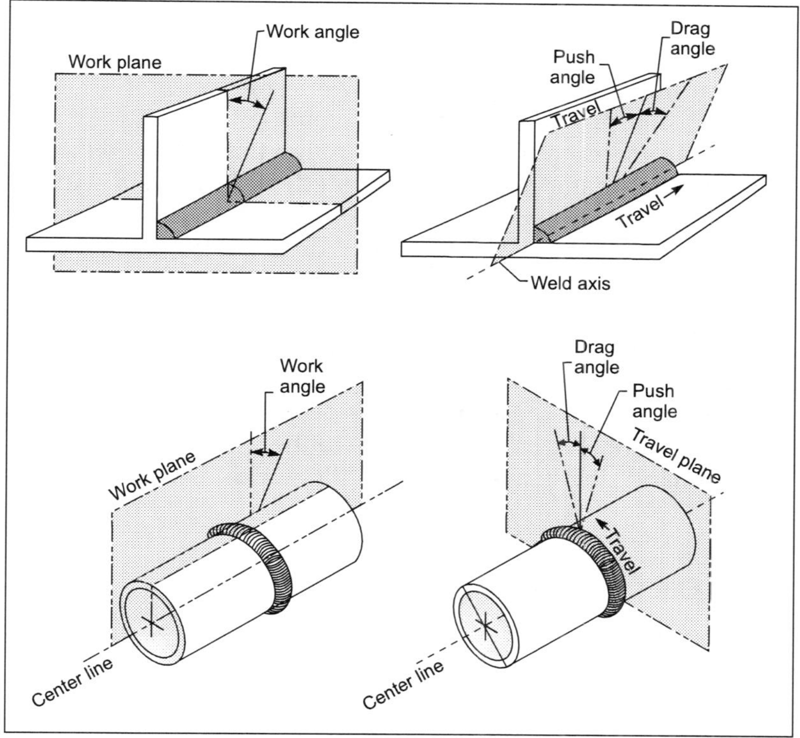

Figure 6.34 Travel angle and work angle

6.8 VARIANTS OF SMAW PROCESS

Standard SMAW processes are carried out by coated electrodes having diameters between 2.5 and 6.3 mm. The variants of this process are:

(i) Touch electrode welding
(ii) Bunched electrode welding
(iii) Multi-arc welding
(iv) Massive electrode welding

6.8.1 Touch Electrode Welding (Figure 6.35)

Touch electrode welding is comparatively a newer technique, improving both the production rate and weld quality. Here, the arc is struck the usual way but as soon as a stable arc is established, the electrode is pushed down towards the workpiece, so that the coating touches it. The electrode is then moved through the intended path while coating remains continuously in touch with the electrode. The rate of melting of the coating material in this type of welding is lower than that of core wire. This provides a barrel of coating around the arc and the material is transferred through this protected passage. Pressure exerted on the electrode manipulates the arc length and the weld width. The rate of electrode advance should be such that the molten metal, the slag and the unburnt crumbled coating are left behind and the electrode is not short circuited to the weld pool.

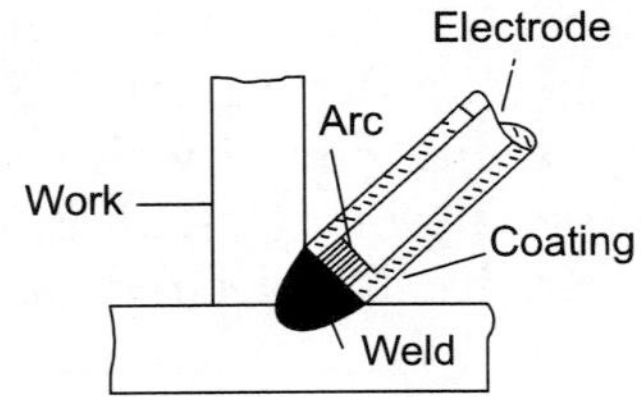

Figure 6.35 Touch electrodes welding of a fillet weld

Electrodes used for touch welding are usually of the rutile and basic coated types of the heavy coated variety. The heavy coating is essential to ensure unburnt lip with which to establish touch technique. Concentration of heat within the small area bounded by the unburnt lip of the coating gives deeper penetration than conventional method. This results in higher production rate to the extent of 50%. Touch welds are applicable to downhand single-layer butt and fillet welds as well as to lap and corner welds.

6.8.2 Bunched Electrode Welding (Figure 6.36)

Welding with bunched electrodes is to enhance the rate of metal deposition. Two to six electrodes may be bunched with a fine wire at three or four places along their lengths and tack welded at the top bare ends. The current is conducted through all the electrodes in the bunch but the arc is established between the work and the nearest electrode. The arc stays on that electrode until the arc length is too long to sustain itself and by now another electrode has moved much closer to the weld pool surface. The arc, therefore jumps to the electrode with the least gap between it and the work. This repeats at regular intervals. It is possible to carry heavier welding current without overheating the electrodes due to cooling intervals, which in fact, enhances their production rate more by about 30% than the normal SMAW techniques. The time of electrode replacement is reduced, and higher heat input efficiency can be achieved.

It is not that popular a technique since it is impossible to maintain uniform arc length due to jumping of arc from one electrode to another. Though used in ship building, this type of welding is not recommended for welding critical components. Sometimes, it becomes difficult to maintain the weld width by this method.

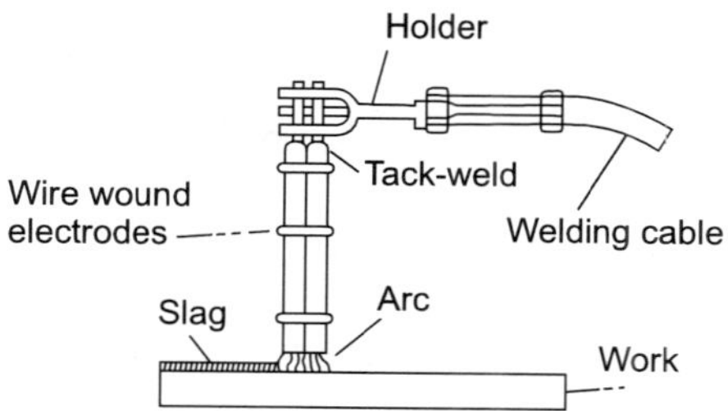

Figure 6.36 Welding with bunched electrodes

6.8.3 Multiple Arc Welding (Figure 6.37)

Twin electrodes which are insulated from each other are used in this type of welding. These are clamped in specifically designed electrode holders which conduct current to each of the electrodes separately. Two phases are connected to the two electrodes and the third phase to the work. Three arcs are maintained at a time, two of which are established between each electrode and the work, while the third is established between the electrodes. As a result, the melting rate of electrodes and production rates are almost doubled in comparison with a single-phase welding. The process is rather cumbersome due to increased weight of the electrodes and the holder. This leads to increased fatigue for the operator.

Three-phase welding transformer is used to supply current for multiple arc welding. The three-phase primary welding is either star or delta connected. The secondary consists of two coils, each wound with bare copper strips on one of the core limbs. The open circuit secondary voltage is 68 V. In addition, the secondary can be tapped at 34 V. When welding is stopped, the electrodes are withdrawn from work and the work is extinguished but the arc between the electrodes remains. To extinguish this third arc, the three-phase welding transformer incorporates a magnetic contactor which disconnects phase two when the arc '*b*' is extinguished. This leads to extinguishing arc '*a*' between the electrodes.

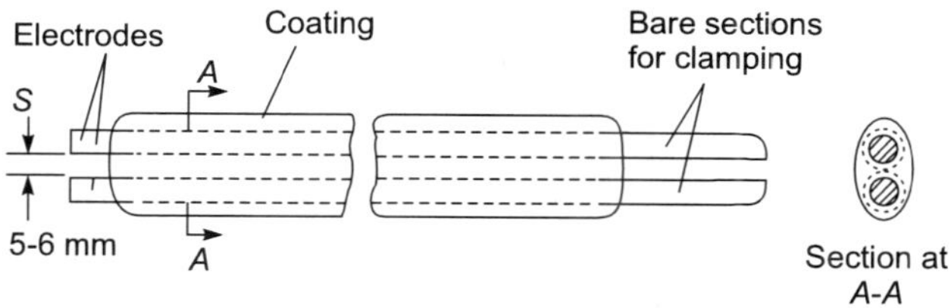

Figure 6.37 Coated electrodes for multiple arc welding

6.8.4 Massive Electrode Welding (Figure 6.38)

Another way to increase the production rate in welding is to use massive electrodes with diameters ranging between 8 and 19 mm and having a maximum length of about 1 m. These electrodes are specifically made for repairing castings and obviously need high welding current. These electrodes are so large and heavy that it is not feasible to use them in normal manual fashion. Instead, they are clamped in manipulators to feed them into work. Data for massive electrode welding is shown in Table 6.5.

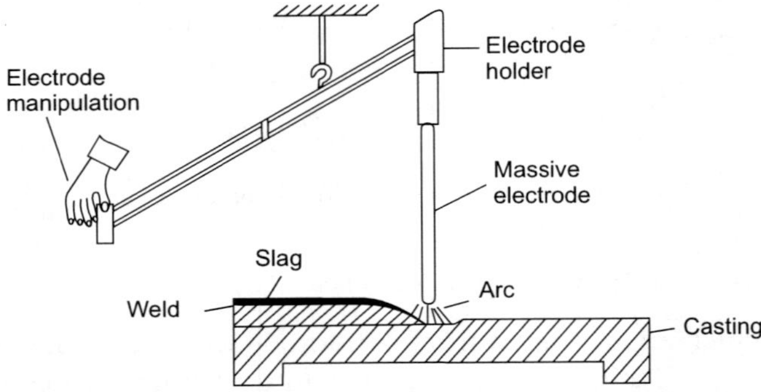

Figure 6.38 Setup for use of massive electrodes

TABLE 6.5 Data for massive electrodes

Electrode Size		Current Range with DCEP (amp)	Weld Pool Size (cm^2)	Weld Metal Deposition rate (kg/hr)
Diameter (mm)	Length (mm)			
8	610	300–500	15–40	4.5
10	610	400–600	40–65	7.0
13	760	600–950	65–130	11.5
16	760	600–1500	130–230	20.5
19	915	1200–2100	230–390	27.0

6.9 WELDING DEFECTS

Shielded metal arc welding, like other welding processes, has welding procedure problems which can cause defects in the weld. Some defects are caused by problems with the materials. Other welding problems may not be foreseeable and may require immediate corrective action. A poor welding technique and improper choice of welding parameters can cause weld defects. Defects that can occur when using the shielded metal arc welding process are slag inclusions, wagon tracks, porosity, piping porosity, undercutting, lack of fusion, overlapping, burn through, arc strikes, craters, and excessive weld spatter. Many of these welding technique problems weaken the weld and can cause cracking. Other problems that can occur and can reduce the quality of the weld are arc blow, finger nailing, and improper electrode coating moisture contents.

Defects Caused by Welding Technique

Defect	Causes	Remedy	Figure
Slag inclusions occur when slag particles are trapped inside the weld metal, which produces a weaker weld.	1. Erratic travel speed. 2. Too wide a weaving motion. 3. Slag left on the previous weld pass. 4. Too large an electrode being used. 5. Letting slag run ahead of the arc.	This defect can be prevented by: 1. A uniform travel speed. 2. A tighter weaving motion. 3. Complete slag removal before welding. 4. Using a smaller electrode. 5. Keeping the slag behind the arc, which is done by shortening the arc, increasing the travel speed, or changing the electrode angle.	
Porosity is gas pockets in the weld metal that may be scattered in small clusters or along the entire length of the weld. Porosity weakens the weld in approximately the same way that slag inclusions do.	1. Excessive welding current. 2. Rust, grease, oil or dirt on the surface of the base metal. 3. Excessive moisture in the electrode coatings. 4. Impurities in the base metal, such as sulfur and base metal phosphorous. 5. Too short an arc length except when using low-hydrogen or stainless-steel electrodes. 6. Travel speed too high, which causes freezing of the weld puddle before gases can escape.	1. Lowering the welding current. 2. Cleaning the surface of the base metal. 3. Redrying electrodes. 4. Changing to a different base metal with a different composition. 5. Using a slightly longer arc length. 6. Lowering the travel speed to let the gases escape. 7. Preheating the base metal, using a different type of electrode, or both.	

Contd.

Undercutting is a groove melted in the base metal next to the toe or root of a weld that is not filled by the weld metal. Undercutting causes a weaker joint and it can cause cracking.	1. Excessive welding current. 2. Too long an arc length. 3. Excessive weaving speed. 4. Excessive travel speed.	1. Choosing proper welding current for the type and size of electrode and the welding position. 2. Holding the arc as short as possible. 3. Pausing at each side of the weld bead when a weaving technique is used. 4. Using a travel speed slow enough so that the weld metal can completely fill all of the melted out areas of the base metal.	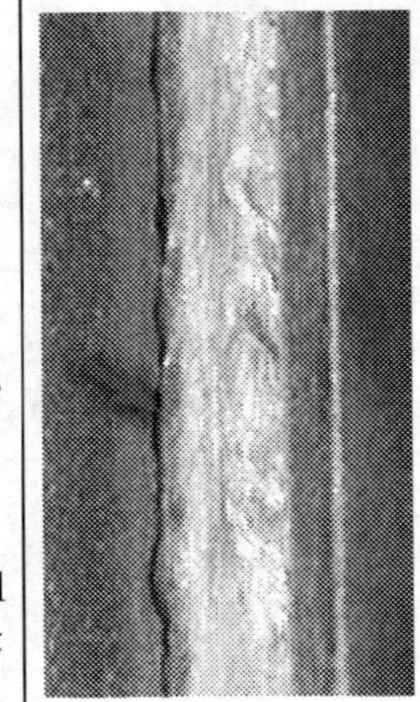
Incomplete fusion is when the weld metal is not fused to the base metal. This can occur between the weld metal and the base metal or between passes in a multiple pass weld.	1. Excessive travel speed. 2. Electrode size too large. 3. Welding current too low. 4. Poor joint preparation. 5. Letting the weld metal get ahead of the arc.	1. Reducing the travel speed. 2. Using a smaller diameter electrode. 3. Increasing the welding current. 4. Better joint preparation. 5. Using a proper electrode angle.	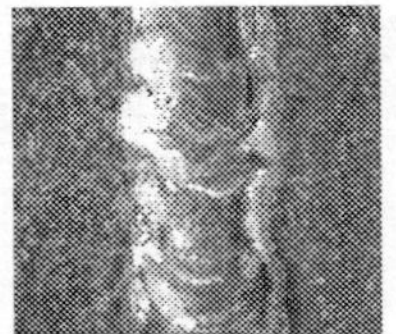 
Overlapping is the protrusion of the weld metal over the edge or toe of the weld bead. This defect can cause an area of lack of fusion and create a notch which can lead to crack initiation.	1. Too slow a travel speed which permits the weld puddle to get ahead of the electrode. 2. An incorrect electrode angle that allows the force of the arc to push the molten weld metal over unfused sections of the base metal. 3. Welding away from the ground connection with large electrodes like the E6020, E6027, E7024, and E7028, which have very fluid weld puddles.	1. A higher travel speed. 2. The electrode angle should be such that the force of the arc does not push the molten metal out of the weld puddle and over the cold base metal. 3. Grinding off excess weld metal.	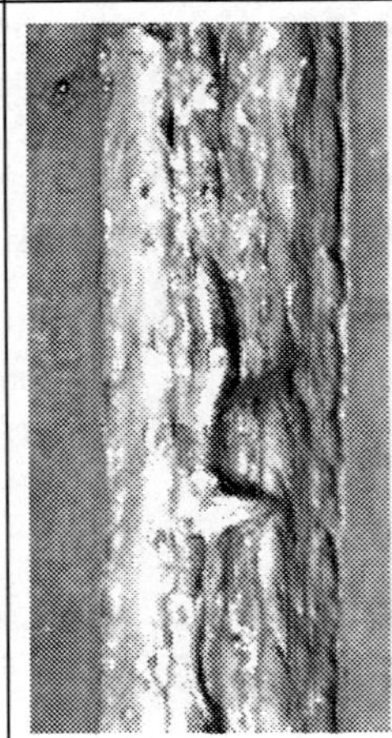

Melt Through

It is when the arc melts through the bottom of the weld.

This can be caused by:

1. Excessive welding current
2. Too slow a travel speed
3. Too wide a root gap

This can be prevented by:

1. Reducing the welding current
2. Increasing the travel speed
3. Reducing the size of the root gap

Arc Strikes

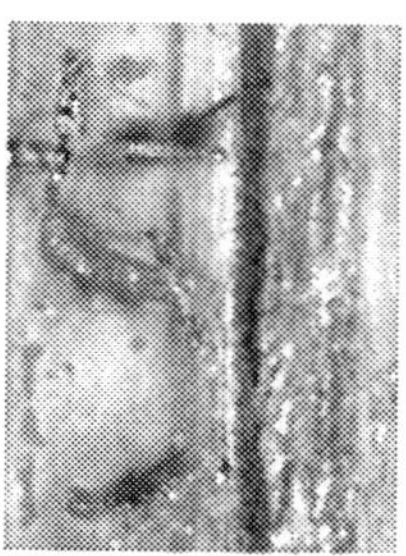

Many codes prohibit striking the arc on the surface of the workpiece. Striking the arc on the base metal outside of the weld joint can produce a hard spot on the base metal surface. Failures can then occur due to the notch effect. The arc strikes may create a small notch on the surface of the metal which can act as an initiating point for cracks.

Craters

Weld craters are a depression on the weld surface at the point where the arc was broken. These are caused by solidification of the metal after the arc has been broken. The weld crater often cracks and can serve as an origin for linear cracking back into the weld metal or into the base metal. These craters can usually be removed by chipping or grinding and the depression can be filled in with a small deposit of filler metal. The best way of preventing weld craters is to reverse the travel of the electrode a little way back into the weld bead from the end of the weld bead before breaking the arc.

Excessive Spatter

Excessive spatter gives the weld a poor appearance, wastes electrodes, difficult slag removal and can lead to lack of fusion in the case of multiple pass welds. If the spatter is coarse, it is usually produced by an arc length that is too long. If the spatter is fine, it is usually caused by too high a welding current. Spatter can be removed by grinding or prevented by using a shorter arc length and lower welding currents.

Distortion and Warpage

Distortion and warpage are caused by the non-uniform expansion and contraction of weld and base metal during the heating and cooling processes of welding. If warpage changes the required dimensions of the total weldment, which may not be acceptable. Methods to reduce distortion and warpage are:

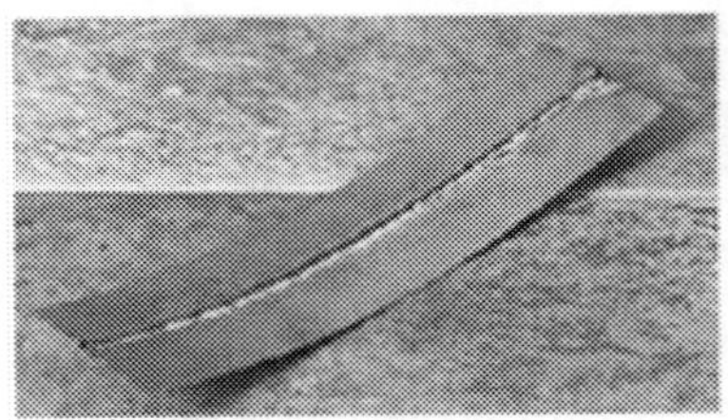

- Deposit only the required amount of weld metal.
- Alternate sides or sequence welds.
- Preset parts to compensate for distortion.

Inadequate Joint Penetration

It is commonly located at the root of the weld and is caused by insufficient heat input while welding. Insufficient heat input can be caused by too low amperage, too much electrode stickout, or too fast a travel speed. Inadequate joint penetration can also be caused by improper joint design or incorrect preparation of the joint edges. A tight root opening or a small groove angle restricts the weld metal, causing lack of penetration at the root. To prevent this problem, insure that enough heat is used on a properly designed joint and that the welder or operator has the skill necessary to produce the desired weld.

Cracking

It may be caused by an improper welding procedure, welder technique, or materials. All types of cracking can be classified as either hot cracking or cold cracking and these cracks can be oriented transversely or longitudinally to the weld. Transverse cracks are perpendicular to the axis of the weld whereas longitudinal cracks are parallel to the axis of the weld. Transverse cracks are often the result of longitudinal shrinkage strains acting on excessively hard and brittle weld metal. Longitudinal cracks are often caused by high joint restraint and high cooling rates.

Hot cracking is a defect that occurs at higher temperatures and generally happens just after the weld metal starts to solidify. This type of cracking is often caused by excessive sulfur, phosphorous, and lead contents in the base metal. It can also occur because of an improper method of breaking the arc or in a root pass, when the cross-sectional area of the weld bead is small as compared to the mass of the base metal. Hot cracking often occurs in deep penetrating welds and it can continue through successive layers if it is not repaired. Hot cracking may be prevented or minimized by:

- Preheating
- Using low-hydrogen electrodes
- Increasing the cross-sectional area of the weld bead
- Changing the contour of the weld bead
- Using base metal with very low sulfur, phosphorous, and lead contents

Crater cracks are shallow hot cracks that are caused by improperly breaking the arc. Several types of these cracks are shown in Figure 6.39.

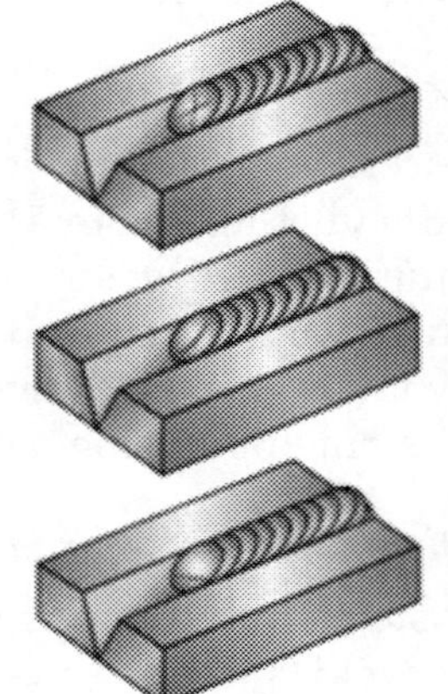

Figure 6.39 Hot cracks

Crater cracks may be prevented the same way that craters are, i.e., by reversing the travel of the electrode a little way back into the weld from the end of the weld before breaking the arc. Cold cracking occurs after the weld metal solidification is complete. Cold cracking may occur several days after welding and is generally caused by hydrogen embrittlement, excessive joint restraint, and rapid cooling. Preheating and the use of low-hydrogen electrodes help reduce this problem.

Center-line cracks are cold cracks that often occur in single-pass concave fillet welds. A longitudinal crack is a center-line crack that runs down the center of the weld, as shown in Figure 6.40.

This problem may be caused by:

1. Too small a weld bead for the thickness of the base metal
2. Poor fit-up
3. High joint restraint
4. Extension of a crater crack

The major methods to prevent center-line cracks are:

1. Increasing the bead size
2. Decreasing the gap width
3. Positioning the joint slightly uphill
4. Preventing weld craters

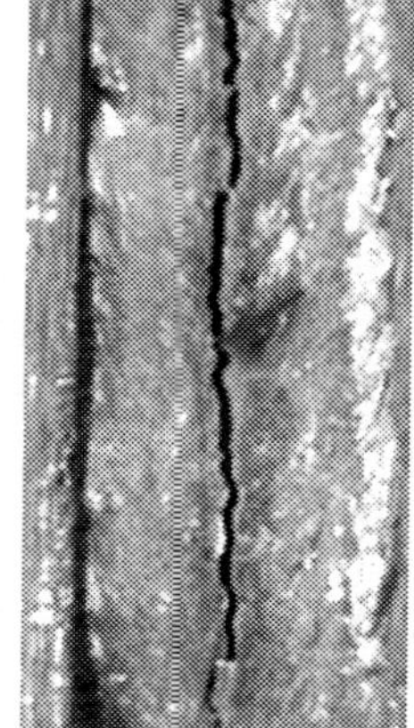

Figure 6.40 Longitudinal crack

Base metal and underbead cracks are cold cracks that form in the heat affected zone of the base metal. Underbead cracks occur underneath the weld bead, as shown in Figure 6.41.

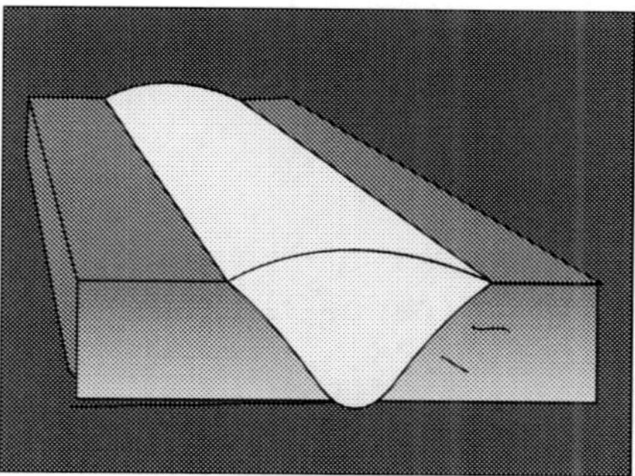

Figure 6.41 Underbead crack

Base metal cracks are those cracks that originate in the heat affected zone of the weld. These are caused by excessive joint restraint, hydrogen, and a brittle microstructure.

A brittle microstructure is caused by rapid cooling or excessive heat input. Underbead and base metal cracking can be reduced or eliminated by using preheat and low-hydrogen electrodes.

6.10 OTHER PROBLEMS

Arc Blow: The electric current that flows through the electrode, workpiece, and ground cable sets up magnetic fields in a circular path that is perpendicular to the direction of the current. When the magnetic fields around the arc are unbalanced, it tends to bend away from the greatest concentration of the field. This deflection of the arc is called arc blow. Deflection is usually in the direction of travel or opposite to it, but it sometimes occurs to the side also. Arc blow can result in excessive weld spatter and lack of fusion. Direct current is highly susceptible to arc blow, especially when welding is done in corners and near the end of joints. Arc blow also occurs when welding complex structures and on massive structures with high currents and poor fit-up. Arc blow occurs with direct current because the induced magnetic field is in one direction. Alternating current rarely is subject to arc blow because the magnetic field builds and collapses all the time due to the reversing current. Forward arc blow is encountered when welding away from the ground connection or at the beginning of the weld joint. Backward arc blow occurs towards the ground connection, into a corner, or towards the end of a weld joint. There are several corrective methods that can be used to correct the arc blow problem, such as:

(a) Changing to alternating current.
(b) Welding towards an existing weld or a heavy tack weld.
(c) Placing the work connection as far as possible from the weld, at the end of the weld or at the start of the weld and weld towards a heavy tack weld.
(d) Reducing the welding current and making the arc length as short as possible.
(e) Wrapping the work lead around the workpiece, so that the magnetic field caused by the current in the ground cable neutralizes the magnetic field causing the arc blow.

Improper Moisture Content

The coatings of all covered electrodes contain a certain amount of moisture. Incorrect moisture content in an electrode coating can cause operating problems with the electrode. The E6010 and E6011 electrodes have relatively high moisture content in their coverings. These electrodes can operate fairly well when the moisture content is above the maximum limit, but an excessive amount of moisture in these electrode coatings can cause blistering of the coatings and poor arc operation. A low moisture content will cause the electrodes to give excessive amount of spatter and possibly porosity. Too much moisture in the coatings of the other types of electrodes can cause blistering of the coating, poor arc operation, and underbead cracking. The

E7015R, E7016R, E7018R, E7028R, and E7048R have relatively low moisture content in their coverings. Covered electrodes should always be stored in dry places. A high moisture content in low hydrogen electrode coatings will damage the quality of the weld deposit. Redrying is often done after a long storage period except on the cellulose electrodes where it is generally not recommended.

Finger nailing is the name given to the burning off of an electrode faster on one side than on the other. This condition can be aggravated if the coating is not concentric with the core wire. Finger nailing is shown in Figure 6.42. Arc blow is a more frequent cause of finger nailing than off-center electrodes. When welding with lower than normal current, finger nailing will appear because there is insufficient arc force to overcome the minor arc blow.

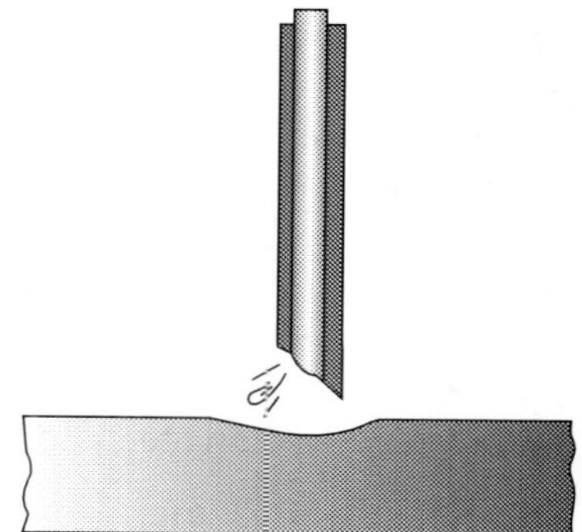

Figure 6.42 Finger nailing

6.11 APPLICATIONS OF SMAW

Of the various arc welding processes, the shielded metal arc welding (SHAW) process is the most extensively used process, because it requires low capital investment. It is easy and economical to use and can be applied on a wide variety of metals, including ferrous and non-ferrous jobs. Nowadays, a variety of coated electrodes are available to practically weld any metal. Some specific applications are given below:

Field Welded Storage Tanks: Field welded storage tanks differ from pressure vessels because they are used to store petroleum, water or other liquids at atmospheric pressure. Shielded metal arc welding is widely used in the fabrication and erection of field welded storage tanks. These tanks are generally constructed of low-carbon and structural steels. Nickel steels are employed when higher toughness is required. This process is used to weld longitudinal and circumferential seams on the tanks as well as the structural support members.

Pressure Vessels: Pressure vessels and boilers are also welded using this process. Shielded metal arc welding is primarily used for welding attachments to the vessel. All sizes of electrodes are commonly used. For applications where the vessels operate at low temperatures, smaller electrodes are used on multiple pass welds. This will produce smaller weld beads that build up the weld in relatively thin layers. The smaller weld beads give a stronger, tougher weld.

Industrial Piping: Shielded metal arc welding is widely used in the industrial piping industry, which includes many types of pressure piping. The types of electrodes most often used are the E6010 and E7018 electrodes for welding low-carbon steel pipes. A common practice is the use of E6010 electrodes to weld in the root passes and the E7018 electrodes are used to weld in the fill and cover passes. Welding on industrial piping is generally done from the bottom to the top, except on small diameter pipes, where it is done both ways. The reason that welding from bottom to top is most often done is because slag is often trapped when welding in the opposite direction. For welding low-carbon steel pipes with a 485MPa tensile strength, E7010 and E7018 electrodes are used. Shielded metal arc welding is often used for welding on other types of industrial piping. EXX15, EXX16, and EXX18 electrodes are used for welding chromium-molybdenum alloy pipes. When welding stainless steel pipes, gas tungsten arc welding is often used to put in the root pass and shielded metal arc welding is used to weld in the fill and cover passes. Medium and high-carbon steel pipes are also welded by this process. For these, smaller diameter electrodes are used than on low-carbon steels to lessen the heat effect on the pipes.

Nuclear Power Plants: The nuclear power industry employs this process for many applications. It is often used in the shop fabrication of low carbon and low alloy steel heavy-walled pressure vessel, for welding longitudinal and circumferential weld seams. Shielded metal arc welding is the best method for welding nozzles and attachments to the vessels. A major application of this process is welding pressure piping for use in the nuclear power facilities. Nuclear power system pressure piping requires stronger quality control than normal pressure piping.

Structures: The construction industry is a major user of shielded metal arc welding. Most of the welding done on buildings and bridges is done in the field at long distances from the power sources, which makes this process popular for these applications. Most types of covered electrodes are used in structural work because of the wide variety in the tensile strength of the steels used.

Ships: Shielded metal arc welding is still the major process used in shipbuilding. It is used for many different applications, including welding in areas where the other processes cannot reach. Most types of low-carbon steel covered electrodes are used, except the EXX12, EXX13, and EXX14 types. These three types of electrodes are not approved for use on the main structural members in the ship because of the relatively low ductility obtained from the weld deposits of these electrodes. The electrodes with large amounts of iron powder in their coatings are popular for many shipbuilding applications because of the high deposition rates obtained. These types of electrodes are especially used on the many fillet welds that are made in a ship structure. Backing tape is often used for backing the weld metal when one side welding is done.

Transportation: Another industry where this welding process is widely used is in the railroad industry. E6OXX and E70XX electrodes are used to weld many parts of the underframe, cab, and engine of the locomotive. The underframe fabrication consists of mostly fillet welds. The frames and brackets for the diesel engines are also welded with these electrodes. Railroad cars are commonly welded together by the shielded metal arc process.

Industrial Machinery: The frames of many types of heavy industrial machinery are welded together using this process. It is the major process used for welding piping associated with this machinery. Shielded metal arc welding is used for welding areas where the other processes cannot reach.

Heavy Equipment: Another major application of this process is in the heavy equipment industries, such as mining, agricultural, and earthmoving equipment. In these industries, shielded metal arc is used for welding structural steels, which are used for the frames, beams, and many other items in the assembly. Most types of covered electrodes are used depending on the type of steel being welded. Stainless steel and non-ferrous metals are also used for some parts.

Miscellaneous: This welding process is commonly used in many general fabrication or small job shops on a wide variety of applications. There are many other applications possible in this area because of the overall versatility of this process, but these are too numerous to mention.

6.12 ADVANTAGES

1. **Low Start-up Costs:** SMAW welding machines are relativity inexpensive.
2. **Portability:** SMAW is a very portable process because it does not require any external shielding gas and equipment due to the ability of the electrode to produce its own shielding gases.
3. **Outdoor Weldability:** SMAW welding can be done outdoors. The shielding of the weld puddle from the coating of the electrode is not susceptible to winds and draft, and therefore, is an excellent choice for outdoor welding.
4. **All Position Welding:** SMAW can be done in all welding positions, which eliminates the need for expensive fixturing or manipulation of the part being welded.
5. **Variety of Materials:** Many base materials and alloys can be welded with the SMAW process. Shielded metal arc welding can be used to weld a wide variety of metal thicknesses. This process is more useful than other welding processes for welding complex structural assemblies because it is easier to use in difficult locations and for multi-position welding. Shielded metal arc welding is a popular process for pipe welding. Weld joints with high quality and strength can be obtained with shielded metal arc welding.

6.13 LIMITATIONS

1. **Operating Factor:** Operating factor can be defined as the amount of an arc on time in a shift of work, labeled as a percentage. SMAW welding has an operating factor of approximately 25% to 30%. Operator duty cycle and overall deposition rates for covered electrodes are usually less than provided with a continuous

electrode process. This is because electrodes have a fixed length and welding must be stopped after each electrode has been consumed. The portion of the electrode that is clamped into the holder must be discarded.

2. **Low Efficiency:** The efficiency of a SMAW electrode can be defined as the percentage of the electrode that is consumed and becomes part of the weld. SMAW electrodes are 60% to 65% efficient. Slag must be removed from the weld after every pass.
3. **Restarts:** SMAW electrodes are only 225 to 450 mm in length. When the electrode is consumed in the arc, it needs to be replaced with the new one. This means that welding stops, and time is taken to replace each electrode.
4. **Operator Skill:** SMAW welding requires a high level of operator skill. This process is one of the hardest arc welding processes to learn.
5. **Slag and Spatter:** Formation of slag and spatter is a serious limitation of this process, especially from surface finish point of view.
6. **Non-ferrous Metals:** Shielded metal arc welding cannot be used to weld some non-ferrous metals.

Comparison of Deposition Rates of the Arc Welding Processes

Figure 6.43 summarizes the deposition rates of the arc welding processes. As shown, the deposition rate increases in the order of GTAW, SMAW, GMAW and FCAW, SAW, and ESW. The deposition rate can be increased by adding iron powder in SAW or using more than one wire in SAW, ESW, and GMAW (not shown).

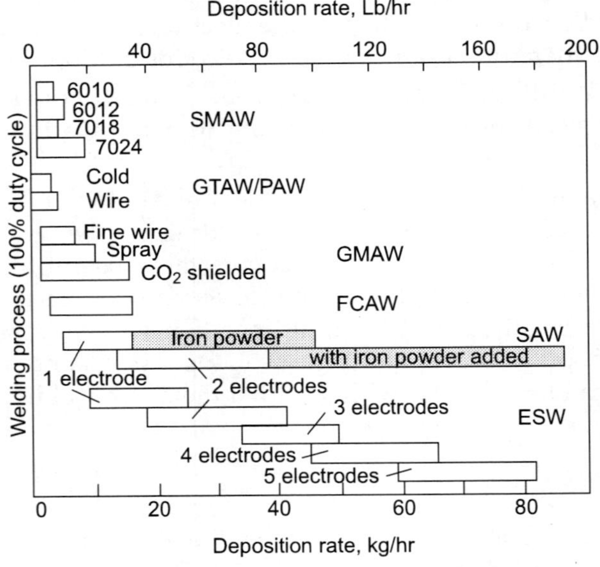

Figure 6.43 Deposition rate in arc welding processes

Worked Examples

Example 1: Manual metal arc welding is used for welding two plates, each measuring $800 \times 200 \times 19$ mm. Given that length of weld = 180 mm, length of electrode is 450 mm, stub length = 50 mm, diameter of electrode is 3 mm, weld reinforcement area is 10%, spatter and other losses 15%, root gap is 2 mm. Estimate the required number of electrodes for the weld. **(GATE 1990)**

Solution: Total area = $A_1 + A_2 + A_3 = A_1 + 2A_2$

$$\text{Total area} = 19 \times 2 + \frac{1}{2}(17 \times \tan 30 \times 17) \times 2 = 204.8 \text{ mm}^2$$

$$\text{Volume of weld bead} = 204.8 \times 180 = 36873 \text{ mm}^3$$

$$h = 19 - 2 = 17 \text{ mm}$$

$$\text{Base} = b = 17 \tan 30$$

Total volume of weld bead = volume of actual weld bead + reinforcement + spatter losses

$$= 1.25 \times \text{volume of weld bead} = 46{,}092 \text{ mm}^3$$

$$\text{Volume electrode} = \frac{\pi d^2 L_{\text{electrode}}}{4} = \frac{\pi \times 3^2 \times (450 - 50)}{4} = 2826 \text{ mm}^3$$

$$\text{No. of electrodes} = \frac{\text{total volume of weld bead}}{\text{volume per electrode}} = \frac{36873}{2826} = 16.3 \cong 17 \text{ electrodes}$$

Example 2: The cross section weld bead is shown in Figure 6.44. The profiles of weld bead and fusion zone are taken circular for convenience. Bead width and radii of curvature of the circular process are given in the figure. What is bead height and depth of penetration? **(GATE 1996)**

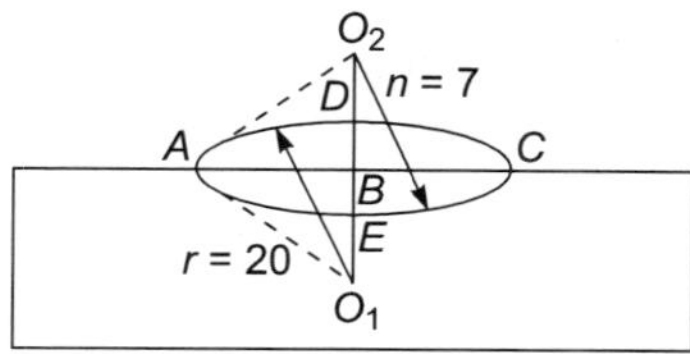

Figure 6.44 Cross section of weld bead

Solution: Given $AC = 10$ mm

$$O_1A = O_1C = 7 \text{ mm}$$

$$O_2A = O_2C = 20 \text{ mm}$$

Height of bead $= BD = O_1D - O_1B = O_1D - \sqrt{O_1A^2 - AB^2} = 20 - \sqrt{20^2 - 5^2} = 0.64$ mm

Depth of penetration

$$DOP = BE = O_1E - O_1B = O_1E - \sqrt{O_2A^2 - AB^2}$$

$$7 - \sqrt{7^2 - 5^2} = 2.10 \text{ mm}$$

Example 3: Welding of C40 steel plates of 10 mm thickness requires a current of 160 A, while it was 360 A when the plate thickness was increased by 50%. Estimate the welding current for 8 mm thickness of the same material.

(GATE 1997)

Solution: Let $t = C_1\sqrt{I} + C_2$

$$t_1 = 10, I_1 = 160 = C_1\sqrt{160} + C_2 \quad (1)$$

$$t_2 = 15, I_2 = 360 = C_1\sqrt{360} + C_2 \quad (2)$$

Solving (1) and (2)

$$C_1 = 0.79 \text{ and } C_2 = 0$$

$$t = C_1\sqrt{t} + C_2 = 0.79\sqrt{I}$$

$$8 = C_1\sqrt{t} + C_2 = 0.79\sqrt{I}$$

Solving we get I = 103 A

Example 4: In butt welding operation on plates, the heat input is given by $Q = 8KT_ct\left[0.2 + \frac{Vb}{4\alpha}\right]$, where, K is the thermal conductivity, T_c is temperature, increasing from room temperature up to melting point, t is the thickness of plate, V is the welding speed, b is the width of weld and α is the thermal diffusivity. Two alloy steel plates as shown in figure are to be welded using a power source rated at 5 kVA, having a duty cycle of 75%. Using the given data, determine the maximum welding speed for the given in mm/s (Figure 6.45).

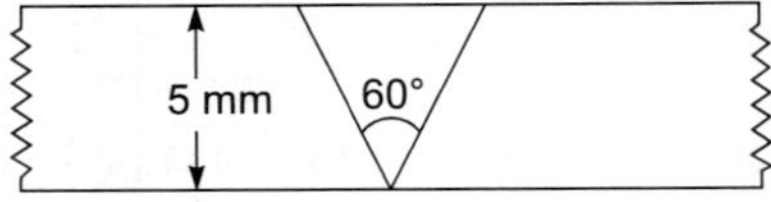

Figure 6.45 Butt weld

Solution: $Q = 8KT_ct\left[0.2 + \frac{Vb}{4\alpha}\right]$

$Q = 5 \text{ kVA} = 5000$ joules

$K = 45, T_c = 1450°C$

$t = 5 \text{ mm} = 5 \times 10^{-3} \text{m}, \alpha = 1.2 \times 10^{-5}$

B = Width of weld

$B = 2 \times 5 \tan 30 = 5.77 \text{ mm} = 5.77 \times 10^{-3} \text{m}$

$$5000 = 8 \times 45 \times 1450 \times 5 \left[0.2 + \frac{V \times 5.77 \times 10^{-3}}{4 \times 1.2 \times 10^{-5}} \right]$$

Solving we get V = 14.2 mm/s

Example 5: Two plates with symmetrical V joint penetration of V angle 60° are to be arc welded in a single pass. The power (kW), arc length L (mm) is given by $P = 4 + 0.8L - 0.1L^2$. Assume that the top surface of the weld is flat and flush with the plate top surface and energy losses are 20%. The density of filler material is 8 gm/cc per cc and energy required to melt 1 gm is 1400 joules. The plate dimensions are 1000 mm length and 5 mm thickness. What is the maximum welding speed?

(GATE 1995)

Solution: $P = 4 + 0.8L - 0.1L^2$

For maximum power

For optimum arc length at maximum power,

$$\frac{dP}{dL} = \frac{d}{dL}(4 + 0.8L - 0.1L^2) = 0$$

Solving we get L = 4 mm

$$P = 4 + 0.8L - 0.1L^2 = 5.6 \text{ kW}$$

Energy losses = 20%, therefore efficiency is 80%

Area of weld bead $= 2 \times \frac{1}{2} \times \text{base} \times \text{heigth} = 5 \tan 30 \times 5 = 14.43$

Volume of weld bead $= 14.43 \times 1000 = 14,433 \text{ mm}^3$

Weight of weld bead $= 14433 \times 10^{-3} \times 8 = 115.5 \text{ gm}$

Heat required for melting weld bead $= 115.5 \times 1400 = 161.66 \text{ kJ}$

$$\text{Time for welding} = \frac{161.66}{0.8 \times 5.6} = 36 \text{ sec}$$

$$\text{Welding speed} = \frac{1000}{36} = 27.78 \text{ mm/sec}$$

Example 6: In an arc welding operation carried out with a power source is maintained at 40 volts and 400 A. The consumable electrode melts and just fills the gap between the metal plates to be butt welded. The heat transfer efficiency for the process is 0.8, melting efficiency for the process is 0.3 and the heat required to melt the electrode is 20 J/m^3. If the travel speed of the electrode is 4 mm/s, what is the cross-sectional area in mm^2 of the weld joint? **(GATE 2014)**

Solution: Heat generated = $VI = 40 \times 400 = 16{,}000$ watts

Total heat required = heat generated × melting efficiency × heat transfer efficiency

Total heat required = $16000 \times 0.8 \times 0.3 = 3840$ watts

Volume filling rate of weld bead = cross-sectional area × velocity

$$= A \times 4\left(\frac{\text{mm}^3}{\text{sec}}\right)$$

Rate of heat required = 20×4 A = 80 A (watts) = heat utilized

$$A = \frac{4800}{80} = 48 \text{ mm}^2$$

Example 7: A butt weld joint is developed on steel plates having yield and ultimate tensile strength of 500 MPa and 700 MPa, respectively. The thickness of plates is 8 mm and width is 20 mm. Improper selection of welding parameter caused an undercut of 3 mm depth along the weld. What is the maximum transverse tensile load in kN carrying capacity of the developed weld joint? **(GATE 2014)**

Solution: b = 20 mm, t = 8 mm, undercut (d) = 3 mm

Due to undercut, the total bearing area = $b \times (t - d) = 20 \times (8 - 3) = 100 \text{ mm}^2$

Maximum load carrying capacity of the weld = ultimate strength × area

= 700 × 100 = 70,000 N

Example 8: Two pipes of inner diameter 100 mm and outer diameter 110 mm each are joined by flash butt welding using 30 V power supply. At the interface, 1 mm of material melts from each pipe which has a resistance of 42 ohms. If the unit melt energy is 64.4 MJ/m^3, what is the time required for welding in seconds? **(GATE 2008)**

Solution: Volume $= \frac{\pi}{4}(10^2 - 100^2) \times 2 = 3298.66 \text{ mm}^3$

Heat required = total volume × heat rate = $3298.66 \text{ mm}^3 \times 64.4 \times 10^6 = 212.4$ joules

$$\text{Power} = VI = V \times \frac{V}{R} = \frac{V^2}{R} = \frac{30^2}{42} = 21.43 \text{ joules/second}$$

$$\text{Heat input} = \frac{\text{power}}{\text{time}}$$

$$212.4 = \frac{21.43}{\text{time}}$$

Welding time = 10 seconds

Example 9: Two 8 mm thick steel plates are placed 5 mm apart and welded by butt joint. Welding is carried out at 20 V and speed is 5 mm/sec. Heat transfer efficiency is 0.80. If the heat required to melt steel is 10 joules/mm^3 and melting efficiency is 0.625 m. What is the weld current?

Solution: Thickness of plate, t = 8 mm

Voltage = 30 volts

Root gap = 5 mm, speed = 5 mm/sec

Efficiency = 0.8, Heat input = 10 joules/mm^3

Melting efficiency is 0.625

$$\text{Volume of weld bead} = t \times \text{root gap} \times L = 8 \times 5 \times L = 40L \text{ mm}^3$$

where, L is the length of the weld bead

$$\text{Heat required} = 40L \times 10 = 400L \text{ (joules)}$$

$$\text{Heat input} = \frac{VI \text{ (in J/sec = watts)}}{\text{speed of welding}\left(\dfrac{\text{mm}}{\text{sec}}\right)} \times \eta \qquad (1)$$

$$\text{Heat input} = \frac{20 \times I \times L}{5} \times 0.8 \times 0.625 \qquad (2)$$

Equating (1) and (2)

Welding current = 200 A

Example 10: A 900 mm long steel plate is welded by manual metal arc welding process using welding current of 150 A, arc voltage of 20 V and welding speed of 300 mm/min. If the process efficiency is 0.8 and surface resistance is 36 micro ohm, what is the heat input? **(GATE 2003)**

Solution: Welding time $= \dfrac{900}{300} = 3 \text{ min} = 180 \text{ sec}$

$$\text{Heat generated} = I^2 R\tau = 150^2 \times 50036 \times 10^{-6} \times 180 \times 0.8$$
$$= 116.64 \text{ J}$$

Example 11: In arc welding of butt joint, the welding speed is to be selected such that highest cooling rate is achieved. Melting efficiency and heat transfer efficiency are 0.5 and 0.7, respectively. The area of the weld cross section is 5 mm^2 and the unit energy required to melt the metal is 10 J/mm^3. If the welding power is 2 kW, what is the welding speed? **(GATE 2008)**

Solution: Net heat input = velocity × unit energy required to melt

$$\text{Power} \times \eta_{\text{melting}} \times \eta_{\text{heat transfer}} = \text{Area} \times \text{velocity} \times \text{unit energy required to melt}$$

$$200 \times 0.5 \times 0.7 = 5 \times \text{velocity} \times 10$$

$$\text{Velocity} = 14 \text{ mm/min}$$

Example 12: A shielded metal arc welding operation is taking place on carbon steel to produce a fillet weld. The desired welding speed is around 25 mm/sec. If the power supply is 10 V, what current is needed if the weld width is to be 7 mm? **(Exercise problem from *Manufacturing Process*: Kalpakjian).**

Solution: Width is given as 7 mm

$$\text{Area of weld} = \frac{7^2}{2} = 24.5 \text{ mm}^2$$

Assume efficiency of 75% for shielded arc welding, specific energy required to melt the metal is 9.7 J/mm^3, welding speed is calculated as

$$\text{Heat input} = VI\left(\text{in} \frac{\text{J}}{\text{sec} = \text{watts}}\right) \times \eta = 10 \times I \times 0.75 \tag{1}$$

$$\text{Heat input} = \text{welding speed} \times \text{area} \times \text{specific energy}$$

$$\text{Heat input} = 25 \frac{\text{mm}}{\text{sec}} \times 24.5 \text{ mm}^2 \times \frac{9.7 \text{ J}}{\text{mm}^3} \tag{2}$$

Equating (1) and (2)
We get current = 792 A

Example 13: Consider the situation where a welding operation is being performed with voltage of 20 volts and current of 200 A and cross-sectional area of weld bead is 30 mm^2. Estimate the welding speed if the workpiece and electrode are made of (a) aluminum, (b) carbon steel, (c) titanium, (d) magnesium, (e) copper, and (f) nickel:

Solution: For aluminum, specific energy required is 2.9 J/mm^3, therefore weld speed is

$$v = \eta \frac{VI}{uA}, \; v = \text{welding speed, } u \text{ is the specific energy and } A \text{ is area}$$

$$v = 0.75 \times \frac{20 \times 200}{2.9 \times 30} = 34.5 \text{ mm/sec}$$

For carbon steel, specific energy required is 12.3 J/mm^3, therefore weld speed is

$$v = 0.75 \times \frac{20 \times 200}{12.3 \times 30} = 8.1 \text{ mm/sec}$$

For titanium, specific energy required is 14.3 J/mm^3, therefore weld speed is

$$v = 0.75 \times \frac{20 \times 200}{14.3 \times 30} = 7 \text{ mm/sec}$$

For magnesium, specific energy required is 2.9 J/mm^3, therefore weld speed is

$$v = 0.75 \times \frac{20 \times 200}{2.9 \times 30} = 34.5 \text{ mm/sec}$$

For copper, specific energy required is 6.1 J/mm^3, therefore weld speed is

$$v = 0.75 \times \frac{20 \times 200}{6.1 \times 30} = 16.4 \text{ mm/sec}$$

For copper, specific energy required is 9.8 J/mm^3, therefore weld speed is

$$v = 0.75 \times \frac{20 \times 200}{9.8 \times 30} = 10.2 \text{ mm/sec}$$

Example 14: A SMAW operation is accomplished in a work cell using a fitter and a welder. The fitter takes 5.5 min to place the unwelded components into the welding fixture at the beginning of the work cycle, and 2.5 min to unload the completed weldment at the end of the cycle. The total length of the several weld seams to be made is 2000 mm, and the travel speed used by the welder averages 400 mm/min. Every 750 mm of weld length, the welding stick must be changed, which takes 0.8 min. While the fitter is working, the welder is idle (resting); and while the welder is working, the fitter is idle. (a) Determine the average arc time in this welding cycle. (b) How much improvement in arc time would result if the welder used FCAW (manually operated), given that the spool of flux-cored weld wire must be changed every five weldments, and it takes the welder 5.0 min to accomplish the change? (c) What are the production rates for these two cases (weldments completed per hour)?

Solution:

(a) SMAW cycle time

$$T_c = 5.5 + 2000/400 + (2000/750)(0.8) + 2.5$$
$$= 5.5 + 5.0 + 2.133 + 2.5 = 15.133 \text{ min}$$

$$\text{Arc time} = 5.0/15.133 = 33.0\%$$

(b) FCAW cycle time

$$T_c = 5.5 + 2000/400 + (1/5)(5.0) + 2.5$$
$$= 5.5 + 5.0 + 1.0 + 2.5 = 14.0 \text{ min}$$

$$\text{Arc time} = 5.0/14.0 = 35.7\%$$

(c) SMAW $R_p = 60/15.133 = 3.96$ pc/hr
FCAW $R_p = 60/14.0 = 4.29$ pc/hr

Example 15: In the previous problem, suppose an industrial robot cell is installed to replace the welder. The cell consists of the robot (using GMAW instead of SMAW or FCAW), two welding fixtures, and the fitter which loads and unloads the parts. With two fixtures, fitter and robot working simultaneously, the robot welding at one fixture while the fitter unloads and loads at the other. At the end of each work cycle, they switch places. The electrode wire spool must be changed every five workparts, which task requires 5.0 minutes and is accomplished by the fitter. Determine (a) arc time, and (b) production rate for this work cell.

Solution:

(a) Fitter: $T_c = 5.5 + 2.5 + (1/5)(5.0) = 9.0$ min
Robot: $T_c = 2000/400 = 5.0$ min
Limiting cycle is the fitter: arc time = 5.0/9.0 = 55.5%

(b) $R_p = 60/9.0 = 6.67$ pc/hr

Example 16: A shielded metal arc welding operation is performed on steel at a voltage = 30 volts and a current = 225 A. The heat transfer factor = 0.85 and melting factor = 0.75. The unit melting energy for steel = 10.2 J/mm^3. Determine (a) the rate of heat generation at the weld, and (b) the volume rate of metal welded.

Solution:

(a) $R_{WR} = f_1 f_2 EI = 0.85 \times 0.75 \times 30 \times 225 = 430.1$ watts

(b) $R_{WV} = (4303.1 \text{ W})/(10.2 \text{ J/mm}^3) = 421.9 \text{ mm}^3\text{/sec.}$

Example 17: In arc welding of a butt joint, the welding speed is to be selected such that highest cooling rate is achieved. Melting efficiency and heat transfer efficiency are 0.5 and 0.7, respectively. The area of the weld cross section is 5 mm^2 and the unit energy required to melt the metal is 10 J/mm^3. If the welding power is 2 kW, what is the welding speed in mm/s? **(GATE Exam)**

Solution: Total energy required to melt material is

$$H = U_m \times A \times v = 10 \times 5 \times v \tag{1}$$

Power supplied to welding

$$H = P \times f_1 \times f_2 = 2000 \times 0.5 \times 0.7 = 700 \text{ watts} \tag{2}$$

Equating (1) and (2) and solving,
We get welding speed = 14 mm/sec

Example 18: In an arc welding process, the voltage and current are 25 V and 300 A, respectively. The arc heat transfer efficiency is 0.85 and welding speed is 8 mm/sec. What is the net heat input (in J/mm)? **(GATE Exam)**

Solution: Power supplied to welding

$$H = \text{power input} = V \times I \times f_1 = 25 \times 300 \times 0.85 = 6375 \text{ joules} \quad (1)$$

$$H\left(\frac{\text{Joules}}{\text{mm}}\right) \times v = \text{power input} \quad (2)$$

$$H\left(\frac{\text{Joules}}{\text{mm}}\right) \times 8 = 6375$$

Solving we get

$$\text{Heat input to workpiece in } \frac{\text{J}}{\text{mm}} = \frac{6375}{8} = 796.9 = 797 \text{ J/mm}$$

Review Questions

1. Explain the principle of operation of SMAW.
2. Describe the differences between AC and DC arc systems?
3. Describe the features of equipment used for SMAW?
4. Describe the features of coated electrodes?
5. Give the classification of coated electrodes.
6. How do we designate the coated electrodes?
7. Describe the advantages of ISO designation of electrodes?
8. Describe the advantages of SMAW?
9. Describe the disadvantages of SMAW?
10. Give applications of SMAW.

7 CHAPTER

Gas Metal Arc Welding

7.0 INTRODUCTION

Gas Metal Arc welding (GMAW) is an arc welding process that uses an arc between a continuous filler metal electrode and the weld pool. The process is used with shielding from an externally supplied gas and without the application of pressure.

The basic concept of GMAW was introduced in 1920, but it was not until 1948 that it was made commercially available. At first it was considered to be, fundamentally, a high-current density, small diameter, bare metal electrode process using an inert gas for arc shielding. Its primary application was for welding aluminum. As a result, the term MZG (Metal Inert Gas) was used and is still a common reference for the process. Subsequent process developments included operation at low-current densities and pulsed direct current, application to a broader range of materials, and the use of reactive gases (particularly CO_2) and gas mixtures. This latter development led to the formal acceptance of the term *gas metal arc welding* (GMAW) for the process because both inert and reactive gases are used.

GMAW may be operated in semiautomatic, machine, or automatic modes. All commercially important metals such as carbon steel, high-strength low alloy steel, stainless steel, aluminum, copper, titanium, and nickel alloys can be welded in all positions with this process by choosing the appropriate shielding gas, electrode, and welding variables.

7.1 OPERATING PRINCIPLE (FIGURE 7.1)

Gas metal arc welding (GMAW) is an arc welding process in which an electric arc is formed and maintained between a continuously fed filler metal electrode wire and the weld pool. In the arc heat, the electrode wire is melted and the molten metal (droplets) is transferred across the arc into the weld pool. The arc and the weld pool is shielded from the atmospheric contamination by an externally supplied gas.

Various materials can be welded by MIG. It includes carbon and low-alloy steel, stainless steel, heat-resistant alloys, copper and its alloys, magnesium alloys, aluminum and its alloys, high strength low alloy steel (HSLA), etc.

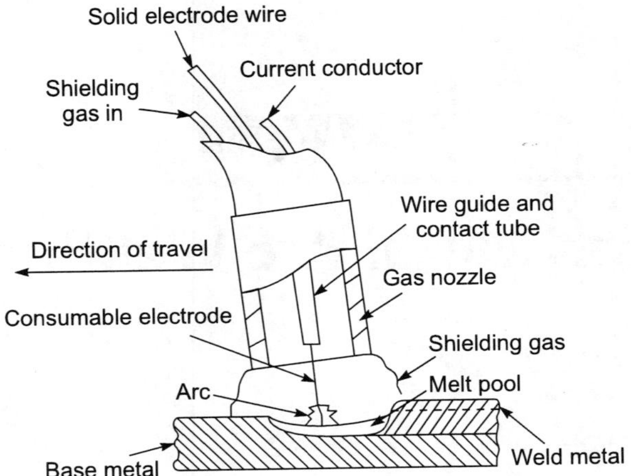

Figure 7.1 Operating principle

7.2 EQUIPMENT

Equipment required for GMAW is shown in Figure 7.2. The system comprises a DC power source with flat characteristics, wire feed unit, welding gun, cable assembly and gas cylinder. The wire electrode is fed continuously into the arc by the wire

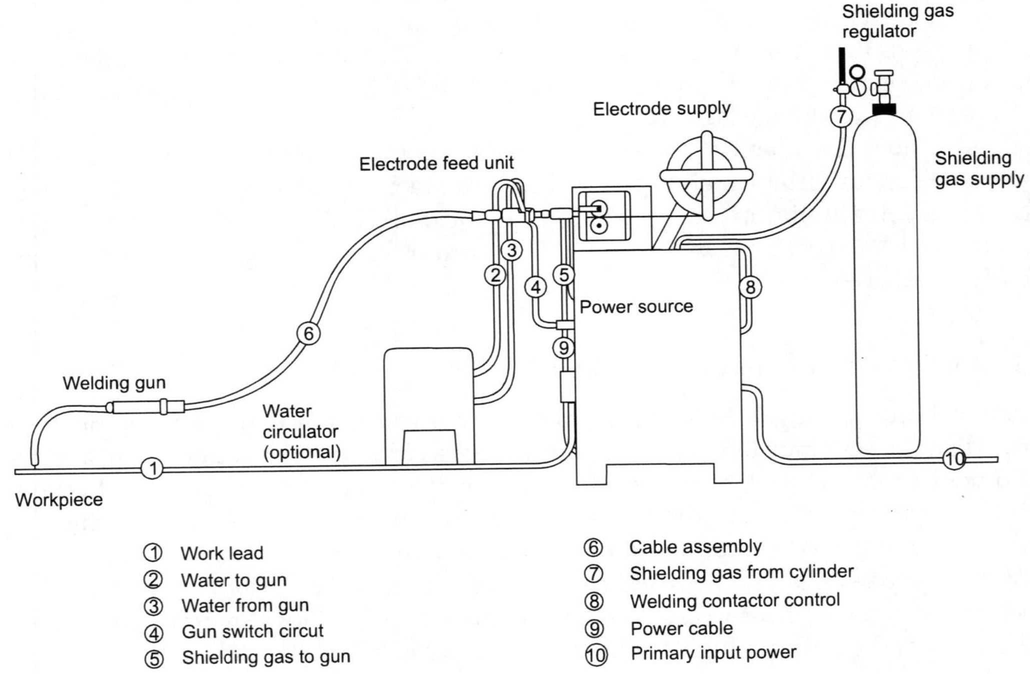

Contd.

1. Power source
2. Wire feed system
3. Welding gun (torch)
4. Gas cylinder/source
5. Ground and weld cables
6. Ground clamp
7. Chiller (refrigeration system)

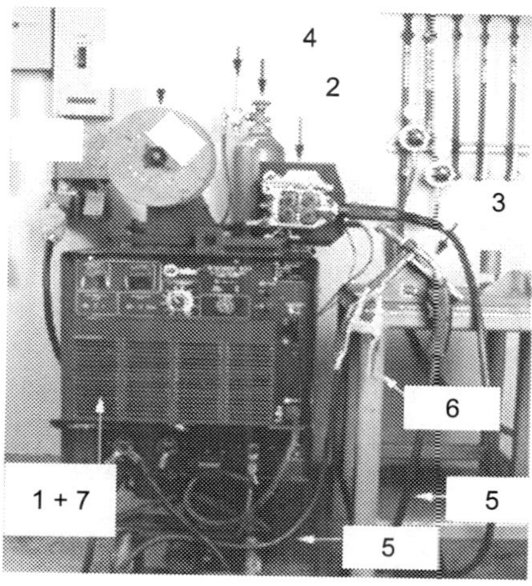

Figure 7.2 Basic GMAW system

feed unit at a speed preset by the operator. The wire feed rate can be varied from 1 m/min to 20 m/min. For a given wire material and diameter, the arc current is determined by the wire feed rate. The required voltage is selected by adjusting the voltage control knobs provided at the power source. The equipment necessary for gas metal arc welding are:

- Power source
- Wire feeder
- Welding gun
- Shielding gas supply
- Solid electrode wire
- Protective equipment

7.2.1 Power Source (Figures 7.3 & 7.4)

Most applications of gas metal arc welding use a constant voltage power supply. As a result, any change in arc length (which is directly related to voltage) results in a large change in heat input and current. A shorter arc length causes a much greater heat input, which makes the wire electrode melt more quickly and thereby, restore the original arc length. This helps operators keep the arc length consistent even when manually welding with hand-held welding guns. To achieve a similar effect, sometimes a constant current power source is

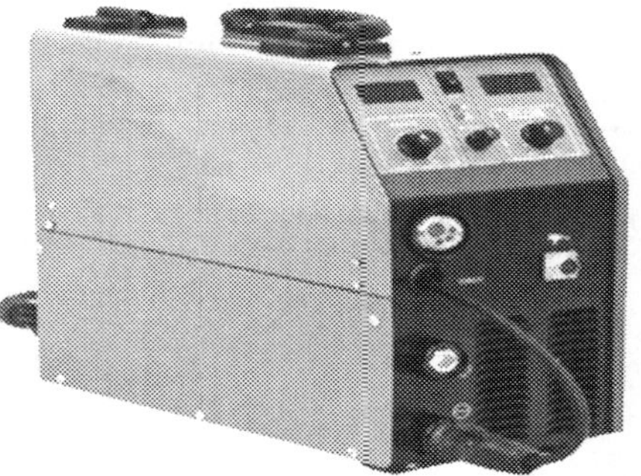

Figure 7.3 Power source

used in combination with an arc voltage controlled wire feed unit. In this case, a change in arc length makes the wire feed rate adjust to maintain a relatively constant arc length. In rare circumstances, a constant current power source and a constant wire feed rate unit might be coupled, especially for welding metals with high thermal conductivities, such as aluminum. This grants the operator additional control over the heat input into the weld, but requires significant skill to perform successfully. Alternating current is rarely used with GMAW; instead, direct current is employed and the electrode is generally positively charged. Since the anode tends to have a greater heat concentration, this results in faster melting of the feed wire, which increases weld penetration and welding speed. The polarity can be reversed only when special emissive coated electrode wires are used, but since these are not popular, a negatively-charged electrode is rarely employed.

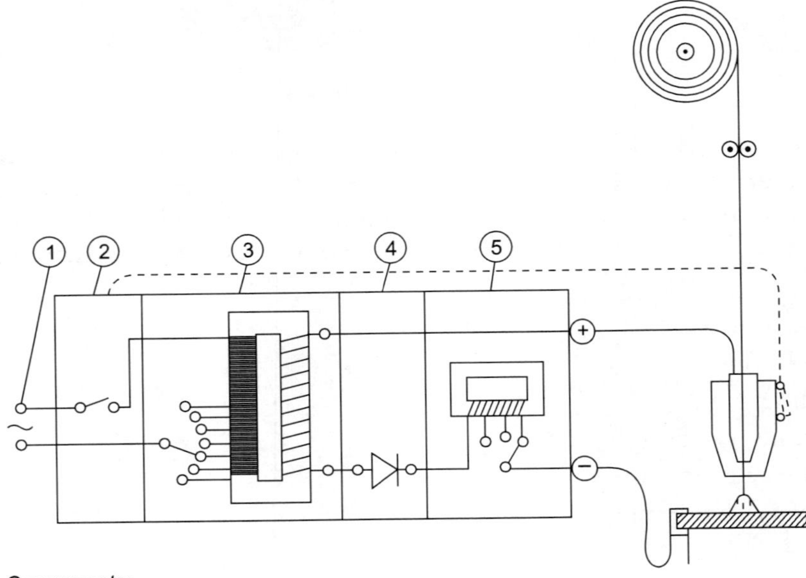

Components:

① Mains supply line

② On/Off swich for welding current

③ Transformer with current switch
Function: To reduce mains voltage to welding voltage and increase mains current to welding current

④ Rectifier
Function: To convert alternating or three-phase current to direct current

⑤ Welding regulator
Function: Smooth out current surges in the welding circuit

Figure 7.4 Power supply for MIG welding

Welding Set Details

- Range of sizes – identified by current

    - Below 200 A – Dip transfer operation only, suitable for steel
    - Larger sets have wire reel and motor as separate unit
        - Can be placed near the job
        - Inductance is controlled by voltage
        - Wire speed control is on wire feed unit
- Electrical input
    - Single-phase 240 V – small sets
    - Three-phase 415 V – medium and large sets
- Output is always DC
    - Flat characteristics for semi-automatic
    - Drooping for mechanized
- Pulse-MIG
    - Sets 40-200 pulse/s give improved results on some jobs
    - Pulse MIG increases number of controls
    - Synergic control system varies all parameters in steps
- Other parts of welding set
    - Built-in holder for gas cylinder
    - Suitable welding gun
        - Water-cooled in case of heavy duty
        - Built-in water tank and radiator in such cases
    - Gun with small reel of wire –'reel-on-gun'
        - For Al welding, to remove the tendency of Al wire to kink through hose.
- Accessories
    - Welding cables – for electrical connections
    - Connectors
    - Clamps or clips
    - Cylinder stand
    - Vaporizer for CO_2 gas
    - Gas regulators and hose, suitable connections
    - Gun and connecting hose
- Welding spares
    - Inlet and outlet guides
    - Drive rolls
    - Contact tip in gun – requires frequent replacement
    - Gas shielding nozzle for gun
    - Wire conduit liner

Power Source Comparison

A power source with a flat characteristic is almost universally used for GMAW process, as it offers several advantages, e.g., latitude in setting the welding conditions, and self-regulated arc. Besides, it also meets the special requirements of dip transfer welding.

The static volt-ampere characteristics of two power sources are shown in Figure 7.5. The curves labelled (*d*) are those of a "drooping arc characteristic" power source and the one labelled (*f*), the "flat characteristic". The lower graph (*b*) represents the typical burn-off characteristics of a wire and depicts the increase in current that results from an increase in wire feed rate, with the wire feed rate set at W_1 m/mm. The lower graph shows that A_l amps is required to burn off this particular wire diameter at this speed. Projecting up to the source curves shows that this current can be supplied at V_3 volts on both power sources, *f* and *d*. If the wire speed is now increased to W_2 m/min, the burn-off characteristic curve (b) indicates that A_2 amps is required to burn off this amount of wire. Projecting up to the source curves shows that the constant voltage power source can supply this higher current at practically the same voltage as V_2 is nearly equal to V_3 but the drooping characteristic source can supply A_2 amps only at V_1 volt. This is an impractical low arc voltage and the arc would be extinguished. A new curve d_2 must be selected on the drooping characteristic power source so that the desired current A_2 can be supplied at the right voltage V_2. Because of this latitude in setting the welding conditions and the independent control of current by wire feed speed, constant voltage power sources are by far the most popular for GMA welding. The control of current with this type of power source merely involves selecting a wire speed from the burn-off characteristic. The current will rise automatically to the required value at a voltage not greatly below the selected open circuit voltage.

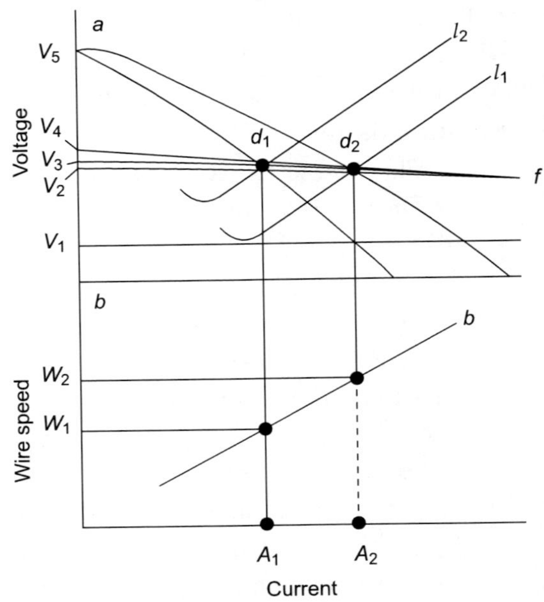

Figure 7.5 Latitude in setting the welding conditions

Another important advantage provided by the flat characteristic power source is its ability to produce self-regulated arc. Self-regulated arc means maintaining the arc length constant. The principle of self-regulation of the arc length in GMAW is illustrated in Figure 7.5. With a constant potential power source, changes in arc voltage will have marked effects on current. Thus, if the arc length is reduced from the set value, there will be an increase in current, resulting in a faster burn-off rate and the arc length will be adjusted back to its original value. On the other hand, an increase in arc length from the set value would increase the arc voltage, resulting in less current and lower burn-off rate. The arc length would be restored to the original length. Self-adjustment will only operate successfully when the change in current produced by voltage fluctuations is sufficiently large to produce a large alteration in burn-off, and rapid response rate and correction that the disturbance cannot beat.

7.2.2 Wire Feeder (Figure 7.6)

When welding with a constant voltage source, the prime function of the wire feeder is to deliver the welding wire to the arc at a constant speed. This is important since varying speeds will produce varying amounts of heat, leading to an inconsistent weld. Advanced electronic equipment make it possible to design motor speed controls that will produce the same speed, even though the load on the motor varies or input voltage to motor fluctuates. When constant current power source is used, the motor speed varies automatically to increase or decrease the wire feed speed as the arc length varies to maintain a constant voltage. Two- or four-roll drive systems deliver the electrode to the welding torch. Two-roll systems are standard with smaller non-industrial systems, but the four-roll system is popular for industrial applications.

Figure 7.6 Wire feeder

7.2.3 Welding Gun (Figures 7.7 and 7.8)

The typical GMAW welding gun has a number of key parts, i.e., a control switch, a contact tip, a power cable, a gas nozzle, an electrode conduit and liner, and a gas hose. The control switch or trigger, when pressed by the operator, initiates the wire feed, electric power, and the shielding gas flow, causing an electric arc to be struck. The contact tip, normally made of copper and sometimes chemically treated to reduce spatter, is connected to the welding power source through the power cable and transmits the electrical energy to the electrode, while directing it to the weld area. It must be firmly secured and properly sized, since it must allow the electrode to pass while maintaining electrical contact. On the way to the contact tip, the wire is protected and guided by the electrode conduit and liner, which help prevent buckling and maintain an uninterrupted wire feed. The gas nozzle directs the shielding gas evenly into the welding zone. Inconsistent flow may not adequately protect the weld area. Larger nozzles provide greater shielding gas flow, which is useful for high current welding operations that develop a larger molten weld pool. A gas hose from the tanks of shielding gas supplies the gas to the nozzle. Sometimes, a water hose is also built in the welding gun for cooling the gun in high heat operations.

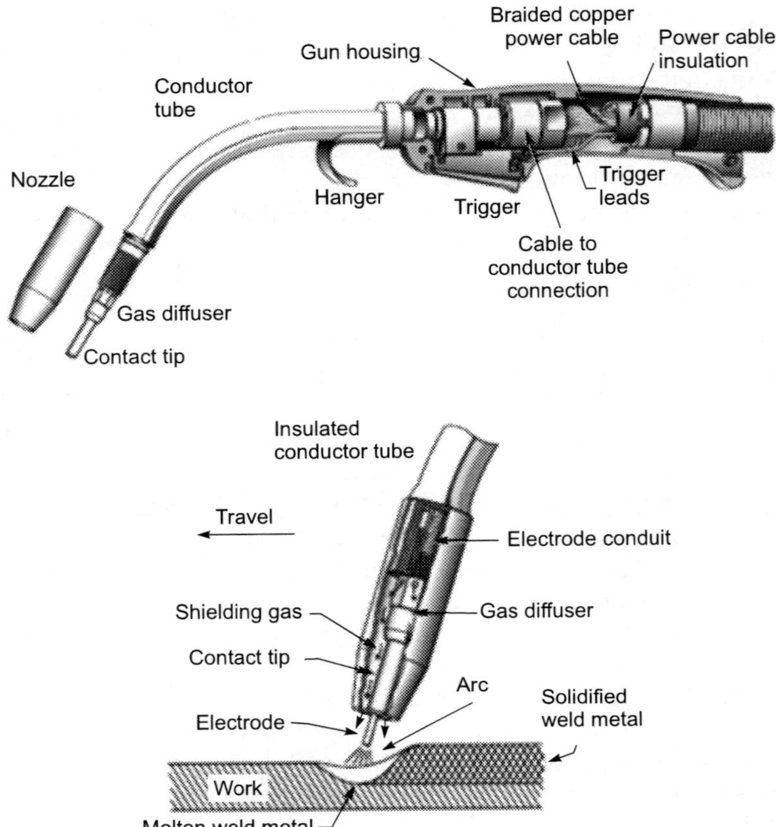

Figure 7.7 Welding gun

7.2.4 Shielding Gases

The primary function of the shielding gas is to exclude the atmosphere from contact with the molten weld metal. This is necessary because most metals, when heated to their melting point in air, exhibit strong tendency to form oxides and to a lesser extent, nitrides. Oxygen will also react with carbon in molten steel to form carbon monoxide and carbon dioxide. These varied reaction products may result in weld deficiencies, such as trapped slag, porosity, and weld metal embrittlement. Reaction products are easily formed in the atmosphere, unless precautions are taken to exclude nitrogen and oxygen. In addition to providing a protective environment, the shielding gas and flow rate also have a pronounced effect on the following:

- Arc characteristics
- Mode of metal transfer

- Penetration and weld bead profile
- Speed of welding
- Undercutting tendency
- Cleaning action
- Mechanical properties of weld metal

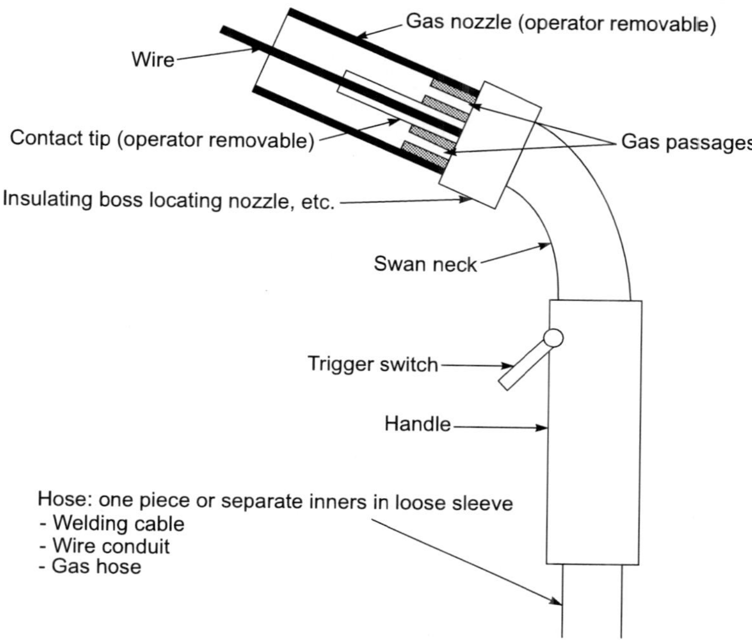

Figure 7.8 Welding gun

The principal gases used in GMAW are shown in Table 7.1. Comparison chart for different inert gases and their effect is given in Table 7.2.

TABLE 7.1 List of recommended gases to be used along with spray transfer

Carbon steel	Argon/CO_2/O_2/He (He less than 50%)
Alloy steel	Argon/CO_2/O_2/He (He less than 50%)
Stainless steel	Argon/O_2/CO_2 (CO_2 max. 2%)
Copper, nickel and Cu-Ni alloys	Argon/helium
Aluminum	Argon/helium

TABLE 7.2 Comparison chart for different inert gases and their effect

Metal	Gas	Gas Effect
Aluminum	Argon Ar 25%, He 75%	Helps to remove surface oxides Oxide removal + porosity control
Mild steel	Carbon dioxide (CO_2)	Low spatter
Magnesium	Argon	Helps to remove surface oxides
Stainless steel	95-99% He Oxygen 1-59%	Oxygen adds to arc stability and reduces under-cutting
Nickel	Argon Ar 25–80% He 50–75%	Helps to control base metal fluidity and provides good welding bead

7.2.5 Electrode

Electrode selection is based primarily on the composition of the metal being welded, the process variation being used, joint design and the material surface conditions. Electrode selection greatly influences the mechanical properties of the weld and is a key factor of weld quality. In general, the finished weld metal should have mechanical properties similar to those of the base material, with no defects such as discontinuities, entrained contaminants or porosity in the weld. To achieve these goals, a wide variety of electrodes exist. All commercially available electrodes contain deoxidizing metals, such as silicon, manganese, titanium and aluminum, in small percentages to help prevent oxygen porosity. Some contain denitriding metals, such as titanium and zirconium, to avoid nitrogen porosity.

Wire Melting Characteristics

A nonlinear relationship exists between the rate at which the wire is fed into the arc and the current required to burn it off to maintain a constant arc length. This is known as the burn-off characteristic and differs for each filler wire composition and diameter. The curves shown in Figure 7.9 illustrate the burn-off characteristics of steel in Ar + O_2 mixture. It will be seen that to alter the welding current, a proportionate increase in the wire feed rate is required. The smaller the diameter of the wire, the faster is the wire feed speed required to reach the same current.

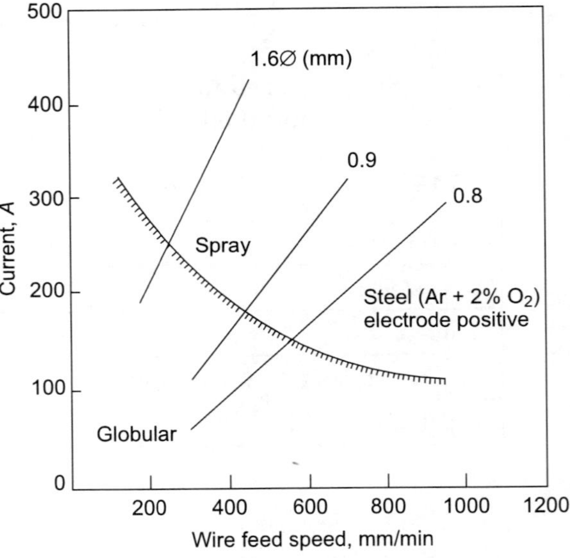

Figure 7.9 Wire burn-off characteristics

7.3 METAL TRANSFER MODES

The type of transfer is determined by a number of factors, the most influential of which are the following:

1. Magnitude and type of welding current
2. Electrode diameter
3. Electrode composition
4. Electrode extension
5. Shielding gas

Metal Transfer

The continuously fed electrode wire is melted in the arc heat and the droplets thus formed at the wire tip are transferred to the weld pool. The three basic modes by which metal is transferred from the electrode to the weld pool are short-circuiting transfer, globular transfer and spray transfer. These basic modes of metal transfers are controlled by the various forces that are acting on the droplets. The electromagnetic pinch force and the surface tension force are considered primary forces responsible for the various modes of metal transfer in GMAW process.

7.3.1 Short-Circuiting Transfer

Short-circuiting transfer occurs when the welding current and arc voltage are low. Metal is transferred from the electrode to the work only when the electrode is in contact with the weld pool. No metal is transferred across the arc gap. The short-circuiting frequency varies from 20 to 200 Hz. The sequence of events in the transfer of metal and the corresponding current and voltage waveform are shown in Figure 7.10. As the droplet on the wire tip makes short-circuit with the weld pool, the current increases until the arc is ignited again. The rate of current rise during short-circuiting must be properly controlled. It must be high enough to maintain a molten electrode tip until filler metal is transferred. Yet, it should not be a sudden surge that causes spatter by disintegration of the transferring drop of the filler metal. The current increase is controlled by adjusting the inductance in the power source. The influence of low and high inductance setting on spattering during short-circuiting transfer welding is illustrated in Figure 7.11. The current setting of the inductance depends on the type of shield gas and the welding parameters used. For short-circuit transfer welding, the open circuit voltage should be low enough so that arc cannot continue under the existing welding conditions. A portion of the energy for arc maintenance is provided by the inductive storage of energy during the period of short circuiting. Short-circuiting transfer offers a number of operating advantages. The small fast-freezing weld pool produced is well suited to join thin sheets, for out of position welding, and for bridging large root openings.

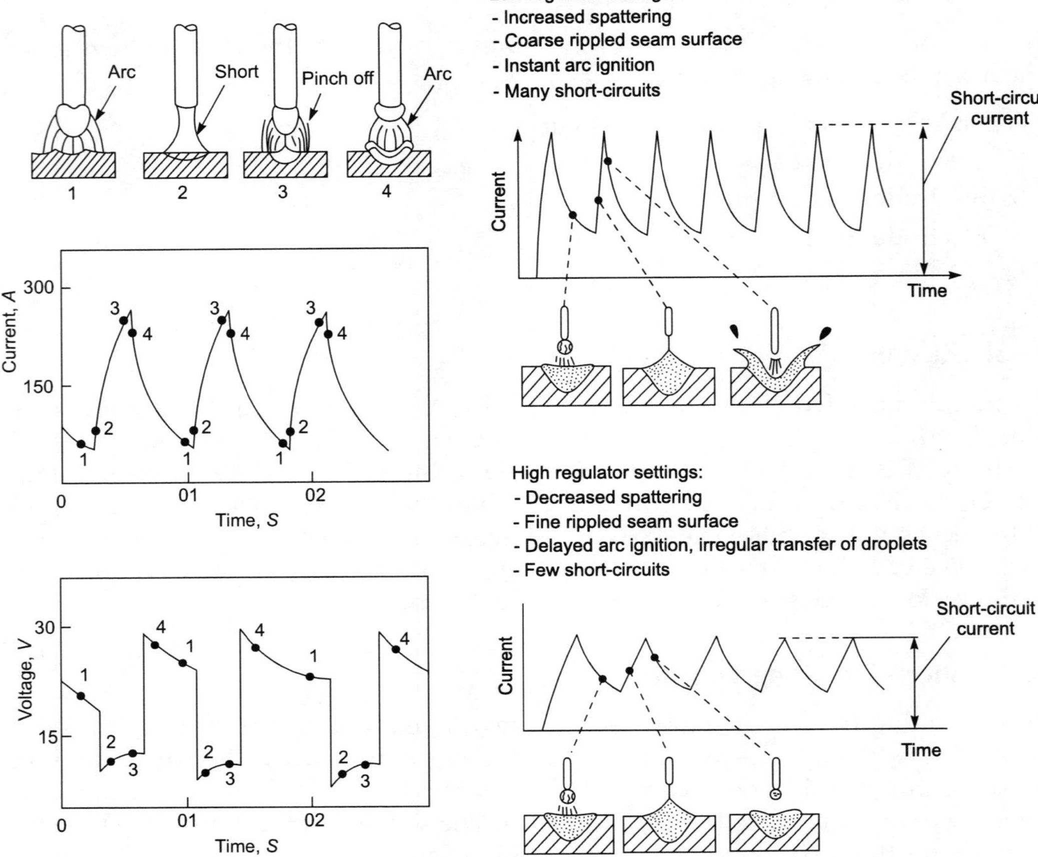

Figure 7.10 Gas metal arc welding using short arc from the electrode to the work

Figure 7.11 Influence on inductance on spatter

Advantages of Short-Circuiting Transfer

- All-position capability, including flat, horizontal, vertical-up, vertical-down and overhead.
- Handles poor fit-up extremely well, and is capable of root pass work on pipe applications.
- Lower heat input reduces weldment distortion.
- Higher operator appeal and ease of use.
- Higher electrode efficiencies, 93% or more.

Limitations of Short-Circuiting Transfer

- Restricted to sheet metal thickness range and open roots of groove joints on heavier sections of base material.

- Poor welding procedure control can result in incomplete fusion. Cold lap and cold shut are additional terms that serve to describe incomplete fusion defects.
- Poor procedure control can result in excessive spatter, and will increase weldment clean-up cost.
- To prevent the loss of shielding gas to the wind, welding outdoors may require windscreen(s).

7.3.2 Globular Transfer

Globular transfer, as shown in Figure. 7.12, is characterized by a drop size of greater diameter than that of the electrode wire. The large drop is easily acted upon by gravity, generally limiting successful to the flat position. Globular transfer occurs in the current range higher than that of short-circuiting transfer. The arc length must be long enough to assure detachment of the drop before it contacts the molten metal. CO_2 shielding results in randomly directed globular transfer. The non-axial transfer is due to electromagnetic repulsive force acting upon the bottom of the molten drop.

Long arc

Description:

- Electrode material transferred in large globules: not free from short-circuits
- Weld pool of low viscosity
- About 100 droplets transferred per second

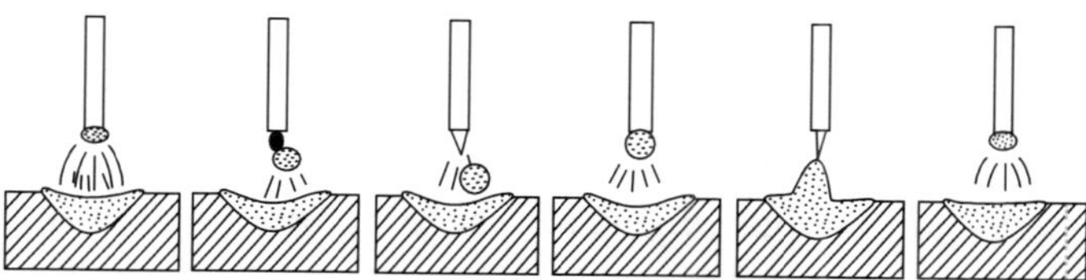

Setting range

High voltage (above 20 V)

Example for: Wire electrode of 1.0 mm diameter
Protective gas in accordance with DIN 32526.C1 (carbon dioxide)

Voltage, V: 10, 20, 30, 40

Current, A: 50, 100, 150, 200, 250

Figure 7.12 Globular transfer

Advantages of Globular Transfer

- Uses inexpensive CO_2 shielding gas, but is frequently used with argon/CO_2 blends.
- Is capable of making welds at very high travel speeds.
- Inexpensive solid or metal-cored electrodes.
- Welding equipment is inexpensive.

Limitations of Globular Transfer

- Higher spatter levels result in costly clean-up.
- Reduced operator appeal.
- Prone to cold lap or cold shut incomplete fusion defects, which results in costly repairs.
- Weld bead shape is convex, and welds exhibit poor wetting at the toes.
- High spatter level reduces electrode efficiency to a range of 87% to 93%.

7.3.3 Spray Transfer

In a gas shield of argon or argon-rich mixture (i.e., Ar + CO_2 gas mixture containing more than 80% Ar), the metal transfer changes from globular to spray type as welding current is increased beyond the transition current. The current value at which the metal transfer mode changes from globular to spray is known as "transition current". Spray-type transfer, as shown in Figure 7.13, has a typical fine arc column and pointed wire tip associated with it. Molten filler metal transfers across the arc as fine droplets. The reduction in droplet size is also accompanied by an increase in the rate of droplet detachment. One important characteristic of spray mode of transfer is the "finger" penetration that it produces. The spray transfer mode can be used to weld any metal. However, applying the process to thin sheets may be difficult because of the high currents needed to produce the spray arc. Also, the large weld pool produced due to the high deposition rate makes it difficult to use it in vertical or overhead positions.

When CO_2 is used for shielding, a globular to spray transition does not occur. Drops become smaller as the current increases, but they are not at all axially directed. The spatter is more than that with argon or argon-rich mixed gas shield. The spatter is minimized if "buried arc" technique is used. Buried arc technique involves using a low voltage for any given current, forcing the arc to locate within the cavity created by the arc force. Under this condition, most of the non-axially directed droplets are trapped in the weld pool.

Spray arc

Description:

- No short-circuit on droplet transfer
- Weld pool of low viscosity
- About 100 to 300 droplets transferred per second

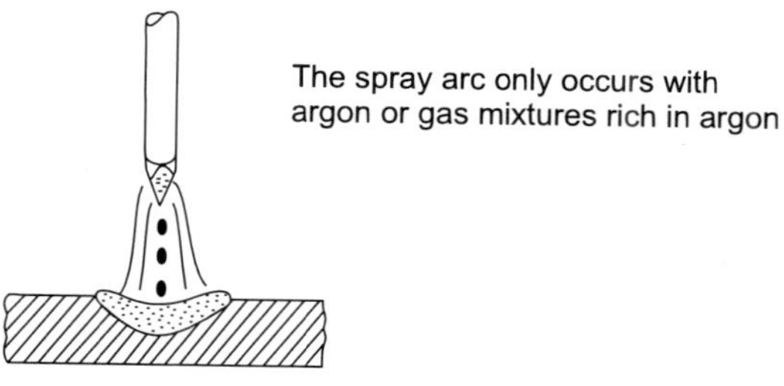

Setting range

High voltage (above 25 V)

Example for : Wire electrode of 1.0 mm diameter
Protective gas in accordance with DIN 32526-M2 (gas mixture)

Voltage, V

40

30

20

10

50 100 150 200 250

Current, *A*

Figure. 7.13 Spray transfer

Advantages of Axial Spray Transfer

- High deposition rates.
- High electrode efficiency of 98% or more.
- Employs a wide range of filler metal types in an equally wide range of electrode diameters.
- Excellent weld bead appearance.
- High operator appeal and ease of use.
- Requires little post-weld clean-up.

- Absence of weld spatter.
- Excellent weld fusion.
- Lends itself to semi-automatic, robotic, and hard automation applications.

Limitations of Axial Spray Transfer

- Restricted to the flat and horizontal welding positions.
- Welding fume generation is higher.
- The higher-radiated heat and the generation of a very bright arc require extra welder and bystander protection.
- The use of axial spray transfer outdoors requires windscreen (s).
- The shielding used to support axial spray transfer costs more than 100% CO_2.

Pulsed Spray Transfer

Pulsed spray metal transfer, known by the acronym GMAW-P, is a highly controlled variant of axial spray transfer, in which the welding current is cycled between a high peak current level to a low background current level. Metal transfer occurs during the high energy peak level in the form of a single molten droplet. GMAW-P was developed for two demanding reasons: control of weld spatter and the elimination of incomplete fusion defects common to globular and short-circuiting transfer. Its earliest application included the welding of high-strength low alloy base material for out-of-position ship hull fabrication. The advantages that it brought to the shipbuilding industry included: higher efficiency electrodes than FCAW, and the ability to deliver lower-hydrogen weld deposits. The mode employs electrode diameters from 0.8 to 1.6 mm solid wire electrodes and metal cored electrodes of 1.1 to 2.0 mm diameter. It is used for welding a wide range of material types. Argon based shielding gas selection with a maximum of 18% CO_2 supports the use of pulsed spray metal transfer with carbon steels.

The welding current alternates between a peak current and a lower background current, and this controlled dynamics of the current results in a lower average current than is found in axial spray transfer. The time, which includes the peak current and the background current, is a period, and the period is known as a cycle (Hz). The high current excursion exceeds the globular to spray transition current, and the low current is reduced to a value lower than that seen in short-circuiting transfer. Ideally, during the peak current, the high point of the period, a single droplet of molten metal is detached and transferred across the arc. The descent to the lower current, known as the background current, provides arc stability and is largely responsible for the overall heat input into the weld. The frequency is the number of times the period occurs per second, or cycles per second. The frequency of the period increases in proportion to the wire feed speed. Taken together, they produce an average current, which leverages its use in a wide material thickness range (Figure 7.14).

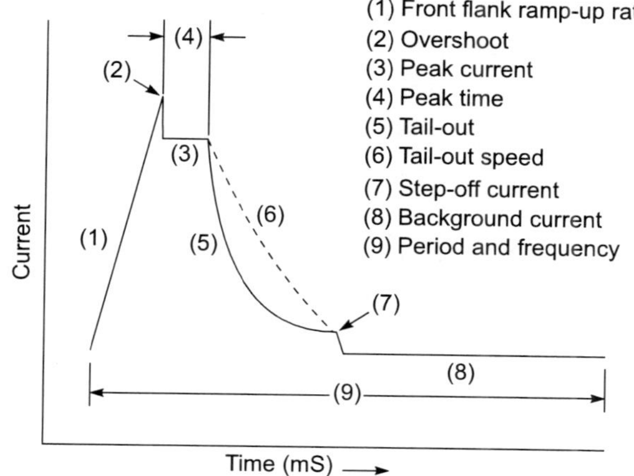

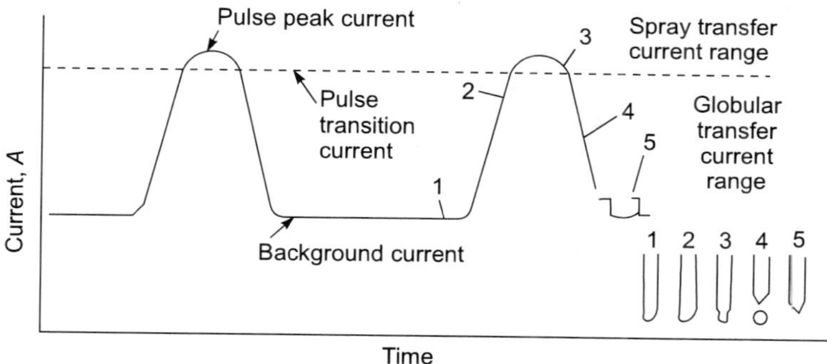

Figure 7.14 Pulsed spray transfer

Advantages of Pulsed Spray Transfer

- Absent or very low levels of spatter.
- More resistant to lack of fusion defects than other modes of GMAW metal transfer.
- Excellent weld bead appearance.
- High operator appeal.
- Offers an engineered solution for the control of weld fume generation.
- Reduced levels of heat induced distortion.
- Ability to weld out-of-position.
- Lower hydrogen deposit.
- Reduces the tendency for arc blow.
- Handles poor fit-up.

- When compared with FCAW, SMAW, and GMAW-S, pulsed spray transfer provides a low cost high-electrode efficiency of 98%.
- Lends itself to robotic and hard automation applications.
- Capable of arc travel speeds greater than 1.2 m/min.

Limitations of Pulsed Spray Transfer

- Equipment to support the process is more expensive than traditional systems.
- Blends of argon based shielding gas are more expensive than carbon dioxide.
- Higher arc energy requires the use of additional safety protection for welders and bystanders.
- Adds complexity to welding.
- Requires windscreens outdoors.

7.4 PROCESS PARAMETERS

The process variables that affect fusion characteristics and weld bead geometry in GMAW are welding current, arc voltage, welding speed, electrode extension, electrode diameter and shielding gas.

Three types of parametric variables that must be considered in MIG welding operations are:

- The variables controlled by an operator
- The machine controlled variables
- The base metal controlled variables

7.4.1 Operator Controlled Variables

Electrode Extension (Stick-Out of the Electrode)

The electrode extension is the distance between the contact tube and the end of the electrode. As this distance increases, the electrical resistance also increases. Therefore, due to resistance heating the electrode temperature rises. Then, only less welding current is required to melt the electrode at a given feed rate. The electrode extension length has to be controlled because too long an extension results in excess weld metal being deposited with low arc heat. This may cause cold lapping and poor bead shape. Also, as the electrode extension increases, the arc becomes less stable and shielding will not be effective. Proper electrode extension would be 6 to 12 mm for short-circuiting transfer welding and 12 to 20 mm for globular and spray type of metal transfer (Figure 7.15).

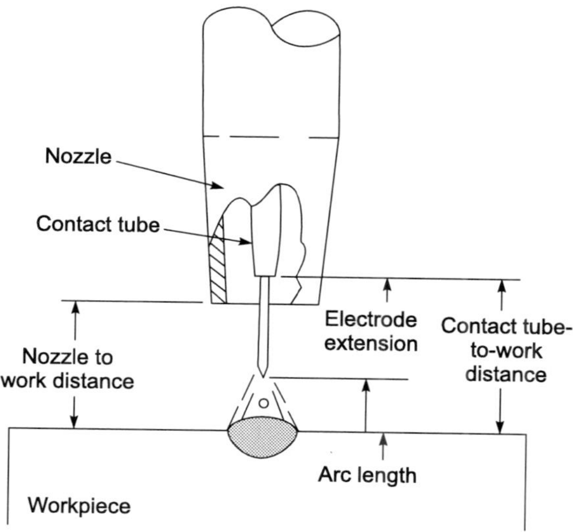

Figure 7.15 Electrical stick-out (ESO)

Each electrode diameter of a given composition has a usable range of currents. The welding amperage range is limited by undesirable effects, such as the absence of proper wetting at very low currents, and spatter porosity and poor bead appearance with excessively high currents. The wire melting rate is a function of current density. If two wires of different diameters are operated at the same amperage, the smaller will have the higher melting rate. Penetration is also a function of current density. Smaller diameter wire will produce deeper penetration than the larger diameter wire when they are used at the same current. However, the weld bead profile will be wider with the larger diameter electrode.

Electrode Orientation

Electrode orientation affects bead shape and penetration to a greater extent than arc voltage or travel speed. The electrode orientation is described in two ways: (1) by the relationship of the electrode axis with respect to the direction of travel (the travel angle), and (2) the angle between the electrode axis and the adjacent work surface (work angle). When the electrode points opposite from the direction of travel, the technique is called backhand welding with a drag angle. When the electrode points in the direction of travel, the technique is called forehand welding with a lead angle. The electrode orientation and its effect on the width and penetration of the weld are illustrated in Figures 7.16 (A), (B), and (C).

When the electrode is changed from the perpendicular to a lead angle technique with all other conditions unchanged, the penetration decreases and the weld bead becomes wider and flatter. Maximum penetration is obtained in the flat position with the drag technique, at a drag angle of about 25° from perpendicular. The drag technique also produces a more convex, narrower bead, a more stable arc, and less spatter on the workpiece. For all positions, the electrode travel angle normally used is a drag angle in the range of 5° to 15° for good control and shielding of the molten weld pool. For some materials, such as aluminum, a lead technique is preferred. This lead technique provides a "cleaning action" ahead of the molten weld metal, which promotes wetting and reduces base metal oxidation. When producing fillet welds in the horizontal position, the electrode should be positioned about 45° to the vertical member (work angle), as illustrated in Figure 7.17.

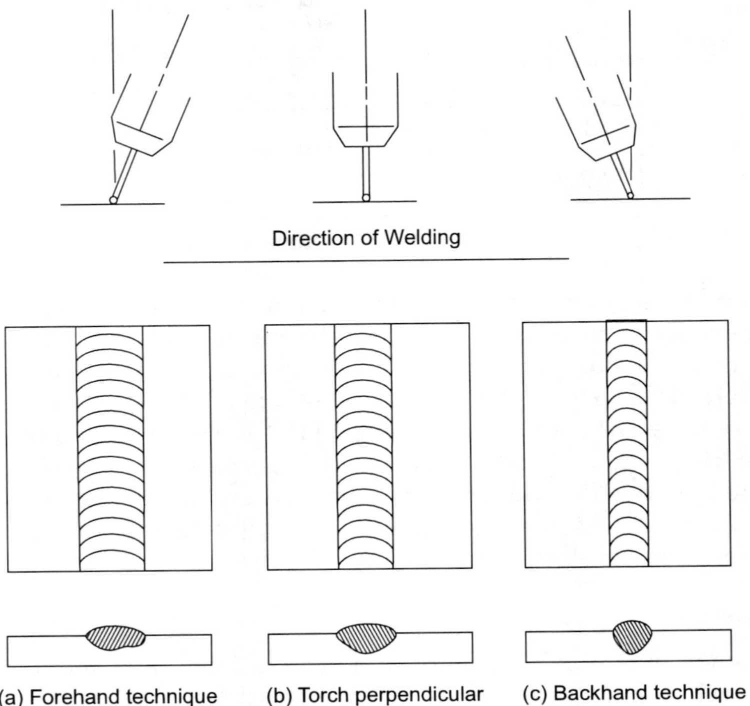

Figure 7.16 Effect of electrode position and welding technique

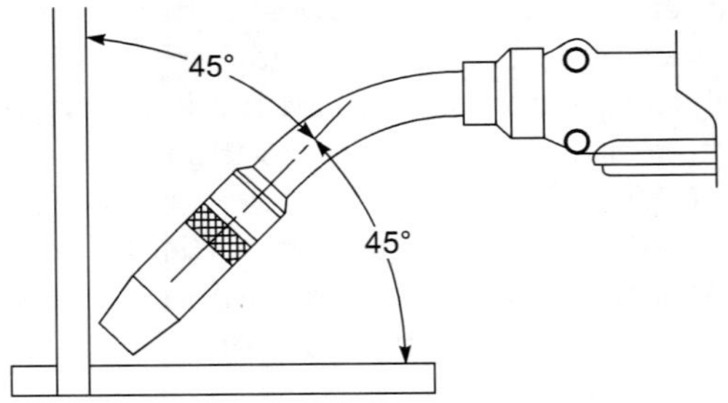

Figure 7.17 Normal work angle for fillet welds

7.4.2 Machine Controlled Variables

The machine controlled variables include the arc voltage, the welding current and the travel speed of the bead. Arc voltage and welding current control will determine whether spray arc welding system, short circuit welding system or pulsed arc welding system is used. These two controls, the voltage and amperage in combination, determine the optimum travel speed of the weld bead.

Welding Current: The welding current is directly related to the electrode feed rate. Welding current has a strong influence on the bead characteristics. With all other variables held constant, an increase in welding current will result in increase in depth and width of penetration, increase in deposition rate, and increase in the size of the weld bead.

Arc Voltage: Arc voltage depends on the arc length as well as many other variables, such as the electrode composition and dimensions, the shield gas, the welding technique and, since it is often measured at the power supply, even the length of the welding cable. Arc voltage is an approximate means of stating the physical arc length in electrical terms, although the arc voltage also includes the voltage drop in the electrode extension beyond the contact tube. With all variables held constant, arc voltage is directly related to arc length. Even though the arc length is the variable of interest and the variable that should be controlled, the voltage is more easily monitored. Arc voltage settings vary, depending on the material, shielding gas, and transfer mode. Typical values are shown in Table 7.3. Trial runs are necessary to adjust the arc voltage to produce the most favourable arc characteristics and weld bead appearance. Trials are essential because the optimum arc voltage depends upon a variety of factors, including metal thickness, type of joint, welding position, electrode size, shielding gas composition, and the type of weld. From any specific value of arc voltage, a voltage increase tends to flatten the weld bead and increase the width of the fusion zone. Excessively high voltage may cause porosity, spatter, and undercut. Reduction in voltage results in a narrower weld bead with a higher crown and deeper penetration. Excessively low voltage may cause stubbing of the electrode.

Welding Speed: Travel speed is the linear rate at which the arc is moved along the weld joint. With all other conditions held constant, weld penetration is maximum at an intermediate travel speed. When the travel speed is decreased, the filler metal deposition per unit length increases. At very slow speeds, the welding arc impinges on the molten weld pool, rather than the base metal, thereby reducing effective penetration. A wide weld bead is also a result.

As the travel speed is increased, the thermal energy per unit length of weld transmitted to the base metal from the arc is at first increased, because the arc acts more directly on the base metal. With further increases in travel speed, less thermal energy per unit length of weld is imparted to the base metal. Therefore, melting of the base metal first increases and then decreases with increasing travel speed. As travel speed is increased further, there is a tendency towards undercutting along the edges of the weld bead because there is insufficient deposition of filler metal to fill the path melted by the arc.

TABLE 7.3 Typical arc voltages for gas metal arc welding of various metals

	Spray[b] Globular Transfer 1/16 in (1.6 mm) Diameter Electrode					Short Circulating Transfer Diameter Electrode			
Metal	**Argon**	**Helium**	**25% Ar-75% He**	**Ar-O_2 (1-5% O_2)**	**CO_2**	**Argon**	**Ar-O_2 (1-5% O_2)**	**75% Ar-25% CO_2**	**CO_2**
Aluminum	25	30	29	–	–	19	–	–	–
Magnesium	26	–	28	–	–	16	–	–	–
Carbon steel	–	–	–	28	30	17	18	19	20
Low alloy steel	–	–	–	28	30	17	18	19	20
Stainless steel	24	–	–	26	–	18	19	21	–
Nickel	26	30	28	–	–	22	–	–	–
Nickel-copper alloy	26	30	28	–	–	22	–	–	–
Nickel-chromium-iron alloy	26	30	28	–	–	22	–	–	–
Copper	30	36	33	–	–	24	22	–	–
Copper-nickel alloy	28	32	30	–	–	23	–	–	–
Silicon bronze	28	32	30	28	–	23	–	–	–
Aluminum bronze	28	32	30	–	–	23	–	–	–
Phosphor bronze	28	32	30	23	–	23	–	–	–

a. Plus or minus approximately ten percent. The lower voltages are normally used on light material and at low amperage; the higher voltages are used on heavy material at high amperage.

b. For pulsed variation of spray transfer, the arc voltage would be from 18 to 28 volts, depending on the amperage range used.

Current density is defined as the current employed with a particular electrode diameter divided by its current carrying cross-sectional area. If the wire feed speed is low, then the current density will be low, and vice versa. From this you can determine that:

- Lower current density applied to a given electrode is associated with the short-circuit mode of metal transfer.
- Higher current density is associated with the higher energy modes of metal transfer: globular, axial spray transfer or the more advanced pulsed spray metal transfer.

The current for a given GMAW solid or metal cored electrode will reach a maximum density level. Once this level of current density is attained, no additional current can be carried by the electrode. In other words, the electrode has reached its maximum current density. In particular, Figure 7.18 demonstrates this phenomenon for 0.9 mm diameter solid wire. It can be seen that the current is relatively linear to approximately 200 A, but as the current reaches just beyond 210 A, the rise in current becomes exponential. At approximately 280 A (18.3 m/min wire feed speed), the electrode reaches its maximum current density. The electrode at this point becomes saturated with current and no more current can be added to the electrode. Therefore, the maximum current density for a given electrode diameter is synonymous with the concept of current saturation. So it can be speculated that this phenomenon occurs for all diameters and material types of electrodes used for GMAW. It is important to note that once the electrode reaches its maximum current density, the saturation point, any added wire feed speed will provide a higher deposition rate with no increase in current.

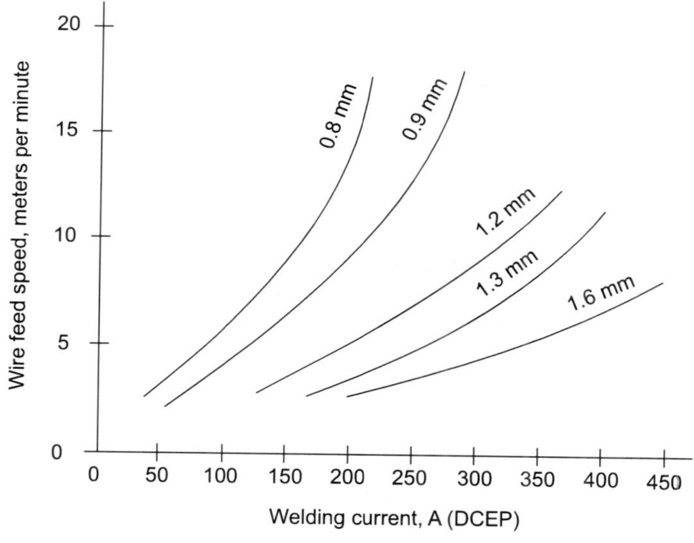

Figure 7.18 Effect of current density

Polarity

The term polarity is used to describe the electrical connection of the welding gun with relation to the terminals of a direct current power source. When the gun power lead is connected to the positive terminal, the polarity is designated as direct current electrode positive (DCEP), arbitrarily called reverse polarity. When the gun is connected to the negative terminal, the polarity is designated as direct current electrode negative (DCEN), originally called straight polarity. The vast majority of GMAW applications use direct current electrode positive (DCEP). This condition yields a stable arc, smooth metal transfer, relatively low spatter, good weld bead characteristics and greatest depth of penetration for a wide range of welding currents.

Direct current electrode negative (DCEN) is seldom used because axial spray transfer is not possible without modifications that have had little commercial acceptance. DCEN has a distinct advantage of high melting rates that cannot be exploited because the transfer is globular. With steels, the transfer can be improved by adding a minimum of 5 percent oxygen to the argon shield (requiring special alloys to compensate for oxidation losses) or by treating the wire to make it thermionic (adding to the cost of the filler metal). In both cases, the deposition rates drop, eliminating the only real advantage of changing polarity. However, because of the high deposition rate and reduced penetration, DCEN has found some use in surfacing applications.

Attempts to use alternating current with the GMAW process have generally been unsuccessful. The cyclic waveform creates arc instability due to the tendency of the arc to extinguish as the current passes through the zero point. Although special wire surface treatments have been developed to overcome this problem, the expense of applying them has made the technique uneconomical.

7.4.3 Base Metal Controlled Variables

The base metal determines the amount and the type of the shielding gas that must be used. It also determines the basic type of joint design, the electrode wire type and the electrode wire size that should be used.

Shielding Gas: The primary function of the shielding gas is to exclude the atmospheric air from the arc zone. However, the shielding gas has significant influence on the arc characteristics, mode of metal transfer, speed of welding, cleaning action and bead geometry. Pure argon is used only for welding of non-ferrous metals. However, pure argon shielding on ferrous alloys causes an erratic arc and a tendency for undercuts to occur. Hence, for welding of ferrous alloys, either CO_2 or Ar+ CO_2 gas mixture is used. Addition of 1 to 5% or 3 to 25% CO_2 produces a noticeable improvement in arc stability and freedom from undercut. Argon-rich gas mixtures produce a finger-shaped penetration whereas CO_2 shielding gas produces wide and deep penetration

The selection of the correct shielding gas for a given application is critical to the quality of the finished weld. The criteria used to make the selection includes, but is not restricted to, the following:

- Alloy of wire electrode
- Desired mechanical properties of the deposited weld metal
- Material thickness and joint design
- Material condition – the presence of mill scale, corrosion, resistant coatings, or oil
- The mode of GMAW metal transfer
- The welding position
- Fit-up conditions

- Desired penetration profile
- Desired final weld bead appearance
- Cost

Shielding Gases: Argon and helium are the two inert shielding gases used for protecting the molten weld pool. The inert classification indicates that neither argon nor helium will react chemically with the molten weld pool. However, in order to become a conductive gas, that is, a plasma, the gas must be ionized. Different gases require different amounts of energy to ionize, and this is measured in terms of the ionization energy. For argon, the ionization energy is 15.7 eV. Helium, on the other hand, has ionization energy of 24.5 eV. Thus, it is easier to ionize argon than helium. For this reason, argon facilitates better arc starting than helium.

The thermal conductivity, or the ability of the gas to transfer thermal energy, is the most important consideration for selecting a shielding gas. High thermal conductivity levels result in more conduction of the thermal energy into the workpiece. The thermal conductivity also affects the shape of the arc and the temperature distribution within the region. Argon has a lower thermal conductivity rate — about 10% of the level for both helium and hydrogen. The high thermal conductivity of helium provides a broader penetration pattern and will reduce the depth of penetration. Gas mixtures with high percentages of argon results in a penetration profile with a finger-like projection into the base material, and this is due to the lower thermal conductivity of argon.

Inert Shielding Gases

Argon is the most commonly used inert gas. Compared to helium, its thermal conductivity is low. The energy required to give up an electron, ionization energy, is low, and this results in the finger-like penetration profile associated with its use. Argon supports axial spray transfer. Nickel, copper, aluminum, titanium, and magnesium alloyed base materials use 100% argon shielding. Argon, because of its lower ionization energy, assists with arc starting. It is the main component gas used in binary (two-part) or ternary (three-part) mixes for GMAW welding. It also increases the molten droplet transfer rate.

Helium is commonly added to the gas mix for stainless and aluminum applications. Its thermal conductivity is very high, resulting in the broad but less deep penetration profile. When in use, arc stability requires additions of arc voltage. Helium additions to argon are effective in reducing the dilution of base material in corrosion-resistant applications. Helium/argon blends are commonly used for welding aluminum more than 25 mm thick.

Reactive Shielding Gases

Oxygen, hydrogen, nitrogen, and carbon dioxide (CO_2) are reactive gases. Reactive gases combine chemically with the weld pool to produce a desirable effect.

Carbon dioxide (CO_2) is a reactive gas widely used in its pure form for gas metal arc welding of carbon and low-alloy steels. It is the only reactive gas suitable for use alone as a shield in the GMAW process. Higher welding speed, greater joint penetration, and lower cost are general characteristics, which have encouraged extensive use of CO_2 shielding gas. CO_2 is inert at room temperature. In the presence of the arc plasma and the molten weld puddle, it is reactive. In the high energy of the arc plasma, the CO_2 molecule breaks apart in a process known as dissociation. In this process, free carbon, carbon monoxide, and oxygen are released from the CO_2 molecule. This occurs at the DC+ anode region of the arc. At the DC – cathode region, which is invariably the workpiece for GMAW, the released elements of the CO_2 molecule undergo the process of recombination. During recombination, higher energy levels exist and are responsible for the deep and broad penetration profile that characterizes the use of carbon dioxide.

Oxygen (O_2) is an oxidizer that reacts with components in the molten puddle to form oxides. In small additions (1-5%), with a balance of argon, it provides good arc stability and excellent weld bead appearance. The use of deoxidizers within the chemistry of filler alloys compensates for the oxidizing effect of oxygen. Silicon and manganese combine with oxygen to form oxides. The oxides float to the surface of the weld bead to form small islands, and are more abundant under CO_2 shielding than with blends of argon and oxygen gas.

Hydrogen (H_2) in small percentages (1-5%), is added to argon for shielding stainless steel and nickel alloys. Its higher thermal conductivity produces a fluid puddle, which promotes improved toe wetting and permits the use of faster travel speeds.

Binary Shielding Gas Blends

Two-part shielding gas blends are the most common and they are typically made up of either argon + helium, argon + CO_2, or argon + oxygen.

Argon + Helium

Argon/helium binary blends are useful for welding nickel-based alloys and aluminum. The mode of metal transfer used is either axial spray transfer or pulsed spray transfer. The addition of helium provides more puddle fluidity and flatter bead shape. Helium promotes higher travel speeds. For aluminum GMAW, helium reduces the finger-like projection found with pure argon. Helium is also linked to reducing the appearance of hydrogen pores in welds that are made using aluminum-magnesium fillers with 5XXX series base alloys. The argon component provides excellent arc starting and promotes cleaning action on aluminum.

Common Argon + Helium Blends

75% Argon + 25% Helium: This binary blend is frequently applied to improve the penetration profile for aluminum, copper, and nickel applications. The puddle is more fluid than with 100% argon.

75% Helium + 25% Argon: The higher helium content increases the thermal conductivity and puddle fluidity. The penetration profile is broad, and it exhibits excellent sidewall penetration.

Argon + CO_2

The most commonly found binary gas blends are those used for carbon steel GMAW welding. All four traditional modes of GMAW metal transfer are used with argon/ CO_2 binary blends.

They have also enjoyed success in pulsed GMAW applications on stainless steel where the CO_2 does not exceed 4%.

Axial spray transfer requires CO_2 contents less than 18%. Argon/CO_2 combinations are preferred where mill scale is an unavoidable welding condition. As the CO_2 percentage increases, so does the tendency to increase heat input and risk burn through.

Argon/CO_2 blends up to 18% CO_2 support pulsed spray transfer.

Short-circuiting transfer is a low heat input mode of metal transfer that can use argon/CO_2 combinations. Optimally, these modes benefit from CO_2 levels greater than or equal to 20%. Use caution with higher levels of argon with short-circuit metal transfer.

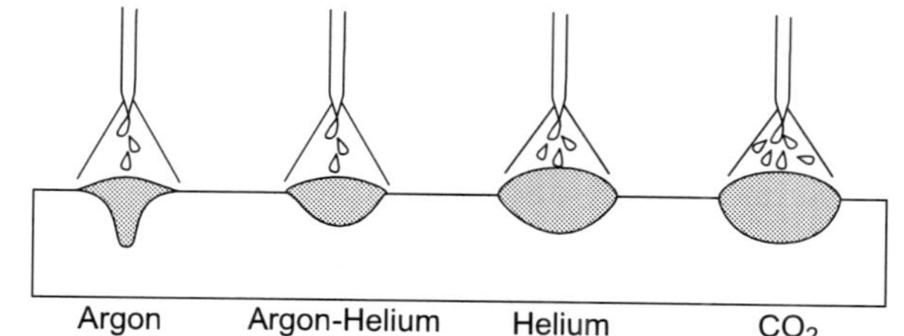

Bead contour and penetration patterns for various shielding gases

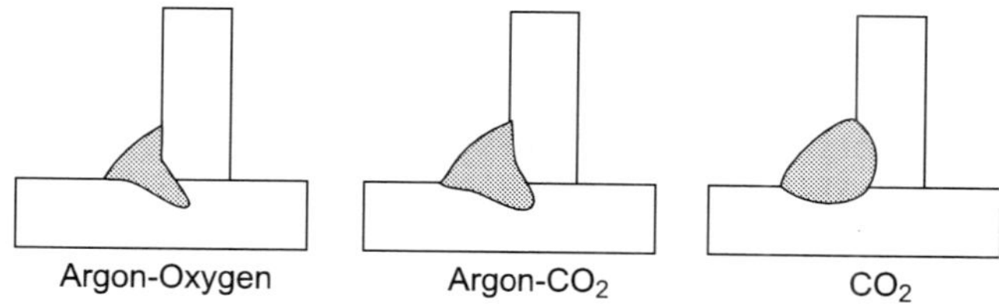

Relative effect of oxygen versus CO_2 additions to the argon shield

Figure 7.19 Effect of gases on bead contours and penetration

Argon-Helium-Carbon Dioxide-Oxygen

This mixture, commonly referred to as *quad mix,* is popular for high-deposition GMAW using the high-current-density metal transfer type arc. This mixture gives good mechanical properties and operability throughout a wide range of deposition rates. Its major application is welding low-alloy, high-tensile base materials, but it has been used on mild steel for high-production welding. Weld economics is an important consideration for using this gas in mild steel welding.

Weld Joint Position

Most spray-type GMAW is done in the flat or horizontal positions, while at low energy levels, pulsed and short-circuiting GMAW can be used in all positions. Fillet welds made in the flat position with spray transfer are usually more uniform, less likely to have unequal legs and convex profiles, and are less susceptible to undercutting than similar fillet welds made in the horizontal position. To overcome the pull of gravity on the weld metal in the vertical and overhead positions of welding, small diameter electrodes are usually used, with either short-circuiting metal transfer or spray transfer with pulsed direct current. Electrode diameters of 1.1 mm and smaller are best suited for out-of-position welding. The low-heat input allows the molten pool to freeze quickly. Downward welding progression is usually effective on sheet metal in the vertical position.

When welding is done in the "flat" position, the inclination of the weld axis with respect to the horizontal plane influences the weld bead shape, penetration, and travel speed. In flat position circumferential welding, the work rotates under the welding gun and inclination is obtained by moving the welding gun in either direction from top dead centre.

By positioning linear joints with the weld axis at 15° to the horizontal and welding downhill, weld reinforcement can be decreased under welding conditions that would produce excessive reinforcement when the work is in the flat position. Also, when traveling downhill, speeds can usually be increased. At the same time, penetration is lower, which is beneficial for welding sheet metal.

Downhill welding affects the weld contour and penetration, as shown in Figure 7.20(a). The weld puddle tends to flow towards the electrode and preheats the base metal, particularly at the surface. This produces an irregularly shaped fusion zone, called a secondary wash. As the angle of declination increases, the middle surface of the weld is depressed, penetration decreases, and the width of the weld increases. For aluminum, this downhill technique is not recommended due to loss of cleaning action and inadequate shielding.

Uphill welding affects the fusion zone contour and the weld surface, as illustrated in Figure 7.20(b). The force of gravity causes the weld puddle to flow back and lag behind the electrode. The edges of the weld lose metal, which flows to the center. As the angle of inclination increases, reinforcement and penetration increase, and the width of the weld decreases. The effects are exactly the opposite of those produced by downhill welding. When higher welding currents are used, the maximum usable angle decreases.

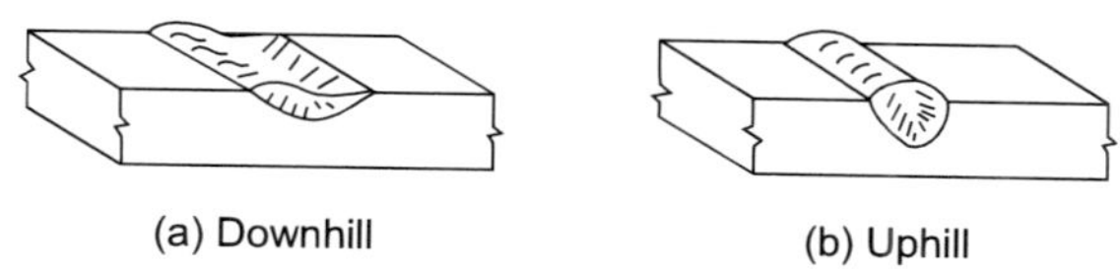

Figure 7.20 Effect of work inclination on weld bead shape

Electrode Size

The electrode size (diameter) influences the weld bead configuration. A larger electrode requires higher minimum current than a smaller electrode for the same metal transfer characteristics. Higher currents, in turn, produce additional electrode melting and larger, more fluid weld deposits. Higher currents also result in higher deposition rates and greater penetration. However, vertical and overhead welding are usually done with smaller diameter electrodes and lower currents.
Summary of welding parameters and variables is given below:

Welding Variables and Parameters

- Electrode/wire extension
    - Affects amperage
    - 10-15 mm stick-out length
- Inductance/choke
    - Smoothens the arc
    - High choke → no penetration
    - Low choke → excessive penetration
- Wire feed speed
    - Affects the amperage
    - Controls fusion and penetration
- Travel speed
    - Controls depth of penetration
- Gas flow rate
    - Protects the weld from atmosphere
- Voltage
    - Controls arc length
    - Set on welding machine
- Tilt angle
    - Not greater than 15° in back/forehand
- Type of weld: Butt, lap, fillet, etc.

7.5 TANDEM GMAW

The tandem GMAW system was developed to take advantage of the potential for higher travel speeds and higher deposition rates when using two electrodes in the same molten puddle. The system employs two power sources, two wire drives, and a system control. It is adapted for either repetitive side-beam type applications or it is employed with a welding robot. This variant of the gas metal arc welding process is capable of higher travel speeds, 1.5 to 2 times the speed of a single electrode. Some travel speeds may exceed 3.81 m/min. Deposition rates to 19.1 kg/hour are achievable for heavier plate welding.

Features of Tandem GMAW

- Capable of higher travel speeds on sheet metal than conventional single electrode GMAW
- Lower hydrogen weld deposit
- Lower spatter levels when compared to other processes and modes of metal transfer
- Capable of high deposition welding for heavy plate fabrication
- May be used for out-of-position welding

Modes of Metal Transfer for Tandem GMAW (Figure 7.21)

The modes of metal transfer used for the tandem GMAW variant are axial spray metal transfer or pulsed spray metal transfer. The combinations of the modes that are popularly employed include:

- **Spray + Pulse:** Axial spray transfer on the lead arc followed by pulsed spray transfer on the trail arc.
- **Pulse + Pulse:** Pulsed spray transfer on both the lead and the trail arc.
- **Spray + Spray:** Axial spray transfer on both the lead and the trail arc. The software programs designed to support the pulse + pulse configuration require that the wire feed speed setting for each, the lead and the trail, is the same. Trim values can be adjusted to account for arc length requirements. The higher energy spray + spray configuration is used for special heavy plate welding where deeper penetration is required. Pulse + pulse allows for heavy welding or high-speed sheet metal welding.

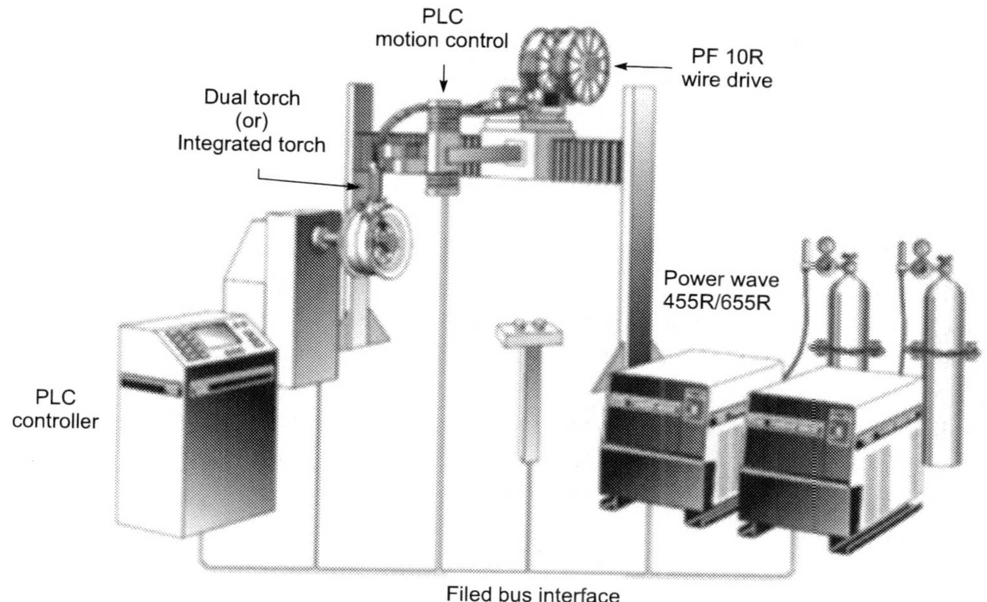

Figure 7.21 Tandem GWAW

7.6 APPLICATIONS

- The standard inert gas welding using a bare electrode and a shielding gas can be used to fuse together various metals, such as aluminum and its alloys, magnesium, copper and nickel based metals, titanium, carbon and low-alloy steels.
- The deciding factor in the use of this process is the shielding gas. It determines the characteristics and penetration of the arc column.
- GMAW is also applied for steel welding as it gives a faster weld time than other weld processes.

7.7 ADVANTAGES AND DISADVANTAGES

Advantages

- GMAW is a versatile process that can be used to weld a wide variety of metals, including aluminum and its alloys, nickel and copper base metals as well as steels.
- It can be used to weld sheet metal or relatively heavy sections. It is the only consumable electrode process that can be used to weld all commercial metals and alloys.

- GMAW overcomes the restriction of limited electrode length encountered with shielded metal arc welding.
- Welding can be done in all positions, a feature not found in submerged arc welding.
- GMAW is easily adapted for high-speed robotic, hard automation and semi-automatic welding applications. In semi-automatic welding, the wire feed speed, voltage, amperage and gas flow are all preset on the equipment. The operator only needs to guide the welding gun along a joint at a uniform speed.
- Practically, all GMAW is done using positive electrode. This polarity provides deep penetration, a stable arc and low spatter levels. GMAW has higher electrode efficiencies, usually between 93% and 98%, when compared to other welding processes.
- It is a continuous welding process and can be mechanized. Because the wire feed is continuous, long welds can be deposited without stops and starts.
- Higher welding speed and metal deposition rate.
- When spray transfer is used, deeper penetration is possible than with shielded metal arc welding, which may permit the use of smaller size fillet welds for equivalent strengths.
- Open arc nature of the process aids monitoring and control of the arc during welding.
- Less distortion due to higher welding speed.
- Avoids the chore of cleaning a solidified slag from the surface of each weld pass.
- Simple equipment components are readily available and affordable.
- Lower hydrogen weld deposit, generally less than 5 mL/100 g of weld metal.
- Lower heat input when compared to other welding processes.
- Minimal post-weld cleaning is required due to the absence of a heavy slag. These advantages make the process particularly well suited to high production and automated welding applications. This has become increasingly evident with the advent of robotics, where GMAW has been the predominant process choice.
- Less welding fumes when compared to SMAW (shielded metal arc welding) and FCAW (flux-cored arc welding) processes.
- Deposition rates are significantly higher than those obtained with shielded metal arc welding. No use of flux during the welding process and deep penetration is possible.
- Welding speeds are higher than those with shielded metal arc welding because of the continuous electrode feed and higher filler metal deposition rates.

Disadvantages

- GMAW requires a wire feeder which is difficult to move and can also be tough to repair.

- The welding arc must be protected against air drafts that disperse the shielding gas. This limits outdoor applications unless protective shields are placed around the welding area.
- GMAW is more difficult to use in hard-to-reach places because the welding gun is larger than a shielded metal arc welding holder, and the welding gun must be close to the joint, between 10 and 19 mm, to ensure that the weld metal is properly shielded.
- Due to the ability to weld at low currents, GMAW has the potential for lack of fusion when running in short-circuit mode.
- GMAW can only handle low to moderate levels of surface contaminants, such as rust, mill scale, dirt, oil and paint. All these have potential to create problems such as porosity, incomplete fusion, bad bead appearance and even cracking.
- The lower heat input characteristic of the short-circuiting mode of metal transfer restricts its use to thin materials.
- The higher heat input axial spray transfer generally restricts its use to thicker base materials.
- Relatively high levels of radiated heat and arc intensity can result in operator-resistance to the process.

7.8 DIFFERENCE BETWEEN MIG AND SMAW

1. In SMAW, electrodes can be changed easily but in MIG electrode, cannot be changed.
2. There is no wastage of wire in the case of MIG, whereas there is a wastage of rod in the case of SMAW.
3. Bare electrode is used in MIG, while coated electrode is used in SMAW.
4. In MIG, mostly DCEP is used.
5. MIG is a semi-automatic process while SMAW is a manual process. Therefore, better control is possible in MIG.

Worked Examples

Example 1: In MIG welding, the power source characteristic is $V_p = 36 - \frac{I}{60}$ and the arc characteristic is $V_a = 2L_a + 27$. Find the change in power of the arc, if the arc length is changed from 2 to 4 mm. If the maximum current capacity of power source is 300 A, what is the maximum sustainable arc length? **(GATE 1992)**

Solution: Welding arc $V = 2L + 27$ (i)

For power source $$V = 36 - \frac{I}{60060} \quad \text{(ii)}$$

For stable arc (i) = (ii)

$$2L + 27 = 36 - \frac{I}{60060}$$

This gives

$$I = 60(9 - 2L)$$

If current is 300 A

$$300 = 60(9 - 2L)$$
$$L = 2 \text{ mm}$$

Welding arc, $V = 2L + 2 = 2 \times 2 + 27 = 31$ V

$$I = 60(9 - 2L) = 60(9 - 2 \times 2) = 300 \text{ A}$$
$$P = VI = 31 \times 300 = 9300 \text{ watts} = 9.3 \text{ kW}$$

Arc length, $L = 4$ mm

$$V = 27 + 2L = 27 + 2 \times 4 = 60 \text{ V}$$
$$P = VI = 60 \times 60 = 2100 \text{ watts} = 2.1 \text{ kW}$$

Change in power = 9300-2100 = 7200 watts

If the maximum current capacity is 300 A, the maximum arc length is 2 mm.

Example 2: A weld is made using MIG welding process with following welding parameters; current = 200 A, voltage = 25 volts, welding speed = 18 cm/min, wire diameter = 1.2 mm, wire feed rate is 4 m/min and thermal efficiency of the process is 65%. What is the heat input per unit length of the weld in kJ/cm. What is the area of cross section of weld bead in mm^2? **(GATE 2001/2003)**

Solution: $I = 200$ amperes, $V = 25$ V, speed = 18 cm/min

$D = 1.2$ mm, $f = 4$ m/min and $\eta = 65\%$

$$\text{Heat input} = \frac{VI}{\text{speed}} \times \eta = \frac{25 \times 200}{18} \times 0.65 \times 60 = 10.83 \text{ kJ/cm}$$

Filing rate of weld bead = filled rate by electrode

$$\text{Area of weld bead} \times \text{speed} = \frac{\pi d^2 f}{4} = \frac{\pi \times 1.2^2 \times 4000}{4 \times 180} = 25.12 \text{ mm}^2$$

$$\text{Heat required} = 114.5 \times 10^{-3} \times 8 \times 1380 = 1264 \text{ joules}$$

Example 3: During the steady state gas metal arc welding with direct current electrode positive polarity, the welding current, voltage and weld speed are 150 A, 30 V and 6 m/min, respectively. A metallic wire electrode of diameter 1.2 mm is fed at a constant rate of 12 m/min. The density, specific heat and melting temperature of the wire electrode are 7000 kg/m^3, 500 J/kg and 1530°C, respectively. Assume the ambient temperature to be 30°C and neglect the latent heat of melting. Also assume that 66.66 % of total electrical power is available for melting of the wire electrode. What is the melting efficiency? **(GATE 2010)**

Solution: $I = 150$ A, $V = 30$ V, speed $= \dfrac{6\text{ m}}{\text{min}}$, wire dia. = 1.2, $f = 12$ m/min

$$\text{Total heat required} = \text{Total volume} \times \text{heat rate}$$

$$\text{Heat required} = \frac{\pi \times D^2 \times f \times \rho \times C_p(T_m - T_a)}{4}$$

$$= \frac{\pi \times 1.2^2 \times 10^{-6} \times 12 \times 7000 \times 500 \times (1530 - 30)}{4 \times 60} = 1187.5 \text{ watts}$$

$$\text{Heat generated} = VI = 150 \times 30 = 4500 \text{ watts}$$

$$\text{Melting efficiency} = \frac{\text{Heat required}}{\text{Heat input}} = \frac{1187.5}{4500 \times \frac{2}{3}} = 0.3958 \cong 40\%$$

Example 4: The accompanying Figure 7.22 shows a metal sheave that consists of two matching pieces of hot rolled, low-carbon steel sheets. These two pieces can be joined either by spot welding or by V-groove welding. Discuss the advantages and limitations of each process for this application.

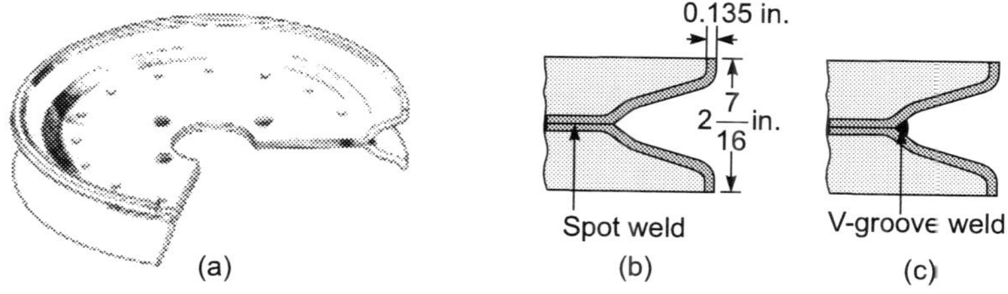

Figure 7.22 Spot welding versus V-goove welding

Solution: The original method of joining the two sheaves was by resistance spot welding, with 16 welds equally spaced around the periphery, as shown in (b). Although the weld quality was satisfactory, the welding time per sheave was 1 minute. In order to increase production rate, an alternative process was chosen (gas-metal-arc welding, GMAW) with a continuous weld around the periphery of the sheave as shown in (c). With an automated welding process, the welding time per sheave was reduced to 40 s.

Review Questions

1. Explain the principle of operation of MIG.
2. Explain the equipment used in MIG.
3. Explain the functions of welding gun.
4. Give brief description of power source for MIG.
5. What are the different metal transfers in MIG?
6. What are the advantages of pulse transfer?
7. Discuss the important variants of MIG.
8. What are the advantages of MIG?
9. What are the disadvantages of MIG?
10. Write a few applications of MIG.

8
CHAPTER

Gas Tungsten Arc Welding

8.0 INTRODUCTION

Gas tungsten arc welding (GTAW), also called tungsten inert gas (TIG) welding, is the welding process performed using the heat of an arc established between a non-consumable electrode and the workpiece. The electrode, the arc and the area surrounding the molten weld puddle are protected from the atmosphere by an inert gas shield, which is supplied directly into the weld zone. The electrode is not consumed in the weld puddle as in shielded metal arc welding. If a filler metal is necessary, it is added to the leading molten puddle. The major inert gases used are argon and helium. TIG welds are stronger, more ductile and more corrosion-resistant than welds made with ordinary shield metal arc welding. The possibility of using helium to shield a welding arc and molten weld pool was first investigated in the 1920s. However, nothing was done with this method until the beginning of World War II when a great need developed in the aircraft industry to replace riveting for joining reactive materials, such as aluminum and magnesium. This process was then called heliarc process, as helium was used for shielding.

In addition, because no granular flux is required, it is possible to use a wider variety of weld designs than in conventional shield arc welding or stick electrode welding. The weld bead has no corrosion because flux entrapment cannot occur and because of gas fluxing, there is little or no post-weld cleaning operation. There is also little weld splatter or weld sparks that damage the surface of the base metal as in conventional shield arc welding. Fusion welds can be made in nearly all commercial metals. The TIG process lends itself ably to the fusion welding of aluminum and its alloys, stainless steel, magnesium alloys, nickel-base alloys, copper-base alloys, carbon steel and low-alloy steels. TIG welding can also be used for combining dissimilar metals, hard facing and the surfacing of metals.

Process of Gas Tungsten Arc Welding (Figure 8.1)

The process uses a non-consumable tungsten (or tungsten alloy) electrode held in a torch. Shielding gas is fed through the torch to protect the electrode, molten weld pool, and solidifying weld metal from contamination by the atmosphere. The electric arc is produced by the passage of current through the conductive, ionized shielding gas. The arc is established between the tip of the electrode and the work. Heat generated by the arc melts the base metal. Once the arc and weld pool are established, the torch is moved along the joint and the arc progressively melts the faying surfaces. Filler wire, if used, is usually added to the leading edge of the weld pool to fill the joint.

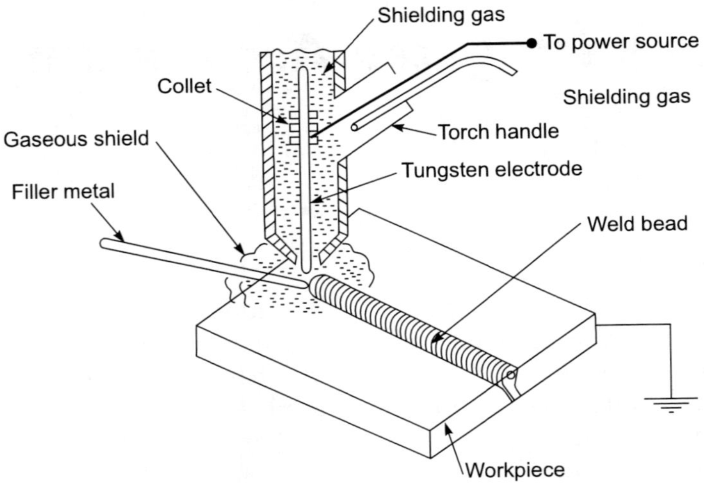

Figure 8.1 Process of gas tungsten arc welding

8.1 MODE OF OPERATION

GTAW lends itself to both manual and automatic operations. In manual operation, the welder holds the torch in one hand and directs the arc into the weld joint. The filler metal is fed manually into the leading edge of the puddle. In automatic applications, the torch may be automatically moved over a stationary workpiece or the torch may be stationary with the work moved or rotated in relation to the torch. Filler metal, if required, is also fed automatically.

In both manual and automatic processes, the TIG arc column is similar to that of the standard shield arc welding arc column. The electron travel is identical to that in shield arc welding. The amount of plasma generated in the center of the arc column, however, is thought to be greater than that generated in the standard shield arc tick electrode welding process because the tungsten electrode has a higher thermionic rating than the mild steel electrode. The thermionic function or emission is the ability of the metal to release its valence electrons.

8.2 EQUIPMENT

The choice of TIG equipment depends on the type of metal being joined, the position of the weld being made and the quality of the weld metal necessary for the application. The basic equipment consists of the following:

1. Power source
2. Electrode holder or electrode torch
3. Shielding gas
4. Tungsten electrode
5. Water supply, if necessary
6. Ground cables
7. Protective equipment

8.2.1 Power Source

TIG power sources typically have either drooping or nearly true constant-current static output characteristics, as shown in Figure 8.2. The static output characteristic is a function of the type of welding current control used in the power source design. A drooping volt-ampere characteristic is typical of magnetically controlled power source designs, including the moving coil, moving shunt, moving core reactor, satiable reactor, or magnetic amplifier designs and also, rotating power source designs. With a drooping characteristic, the welder may vary the current level slightly by changing the arc length. The advantages of magnetically controlled power sources are that they are simple to operate, require little maintenance and are relatively inexpensive. The disadvantages are that they are large in size and weight and have a lower efficiency compared to electronically controlled power sources. Also, most magnetically controlled techniques are open loop, which limit repeatability, accuracy, and response.

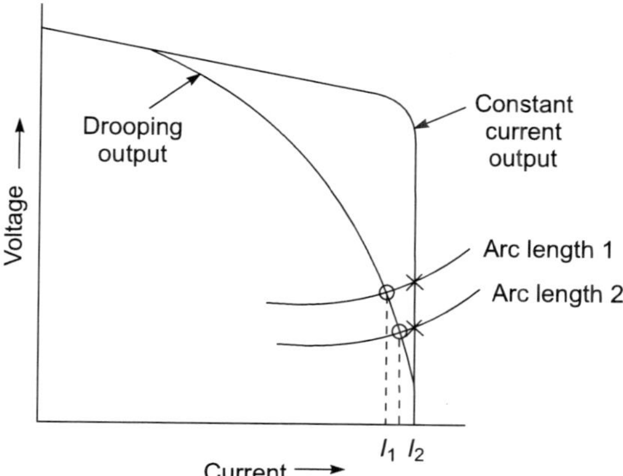

Figure 8.2 Static VA curves for constant current power supplies

An essentially constant-current volt-ampere characteristic can be provided by electronically controlled power sources, such as the series linear regulator, silicon controlled rectifier, secondary switches, and inverter designs. Inverter power sources are the most versatile, offering multiprocess capabilities and variable welding current waveform output. Inverters are also lighter and more compact than other power source designs of equivalent current rating. The advantages of electronically controlled power sources are that they offer rapid dynamic response and provide variable current waveform output. Most truly constant current power sources are closed loop controlled, in which the actual current is measured and compared to the desired current setting. Adjustments are made electronically within the power source to maintain the desired current, as welding conditions change. The disadvantages are that they are more complex to operate and maintain and are relatively expensive. It is important to select a TIG power source based on the type of welding current required for a particular application. The types of welding current include AC sine wave, AC square wave, DC, and pulsed DC.

Direct Current

Using direct current, the tungsten electrode may be connected to either the negative or the positive terminal of the power supply. In almost all cases, electrode negative (cathode) is chosen. With that polarity, electrons flow from the electrode to the work and positive ions are transferred from the work to the electrode and the connection is known as DCEN (straight polarity) connection. When the electrode is positive (anode), the directions of electron and positive ion flow are reversed, the connection is known as DCEP (reverse polarity) connection.

Current Type	DCEN	DCEP	AC (Balanced)
Electrode polarity	Negative	Positive	
Electron and ion flow penetration characteristics	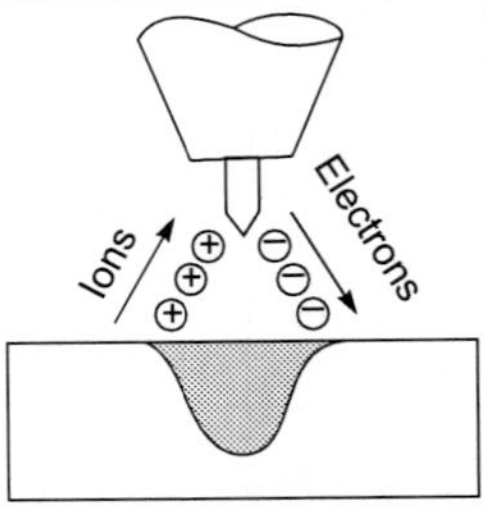 Ions / Electrons	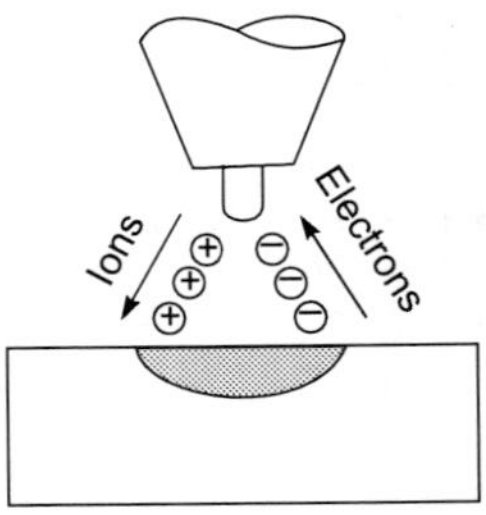 Ions / Electrons	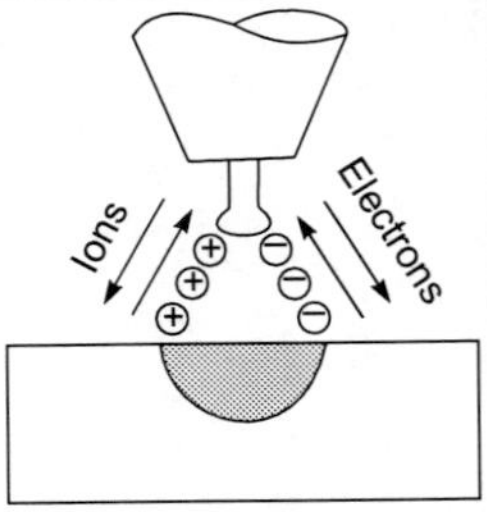 Ions / Electrons
Oxide cleaning action	No	Yes	Yes-once every half cycle
Heat balance in the arc (approx.)	70% at work end 30% at electrode end	30% at work end 70% at electrode end	50% at work end 50% at electrode end
Penetration	Deep; narrow	Shallow wide	Medium
Electrode capacity	Excellent e.g., 1/8 in. (3.2 mm) 400 A	Poor e.g., 1/4 in. (6.4 mm) 120 A	Good e.g., 1/8 in. (3.2 mm) 225 A

Figure 8.3 Welding characteristics of DCEN, DCEP and AC arcs

With DCEN and a thermionic electrode, such as tungsten, approximately 70%, of the heat is generated at the anode and 30% at the cathode. Since DCEN produces the greatest amount of heat at the workpiece, for a given welding current DCEN provides deeper weld penetration than DCEP (Figure 8.3). DCEN is the most common configuration used in TIG welding and is used with argon, helium, or a mixture of the two, to weld most metals.

When the tungsten electrode is connected to the positive terminal (DCEP), a cathodic cleaning action is created at the surface of the workpiece. This action occurs with most metals but is most important when welding aluminum and magnesium because it removes the refractory oxide surface that inhibits arc stability and wetting by the weld metal.

Unlike DCEN, in which the electrode tip is cooled by the evaporation of electrons, when the electrode is used as the positive pole, its tip is heated by the bombardment of electrons as well as by its resistance to their passage through the electrode. Therefore, to reduce resistance heating and increase thermal conduction into the electrode collet, a large diameter electrode is required for a given welding current when reverse polarity is used. The current carrying capacity of an electrode connected to the positive terminal is approximately one-tenth that of an electrode connected to the negative terminal. DCEP is generally limited to welding thin sheet metal.

Pulsed DC Welding

Pulsed DC involves the repetitive variation in arc current from a background (low) value to a peak (high) value. Adjustments of the pulse current time, background current time, peak current level, and background current level can be made to provide a current output waveform fitted to a particular application (Figure 8.4). Generally, pulse frequencies are adjustable from 0.5 to 20 pulses per second. Pulsed DC is usually applied with the electrode negative (DCEN).

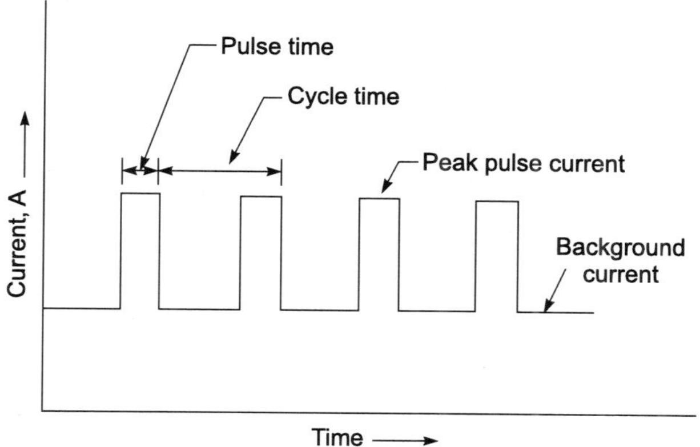

Figure 8.4 Pulsed DC waveform characteristics

In pulsed DC welding, the pulse current level is typically set at 2 to 10 times the background current level. This combines the driving, forceful arc characteristics of high current with the low-heat input of low current. The pulse current achieves good fusion and penetration, while the background current maintains the arc and allows the weld area to cool.

There are several advantages of pulsed current. For a given average current level, greater penetration can be obtained than with steady current, which is useful on metals sensitive to heat input and minimizes distortion. Very thin metals can be joined with pulsed DC. In addition, one set of welding variables can be used on a joint in all positions, such as a circumferential weld in a horizontal pipe. Pulsed DC is also useful for bridging gaps in open root joints.

Alternating Current

Alternating current undergoes periodic reversal in polarity from electrode positive to electrode negative. Thus, AC can combine the work cleaning action of electrode positive (reverse polarity) with the deep penetration characteristic of electrode negative (straight polarity).

Conventional AC welding power sources produce a sinusoidal open circuit voltage output. The frequency of voltage reversal is typically fixed at the standard 50/60 Hz frequency of the primary power. When the current decays to zero, various effects occur, depending on the polarity. When the thermionic tungsten electrode becomes negative, it supplies electrons immediately to reignite the arc. However, when the weld pool becomes negative, it cannot supply electrons until the voltage is raised sufficiently to initiate cold-cathode emission. Without the voltage, the arc becomes unstable.

Some means of stabilizing the arc during voltage reversal are required with conventional sinusoidal welding power sources. This has been done by using high voltage high-frequency sparks in parallel with the arc; and by using power supplies with a square wave output. Square wave AC welding power sources can change the direction of the welding current in a short period of time. The presence of high voltage, coupled with high electrode and base metal temperature at current reversal, allows the arc to be reignited without the need for an arc stabilizer. Also, the lower "peak" current of the square waveform tends to increase the usable current range of the electrode. Since it is easier to provide electrons needed to sustain an arc when the electrode is negative, the voltage required is also less. The result is a higher welding current during the DCEN interval than during DCEP. In effect, the power supply produces both direct current and alternating current. Such rectification can cause damage to the power supply due to overheating or, with some machines, decay in the output. Rectification is eliminated by wave balancing AC waveforms, as shown in Figure 8.5.

Earlier balanced current power supplies involved either series connected capacitors or a DC voltage source (such as a battery) in the welding circuit. Modern power supply circuits use electronic wave balancing. Balanced current flow is not essential for most manual welding operations. It is, however, desirable for a high-speed machine for automatic welding.

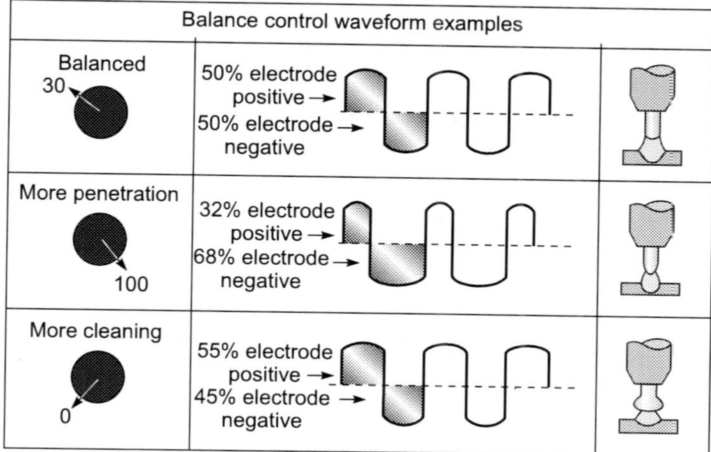

Figure 8.5 Typical AC waveforms used in GTAW

Advantages of balanced current flow

1. Better oxide removal
2. Smoother, better welding
3. No requirement for reduction in output rating of a given size of conventional welding transformer (the unbalanced core magnetization that is produced by the DC component of an unbalanced current flow is minimized).

Disadvantages of balanced current flow

1. Larger tungsten electrodes are needed
2. Higher open-circuit voltages, generally associated with some wave balancing means, may constitute a safety hazard
3. Balanced wave welding power sources are more costly

Some square wave AC power sources adjust the current level during the electrode positive and electrode negative cycles at our standard 60 Hz frequency. More expensive power sources adjust the time of each polarity half cycle as well as the current level during that half cycle. Such variable waveforms adjust the welding current to suit a particular application. The characteristics of variable square wave alternating current are shown in Figure 8.6.

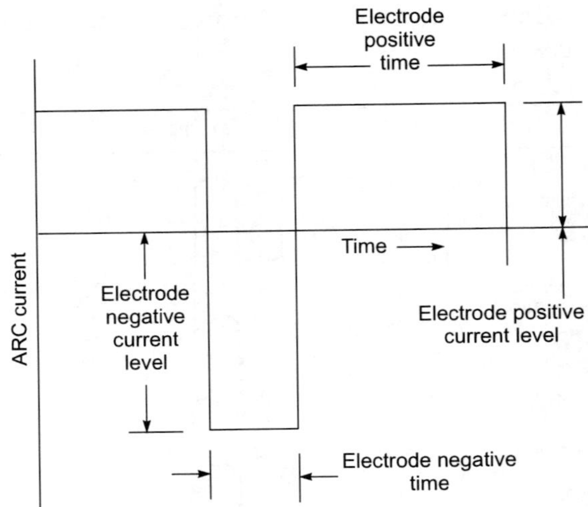

Figure 8.6 Characteristics of variable square wave AC

8.2.2 Welding Torch

GTA welding torches hold the tungsten electrode, which conducts welding current to the arc, and provide a means for conveying shielding gas to the arc zone. Torches are rated in accordance with the maximum welding current that can be used without overheating. A range of electrode sizes and various types and sizes of nozzles can be used in the same torch. Manual GTAW torches often have auxiliary switches and valves built into their handles for controlling current and gas flow. Torches for machine or automatic GTAW are typically mounted on a device which centers the torch over the joint, may move the torch along the joint, and may change or maintain the torch-to-work distance.

The heat generated in the torch during welding is removed either by gas cooling or water cooling. Gas cooled torches provide cooling by the flow of the relatively cool shielding as through the torch. Gas cooled torches are used for maximum welding current of about 100 A. In water cooled torches, cooling water enters the torch through the inlet hose, circulates through the torch, and exits through an outlet hose. The power cable from the power supply to the torch is enclosed within the cooling water outlet hose. Water cooled torches are designed for use at higher welding currents in the range of 200 to 500 A. Most machines for automatic welding applications use water cooled torches.

Tap water can be used to cool water cooled torches. In this case, water flows through the torch and then down a drain, or a closed system involving a reservoir, pump, and radiator or water chiller to disperse heat from the system can be used.

Collets

Collets are used to grip the tungsten electrodes of various diameters in the electrode holder. Good contact between the electrode and the inside diameter of the collet is essential for proper current transfer and electrode cooling. Tungsten electrode of any particular diameter requires a collet of matching size.

Filler Wire

Filler metal is generally not used for welding thin sheet by GTAW. Welding of thick steel plates by GTAW to produce high quality welds for critical applications such as joining of nuclear and aerospace components, requires addition of filler metal to fill the groove. The filler wire can be fed manually or by some wire feed mechanism. For feeding small diameter filler wires (0.8 to 2.4 mm), usually push-type wire feed mechanism with speed control device is used.

Nozzles

Shielding gas is directed to the weld zone by gas nozzles or cups fitted onto the head of the torch.

Nozzle Materials: Nozzles are made of ceramic, metal, metal jacketed ceramic, fused quartz, or other materials. Ceramic nozzles are the least expensive and most popular, but are brittle and must be replaced often. Fused quartz nozzles are transparent and allow better vision of the arc and electrode. However, contamination from metal vapors from the weld can cause them to become opaque, and they are also brittle. Water cooled metal nozzles have longer life and are used mostly for machine and automatic welding applications, where welding currents exceed 250 A.

Sizes and Shapes of Nozzles: The gas nozzle or cup must be large enough to provide shielding gas coverage of the weld pool area and surrounding hot base metal. The nozzle diameter must be appropriate for the volume of shield gas needed to provide protection and the stiffness needed to sustain coverage in drafts. A delicate balance exists between the nozzle diameter and the flow rate. If the flow rate for a given diameter is excessive, the effectiveness of the shield is destroyed because of turbulence. High flow rates without turbulence require large diameters; this is an essential condition at high currents. Size selection depends on electrode size, type of weld joint, weld area to be effectively shielded, and access to the weld joint.

Gas Lenses: One device used for assuring a laminar flow of shielding gas is an attachment called gas lens. Gas lenses contain a porous barrier diffuser and are designed to fit around the electrode or collet. Gas lenses produce a longer, undisturbed flow of shielding gas. They enable operators to weld with the nozzle of 25.4 mm or more from the work, improving their ability to see the weld pool and allowing them to reach places with limited access, such as inside corners.

8.2.3 Shielding Gases

Shielding gas is directed by the torch to the arc and weld pool to protect the electrode and the molten weld metal from atmospheric contamination. Uniformity of root bead contour, freedom from undercutting, and the desired amount of root bead reinforcement are more likely to be achieved when using gas backup under controlled conditions. In some materials, gas backup reduces root cracking and porosity in the weld.

Table 8.1 lists the properties of some shielding gases. As shown, the *ionization potentials* for argon and helium are 15.7 and 24.5 eV (electron volts), respectively. Since it is easier to ionize argon than helium, arc initiation is easier and the voltage drop across the arc is lower with argon. Also, since argon is heavier than helium, it offers more effective shielding and greater resistance to cross draft than helium. With DCEP or AC, argon also has a greater oxide cleaning action than helium. These advantages, plus the lower cost of argon, make it more attractive for GTAW than helium.

TABLE 8.1 Properties of shielding gases used for welding

Gas	Chemical Symbol	Molecular Weight (g/mol)	Specific Gravity with Respect to Air at 1 atm and 0°C	Density (g/L)	Ionization Potential (eV)
Argon	Ar	39.95	1.38	1.784	15.7
Carbon dioxide	CO_2	44.01	1.53	1.978	14.4
Helium	He	4.00	0.1368	0.178	24.5
Hydrogen	H_2	2.016	0.0695	0.090	13.5
Nitrogen	N_2	28.01	0.967	1.25	14.5
Oxygen	O_2	32.00	1.105	1.43	13.2

Because of the greater voltage drop across a helium arc than an argon arc, however, higher power inputs and greater sensitivity to variations in the arc length can be obtained with helium. The former allows welding of thicker sections and the use of higher welding speeds. The latter, on the other hand, allows a better control of the arc length during automatic GTAW.

Details about Shielding Gases

Argon and helium or mixtures of the two, are the most common types of inert gas used for shielding. Argon-hydrogen mixtures are used for special applications. Depending on the volume of usage, these gases may be supplied in cylinders or as a liquid in insulated tanks. The liquid is vaporized and piped to points within the plant, thus eliminating cylinder handling.

Argon

It is obtained from the atmosphere by the separation of liquified air. Welding grade argon is refined to a minimum purity of 99.95%. This is acceptable for GTAW of most metals, except the reactive and refractory metals, for which a minimum purity of 99.997% is required. Often, such metals are fabricated in chambers from which all traces of air have been purged prior to initiating the welding operation.

Argon is used more extensively than helium because of the following advantages:

1. Smoother, quieter arc action
2. Reduced penetration
3. Cleaning action when welding materials such as aluminum and magnesium
4. Lower cost and greater availability
5. Lower flow rates for good shielding
6. Better cross-draft resistance
7. Easier arc starting

Helium

It is obtained by separation from natural gas. Welding grade helium is refined to a purity of at least 99.99%. For given values of welding current and arc length, helium transfers more heat into the work than argon. The greater heating power of the helium arc can be advantageous for joining metals of high thermal conductivity and for high speed mechanized applications. Also, helium is used more often than argon for welding heavy plates. Mixtures of argon and helium are useful when some balance between the characteristics of both is desired.

Characteristics of Argon and Helium

The chief influencing shielding effectiveness is the gas density. Argon is approximately one and one-third times as heavy as air and ten times heavier than helium. Argon, after leaving the torch nozzle, forms a blanket over the weld area. Helium, because it is lighter, tends to rise around the nozzle. Experimental work has consistently shown that to produce equivalent shielding effectiveness, the flow of helium must be two to three times that of argon.

The important characteristics of these gases are the voltage-current relationships of the tungsten arc in argon and in helium that are illustrated in Figure 8.7.

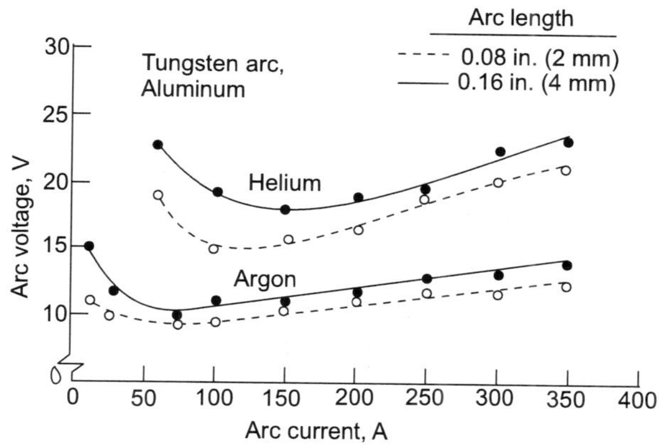

Figure 8.7 Voltage-current relationship with argon and helium shielding

At all current levels, for equivalent arc lengths, the arc voltage obtained with helium is appreciably higher than that with argon. The higher available heat for helium favors its selection when welding thick materials and metals having high thermal conductivity or relatively high melting temperatures.

Comparison between argon, helium and mixtures is given in Table 8.2.

TABLE 8.2 Behavior of gas mixtures

Characteristics	Argon	Helium	Argon/Helium Mixes
Travel speed	Reduced travel speeds	Faster travel speeds	Improved travel speeds over 100% argon
Penetration	Reduced penetration	Increased penetration	Improved penetration over 100% argon
Cleaning	Good cleaning action	Less cleaning action	Cleaning properties closer to argon
Arc starting	Easier arc starting	Difficult arc starting	Improved arc starting over 100% helium
Arc stability	Good arc stability	Less low amperage stability	Improved arc starting over 100% helium
Arc cone	Focused arc cone	Flared arc cone	Arc cone shape more focused than w/helium
Arc voltage	Lower arc voltages	Higher arc voltages	Arc voltages between 100% Argon and helium
Flow rate	Lower flow rates 10-30 CFH	Higher flow rates (2 times)	Higher flow rates than argon
Cost	Lower cost and greater availability	Higher cost than argon	Cost higher than argon

Argon-Hydrogen Mixtures

Argon-hydrogen mixtures are employed in special cases, such as mechanized welding of light gage stainless steel tubing, where hydrogen does not cause adverse metallurgical effects such as porosity and hydrogen induced cracking. Increased welding speeds can be achieved in almost direct proportion to the amount of hydrogen added to argon because of the increased arc voltage. Excessive hydrogen causes porosity. Hydrogen concentrations up to 35% have been used on all thicknesses of stainless steel where a root opening of approximately 0.010 to 0.020 inches (0.25 to 0.5 mm) is used. Argon-hydrogen mixtures are limited to use on stainless steel, nickel-copper, and nickel-base alloys.

Selection of Shielding Gas

Either argon, helium, or a mixture of argon and helium may be used successfully for most applications, with the possible exception of manual welding on extremely thin material, for which argon is essential. Argon generally provides an arc, operates more smoothly and quietly, is handled more easily, and is less penetrating than an arc shielded by helium. In addition, the lower unit cost and the lower flow rate requirements of argon make argon preferable from an economic point of view. Argon is preferred for most applications, except where helium's higher heat penetration is required for welding thick sections of metals with high heat conductivity, such as aluminum and copper. A guide to the selection of gases is provided in Table 8.3.

TABLE 8.3 Recommended types of current, tungsten electrodes and shielding gases for welding different metals

Type of Metal	Thickness	Type of Current	Electrode"	Shielding Gas
Aluminum	All over 1/8 in. under 1/8 in.	Alternating current DCEN DCEP	Pure or zirconium Thoriated Thoriated or zirconium	Argon or argon-helium Argon-helium or argon Argon
Copper, copper alloys	All under 1/8 in.	DCEN Alternating current	Thoriated Pure or zirconium	Helium Argon
Magnesium alloys	All under 1/8 in.	Alternating current DCEP	Pure or zirconium Zirconium at thoriated	Argon Argon
Nickel, nickel alloys	All	DCEN	Thoriated	Argon
Plain carbon, low-alloy steels	All under 1/8 in.	DCEN Alternating current	Thoriated Pure or zirconium	Argon or argon-helium Argon
Stainless steel	All under 1/8 in.	DCEN Alternating current	Thoriated Pure or zirconium	Argon or argon-helium Argon
Titanium	All	DCEN	Thoriated	Argon

" Where thoriated electrodes are recommended, ceriated or lanthanated electrodes may also be used.

Recommended Gas Flow Rates

Shielding gas flow requirements are based on cup or nozzle size, weld pool size, and air movement. In general, the flow rate increases in proportion to the cross-sectional area at the nozzle. The minimum flow rate is determined by the need for a stiff stream to overcome the heating effects of the arc and cross-drafts. With the more commonly used torches, typical shielding gas flow rates are 7 to 16 L/min for argon and 14 to 24 L/min for helium. Excessive flow rates cause turbulence in the gas stream which may aspirate atmospheric contamination into the weld pool.

A crosswind or draft moving at five or more miles per hour can disrupt the shielding gas coverage. The stiffest, non-turbulent gas streams (with high-stream velocities) are obtained by incorporating gas lenses in the nozzle and by using helium as the shield gas. However, in the interest of cost, protective screens to block air flow are preferred to increase shielding gas flow.

Backup Purge

When making the root passes of welds, the air contained on the back side of the weldment can contaminate the weld. To avoid that problem, the air must be purged from this region. Argon and helium are satisfactory for the backup purge when welding all materials. Nitrogen may be used satisfactorily for backing up welds in austenitic stainless steel, copper, and copper alloys. Gas flow requirements for the backup purge range from 1 to 90 CFH (0.5 to 42 L/min), based on the volume to be purged.

As a rule of thumb, a relatively inert atmosphere will be obtained by flushing with four times the volume to be purged. Several devices are available to contain shielding gas on the back side of plate and piping weldments. One of these is shown in Figure 8.8.

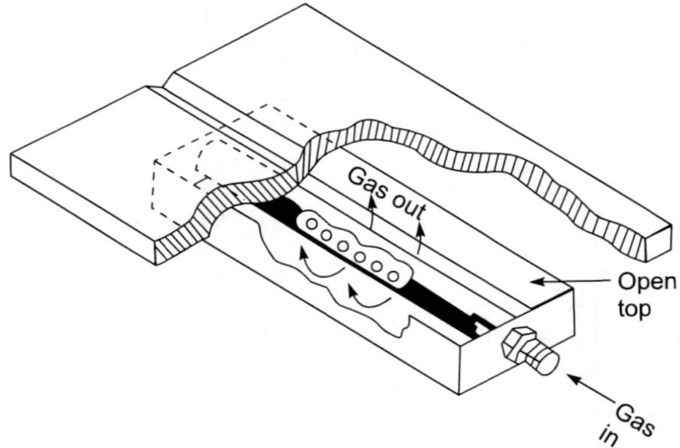

Figure 8.8 Backup purge gas channel

When purging piping systems, provisions for an adequate vent or exhaust, as shown in Figure 8.9, are important to prevent excessive pressure build up during welding. The area of vents through which the backup gas is exhausted to the atmosphere should be at least equal to the area of the opening through which the gas is admitted to the system. Extra care should be taken to ensure that the backup purge pressure is not excessive when welding the last inch or two on the root pass, to prevent weld pool blow-out or root concavity.

When using argon or nitrogen, the backup gas should preferably enter the system at a low point, to displace the atmosphere upwards, and be vented at points beyond the joint to be welded.

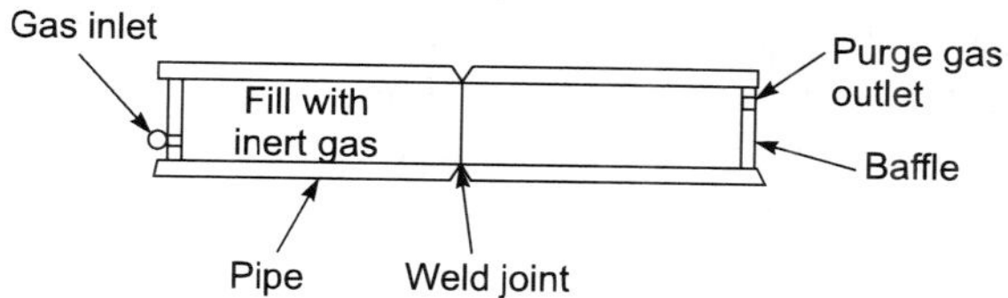

Figure 8.9 Backup gas purge arrangement for circumferential pipe joint (*Note:* Baffles used to contain pure gas)

8.2.4 Tungsten Electrodes

Tungsten electrodes are non-consumable because they do not melt or transfer to the weld. The function of a tungsten electrode is to serve as one of the electrical terminals of the arc that supplies the heat required for welding. Its melting point is 3410°C. Approaching this high temperature, tungsten becomes thermionic; it is a ready source of electrons. It reaches this temperature by resistance heating and, were it not for the significant cooling effect of electrons boiling from its tip, resistance heating would cause the tip to melt. In fact, the electrode tip is much cooler than that part of the electrode between the tip and the externally cooled collet.

The shape of the tip of the electrode is an important factor in GTAW. When welding with DCEN, the tip must be ground to a point. The included angle at which the tip is ground varies with application, the electrode diameter and the welding current. Narrow joints require a relatively small included angle. When welding very thin material at low currents, a needle-like point ground onto the smallest available electrode may be necessary to stabilize the arc. Properly ground electrodes assure easy starting, good arc stability and proper bead width.

The American Welding Society (AWS) has published Specification AWS A5.12-80 for tungsten arc welding electrode which classifies the electrodes on the basis of their chemical composition, size and finish. They are:

1. **Pure Tungsten:** Used for less critical applications. The cost is low and they give good results at relatively low currents on a variety of metals. Most stable arc when used on AC, either balanced wave or continuous high frequency.
2. **1% Thoriated Tungsten:** Good current carrying capacity, easy arc starting and they provide a stable arc. Less susceptible to contamination. Designed for DC applications of non-ferrous materials.
3. **2% Thoriated Tungsten:** Longer life than 1% thoriated electrodes. Maintain the pointed end longer, used for light gage critical welds in aircraft work. Like 1%, designed for DC applications for non-ferrous materials.
4. **5% Thoriated Tungsten:** Sometimes called "striped" electrode because it has 1.0 to 2.0%. Combines the good properties of pure and thoriated electrodes. Can be used on either AC or DC applications.

5. **Zirconia Tungsten:** Longer life than pure tungsten. Better performance when welding with AC. Melts more easily than thorium-tungsten when forming rounded or tapered tungsten end. Idea for applications where tungsten may contaminate must be minimized.

Electrode Tip Configuration: The shape of the tungsten electrode tip is an important process variable in TIG welding. Tungsten electrodes may be used with a variety of tip preparations. With AC welding, pure or zirconiated tungsten electrodes form a hemispherical ball end. The size of the hemisphere should not exceed 1 to 1.5 times the electrode diameter, otherwise it may fall off while it is molten (Figure 8.10). For DC welding with thoriated, ceriated or lanthanated tungsten electrodes, the end is typically ground to a specific included angle, often with a truncated end. In general, as the included angle increases, the weld bead decreases. Tips are made by grinding and chemical treatment.

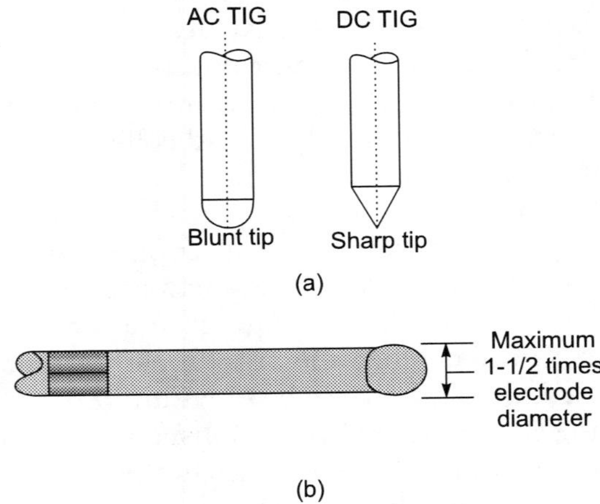

Figure 8.10 (a) Electrode tips for AC and DC (b) Ball end of the tungsten before welding by applying either an AC amperage slightly higher than what is recommended for a given electrode diameter or a DC electrode positive amperage

Grinding: To produce optimum arc stability, grinding of tungsten electrodes should be done with the axis of the electrode perpendicular to the axis of the grinding wheel. The grinding wheel should be reserved for grinding only tungsten to eliminate possible contamination of the tungsten tip with foreign matter during the grinding operation.

Chemical Sharpening: Chemical sharpening consists of submerging the red-hot end of a tungsten electrode into a container of sodium nitrate. The chemical reaction between the hot tungsten and the sodium nitrate will cause the tungsten to erode at a uniform rate all around the circumference and at the end of the electrode. Repeated heating and dipping of the tungsten into the sodium nitrate will form a tapered tip.

Electrode Contamination: Contamination of the tungsten electrode is most likely to occur when a welder accidentally dips the tungsten into the molten weld pool or touches the tungsten with the filler metal. The tungsten electrode may also become oxidized by an improper shielding gas or insufficient gas flow, during welding or after the arc has been extinguished. Other sources of contamination include: metal vapors from the welding arc, weld pool eruptions or spatter caused by gas entrapment, and evaporated surface impurities. The contaminated end of the tungsten electrode adversely affects the arc characteristics and may cause tungsten inclusions in the weld metal. If this occurs, the welding operation should be stopped and the contaminated portion of the electrode removed. Contaminated tungsten electrodes must be properly dressed by breaking off the contaminated section and grinding to shape according to the manufacturer's suggested procedure.

8.3 ARC INITIATION METHODS

8.3.1 Scratch or Touch Start

With the power supply energized, and the shielding gas flowing from the cup, the torch is lowered towards the workpiece until the tungsten electrode makes contact with the workpiece. The torch is quickly withdrawn a short distance to establish the arc.

The advantage of this method of arc initiation is its simplicity of operation for both manual and machine welding. The disadvantage of touch starting is the tendency for the electrode to stick to the workpiece, causing electrode contamination and transfer of tungsten to the workpiece.

8.3.2 High Frequency Start

High frequency starting can be used with DC or AC power sources for both manual and automatic applications. High-frequency generators usually have a spark-gap oscillator that superimposes a high voltage AC output at radio frequencies in series with the welding circuit. The circuit is shown in Figure 8.11. The high voltage ionizes the gas between the electrode and the work, and the ionized gas then conducts welding current that initiates the welding arc.

8.3.3 Pulse Start

Application of a high voltage pulse between the tungsten electrode and the work will ionize the shielding gas and establish the welding arc. This method is generally used with DC power supplies in machine welding applications.

8.3.4 Pilot Arc Start

Pilot arc starting may be used with DC welding power sources. The pilot arc is maintained between the welding electrode and the torch nozzle. The pilot arc supplies

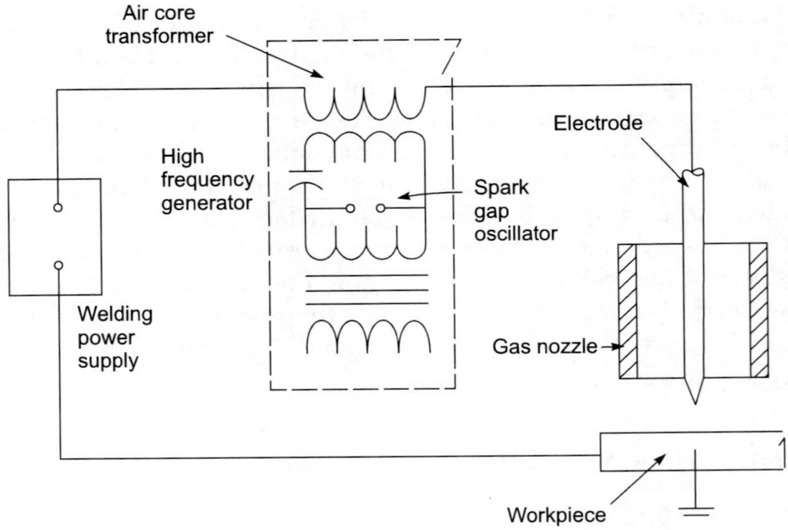

Figure 8.11 High frequency arc starting

the ionized gas required to establish the main welding arc, as shown in Figure 8.12. The pilot arc is powered by a small auxiliary power source and is started by high-frequency initiation.

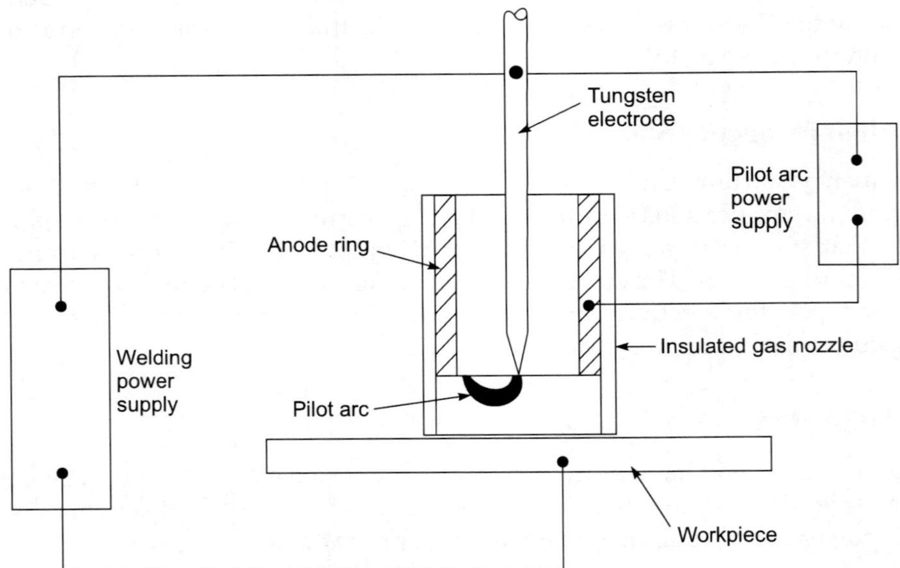

Figure 8.12 Pilot arc starting circuit used for gas tungsten arc spot welding

8.4 GTAW WELDING TECHNIQUES (FIGURE 8.13)

8.4.1 Manual Welding

The word "manual" in the GTAW process implies that a person controls all the functions of the welding process. The functions include manipulation of the electrode holder, control of filler metal additions, welding current, travel speed, and arc length.

Manual Welding Equipment

In addition to an appropriate power supply and a source of shielding gas, manual GTAW equipment include the welding torch, hoses and electrical conductors, foot pedal (or switch on torch) for controlling welding current levels during the welding cycle, and gas flow controls.

Manual Welding Techniques

The torch is then held at an angle of 15° from the vertical and is moved along the joint to progressively melt the faying surfaces. Filler metal, if used, is added to the leading edge of the pool.

The electrode holder and welding rod must be moved progressively and smoothly so that the weld pool, the hot welding rod end, and the hot solidified weld are not exposed to air that will contaminate the weld metal area or heat affected zone. Generally, a large shielding gas envelope will prevent exposure to air.

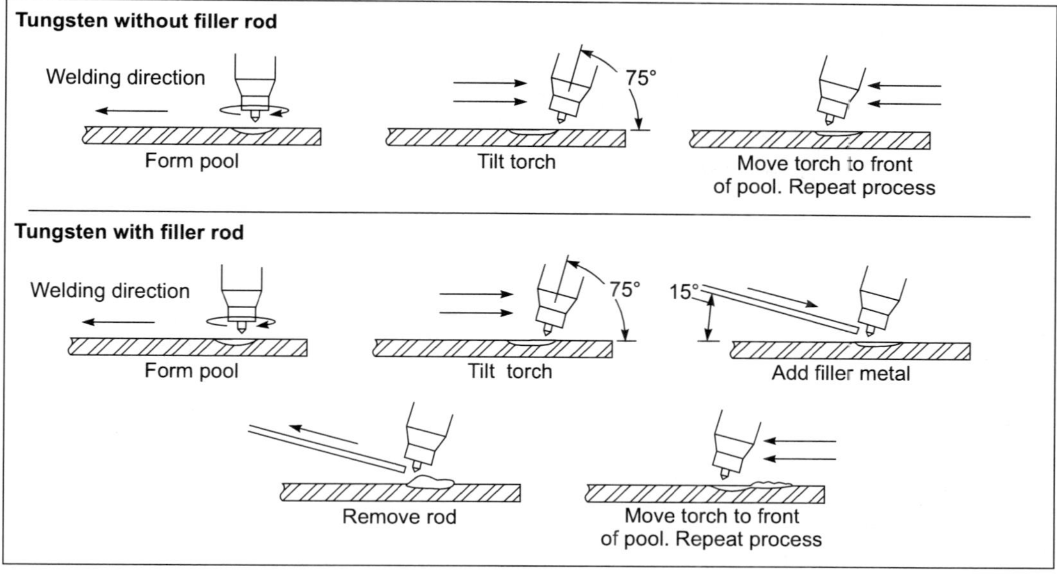

Figure 8.13 GTAW welding techniques

8.4.2 Machine Welding

Machine welding is done with equipment that performs the welding operation under the constant observation and control of a welding operator. Machine GTAW provides greater control over travel speed and heat input to the workpiece. Weld sequencers operate in an open loop control mode: variables are maintained at preset levels and no attempt is made to adjust them in response to changing weld quality. High precision parts and sturdy fixturing increase production cost, but welding sequencers usually cost less than more sophisticated automatic controllers.

8.4.3 Arc Spot Welding

Gas tungsten arc spot welding is often done manually with a pistol-like holder that has a vented, water cooled gas nozzle, a tungsten electrode that is concentrically positioned with respect to the gas nozzle, and a trigger switch for controlling the operation. Figure 8.14 illustrates such arrangement. Gas tungsten arc spot welding electrode holders are also available for automatic applications. The nozzle is often used to press against the workpiece to assure tight fit-up of the faying surfaces. This technique also controls the electrode-to-work distance. Automatic sequencing controls are generally used because of the relatively complex cycles involved. The controls automatically establish the pre-weld gas and water flow, start the arc, time the arc duration, and provide the required post-weld gas and water flow.

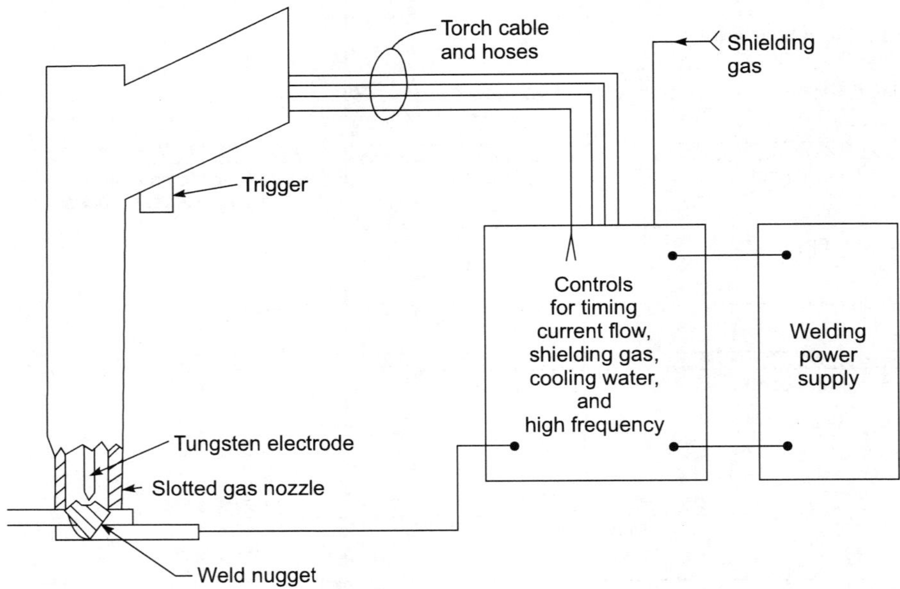

Figure 8.14 Schematic of manual gas tungsten arc spot welding

8.5 WELDING PREPARATIONS

8.5.1 Joint Design

Due to the variety of base metals and their individual characteristics (such as surface tension, fluidity, melting temperature, etc.), joint geometries or designs that provide optimum welding conditions should be used. Factors affecting joint designs include metal composition and thickness, weld penetration requirements, joint restraint and joint efficiency requirements.

Basic Joint Configuration

The five basic joints (butt, lap, T, edge, and corner), shown in Figure 8.15, may be used for virtually all metals. Many variations are derived from these basic joints. In all instances, the primary objective is to minimize welding cost while maintaining the desired weld quality and performance level for the design.

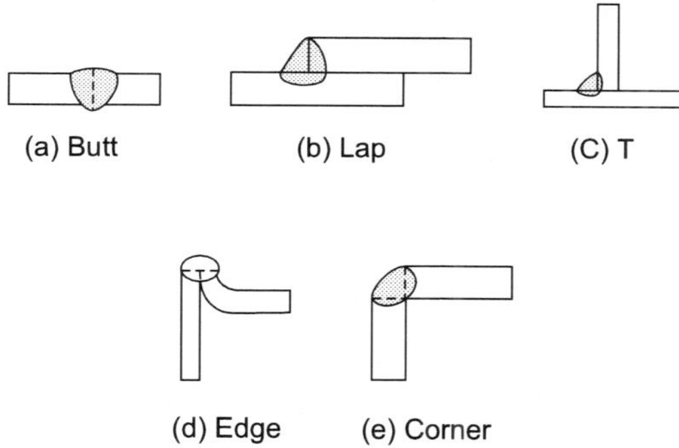

Figure 8.15 Five basic weld joints

8.5.2 Joint Preparation

There are many ways to remove metal to prepare a given joint configuration. However, many GTAW problems, or supposed problems, are a direct result of using improper methods to prepare the joint. Chief among these is the improper use of grinding wheels to prepare joints. Soft materials, such as aluminum, become impregnated with micro-sized abrasive particles which, unless subsequently removed, result in excessive porosity. Grinding wheels should be cleaned and dedicated exclusively to the material being welded. The ideal joint preparation is obtained with cutting tools such as a lathe for round or cylindrical joints, or a milling cutter for longitudinal preparations.

8.5.3 Joint Tolerance

The allowable tolerance of joint dimensions depends upon whether GTAW is to be done manually or by mechanized means. Manual welding applications can tolerate greater irregularities in joint fit-up than mechanized welding, particular tolerance for a given application can be determined only by actual testing, and this tolerance should be specified for future work.

8.5.4 Cleaning

Cleanliness of both the weld joint areas and the filler metal is an important consideration when welding with the gas tungsten arc process. Oil, grease, shop dirt, paint, marking crayon, and rust or corrosion deposits all must be removed from the joint edges and metal surfaces, to a distance beyond the heat affected zone. Their presence during welding may lead to arc instability and contaminated welds.

8.5.5 Fixturing

Fixturing may be required if the parts to be welded cannot be self-supported during welding or if any resultant distortion cannot be tolerated or corrected by straightening. The fixturing should be massive enough to support the weight of the parts and weldment and to withstand welding stresses caused by thermal expansion and contraction. The fixtures must also handle the normal wear and tear that occurs during production.

The following are the primary functions of fixturing:

1. Locate parts precisely within the assembly
2. Maintain alignment during welding
3. Minimize distortion in the weldment
4. Control heat build up

8.6 PROCESS PARAMETERS

8.6.1 Process Variables

The primary variables in GTAW are arc voltage (arc length), welding current, travel speed, and shielding gas. The amount of energy produced by the arc is proportional to the current and voltage. The amount transferred per unit length of weld is inversely proportional to the travel speed. The arc in helium is more penetrating than that in argon. However, because all of these variables interact strongly, it is impossible to treat them as truly independent variables when establishing welding procedures for fabricating specific joints.

Arc Current

As a general statement, arc current controls the weld penetration, the effect being directly proportional, if not somewhat exponential. Arc current also affects the voltage, with the voltage at a fixed arc length increasing in proportion to the current.

For this reason, to keep a fixed arc length, it is necessary to change the voltage setting when the current is adjusted.

The process can be used with either direct or alternating current, the choice depending largely on the metal to be welded. Direct current with the electrode negative offers the advantages of deep penetration and fast welding speeds, especially when helium is used as the shield. Helium is the gas of choice for mechanized welding. Alternating current provides a cathodic cleaning (sputtering) which removes refractory oxides from the joint surfaces of aluminum and magnesium, allowing superior welds to be made. In this case, argon must be used for the shield because sputtering cannot be obtained with helium. Argon is the gas of choice for manual welding whether used with direct current or alternating current. A third power option is also available, that of using direct current with the electrode positive. This polarity is used only rarely because it causes electrode overheating. These effects of polarity are explained in more detail in the following discussion of direct current.

Arc Voltage

The voltage measured between the tungsten electrode and the work is commonly referred to as the arc voltage. Arc voltage is a strongly dependent variable, affected by the following:

1. Arc current
2. Shape of the tungsten electrode tip
3. Distance between the tungsten electrode and the work
4. Type of shielding gas

The arc voltage is changed by the effects of the other variables, and is used in describing welding procedures only because it is easy to measure. Since the other variables such as the shield gas, electrode, and current have been predetermined, arc voltage becomes a way to control the arc length, a critical variable that is difficult to monitor. Arc length is important with this process because it affects the width of the weld pool; pool width is proportional to arc length. Therefore, in most applications other than those involving sheet, the desired arc length is as short as possible. Of course, recognition needs to be given to the possibility of short-circuiting the electrode to the pool or filler wire if the arc is too short. However, with mechanized welding, using a helium shield, DCEN power, and a relatively high current, it is possible to submerge the electrode tip below the plate surface to produce deeply penetrating but narrow welds at high speeds. This technique is called buried arc.

When arc voltage is used to control arc length in critical applications, care must be taken to observe the other variables which affect arc voltage. Among them are electrode and shielding gas contaminants, improperly fed filler wire, temperature changes in the electrode, and electrode erosion. Should any of these change enough to affect the arc voltage during mechanized welding, the arc length must be adjusted to restore the desired voltage.

Travel Speed

Travel speed affects both the width and penetration of a gas tungsten arc weld. However, its effect on width is more pronounced than that on penetration. Travel speed is important because of its effect on cost. In some applications, travel speed is defined as an objective, with the other variables selected to achieve the desired weld configuration at that speed. In other cases, travel might be a dependent variable, selected to obtain the weld quality and uniformity needed under the best conditions possible with the other combination of variables. Regardless of the objectives, travel speed generally is fixed in mechanized welding while other variables, such as current or voltage, are varied to maintain control of the weld.

Wire Feed

In manual welding, the way filler metal is added to the pool influences the number of passes required and the appearance of the finished weld. In machine and automatic welding, wire feed speed determines the amount of filler deposited per unit length of weld. Decreasing wire feed speed increases penetration and flattens the bead contour. Feeding the wire too slowly can lead to undercut, center-line cracking, and lack of joint fill. Increasing wire feed speed decreases weld penetration and produces a more convex weld bead.

Table 8.4 gives the summary of control of welding parameters in TIG, MIG and SMAW welding.

TABLE 8.4 Control of welding parameters in TIG, MIG and SMAW processes

Welding Processes	Arc Length	Voltage	Electrode Feed Rate	Current
TIG	Welder	Welder	Not applicable	Power supply
MIG	Power supply via voltage	Power supply	Wire feed	Electrode speed via wire feed motor
MMA	Welder	Welder via arc length	Welder	Power supply

8.7 WELD QUALITY

8.7.1 Discontinuities

Discontinuities are interruptions in the typical structure of a weldment, and they may occur in the base metal, weld metal, and heat affected zones. Discontinuities in a weldment that do not satisfy the requirements of an applicable fabrication code or specification are classified as defects, and are required to be removed because they could impair the performance of that weldment in service.

8.7.2 Tungsten Inclusions

One discontinuity found only in gas tungsten arc welds is tungsten inclusions. Particles of tungsten from the electrode can be embedded in a weld when improper welding procedures are used with the GTAW process.

Typical causes are the following:

1. Contact of electrode tip with molten weld pool
2. Contact of filler metal with hot tip of electrode
3. Contamination of the electrode tip by spatter from the weld pool
4. Exceeding the current limit for a given electrode size or type
5. Extension of electrodes beyond their normal distances from the collet (as with long nozzles), resulting in overheating of the electrode
6. Inadequate tightening of the holding collet or electrode chuck
7. Inadequate shielding gas flow rates or excessive wind drafts resulting in oxidation of the electrode tip
8. Defects such as splits or cracks in the electrode
9. Use of improper shielding gases such as argon-oxygen or argon-COz mixtures that are used for gas metal arc welding

8.7.3 Lack of Shielding

Discontinuities related to the loss of inert gas shielding are the tungsten inclusions previously described; porosity, oxide films and inclusions, incomplete fusion, and cracking. The extent to which they occur is strongly related to the characteristics of the metal being welded. In addition, the mechanical properties of titanium, aluminum, nickel, and high-strength steel alloys can be seriously impaired with loss of inert gas shielding. A bright, silvery spot will be evident if shielding is effective.

8.8 SAFE PRACTICES

Safe Handling of Gas Cylinders and Regulators: Compressed gas cylinders should be handled carefully. Knocks, falls, or rough handling may damage cylinders, valves, or safety devices and cause leakage or explosive rupture accidents.

Gas Hazards: The major toxic gases associated with GTAW are ozone, nitrogen dioxide, and phosgene gas. Degreasing or other cleaning operations involving chlorinated hydrocarbons should be performed where vapors from these operations are not exposed to radiation from the welding arc.

Ozone: Average concentration of ozone generated in GTAW does not pose any health hazard, if good ventilation is provided and good welding practice is followed.

Nitrogen Dioxide: Natural ventilation quickly reduces these concentrations to safe levels in the welder's breathing zone.

Inert Shielding Gases: Provision for adequate ventilation should be made when inert gas shielding and purging gases are used. Accumulation of these gases could cause suffocation of welding and inspection personnel.

Welding Equipment Safety: All welding equipment should be on an approved list from an NFPA recognized testing agency. No welding should be done until all electrical connections, power supply, welding leads, welding machines, and work clamps are secure, and the welding power source frame is well grounded.

8.9 APPLICATIONS

Gas tungsten arc welding has become indispensable as a tool for many industries because of the high quality welds produced at relatively low equipment cost. The applications of GTAW are as follows:

1. Due to the flexibility and repair work it provides, it is used extensively in the aerospace and nuclear industries.
2. It provides precise control of heat input, for that reason it is preferred for joining thin gage metals and to produce weld close to heat-sensitive components.
3. It is specially used for joining aluminum and magnesium which form refractory oxides, and for reactive metals like titanium and zirconium that can become embrittled.
4. It is used extensively in the manufacture of space vehicles, and is also frequently employed to weld small-diameter, thin-wall tubing, such as those used in the bicycle industry.
5. GTAW is often used to make root or first-pass welds for piping of various sizes. In maintenance and repair work, the process is commonly used to repair tools and dies, especially components made of aluminum and magnesium.

8.10 ADVANTAGES AND LIMITATIONS

The GTAW process is applicable when highest weld quality is required. It can be used to weld almost all types of metals. The operator has excellent control of heat input and vision is not limited by fumes or slag from the process.

Advantages of Gas Tungsten Arc Welding

1. GTAW produces high quality welds, generally free of defects
2. The weld is automatically protected by the inert gas during the welding process
3. No slag is produced
4. TIG welding can be done in any position (1G, 2G, 3G, 4G, 6G)

5. Produces high-quality, low-distortion welds
6. It is free of the spatter which occurs with other arc welding processes
7. It can be used with or without filler metal as required for the specific application
8. Welds almost all metals. Very suitable for non-ferrous metals and stainless steel
9. It gives precise control of welding heat and welding variables
10. It allows excellent control of root pass weld penetration
11. It can produce inexpensive autogenous welds at high speeds
12. It can use relatively inexpensive power supplies
13. It can be used for high quality welding of thin materials

Limitations of Gas Tungsten Arc Welding

1. There is a chance for tungsten electrode contamination and shielding becomes difficult on drafty environments
2. Produces lower deposition rates than consumable electrode arc welding processes
3. Requires slightly more skill as compared to GMAW or shielded metal arc welding (SMAW)
4. Less economical for section thickness greater than 9.5 mm
5. Problematic in drafty environments because of difficulty in shielding the weld zone properly
6. Tungsten inclusions if the electrode is allowed to contact the weld pool
7. Contamination of the weld metal, if proper shielding of the filler metal by the gas stream is not maintained
8. Low tolerance for contaminants on filler or base metals
9. Arc blow or arc deflection, as with other processes

Potential problems with the process

1. Tungsten inclusions can occur if the electrode is allowed to contact the weld pool
2. Contamination of the weld metal can occur if proper shielding of the filler metal by the gas stream is not maintained
3. There is low tolerance for contaminants on filler or base metals
4. Possible contamination or porosity is caused by coolant leakage from water-cooled torches
5. Arc blow or arc deflection, as with other processes

8.11 COMPARISON BETWEEN TIG AND MIG

MIG weld is not considered as clean as TIG welding. Difference in cleanliness of the weld produced by MIG and TIG welding is primarily attributed to the variation in effectiveness of shielding gas to shield the weld pool in two processes. Effectiveness of shielding in two processes is mainly determined by two characteristics of the welding arc, namely, stability of welding arc and length of arc. The MIG arc is relatively longer and less stable than TIG arc. Differences in stability of two welding arcs is primarily due to (a) the type of electrode consumable/non-consumable electrode, and (b) electrode material. Consumption of the electrode during welding slightly decreases the stability of the arc. Summary of comparison between GTAW (TIG) and GMAW (MIG) is given in Table 8.5.

TABLE 8.5 Comparison between GTAW (TIG) and GMAW (MIG)

S. No.	Aspects	TIG Trading	MIC Welding
1.	Name of the process	Tungsten inert gas welding	Metal inert gas welding
2.	Type of electrode used	Non-consumable tungsten electrode	Consumable metallic electrode
3.	Electrode feed	Electrode feed not required.	Electrode need to be fed at a constant speed from a wire reel
4.	Electrode holder	It is called welding torch and has got a cap filled on the back to cover the tungsten electrode. It also has connections for shielding gas, cooling water and a control cable. It may be air-cooled also.	It is called welding gun or torch. It has facility to continuously feed wire electrodes; shielding inert gas, cooling water and control table
5.	Welding current	Both A.C. and D.C. can be used	D.C. with reverse polarity is used
6.	Feed metal	Filler metal may or may not be used	Filler metal in the form of fire wire is used
7.	Bases metal thickness	Metal thickness which can be welded is limited to about 5 mm	Thickness limited to about 40 mm
8.	Welding speed	Slow	Fast

Worked Examples

Example 1: In linear gas tungsten arc welding of two plates of the same material, the peak temperature T 9 in K is given as $T = \frac{C_1 q}{\propto}$, where, q is the heat input per unit length (in J/m) of weld, $\propto$ is the thermal diffusivity in m^2/s of the plate material and

C is a constant independent of process parameters and material types. Two welding cases are given below:

Case 1: Voltage $V = 15$ V, $I = 200$ V, $v = 5$ mm/s, $k = 150$ W/mK, $\rho = 3000$ kg/m^3. C = 900 J/kg-k

Case 2: Voltage $V = 15$ V, $I = 300$ V, $v = 10$ mm/s, $k = 50$ W/mK, $\rho = 8000$ kg/m^3. C = 450 J/kg-k

where, V is welding voltage, I is welding current, v is the welding speed and k, ρ and C refers to thermal conductivity, the density and specific heat of the plate material, respectively. Consider that the electrical energy is completely converted to thermal energy, all other conditions remain same, what is the ratio of peak temperature in Case I to that in Case II? **(GATE 2016)**

Solution: Thermal diffusivity is given by

$$\propto = \frac{k}{\rho C}$$

$$\propto_1 = \frac{150}{3000 \times 900} = 55.5 \times 10^{-6}$$

$$\propto_2 = \frac{50}{8000 \times 450} = 13.88 \times 10^{-6}$$

$$q_1 = \text{heat input} = \frac{VI}{\text{welding speed}} = \frac{15 \times 200}{5 \times 10^{-3}} = 6 \times 10^5$$

$$q_2 = \text{heat input} = \frac{VI}{\text{welding speed}} = \frac{15 \times 300}{10 \times 10^{-3}} = 4.5 \times 10^5$$

$$\frac{T_1}{T_2} = \frac{q_1 \times \propto_2}{q_2 \times \propto_1} = \frac{6 \times 10^5 \times 13.88 \times 10^{-6}}{4.5 \times 10^5 \times 55.5 \times 10^{-6}} = \frac{1}{3}$$

$$T_2 = 3T_1$$

Example 2: Two aluminum alloy plates each 10 mm thick and 1 m long were welded without crowning by multipass tungsten inert gas butt welding. The joint configuration is V type with 60° angle and root gap is maintained at 5 mm. If electrode of 5 mm diameter with 500 mm length is used for welding, what is the number of electrodes required? **(GATE 2015)**

Solution: Area of weld bead is given by

$$5 \times 10 + 2 \times 0.5 \times 10 \tan 30 \times 10 = 107.73 \text{ mm}^2$$

Volume of weld bead is given by

$$V = \text{area} \times \text{lenght} = (5 \times 10 + 2 \times 0.5 \times 10 \tan 30 \times 10)1000 = 107735 \text{ mm}^2$$

$$\text{Volume per electrode} = \frac{\pi \times 5^2 \times 500}{4} = 9812.5 \text{ mm}^3$$

No. of electrodes required = 107735/9812 = 10.97 = 11 electrodes

Example 3: Autogeneous gas tungsten arc welding of a steel plate is carried out with welding current of 500 A, voltage of 20 V m, and weld speed of 20 mm/min. Consider the heat transfer efficiency from the arc to the weld pool as 90%, what is the heat input per unit length in kJ/mm? **(GATE 2009)**

Solution: $\text{Heat input} = \text{power} = \dfrac{VI\,(\text{in J/sec} = \text{watts})}{\text{speed of welding}\left(\dfrac{\text{mm}}{\text{sec}}\right)} \times \eta$

$$\text{Heat input} = \text{power} = \frac{20 \times 500}{20/60} \times 0.9 = 27 \text{ kJ/mm}$$

Example 4: In argon shielded tungsten arc welding (TIG), the cathode drop was found to be 10 volts for a welding current of 120 volts and an arc voltage of 18 volts. Determine (a) the arc length, if the arc efficiency be 55% with an arc temperature of 10,000 K. Assume column voltage drop is 1.2 volt per mm. and that 20% of the heat of the column is transferred to the anode. (b) The arc efficiency if the same process parameters are applicable to GMAW process and the wire electrode is made the anode. Take work function for tungsten at 4.5 eV.

Solution: $V_c I = \phi I + q_e$

where, ϕ work function at temperature T degrees kelvin

$$\phi = \phi_0 + \frac{3}{2} kT$$

where, ϕ_0 = work function at 0 degree kelvin,

k = Boltzmann constant = 8.620×10^{-5} eV/K = 1.381×10^{-23} joule/K

T = temperature in degrees kelvin

q_e, heat going to rod electrode

$$V_c I = \phi I + q_e = \left(\phi_0 + \frac{3}{2} kT\right) I + q_e$$

$$10 \times 120 = \left(4.5 + \frac{3}{2} \times 10000\right)120 + q_e$$

Solving, we get $q_e = 505.2$ joules

$$\eta = \frac{h_a}{VI} = \frac{VI - (q_e + (1-m)V_p I)}{VI} = 1 - \frac{505.2 + 96V_p \times 120}{2160} = 0.55$$

Solving, we get $V_p = 4.8625$ volts

Arc length = 4.8625/1.2 = 4 mm

$$\eta = \frac{h_a}{VI} = \frac{VI - (q_e + (1-m)V_p I)}{VI} = 1 - \frac{(1-m)V_p}{V}$$

$$\eta = 1 - \frac{(1-m)V_p}{V} = 1 - \frac{(1-0.2) \times 4.8625}{18} = 0.7839 = 78.39\%$$

Example 5: A TIG process uses DCEN polarity, arc voltage of 30 V and welding current of 120 A for the welding of 3 mm thin plates. Assuming the voltage drop in anode, cathode and plasma regions is 16 V, 10 V and 4 V, respectively and 20 % heat generated in plasma region is used for melting the base metal and all the heat generated in anode drop zone is used for welding. Neglecting the voltage drop on account of work function of metal during welding, calculates the arc efficiency.

Solution: Arc efficiency = $\dfrac{\text{Heat generated in anode drop zone + heat generated in plasma used welding}}{\text{Total heat produced by welding arc}}$

$$\text{Arc efficiency} = \frac{V_a \times I + m(V_p \times I)}{VI} = \frac{V_a + m \times V_p}{V}$$

$$\text{Arc efficiency} = \frac{V_a + m \times V_p}{V} = \frac{16 + 0.2 \times 4}{30} = 0.56 = 56\%$$

Example 6: A welding operation takes place on an aluminum alloy plate. A pipe 50 mm in diameter with a 4 mm wall thickness and a 60 mm length is butt-welded onto a section of 15 × 15 × 5 mm angle iron. The angle iron is of an L-shape and has a length of 0.3 m. If the weld zone in a gas tungsten arc welding process is approximately 8 mm wide, what would be the temperature increase of the entire structure due to the heat input from welding only? What if the processes were an electron beam welding operation with a bead width of 6 mm? Assume that the electrode requires 1500 J and the aluminum alloy requires 1200 J to melt one gram. Assume that the electrode is placed around the entire pipe and weld cross section is triangular **(Exercise problem from *Manufacturing Process*: Kalpakjian)**.

Solution: Case A: If the welding is done using TIG

Since the electrode is placed around the entire pipe, we can find out the weld length as $D = \pi \times 50\text{ mm} = 0.157\text{ m}$. The weld cross section is triangular, its volume is approximately

$$\text{Volume} = \frac{bhL}{2} = \frac{0.008 \times 0.008 \times 0.157}{2} = 5.02 \times 10^{-6}\,\text{m}^3 = 5020\text{ mm}^3$$

The electrode material should have similar properties as that of aluminum. The density $\rho = 2700\text{ kg/m}^3$ and $C = 900\frac{\text{J}}{\text{kg-k}}$. Specific heat to melt aluminum is 2.9 J/mm^3.

Heat energy input = 2.9(5020) = 14.5 kJ. The total volume of aluminum is

$$V = \frac{\pi}{4}(d_0^2 - d_i^2)L + 2btl$$

$$V = \frac{\pi}{4}(50^2 - 42^2)60 + 2 \times 15 \times 5 \times 30 = 79{,}683\text{ mm}^3 = 7.968 \times 10^{-5}\text{ m}^3$$

The temperature rise is then calculated as

$$E = \rho V C \Delta T$$

$$14500 = 2700 \times 7.968 \times 10^{-5} \times 900 \times \Delta T$$

Solving we get

$$\Delta T = 75°\text{C}$$

Case B: If the welding is done using EBM

$$\text{Volume} = \frac{bhL}{2} = \frac{0.006 \times 0.006 \times 0.157}{2} = 2.826 \times 10^{-6}\,\text{m}^3 = 2826\text{ mm}^3$$

Heat energy input = 2.9(2856) = 8.20 kJ

The temperature rise is then calculated as

$$E = \rho V C \Delta T$$

$$8200 = 2700 \times 7.968 \times 10^{-5} \times 900 \times \Delta T$$

Solving we get

$$\Delta T = 42°\text{C}$$

Example 7: A GTAW operation is performed on low-carbon steel, whose unit melting energy is 10.3 J/mm^3. The welding voltage is 22 volts and the current is 135 A. The heat transfer factor is 0.7 and the melting factor is 0.65. If filler metal wire of 3.5 mm diameter is added to the operation, the final weld bead is composed of 60%

volume of the filler and 40% volume base metal. If the travel speed in the operation is 5 mm/sec, determine (a) cross-sectional area of the weld bead, and (b) the feed rate (mm/sec) at which the filler wire must be supplied.

Solution:

(a) $R_{HW} = f_1 f_2 EI = U_m A_w v$

where, R_{HW} = rate of heat used for welding

$f_1 f_2$ = heat transfer coeff. and melting efficiency factor, respectively

U_m = unit energy required to melt material

v = welding speed

$$0.7 \times 0.62 \times 22 \times 135 = 10.3 \times A_w \times 5$$

Solving we get

$$A_w = 26.24 \text{ mm}^2$$

(b) Volume of weld = $A_w \times 5 = 26.24 \times 5 = 131.2 \dfrac{\text{mm}^3}{\text{s}}$

Filler wire, $A = \pi D^2/4 = \pi \times 3.5^2/4 = 9.62 \text{ mm}^2$

At 60% filler metal, feed rate of filler wire = 131.2(0.60)/9.62 = 8.18 mm/s

Example 8: A gas tungsten arc welding operation is performed using a current of 250 amperes and an arc voltage of 20 volts at a welding speed of 5 mm/s. Assuming the arc efficiency is 70%. The net input per unit length of the weld will be __________ kJ/min (round off to one decimal place). **(GATE 2019-S2)**

Solution: V = 20 volts

I = 250 amperes

Arc efficiency = 0.7

Welding speed = 5 mm/sec

Since arc efficiency = 70%

The net heat input = 0.7 × 5000 = 3500 joules/sec

$$\text{Net heat input/length} = \frac{\text{Net heat input}}{\text{Welding speed}}$$

$$\text{Net heat input/length} = \frac{3500\left(\frac{\text{joules}}{\text{sec}}\right)}{5\left(\frac{\text{mm}}{\text{sec}}\right)} = 700 \frac{\text{joules}}{\text{sec}} = 0.7 \text{ kJ/min}$$

Review Questions

1. Explain the principle of operation of TIG.
2. Explain the equipment used in TIG.
3. Explain differences between TIG and MIG.
4. Give brief descriptions of power source for TIG.
5. What are the different welding techniques for TIG?
6. What are the advantages of tungsten arc spot welding?
7. Discuss the important variants of TIG.
8. What are the advantages of TIG?
9. What are the disadvantages of TIG?
10. Write some few applications of TIG.

Submerged Arc Welding

9.0 INTRODUCTION

It is one of the arc welding processes wherein coalescence is produced by heating with an electric arc setup between the bare electrode and workpiece under a blanket of granular flux, which is continuously fed. No pressure is applied in this process.

Submerged arc welding (SAW) is a common arc welding process. The first patent on the SAW process was taken out in 1935 and covered an electric arc beneath a bed of granulated flux. Originally developed and patented by Jones, Kennedy and Rothermund, the process requires a continuously fed consumable solid or tubular (metal cored) electrode. The molten weld and the arc zone are protected from atmospheric contamination by being "submerged" under a blanket of granular fusible flux consisting of lime, silica, manganese oxide, calcium fluoride, and other compounds. When molten, the flux becomes conductive, and provides a current path between the electrode and the work. This thick layer of flux completely covers the molten metal, thus preventing spatter and sparks as well as suppressing the intense ultraviolet radiation and fumes.

SAW is normally operated in the automatic or mechanized mode, however, semi-automatic (hand-held) SAW guns with pressurized or gravity flux feed delivery are also available. The process is normally limited to the flat or horizontal fillet welding positions (although horizontal groove position welds have been done with a special arrangement to support the flux).

Factors that determine whether to use submerged arc welding include:

1. The chemical composition and mechanical properties
2. Thickness of base metal to be welded
3. Joint accessibility
4. Position in which the weld is to be made
5. Frequency or volume of welding to be performed

9.1 PROCESS DETAILS (FIGURES 9.1 AND 9.2)

Submerged arc welding (SAW) process uses heat generated by an electric arc established between a bare consumable electrode wire and the workpiece. Since in this process, welding arc and the weld pool are completely submerged under cover of granular fusible and molten flux, therefore it is called submerged arc welding. During welding, granular flux is melted using heat generated by arc and forms cover of molten flux layer, which in turn, avoids spatter tendency and prevents accessibility of atmospheric gases to the arc zone and the weld pool. The molten flux reacts with the impurities in the molten weld metal to form slag which floats over the surface of the weld metal. Layer of slag over the molten weld metal results in (a) increased protection of weld metal from atmospheric gas contamination and so improved properties of weld joint, and (b) reduced cooling rate of weld metal and heat affected zone (HAZ), owing to shielding of the weld pool by molten flux and the solidified slag, in turn, leads to smoother weld bead and reduced cracking tendency of hardenable steel.

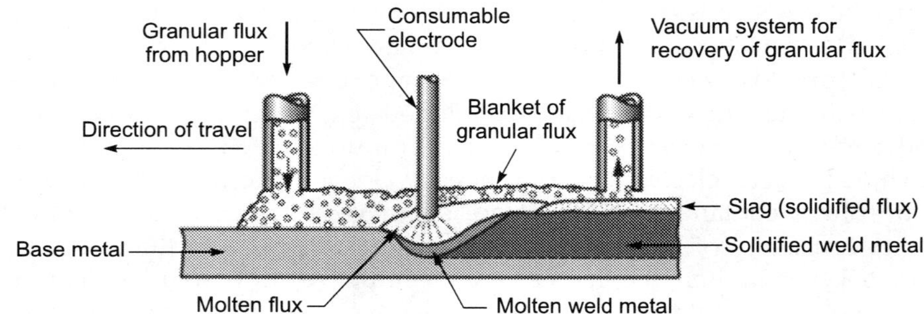

Figure 9.1 Schematic of submerged arc welding

This type of arc welding involves the formation of an arc between a continuously fed bare wire electrode and the workpiece. A flux is generated which makes a protective gas and slag and also helps in controlling the composition of the deposited metal by providing alloying elements to the weld pool. A thin layer of flux powder is placed on the workpiece surface before the welding takes place. The arc, which is completely covered by flux layer, thereby having lesser heat loss, moves along the joint line. It is capable of a thermal efficiency in the range of 80% to 90% and produces no visible light, a splatter free welding, and no need for fume extraction.

Submerged arc welding is a high deposition, downhand welding process capable of making welds with currents up to 2000 A, AC or DC, using single or multiple wires or strips of filler metal. The submerged arc process is also capable of welding fairly thin gage material. However, because of the greater welding current densities, the electrode many times is pointed or slanted in a particular direction. This wire slant determines to a great extent the bead appearance. The forward pointing wire points in the direction of the weld travel and is similar to forehand welding in the gas metal arc welding process, which deflects much of the heat of the arc column, yielding a decrease in the penetration, a wider and flatter crown to the bead, and a smaller heat affected zone. The backward slanting point, however, does just the opposite and is comparable to the backhand welding technique in the GMAW process. The backward pointing electrode yields greater penetration, a larger heat affected zone, and a narrower weld bead.

Normally, the submerged arc welding process is known as a process that achieves deep penetration. However, penetration of the series dual electrode is based on the magnetic lines of force opposing each other, resulting in a force being exerted at an angle of 90° to the electron flow, which accounts for the shallow penetration.

This process is applicable to the surfacing of a metal with a dissimilar metal, such as in hard surfacing. The submerged arc process produces high-quality welds having excellent bead profile. The elimination of fumes, smoke, and any visible arc column adds to its ease of operation and efficiency, which, in turn, encourages its application in the welding industry.

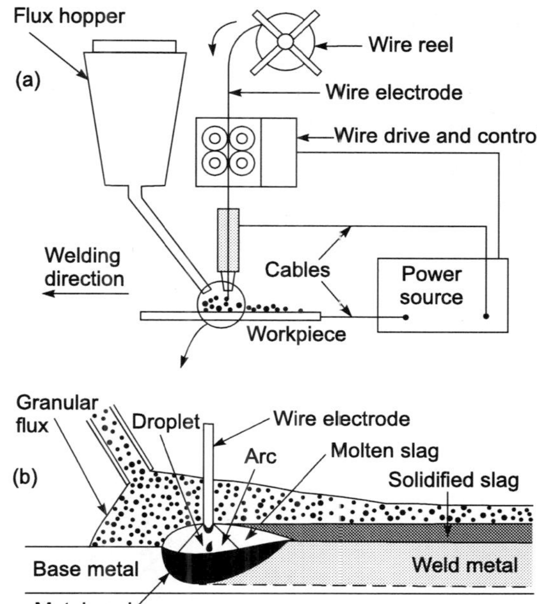

Figure 9.2 Submerged arc welding details

9.2 EQUIPMENT (FIGURE 9.3)

Submerged arc welding can be semi-automatic but is usually operated as a fully mechanized or automatic process. The electrode which acts as a filler wire is continuously fed by a wire feeding mechanism, after the initiation of the arc, towards the joint at a controlled rate and the welding torch is moved manually or automatically along the weld seam.

Flux is continually fed in front of and around the electrode from a hopper as it keeps moving over a seam. The heat generated by the electric arc progressively melts some of the flux, the end of the electrode wire and the adjacent edges of the base metal the creating a pool of molten metal below the layer of liquid slag. The flux which floats on the molten metal completely shields the molten zone from the atmosphere. It also dissolves impurities in the base metal and electrode and makes them float up to the surface.

As the welding arc advances along the seam, the weld metal and then the liquid flux cools down and solidifies, forming a weld bead and a protective slag shield over it. This slag shield results in a slower cooling rate for the deposited weld metal and thus, provides an annealing effect to the weld deposit.

Figure 9.3 shows a schematic representation of the submerged arc welding process.

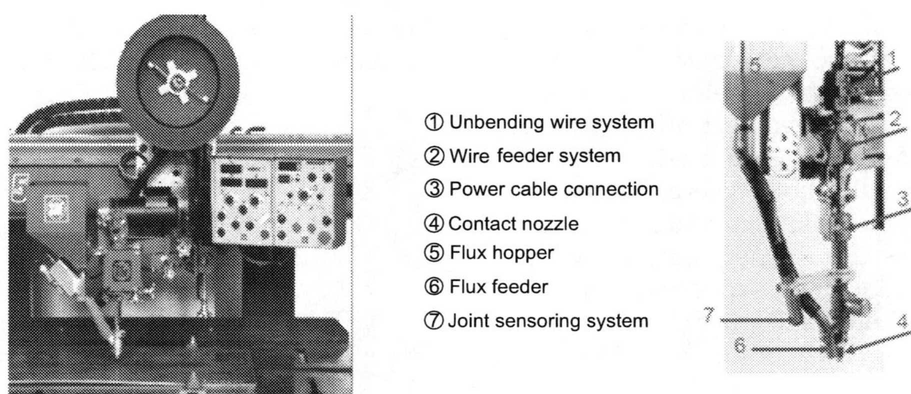

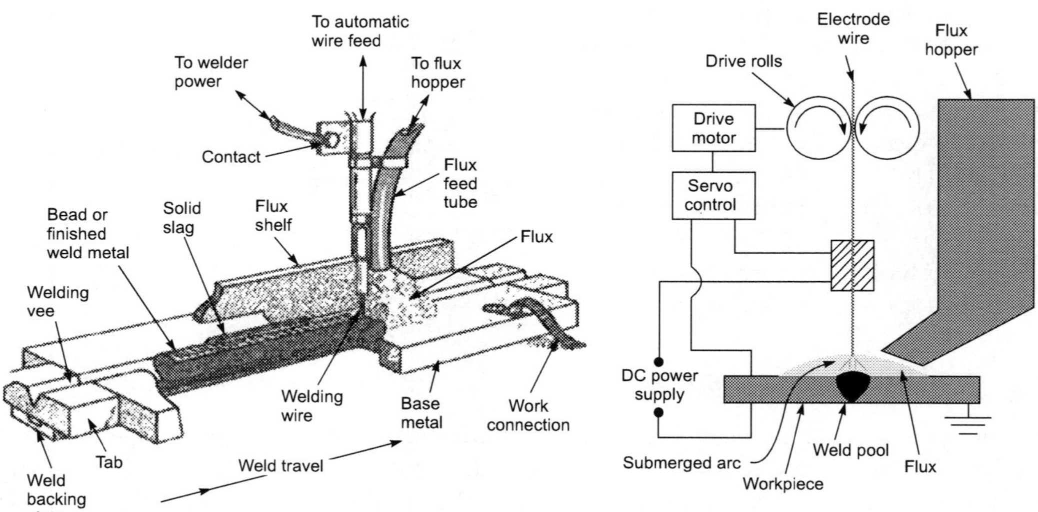

Figure 9.3 Equipment for SAW

9.2.1 Power Sources

Several types of power supply are suitable for submerged arc welding. This being generally a high current process with high duty cycle, a power supply capable of providing high amperage at 100% duty cycle is recommended. A constant voltage (CP) power supply being self-regulating, can be used with a constant speed wire feeder. The wire feed speed and wire diameter control current and the power supply controls the voltage. The CP, DC power supply gives the best result for high-speed welding of thin steels. In DC operation, the electrode is normally connected to the positive terminal.

AC power sources usually have a constant current output characteristics and are therefore, not self-regulating. The arc with this type of power supply is controlled by sensing the arc voltage and using the signal to control the wire feed speed. In practice, for a given welding current level, arc length is determined by wire burn-off rate, i.e., the balance between the welding current setting and wire feed speed, which is under feedback control. AC square-wave power sources have a constant voltage output characteristic. It provides constant wire feed speed control and easier arc ignition (Figure 9.4).

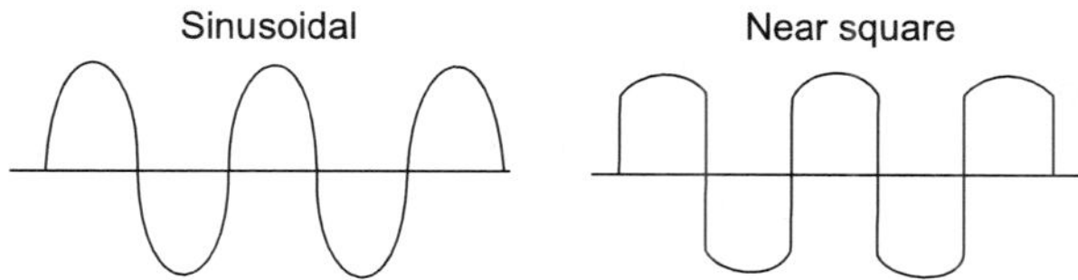

Figure 9.4 AC power sources used for intermediate deposition and/or penetration efficiency. It is less sensitive to arc blow effects

9.2.2 Electrodes

SAW produces weld deposits matching carbon steels, low-alloy steels, high-carbon steels, special alloy steels, stainless steels, nickel alloys and special alloys for surfacing applications. These electrodes are supplied as bare solid wire and as composite metal cored electrodes.

Electrodes are normally packaged as coils or drums, ranging in weight from 11 to 454 kg. Large electrode packages are economical.

Steel electrodes are usually copper coated, except those for welding corrosion-resisting materials or for certain nuclear applications. The copper coating provides good shelf life, decreases contact tube wear, and improves electrical conductivity.

Submerged arc welding electrodes vary in size from 1.6 to 6.4 mm in diameter, as shown in Table 9.1. The wide amperage ranges are typical of submerged arc welding.

TABLE 9.1 Wire diameter versus current range

Wire Diameter		Current Range (Amperes)
Inches	mm	
5/64	2.3	200-500
3/32	2.4	300-600
1/8	3.2	300-800
5/32	4.0	400-900
3/16	4.8	500-1200
1/32	5.6	600-1300
1/4	6.4	600-1600

9.2.3 Flux

The flux protects the weld fusion zone and molten drops transferred from the filler metal. Commonly used fluxes in submerged arc welding are granular fusible minerals containing oxides of manganese, silicon, titanium, aluminum, calcium, zirconium, magnesium and other compounds, such as calcium fluoride. The flux is specially formulated to be compatible with a given electrode wire type so that the combination of flux and wire yields desired mechanical properties. All fluxes react with the weld pool to produce chemical composition and mechanical properties of the weld metal. Fluxes are commonly referred to as 'active' if they add manganese and silicon to the weld. The amount of manganese and silicon that gets added is influenced by the arc voltage and the welding current level. Fluxes, used in submerged arc welding, are primarily of two types. The flux plays the following roles, apart from shielding the arc and the molten pool from atmospheric contamination:

1. The stability of the arc depends on the flux
2. Chemical and thus the mechanical properties of the weld metal can be controlled by flux
3. The quality of the weld may be affected by the quality and quantity of the flux used over the arc
4. Flux creates an appropriate easy way to ionize gas environment to form the plasma for the electric arc
5. It generates slag to cover the molten metal during solidification and cooling
6. Improves chemical composition of weld metal
7. Promoting a stable transference of filler metal into the weld bead

(a) Fused Flux: To manufacture a fused flux, the raw materials are dry mixed and melted in an electric furnace. After melting and any final additions, the furnace charge is poured and cooled. Cooling may be accomplished by shooting the melt through a stream of water or by pouring it onto large chill blocks. The result is a product with a glassy appearance which is then crushed, screened for size, and packaged.

The advantages of fused fluxes are as follows:

(a) Good chemical homogeneity
(b) Easy removal of the fines without affecting the flux composition
(c) Not hygroscopic normally, which simplifies handling, storage, and welding problems
(d) Readily recycled through feeding and recovery systems without significant change in particle size or composition
(e) The main advantage of this flux is the production of smooth stable arcs, with welding currents up to 2000 A and consistent weld metal properties

The disadvantages are:

- Difficulty of adding deoxidizers and ferroalloys to them during manufacture without segregation or extremely high losses.
  The high temperatures needed to melt the raw ingredients limit the range of flux compositions.

(b) Bonded Flux: Bonded fluxes are made from raw materials that are powdered, dry mixed and bonded together with either a potassium silicate or sodium silicate binder or a combination of both. The wet bonded mixture is palletized and baked at low temperature. When drying is complete, the pellets are broken up, screened for particles sizing and packaged for shipment. Bonded fluxes contain metallic deoxidizers, which help to prevent weld porosity. These fluxes are effective over rust and millscale.

The advantages of bonded fluxes include the following:

- Easy addition of deoxidizers and alloying elements. A alloying elements are added as ferroalloys or as elemental metals to produce alloys not readily available as electrodes, or to adjust weld metal compositions.
- Allows thicker layer of flux during welding
- Can be easily identified by color

The disadvantages of bonded fluxes are the following:

- Tendency for some fluxes to absorb moisture in a manner similar to coatings on some shielded metal arc electrodes
- Possible gas evolution from the molten slag
- Possibility of change in flux composition due to loss of fine particle size

(c) Agglomerated Flux: These are easily manufactured as compared to fuse type flux at low temperature. They are homogeneous and produced by agglomeration of grain with water glass as binding material.

Agglomerated fluxes are light in color and roughly spherical in shape. They give the best mechanical properties and low hydrogen potential possible, requiring the flux to be preheated (baked). Agglomerated fluxes absorb moisture, so at the end of work they must be removed and dried. Agglomerated fluxes are produced similar to bonding fluxes, except that a ceramic binder is used which limits the use of deoxidisers and alloy elements similar to those of bonded fluxes.

Many elements are added to a flux to produce desired properties:

1. Deoxidation element added (FeSi, FeMn or FeSiMn), so that the process does not depend upon the reduction of MnO_2 to MnO
2. Alloying element added to give the flux varying properties
3. Easy slag removal and good weld appearance
4. High resistance to porosity
5. Low specific flux consumption

Basicity of fluxes used in submerged arc welding can be expressed by the following equation:

$$\text{Basicity(B)} = \frac{CaO + MgO + Na_2O + K_2O + CaF_2 + \dfrac{(MnO + FeO)}{2}}{SiO_2 + \dfrac{(Al_2O_3 + TiO_2 + ZnO_2)}{2}}$$

(d) Mechanical Flux: Mechanical flux is achieved by mixing two or more fluxes, bonded or agglomerated fluxes, to get the desired properties.

The advantage is:

- Several commercial fluxes which are readily available can be mixed in critical welding situations.

The disadvantages are:

- Segregation of the combined fluxes during shipment, storage, and handling
- Segregation occurring in the feeding and recovery systems during the welding operation
- Inconsistency in the combined flux from mix to mix

Particle Size and Distribution

Flux particle sizes and their uniform distribution within the bulk flux influence feeding and recovery, amperage level, and weld bead smoothness and shape. As amperage increases, the average particle size for fused fluxes should be decreased and the percentage of small particles should be increased. If the amperage is too high with a given particle size, the arc may be unstable and leave ragged, uneven bead edges.

Some flux manufacturers may mark their packages with sizing information presented in the form of two mesh numbers. The numbers represent the largest and smallest particle sizes present when standardized screens are used to measure them. The first number identifies the mesh (screen) through which essentially all of the particles will pass, and the second number identifies the mesh through which most or all of the particles will not pass.

The particle size does not provide all the information that may be needed. Some flux manufacturers offer each of their fluxes in only one particle size range, customized for a general area of flux application.

Flux Usage

If the flux is too fine, it will pack and not feed properly. If a fine flux or a flux with small amounts of fine particles is recovered by a vacuum system, the fine particles may be trapped by the system. Only the coarser particles will be returned to the feeding system for reuse, which may cause welding problems.

When alloy bearing fluxes are used, it is necessary to maintain a fixed ratio between the quantities of flux and electrode melted, to obtain a consistent weld metal composition. This ratio is actually determined by the variables of the welding procedure.

For example, deviation from an established volt-ampere relation will change the alloy content of the weld metal by changing the flux-electrode melting ratio. Fluxes are also identified as chemically basic, chemically acidic, or chemically neutral. The basicity or acidity of a flux is related to the ease with which the component oxides of the flux ingredients dissociate into a metallic cation and an oxygen anion. Chemically basic fluxes are normally high in MgO or CaO, while chemically acidic fluxes are normally high in SiO_2.

Flux Recovery Units

Flux recovery units are frequently used to maximize flux utilization and minimize manual clean-up. Flux recovery units may do any combination of the following:

1. Remove unfused flux and fused slag behind the weld
2. Screen out fused slag and other oversized material
3. Remove magnetic particles
4. Remove fines
5. Recirculate flux back to a hopper for reuse
6. Heat flux in a hopper to keep it dry

Pneumatic flux feeding is commonly used in semi-automatic SAW and frequently in automatic SAW.

9.2.4 Welding Guns (Figure 9.5)

There are three types of guns used in submerged arc welding process: the side flux delivery, the concentric flux delivery and the deep groove type of gun. When the side delivery flux gun or deep groove gun is used, the flux is fed from an overhead gravity hopper to the flux shut-off assembly on the gun. The flux is deposited in front of the wire normally to a depth to prevent visible arcing or flashing.

Figure 9.5 Welding gun

9.3 OPERATING CHARACTERISTICS

Selection of the correct welding conditions for the plate thickness and joint preparation to be welded is very important, if satisfactory joints free from defects, such as cracking, porosity and undercut, are to be obtained. The process variables, which have to be considered, are:

1. Welding current
2. Welding voltage
3. Welding speed
4. Electrode diameter
5. Electrode extension (length of stick-out)
6. Type of flux
7. Width and depth of flux
8. Polarity

These variables determine bead size, bead shape, and depth of penetration and in some circumstances, metallurgical effects such as incidence of cracking, porosity and weld metal composition.

9.3.1 Welding Current (Figure 9.6)

Welding current is the most important parameter of submerged arc welding. Since the amount of heat generated is directly proportional to the square of the current, it controls the melting rate of the electrode and thereby, controls the deposition rate. It also directly influences the depth of penetration and the extent of base metal fusion.

Following are the three rules concerning welding current:

(a) Increasing current increases penetration and melting rate
(b) Excessively high current produces a digging arc and undercut, or a high, narrow bead
(c) Too low welding current produces an unstable arc

If the welding current is too high at a given welding speed, it will lead to:

1. Excessive depth of penetration resulting into melt through the metal being joined
2. Excessive reinforcement beading to wastage of electrode
3. Increased heat affected zone
4. Increased weld induced distortion

If the current is too low, it may result in:

1. Inadequate penetration or incomplete fusion
2. Unstable arc

The metal transfer process from the electrode in submerged arc welding is much less known as compared to that of gas metal arc welding. Different forms of transfer are known to occur with different welding currents and other process variables.

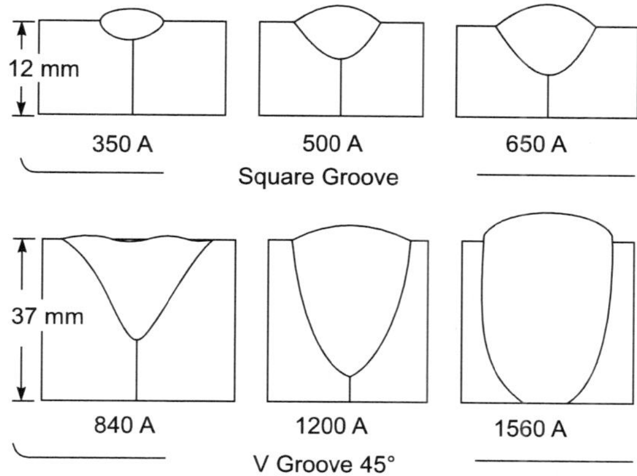

Figure 9.6 Influence of current on weld bead morphology

9.3.2 Welding Arc Voltage (Figure 9.7)

Welding voltage adjustment varies the length of the arc between the electrode and the molten weld metal. If the overall voltage is increased, the arc length increases; if the voltage is decreased, the arc length decreases. Voltage has little effect on the electrode deposition rate, which is determined by welding current. The voltage principally determines the shape of the weld bead cross section and its external appearance.

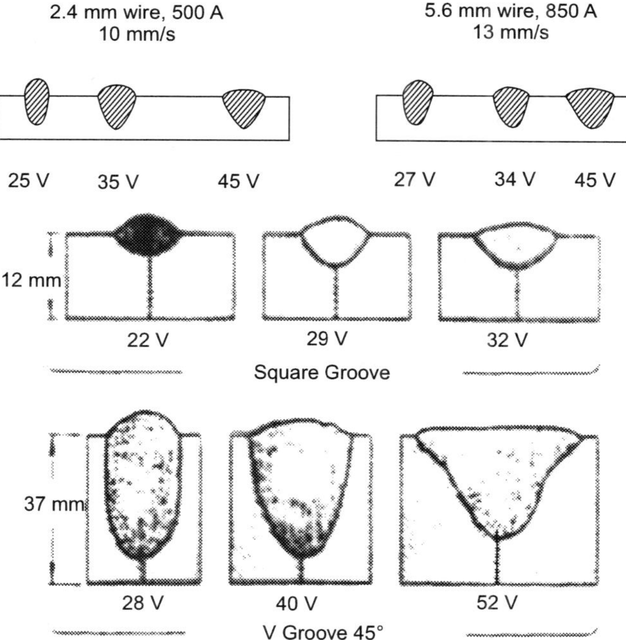

Figure 9.7 Effect of arc voltage variations on weld bead shape and penetration

If the welding voltage is increased, keeping current and welding speed constant, it will result in:

(a) Flatter and wider bead
(b) Increased flux consumption
(c) Reduction of porosity, due to rust or scale on the base metal
(d) Increased tolerance of root opening arising out of poor fit-up. It will help to bridge an excessive root opening
(e) Increase in pickup of alloying elements from flux

If the arc voltage is increased excessively, it will:

(a) Produce a very wide weld bead, which is more prone to cracking
(b) Make slag removal difficult in groove welds
(c) Increase undercut along the edge (s) of fillet welds

If the arc voltage is low, it will:

(a) Improve penetration in a deep weld groove
(b) Resist arc blow

However, excessive reduction in arc voltage will produce a high and narrow bead and slag removal along the bead edges becomes difficult.

9.3.3 Welding Speed (Figure 9.8)

The welding speed is another important variable, which has a pronounced effect on weld size and penetration for a given combination of welding current and arc voltage.

If the welding speed is increased:

(a) Heat input per unit length decreases
(b) Less filler metal is applied per unit length of weld, leading to less weld reinforcement
(c) Weld bead becomes smaller

Excessively high welding speeds may lead to:

(a) Undercut, porosity and uneven bead shape
(b) Arc blow

If the welding speed is decreased:

(a) Heat input per unit length of weld is increased
(b) More filler metal is consumed
(c) Leads to more weld reinforcement increased penetration
(d) Provides more time for gases to escape, thus reducing porosity

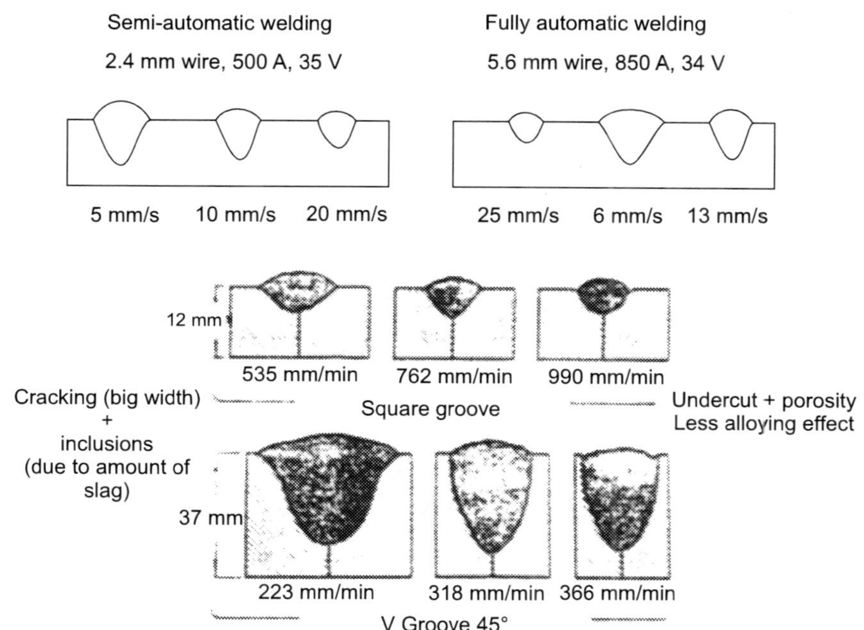

Figure 9.8 Effect of travel speed variation on weld bead shape and penetration

If the welding speed is further reduced, i.e., excessively slow speed will produce:

(a) A convex bead shape, more prone to cracking
(b) A large molten weld pool, resulting in a rough bead and slag inclusions
(c) Weld penetration is not further increased because much of the molten weld pool will be beneath the arc. The penetrating force of the arc gets cushioned by the molten pool, thus restricting further penetration

9.3.4 Wire Feed Speed

Wire feed speed is directly proportional to the current. It controls the wire melt-off rate, depth of penetration and the amount of base metal dilution. Increased current/wire feed rate increases penetration, wire melt-off rate, deposition, results in an unstable arc, undercut, excessive weld reinforcement, increased weld shrinkage, added stress and a greater chance of distortion. Low wire speed produces an unstable arc, lack of proper fusion and low deposition.

9.3.5 Electrode Diameter (Figure 9.9)

The electrode diameter at a given current affects the weld bead shape and the depth of penetration. In case of poor fit-up, a large diameter electrode is better suited compared to thinner one for bridging large root openings. The metal deposition rate is also influenced by electrode diameter. At any given current, a smaller diameter electrode will have a higher current density and will provide a higher deposition rate as compared to that of a larger diameter electrode. However, a higher diameter electrode can carry much higher current than thinner electrode, thus it can produce higher deposition rate at higher amperage.

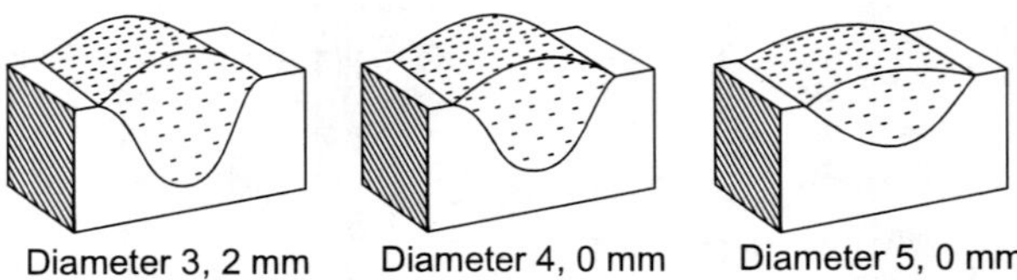

Figure 9.9 Effect of electrode size on weld bead shape and penetration

9.3.6 Electrode Extension

Electrode extension or length of stick-out becomes an important operating variable for current densities above 125 A/m². The melting rate increases because of resistance heating of the electrode between the contact tube and the arc at higher current densities. Thus, deposition rates can be increased from 25% to 50% by using long electrode extensions with no change in welding current. For procedure development, an electrode extension of about eight times the electrode diameter is a good starting point. The length is to be suitably adjusted to achieve the desired melting rate. Melting rate depends on the heat generated by the arc when there is no electrode extension and due to resistance heating of the electrode. The total electrode melting rate m_r, (g/s) is given as:

$$m_r = \frac{I}{60}\left[0.35 + \frac{d^2}{645} + 2.08 \times 10^{-7}\left(\frac{I \times L_e \times \pi}{d^2}\right)^{1.22}\right]$$

where,

d – electrode diameter in mm.

L_e – length of stick out in mm.

Since increase in electrode extension adds to the resistance component of the electrode in the welding circuit, it causes an increased voltage drop across the electrode, thus reducing the arc voltage. Hence, the bead shape changes with change in electrode extension. It is increased to take advantage of the higher melting rate, however the voltage setting of the power supply should be accordingly increased to maintain proper arc voltage to achieve proper bead shape.

However, an increase in deposition rate is accompanied by a decrease in penetration. Therefore, where deep penetration is required it is not recommended to have a long extension of electrode. On the other hand, where melt through may occur as in case of thin plates, a longer electrode extension is preferable. A long stick-out is normally only used in cladding and surfacing applications where there is greater emphasis on deposition rate and control of penetration, rather than accurate positioning of the wire (Figure 9.10).

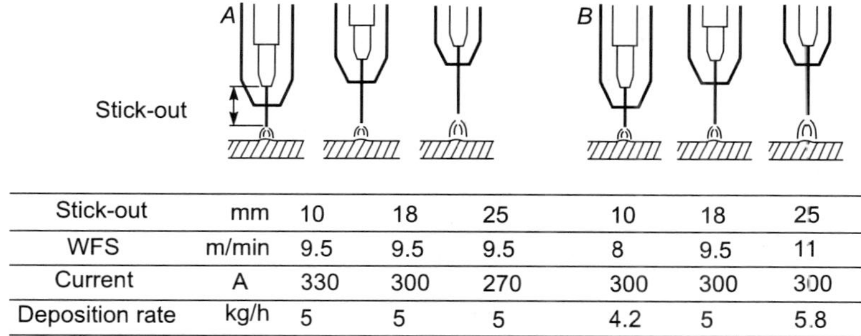

		A			B		
Stick-out	mm	10	18	25	10	18	25
WFS	m/min	9.5	9.5	9.5	8	9.5	11
Current	A	330	300	270	300	300	300
Deposition rate	kg/h	5	5	5	4.2	5	5.8

Figure 9.10 Influence of stick-out on deposition rate

The recommended and maximum stick-out lengths are given in Table 9.2.

TABLE 9.2

Wire Diameter (mm)	Current Range (A)	Wire Stick-out	
		Normal (mm)	Maximum (mm)
0.8	100 to 200	12	–
1.2	150 to 300	20	–
1.6	200 to 500	20	–
2.0	250 to 600	25	63
3.2	350 to 800	30	76
4.0	400 to 900	32	128
4.75	450 to 1000	35	165

9.3.7 Width and Depth of Flux

The bead appearance and soundness of the finished weld also depend on the width and depth of the granular flux layer. If the layers are too deep, the gases generated during welding cannot readily escape and thus, results in a distorted weld surface. If the layer is too shallow, the arc may not get fully submerged, causing flashing and spattering of the molten metal. It will result in a poor bead appearance and may cause porosity. The optimum flux depth can be achieved by slowly increasing the flow of the flux until the arc is submerged and flashing no longer occurs. The unburnt flux can be collected back and recycled after straining through a screen to remove any coarse particle, if present. The flux prior to use should be baked to remove moisture. Moisture, if present in the flux, will cause porosity in the weld deposit.

9.3.8 Polarity (Figure 9.11)

Direct current is the most commonly used but base metal type, joint design and thickness are the factors when choosing polarity. DC usually provides more control over the bead shape, penetration, deposition and welding speed. DC makes starting easier than conventional AC power supplies. AC power sources have an added advantage of reduced possibility of arc blow problems.

DCRP (electrode positive) provides the best control of bead shape, maximum penetration and has a tendency to produce less weld porosity. DCSP (electrode negative) is used on occasions where deep penetration is not required; and on hard surfacing applications where higher melt-off rate is desired with less base metal dilution.

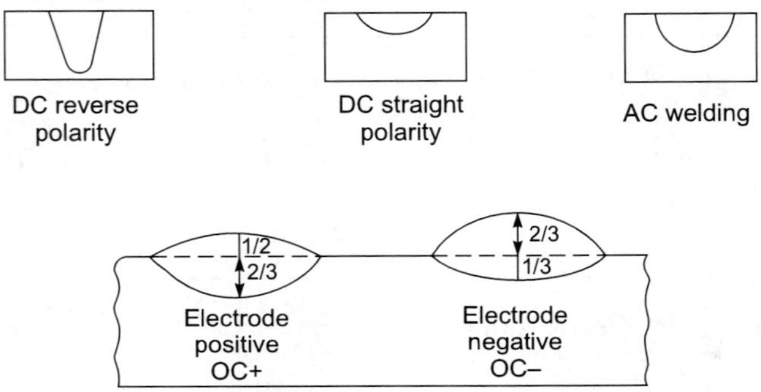

Figure 9.11 Comparison of penetration contours

Electrode Angle (Figure 9.12)

Since the angle between the electrode and the plate determines the point of application and direction of the arc force, it has a profound effect on both penetration and undercut. The first figure shows the effect on horizontal/vertical fillet welds and the second figure compares the effect obtained with a vertical arc with those obtained with leading and trailing arcs. The effect on undercutting can be particularly marked.

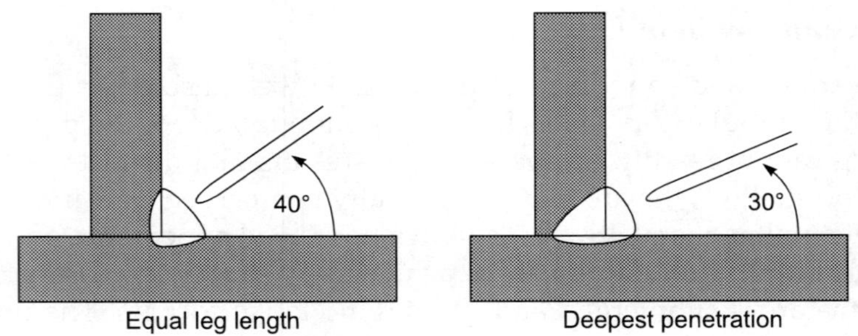

Figure 9.12 (a) Effect of electrode angle on HV fillet welds

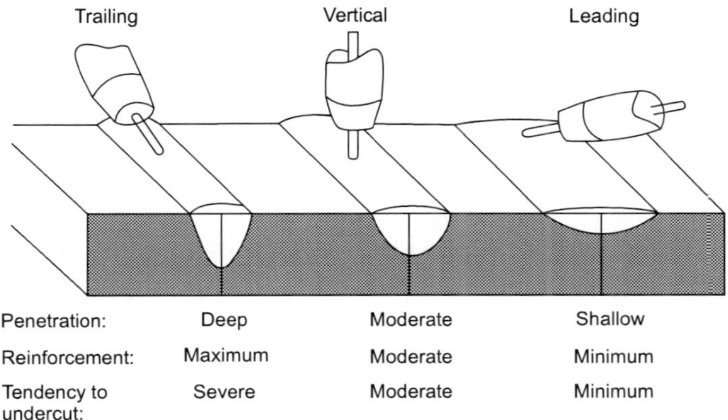

Figure 9.12 (b) Effect of electrode angle on butt welds

Flux Depth

The depth of the flux, or flux burden, is often ignored and the powder heaped around the wire until the arc is completely covered. If optimum results are to be obtained the flux depth should be just sufficient to cover the arc, although the point where the electrode enters the flux bed, light reflected from the arc should be just visible. Too shallow a flux bed gives flash-through and can cause porosity because of inadequate metallurgical protection of the molten metal. Too deep a flux bed gives a poor bead appearance and can lead to spillage on circumferential welds. On deep preparations in thick plate, it is particularly important to avoid excessive flux depth otherwise, the weld bead shape and slag removal can be unsatisfactory.

Penetration

Deep penetration is obtained by this process. Narrow welding groove can be used. Also because of this, small addition of filler material is required. Approximately two volumes of base metal are fused for each volume of filler metal added.

To get good quality welds, the weld area must be free from scale, rust, oil, grease and other gas involving substances. The welding zone should be free from moisture. Current as high as 1200 A can be used with a single electrode. Current densities commonly range from 20,000 to 70,000 A/inch^2. The preparation of face to be welded by this method is very similar to the edge preparation for other fusion welding processes. Due to deep penetration, plates of considerable thickness can be welded without bevelling. Also, to prevent the molten metal from running out the opposite side of the weld may have to be backed up adequately. Water cooled copper plates, steel plates or the powdered flux, itself may be used as the backing.

It is found that the penetration depends on the following factors:

1. Arc voltage has no appreciable influence on penetration
2. It is directly proportional to the current

3. Inversely proportional to the electrode size
4. Inversely proportional to the speed of the travel
5. Directly proportional to the spacing of the plates

9.4 JOINT PREPARATION

- Joint preparation depends on plate thickness, type of joint, e.g., circumferential or longitudinal and to some extent, on the standards to which the structure is being made.
- Plates of up to 14 mm thickness can be butt welded without preparation with a gap not exceeding 1 mm or 10% of the plate thickness, whichever is the greater. Thicker plates need preparation if full penetration is to be obtained. Variable fit-up cannot be tolerated.
- A welder using stick electrodes can adjust his technique to cope with varying joint gaps and root faces or varying dimensions. If conditions are set up for a root gap of 0.5 mm and this increases to 2 or 3 mm, burn-through will occur unless an efficient backing strip is used. In such circumstances, a hand-welded root run using MIG or MMA is advisable. All plate edges must be clean and free from rust, oil, millscale, paint, etc. If impurities are present and are melted into the weld, porosity and cracking can easily occur.
- Time spent in minimising such defects by careful joint preparation and thorough inspection prior to welding is time well spent, since cutting out weld defects and rewarding is expensive and time consuming.

9.5 WELD QUALITY

9.5.1 Porosity Problems

Submerged arc deposited weld metal is usually clean and free of injurious porosity because of the excellent protection provided by the blanket of molten slag. When porosity does occur, it may be found on the weld bead surface or beneath a sound surface. Various factors that may cause porosity are the following:

1. Contaminants in the joint
2. Electrode contamination
3. Insufficient flux coverage
4. Contaminants in the flux
5. Entrapped flux at the bottom of the joint
6. Segregation of constituents in the weld metal
7. Excessive travel speed
8. Slag residue from tack welds made with covered

Electrodes

As with other welding processes, the base metal and electrode must be clean and dry. High travel speeds and associated fast weld metal solidification do not provide time for gas to escape from the molten weld metal. The travel speed can be reduced, but other solutions should be reinvestigated to avoid higher welding costs. Porosity from covered electrode tack welds can be avoided by using electrodes that do not leave a porosity-causing residue. Recommended tack weld electrodes are E6010, E6011, E7016, and E7018.

9.5.2 Cracking Problems

Cracking of welds in steel is usually associated with liquid metal cracking (center bead cracking). This cause may be traced to the joint geometry, welding variables, or stresses at the point where the weld metal solidifies. This problem can occur in both butt welds and in fillet welds, including grooves and fillet welds simultaneously welded from two sides. One solution to this problem is to keep the depth of the weld bead less than or equal to the width of the face of the weld. Weld bead dimensions may best be measured by sectioning and etching a sample weld. To correct the problem, the welding variables or the joint geometry must be changed. To decrease the depth of penetration compared to the width of the face of the joint, the welding travel speed as well as the welding current can be reduced. Cracking in the weld metal or the heat affected zone may be caused by diffusible hydrogen in the weld metal. Hydrogen may enter the molten weld pool from the following sources: flux, grease or dirt on the electrode or base metal, and hydrogen in the electrode or base metal. Cracking due to diffusible hydrogen in the weld metal is usually associated with low-alloy steels and with increasing tensile and yield strengths. It sometimes can occur in carbon steels. There is always some hydrogen present in deposited weld metal, but it must be limited to relatively small amounts. As tensile strength increases, the amount of diffusible hydrogen that can be tolerated in the deposited weld decreases. Cracking due to excessive hydrogen in the weld is called delayed cracking, it usually occurs several hours, up to approximately 72 hours, after the weld has cooled to ambient temperature. Hydrogen will diffuse out of the base metal at elevated temperatures [above approximately 200°F (93°C)] without resulting in cracking. It is at ambient temperatures that hydrogen accumulated at small defects in the weld metal or base metal results in cracking. To keep the hydrogen content of the weld metal low:

1. Remove moisture from the flux by baking in an oven (follow the manufacturer's recommendations).
2. Remove oil, grease, or dirt from the electrode and base material.
3. Increase the work temperature to allow more hydrogen to escape during the welding operation. This may be done by continuing the "preheat" until the seam is completely welded, or by post-heating the weld joint for several hours before letting it cool to ambient temperature.

9.6 APPLICATIONS

General Process Application

Submerged arc welding (SAW) is used in a wide range of industrial applications. High weld quality, high deposition rates, deep penetration, and adaptability to automatic operation make the process suitable for fabrication of large weldments. It is used extensively in pressure vessel fabrication, ship and barge building, railroad car fabrication, pipe manufacturing, and the fabrication of structural members where long welds are required. Automatic SAW installations manufacture mass produced assemblies joined with repetitive short welds. The process is used to weld materials ranging from 1.5 mm sheet to thick, heavy weldments. Submerged arc welding is not suitable for all metals and alloys. It is widely used on carbon steels, low-alloy structural steels, and stainless steels. It joins some high-strength structural steels, high-carbon steels, and nickel alloys. However, better joint properties are obtained with these metals by using a process with lower heat input to the base metal, such as gas metal arc welding. Submerged arc welding is used to weld butt joints in the flat position, fillet welds in the flat and horizontal positions, and for surfacing in the flat position. With special tooling and fixturing, lap and butt joints can be welded in the horizontal position.

1. Welds made by the submerged arc welding process have high strength and ductility with low hydrogen or nitrogen contents. This process is suitable for welding low-alloy, high-tensile steels as well as the mild, low-carbon steel.
2. The submerged arc process is also capable of joining medium carbon steel, heat-resistant steel, corrosion-resistant steels, and many of the high-strength steels.
3. Also, the process is adaptable to nickel, monel and other non-ferrous metals.
4. It is used in the fabrication of pipes, penstocks, boiler pressure vessels, railroad tank cars and structural shapes.
5. The submerged arc process lends itself to the production of high quality welds with uniform weld surface. The elimination of fumes, smoke and any visible arc column add to its ease of operation and efficiency, which in turn, encourages its application in the welding industry.
6. A major advantage of the submerged arc welding process is that joint designs can be kept simple, yet the process can ensure adequate penetration, even in thicker cross sections.
7. Submerged arc welding is widely used in steel fabrication for structural shapes (e.g., welded I-beams); longitudinal and circumferential seams for large diameter pipes, tank, and pressure vessels; and welded components for heavy machinery. In these kinds of applications, steel plates of 25 mm thickness and heavier are routinely welded by this process. Low-carbon, low-alloy, and stainless steels can be readily welded by SAW; but not high-carbon steels, tool steels, and most non-ferrous metals. Because of the gravity feed of the granular flux, the parts must always be in a horizontal orientation, and a back-up plate is often required beneath the joint during the welding operation.

8. It is mainly used in shipbuilding, heavy fabrication, circumferential welds (e.g., rotated vessels), longitudinal seams (e.g., for pipelines), cladding applications.
9. Strip cladding: Although most applications of submerged arc process make use of single or multi-wire systems using round wires, electrode in the form of a strip are often used for cladding purposes. Strips are usually 0.5 mm thick, the commonest strip width being 60 mm, but wider strips, e.g., 100 mm, can be used without any loss of quality. The big advantage of strip cladding is that the penetration is low, particularly with DC electrode negative polarity, and deposition rate is relatively high. Modern fluxes designed for strip cladding have greater current tolerance than earlier types and use of currents up to 1200 A with austenite polarity. Inconel can also be deposited and, provided the flux is low silica type and the inconel strip used contains 2% to 3% Nb, good quality crack-free deposits can be obtained. Monel, aluminum, bronze, nickel and 13% Cr strips have also been successfully used as strip cladding electrodes. Good electrical contact between the strip and feed nozzle is essential.

9.7 ADVANTAGES AND LIMITATIONS

The advantages of submerged arc welding are:

1. Since the radiance of the arc is blanketed by the loose flux, there is no need of a protective welding hood, there is no spatter and only a very minimal amount of fumes escape from under the blanket.
2. High-welding currents, quite commonly in the 3000 to 1600 A range, are used. These high currents, combined with fast travel speeds, make SAW a high deposition process that is especially suitable for applications that require a series of repetitious welds.
3. Some setups allow two or more electrodes to be fed simultaneously into the joint, further increasing the deposition rate and speed.
4. Smooth, uniform finished weld with no spatter.
5. Minimal need for protective clothing as there is no arc flash.
6. High utilization of electrode wire.
7. Easy automation for high-operator factor.
8. Normally, no involvement of manipulative skills.
9. Thicker plates can be welded.
10. Efficiency of the process is high.
11. Less probability of spark, smoke, flash or spatters.
12. No wastage of electrode in this process.

The disadvantages of submerged arc welding are:

1. The flux must be deposited and collected for every welding pass. This requires additional equipment and handling.

2. Also, because of the loose flux, the process is limited to the flat and horizontal positions.
3. The equipment for SAW is commonly quite bulky which limits its nobility, and although the process works well on thick materials, usually is not satisfactory for thin gage material.
4. The amperages commonly used may cause excessive heat build-up in the base metal that may result in distortion or brittleness.
5. Large HAZ which affects the mechanical and metallurgical properties of parent metal.
6. Edge preparation is must.
7. Cast iron, Al alloys, Mg alloys, Pb and Zn cannot be welded by the process.
8. If backing plate is not used, the weld may burn off.
9. Flux cannot be reused.
10. Skilled operator is required.

Worked Example

Example 1: A submerged arc welding operation takes place on 10 mm thick stainless steel, producing a butt weld as shown in Figure 9.13. The weld geometry can be approximated as a trapezoid with 15 mm and 10 mm as the top and bottom dimensions, respectively. If the voltage provided is 40 V at 400 A, estimate the welding speed if a stainless steel filler wire is used.

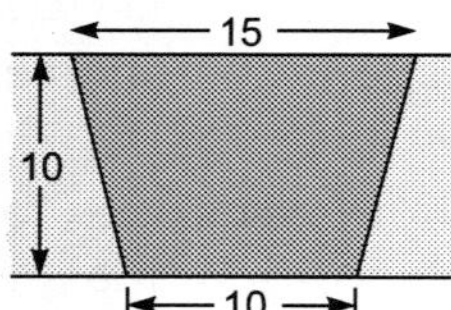

Figure 9.13 Cross section of weld

Solution: Area of trapezoid $= 10 \times 10 + 2 \times \frac{1}{2} \times 2.5 \times 10 = 125 \text{ mm}^2$

Assume efficiency of 90 % for submerged arc welding, specific energy required to melt the metal is 9.4 J/mm^3

$$\text{Heat input} = VI\left(\text{in}\frac{\text{J}}{\text{sec-watts}}\right) \times \eta = 40 \times 400 \times 0.90 \tag{1}$$

$$\text{Heat input} = \text{Welding speed} \times \text{area} \times \text{specific energy}$$

$$\text{Heat input} = v \times 125 \text{ mm}^2 \times \frac{9.4 \text{ J}}{\text{mm}^3} \tag{2}$$

Equating (1) and (2),
We get welding speed = 12.2 mm/s

Review Questions

1. Explain the principle of operation of SAW.
2. Explain the equipment used in SAW.
3. Explain differences between MIG and SAW.
4. Give brief descriptions of power source for SAW.
5. What are the functions of flux in SAW?
6. What are the advantages of wire feed mechanism in SAW?
7. Discuss the important advantages of bonded flux in SAW.
8. What are the disadvantages of SAW?
9. Write a few applications of SAW.

10 CHAPTER

Electroslag Welding

10.0 INTRODUCTION

Electroslag welding (ESW) was discovered in Russia (Ukraine). In this welding process fusion of base metal as well as filler metal is achieved from the heat generated by the electrical resistance of the molten slag. The weld pool is shielded by this slag, which moves along the full cross section of the joint as welding progresses. The process is initiated by an arc that heats up the granulated flux and melts it to form the slag. The arc then gets extinguished because the molten slag provides the conducting path to the current. The slag remains in the molten state by the heat generated by its resistance to electric current passing through it between electrode and the base metals.

Thus, the electrode remains dipped in the molten slag and gets melted and deposited as the welding progresses. Electroslag welding is a vertical position machine welding process and continues to completion. Since during welding no arc is involved, the welding action is quiet and spatters free. It is essentially a single-pass welding process, irrespective of plate thickness. Extremely high deposition rates can be achieved with ESW and thus, very thick sections can be welded in a single pass. A high quality weld deposit results from the nature of the melting and solidification during welding. Because of even distribution of heat across the thickness, no angular distortion of plates results. The molten slag bath over the weld pool acts both as the heat source as well as a shielding medium.

The molten slag bath over the weld pool acts both as the heat source as well as a shielding medium. The molten slag with a temperature up to 2000°C washes the joint edges, melting into them to produce a weld with a dilution as high as 50%. The melted electrode and base metals form a molten pool beneath the molten slag and slowly solidify to from the weld. A progressive solidification takes place from bottom upwards, and there is always a molten metal pool over the solidifying weld metal. There are two types of electroslag welding. One uses a wire electrode with a non-consumable guide-cum-contact tube and the other uses a electrode with a consumable guide extending down to the molten slag.

10.1 ELECTRODE WITH NON-CONSUMABLE GUIDE

In this method, one or more electrodes are fed to the molten slag pool depending upon the thickness of the metals being joined (Figure 10.1). Water cooled copper shoes are normally used on both sides of the plates being joined to contain the molten metal and the molten slag. These shoes are attached to the welding machine and move vertically up with the machine. The speed of this vertical movement determines the welding speed and it is consistent with the electrode melting and deposition rate. The contact surfaces of the shoes are shaped concave, which provides the required amount of reinforcement on the weld deposit. As the shoes move up, the solidified weld deposits get exposed. The exposed surfaces are covered with a fine layer of slag. Hence, small amounts of flux must be added at intervals to compensate for consumption of slag. In this method, the metal deposition is from both electrode and the consumable guide. The consumable guide provides approximately 5 to 15% of the filler metal. As the welding proceeds and the molten slag pool rises, the consumable guide melts and becomes part of the weld deposit. The electrode wire is directed to the bottom of the joint by the guide tube extending the entire length of the joint.

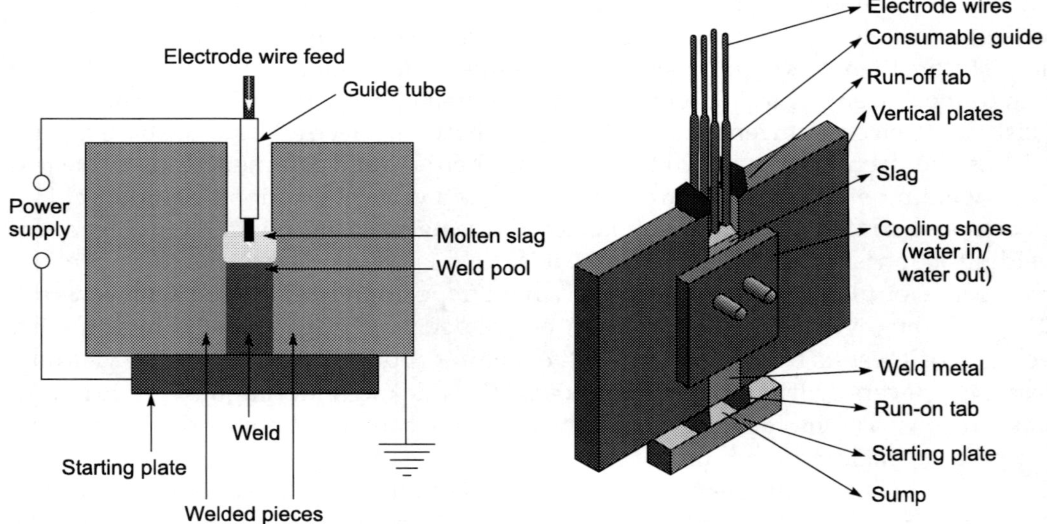

Figure 10.1 Non-consumable guide method of electroslag welding

The welding machine, in this method remains stationary and non-sliding shoes are used. On short length of welds, the entire joint height is covered by the shoes that means, the shoes are of same length as the joint. For longer joint, several sets of shoes are used. With the progress of the weld as the metal solidifies, the shoes from below are removed and placed above the top shoes. This process is repeated till completion of the weld.

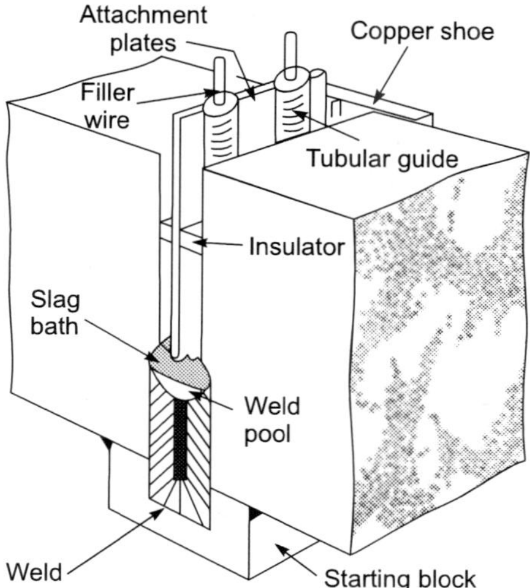

Figure 10.2 Electroslag welding using a plate crawler & electrode with consumable guide

This method can be effectively used to weld plates ranging in thickness from approximately 19 to 460 mm. As the welding current is carried by the tube, it needs to be insulated. The required insulation may be provided by one of the following means:

- Flux coating of the outer surface guide of the tube
- Doughnut-shaped insulators
- Fiberglass sleeves
- Insulating tape

This method can be used to weld sections of virtually unlimited thickness. Summary of these two processes is given in Table 10.1.

TABLE 10.1 Comparison between consumable and non-consumable guide tubes

Non-Consumable Guide Tube ESW (Figure 10.1)	Consumable Guide ESW (Figure 10.2)
Conventional electroslag welding in which a non-consumable guide tube is used to direct the electrode into the molten slag bath. Current is conducted to the electrode wire by the guide tube. Moreover, the welding head moves progressively upwards as the weld is deposited at a rate consistent with the electrode deposition rate. Thicknesses from 18 to 460 mm are most commonly welded by the conventional process.	In consumable guide electroslag welding a consumable guide tube is used which means, both the electrode wire and its guiding member melt and supply the filler metal. Welding current is carried by the guide tube. The electrode wire melts inside the molten flux whereas the guide tube melts off just above the surface of the slag bath.

10.2 ELECTROSLAG WELDING EQUIPMENT (FIGURE 10.3)

Power Supply

A constant potential DC power source commonly used is transformer rectifier set. Motor generator sets are much more costly than transformer rectifiers of the capacity required for electroslag welding. Current rating for electroslag welding varies from 750 to 1000 A and load voltages generally range from 30 to 55 V.

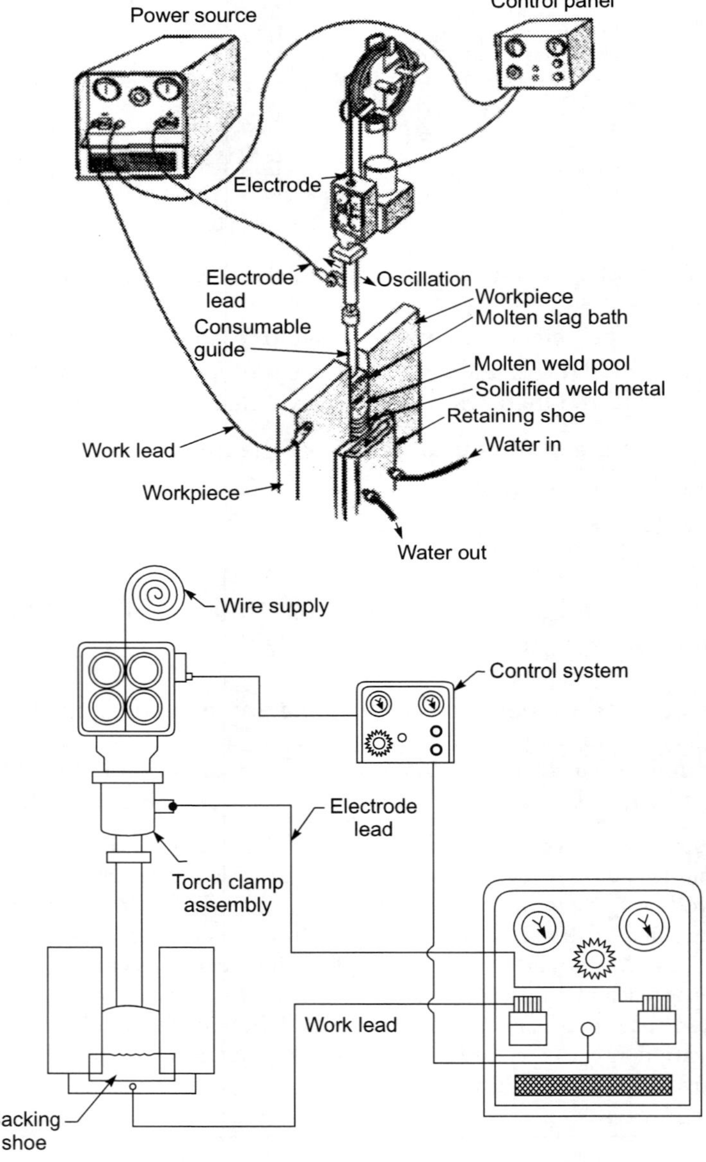

Figure 10.3 Equipment setup

Welding Head: In case of conventional electroslag welding process, a motorized vertical traversing welding head incorporates an oscillating traverse for the wire guide tubes, the water cooled copper shoes and the wire feeding unit. The non-consumable electrode guide tube is made of beryllium copper alloy. It energizes the electrode as well as guides the same from the wire feed rolls into the molten slag path. Since the guide is positioned close to the molten slag, its end will deteriorate. The end may then be cut off and the guide used where a shorter one will suffice.

Electrode Wire: Both solid and tubular electrode wires can be used as the filler metal in both systems of electroslag welding. It is customary to use a standard wire size of 3 mm. The wire is normally copper coated. Tubular electrode has advantages in special situations where alloying elements need to be added to produce a desired composition of weld deposit. Tubular electrodes, however, result in excessive buildup of flux in the pool and good fusion is impaired.

Fluxes: The molten flux is the slag that gives the electroslag process its name. An electroslag welding flux must:

1. be conductive,
2. possess medium viscosity,
3. have a melting range lower than that of the weld metal,
4. be metallurgically compatible with tile alloy being welded, and
5. be easily removed from the surface of the weld when solidified.

10.3 OPERATING VARIABLES

In electroslag welding, it is essential to understand the effect of each variable since they differ from those with conventional arc welding processes.

Form Factor

In electroslag welding, solidification of the molten metal takes place at a slow rate. The solidification progresses from the bottom of the joint upwards. The angle at which the grains meet in the center is determined by the shape of the molten weld pool. Weld pool shape can be expressed by a "form factor". The form factor is the ratio of the weld pool width to its maximum depth. Width is the root opening plus the total penetration into the base metal. Depth is the distance from the top of the molten weld pool to the lowest level of the liquid-solid interface. A high form factor weld deposit tends to solidify with the grains meeting at an acute angle, whereas in case of low form factor, the grains meet at an obtuse angle. The cracking resistance is high in case of acute angles and low in case of obtuse angles. Therefore, maximum resistance to cracking is obtained with a high form factor. However, form factor alone does not control cracking. The base metal composition (especially the carbon content), the tiller metal composition and joint restraint have a significant effect on crack formation. The shape of the molten pool and hence, the form factor depends on the other welding variables.

Welding Current

Increasing welding current implies increased heat input and results in increased depth of molten weld pool. Thus, the effect of increasing current is to decrease the form factor and thereby, lowering the resistance to cracking. Generally, currents of 500 to 700 A are used with 3.2 mm electrode. For metals more prone to cracks, current below 500 A is used.

Welding Voltage

In ESW, welding voltage is an extremely important variable. It has a major effect on the depth of fusion into the base metal and on the stable operation of the process. Increasing the voltage increases the sideways depth of fusion in the base metal, resulting in increase in the width of the weld. Therefore, it also increases the form factor and thereby, increases the resistance to cracking. However, for stable operation voltages within 32 to 55 V per electrode are used. Too high a voltage may lead to slag spatter and arcing on top of the slag bath. If the voltage is too low, short-circuiting or arcing to the weld pool may occur. However, with thicker sections, higher voltages are used.

Electrode Extension

Electrode extension is the distance between the slag bath surfaces to the end of the guide tube. Using constant voltage power and constant electrode feed speed, increasing the electrode extension will increase the resistance. This causes the power supply to reduce its current output, which slightly increases the form factor. Electrode extensions of 50 to 75 mm are generally used. Less than 50 mm electrode extensions result in overheating of the guide tube, while extensions higher than 75 mm cause overheating of the electrode because of increased electrical resistance. Hence, with long electrode extensions, the electrode may melt at the slag bath surface instead of inside the bath. This will lead to improper slag bath heating and resulting instability.

Slag Bath Depth

A minimum slag bath depth is necessary to ensure electrode immersion and melting beneath the surface. A shallow bath will cause slag spitting and surface arcing, whereas too large a bath depth provides a larger area for heat transfer into the retaining shoes and base metal. This reduces the overall temperature of the slag bath, thereby reducing the weld width and hence, the form factor. Higher slag bath depth also leads to poor circulation of the cooler slag at the bottom which may tend to solidify on the surface of the base metal, resulting in slag inclusion. Optimum bath depth is 40 mm, but it can also be as low as 25 mm or as high as 50 mm, without much significant effect.

Root Opening

Root opening should be enough to provide clearance for the guide tube and its insulation. However, for sufficient slag bath size and good slag circulation, a minimum root opening is required. Increased root opening does not affect the weld pool depth.

However, it increases the weld width and hence, the form factor. Excessive root opening will require extra amount of filler metal, which may become uneconomical. Also, it may give rise to lack of edge fusion. Depending on base metal thickness, number of electrodes and the use of electrode oscillation, generally 20 to 40 mm of root opening is used. As welding progresses up the joint, the parts are drawn together by weld shrinkage. Therefore, for long welds, the root opening at the top of the joint should be approximately 3 to 6 mm more than that at the bottom to allow for this shrinkage. Factors influencing shrinkage allowance include material type, joint thickness and joint length. The use of proper root opening and shrinkage allowance is important for maintaining the dimension of the weldment. However, if the root opening turns out to be incorrect, the same may be rectified within limits by changing the welding parameters. For example, if the initial root opening is too small, the wire feed speed may be reduced, to reduce deposition rate and to increase penetration. If the root opening is too large, the wire feed speed may be increased. Regardless of the type of base metal, the voltage should be increased for wider root openings. For excessive root openings, which are beyond the capacity of the normal number of electrodes, an additional electrode may be added if space is available. Also, electrode oscillation may become necessary to compensate for wide root openings. On the other hand, when the root openings are too small, the joint may fill too fast, causing weld cracks or lack of edge fusion. In case of small root openings, it may further lose up due to weld shrinkage and may lead to stoppage of oscillating guide tube from traversing.

Electrode Oscillation

Oscillation of electrode is generally carried out horizontally across the thickness. The oscillation of electrode distributes the heat and thus, helps to obtain better edge fusion. Oscillation speeds vary between 8 and 40 mm/s, with the speed increasing to match the plate thickness. Generally, the oscillation speeds are based on a traverse time of 3 to 5 5 sec. Increase in oscillation speed reduces the weld width and hence, the form factor. Thus, the oscillation speed must be balanced with other parameters. A dwell period of 2 to 7 sec. is used at each end of the oscillation travel to obtain complete fusion at each end and to overcome the chilling effect of the retaining shoes.

10.4 MATERIALS

- Carbon steel: Many types of carbon steels can be electroslag welded in production, such as AISI 1020, AISI 1045, AISM A36, AISM A441, AISM A515. They generally can be welded without post-weld heat treatment. In addition to carbon steels, other steels are successfully electroslag welded, they include AISI 4130, AISI 8620, ingot iron and AISM A514. Most of these steels require special electrodes and post-weld heat treatment to develop required weld or weld heat affected zone properties.
- Low-alloy steels.
- Stainless steels (less common).

10.5 SAFETY

Hot molten metal may spill out of the joint, or leak around the cooling shoe. Poor fit-up of the cooling shoe or misalignment of sumps and runoffs also may open gaps, where leaks can occur. Slag leakage can be especially dangerous because of the large volumes of slag bath associated with the ESW process. For the same reason, special attention should be given to shoe fit-up and alignment, and the cooling shoes should be left in places after completion of the weld until the slag bath has solidified. Water cooled shoes can be much safer than solid shoes, but both types can cause severe burns.

10.6 APPLICATIONS

Electroslag welding is used in shipbuilding, mainly for joining the vertical side shells of two adjacent blocks. Plate thicknesses of 12 to 32 mm are commonly found in side shells and the weld length varies from 12 to 20 m, depending on the size of the ship.

- Thick pressure vessels for the chemical, petroleum, marine, and power generating industries.
- ESW is often used to weld stiffeners in structural box columns and wide flanges.
- Manufacture of large presses and machine tools work with large heavy plates.
- Other machinery applications include kilns, gear blanks, motor frames, press frames, turbine rings, shrink rings, crusher bodies, rebuilding metal mill rolls and rims for road rollers.
- Electroslag welding is used for welding of low-carbon steel, medium-carbon steel, high-strength alloys steel, stainless steel and Ni-based alloys which is employed for welding of plates of thickness ranging from 35 to 300 mm thick.
- Heavy forging and heavy casting can be welded using this process.

10.7 ADVANTAGES

- Much thicker steels can be welded in single pass and more economically. Thicknesses up to 450 mm in plain and alloy steels can be welded without difficulty.
- Electroslag welding can have extremely high deposition rates, but only one single pass is required, no matter how thick the workpiece is. It may have 15 to 25 kg per hour per electrode deposition rate.
- Preheating, even for materials with high hardenability, is normally not required.
- Produces high quality weld deposit.

- Minimum fit-up requirements.
- High-duty cycle and high deposition rate.
- Minimum materials handling, once started it continues to completion.
- Elimination of weld spatter, which results in 100% filler metal deposition efficiency.
- Low flux consumption; 1 kg of flux consumed for each 20 kg of weld metal.
- ESW is the fastest welding process for large and thick materials.
- Unlike SAW or other arc welding processes, there is no angular distortion in ESW because the weld is symmetrical with respect to its axis.
- High welding speed and good stress distribution across the weld.
- Joint preparation is often much simpler than other arc welding processes. No special edge preparation is required on plates to be welded and in most cases, a square head is adequate.
- Residual stresses and distortion produced are low and no angular distortion.
- Flux composition and consumption as compared to submerged arc welding (SAW) is very low.
- During the electroslag process, since no arc exists, no spattering or intense arc flashing occurs.

10.8 LIMITATIONS

- Only carbon and low-alloy steels and some stainless steel can be welded by ESW.
- Once the welding has started, it must continue to completion otherwise, defects are likely to occur.
- ESW cannot generally be used on materials thinner than about 12 mm.
- Electroslag welding is restricted to vertical position welding because of the large pools of the molten metal and slag.
- The heat input is very high and the weld quality can be rather poor, including low toughness caused by the coarse grains in the fusion zone and the heat affected zone.
- In electroslag welding, there is some tendency towards hot cracking and notch sensitivity in the heat affected zone.
- It is difficult to close cylindrical welds. Complex shape joints cannot be welded.
- Electroslag welding tends to produce large grain sizes.
- Submerged arc welding is more economical than electroslag welding for joints below 60 mm.

Review Questions

1. Explain the principle of operation of electroslag welding.
2. Explain the equipment used in electroslag welding.
3. Explain differences between electroslag welding and thermit welding.
4. Give brief descriptions of power source for electroslag welding.
5. What are the functions of flux in electroslag welding?
6. What are the important operating variables in electroslag welding?
7. Discuss the important safety precautions to be taken in electroslag welding.
8. What are the advantages of electroslag welding?
9. What are the disadvantages of electroslag welding?
10. Write a few applications of electroslag welding.

11 CHAPTER

Electrogas Welding

11.0 INTRODUCTION

The first available thick-plate single-pass vertical welding method was electroslag welding. Demand arose immediately for equipment that would apply the process to thinner sections. Then in 1961, laboratory studies with an electroslag welding machine adapted to feed auxiliary gas shielding around a flux cored electrode demonstrated that plate as thin as 13 mm could be satisfactorily welded in the vertical position in a single pass. This technique is called electrogas welding (EGW). Electrogas welds normally exhibit excellent weld metal soundness. Thus, the process offers economic benefits and quality enhancements. The mechanical aspects of electrogas welding are strikingly similar to those of the electroslag process from which it was developed. Applications involving aluminum alloys and stainless steel will not be discussed, although successful examples have been reported. Some of the advantages associated with EGW have resulted in considerable cost savings, particularly in joining thicker materials. Savings have been achieved where components can be joined in the vertical position with a continuous vertical weld. In thicker materials, EGW is often less expensive than the more conventional joining methods, such as submerged arc welding and flux cored arc welding. Even in some applications involving thinner base materials, EGW may result in cost savings because of its efficiency and simple joint preparation.

11.1 COMPARISON BETWEEN ELECTROSLAG AND ELECTROGAS WELDING

The electrogas welding is similar to the electroslag welding but for the following differences:

1. The heat for welding is furnished by a continuous arc between the electrode and the molten pool and not by the electrical resistance of slag as in electroslag welding.
2. The shielding is provided mainly by a gas like argon or carbon dioxide.
3. No flux is fed into the weld zone as in electorslag welding but a coated electrode is used.

4. Only DC can be used whereas both AC and DC can be employed for electroslag welding.

TABLE 11.1 Comparison between electroslag welding and electrogas welding

Electroslag Welding	Electrogas Welding
No electric arc is there during welding.	Heat required for welding is obtained by electric arc.
Weld pool is shielded by molten slag.	An inert gas shields the molten weld pool.
Either AC or DC can be used for electroslag welding.	Only DC can be used for electrogas welding.
Chance of flux entrapment is more.	No flux is used, therefore there is no chance of flux entrapment (inert gas is used).
Energy required for welding is generated by electrical resistance of a molten slag bath, which converts electrical energy into thermal energy.	Heat required for welding purpose is obtained by maintaining a continuous electric arc between the electrode and molten metal weld pool.

11.2 FUNDAMENTALS OF THE PROCESS

Process Description

Electrogas welding is an arc welding process that uses an arc between a continuous filler metal electrode and the weld pool, employing vertical position welding with backing to confine the molten weld metal. The process is used with or without an externally supplied shielding gas and without the application of pressure.

A square groove or single V-groove joint is positioned so that the axis or length of the weld is vertical. There is no repositioning of tie joint once welding has started; welding continues to completion, so that the weld is made in one pass. Electrogas welding is a machine welding process. The nature of the melting and solidification during welding results in a high quality weld deposit. There is little or no angular distortion of the base metal with single-pass welds. The welding action is quiet, with little spatter.

Principles of Operation (Figure 11.1)

The consumable electrode, either solid or flux cored, is fed downwards into a cavity formed by the base metals to be welded and the retaining shoes. A sump (starting tab) is used at the beginning of the weld to allow the process to stabilize before the molten weld metal reaches the work. An arc is initiated between the electrode and the sump. Heat from the arc melts the continuously fed electrode and the groove faces. Melted filler metal and base metal collect in a pool beneath the arc and solidify to form the weld. The electrode may be oscillated horizontally through the joint for uniform distribution of heat and weld metal. As the cavity fills, one or both shoes may move upwards. Although the weld travel is vertical, the weld metal is actually deposited in the flat position at the bottom of the cavity.

Figure 11.1 Principle of operation

11.3 VARIATIONS OF ELECTROGAS WELDING

Solid Electrode

A schematic view of a typical electrogas welding installation using a solid electrode is shown in Figure 11.2. The electrode is fed through a welding gun, called a non-consumable guide. The electrode may be oscillated horizontally to weld thicker materials. Gas shielding, normally carbon dioxide (CO_2) or an argon-carbon dioxide

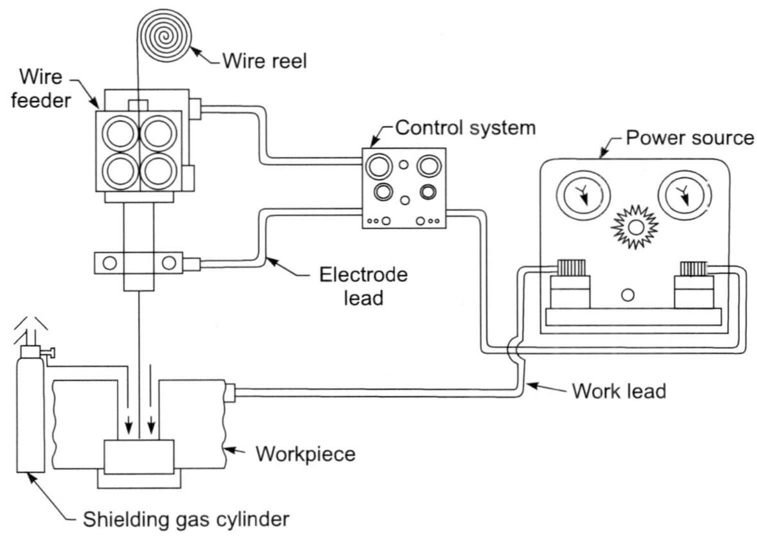

Figure 11.2 Electrogas welding with a solid electrode

($Ar-CO_2$) mixture, is provided to the weld cavity through gas ports, boxes, or nozzles. Water cooled copper retaining shoes are normally used on each side of the joint to retain the molten weld metal. The shoes are usually attached to the welding machine and move vertically as the machine moves. Vertical movement of the welding machine must be consistent with the deposition rate. Vertical movement may be automatic or controlled by the welding operator. Electrogas welding with solid electrodes can be used to weld base metals ranging in thickness from approximately 10 mm to 100 mm. Base metal thicknesses most commonly welded are between 13 mm and 76 mm. Electrode diameters most commonly used are 1.6, 2.0, 2.4, and 3.2 mm.

Flux Cored Electrode

Figure 11.3 illustrates electrogas welding using a self-shielded flux cored electrode. The principles of operation and characteristics are identical to the solid electrode variation, except that no separate gas shielding is needed. The flux cored electrode creates a thin layer of slag between the weld metal and copper shoes to provide a smooth weld surface. Electrogas welding with a flux cored electrode may be done with an external gas shield or a self-shielding electrode. Self-shielded electrodes operate at higher current levels and deposition rates than gas-shielded types. Diameters of flux cored electrodes commonly vary from 1.6 mm to 3.2 mm. The wire (electrode) feeder must be capable of smooth continuous feeding of small diameter wires at high speeds and larger diameter wires at slower speeds.

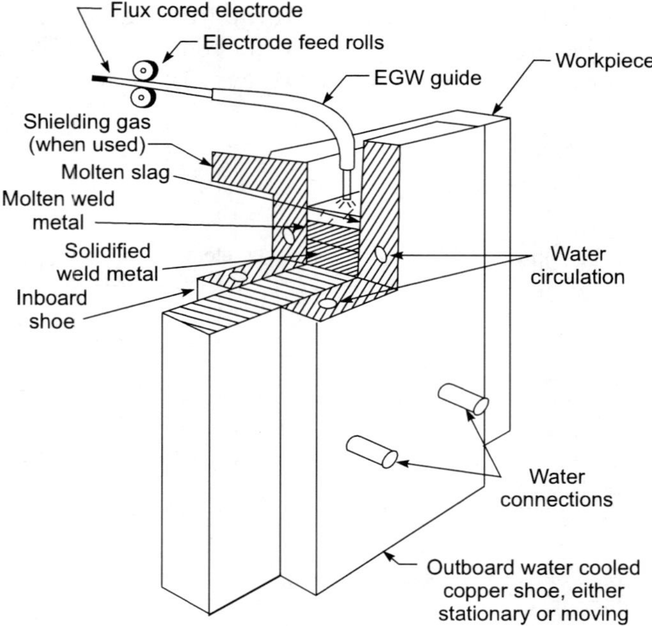

Figure 11.3 Electrogas welding with a self-shielded flux cored electrode

Consumable Guide Process

EGW with a consumable guide is similar to consumable guide electroslag welding. This variation of EGW is primarily used for short weldments in shipbuilding, and in column and beam fabrication. Consumable guide EGW uses relatively simple equipment, as shown in Figure 11.4. The main difference is that none of the equipment moves vertically during consumable guide welding. Instead, the electrode is fed through a hollow consumable guide tube which extends to about 25 mm from the bottom of the joint. As the weld progresses vertically, the electrode melts back to the guide tube. Initially, the wire electrode penetrates about an inch beyond the end of the guide tube. Then a steady-state relationship develops between melting of the end of guide tube and the electrode wire, as shown in Figure 11.5.

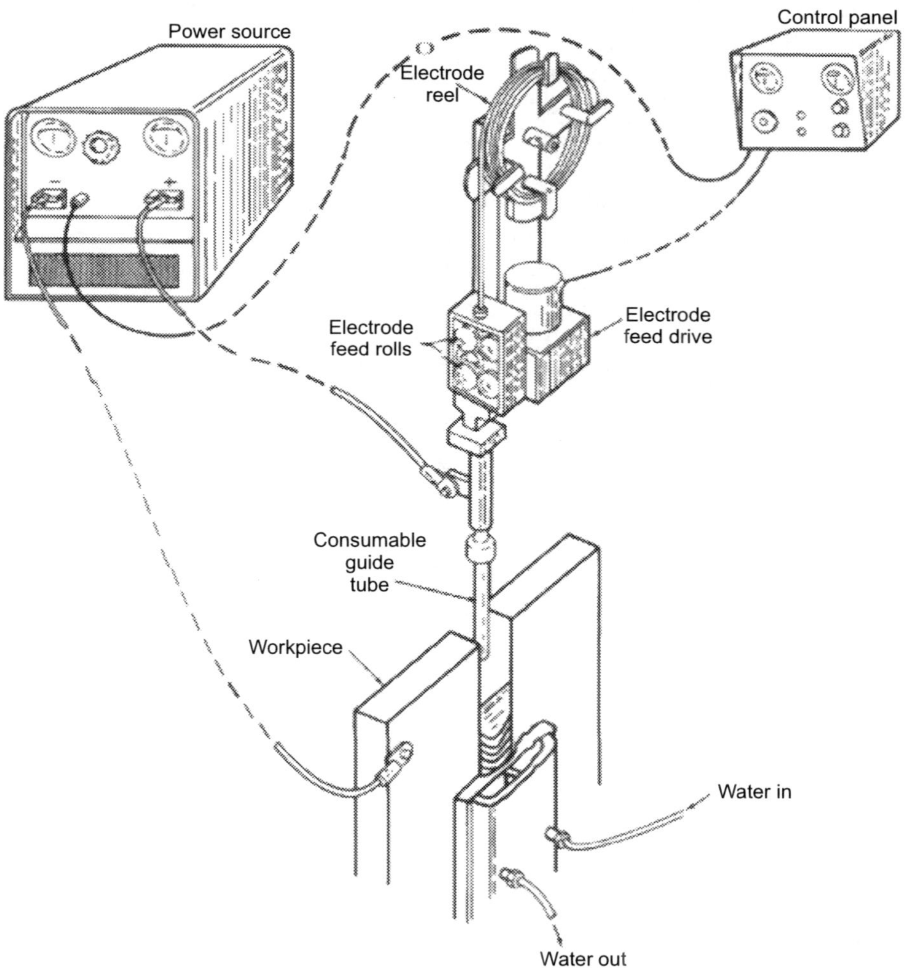

Figure 11.4 Consumable guide electrogas welding equipment

The consumable guide provides approximately 5 to 10 percent of the deposited metal, with the balance supplied from the wire electrode. Arc initiation begins in the same general fashion as in the conventional process. A collar or clamp device, usually made of copper, holds the guide tube in position in the consumable guide process. The consumable guide is selected so that its chemical composition is compatible with the base material. Guide tubes are usually 13 to 16 mm O.D. and 3 to 5 mm I.D. Guide tubes are available in various lengths and for some applications, may be welded together to accommodate longer joints. Oscillation with a consumable guide tube over 1 m long becomes increasingly more difficult as weld length increases because the bottom end of the tube starts a "whipping" motion and may distort due to heating. This tendency can be eliminated by adding one or more additional electrodes and thereby, reducing or eliminating the oscillation. Circular insulators may be used approximately every 300 to 450 mm along the length of the tube to prevent shorting of the tube to the wall of the groove. In consumable guide EGW, the retaining shoes are independent of the welding equipment, as indicated in Figure 11.3. The shoes may be held in place with wedges placed between the shoes and strong backs welded to the workpiece. On short welds, shoes may be of same length as the joint. Several sets of shoes may be stacked one on top of the other for longer joints. One complication in consumable guide EGW is the buildup of slag on top of the molten weld pool. Excessive slag buildup can lead to poor penetration or non-fusion. Slag buildup can be overcome by either providing weep holes in the retaining shoes or by shimming the retaining shoes intermittently along the length of the weld. Either method will permit the excess slag to escape from the weld groove. Consumable guide tubes may sometimes be used for complex shapes. The guides may be performed to fit along the length of a curved joint.

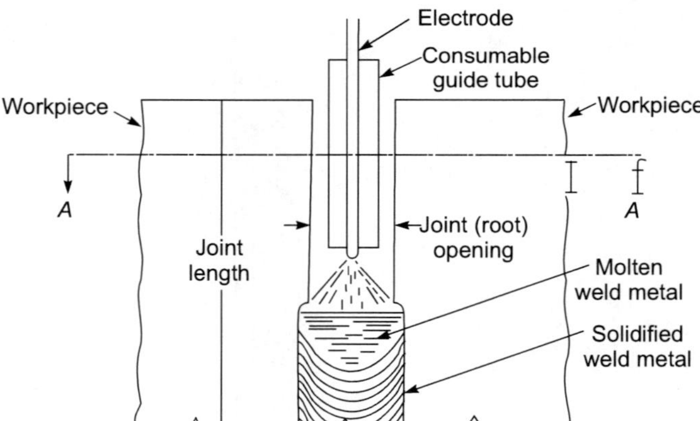

Figure 11.5 Schematic diagram of consumable guide electrogas welding process

11.4 EQUIPMENT

The basic mechanical equipment for electrogas welding consists of a direct current power supply, a device for feeding the electrode, shoes for retaining molten metal, an electrode guide, a mechanism for oscillating the electrode guide, and equipment needed for supplying shielding gas, when used. In a typical electrogas welding system, the essential components (with the exception of the power supply) are incorporated in an assembly that moves vertically as welding progresses.

Power Supply

Direct current electrode positive (reverse polarity) is normally used for EGW, with the power supply being either constant voltage or constant current. The power source should be capable of delivering the required current without interruption during the welding of a seam that may be of considerable length. Power sources used for electrogas welding usually have capacities of 750 to 1000 A at 30 to 55 V and 100% duty cycle. Direct current is usually supplied by transformer-rectifier power sources, although motor-driven and engine-driven generators may be used.

Electrode Feeder

The wire feeder for the electrode is of the push-type, such as used with automatic GMAW or FCAW. The wire feeder is normally mounted as an integral part of the vertical moving welding machine. Wire feed speeds may vary up to 230 mm/s. The wire feed system may include a wire straightener to eliminate the cast and helix set in the electrode and thus minimize electrode wander at the joint. Since the electrode extension in EGW is relatively long, 40 mm or more, a straight length of electrode projecting from the guide is necessary to assure positive arc location in the joint.

Electrode Guide

Electrode guides, in many respects, are similar to the welding guns used for semi-automatic GMAW or for FCAW. The guide may have a shielding gas outlet to deliver gas around the protruding electrode (usually supplementary to the shielding gas supplied by ports in the shoes), and it may or may not be water cooled. The electrode guide (or at least part of it, in the case of a curved guide) rides inside the gap directly above the molten weld pool. It must be narrow enough to be contained in the gap with clearance to permit horizontal oscillation, if used, between the two shoes. For this reason, the width of an electrode guide is often limited to 10 mm, so as to accommodate the 18 mm minimum gaps.

Electrode Guide Oscillators

With base metals 30 mm to 100 mm thick, it is necessary to move the arc back and forth between the shoes and over the molten weld pool to achieve uniform metal deposition and ensure fusion to both workpieces. Horizontal oscillation of the arc is usually not needed for joints in base metal less than 30 mm thick, but is sometimes used with thinner base metals to minimize penetration into the work and to improve weld properties. This horizontal motion is accomplished by a system that oscillates the electrode guide and provides adjustable dwell times at either end of the oscillation. The arc is normally oscillated to about 7 mm, from the retaining shoe on each side of the joint. Dwell time may be adjusted to produce good fusion at the base metal surface.

Retaining Shoes

Retaining shoes, also called dams, are pressed against each side of the gap between the base metals to be welded to retain (dam) the molten weld metal in the groove. One or both shoes may move upwards as welding progresses. In some weldments, a steel backing bar, which fuses to the weld, may replace the outboard shoe. Non-fusing ceramic backups have also been used. Retaining shoes are usually water cooled; moving shoes must be water cooled. The shoes are usually grooved to develop the desired weld reinforcement. Sliding shoes may or may not contain gas ports for supplying shielding gas directly into the cavity formed by the shoes and the weld groove. When gas ports are not used in the shoes, a "gas box" arrangement may be mounted on the shoes so as to surround the electrode and the welding arc with shielding gas. Gas ports in the shoes and "gas boxes" are not required when using self-shielded flux cored electrodes.

Controls

With the exception of the vertical travel control, EGW controls are primarily adaptations of the devices used with GMAW and FCAW. Vertical travel controls, electrical, optical, or manual, maintain a given electrode extension, with the top of the movable shoe a specific distance above the molten weld pool. On equipment using a self-shielded flux cored electrode, starting procedures must be carefully controlled to minimize porosity in the starting sump area. Welds are usually started at lower wire feed speeds and high voltages. These lower start settings must be maintained until the arc stabilizes and the starting area is heated. After a preset time, the equipment automatically changes the feed speed and voltage to the operating welding speed and voltage. Automatic, rather than manual changing, reduces the chance that operator error will cause porosity.

11.5 CONSUMABLES

Electrodes

Both flux cored and solid electrodes are used with the electrogas welding process. Flux cored EGW electrodes contain a lower percentage of slagging compounds than typical electrodes used in the FCAW process. These special electrodes allow a thin slag layer to form between the shoes and the weld to provide a smooth weld surface. Flux cored electrodes arc supplied in sizes from 1.6 to 3.2 mm. Only flux cored electrodes specifically designed for EGW should be used. Solid electrodes are generally identical to those used for GMAW. They are supplied in sizes from 0.8 to 3.2 mm. Both flux cored and solid electrodes are available in various chemical compositions designed to introduce the necessary alloying elements to achieve strength, impact properties, or appropriate combinations of these and other properties in the deposited weld metal. For steel welding, these additional elements include manganese, silicon, and nickel. The required properties are normally obtained in the as-welded condition.

Shielding

Self-shielded flux cored electrodes for EGW contain core materials that shield the molten weld metal. Other flux cored electrodes require an additional external shielding gas (normally carbon dioxide, but mixtures of argon and carbon dioxide are also used). Recommended gas flow rates range from 30 to 140 CFH (14 to 66 L/min). Gas flow rates depend upon the equipment design, and manufacturer's recommendations should be followed. Mixtures of argon and carbon dioxide are normally used for welding steel with solid electrodes and may be used with flux cored electrodes.

11.6 WELD PROPERTIES

Post-weld Heat Treatment

Most applications of electrogas welding, particularly the welding of field erected structures, require no post-weld heat treatments. Stress relief heat treatment of electrogas weldments typically results in a slight drop in the yield and tensile strength and a slight improvement in weld and heat affected zone notch toughness. Electrogas welded quenched and tempered steels will exhibit lower strength in the weld metal and the heat affected zone. Matching strength may be restored by a full heat treatment.

Residual Stresses and Distortion

The residual stress pattern is just the reverse of the residual stresses that are likely to occur with multipass arc welding processes. Angular distortion in the horizontal plane is virtually non-existent in single-pass electrogas welds. This is due to the symmetry of most electrogas joint designs about the mid-thickness of the base material, which results in uniform shrinkage in the thickness direction. Factors influencing shrinkage allowance include material type, joint thickness, joint length, and degree of restraint of the parts being joined. Angular distortion can occur during multiple-pass applications of electrogas welding. It may be compensated by presetting the parts.

Mechanical Properties

EGW involves high base metal penetration and therefore, melted base metal may contribute up to 35% of the total weld metal. Thus, weld composition and mechanical properties can vary substantially with different types of steel and even with different heats of the same steel grade. Mechanical properties also vary with welding conditions, such as current and voltage settings, joint design, type of backup, and cooling rate.

One problem encountered in electrogas welding has been low notch toughness properties of welds and heat affected zones in the as-welded condition. By using suitably alloyed electrodes and controlled welding conditions, the notch toughness properties of the weld metal can be equal to or better than those of the base metal for most structural and pressure vessel steels.

11.7 PROCESS VARIABLES

Process variables for electrogas welding include arc voltage, current, electrode feed speed, electrode extension, electrode oscillation and dwell, and joint opening. The proper selection of these variables affects the operation and economics of the process and the resultant weld quality.

Welding Voltage (Figure 11.6)

The welding voltage is the major variable affecting weld width and base metal melting in electrogas welding. Welding voltage of 30 to 35 V are normally used. Increasing the voltage increases the sidewall penetration and width of the weld as illustrated in the figure. For thicker base metals or for higher deposition rates, the voltage should be increased. However, if excessive high voltage is used, then electrode may arc to the joint sidewalls above the molten weld pool, causing unstable operation.

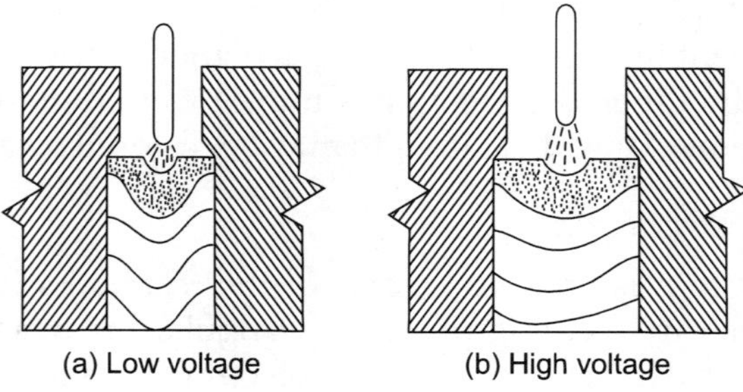

Figure 11.6 Effect of welding voltage on the shape of molten weld pool

Welding Current and Electrode Feed Speed (Figure 11.7)

Welding current and electrode feed speed are proportional for a given electrode diameter and extension. Increasing the electrode feed speed increases the deposition rate, welding current, and travel speed. For a given set of conditions, increasing the current decreases the sidewall penetration and width of the weld, as shown in Figure 11.7. Low current levels result in slow travel speeds and wide welds. Excessive welding current may cause a severe reduction in weld width and sidewall penetration. Excessive welding current also causes a low form factor which contributes to center-line weld crack susceptibility.

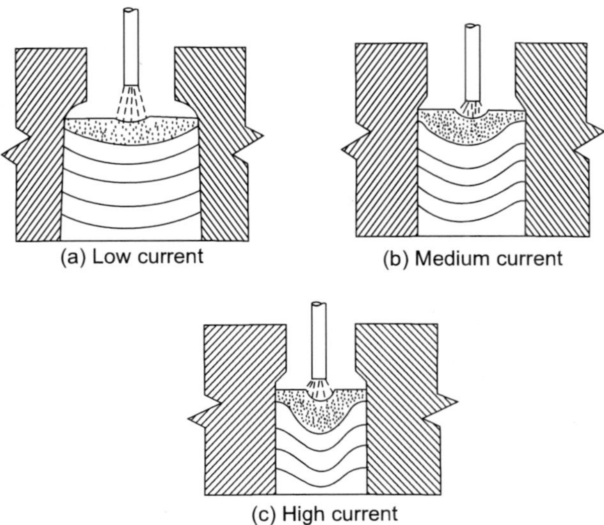

Figure 11.7 Effect of weld current on molten weld pool

Form Factor (Figure 11.8)

The resistance of the weld to centerline cracking is greatly influenced by the manner in which it solidifies. Heat is removed from the molten weld metal by the base metal and retaining shoes. Solidification begins at these cooler areas and progresses towards the center of the weld. The resulting solidification pattern is shown by the sketch of a vertical section of an electrogas weld in Figure 11.8. The molten weld pool solidification pattern can be expressed by the term "form factor". It is the ratio of the maximum molten pool width (gap opening plus sidewall penetration) to its maximum depth. The form factor is an empirical number that provides a relative measure of the resistance to centerline cracking and describes the shape of the weld metal solidification pattern. The shape of the molten weld pool and its resultant form factor are controlled by the welding variables. In general, increasing root opening or the voltage increases the form factor, while increasing the current or decreasing the root opening decreases the form factor.

Electrode Extension (Figure 11.9)

When using a constant potential power source, increasing the electrode extension by jogging the welding head up will reduce the arc voltage. The width of the resultant weld will be reduced. Increasing the electrode extension by increasing the electrode feed speed will increase the deposition rate and decrease the sidewall penetration and weld width. For a constant current source, increasing the electrode extension, by jogging the welding head up, will cause an increase in the voltage. This will result in a wider weld. Increasing the electrode extension by increasing the wire feed speed will decrease the arc voltage and result in a higher deposition rate.

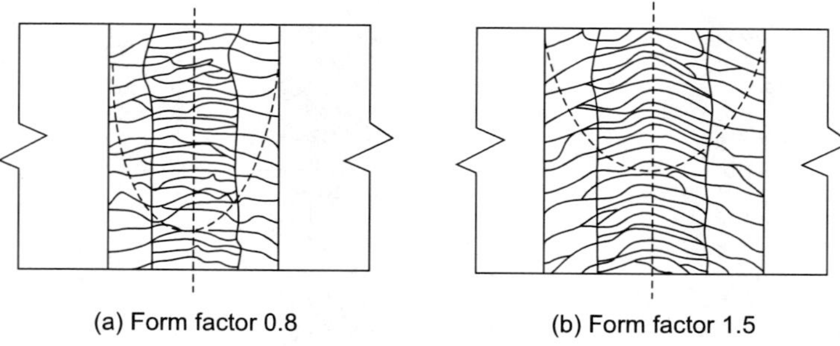

Figure 11.8 Effect of form factor on molten weld pool

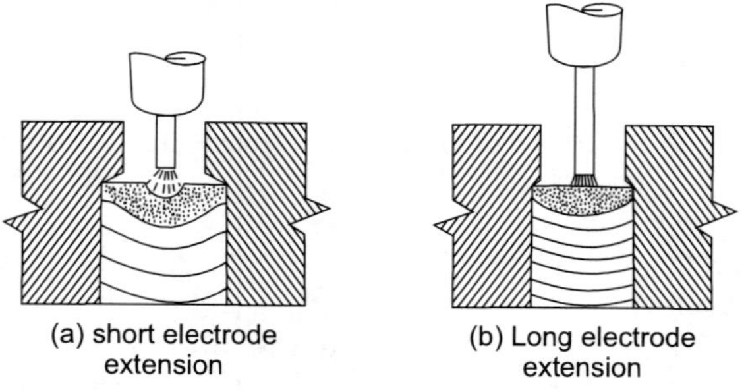

Figure 11.9 Effect of electrode extension on molten weld pool

Electrode Oscillation

The necessity for electrode oscillation depends upon the welding conditions including the current, voltage, and electrode diameter. Base metals up to 3/4 inches (19 mm) thick are commonly welded with a stationary electrode, and base metals between 3/4 and l-1/4 inches (32 mm) may be welded with stationary electrodes, but oscillation is commonly employed for greater thicknesses. Base metals with a thickness between 1-1/4 and 1-1/2 inches (40 mm) have been welded with a stationary electrode, but high voltage and special techniques are required to prevent lack of edge fusion. A dwell period at each end of the oscillation travel is often needed to assure complete fusion at the joint corners.

Joint Opening

A minimum joint opening is needed to give sufficient clearance for the electrode guide, which usually reaches inside the joint. Increasing the joint opening increases the weld width. Excessive joint openings will increase the welding time and consume extra filler metals and gases, which will increase welding costs. Excessive joint openings may also cause lack of edge fusion if the voltage is not increased.

Work Lead

The work lead is usually connected to the starting sump, but it can be split and attached to each side of the joint at the bottom. Although this work lead location has proved satisfactory in most applications, severe arc blow can cause excessive starting porosity, lack of fusion on one side of the joint, or both. The final work lead location must be determined by the user to suit the specific conditions of each application.

11.8 METHODS OF INSPECTION

Conventional non-destructive (NDE) methods can be used to inspect electrogas welded joints. Non-destructive examination techniques normally used with EGW include:

1. VT-visual
2. RT-radiography (isotope or X-ray)
3. MT-magnetic particle
4. PT-liquid penetrant
5. UT-ultrasonic (the large grain size may require special techniques)

All NDE should be performed in accordance with qualified procedures by qualified technicians. The inspection method used will depend not only upon the requirements of the governing code but also the owner's or purchaser's contract specifications. Every joint should have thorough visual examination (VT) to reveal incomplete edge fusion, under fill, and in some cases, cracks. When the sump and runoff tabs are removed, defects such as cracking and porosity may be detected. However, to find internal defects (such as porosity, inclusions, cracking, and in rare instances, lack of internal fusion), RT and UT testing are the most effective means of examination. Magnetic particle (MT) testing may also be used to detect cracks and lack of fusion.

11.9 TROUBLESHOOTING GUIDE

Porosity at Start

Several factors contribute to starting porosity in electrogas welds. Among them are the following:

1. Low electrode feed speed, high voltage, short electrode extension, insufficient time in sump
2. Contaminants in sump or between sump and base
3. Condensation on shoes or between shoes and base
4. Fast cooling rates caused by thick plates
5. Sump too shallow

6. Arc location too close to shoes
7. Water leakage in sump

Porosity in Weld (Figure 11.10)

Weld metal porosity, as illustrated in the figure, is generally caused by expanding gases that are insoluble in the freezing metal, but gas-like voids can also be caused by mechanical problems. Some of the causes are the following:

1. Starting porosity that extends into the production
2. Excessive voltage
3. Low electrode feed speed
4. Electrode extension too short
5. A "cold" weld (low electrode feed speed and voltage)
6. Contaminants in weld area

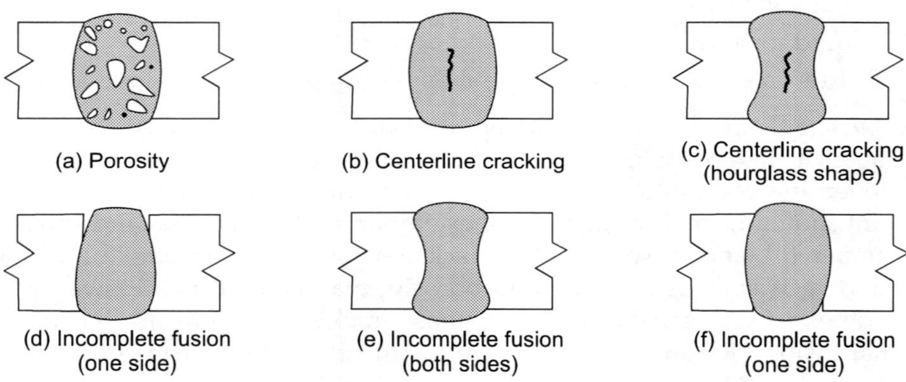

Figure 11.10 Common weld defects in electrogas welding

Porosity at Weld Termination

Porosity at the end of the weld can be caused by the same factors as porosity within the weld. In addition, the following factors can cause porosity:

1. Runoff tabs or stationary shoe too short
2. Slag leaking due to improperly attached runoff tabs
3. Arc blow caused by improper location of work lead

Centerline Weld Cracking

Centerline cracking, as shown in Figure 11.10, may be related to procedure variables, thermal conditions, and weld restraint. Some steel grades are more susceptible to centerline cracking than others. The procedure variables that contribute to a low form factor contribute to centerline cracking:

1. Electrode feed speed too high
2. Arc voltage too low
3. Gap too narrow
4. Dwell time too long

Incomplete Fusion to Both Sidewalls

Incomplete fusion at both sidewalls, as shown in Figure 11.10, is caused by unsatisfactory thermal conditions that prevent the sidewalls from melting. Poor heat distribution as well as insufficient heat can result in this defect.

Incomplete Fusion to One Sidewall

Incomplete fusion to one sidewall; as illustrated in the figure, is caused by asymmetric thermal conditions.

These can result from the following:

1. Arc location off-center
2. Electrode angled towards one sidewalls
3. Arc blow caused by improper location of work lead

Overlap

Overlap is caused by weld metal flow out of the joint without melting the base metal. It is often caused by a poor fit of the base metal and it can also result if the sliding shoe is lifted from contact with the base metal by foreign material, such as spatter. This allows weld metal to leak and freeze against the plate.

1. Arc location too far back
2. Bevel angle too large
3. Cold weld

Underfill

Underfill, under certain conditions, may be acceptable but at best it represents marginal workmanship. Moreover, it is fairly easy to prevent. It can be caused by excessive melting of base metal beyond the shoe or too narrow a groove in the shoe.

Hot Cracking

Hot cracking can be caused by the partial dissolution of the copper molding shoes. The cracks are generally at or near the surface. Electrogas welds of consistent quality can be fabricated by maintaining the equipment and following the weld procedure exactly.

11.10 APPLICATIONS

1. Applications of electrogas welding include storage tanks, ship hulls, structural members, and pressure vessels. EGW should be considered for any joint to

be welded in a vertical position in materials ranging in thickness from 310 to 100 mm.

2. For joining thick plates in vertical position.
3. Ships, bridges, large tanks, pressure vessels, offshore drilling rigs and some parts for high-rise buildings are profitably fabricated by the electrogas welding process.

11.11 ADVANTAGES OF ELECTROGAS WELDING

For welding job thicknesses 20 to 75 mm, the electrogas welding has the following advantages over electroslag process:

1. Weld is much more visible to the welder
2. Restarting the weld is much easier
3. Welded impact properties are better
4. Welding is usually done in one pass

11.12 DISADVANTAGES OF ELECTROGAS WELDING

1. For job thicknesses above 75 mm, electroslag welding is usually more practical than electrogas welding
2. As compared to electrogas welding, electroslag welding usually produces welds that are cleaner and nearly crack free
3. In electrogas process, porosity generally increases as job thickness increases because gas shielding becomes less effective

Review Questions

1. Explain the principle of operation of electrogas welding.
2. Explain the equipment used in electrogas welding.
3. Explain the differences between electroslag welding and electrogas welding.
4. Give brief descriptions of power source for electrogas welding.
5. What are the functions of flux in electrogas welding?
6. What are the important operating variables in electrogas welding?
7. Discuss the important weld defects in electrogas welding.
8. What are the advantages of electrogas welding?
9. What are the disadvantages of electrogas welding?
10. Write a few applications of electrogas welding.

Section B

Resistance Welding Processes

12
CHAPTER

Resistance Spot Welding

12.0 INTRODUCTION

Large scale and high rate production of components using sheet metal have been possible only due to the resistance welding processes, which have the dual advantage of making joints quickly and with minimum distortion. Resistance welding is based on the principle that heat is generated by the resistance offered to the flow of current. There are at least eight important resistance welding processes, which include spot welding, upset butt welding, seam welding, projection welding, flash welding, percussion welding, high-frequency resistance welding and high-frequency induction welding.

12.1 PRINCIPLES OF OPERATION (FIGURE 12.1)

Spot welding operations involve a coordinated application of electric current and mechanical pressure of proper magnitudes and durations. The welding current must pass from the electrodes through the work. Its continuity is assured by forces applied to the electrodes, which are shaped to provide the necessary current density and pressure. The sequence of operation must first develop sufficient heat to raise a confined volume of metal to the molten state. This metal is then allowed to cool while under pressure until it has adequate strength to hold the parts together. The current density and pressure must be such that a nugget is formed, but not so high that molten metal is expelled from the weld zone. The duration of weld current must be sufficiently short to prevent excessive heating of the electrode faces. Such heating may bond the electrodes to the work and greatly reduce their life.

The heat required for these resistance welding processes is produced by the resistance of the workpieces to an electric current passing through the material. Because of the short electric current path in the work and limited weld time, relatively high welding currents are required to develop the necessary welding heat.

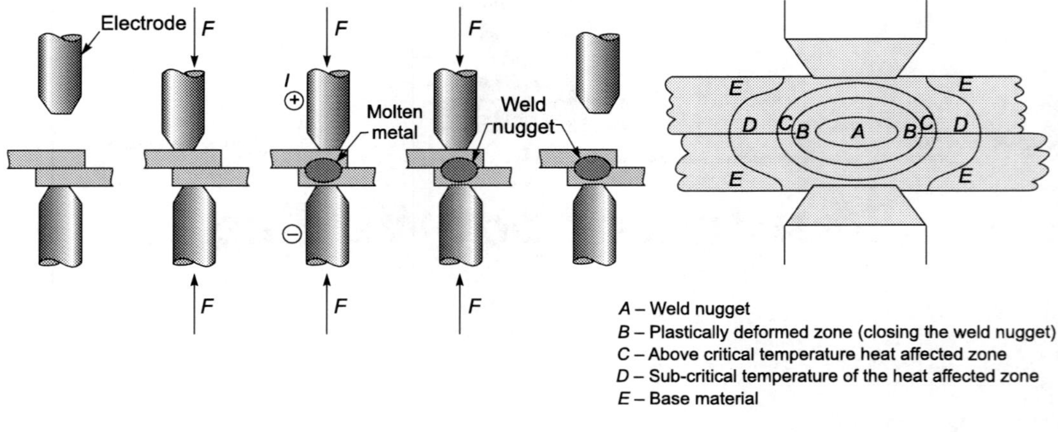

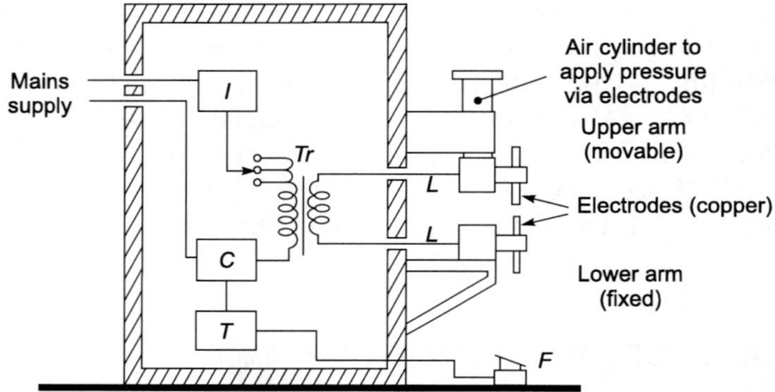

Figure 12.1 Principal components of a resistance spot welding

12.2 HEAT GENERATION

In an electrical conductor, the amount of heat generated depends upon three factors: (1) the amperage, (2) the resistance of the conductor (including interface resistance), and (3) the duration of current. These three factors affect the heat generated as expressed in the formula:

$$Q = I^2 R t$$

where,

Q = heat generated, joules

I = current, amperes

R = resistance of the work, ohms
t = duration of current, seconds

The heat generated is proportional to the square of the welding current and directly proportional to the resistance and the time. Part of the heat generated is used to make the weld and part is lost to the surrounding metal. The welding current required to produce a given weld is approximately inversely proportional to the square root of the time. Thus, if the time is extremely short, the current required will be very high. A combination of high current and insufficiently short time may produce an undesirable distribution of heat in the weld zone, resulting in severe surface melting and rapid electrode deterioration. The secondary circuit of a welding machine and the work being welded constitute a series of resistances. The total resistance of the current path affects the current magnitude. The current will be same in all parts of the circuit, regardless of the instantaneous resistance at any location in the circuit, but the heat generated at any location in the circuit will be directly proportional to the resistance at that point. An important characteristic of resistance welding is the rapidity with which welding heat can be produced. The temperature distribution in the work and electrodes, in the case of spot welding, is illustrated in Figure 12.2. There are, in effect, at least seven resistances connected in series in a weld that account for the temperature distribution. For a two-thickness joint, these are the following:

(a) **1** and **7,** the electrical resistance of the electrode material.

(b) **2** and **6**, the contact resistance between the electrode and the base metal. The magnitude of this resistance depends upon the surface condition of the base metal and the electrode, the size and contour of the electrode face, and the electrode force. (Resistance is roughly inversely proportional to the contacting force.) This is a point of high heat generation, but the surface of the base metal does not reach its fusion temperature during the current passage, due to the high thermal conductivity of the electrodes (**1** and **7**) and the fact that they are usually water cooled.

(c) **3** and **5,** the total resistance of the base metal itself, which is directly proportional to its resistivity and thickness, and inversely proportional to the cross-sectional area of the current path.

(d) **4,** the base metal interface resistance at the location where the weld is to be formed. This is the point of highest resistance and therefore, the point of most heat generation. Since heat is also generated at points **2** and **6,** the heat generated at interface **4** is not readily lost to the electrodes. Heat is generated at all of these locations, not at the base metal interface alone. The flow of heat to or from the base metal interface is governed by the temperature gradient established by the resistance heating of the various components in the circuit. This, in turn, assists or retards the creation of the proper localized welding heat.

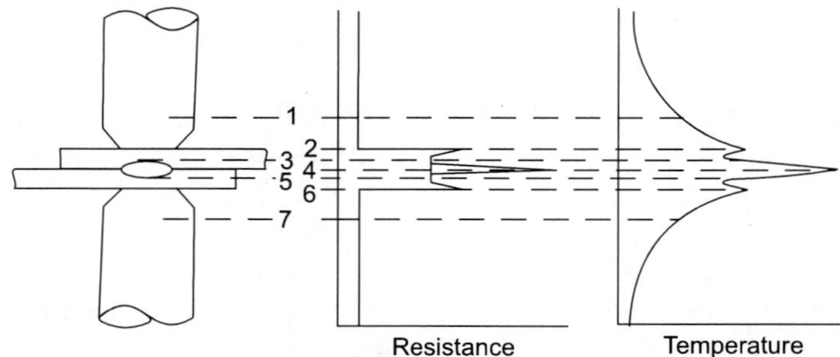

Figure 12.2 Graph of resistance and temperature as a function of location in the diagrammed circuit

Heat is generated in each of the seven locations in Figure 12.2 in proportion to the resistance of each. Welding heat, however, is required only at the base metal interface, and the heat generated at all other locations should be minimized. Since the greatest resistance is located at **4**, heat is most rapidly developed at that location. Points of next lower resistance are **2** and **6**. The temperature rises rapidly at these points also, but not as fast as at **4**. After about 20 percent of the weld time, the heat gradient may conform to the profile shown in Figure 12.2. Heat generated at **2** and **6** is rapidly dissipated into the adjacent water cooled electrodes, **1** and **7**. The heat at **4** is dissipated much more slowly into the base metal. Therefore, while the welding current continues, the rate of temperature rise at plane **4** will be much more rapid than at **2** and **6**. The welding temperature is indicated on the chart at the right of the figure, number of dots within the drawing leading to the matching curve.

In a well controlled weld, the welding temperature will first be reached at numerous point contacts at the interface that melt and with time quickly grow into a nugget.

Factors that affect the amount of heat generated in the weld joint by a given current for a unit of weld time are: (a) the electrical resistances within the metal being welded and the electrodes, (b) the contact resistances between the workpieces and between the electrodes and the workpieces, and (c) the heat lost to the workpieces and the electrodes.

12.2.1 Effect of Welding Current (Figure 12.3)

In the formula, $Q = I^2RT$, current has a greater effect on the generation of heat than either resistance or time. Therefore, it is an important variable to be controlled. Two factors that cause variation in welding current are fluctuations in power line voltage and variations in the impedance of the secondary circuit with AC machines. Impedance variations are caused by changes in circuit geometry or by the introduction of varying masses of magnetic metals into the secondary loop of the machine. Direct current machines are not significantly affected by magnetic metals in the secondary loop and are little affected by circuit geometry.

In addition to variations in welding current magnitude, current density varies at the weld interface. This can result from shunting of current through preceding welds and contact points other than those at the weld. An increase in electrode face area will decrease current density and welding heat. This may cause a significant decrease in weld strength. A minimum current density for a finite time is required to produce fusion at the interface. Sufficient heat must be generated to overcome the losses to the adjacent base metal and the electrodes.

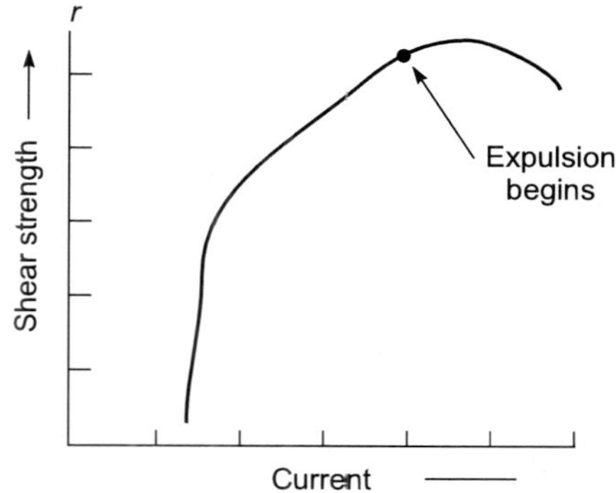

Figure 12.3 Effect of welding current on spot weld shear strength

Weld nugget size and strength increase rapidly with increasing current density. Excessive current density will cause molten metal expulsion (resulting in internal voids), weld cracking, and lower mechanical strength properties. Typical variations in shear strength of spot welds as a function of current magnitude are shown in Figure 12.3. In the case of spot welding, excessive current will overheat the base metal and result in deep indentations in the parts and, it will cause overheating and rapid deterioration of the electrodes.

12.2.2 Effect of Weld Time

The rate of heat generation must be such that welds with adequate strength will be produced without excessive electrode heating and rapid deterioration. The total heat developed is proportional to weld time. Essentially, heat is lost by conduction into the surrounding base metal and the electrodes; a very small amount is lost by radiation. These losses increase with increase in weld time and in metal temperature, but they are essentially uncontrollable.

During a spot welding operation, some minimum time is required to reach melting temperature at some suitable current density. If current is continued, the temperature at plane **4** in the weld nugget will far exceed the melting temperature, and the internal pressure may expel molten metal from the joint. Generated gases or metal vapor may be expelled together with minute metal particles. If the work surfaces are scaly or pitted, gases and particles may also be expelled at planes **2** and **6**. Excessively long weld time will have the same effect as excessive amperage on the base metal and electrodes. Furthermore, the weld heat affected zone will extend farther into the base metal. In most cases, the heat losses at some point during an extended welding interval will equal the heat input; temperatures will stabilize. An example of the relationship between weld time and spot weld shear strength is shown in Figure 12.4, assuming all other conditions remain constant.

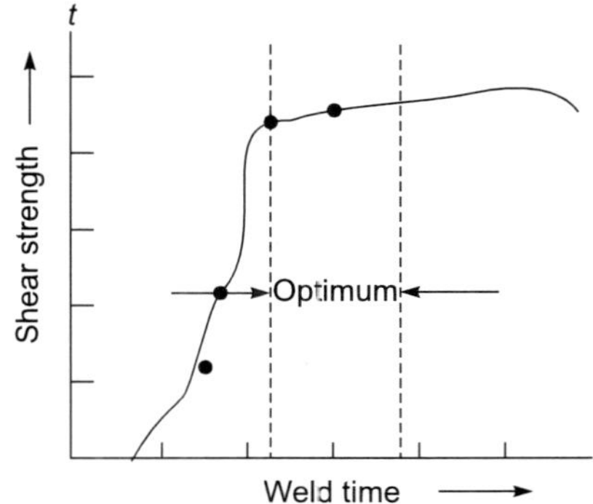

Figure 12.4 Tensile-shear strength as a function of weld time

To a certain extent, weld time and amperage may be complementary. The total heat may be changed by adjusting either the amperage or the weld time. Heat transfer is a function of time and the development of the proper nugget size requires a minimum length of time, regardless of amperage. When spot welding heavy plates, welding current is commonly applied in several relatively short impulses without removing the electrode force. The purpose of pulsing the current is to gradually build up the heat at face between the workpieces. The amperage needed to accomplish welding can rapidly melt the metal if the heat pulse time is too long, resulting in expulsion.

12.2.3 Effect of Welding Pressure

The resistance in the heat formula ($Q = I^2RT$) is influenced by welding pressure through its effect on contact resistance at the interface between the workpieces. Welding pressure is produced by the force exerted on the joint by the electrodes. Electrode force is considered to be the net dynamic force of the electrodes upon the work, and it is the resultant pressure produced by this force that affects the contact resistance. Pieces to be spot welded must be clamped tightly together at the weld location to enable the passage of the current. Everything else being equal, as the electrode force or welding pressure is increased, the amperage will also increase up to some limiting value. The effect on the total heat generated, however, may be the reverse. As the pressure is increased, the contact resistance and the heat generated at the interface will decrease. To increase the heat to the previous level, amperage or weld time must be increased to compensate for the reduced resistance.

The surfaces of metal components, on a microscopic scale, are a series of peaks and valleys. When they are subjected to light pressure, the actual metal-to-metal contact will be only at the contacting peaks, a small percentage of the area. Contact resistance will be high. As the pressure is increased, the high spots are depressed and the actual metal-to-metal contact area is increased, thus decreasing the contact resistance. In most applications, the electrode material is softer than the workpieces; consequently, the application of a suitable electrode force will produce better contact at the electrode-to-work interfaces than at the interface between the workpieces.

12.2.4 Influence of Electrodes (Figure 12.2)

Electrodes play a vital role in the generation of heat because they conduct the welding current to the work. In the case of spot, the electrode contact area largely controls the welding current density and the resulting weld size. Electrodes must have good electrical conductivity, but they must also have adequate strength and hardness to resist deformation caused by repeated applications of high electrode force. Deformation or "mushrooming" of the electrode face increases the contact area and decreases both current density and welding pressure. Weld quality will deteriorate as tip deformation proceeds; consequently, the electrodes must be reshaped or replaced at intervals to maintain adequate heat generation for acceptable weld properties.

When the electrodes are slow in following a sudden decrease in total work thickness, a momentary reduction in pressure will occur. If this happens while the welding current is on, interface contact resistance at locations **2**, **4** and **6** and the rate of heat generation will increase. An excessive heating rate at the three contacting surfaces tends to cause overheating and violent expulsion of molten metal. Molten metal is retained at each interface by a ring of fused metal surrounding the weld nugget. A momentary reduction in electrode force permits the internal metal pressure to rupture this surrounding ring of unfused metal. Internal voids or excessive electrode indentation may result. Weld properties may fall below acceptable levels, and electrode wear will be greater than normal.

12.2.5 Influence of Surface Condition

The surface condition of the parts influences heat generation because contact resistance is affected by oxides, dirt, oil, and other foreign matter on the surfaces. The most uniform weld properties are obtained when the surfaces are clean.

The welding of parts with a non-uniform coating of oxides, scale, or other foreign matter on the surface causes variations in contact resistance. This produces inconsistencies in heat generation. Heavy scale on the work surfaces may also become embedded in the electrode faces, causing rapid electrode deterioration. Oil and grease pick up dirt which will also contribute to electrode deterioration.

12.2.6 Influence of Metal Composition

The electrical resistivity of a metal directly influences resistance heating during welding. In high-conductivity metals, such as silver and copper, little heat is developed even under high-current densities. The small amount of heat generated is rapidly transmitted into the surrounding work and the electrodes.

The composition of a metal determines its specific heat, melting temperature, latent heat of fusion, and thermal conductivity. These properties govern the amount of heat required to melt the metal and produce a weld. However, the amount of heat necessary to raise unit masses of most commercial metals to their fusion temperatures are nearly the same. For example, stainless steel and aluminum require the same joules per gram to reach fusion temperature, even though they differ widely in spot welding characteristics. As a result, the electrical and thermal conductivities become dominant. The conductivities of aluminum are about ten times greater than those of stainless steel. Consequently, the heat lost into the electrodes and the surrounding metal is greater with aluminum. Accordingly, the welding current for aluminum must be considerably greater than that for stainless steel.

12.3 HEAT BALANCE

Heat balance occurs when the depths of fusion (penetration) in the two workpieces are approximately the same. The majority of spot welding applications are confined to the welding of equal thicknesses of the same metal, with electrodes of the same alloy, shape, and size. Heat balance in these cases is automatic; however, in many applications, the heat generated in the parts is unbalanced.

Heat balance may be affected by the following:

1. Relative electrical and thermal conductivities of the metals to be joined
2. Relative geometry of the parts at the joint
3. Thermal and electrical conductivities of the electrodes
4. Geometry of the electrodes

Heating will be unbalanced when pieces to be welded have significantly different compositions, different thicknesses, or both. The unbalance can be minimized in many cases by part design, electrode material and design. Heat balance can also be improved by using the shortest weld time and lowest current that will produce acceptable welds.

12.4 HEAT DISSIPATION

During welding, heat is lost by conduction into the adjacent base metal and the electrodes, as shown in Figure 12.5. This heat dissipation continues at varying rates during current application and afterward, until the weld has cooled to room temperature. It may be divided into two phases: (1) during the time of current application, and (2) after the cessation of current. The extent of the first phase depends upon the composition and mass of the workpieces, the welding time, and the external cooling means. The composition and mass of the workpieces are determined by the design. External cooling depends upon the welding setup and the welding cycle.

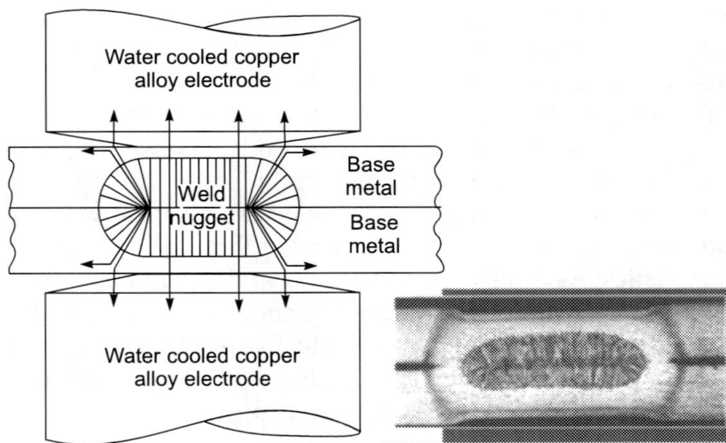

Figure 12.5 Heat dissipation during resistance welding into the surrounding base metal and electrodes

The heat generated by given amperage is inversely proportional to the electrical conductivity of the base metal. The thermal conductivity and temperature of the base metal determine the rate at which heat is dissipated or conducted from the weld. In most cases, the thermal and electrical conductivities of a metal are similar. In a high-conductivity metal, such as copper or silver, high amperage is needed to produce a weld and compensate for the heat that is dissipated rapidly into the adjacent base metal and the electrodes. Spot welding of these metals is very difficult.

If the electrodes remain in contact with the work after weld current ceases, they rapidly cool the weld nugget. The rate of heat dissipation into the surrounding base metal decreases with longer welding times because a larger volume of base metal would have been heated. This reduces the temperature gradient between the base metal and the weld nugget. For thick sheets of metal where long welding times are generally required, the cooling rates are slower than with thin sheets and short weld times. If the electrodes are removed from the weld too quickly after the welding current is turned off, problems may result. With thin sheets, this procedure may cause excessive warpage. With thick sheets, adequate time is needed to cool and solidify the large weld nugget while under pressure. It is usually best, therefore, to have the electrodes in contact with the work until the weld cools to a temperature where it is strong enough to sustain any loading imposed when the pressure is released. The cooling time for a weld nugget is short when the electrodes are rotated continuously. Therefore, welding is commonly done with water flowing over the workpieces to remove the heat as rapidly as possible. It is not always a good practice to cool the weld zone rapidly. With quench hardenable alloy steels, it is usually best to retract the electrodes as quickly as possible to minimize heat dissipation to the electrodes, and thus, to retard the cooling rate of the weld.

12.5 WELDING CYCLE

The welding cycle for spot welding consists basically of four phases:

1. Squeeze time – The time interval between timer initiation and the first application of current; the time interval is to assure that the electrodes contact the work and establish the full electrode force before welding current is applied.
2. Weld time – The time that welding current is applied to the work in making a weld in single-impulse welding.
3. Hold time – The time during which force is maintained to the work after the last impulse of current ends; during this time, the weld nugget solidifies and is cooled until it has adequate strength.
4. Off time – The time during which the electrodes are off the work and the work is moved to the next weld location; the term is generally applied where the welding cycle is repetitive.

Figure 12.6 shows a basic welding cycle. One or more of the following features may be added to this basic cycle to improve the physical and mechanical properties of the weld zone:

1. Precompression force to seat the electrodes and workpieces together
2. Preheat to reduce the thermal gradient in the metal at the start of weld time
3. Forging force to consolidate the weld nugget

4. Quench and temper times to produce the desired weld strength properties in hardenable alloy steels
5. Postheat to refine the weld grain size in steels
6. Current decay to retard cooling on aluminum

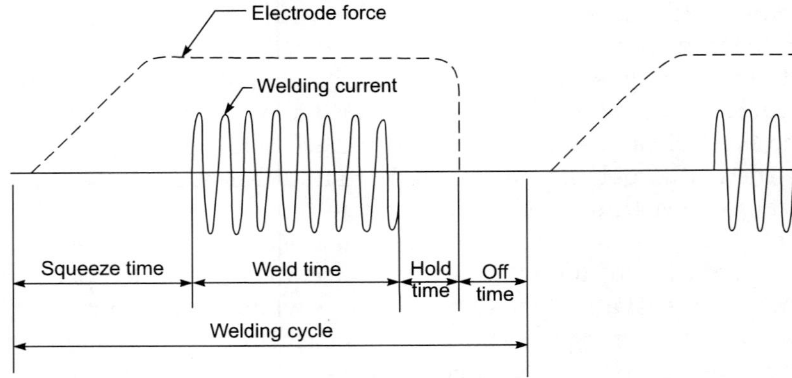

Figure 12.6 Basic single impulse welding cycle for spot welding

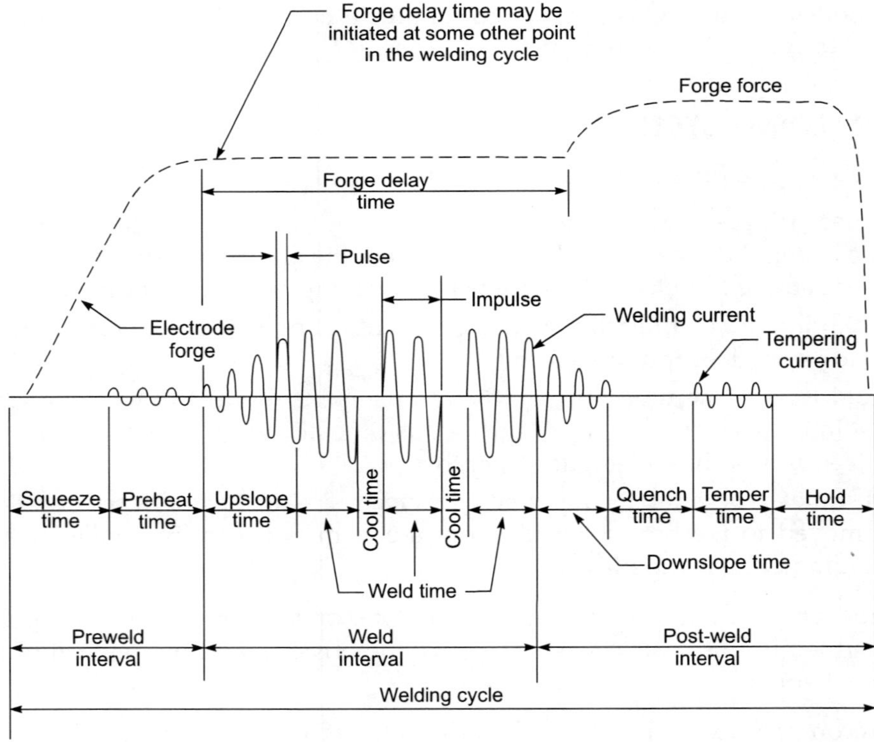

Figure 12.7 Enhanced welding cycle which includes: preheat time, upslope time, downslope time, quench time, temper time, and forging force

In some applications, the welding current is supplied intermittently during a weld interval time; it is on during heat time and ceases during cool time. Figure 12.7 shows the sequence of operations in a more complex welding cycle.

12.6 WELDING CURRENT

Both alternating current (AC) and direct current (DC) are used to produce spot welds. The welding machine transforms line power to low voltage, high amperage welding power, some applications use single phase AC of the same frequency as the power line, usually 60 Hz. Direct current is used for applications that require high amperage because the load can be balanced on a 3-phase power line. Its use also reduces the power losses in the secondary circuit. Direct current may be essentially constant for a timed period or in the form of a high-peaked pulse. The latter is normally produced from stored electrical energy.

Current Programming

With direct energy machines, the rate of rise and fall of current can be programmed. The current rise period is commonly called upslope time and the current fall period is called downslope time (Figure 12.7). These features are available on machines equipped with electronic control systems. Upslope is generally used to avoid overheating and expulsion of metal at the beginning of weld time, when the base metal interface resistance is high. Downslope is used to control weld nugget solidification, to avoid cracking in metals that are quench-hardenable or subject to hot tearing. Prior to welding, the base metal can be preheated using a low current. Following the formation of the weld nugget, the current can be reduced to some lower value for postheating of the weld zone. This may be part of the weld interval, as shown in Figure 12.7, or a separate application of current following a quench time period.

Weld Time

The time of current application, or weld time for other than stored energy power, is controlled by electronic, mechanical, manual, or pneumatic means. Time commonly ranges from one half cycle (1/200 sec) for very thin sheets to several seconds for thick plates. For the capacitor or magnetic type of stored energy machines, the weld time is determined by the electrical constant of the system.

Single Impulse

The use of one continuous application of current to make an individual weld is called single impulse welding (Figure 12.6). Up or down current slope may be included in the time period.

Multiple Impulse Welding

Multiple impulse welding consists of pulses of current separated by a preset cool time (Figure 12.7). This sequence is used to control the rate of heating at the interface while spot welding of relatively of thick steel sheet.

12.7 ELECTRODE FORCE

Completion of electrical circuit through the electrodes and the work is assured by the application of electrode force. This force is produced by hydraulic, pneumatic, magnetic, or mechanical devices. The pressure developed at the interfaces depends upon the area of the electrode faces in contact with the workpieces. The functions of this force or pressure are to: (1) bring the various interfaces into intimate contact, (2) reduce initial contact resistance at the interfaces, (3) suppress the expulsion of molten weld metal from the joint, and (4) consolidate the weld nugget.

Forces may be applied during the welding cycle as follows:

1. A constant weld force
2. Precompression and weld forces – a high initial level to reduce initial contact resistance and bring the parts into intimate contact, followed by a lower level for welding
3. Precompression, weld, and forging forces – the first two levels as described are followed by a forging force near the end of the weld time; forging is used to reduce porosity and hot cracking in the weld nugget
4. Weld and forging forces

12.8 WELDABILITY

Most of the industrial metals can be welded by one or the other resistance welding process. Thicker sections are more difficult to weld and some metals may need postweld heat treatment to attain the desired properties. Weldability depends on three factors; electrical resistivity, thermal conductivity and melting point of the metal. Ferrous metals are easily welded. Metals with low resistivity and high thermal conductivity like aluminum and magnesium alloys and refractory metals like tungsten and tantalum are difficult to weld. Weldability for resistance welding, in general, and spot welding in particular, can be expressed by the following relationship:

$$\text{Percentage weldability} = \frac{\text{Resistivity}}{\text{Thermal conductivity} \times \text{melting point}} \times 100$$

$$W = \frac{\rho}{k \times t} \times 100$$

W = percentage weldability

ρ = electrical resistivity of work material ($\mu\Omega$/cm),

k = relative thermal conductivity with copper
t = melting point, °C.

From the above relationship, weldability above 2.0 is excellent, between 0.75 and 2.0 is good and below 0.25 is poor. Weldability rating for mild steel is above 10 while for aluminum alloys, it ranges from 1 to 2. Copper and brass have poor weldability and are difficult to weld.

12.9 EQUIPMENT

Spot welding equipment consists of three basic elements: an electrical circuit, the control equipment, and a mechanical system.

12.9.1 Electrical Circuit

This circuit comprises a welding transformer, a primary contactor, and a secondary circuit. The secondary circuit includes the electrodes that conduct the welding current to the work. In some cases, a means of storing electrical energy is also included in the circuit. Both alternating current and direct current are used for resistance welding. The welding machine converts 50 Hz power to low voltage, high amperage power in the secondary circuit of the welding machine.

Alternating Current

Some resistance welding machines produce single-phase alternating current (AC) of the same frequency (50 Hz) as the power line, these machines contain a single-phase transformer that provides the high welding currents required at low voltage. Depending upon the thickness and type of material to be welded, currents may range from 1000 to 1,00,000 amperes. A typical electrical circuit designed for this type of machine is shown in Figure 12.8.

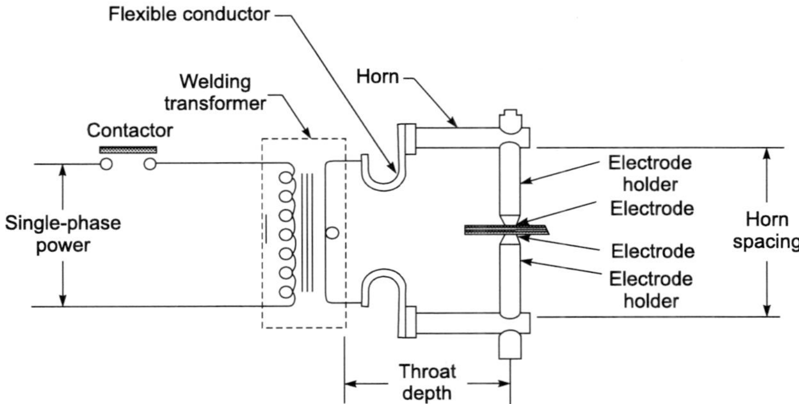

Figure 12.8 Electrical circuit, which consists of a welding transformer, tap switch and a secondary circuit

Direct Current

Welding machines may produce direct current of continuous polarity, pulses of current of alternating polarity, or high peaked pulses of current. The latter type is produced by stored electrical energy.

Rectifier-Type Machines: These machines are direct energy types, in that AC power from the plant distribution system passes through a welding transformer and is then rectified to DC power. Silicon diode rectifiers are widely used in secondary circuits because of their inherent reliability and efficiency. The system can be single phase; however, one of the advantages of direct current systems is the ability to use a three-phase transformer to feed the rectifier system in the secondary circuit. This makes it possible to use balanced three-phase line power.

Frequency Converter Machines: This type of machine has a special welding transformer with a three-phase primary and a single-phase secondary. The primary current is controlled by ignitron tubes or silicon controlled rectifiers (SCRs). Half cycles of three-phase power, either positive or negative, are conducted to the transformer for a timed period that depends upon the transformer design. The transformer output is a pulse of direct current. By switching the polarity of the primary half cycles, the polarity of the secondary current is reversed. A weld may be made with one or more DC pulses.

Energy Machines: Stored energy machines are of electrostatic design. They draw power from a single-phase system, store it, and then discharge it in a very short pulse period to make the weld. These machines draw power from the supply line over a relatively long time between welds, accumulating power to deliver to the electrodes during a short weld time.

The equipment for electrostatic stored energy welding consists primarily of a bank of capacitors, a circuit for charging these capacitors to a predetermined voltage, and a system for discharging the capacitors through a suitable welding transformer. High-voltage capacitors are generally used, the most common varying from 1500 to 3000 volts.

The different welding current and time for various electrical circuit types is given in Figure 12.9.

Electrodes

Resistance welding electrodes perform four functions:

1. Conduct the welding current to the work and for spot welding, fix the current density in the weld zone.
2. Transmit a force to the workpieces.
3. Dissipate part of the heat from the weld zone.
4. Maintain relative alignment and position of the workpieces.

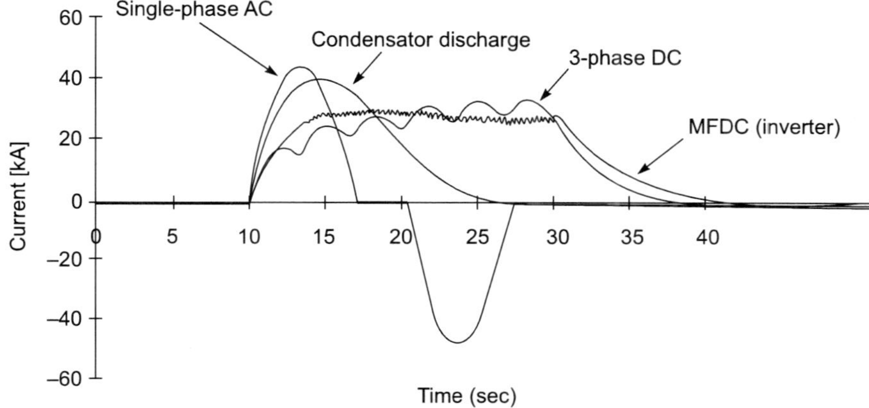

Figure 12.9 Different current/time welding machines

If the applications of pressure were not involved, electrode material selection could be made almost entirely on the basis of electrical and thermal conductivity. Since the electrodes are subjected to forces that are often of considerable magnitude, they must be capable of withstanding the imposed stresses at elevated temperatures without excessive deformation. Proper electrode shape is important because the current must be confined to a fixed area to achieve needed current density.

When only one spot is to be made at a time, only one pair of electrodes is required. In this case, the force and current are applied to each weld by shaped electrodes. Electrodes of several copper alloys with satisfactory physical and mechanical properties are available. Generally speaking, the harder the alloy, the lower are its electrical and thermal conductivities. The choice of a suitable alloy for any application is based on a compromise of its electrical and thermal properties with its mechanical qualities. Electrodes selected for aluminum welding, for instance, should have high conductivity at the expense of high compressive strength to minimize sticking of electrodes to the work. Electrodes for welding stainless steel, on the other hand, should sacrifice high conductivity to obtain good compressive strength, to withstand the required electrode force.

Resistance to deformation or mushrooming depends upon the proportional limit and hardness of the electrode alloy. The proportional limit is largely established by heat treatment. The temperature of the electrode face is the governing factor, because this is where softening takes place. The sizes and shapes of electrodes are usually determined by the sheet thickness and the metal to be welded.

Types of Electrodes (Figure 12.10)

To achieve the desired current density, it is important to have a proper electrode shape, for which three main types of electrodes are used. These are truncated, dome/ball-shaped, and flat tip.

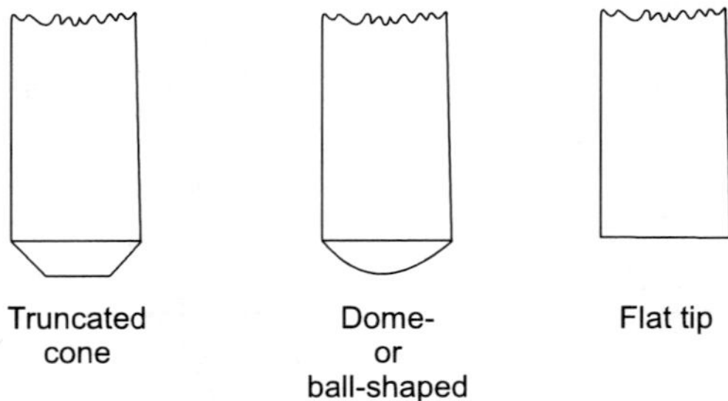

Figure 12.10 Different types of electrodes

Pointed tips are most widely used particularly for ferrous materials; with continued wear they mushroom uniformly. The pointed electrodes are truncated cone electrodes, with an angle of 120° to 140°. Contact area can be controlled more accurately and any wear in service can be easily seen. The pointed tip electrodes are more obvious surface marking on the workpiece and require more accurate alignment.

Domed electrodes are characterized by their ability to withstand heavier pressure and severe heating without mushrooming, which makes them particularly useful for welding non-ferrous metals. The radius of the dome varies but 50 to 100 mm radius is commonly used. Filing of electrodes should not be permitted as it is not possible to reproduce their original shape by that method. When inconspicuous or invisible welds are desired or where the weld indentation is to be minimum, a flat tip electrode is used. A combination of one flat and one domed electrode is usually used in such cases.

Offset electrodes can be used to make spot welds in places that are inaccessible to weld by conventional electrodes, e.g., for making corner welds and for parts with overhanging flanges.

Electrodes for aluminum welding are required to have high electrical conductivity, and lower compressive strength so as to minimize electrode sticking to work.

Electrode Size

The size of the tip of the electrode depends upon the size of the spot weld or the required nugget diameter. The spot weld diameter is decided by the sheet thickness to be welded. Often the diameter of the nugget is selected as equal to the diameter of the corresponding rivet used for joining the sheet of same thickness. Using Unwin's formula, nugget size is given by $d_n = 6\sqrt{t}$ where, d_n = diameter of the nugget and t is single sheet thickness (mm). The efficiency of a spot weld is considered to be higher than that of a riveted joint designed with the same design formula. The electrode tip size is considered nearly equal to the nugget size and is often taken as $d_e = 5\sqrt{t}$ where, d_e is electrode tip size. It is an empirical relationship but gives satisfactory results. Another empirical formula used for designing electrode tip size is $d_e = 2.5 + 2t$

where, t is single sheet thickness (mm). It gives nearly the same result as the first formula except for very thick or thin workpieces. If full length of the electrode is made of that same size, it will be too weak to withstand the pressure to be exerted through it and also cause high resistance to the electric flow, resulting in overheating. The practical electrodes made of copper alloys are of substantial diameter machines to a truncated cone with an angle of 120° to 140°. Radius of the dome is used to control the area of contact when domed electrodes are used. The electrode pressure and the sheet thickness are other factors that decide the area of contact in domed electrodes.

Electrode Maintenance

Maintenance of electrodes is necessary for the production of consistent welds. An abnormal increase in the size of the electrode faces contacting the work is detrimental to strength and quality. For example, if a 6.35 mm diameter electrode face is allowed to increase to 7.94 mm diameter by mushrooming, the contact area will increase 50% with a corresponding decrease in current density and pressure. Depending somewhat upon the weld schedule, the result may be weak or defective welds. A danger sign is the production of poorly shaped spots, which may be caused by the following:

1. Non-circular electrode faces
2. Too large a flat face on the electrode
3. Concavity or convexity of the electrode face
4. Misalignment of the electrodes with respect to the work

Correct electrode alignment is relatively easy to maintain with stationary welding machines and proper supporting fixtures; however, misalignment is common with portable gun type machines. The seriousness of this condition is dependent upon the ease with which the equipment can be manipulated and correctly positioned for welding. It is likely that the electrodes will have longer life between dressings on positioned work (stationary machines) than on non-positioned work (portable welding guns).

12.9.2 Control Equipment

Welding controls may provide one or more of the following principal functions:

1. Initiate and terminate current to the welding transformer
2. Control the magnitude of the current
3. Actuate and release the electrode force mechanism at the proper times

They may be divided into three groups based on their purposes: welding contactors, timing and sequencing controls, other current controls and regulators. A welding contactor connects and disconnects the primary power and the welding transformer. Electronic contactors use silicon-controlled rectifiers (SCRs), ignitron tubes, or thyratron tubes to interrupt the primary current.

The timing and sequence control establish the welding sequence and the duration of each function of the sequence. This includes application of electrode force and current, as well as the time intervals following each function. The welding current output of a machine is controlled by transformer taps, or an electronic heat control, or both. An electronic heat control is used in conjunction with ignitron tubes or SCRs. It controls current by delaying the firing of the ignitron tubes or SCRs during each of half cycle. Varying the firing delay time can be used to gradually increase or decrease the primary (RMS) amperage. This provides upslope and downslope control of welding current.

Transformer taps are used to change the number of primary turns connected across the AC power line. This changes the turns ratio of the transformer, with an increase or decrease in open circuit secondary voltage. Decreasing the turns ratio will increase the open circuit secondary voltage, the primary current, and the welding current.

12.9.3 Mechanical Systems

In spot welding machines, electrodes approach and retract from the work at controlled times and rates. Electrode force is applied by hydraulic, pneumatic, magnetic, or mechanical means. The rate of electrode approach must be rapid, but controlled, so that the electrode faces are not deformed from repeated blows. The locally heated weld metal expands and contracts rapidly during the welding cycle and the electrodes must follow this movement to maintain welding pressure and electrical contact. The ability of the machine to follow motion is influenced by the mass of the moving parts, or their inertia, and by friction between the moving parts and the machine frame.

If the pressure between the electrodes and work drops rapidly during weld time, the contact surfaces of the electrodes and workpieces may overheat and result in burning or pitting of the electrode faces. The electrodes may stick to the work and, in some cases, the surfaces of the parts that are being welded may vaporize from the very high energy. The electrode force used during the melting of the weld nugget may not be adequate to consolidate the weld metal and to prevent internal porosity or cracking. Multilevel force machines may be employed to provide a high forging pressure during weld solidification. The magnitude of this pressure should suit the composition and thickness of the metal and the geometry of the parts. The forging pressure is often two to three times the welding pressure. Since the weld cools from the periphery inward, the forging pressure must be applied at or close to the current termination time.

12.10 TYPES OF EQUIPMENT

The machine used for resistance welding can be either stationary or portable. Such machines may range from small inexpensive units for welding thin sheets to large multiple spot welding machines having high current outputs. Two types of stationary single spot welding machines are commonly used. They are rocker arm type, and press type.

12.10.1 Rocker Arm Type Machines (Figure 12.11)

In its simple form, it has a single pair of electrodes to which the work is offered manually. With small units, a foot operated pedal is linked to a pivoted lever which holds the upper arm. The machines are available with pneumatic, and cam operated arms. Electrode force depends upon the mechanical advantage of the lever arm and setting of spring and weights, etc. They are low cost, simple machines suitable for welding thin low-carbon sheets with simple welding cycles. A major disadvantage of this type of machines is that the upper electrode is moved in an arc with consequential difficulty in electrode alignment. This results in uneven wearing of the electrode tips which may produce an unsatisfactory weld. The power of these machines is usually limited to 50 kVA.

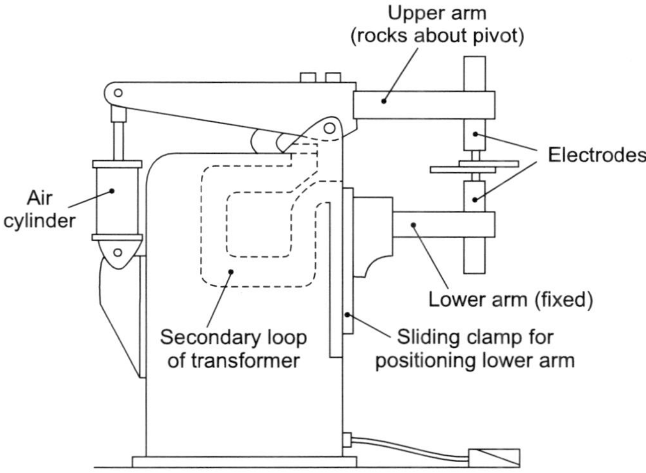

Figure 12.11 Rocker arm welder

12.10.2 Press Type Machines (Figure 12.12)

High force cannot be exerted with rocker arm type machines and so for welding thick metals requiring heavy force and precise control, the press type machines are used. The upper electrode moves in a slide using hydraulic or pneumatic cylinders. Proper alignment of this is easy. Machines can be used with mechanized work feeds. Medium to the heaviest thickness weldable can be welded. Rating varies between 5 and 500 kVA.

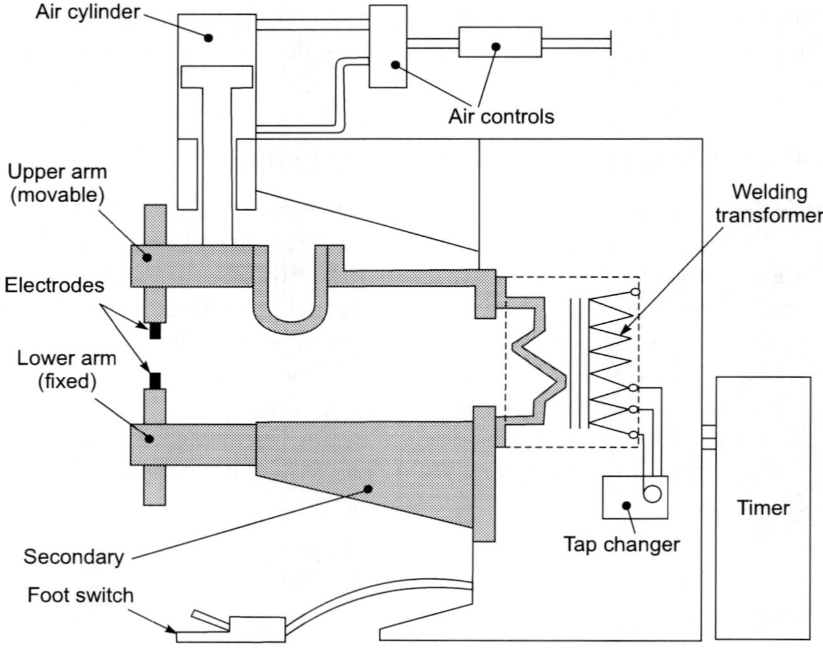

Figure 12.12 Press type spot welder

12.10.3 Portable Welder (Figure 12.13)

When the work is too bulky, a portable welder can be employed. Different designs of portable welders are available to suit the job. The welding transformer is separated

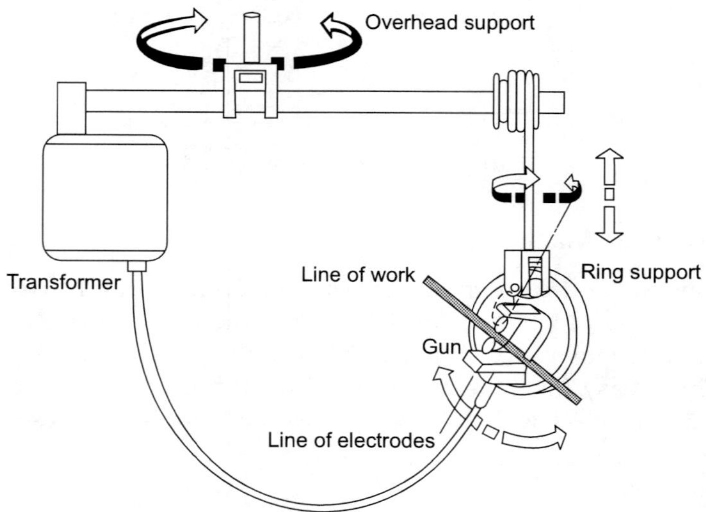

Figure 12.13 Portable welder

from the welding gun to facilitate ease in handling but necessitates the use of heavy cables to reduce losses in the secondary circuit to a minimum. The circuit reactance is kept low by running equal number of closely spaced conductors in both directions. It ensures minimum reactance and prevents kicking of the cable, which otherwise would occur due to the passage of heavy current. Such cables are also called kickless cables. Current control is usually through primary tap switching as the transformers are of single phase and work is of limited thickness.

There are two types of portable welders depending upon the type of clamping:

(i) The **pincer type gun,** uses levers as in arm or horn type machines.

(ii) In the **C-clamp gun,** the force is applied through a piston as in press type welder.

These machines are mostly slung from a balancing boom to reduce operator fatigue. These are widely used in the car industry and for manufacture of refrigerator shell bodies.

12.11 MACHINE RATING

The actual welding is done in the secondary circuit of the resistance welding machine but the voltage is not known and the current flowing, for a fraction of second for each weld cycle, is heavy. The machine rating is based upon the current drawn by the system at main supply and is in kVA unit. Water cooling of the system is necessary as the heat loss rises the temperature of the transformer. kVA ratings are quoted based upon the power that can be drawn for thirty seconds in each minute, which accounts for cooling requirements of the system. The kVA ratings of the welding machines for spot, steam and projection welding, normally varies between 5 and 500 kVA.

12.12 SURFACE PREPARATION

For all types of resistance welding, the condition of the surfaces of the parts to be welded largely controls how consistent the weld quality will be. The contact resistance of the faying surfaces has a significant influence on the amount of welding heat generated; hence, the electrical resistance of these surfaces must be highly uniform for consistent results. They must be free of high-resistance materials, such as paint, scale, thick oxides, and heavy oil and grease. If it is necessary to use a primer paint on the faying surfaces prior to welding, as is sometimes the case, the welding operation must be performed immediately after applying the primer, or special conducting primers must be used. For best results, the primer should be as thin as possible so that the electrode force will displace it and give metal-to-metal contact. Paint should never be applied to the outside base metal surfaces before welding, because it will reduce electrode life and produce poor surface appearance. Heavy scale should be removed by mechanical or chemical methods. Light oil on steel is not harmful unless it has picked up dust or grit. Drawing compounds containing mineral fillers should be removed before welding. The methods used for preparing surfaces for resistance welding differ for various metals and alloys. A brief description of surface conditions and methods of cleaning follows.

Aluminum

The chemical affinity of aluminum for oxygen causes it to become coated with a thin film of oxide whenever it is exposed to air. The thin oxide film that forms on a freshly cleaned aluminum surface does not cause sufficient resistance to be troublesome for resistance welding. The permissible holding period or elapsed time between cleaning and welding, may vary from 8 to 48 hours or more, depending upon the cleaning process, cleanliness of the shop, the particular alloy, and the application.

An aluminum surface may be mechanically cleaned for resistance welding with a fine grade of abrasive cloth, fine steel wool, or a fine wire brush. Clad aluminum may also be cleaned by mechanical means, but care must be taken not to damage the cladding. Numerous commercial chemical cleaners are available for aluminum. Chemical cleaning is usually preferred in large volume production for reasons of economy as well as uniformity and control.

Magnesium

Cleaning magnesium alloys is particularly important since they readily alloy with copper at elevated temperatures. The contact resistance between the electrode and the work must be kept as low as possible. Magnesium alloys are supplied coated with an oil or are chrome-pickled, to protect the metal from oxidation during shipment and storage. To obtain sound and consistent welds, the protective coating must be removed to facilitate the removal of residual magnesium oxide.

Copper

Cleaning of copper alloys is important. The coppers and aluminum-bronzes are particularly difficult to clean by chemical means and thus, mechanical means are preferred. In some instances, a flash coating of tin is applied to produce a uniformly higher surface resistance than pure copper would have.

Nickel

Nickel and its alloys demand high standards of material cleanliness for successful resistance welding. The presence of grease, dirt, oil and paint increases the probability of sulphur embrittlement during welding, and will result in defective welds. Oxide removal is necessary if heavy oxides are present from prior thermal treatments. Machining, grinding, blasting, or pickling may be employed. Wire brushing is not satisfactory.

Titanium

Before welding, the surfaces of titanium parts should be scrupulously clean. Materials such as oil, grease, dirt, oxides and paint can affect both weld consistency and chemical composition. Titanium and its alloys react with many elements and compounds at welding temperatures. Contamination by oxygen, hydrogen, nitrogen, and carbon, which enter the microstructure interstitially, can significantly reduce weld ductility and toughness. Scale-free surfaces may be welded either after degreasing or after degreasing plus acid pickling. The surfaces may be degreased with acetone, methyl ethyl ketone, or a dilute solution of sodium hydroxide. Chlorinated solvents should not be used. Titanium and its alloys are susceptible to stress corrosion.

Pickling may be used to remove light oxide scale before welding. Pickling is usually performed with solutions containing 2 to 5% HF and 30 to 40% HNO_3 by volume, balance water.

Steels

Plain carbon and low-alloy steels have relatively low resistance to corrosion in ordinary atmosphere; hence, these metals are usually protected by thin oil during shipment, storage, and processing. This oil film has no harmful effects on the weld, provided the oily surfaces are not contaminated with shop dirt or other poorly conductive or dielectric materials.

Steels are supplied with various surface finishes. Some of the most common are (1) hot-rolled, unpickled; (2) rolled, pickled, and oiled; and (3) cold-rolled with or without an anneal. Unpickled hot-rolled steel must be pickled or mechanically cleaned prior to welding. Hot-rolled pickled steel is weldable in the as-received condition, except for possible wiping to remove loose dirt. Cold-rolled steel presents the best welding surface and, if properly protected by oil, requires no cleaning prior to welding other than wiping to remove loose dirt.

High-alloy steels and stainless steels are non-corrosive and usually do not require cleaning before resistance welding. When exposed to elevated temperatures, stainless steels acquire an oxide film; the thickness depends upon the temperature and time of exposure. The scale is an oxide of chromium which is effectively removed by pickling. Oil and grease should be removed with solvents or by vapor degreasing prior to welding.

Coated Steels

The coatings and plantings applied to carbon steel to provide corrosion resistance or for decoration, lend themselves satisfactorily to resistance welding with a few exceptions. In general, good results may be obtained without special cleaning processes. Welding of aluminized steel results in less expulsion and pickup, if the surfaces are wire brushed. Phosphate coatings increase the electrical resistance of the surfaces to a degree that welding current cannot pass through the sheets with low welding pressures. Higher pressures produce welds, but slight variations in coating thickness may prevent welding.

Surface Preparation Control

Surface preparation control can be maintained by periodically measuring the room temperature and contact resistance of the workpiece, immediately after cleaning. The measurement is most readily taken tip to tip between two RW electrodes, through two or more thicknesses of metal. Unit surface resistance varies inversely with pressure, temperature, and area of contact. The test conditions must be specified to make the measurements significant in control of surface cleanliness.

12.13 APPLICATIONS OF RESISTANCE SPOT WELDING

Resistance spot welding (RSW) is used to fabricate sheet metal assemblies up to about 3.2 mm thickness, when the design permits the use of lap joints and leak-tight joints are not required. Occasionally, the process is used to join steel plates 6.35 mm thick or thicker; loading of such joints is limited and the joint overlap adds weight and cost to the assembly when compared to the cost of an arc welded butt joint. The process is used in preference to mechanical fastening, such as riveting or screwing, when disassembly for maintenance is not required. It is much faster and more economical because separate fasteners are not needed for assembly.

Spot welding is used extensively for joining low-carbon steel sheet metal components for automobiles, cabinets, furniture, and similar products. Stainless steel, aluminum, and copper alloys are commonly spot welded commercially.

Resistance welding is used for joining sheets, tubes, brass for making furniture. It also finds its applications in aircraft and automobile industries.

12.14 ADVANTAGES AND LIMITATIONS

The major advantages of resistance spot welding are its high speed and adaptability for automation in high-rate production of sheet metal assemblies. Spot welding is also economical in many job shop operations, because it is faster than arc welding or brazing and requires less skill to perform.

Advantages of Resistance Welding

1. No filler rod is required
2. High rate of production
3. Semi-skilled operator can perform the welding easily
4. Warping and distortion is minimum
5. No edge preparation is required
6. Less chance of workpiece burn-through
7. Process is fast and less time consuming
8. There is not much change in the metallurgy of workpiece unlike in other processes

Limitations of Resistance Welding

1. Disassembly for maintenance or repair is difficult
2. A lap joint adds weight and material cost to the product, when compared to a butt joint
3. The equipment costs are generally higher than the costs of most arc welding equipment
4. The short time, high-current power requirement produces unfavorable line power demands, particularly with single-phase machines
5. Spot welds have low tensile and fatigue strengths, because of the notch around the periphery of the nugget between the sheets
6. The full strength of the sheet cannot prevail across a spot welded joint, because fusion is intermittent and loading is eccentric due to the overlap
7. All types of metal cannot be welded
8. Complex shapes are difficult to join
9. Welding joint is not so strong
10. Thicker plates cannot be welded easily

12.15 PROCESS VARIATIONS

Variation of resistance spot welding process differs in the application of welding current and pressure, and the arrangement of the secondary circuit.

12.15.1 Direct and Indirect Welding (Figure 12.14)

Direct welding is a resistance welding secondary circuit variation in which welding current and electrode force are applied to the workpieces by directly opposing electrodes. Indirect welding is a variation in which the welding current paths through the workpieces are away from, as well as at, the spot weld locations.

Typical arrangements for direct and indirect spot welding are shown schematically in Figure 12.14. Figures 12.14(a), (b), and (c) apply to direct resistance welding. In Figure 12.14(a), direct welding is done using electrodes with similar geometries. A larger electrode against one workpiece in Figure 12.14(b) provides an increased contact surface area for applications requiring better welding heat balance, or in order to reduce the marking on the lower sheet by the electrodes. The larger electrode surface area will conduct heat away from the weld joint more rapidly. Figure 12.14(c) is a schematic of two or more electrodes connected in series to a single transformer. Electrode arrangements such as this can make spot welds in rapid succession with one set of electrodes in contact with the work at a time. This arrangement is economical with respect to equipment costs. Figures 12.14(d) through (g) represent indirect resistance welding arrangements. A backing plate arrangement in Figure (d) provides a current path and pressure when the backing plate is made of a conducting material.

If the backing plate is non-conductive, the plate only provides welding pressure, and the current path will be from the top electrode, through the faying surface location, along the lower workpiece to the return connection further down the joint. Figure 12.14(e) is similar to (d) with the exception that the pressure electrode is away from the lap joint. Current may be through a conducting backing plate between the electrodes, or through the base material between the electrodes. Figures 12.14(f) and (g) are similar to the joints in (e) and (d), but are for spot welding high-resistance materials which require higher voltages. The two secondary circuits are in series, and are connected to two transformers. The primary circuits may be connected either in series or in parallel.

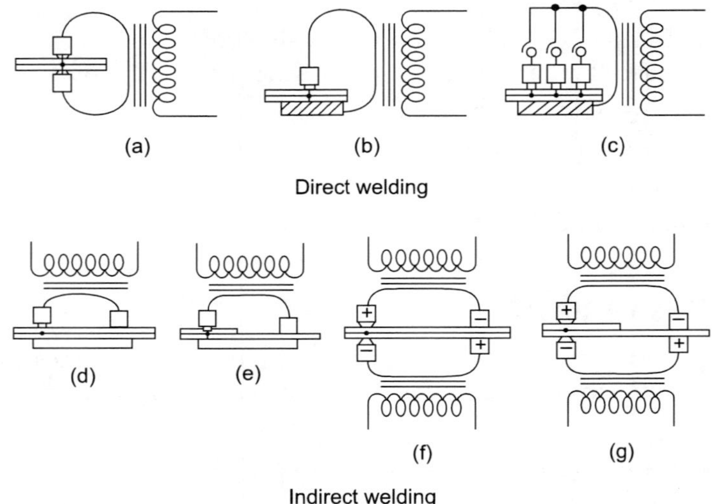

Figure 12.14 Typical secondary circuit arrangements for direct and indirect resistance spot welding

12.15.2 Parallel and Series Welding (Figure 12.15)

Parallel and series resistance welding are secondary circuit variations used for multiple spot welding applications. Examples of parallel and series welding arrangements are shown schematically in Figure 12.15. Parallel welding arrangements divide and conduct the secondary current through the workpieces and electrodes in parallel electrical paths, simultaneously forming spot welds. Figures 12.15(a) and (b) represent parallel welding arrangements. In Figure (a), the welding current is from a single transformer with multiple electrodes in the secondary circuit in parallel. Figure (b) represents a parallel welding system operating with a three-phase primary. This particular system is limited to three work stations.

In series welding (Figures 12.15(c) and (d), the secondary circuit current is conducted through the workpieces and the electrodes in a series electrical path, simultaneously forming multiple spot welds at the electrode locations. A series welding arrangement requires equal resistance values at the faying surfaces in order to obtain uniform heating at each spot weld. When spot welding with two electrodes is in series, a portion of the current travels through the adjacent workpiece from one electrode to the other, bypassing the faying surfaces. This shunted current does not contribute to the spot weld, and must be taken into account when developing a series spot welding procedure.

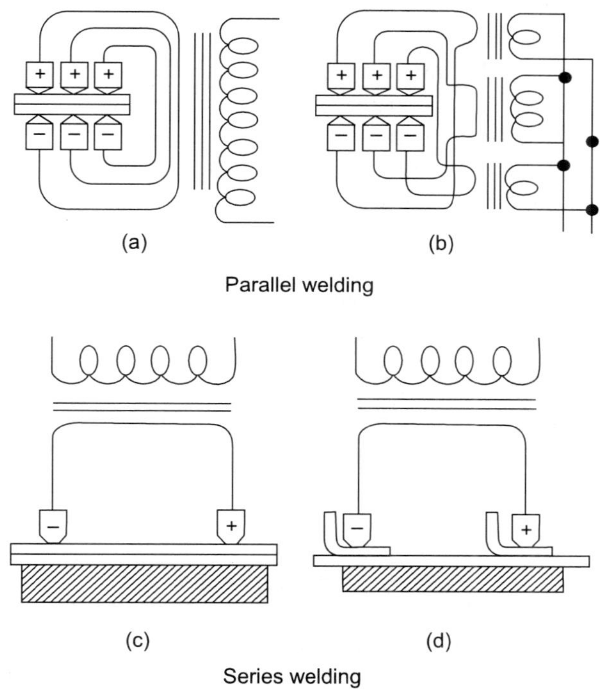

Figure 12.15 Typical secondary arrangements for direct multiple spot welding

12.15.3 Weld Bonding

Weld bonding is a combination of resistance spot welding and adhesive bonding. Paste or film adhesive is placed between the members to be joined, and resistance welds are then made through the adhesive layer. The adhesive is allowed to cure either at ambient temperature or heated in an oven, as required by the adhesive manufacturer. The spot welds principally hold the joint together during curing; they are fewer in number than otherwise required and so do not contribute greatly to the strength of the joint. Common structures joined by weld bonding are found in the aerospace and transportation industries. Weld bonding is used to attach beaded panels to aircraft skins, and aircraft or truck skins to channels, angles, and other types of reinforcement.

The adhesive, whether paste or film type; can be applied to one or both joint surfaces. The electrode force during welding squeezes out the adhesive at the spot weld locations, creating a current path through the sheets. The adhesive must have good wetting and flow characteristics, in order to bond the faying surfaces securely. Premature curing of the adhesive, during or prior to the spot welding, may hamper proper adhesive movement and result in high resistance between the faying surfaces. High resistance may impede the weld current, or result in excessive heating and subsequent metal expulsion. Application of a precompression electrode force prior to the welding cycle may help displace the adhesive at each weld site.

Weld bonding improves the fatigue life and durability of the joint over that obtained with spot welding alone. The process may also improve stress distribution, joint rigidity, and buckling resistance in thin sheets. The adhesive in the joint dampens vibration and noise, and provides some corrosion resistance. In some aircraft components, greater cost effectiveness is obtainable with weld bonding than with mechanical fastening or adhesive bonding alone.

Disadvantages include the additional costs of the adhesive, the added curing operation, and the time and labor costs for cleaning the components, treating the surfaces, and applying the adhesive. In addition, operating temperatures for the component are limited to the effective service temperature of the adhesive.

The presence of adhesive in the joint makes welding more difficult, and may contribute to significant variations in weld quality. Regardless of welding conditions, not all of the adhesive will be displaced from the sheets and therefore, the contact resistance will be higher than with clean sheets.

Worked Examples

Example 1: Aluminum strips of 2 mm thickness are joined together by resistance spot welding process by applying an electric current of 6000 A for 0.15 seconds. The heat required for melting aluminum is 2.9 J/mm^3. The diameter and thickness of weld nugget are found to be 5 mm and 2.5 mm, respectively. Assume the electrical resistance to be 75 micro-ohms. What is the percentage of total energy used in forming the weld nugget? **(GATE 2008)**

Solution: Given

$$I = 6000\text{ A},\ \tau = 0.15\text{ sec},\ t = 2\text{ mm}$$

$$\text{Heat rate} = 2.9\text{ J/mm}^3$$

$$\text{Diameter } (D) = 5\text{ mm, thickness } (t) = 2.5\text{ mm}$$

$$\text{Total heat required} = \text{Total volume} \times \text{heat rate}$$

$$\text{Heat required} = \frac{\pi D^2 h}{4} = \frac{\pi \times 5^2 \times 2.5 \times 2.9}{4} = 142.35\text{ joules}$$

$$\text{Heat generated} = I^2 R\tau = 6000^2 \times 75 \times 10^{-6} \times 0.15 = 400 \text{ J}$$

$$\text{Melting efficiency} = \frac{\text{Heat required}}{\text{Heat input}} = \frac{142.28}{405} = 0.35 = 35\%$$

Example 2: For spot welding of two steel sheets (base metal), each of 3 mm thickness, welding current of 10,000 A is applied for 0.2 sec. The heat dissipated to the base metal is 1000 J. Assuming that the heat required for melting 1 mm^3 volume of steel is 20 J and interfacial contact resistance between sheets is 0.0002 ohms. What is the volume of weld nugget in mm^3? **(GATE 2014)**

Solution: I = 1000 A, τ = 0.2 sec, t = 3 mm

$$\text{Heat rate} = 20 \text{ J/mm}^3$$
$$\text{Total heat required} = \text{Total volume} \times \text{heat rate}$$
$$\text{Heat required} = \text{Total volume} \times 20$$
$$\text{Heat generated} = I^2 R\tau = 10000^2 \times 0.0002 \times 0.2 = 4000 \text{ J}$$

Heat dissipated to base plate is 1000 J

Net heat available for welding is 4000-1000 J

$$\text{Heat generated} = \text{Heat available for welding}$$
$$\text{Total volume} \times 20 = 3000$$

Solving, we get total volume = 150 mm^3.

Example 3: Two flat steel plates, each of 2.5 mm thickness are being resistance spot welded using a current of 6000 A and weld time of 0.2 sec. The contact resistance at the interface between the two sheets is 200 micro-ohms and specific energy to melt steel is 10 × 109 J/m^3. A spherical melt pool of diameter 4 mm is formed at the interface due to the current flow. Consider that electrical energy is completely converted to thermal energy. What is the ratio of heat used for melting to the total resistive heat generated? **(GATE 2016)**

Solution: Given

$$I = 6000 \text{ A}, \ \tau = 0.2 \text{ sec}$$
$$\text{Resistance} = 200 \text{ micro-ohm}$$

Heat required = Volume of nugget × Heat required to melt unit volume of material

$$\text{Volume} = \frac{4\pi r^3}{3} = \frac{4 \times \pi \times (2 \times 10^{-3})^3}{3}$$

$$\text{Heat required} = \frac{4 \times \pi \times (2 \times 10^{-3})^3}{3} \times 10 \times 10^9 = 335.1032 \text{ J}$$

$$\text{Heat generated} = I^2R\tau = 6000^2 \times 200 \times 10^{-6} \times 0.2 = 1440 \text{ J}$$

$$\text{Melting efficiency} = \frac{\text{Heat required}}{\text{Heat input}} = \frac{335.1032}{1440} = 0.2327 = 23.27\%$$

Example 4: Two different pairs of sheets of same material are welded by resistance spot welding. In one pair, the average radius of each spherical bridge is 0.2 mm and the number of bridges per cm^2 is 25. In another pair, the number of bridges per cm^2 is 50 with same radius of bridge, the contact resistance per unit area is given by $R = 0.85\frac{\rho}{n\pi r}$, where, r is the radius of spherical bridge and ρ is the resistivity of the metal. It the voltage applied is 5 volts and the resistivity of metal is 2×10^{-5} ohm cm, what is the heat generated per cm^2 in each case? **(GATE 2000)**

Solution: $R_1 = 0.85\frac{\rho}{n\pi r} = 0.85\frac{2\times 10^{-5}}{25\times \pi \times 0.02} = 1.082\times 10^{-5}$

$$\text{Power, } VI = V \times \frac{V}{R} = \frac{V^2}{R} = \frac{5^2}{1.082\times 10^{-5}} = 2.3 \text{ MW}$$

$$R_1 = 0.85\frac{\rho}{n\pi r} = 0.85\frac{2\times 10^{-5}}{50\times \pi \times 0.02} = 0.541\times 10^{-5}$$

$$\text{Power, } VI = V \times \frac{V}{R} = \frac{V^2}{R} = \frac{5^2}{0.541\times 10^{-5}} = 4.6 \text{ MW}$$

Example 5: Two sheets of low-carbon steel, each 1.5 mm thick, are spot welded by passing a current of 10,000 A for 5 cycles in 50 Hz main supply. What is the heat generated due to supply of power if the effective resistance is 200 micro-ohms? The maximum indentation is 10% of the sheet thickness and density of spot weld nugget is 8 grams/cc. If 1380 joules are required to melt 1 gram of steel, determine the heat required for welding operation, if $D - 6\sqrt{t}$ is used to determine nugget diameter D and assume the nugget size to be equal to metal between the two electrodes.

(GATE 1992)

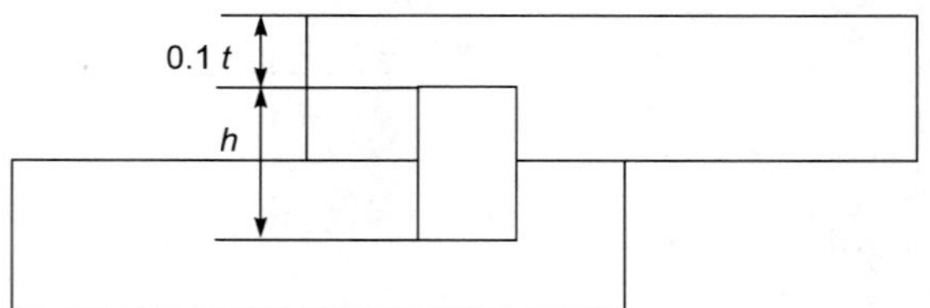

Figure 12.16

Solution: Heat generated $= I^2R\tau = 10000^2 \times 200 \times 10^{-6} \times \frac{5}{50} = 2000$ joules

$$h = 2t - 2(0.1t) = 1.8t = 1.8 \times 1.5 = 2.7 \text{ mm}$$

$$D = 6\sqrt{t} = 6\sqrt{1.5} = 7.35 \text{ mm}$$

$$\text{Volume of nugget} = \frac{\pi \times 7.35^2}{4} \times 2.7 = 114.5 \text{ mm}$$

$$\text{Heat required} = \text{Volume of nugget} \times \text{density} \times \text{heat generated per gram}$$

$$\text{Heat required} = 114.5 \times 10^{-3} \times 8 \times 1380 = 1264 \text{ joules}$$

Example 6: Resistance spot welding of two sheet sheets is carried out in lap joint configuration by using a welding current of 3 kA and a welding time of 0.2 sec. A molten weld nugget of volume 20 mm^2 is obtained. The effective contact resistance is 200 μΩ. The material properties of steel given are: latent heat of melting is 1400 kJ/kg, density is 8000 kg/m^3, metling temperature is 1520°C, specific heat is 0.5 kJ/kg°C and the ambient temperature is 20°C. What is heat in joules used for producing weld nugget, assuming 100% heat transfer efficiency. What is the heat dissipated to the base metal, neglecting all other heat losses? **(GATE 2009)**

Solution: Given

$$I = 3000 \text{ A}, \tau = 0.2 \text{ sec}$$

$$\text{Resistance} = 200 \ \mu\Omega$$

$$\text{Heat generated} = I^2R\tau = 3000^2 \times 200 \times 10^{-6} \times 0.2 = 360 \text{ J}$$

$$\text{Heat required} = mh_{fg} + mC_p(T_m - T_a)$$

$$m = \text{Volume} \times \rho = 20 \times 10^{-9} \times 8000$$

$$\text{Heat required} = 20 \times 10^{-9} \times 8000\,[1400 \times 1000 + 500(1520 - 20)] = 344 \text{ joules}$$

$$\text{Heat dissipated} = 360 - 344 = 16 \text{ joules}$$

Example 7: A resistance welder is rated at 50 kVA at 50% duty cycle. Secondary voltage is always constant at 25 volts. If a job requires 1500 A current, determine the maximum welding time in a total cycle time of 30 seconds. **(GATE 1991)**

Solution: Rated current, $I_R = \frac{50 \times 1000}{25} = 2000$ A

$$\text{Rated duty cycle, } D_R = 50\%$$

$$\text{Desired duty cycle} = 100\%$$

$$D_R \times I_R^2 = I_{100}^2 \times D_{100}$$

where,

I_{100} is current at 100% duty cycle
D_{100} is 100% duty cycle
I_R is current at required duty cycle = 1500 A
D_R is required duty cycle

$$50 \times 2000^2 = 1500^2 \times D_{100}$$

Solving we get,

$$D_{100} = 88.8$$

$$D_{100} = \frac{\text{Arc on time}}{\text{Total welding time}}$$

$$\frac{88.8}{100} = \frac{\text{Arc on time}}{30}$$

Arc on time = 26.67 seconds

Example 8: Spot welding of two steel sheets, each 2 mm thick, carried out successfully by passing 4 kA of current for 0.2 seconds through the electrodes. The resulting weld nugget formed between the sheet is 5 mm in diameter. Assuming cylindrical shape for the nugget, what is the thickness of the nugget in mm. Use the following data.

(GATE 2016)

Latent heat of fusion for steel	1400 kJ/kg
Effective resistance of the weld joint	200 micro-ohm
Density of steel	8000 kg/mm^3

Solution: Heat generated $= I^2 R\tau = 4000^2 \times 200 \times 10^{-6} \times 0.2$ (1)

Heat used = Volume of nugget × density × latent heat

= Volume of nugget × 8000 × 1400 × 1000 (2)

$$\text{Volume of nugget} = \frac{\pi \times 5^2}{4} \times h \times 10^{-6} \quad (3)$$

Using (3) in (2) and equating to (1) we get

$$I^2 R\tau = 4000^2 \times 200 \times 10^{-6} \times 0.2 = \frac{\pi \times 5^2}{4} \times h \times 10^{-6}$$

Solving we get thickness of the nugget = 2.91 mm

Example 9: In spot pulsed laser welding of aluminum plates of density 2700 kg/m^3, specific heat is 896 J/kg, melting temperature is 933 K, latent heat of melting is 398 kJ/kg at a temperature of 30°C pulse with energy of 0.5 J is focused onto an area of 0.05 mm^2. If the entire energy is coupled into the material, what will be the depth of the weld, assuming the cross-sectional area of the weld is circular and is uniform throughout its depth, and heat conduction only in the direction of penetration?

Solution: Heat supplied = Heat utilized

$$0.5 \text{ J} = m(\text{sensible heat} + \text{latent heat})$$

$$\text{Heat required} = mh_{fg} + mC_p(T_m - T_a)$$

$$m = a \times h \times \rho = 0.05 \times 10^{-6} \times h \times 2700 \times 398000$$

$$0.5 = 0.05 \times 10^{-6} \times h \times 2700 \times 398000 \times + 0.05 \times 10^{-6} \times h \times 896(1793 - 292)$$

Solving we get, h = 0.00385 m = 3.85 mm

Example 10: Two metallic sheets, each of 2 mm thickness are welded in lap joint configuration by resistance spot welding at a welding current of 10 kA and welding time of 10 milliseconds. A spherical fusion zone extending up to the full thickness of each sheet is formed. The properties of the metallic sheet are given below:

Ambient temperature = 293 K, melting temperature = 1793 K, density = 7000 kg/m^3, latent heat of fusion = 300 kJ/kg and specific heat is 800 J/kg K. Assume (i) Contact resistance along sheet-sheet interface is 500 micro-ohm and along the electrode interface is zero. (ii) No conductive heat loss through the bulk sheet material. (iii) The complete weld fusion zone is at melting temperature. What is the melting efficiency? **(GATE 2007)**

Solution: Given

$$I = 10{,}000 \text{ A}, \tau = 0.01 \text{ sec}$$

$$h_{fg} = 300 \frac{\text{kJ}}{\text{kg}} \; C_p = 800 \frac{\text{J}}{\text{kg}} \text{K}$$

$$\text{Ambient temperature, } T_m = 293 \text{ K}$$

$$\text{Mass of nugget} = \text{Volume of nugget} \times \text{density}$$

$$m = \frac{4\pi r^3 \rho}{3} = \frac{4 \times \pi \times 0.02^3 \times 7000}{3} = 2.346 \times 10^{-4} \text{ kg}$$

$$\text{Heat required} = mh_{fg} + mC_p(T_m - T_a)$$

$$= 2.346 \times 10^{-4} \times 300 + 2.346 \times 10^{-4} \times 0.8(1793 - 292) = 351.9 \text{ kJ}$$

$$\text{Heat generated} = I^2 R\tau = 10000^2 \times 500 \times 10^{-6} \times 0.01 = 500 \text{ kJ}$$

$$\text{Melting efficiency} = \frac{\text{Heat required}}{\text{Heat input}} = \frac{351.9}{500} = 70.38\%$$

Example 11: Resistance welding is performed on two plates of 1.5 mm thickness with 6 mm diameter electrode, using 15,000 A current for a time duration of 0.25 seconds. Assuming the interface resistance to be 0.0001 ohms, what is the heat generated to form the weld? **(GATE 2001)**

Solution: Heat generated $= I^2R\tau = 15000^2 \times 0.0001 \times 0.25 = 5625\text{ J}$

Example 12: Two 1 mm thick steel sheets are to be spot welded at a current of 5000 A. Assuming effective resistance to be 200 micro-ohms and current flow of 0.2 sec. What is the heat generated during the process? **(GATE 2004)**

Solution: Heat generated $= I^2R\tau = 5000^2 \times 200 \times 10^{-6} \times 0.2 = 1000\text{ J}$

Example 13: Two flat copper sheets (each 1.5 mm thick) are spot welded by the use of a current of 7000 A and a current flow time of 0.3 sec. The electrodes are 5 mm in diameter. Estimate the heat generated in the weld zone. Assume that the resistance is 200 μ. Specific energy needed to melt copper is 6.2 J/mm^3. Calculate the temperature rise, assuming that the heat generated is confined to the volume of material directly between the two electrodes and that the temperature distribution is uniform. The specific heat for copper is 385 J/kg K.

(*Manufacturing Process*: Kalpakjian)

Solution: Heat generated $= I^2R\tau = 7000^2 \times 200\ \ 10^{-6} \times 0.3 = 2940\text{ J}$

$$\text{Also volume of nugget} = \frac{\pi \times 5^2}{4} \times 3 = 58.9\text{ mm}^3$$

$$\text{Heat needed} = \text{Volume} \times \text{specific energy to melt copper}$$
$$= 58.9 \times 6.2 = 359\text{ J}$$

The remaining heat (that is, 2940 – 359 J = 2581 J) is dissipated into the volume of metal surrounding the weld nugget.

The volume of metal directly under the 5 mm electrodes is 58.9 mm^3.

Mass of the metal = Volume × density = 58..9 mm^3 × 0.00897 = 00.53 g

The specific heat for copper is 385 J/kg K. Therefore, the theoretical temperature rise is

$$\Delta T = \frac{2940\text{ Joules}}{385\,\dfrac{\text{joules}}{\text{kg K}} \times 0.00053\text{ kg}} = 14,400\text{ K}$$

Melting point of copper is 1352 K and much energy is needed to melt this small amount of volume of material.

Example 14: An RSW operation is used to make a series of spot welds between two pieces of aluminum, each 2.0 mm thick. The unit melting energy for aluminum = 2.90 J/mm^3, welding current = 6000 A, and time duration = 0.15 sec. Assume that the resistance = 75 micro-ohms. The resulting weld nugget measures 5.0 mm in diameter by 2.5 mm thick. How much of the total energy generated is used to form the weld nugget? **(*Fundamentals of Modern Manufacturing*: Groover)**

Solution: $H = I^3R\tau = 600^2(75 \times 10^{-6}) \times 0.15 = 405$ J

$$\text{Weld nugget volume} = \frac{\pi D^2 d}{4} = \frac{\pi \times 5^2 \times 2.5}{4} = 49.1\,\text{mm}^3$$

$$\text{Heat required for melting} = U_m \times \text{V} = (2.9\,\text{J}/\text{mm}^3)\,(49.1\,\text{mm}^3) = 142.4\,\text{J}$$

Proportion of heat used for welding = 142.4/405 = 0.351 = 35.1%

Example 15: The unit melting energy for a certain sheet metal is 9.5 J/mm^3. The thickness of each of the two sheets to be spot welded is 3.5 mm. To achieve required strength, it is desired to form a weld nugget that is 5.5 mm in diameter and 5.0 mm thick. The weld duration will be set at 0.3 sec. If it is assumed that the electrical resistance between the surfaces is 140 micro-ohms, and that only one-third of the electrical energy generated will be used to form the weld nugget (the rest being dissipated), determine the minimum current level required in this operation.

(*Fundamentals of Modern Manufacturing*: Groover)

Solution: $\text{Weld nugget volume} = \dfrac{\pi D^2 d}{4} = \dfrac{\pi \times 5.5^2 \times 5}{4} = 118.8\,\text{mm}^3$

$$\text{Heat required for melting} = U_m \times \text{V} = (9.5\,\text{J}/\text{mm}^3)(118.8\,\text{mm}^3) = 1129\,\text{J}$$

$$\text{Required heat for RSW operation, } H = \frac{1129}{\frac{1}{3}} = 3386\text{ joules}$$

$$H = I^2R\tau = 6000^2(140 \times 10^{-6}) \times 0.3 = 3386\text{ joules}$$

Solving for I, we get current = 8980 A

Example 16: An experimental power source for spot welding is designed to deliver current as a ramp function of time: I=100,000t, where I = amp and t = sec. At the end of the power-on time, the current is stopped abruptly. The sheet metal being spot welded is low-carbon steel whose unit melting energy = 10 J/mm^3. The resistance R

= 85 micro-ohms. The desired weld nugget diameter = 4 mm and thickness = 2 mm (assume a disc-shaped nugget). It is assumed that 1/4 of the energy generated from the power source will be used to form the weld nugget. Determine the power-on time, the current must be applied in order to perform this spot welding operation.

(*Fundamentals of Modern Manufacturing*: Groover)

Solution: Weld nugget volume $= \dfrac{\pi D^2 d}{4} = \dfrac{\pi \times 4^2 \times 2}{4} = 25.14 \text{ mm}^3$

Heat required for melting $= H_m = U_m \times V = (10 \text{ J/mm}^3)(25.14 \text{ mm}^3) = 251.4 \text{ J}$

$$H = \frac{251.4}{0.25} = 1005.6 \text{ J} \qquad (1)$$

$$\text{Power} = \int I^2 R dt = \int (100{,}000t)^2 \times (75 \times 10^{-6}) \times dt$$

$$\text{Power} = \int_o^t (100{,}000t)^2 (75 \times 10^{-6}) \times dt = 31481.5t^3 \qquad (2)$$

Equating (1) and (2) and solving it, we get

$$t = 0.317 \text{ sec}$$

Example 17: Spot welding of two 1 mm thick sheets of steel (density = 8000 kg/m^3) is carried out successfully by passing a certain amount of current for 0.1 second through the electrodes. The resultant weld nuggets formed are 5 mm in diameter and 1.5 mm thick. The latent heat of fusion of steel is 1400 kJ/kg and the effective resistance in welding operation is 2000 micro-ohms. What is the current passing through the electrode. **(GATE 2005)**

Solution: Density = 8000 kg/m^3

$$t = 0.1 \text{ sec}$$
$$d = 5 \text{ mm}$$
$$w = 1.5 \text{ mm}$$
$$L = 1400 \text{ kJ/kg}$$
$$R = 200 \ \mu\Omega$$

Mass can be calculated using the equation

Mass = density × volume

$$\text{Mass} = 8000 \times \frac{\pi}{4}\left(\frac{5}{1000}\right)^2 \times \frac{1.5}{1000} = 2.35 \times 10^{-4} \text{ kg}$$

Total heat of fusion can be found out by

$$Q = \text{mass} \times \text{latent heat of fusion}$$

$$Q = 2.35 \times 10^{-4} \times 1400 \times 1000 = 329 \text{ joules} \quad \ldots(1)$$

We know that in spot welding the amount of the heat developed in contact area,

$$Q = I^2Rt = I^2 200 \times 10^{-6} \times 0.1 \quad \ldots(2)$$

Equating (1) and (2)

$$I^2 200 \times 10^{-6} \times 0.1 = 329$$

Solving we get current, $I = 4060$ amperes.

Example 18: A 900 mm long steel plate is welded by manual metal arc welding process using welding current of 150 A, arc voltage of 20 V and welding speed of 300 mm/min. If the process efficiency of is 0.8 and surface resistance is 36 micro-ohms, the heat input in J/mm. **(GATE 2003)**

(a) 600 kJ/mm (b) 480 kJ/mm (c) 146 kJ (d) 116.5 kJ

Solution: Welding time = length/speed = 900/300 = 3 minutes = 1800 seconds

Heat generated = $(I^2Rt) \times$ efficiency

$$Q = 150^2 \times 36 \times 10^{-6} \times 180 \times 0.80 = 116 \text{ kJ}$$

Therefore, option (d) is correct.

Example 19: In spot pulsed laser welding of aluminum plates (density = 2700 kg/m^3, specific heat = 896 J/kg, melting temp = 933 K, latent heat of melting = 398 kJ/kg at a temp of 30 deg, pulse with energy of 0.5 J) is focused onto an area of 0.05 mm^2. If the entire energy is coupled into the material, what will be the depth of the weld assuming the cross-sectional area of the weld is circular and is uniform throughout its depth and only heat conduction in the direction of penetration. **(GATE 2002)**

(a) 5.34 mm (b) 2.15 mm (c) 4.23 mm (d) 3.95 mm

Solution: Heat supplied = heat utilized

0.5 joules = mass (sensible heat + latent heat)

Mass can be calculated using the equation

$$\text{Mass} = \text{density} \times \text{volume}$$

$$\text{Mass} = 2700 \times 0.05 \times 10^{-6}\, h = 1.35 \times 10^{-4}\, (h) \text{ kg}$$

where, h is the depth of the weld.

Total heat of fusion can be found out by

$$Q = \text{mass} \times \text{latent heat of fusion}$$

$$Q_1 = 1.35 \times 10^{-4}\, (h) \times 398 \times 1000 = 53.73\, h \text{ joules} \quad \ldots(1)$$

$$Q = \text{mass} \times \text{sensible heat}$$

$$Q_2 = 1.35 \times 10^{-4}\, h \times 896 \times (933 - 273) = 72.576\, h \text{ joules} \qquad \ldots(2)$$

$$Q = Q_1 + Q_2 = 53.73\, h + 72.576\, h = 0.5$$

$$h = \frac{0.5}{126.306} = 3.95 \times 10^{-3} \text{ mm} = 3.95 \text{ mm}$$

Review Questions

1. What is spot welding?
2. Explain with a neat sketch the process of resistance spot welding.
3. How does a spot welding machine work?
4. What is a spot welding gun?
5. What are the applications of resistance spot welding?
6. What are the advantages of resistance spot welding?
7. What are the limitations of resistance spot welding?
8. What are the spot weld gun types?
9. Explain the process parameters of resistance spot welding.
10. What are the types of spot welding electrodes?

13

CHAPTER

Resistance Seam Welding

13.0 INTRODUCTION

Resistance seam welding is a resistance welding process which produces coalescence at the flying surface using the resistive heating obtained by contact resistance at frying surfaces, which are held together under pressure by two electrodes. Seam welding is also known as continuous spot welding process as the spot is produced in the work continuously.

The resulting weld is a series of overlapping resistance spot welds made progressively along a joint rotation under pressure by electrodes. When the spots are not overlapped enough to produce gaslight welds, it is a variation known as roll resistance spot welding. This process differs from spot welding since the electrodes are wheels. Both lower and upper electrode wheels are powered. Pressure is applied in the same manner as a press type welder. The wheels can be either in line with the throat of the machine or transverse. If they are in line, it is normally called a seam welding machine. Welding current is transferred through the bearing of the roller electrode wheels. Water cooling is not provided internally and therefore, the weld area is flooded with cooling water to keep the electrode wheels cool.

In seam welding, a rather complex control system is required. This involves the travel speed as well as the sequence of current flow to provide for overlapping welds. The welding speed, the spot per inch, and the timing schedule are dependent on each other. Welding schedules provide the pressure, the current, the speed, and the size of electrode wheels.

This process is quite common for making flange welds, watertight joints for tanks, etc. Another variation is so called mash seam welding, where the lap is fairly narrow and the electrode wheel is at least twice as wide as used for standard seam welding. The pressure is increased to approximately 300 times normal pressure. The final weld mash seam thickness is only 25% greater than the original single sheet (Figure 13.1).

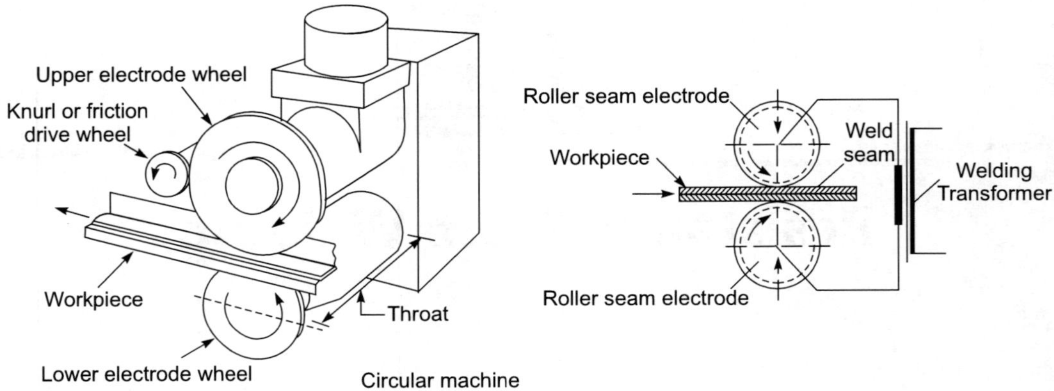

Figure 13.1 Position of electrode wheels on resistance seam welder

13.1 TYPES OF SEAM WELDING MACHINE

Circular: This is common type of seam welding machine in which face of the electrode wheels are parallel to the throat of the machine and throat depth is typically 30 to 90 cm.

Longitudinal: This type of machines are used for welding short seam in containers.

Universal: Seam welding replaces spot welding for producing continuous leak-proof joint for use in sheet metal item (containers), which are frequently used in automobile industries.

13.2 CLASSIFICATION OF SEAM WELDING

13.2.1 Lap Seam Welding

Lap joints can be seam welded using two wheel electrodes (Figure 13.2(a)) or with one wheel and a mandrel. The minimum joint overlap is the same as for spot welding, namely, twice the minimum edge distance (distance from the center of the weld nugget to the edge of the sheet).

13.2.2 Mash Seam Welding

Mash seam welding is a resistance welding variation that makes a lap joint primarily by high-temperature plastic forming and diffusion, as opposed to melting and solidification; the joint thickness after welding is less than the original assembled thickness.

Mash seam welding (Figure 13.2(b)) requires considerably less overlap than the conventional lap joint. The overlap is about 1 to 1.5 times the sheet thickness, with proper welding procedures. Wide, flat faced wheel electrodes, which completely cover the overlap, are used. Mash seam welding requires high electrode force, continuous welding current and accurate control of force, current, welding speed, overlap, and joint thickness in order to obtain consistent welding characteristics. Overlap is maintained at close tolerances, usually by rigidly clamping or tack welding the pieces.

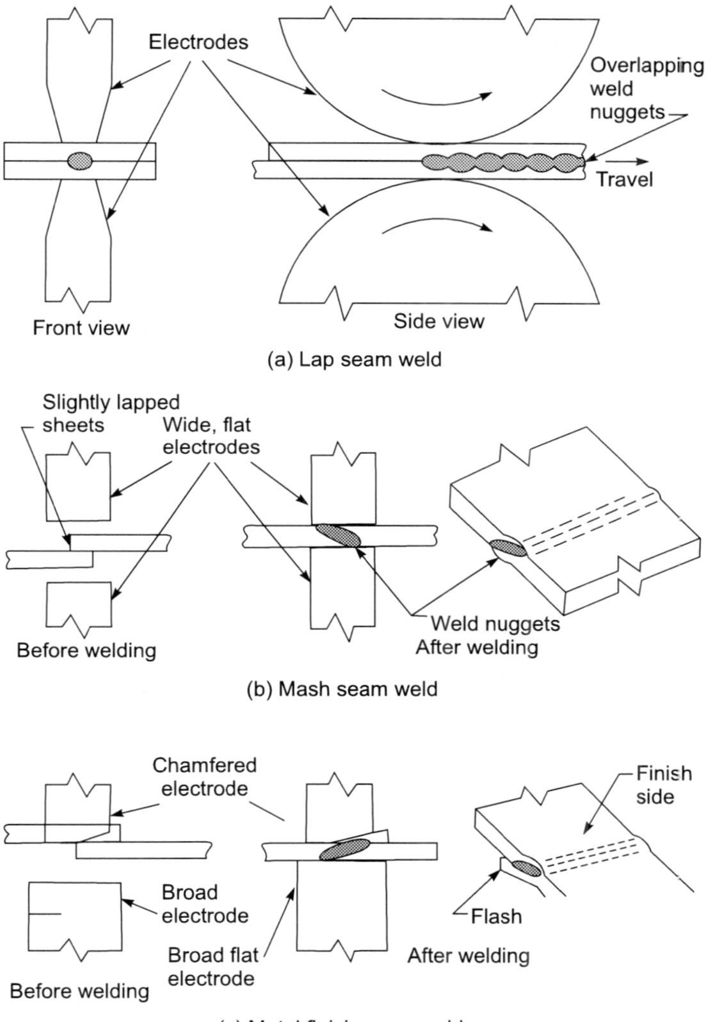

Figure 13.2 Resistance seam welding variations

Typically, the exposed or show side of the welded component is placed against a mandrel which acts as an electrode and supports the members to be joined. A welding wheel electrode is applied to the side of the joint that does not show. The show surface of the joint must be mashed as nearly flat as possible so that it presents a good appearance. Proper positioning of the wheel with respect to the joint is required to obtain a smooth weld face. Some polishing of the weld area may be required before painting or coating, when the appearance of the finished product is important.

Mash seam welding produces continuous seams, which have good appearance and are free of crevices. Crevice-free joints are necessary in applications having strict contamination or cleanliness requirements, such as joints in food containers or refrigerator liners.

Disadvantages of the mash seam welding process include the following:

1. Offset at the joint, due to the inability of the process to completely flatten the seam.
2. Distortion: the inherent lateral flow of metal, as it is welded, is restrained by fixturing or by tack welds.
3. Very rigid fixturing, required to resist weld distortion.

In order to obtain acceptable welds, the materials to be joined by mash seam welding must have wide plastic temperature ranges. Low-carbon steel and stainless steel can be mash seam welded for certain applications.

13.2.3 Metal Finish Seam Welding

Lap and mash seam welds differ with respect to the amount of forging or, as the name implies, mash down. The lap weld has practically no mash down, while the thickness of a mash seam weld approaches that of one sheet thickness. In metal finish seam welding, mash down occurs on only one side of the joint (Figure 13.2(c)), and is a compromise between lap and mash seam welding.

The amount of deformation or mash, is affected by the geometry of one electrode wheel face and the position of the joint with respect to that face. The wheel face is beveled on one side of the midpoint (Figure 13.3). This varies the amount of deformation across the joint. Good surface finish can be produced on the side of the joint against the flat wheel by using proper welding procedures.

The location of the edge of the sheet contacting the flat faced electrode, relative to the bevel on the other electrode, must be held within close tolerance. With 0.031 inch (0.8 mm) thick low-carbon steel sheet, for example, the edge must be within 0.4 mm of the center. The overlap distance is not critical.

Higher amperage and electrode force are required than those for mash seam welding, because of the greater overlap distance. Materials that are easily mash seam welded (those with wide plastic temperature ranges) are also easily welded using the metal finish seam welding variation.

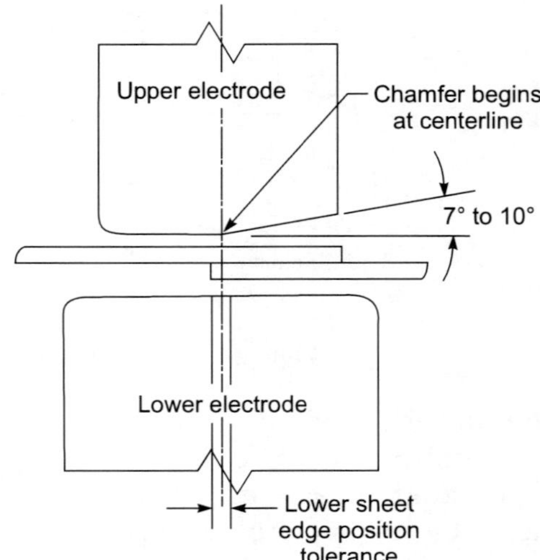

Figure 13.3 Electrode face contour and joint position for metal finish seam welding

13.2.4 Electrode Wire Seam Welding

The electrode wire seam welding process uses an intermediate wire electrode between each wheel electrode and the workpiece (Figure 13.4). The electrode wire seam welding process is used almost exclusively for seam welding of tin mill products to fabricate cans. The copper wire travels around the wheel electrodes at the welding speed. The copper wire electrode provides a continuously renewed surface, but it is not consumed in the welding operation. The tin build-up which would occur on a copper wheel electrode is avoided. The copper wire electrode may have a circular or flat cross section.

The process requires specially designed welding systems. The seam welds may be made with two wheel electrodes, or one wheel electrode and a mandrel electrode. The temperature range of an electrode wire seam weld should provide good solid phase bonding and not exceed the melting point of the base material. If the seam weld reaches temperatures greater than the melting point of the base metal, spikes or splashes of molten metal are expelled from the seam weld. Splashes of material can lead to corrosion of the welded component and are, therefore, undesirable. Electrode wire seam welding has small tolerance for temperature variation. Variations in welding temperature resulting from fluctuations in electrical power or electrode pressure, and changes in overlap distances are usually acceptable.

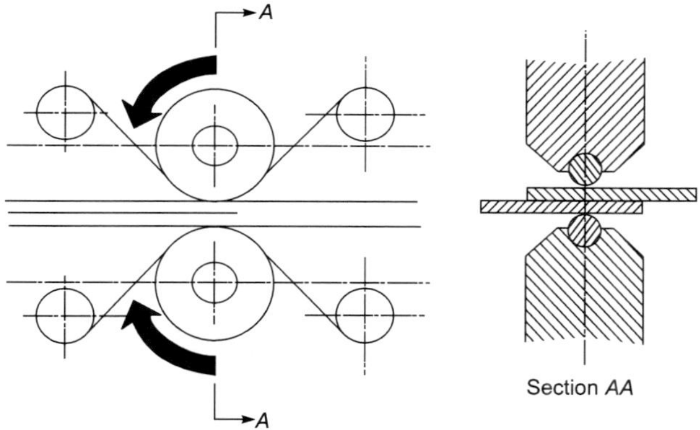

Figure 13.4 Electrode wire seam weld butt joint seam welding

Butt joint seam welding is done with the edges of the sheets forming a butt joint. A thin, narrow strip of metal, fed between the workpieces and the wheel electrode, is welded to one or both sides of the joint. The metal strip bridges the gap between the workpieces, distributes the welding current to both sheet edges, offers added electrical resistance, and contains the molten weld nugget as the nugget forms. The strip serves as a filler metal and produces a flush or slightly reinforced weld joint (Figure 13.5).

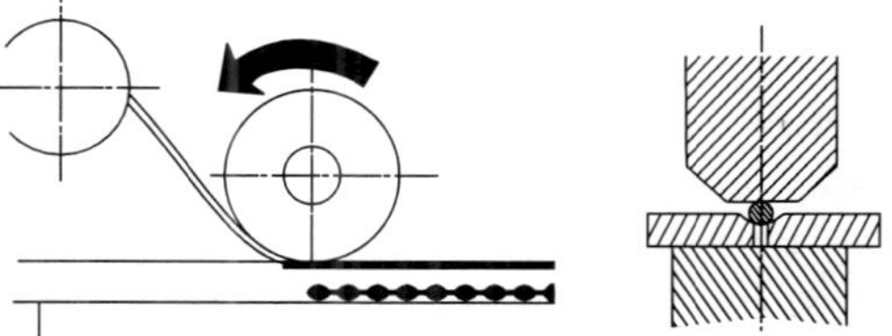

Figure 13.5 Butt joint seam weld

Strip electrode configurations can be circular, triangular, or flat. The metal strip must be guided accurately and centered on the joint to secure even current distribution to both sheet edges. The strip may be roll spot welded to the sheets at low power before the joint is scam welded. Very little forging of the joint during welding is required; therefore, there is lower distortion of the joint compared to lap seam welding. The travel speed for butt joint seam welding of low-carbon steel is comparable to that for lap seam welding. Welding procedures must ensure welding of the strips over their entire width, in order to avoid reduced corrosion resistance.

13.2.5 Roll Spot Welding (Figure 13.6)

Roll spot welding consists of making a series of spaced spot welds in a row with a seam welding machine without retracting the electrode or removing the electrode force between welds. Electrode wheel rotation may or may not be stopped during the welding cycle. The radius of the wheel electrode, the contour of its face, and the weld time influence the shape of the nugget. The nugget is usually oval shaped.

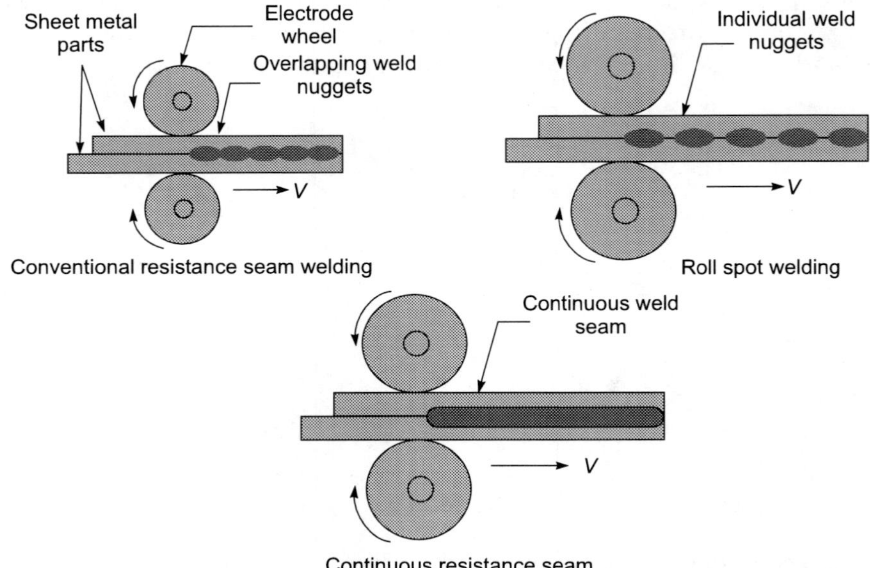

Figure 13.6 Roll spot welding

The weld spacing is obtained by adjusting of cool time with the wheel electrodes continuously rotating at a fixed speed. Hold time is effectively zero. Roll spot welding may also be done with interrupted electrode rotation when a hold time period is needed to consolidate the weld nugget as it cools. When continuously moving electrodes are employed, as is commonly the case, weld time is usually shorter and welding amperage higher than those used for conventional spot welding. The higher amperage employed, may sometimes, require the use of a higher electrode force. Otherwise, recommended practices for spot welding apply.

13.3 HEAT BALANCE

Seam welding of dissimilar metals or unequal thicknesses presents the same heat balance problems as the spot welding. The techniques for improving joint heat balance are similar. On workpieces requiring a lower current density or faster cooling, the contact area between the work and the electrode can be enlarged by increasing the diameter or width of the wheel electrode. As an alternative, one of the electrode wheels or mandrels may be made of an alloy of higher thermal conductivity in order to facilitate heat conduction from the workpiece via the electrode.

13.4 WELDING CYCLE

Resistance seam welds typically require higher welding currents than resistance spot welds, due to the shunting of the welding current through previously made welds. The resistance welding current is normally supplied in timed pulses (heating times), which are separated by periods of cool times (hold times). A weld nugget is produced during each pulse of current. For a given welding speed and heating-cooling cycle, the welding current determines the depth of weld penetration. The welding current schedule and weld speed control the weld nugget overlap. As the welding speed increases, the ratio of heating time to cool time must be increased in order to maintain weld nugget overlapping. The heat time controls the size of the weld nugget. Figure 13.7 shows the effect that cool time has on weld nugget penetration (macro-section perpendicular to weld centerline) and weld overlap (macro-section parallel to weld centerline).

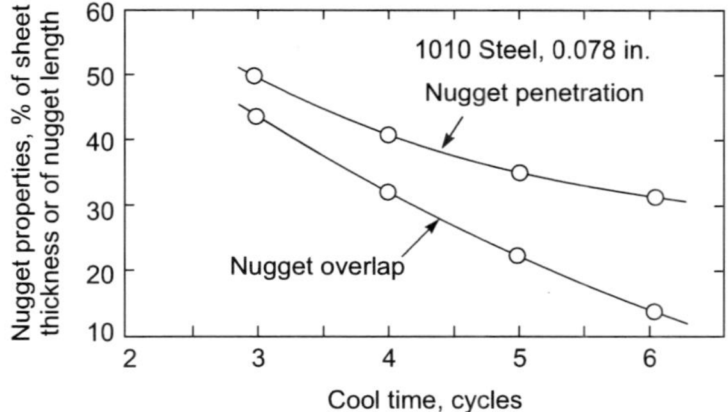

Note: Welding conditions: Heat time, 6 cycles; electrode force, 1500 Lb; welding current, 18950 A; welding speed, 55 in./min.

Figure 13.7 Effects of cool time on nugget penetration and nugget overlap in seam welding

Welding currents higher than those required to obtain proper joint properties may result in excessive indentation, or burning of the welded workpieces. Short heating times, or fast weld speeds, require higher current levels for proper weld nugget formation, but may produce greater electrode wear.

Either pulsed (intermittent) current or continuous alternating current is used for resistance seam welding. As an example of the lap seam welding process, Table 13.1 provides suggested resistance seam welding conditions for welding uncoated low-carbon steel sheet.

TABLE 13.1 Resistance seam welding conditions for welding uncoated low-carbon steel sheet

Electrode width and shape (in.) (W; R = 3″; E)					Minimum contact overlap (L)				
Thickness	**W_x Min.**	**E_l Max.**	**Force (in.)**	**On Time (Cycles) (60 per Second)**	**Off Time (Cycles)**	**Weld Speed (in./min.)**	**Welds Per inch**	**Current (amp.)**	**(in.)**
0.010	0.380	0.18	400	2	1	80	15.0	8000	0.38
0.021	0.380	0.19	550	2	2	75	12.0	11000	0.44
0.031	0.500	0.25	900	3	2	72	10.0	13000	0.50
0.040	0.500	0.25	980	3	3	67	9.0	15000	0.50
0.050	0.500	0.31	1050	4	3	65	8.0	16500	0.56
0.062	0.500	0.31	1200	4	4	63	7.0	17500	0.62
0.078	0.062	0.38	1500	6	5	55	6.0	19000	0.69
0.094	0.062	0.44	1700	7	6	50	5.5	20000	0.75
0.109	0.750	0.50	1950	9	6	48	5.0	21000	0.81
0.125	0.750	0.50	2200	11	7	45	4.5	22000	0.88

Notes:
1. Type of steel – SAE 1010.
2. Material should be free from scale, oxides, paint, grease and oil.
3. Welding conditions determined by thickness outside piece "T".
4. Data for total thickness of pile-up, not exceeding 4 "T".
   Maximum ratio between thicknesses: 3 to 1.
5. Electrode material: Class 2.
   Minimum conductivity – 75% of copper.
   Minimum hardness – 75 Rockwell "B".
6. For large assemblies, minimum contacting overlap indicated should be increased by 30%.

13.5 PULSED CURRENT

Pulsed current is usually desirable for most seam welding operations for the following reasons:

1. Good control of heat is obtained.
2. Each weld nugget in the seam is allowed to cool under pressure.
3. Distortion of the work parts is minimized.
4. It is easy to control expulsion or burning.
5. Sound welds with better surface appearance are possible.

To produce a leak-tight seam, the nuggets should overlap 15 to 20% of the nugget diameter. For maximum strength, the overlap should be 40 to 50%. The size of the nugget depends upon the heat time for a given welding speed and current. The amount of overlap depends upon the cooling time.

For a particular metal and sheet thickness, the number of welds (nuggets) per inch that can be produced economically will fall within a range. In general, as the sheet thickness decreases, the number of welds per unit of length must increase to obtain a strong, leak-tight seam weld. The ratio of welds per inch to welds per minute will establish the welding speed in inches per minute. The number of welds per minute is the number of cycles of AC per minute, divided by the sum of the heat and cooling time (in cycles) for a single weld.

To obtain the minimum number of welds per inch that will produce the required seam at a given welding speed, the heat time and welding current should be adjusted to give the required weld nugget geometry. The cooling time should then be set to give the necessary nugget overlap. Since decreasing the cooling time may increase the heat buildup, nugget penetration may increase.

13.6 CONTINUOUS CURRENT

With low-carbon steel, welding current can be applied continuously along the length of the seam with high travel speeds, if the current waveform available will produce the proper nugget size and spacing. In this case, weld quality is secondary to high production requirements. Continuous current can be used for sheet up to and including 1 mm thickness. Above this thickness, surface condition has a significant effect on welding, and electrode life is short. Continuous current welds in a particular thickness can be made over a wide range of speeds. For example, two thicknesses of 1 mm. Steel stock can be welded at speeds ranging from 44 to 131 mm/s. The required amperage increases with speed.

A problem that may arise when using continuous alternating current operations is arcing between the wheel electrode and a localized region of the weld assembly on the exit side of the electrode. The arcing may produce superficial melting of the sheet surface and the electrode. In steels, rapid cooling of molten metal that results from workpiece-to-electrode arcing, can result in brittle martensite formation, the localized martensitic microstructure may provide initiation sites for crack formation.

The speed of welding depends upon the metal being welded, stock thickness, and the weld strength and quality requirements. In general, permissible welding speeds are much lower with stainless steels and non-ferrous metals, because of restrictions on the heating rate to avoid weld metal expulsion.

In some applications, it is necessary to stop the movement of the electrodes and work as each weld nugget is made. This is usually the case for sections over 4.78 mm thick, and for metals that require postheating or forging cycles to produce the desired weld properties. Interrupted motion significantly reduces welding speed because of the relatively long time required for each weld.

With continuous motion, the welding current must be increased and heat time decreased as welding speed is increased to maintain weld quality and joint strength. There is a speed beyond which the required welding current may cause undesirable surface burning and electrode pickup. This will accelerate electrode wear.

13.7 ELECTRODES

Seam welding electrodes are normally wheels with diameters ranging from 50 to 600 mm. Common sizes have diameters of 175 to 300 mm and widths of 10 to 19 mm.

The width of the weld cross section at the interface of the two workpieces should range from 1.5 to 3 times the thickness of the thinner member. The ratio of weld width to sheet thickness normally decreases as the thickness increases. The weld width is always slightly less than the electrode's face width when commercial welding schedules are used.

13.8 EXTERNAL COOLING

Flood cooling, immersion, or mist cooling is commonly used with seam welding. This is generally in addition to any internal cooling of the components in the secondary circuit of the welding machine. When external cooling is not used, electrode wear and distortion of the work may be excessive. For welding non-ferrous metals and stainless steel, clean tap water is satisfactory. For ordinary steels, a 5% borax solution is commonly used to minimize corrosion.

13.9 JOINT DESIGN

The various requirements that must be met in designing spot welded joints apply to seam welded joints. With seam welding, electrode design together with mounting and leak-tightness requirements place some limitations on design.

The wheel type electrodes are relatively large and require unobstructed access to the joint. Since the electrodes rotate during welding, they cannot be inserted into small recesses or internal corners. External flanges must change direction over large radii in order to produce a strong, leak-tight seam weld. Joint designs that incorporate corners having small radii may result in welding problems when resistance seam welding is used. Decreased welding speeds are sometimes required in order to maintain weld quality.

Figure 13.8 presents some common designs of seam welded joints. They are similar to those used for spot welding applications. The lap joint (Figure 13.8(a)) is the most common design. The workpiece edges must overlap sufficiently to prevent expulsion of the weld metal from the edges of the workpiece. Excessive overlapping however, may entrap dirt or moisture within the joint, and may cause subsequent manufacturing or service problems. Lap seam welds are used for the longitudinal seams in cans, buckets, water tanks, mufflers, and large diameter, thin-walled pipes.

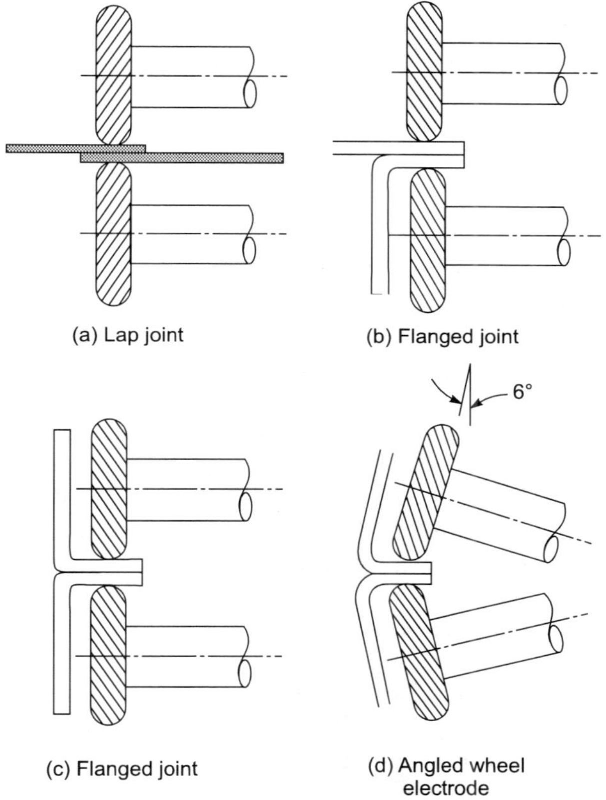

Figure 13.8 Examples of resistance seam welded joints

Flange joints are forms of lap joints. The design in Figure 13.8(b), in which one of the pieces is straight, is commonly used to weld flanged ends to containers of various types. In Figure 13.8(c), both pieces are flanged. This design is used to join the two sections of automotive gasoline tanks. Often, the flanged pieces are dished to obtain added strength, in which case it is necessary to mount one or both wheels at an angle to clear the work, as shown in Figure 13.8(d). A practical limit is 6^c because greater angles cause excessive bearing thrust.

Specialized workpiece designs may require the adjustment of the shape and contour of the wheel electrode. Workpieces that contain regularly spaced contours may be welded with notched or segmented wheel electrodes (Figure 13.9).

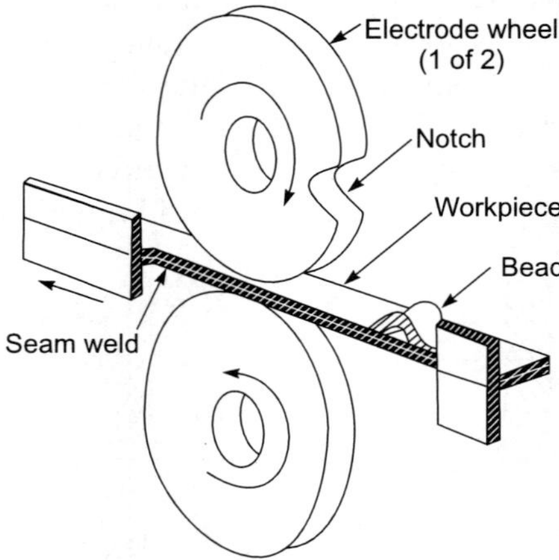

Figure 13.9 Notched electrode wheel for seam welding of a workpiece having obstructions in the path of the wheel

13.10 APPLICATIONS OF RESISTANCE SEAM WELDING

A resistance seam weld (RSEW) is made on overlapping workpieces and is a continuous weld formed by overlapping weld nuggets, by a continuous weld nugget, or by forging the joint as it is heated to the welding temperature by its resistance to the welding current. Seam welds are typically used to produce continuous gas- or liquid-tight joints in sheet assemblies, such as automotive gasoline tanks. The process is also used to weld longitudinal seams in structural tubular sections that do not require leak-tight seams. In most applications, two wheel electrodes, or one translating wheel and a stationary mandrel, are used to provide the current and pressure for resistance seam welding. Seam welds can also be produced using spot welding electrodes; this requires the purposeful overlapping of the spot welds in order to obtain a leak-tight seam weld. Overlapping spot welding requires an increase in power after the first spot weld, to offset the shunting effect in order to obtain adequate nugget formation as welding progresses. Seam welding is used for welding carbon steels, alloy steels, stainless steel, alloy of aluminum, nickel and magnesium. Copper and its alloys cannot be welded by seam welding.

13.11 ADVANTAGES AND LIMITATIONS

Resistance seam welding has the same advantages and limitations as resistance spot welding. An additional advantage is the ability to produce a continuous leak-tight weld.

Seam welds must be made in a straight or uniformly curved path. Abrupt changes in welding direction or in joint contour along the path cannot be welded leak-tight. This limits the design of the assembly.

Strength properties of seam welded lap joints are generally lower than those of fusion welded butt joints, due to the eccentricity of loading on lap joints and the built-in notch along the nugget at the sheet interface.

Advantages of Seam Welding

1. Capable of producing gas/liquid-tight joint
2. Semi-skilled worker is required
3. Overlap can be less than the spot or projection weld
4. It is suitable for mass production since welding can be done quickly

Disadvantages

1. Process is limited to lap joint
2. It is difficult to weld complex shapes
3. Thicker plates (>3 mm) cannot be welded

Worked Examples

Example 1: A resistance seam welding operation is performed on two pieces of 2.5 mm thick austenitic stainless steel to fabricate a container. The weld current in the operation is 10,000 A, the weld duration = 0.3 sec, and the resistance at the interface is 75 micro-ohms. Continuous motion welding is used, with 200 mm diameter electrode wheels. The individual weld nuggets formed in this RSEW operation have diameter = 6 mm and thickness = 3 mm (assume the weld nuggets are disc shaped). These weld nuggets must be contiguous to form a sealed seam. The power unit driving the process requires an off-time between spot welds of 1.0 sec. Given these conditions, determine (a) the unit melting energy of stainless steel using the methods of the previous chapter, (b) the proportion of energy generated that goes into the formation of each weld nugget, and (c) the rotational speed of the electrode wheels.

(Fundamentals of Modern Manufacturing: Groover***)***

Solution: Part (a)

Melting temperature for austenitic stainless steel = 1670 K

$$U_m = 3.33 \times 10^{-6} 1670^2 = 9.29 \text{ J/mm}^3$$

Part (b)

$$\text{Heat required for welding, } H_w = U_m \times V$$

$$\text{Weld nugget volume} = \frac{\pi D^2 d}{4} = \frac{\pi \times 6^2 \times 3}{4} = 84.82 \text{ mm}^3$$

Heat required for welding, $H_w = U_m \times V = (9.2\ \text{J/mm}^3) \times 84.82\ \text{mm}^3 = 788\ \text{J}$

$$H = I^2R\tau = 10{,}000^2\,(75 \times 10^{-6}) \times 0.3 = 2225\ \text{J}$$

Proportion of heat used for welding = 788/2225 = 0.354 = 35.4%

Part (c)

$$T_c = 0.3 + 1.0 = 1.3\ \text{sec.}$$

Distance moved per spot weld in order to have contiguous spot weld for the seam, $D = 0.6$ mm.

Therefore, surface speed of electrode wheel, $v = \dfrac{6}{1.3} = 4.61\dfrac{\text{mm}}{\text{s}} = 276.9\ \text{mm/min}$

$$N = \frac{v}{\pi D} = \frac{276.9}{\pi \times 200} = 0.441\ \text{rev/min}$$

Example 2: Suppose in the previous problem that a roll spot welding operation is performed instead of seam welding. The interface resistance increases to 100 micro-ohms, and the center-to-center separation between weld nuggets is 25 mm. Given the conditions from the previous problem, with the changes noted here, determine: (a) the proportion of energy generated that goes into the formation of each weld nugget; (b) the rotational speed of the electrode wheels, and (c) at this higher rotational speed, how much does the wheel move during the current on-time, and might this have the effect of elongating the weld nugget (making it elliptical rather than round)?

Solution:

(a) $U_m = 3.33 \times 10^{-6}\,(1670)^2 = 9.29\ \text{J/mm}^3$ from previous problem.
$H_w = (9.29\ \text{J/mm}^3)\,(84.82\ \text{mm}^3) = 788$ J from previous problem.
$H = I^2Rt = (10{,}000)^2\,(100 \times 10^{-6})\,(0.3) = 300\ \text{J}$
Proportion of heat for welding = 788/3000 = 0.263 = 26.3%

(b) Total cycle time per spot weld $T_c = 1.3$ sec as in previous problem.
Distance moved per spot weld = 25 mm as given.
Surface speed of electrode wheel, v = 25 mm/1.3 sec = 19.23 mm/s = 1153.8 mm/min
$N = v/\pi D = (1153.8\ \text{mm/min})/(200\pi\ \text{in/rev}) = 1.836\ \text{rev/min}$

(c) Power-on time during cycle = 0.3 sec.
Movement of wheel during 0.3 sec = (0.3 sec) (19.23 mm/s) = 5.77 mm. This movement is likely to make the weld spot elliptical in shape.

Review Questions

1. What is resistance seam welding?
2. Explain with a neat sketch the process of resistance seam welding.
3. How does a seam weld machine work?
4. What are the types of seam welding machines?
5. What are the applications of resistance seam welding?
6. What are the advantages of resistance seam welding?
7. What are the limitations of resistance seam welding?
8. What is roll spot welding?
9. Explain the process parameters of resistance seam welding.
10. What is pulsed current and continuous current?

14

CHAPTER

Projection Welding

14.0 INTRODUCTION

Projection welding is a modification of spot welding. It is a resistance welding process which produces coalescence of metals with the heat obtained from resistance to electrical current through the work parts held together under pressure by electrodes. The resulting welds are localized at predetermined points by projections, embossment, or intersections. Localization of heating is obtained by a projection or embossment on one or both of the parts being welded. There are several types of projections: the button or dome type, usually round; elongated projections; ring projections; shoulder projections; cross wire welding and radius projection. The major advantage of projection welding is that electrode life is increased because larger contact surfaces are used. A very common use of projection welding is the use of special nuts that have projections on the portion of the part to be welded to the assembly.

14.1 PROCESS OF PROJECTION WELDING (FIGURE 14.1)

In projection welding, one or both of the workpieces are provided with small projections so that current flow and heating are localized at those spots. The projections are usually produced by die pressing or any other similar method. When high density current (less than spot welding current) is passed, localized heating takes place at the points of contact. Projections collapse under the external applied force, forming a close surface-to-surface contact, yielding a well defined, finished weld similar to multi-spot welding. When the current is switched off, the weld cools down and solidification takes place under the applied force. The electrode force is then released, and the welded workpiece is removed. As in the case with spot welding, the entire projection welding process takes only a fraction of second.

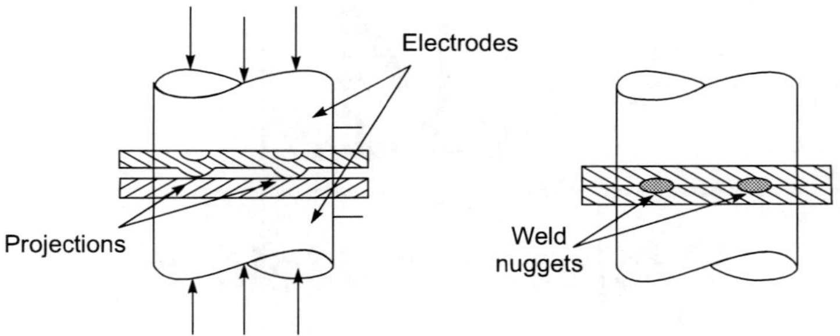

Figure 14.1 Projection welding

14.2 EQUIPMENT (FIGURE 14.2)

Equipment used for projection welding is similar to that used for spot welding except that the rod electrodes are replaced by flat copper platens or dies as shown in Fig. 2.20. Usually, two or three projections are welded at the same time. With more than three simultaneous welds, however, there is a tendency for lack of consistency and the welds may not have same shape or strength. When long seam is to be projection welded then the limit is usually 250 mm length.

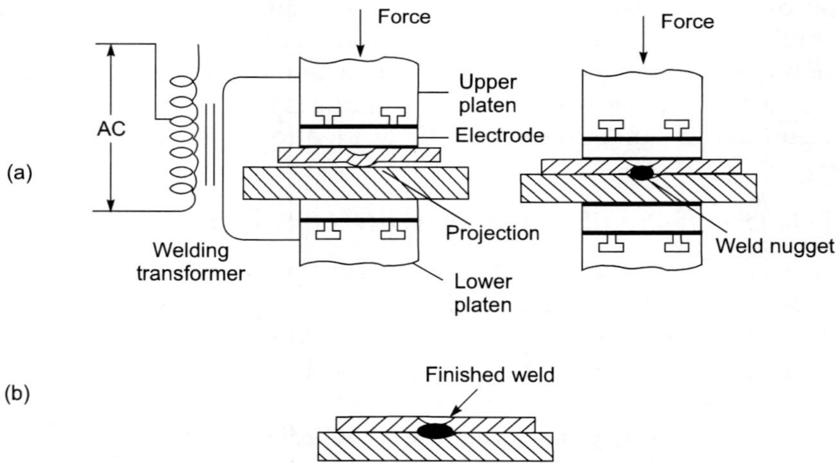

Figure 14.2 Projection welding (a) Setup (b) Finished weld

14.3 HEAT BALANCE

Factors that affect the heat balance of welding are as follows:

1. Projection design and location
2. Thickness of the sections
3. Thermal and electrical conductivities of the metal being welded

4. Heating rate
5. Electrode alloy

The distribution of heat in the two sections being projection welded together must be reasonably uniform to obtain strong welds, as in spot welding. The major portion of the heat develops in the projections during projection welding. Consequently, heat balance is generally easier to obtain in projection welding than in spot welding; however, it may be complicated by simultaneously making multiple projection welds. Uniform division of welding current and electrode force is necessary to obtain even heating of all projections. Since the current paths through the projections are in parallel, any variation in resistance between the projections will cause the current to be distributed unequally.

Projections must be designed to support the electrode force needed to obtain good electrical contact with the mating part, and to collapse when heated. With multiple projections, slight variation in projection heights can affect heat balance. This may occur as a result of wear of the projection forming punches. Heat balance in dissimilar thickness materials is maintained by placing the projection in the thicker pieces to be welded. The size of the projection is based upon the requirements for heating the thinner section. Similarly, to maintain heat balance in materials of dissimilar conductivity, the projection is located in the piece of higher conductivity (lower resistivity). The choice of electrode alloy determines the conductivity of the electrode, which can also affect heat balance.

14.4 WELDING CYCLE

Current

The current for each projection is generally less than that required to produce a spot weld in the same thickness of that same metal. The projection will heat rapidly and excessive current will melt it and result in expulsion; however, the current must be at least high enough to create fusion before the projection has completely collapsed. For multiple projections, the total welding current will approximately equal the current for one projection multiplied by the number of projections. Some adjustment may be required to account for normal projection tolerances, part designs, and the impedance of the secondary circuit.

Weld Time

Weld time is about the same for single or multiple projections of the same design. Although a short weld time may be desirable from a production standpoint, it will require correspondingly higher amperage. This may cause overheating and metal expulsion. In general, longer weld times and lower amperages are used for projection welding than those for spot welding. In some cases, multiple impulse welding may be advantageous to control the heating rate. This is helpful with thick sections and with metals of low thermal conductivity.

Electrode Force (Figure 14.3)

The electrode force used for projection welding depends upon the metal being welded, the projection design, and the number of projections in the joint. The force should be adequate to flatten the projections completely when they reach welding temperature, and to bring the workpieces in contact. Excessive force will prematurely collapse the projections and the weld nuggets will be ring shaped, with incomplete fusion in the center.

The welding machine must be capable of mechanically following the work with the electrodes as the projections collapse. Slow followup will permit metal expulsion before the workpieces are together. The sequence of events during the formation of a projection weld is shown schematically in Figure 14.3. In Figure 14.3(a), the projection is shown in contact with the mating sheet. In Figure 14.3(b), the current has started to heat the projection to welding temperature. The electrode force causes the heated projection to collapse rapidly and then fusion takes place as shown in Figure 14.3(c). The completed weld is shown in Figure 14.3(d).

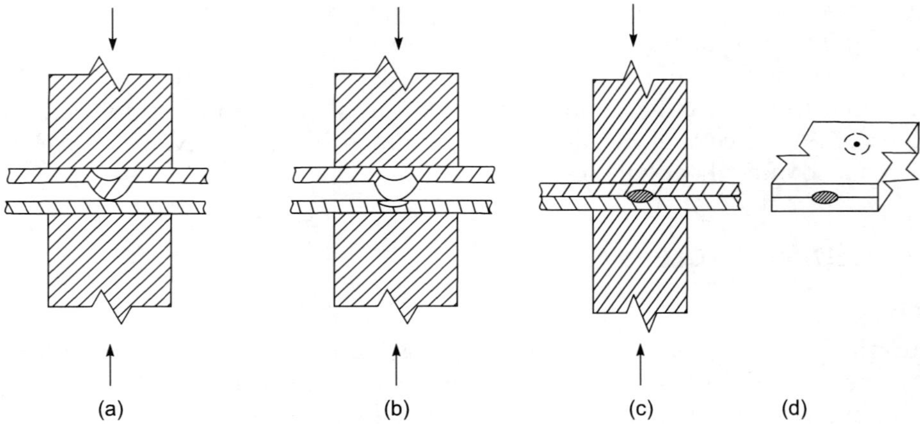

Figure 14.3 Sequence of events during the formation of a projection weld

Electrodes and Welding Dies

The areas of parts to be joined are frequently flat except for the projections. In such cases, large flat faced electrodes are used. When the surfaces to be contacted are contoured, the electrodes are fitted to them. With such electrodes, the electrode force can be applied without distorting the parts, and the welding current can be introduced without overheating the contact areas.

For a single projection, the electrode face diameter should be at least twice the diameter of the projection. With multiple projections, the electrode face should extend a minimum of one projection diameter beyond the boundary of the projection pattern. The best electrode material is one that is sufficiently hard (to minimize wear) but does not crack or cause surface burning of the part. If burning or cracking is encountered, a softer alloy of higher conductivity should be used. With multiple projections, electrode wear can upset the balance of welding current and electrode force on the projections. Then, the strength and quality of the welds may become unacceptable.

Welding electrodes and locating dies for projection welding are usually combined. With the proper dies, it is possible to attain accuracy with projection welding equal to that of any other assembly process. The welding dies should meet the following requirements:

1. Provide accurate positioning of the parts
2. Permit rapid loading and unloading
3. Have no alternative path for the welding current
4. For AC welding, it should be made of non-magnetic materials
5. It should be properly designed for operator safety

The dies must be mounted solidly on the welding machine. The parts are mated in one die and all the welds are made at once with one operation of the machine. One part may be located in relation to the other by punching holes in one, with semi-punchings in the other to match. The projections can usually be embossed or forged in the same operation. In some designs, insulated pins or sleeves may be used in the electrode or dies to position and align the parts. Simple examples are shown in Figures 14.4.

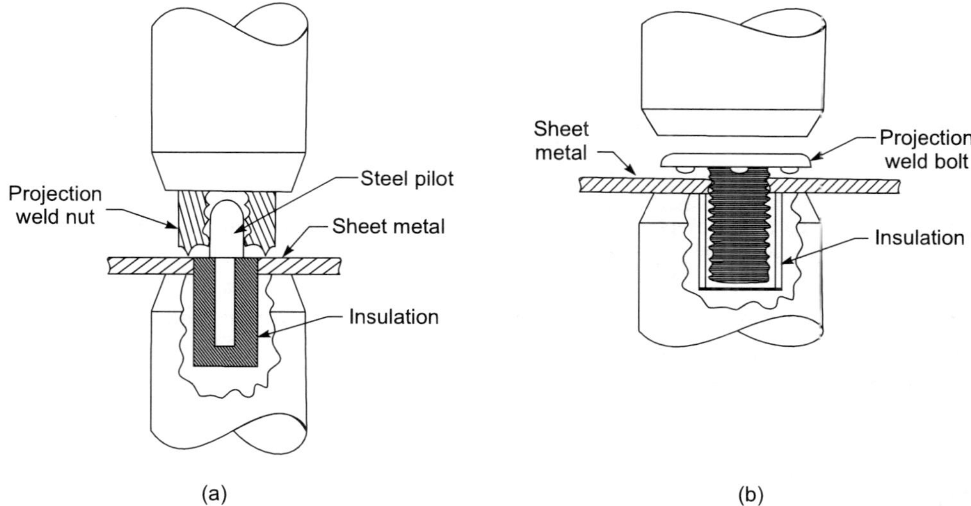

Figure 14.4 (a) Use of an insulating pin to locate a weld nut & (b) Positioning a weld bolt with an insulating sleeve

When the small part of an assembly can be placed on the bottom and the large part on top, it is a simple matter to hold the small part in a recessed lower electrode, such as shown in Figure 14.5(a). When it is desired to locate a small part on top of a larger part, problem exists. Sometimes the small part can be located and held by a removable device and then welded with a flat upper electrode. Parts that nest into the upper electrode may be held by spring clips attached to the electrode. Figure 14.5(b) shows spring loaded dies. Check for parallelism of the platens should be made under intended operating forces. This can best be done by placing a steel block with smooth parallel faces between the platens, applying the intended electrode force, and then checking for gap with feeler gages.

The next step is to check the bases of the die blocks. They must be clean, smooth, flat, and free from burrs and nicks. If not, they should be machined.

The dies are then installed on the machine. Most machines have 'tee' slots at right angles to one another in the two platens to permit universal alignment of the dies. After the dies are properly lined up, they should be clamped securely to the platens. With the work in place in the dies, the position of the ram or knee of the machine should be adjusted for the proper stroke, including the necessary allowance for upset of the projections.

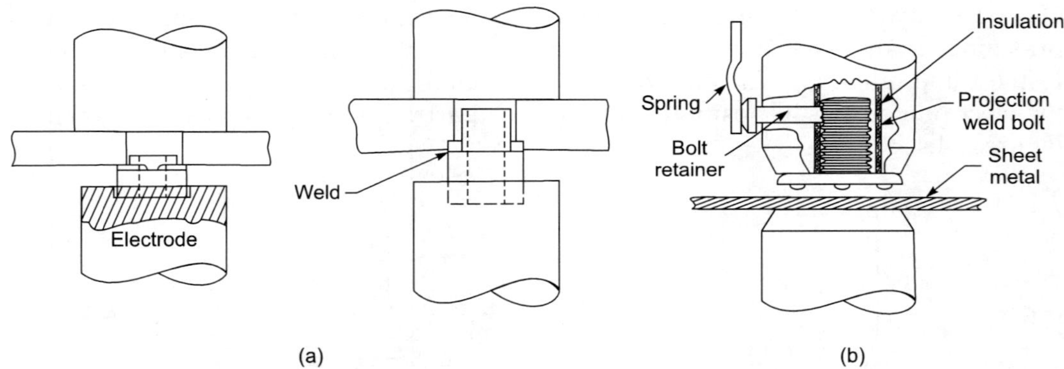

Figure 14.5 (a) Recessed electrode to position a weld nut & (b) Holding a bolt in the upper electrode with a spring retainer

If the tips of the projections are in one plane and of uniform height, the setup is ready for trial welds. Non-uniformity of current or force on the projections may be caused by the following:

1. Shunting of current through locators
2. Unequal secondary circuit path lengths
3. Excessive play in the welding head
4. Too much deflection in the knee of the machine

The use of shims between die components or between dies and platens should be avoided. If shims must be used, they should be clean, annealed pure copper sheets of sufficient area to carry the secondary current.

14.5 JOINT DESIGN

Lap joint designs for projection welding are similar to those for spot welding. In general, joint overlap and edge distances for projection welding can be less than for spot welding. Most applications use multiple projections where the minimum distance between projections should be twice the projection diameter.

Fit-up is important with multiple projection welding. Each projection must be in contact with the mating surface to accomplish a weld. Uniformity of projection heights is a factor in good fit-up. The welding dies must be carefully designed and accurately manufactured to mate with the parts at the weld locations. They should not need to deform the parts to obtain good fit-up.

When projection welds are used to attach other fasteners, such as weld nuts and bolts, they must contain a sufficient number of projections to carry the design load. The design should be proven by applicable mechanical testing. Production quality control should be programmed to ensure that weld quality does not drop below the design standards.

Where surface marking of one part must be minimized, the projections should be placed in the other part. A large, flat electrode on the show side of the joint should prevent electrode marking, although slight shrinkage may occur at each projection weld. This may be visible after some finishing operations.

14.6 APPLICATIONS OF PROJECTION WELDING

Projection welding is primarily used to join a stamped, forged, or machined part to another part. One or more projections are produced on the parts during the forming operations. Fasteners or mounting devices, such as bolts, nuts, pins, brackets, and handles, can be projection welded to a sheet metal part. Projection welding is especially useful for producing several weld nuggets simultaneously between two parts. Marking of one part can be minimized by placing the projections on the other part.

The process is generally used for section thicknesses ranging from 0.02 to 0.125 inch (0.5 to 3.2 mm) thick. Thinner sections require special welding machines capable of following the rapid collapse of the projections. Various carbon and alloy steels and some nickel alloys can be projection welded.

Applications of Annular Welding (Figure 14.6)

The applications of projection welding using annular projections are explained in the following examples. The summit of the circular ridge should be rounded, particularly with heavy sections, to improve heat balance. Relief should be provided at the base of projection, for the upset material to fill as projection collapses to ensure a tight joint without gap.

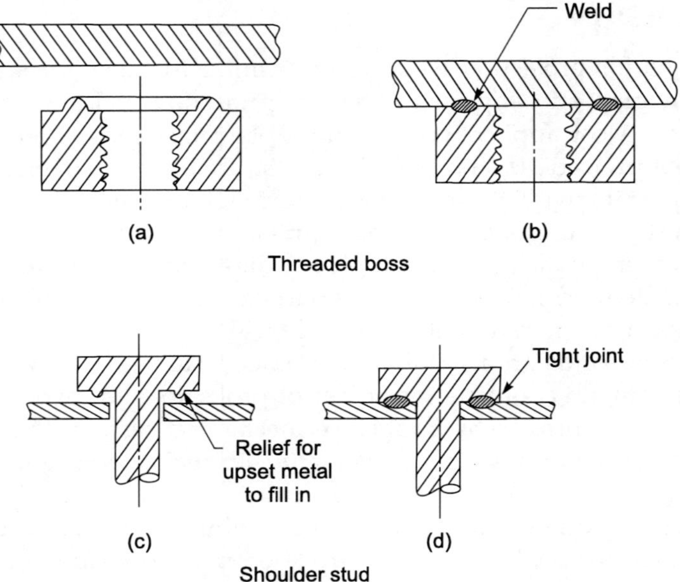

Figure 14.6 Application of projection welding using annular projections

14.7 ADVANTAGES AND LIMITATIONS

In general, projection welding can be used instead of spot welding to join small parts to each other and to larger parts. Selection of one method over another depends upon the economics, advantages, and limitations of the two processes.

Advantages of Projection Welding

1. A number of welds can be made simultaneously in one welding cycle of the machine. The limitation on the number of welds is the ability to apply uniform electrode force and welding current to each projection.
2. Less overlap and closer weld spacings are possible, because the current is concentrated by the projection, and shunting through adjacent welds is not a problem.
3. Thickness ratios of at least 6 to 1 are possible, because of the flexibility in projection size and location. The projections are normally placed on the thicker section.
4. Projection welds can be located with greater accuracy and consistency than spot welds, and the welds are generally more consistent because of the uniformity of the projections. As a result, projection welds can be smaller in size than spot welds.

5. Projection welding generally results in better appearance, on the side without the projection, than spot welding can produce. The most deformation and greatest temperature rise occur in the part with the projection, leaving the other part relatively cool and free of distortion, particularly on the exposed surface.
6. Large, flat faced electrodes are used; consequently, electrode wear is much less than that with spot welding and this reduces maintenance costs. In some cases, the fixturing or part locators are combined with the welding dies or electrodes when joining small parts together.
7. Oil, rust, scale, and coatings are less of a problem than with spot welding, because the tip of the projection tends to break through the foreign material early in the welding cycle; however, weld quality will be better with clean surfaces.

Limitations of Projection Welding

1. The forming of projections may require an additional operation unless the parts are press-formed to design shape.
2. With multiple welds, accurate control of projection height and precise alignment of the welding dies are necessary to equalize the electrode force and welding current.
3. With sheet metal, the process is limited to thicknesses in which projections with acceptable characteristics can be formed, and for which suitable welding equipment is available.
4. Multiple welds must be made simultaneously, which requires higher capacity equipment than spot welding does. This also limits the practical size of the component that contains the projections.

Worked Example

Example 1: Resistance projection welding is used to simultaneously weld two thin, steel plates together at four locations. One of the pieces of steel plate is performed with projections that have a diameter of 0.25 inch and a height of 0.20 inch. The duration of current flow during the weld is 0.30 sec and all four projections are welded simultaneously. The plate steel has a unit melting energy of 140 Btu/in^3 and a resistance between plates of 90.0 micro-ohms. Experience has shown that 55% of the heat is dissipated by the metal and 45% melts the weld nugget. Assume the volume of the nuggets will be twice the volume of the projections because metal from both plates is melted. How much current is required for the process?

*(**Fundamentals of Modern Manufacturing:** Groover)*

Solution: Volume of single projection $= \frac{\pi D^2 d}{4} = \frac{\pi \times 0.25^2 \times 0.2}{4} = 0.00982 \text{ in}^3$

Volume of one nugget $= 0.00982 \text{ in}^3 \times 2 = 0.01964 \text{ in}^3$

Volume of four nuggets $= 0.00982 \text{ in}^3 \times 8 = 0.07854 \text{ in}^3$

Heat required for melting $= U_m \times V = (140 \text{ Btu/in}^3)(0.07854 \text{ in}^3) = 11 \text{ Btu}$

Required heat for operation, $H = \frac{11}{0.45} = 24.43 \text{ Btu}$

Total heat for joules $H = 24.43 \text{ Btu} \times 1055 = 25780 \text{ J}$

$$H = I^2 R\tau = I^2(90 \times 10^{-6}) \times 0.3 = 25780 \text{ J}$$

Solving for I, we get current = 30900 A

Review Questions

1. What is projection welding?
2. Explain with a neat sketch the process of projection welding.
3. How does a projection welding machine work?
4. How is heat balance obtained in projection welding?
5. What are the applications of projection welding?
6. What are the advantages of projection welding?
7. What are the limitations of projection welding?
8. What are the requirement of welding dies?
9. Explain the process parameters of projection welding.
10. What are the causes of non-uniform pressure in projection welding?

15 CHAPTER

Flash and Upset Welding

15.0 INTRODUCTION: FLASH

Flash welding (FW) is a resistance welding process that produces a weld at the faying surfaces of a butt joint by a flashing action and by the application of pressure after the heating is substantially completed. The flashing action, caused by the very high current densities at small contact points between the workpieces, forcibly expels material from the joint as the workpieces are slowly moved together. The weld is completed by a rapid upsetting of the workpieces.

Two parts to be joined are clamped in dies (electrodes) connected to the secondary of a resistance welding transformer. Voltage is applied as one part is advanced slowly towards the other. When contact occurs at surface irregularities, resistance heating occurs at these locations. High amperage causes rapid melting and vaporization of the metal at the points of contact, and then minute arcs form. This action is called "flashing". As the parts are moved together at a suitable rate, flashing continues until the faying surfaces are covered with molten metal and a short length of each part reaches forging temperature. A weld is then created by the application of an upset force to bring the molten faying surfaces in full contact and forge the parts together. Flashing voltage is terminated at the start of upset. The solidified metal expelled from the interface is called "flash".

Note: The major difference between friction welding and upset and flash welding is that the heat for friction welding is developed by rubbing friction between the faying surfaces, rather than from electrical resistance. Upset welding is similar to flash welding except that no flashing action occurs.

15.1 PRINCIPLES OF OPERATION

The basic steps in a flash welding sequence are as follows:

- Position the parts in the machine.
- Clamp the parts in the dies (electrodes).
- Apply the flashing voltage.

- Start platen motion to cause flashing.
- Flash at normal voltage.
- Terminate flashing.
- Upset the weld zone.
- Unclamp the weldment.
- Return the platen and unload.

Figure 15.1 illustrates these basic steps. Additional steps such as preheat, dual voltage flashing, postheat, and trimming of the flash may be added as the application dictates. Flashing takes place between the faying surfaces as the movable part is advanced towards the stationary part. Heat is generated at the joint and the temperature of the parts increases with time. Flashing action (metal loss) increases with part temperature.

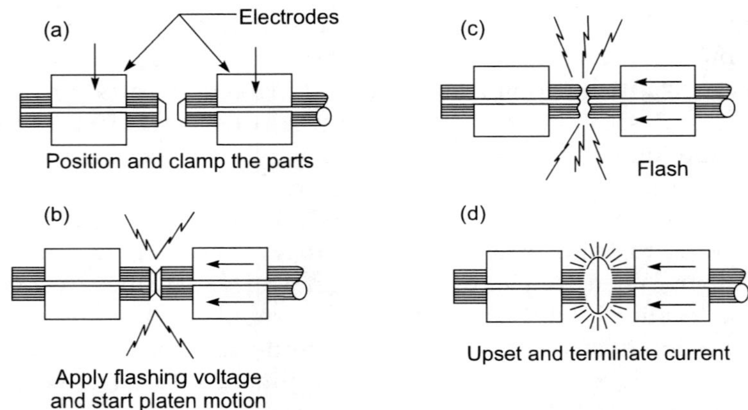

Figure 15.1 The basic steps in flash welding (a) Position and clamp the parts (b) Apply flashing voltage and start platen motion (c) Flash (d) Upset and terminate current

A graph relating part motion with time is known as the flashing pattern. In most cases, a flashing pattern should show an initial period of constant velocity motion of one part towards the other to facilitate the start of flashing. This and the average temperature of the faying surface should be the melting temperature of the metal. Once these conditions are reached, further flashing is not necessary. The steepness of the temperature gradient corresponding to a stable temperature distribution is a function of the part acceleration during parabolic flashing. In general, the higher the rate of part acceleration, the steeper is the stable temperature gradient produced. Thus, the shape of the temperature distribution curve in a particular application can be controlled by appropriate choice of flashing pattern. Since the compressive yield strength of a metal is temperature sensitive, the behavior of the metal during the upsetting portion of the welding cycle markedly depends upon the flashing pattern. Therefore, the choice of flashing pattern is extremely important for the production of sound flash welds.

The minimum flashing distance is the amount of flashing required to produce a stable temperature distribution. From a practical standpoint, the flashing distance should be slightly greater than the minimum acceptable amount, to ensure that a stable temperature distribution is always achieved. Upset occurs when a stable temperature distribution is achieved by flashing and the two parts are brought together rapidly. The movable parts should be accelerated rapidly so that the molten metal on the flashing surfaces will be extruded before it can solidify in the joint. Motion should continue with sufficient force to upset the metal and weld the two pieces together. Upset current is sometimes applied as the joint is being upset to maintain temperature by resistance heating. This permits upset of the joint with lower force than would be required without it. Upset current is normally adjusted by electronic heat control on the basis of either experience or welding tests.

15.2 EQUIPMENT (FIGURE 15.2)

Typical Machines

A typical flash welding machine consists of six major parts:

- The machine bed which has platen ways attached
- The platens which are mounted on the ways
- Two clamping assemblies, one of which is rigidly attached to each platen to align and hold the parts to be welded
- A means for controlling the motion of the movable platen
- A welding transformer with adjustable taps
- Sequencing controls to initiate part motion and flashing current

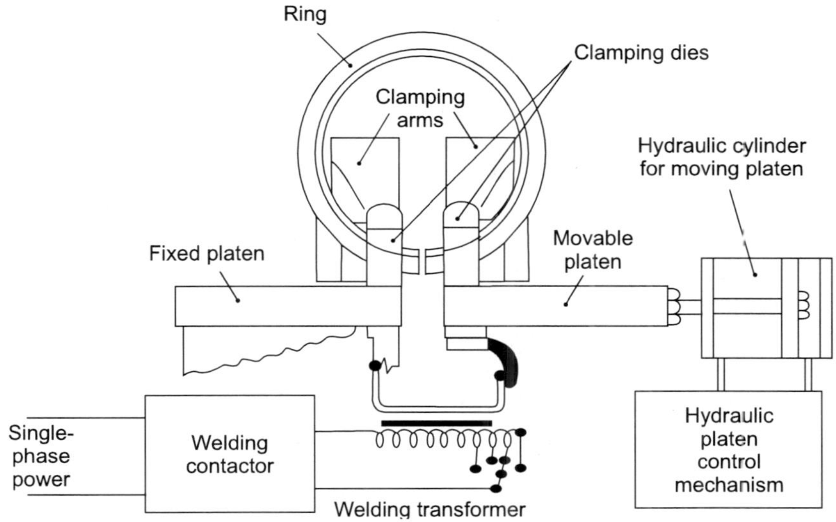

Figure 15.2 Equipment

Flash welding machines may be manual, semi-automatic, or fully automatic in their operations; however, most of them are either semi-automatic or fully automatic. With manual operation, the operator controls the speed of the platen from the time that flashing is initiated until the upset is completed. In semi-automatic operation, the operator usually initiates flashing manually and then energizes an automatic cycle that completes the weld. In fully automatic operation, the parts are loaded into the machine and the welding cycle is then completed automatically. The platen motion of many small flash welding machines is provided mechanically by a cam that is driven by an electric motor through a speed reducer. Large machines may be hydraulically or pneumatically operated. Operating personnel should be given instructions on how to operate the machinery in a safe manner. Hands must be kept clear of moving machinery, and contact with electrically charged surfaces must be avoided.

Controls and Auxiliary Equipment

Electrical controls on flash welding machines are integral types designed to sequence the machine, control the welding current, and precisely control the platen position during flashing and upsetting. Silicon controlled rectifier (SCR) contactors are widely used on machines drawing up to 1200 A from the power lines. Ignitron contactors are common on larger machines. Preheat and postheat cycles are normally controlled with electronic timers and phase-shift heat controls. Timers for these functions may be initiated manually or automatically in proper order during the welding period.

Dies

Flash welding dies, compared to spot and seam welding electrodes, are not in direct contact with the welding area. Dies may be considered work holding and current conducting clamps. Since the current density in these dies is normally low, relatively hard materials with low electrical conductivity may be used. However, water cooling of the dies may be necessary in high production to avoid overheating. There are no standardized designs for these dies since they must fit the contour of the parts to be welded. The size of the dies depends largely upon the geometry of the parts to be welded and the mechanical rigidity needed to maintain proper alignment of parts during upsetting. The dies are usually mechanically fastened to the welding machine platens. Electrode contact area should be as large as possible to avoid local die burns. The contact surfaces may be incorporated in small inserts attached to larger dies for low-cost replacement and convenient detachment for redressing. A facing insert of special material which is brazed to the die is frequently used for maximum wear resistance. If the parts are backed up so that the clamping dies do not need to carry the upset force, clamping pressures need only be sufficient to provide good electrical contact. If the work cannot be backed up, it may be necessary to use serrated clamp inserts. In this case, the inserts are usually made of hardened tool steel. Flash welding dies tend to wear but do not mushroom. As wear takes place, the contact area may decrease and cause local hot spots (die burns). The dies should be kept clean. Flash and dirt will tend to embed in the dies and cause hot spots and die burns. All bolts, nuts, and other die-holding devices should be tight.

Fixtures and Backups

The functions of fixtures for flash welding are: (1) to rapidly and accurately locate two or more parts relative to each other, (2) to hold them in proper location while they are being welded, and (3) to permit easy release of the welded assembly. A fixture is either fastened to the machine or built into it. Parts are loaded directly into the fixture and welded. Resistance welding processes are very rapid compared to other methods of joining. If maximum production is to be attained, fixtures must be easily loaded and unloaded. The following factors should be considered when designing a fixture:

- Quick-acting clamps, toggles, and other similar devices should be employed. Sometimes ejector pins are used to facilitate removal of the finished assembly.
- The fixture must be designed so that welding current is not shunted through any locating devices. This may require insulation of pins and locating strips.
- Non-magnetic materials are usually preferred, because any magnetic material located in the throat of the machine will increase the electrical impedance and limit the maximum current which the machine can deliver.
- The operator should be able to load and unload the parts safely. This may require the use of swivel devices or slides, so that the fixture can be moved out of the machine. A guard should swing in place to prevent the operator from reaching between the platens. The guard can also act as the flash shield.
- A fixture must provide for movement of the parts as they are being clamped in the dies.
- All bearings, pins, slides, etc., should be protected from spatter and flash.

Backups are needed if the clamping dies cannot prevent slippage of the parts when the upsetting force is applied. Slippage usually occurs when the section of the part in the die is too small for effective clamping, or the part is unable to withstand sufficient clamping force without damage. A backup often consists of a steel bracket that can be bolted in various positions to the platen. Brackets can have either fixed or adjustable stops against the parts.

15.3 WELDING PROCEDURES

Every welding operation involves numerous variables that affect the quality of the resulting weld. For this reason, a welding procedure should be developed that prescribes the settings for the welding variables to ensure consistent weld quality. Flash welding involves dimensional, electrical, force, and time variables. The dimensional variables are shown in Figures 15.3(a) and (b) and 15.4. The paths of the movable platen and the faying surfaces during flashing and upsetting are also shown in Figure 15.4. The current, force, and time variables are shown in Figure 15.5. Most operations do not involve all of the variables shown. A simple flash welding cycle involves flashing at one voltage setting followed by upset. Table 15.1 gives the recommended maximum joint lengths for flash welding.

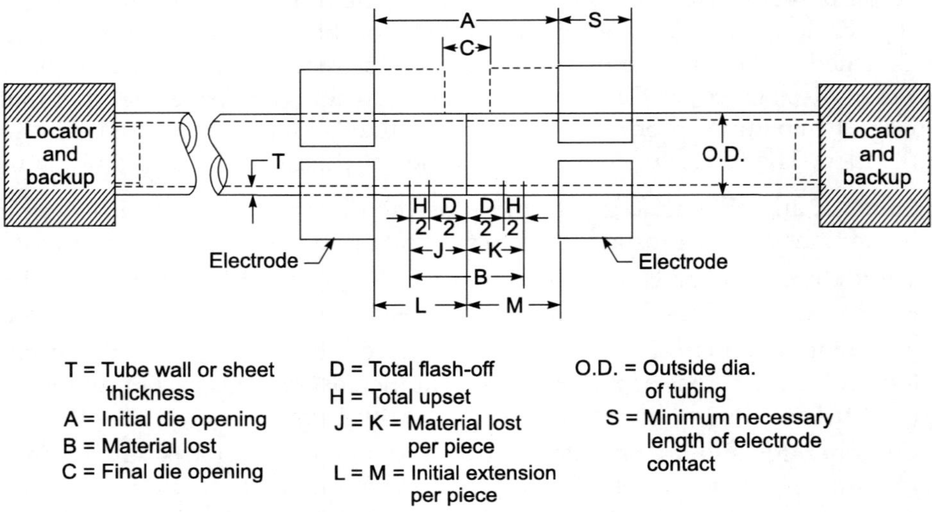

Figure 15.3 (a) Flash welding of tubing and flat sheets

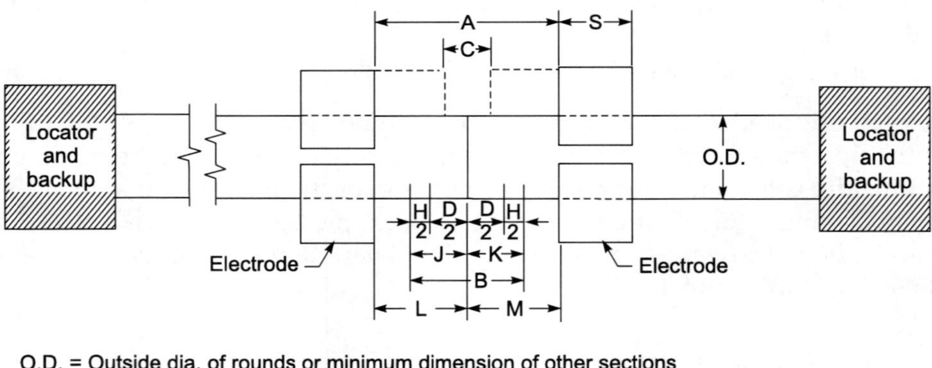

O.D. = Outside dia. of rounds or minimum dimension of other sections

A = Initial die opening
B = Material lost
C = Final die opening
D = Total flash-off
H = Total upset
J = K = Material lost per piece
L = M = Initial extension per piece
S = Minimum necessary length of electrode contact

Figure 15.3 (b) Flash welding of solid round, hex, square and rectangular bars

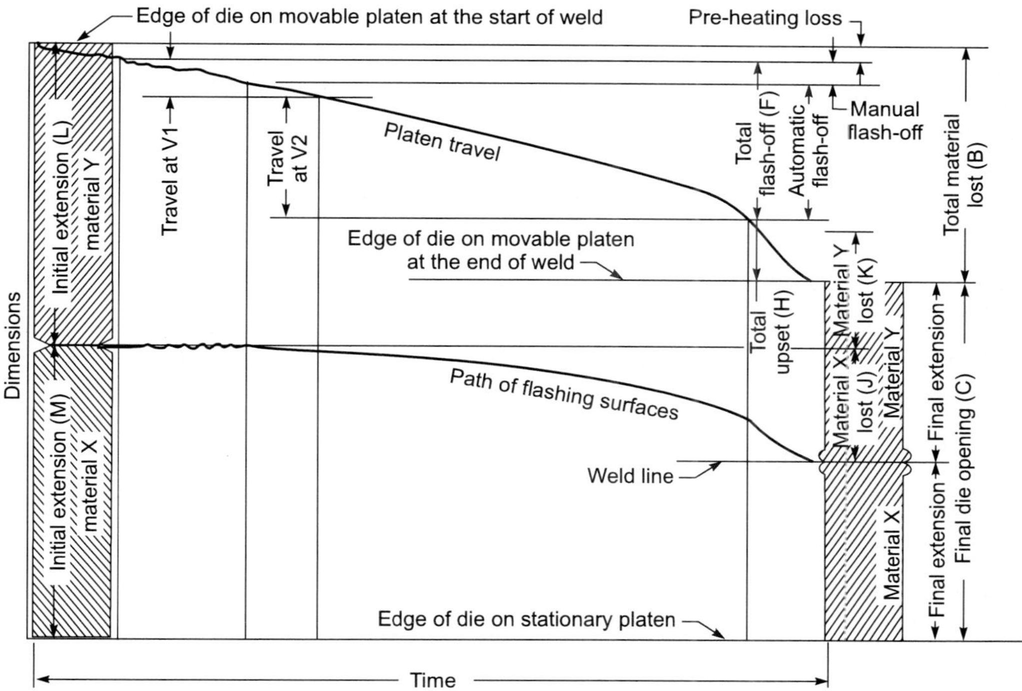

Figure 15.4 Flash welding: dimensional variables and motions

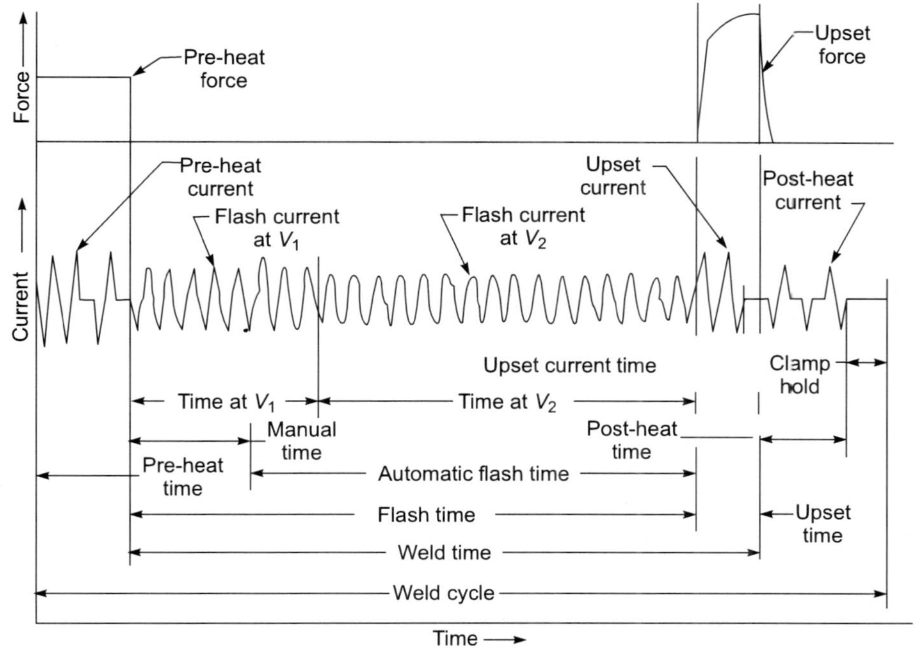

Figure 15.5 Flash welding: dimensional variables and motions

TABLE 15.1 Recommended max. joint lengths for flash welding

Sheet Thickness		Max. Joint Length		Sheet Thickness		Max. Joint Length	
inch(s)	mm	inches	mm	inch(s)	mm	inches	mm
0.010	0.25	1.00	25	0.060	1.5	25.00	635
0.020	0.50	5.00	125	0.080	2.0	35.00	890
0.030	0.75	10.00	250	0.100	2.5	45.00	1145
0.040	1.0	15.00	375	0.125	3.2	57.00	1450
0.050	1.3	20.00	500	0.187	4.8	88.00	2235

15.4 JOINT DESIGN

Three common types of welds made by flash welding are shown in Figure 15.6. Several basic design rules for flash welding are as follows:

1. The design should provide for an even heat balance in the parts so that the ends to be welded will have nearly equal compressive strengths at the end of the flashing time.
2. The metal lost during flashing (flash loss) and upset must be included in the initial length when designing the part. With miter joints, the angle between the two members must be taken into account in the design.
3. The parts must be designed so that they can be suitably clamped and held in accurate alignment during flashing and upset, with the joint perpendicular to the upset force direction.
4. The end preparation should be designed so that the flash material can escape from the joint, and that flashing starts at the center or the central area of the parts.

In general, the two parts to be welded should have the same cross section at the joint. Bosses may have to be machined, forged, or extruded on parts to meet this requirement. In the flash welding of extruded or rolled shapes with different thicknesses within the cross section, the temperature distribution during flashing will vary with section thickness. This tendency can often be counteracted by proper design of the clamping dies, provided the ratio of the thicknesses does not exceed about 4 to 1. The recommended maximum joint lengths for several thicknesses of steel sheet are given in Table 15.1.

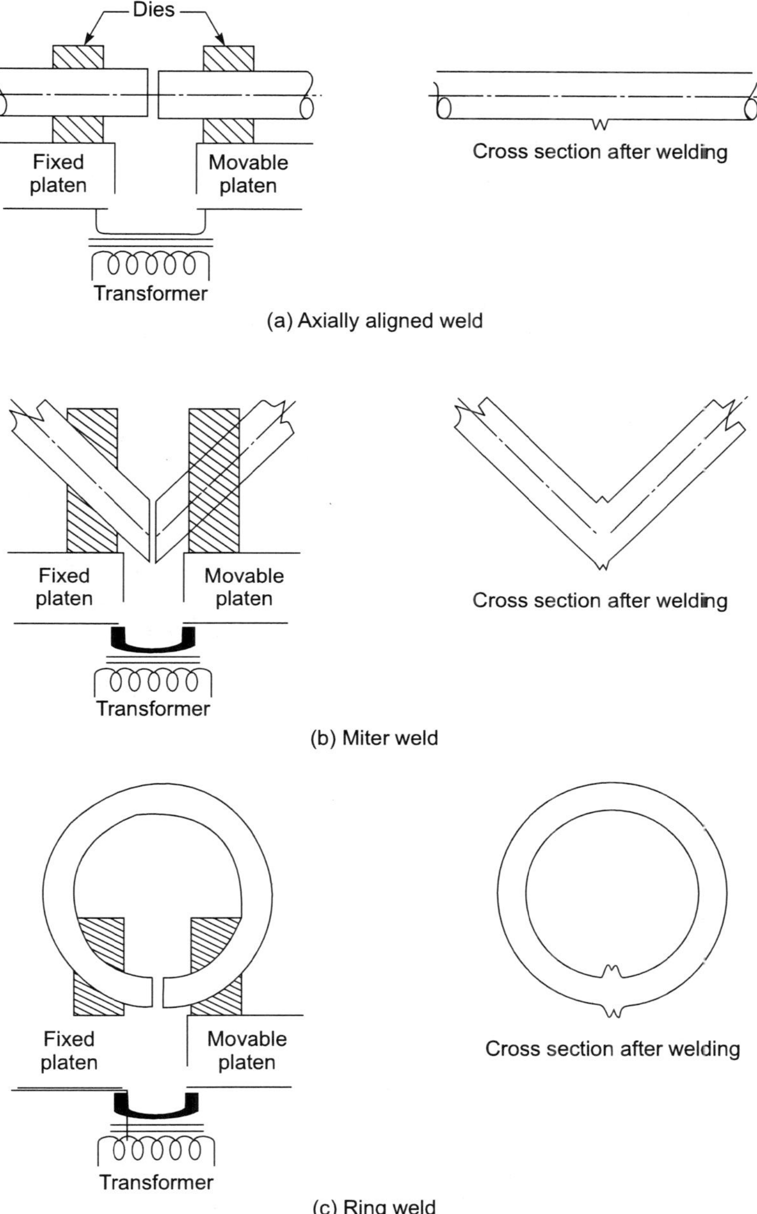

Figure 15.6 Common types of flash welds

15.5 PROCESS VARIABLES

Flashing Voltage

Flashing voltage is determined by the welding transformer tap setting. It should be selected to be as low as possible consistent with good flashing action. Electronic phase-shift heat control is not an effective means for reducing the flash voltage. The secondary voltage waveform produced by this means is incompatible with good flashing action. Changes in flashing voltage should be made only by changing the tap setting of the transformer. One system for providing two voltage ranges uses two primary contactors, each of which is connected to separate transformer taps. One contactor is energized to provide a high secondary voltage (V_1 in Figure 15.5) during the initial stages of flashing. The high voltage assists in starting the flashing action. The other contactor is energized after a predetermined time in the flashing operation to provide a normal secondary voltage (V_2 in Figure 15.5). The first contactor is de-energized at the same time. The best flashing action is achieved with this arrangement.

Flashing Time

Flashing is carried out over a time interval to obtain the required flash loss of metal. The time required will be related to the secondary voltage and the rate of metal loss as flashing progresses. Since a flashing pattern is generally parabolic, the variables are interrelated. In any case, smooth flashing action for some minimum flashing distance during some time interval is necessary to produce a sound, and strong weld.

Upset

In the production of a satisfactory flash weld, the flash and upset variables must be considered together since they are interrelated. The upset variables include the following:

- Flashing voltage cut-off
- Upset rate
- Upset distance
- Upset current magnitude and duration

Flashing Voltage Cut-off

Flashing voltage should be terminated at the moment when upset of the weld commences. Adjustments should be made during actual welding tests to ensure that voltage termination does not take place before the faying surfaces make full contact.

Upset Rate

Upset is initiated by increasing the acceleration of the parts to bring the faying surfaces together quickly. The molten metal and oxides present on the surfaces are forced out of the joint as this occurs. The hot weld zone is upset. The upset rate must be sufficient to expel the molten metal before it solidifies and to produce the optimum upset while the metal has adequate plasticity. The welding machine must apply a force to the movable platen to properly accelerate the part and overcome the resistance of the parts to plastic deformation. The force required depends upon the cross-sectional area of the joint, the yield strength of the hot metal to be welded, and the mass of the movable platen. Table 15.2 gives the approximate minimum upset pressures for flash welding typical alloys. These values may be used for a first approximation in determining the welding machine size required to flash weld a particular joint area on one of these alloys.

TABLE 15.2 Upsetting pressures for various classes of alloys

Strength Classification	Examples	Upset Pressure (ksi)	MPa
Low forging	SAE 1020, 1112, 1315 and those steels commonly designated as high-strength low alloy	10	69
Medium forging	SAE 1045, 1065, 1335, 3135, 4130, 4140, 8620, 8630	15	103
High forging	SAE 4340, 4640, 300M, tool steel, 12% Cr and 18-8 stainless steel, titanium, aluminum	25	172
Extra high forging	Materials exhibiting extra high compressive strength at elevated temperature such as A286, 19-9 DL, nickel and cobalt based alloys	35	241

Upset Distance

The magnitude of the upset distance must be sufficient to accomplish two actions:

- The oxides and molten metal must be expelled from the faying surfaces.
- The two faying surfaces must be brought into intimate metal-to-metal contact over the entire cross section.

The amount of upset required to obtain a sound flash weld depends upon the metal and the section thickness. If the flashing conditions produce relatively smooth flashed surfaces, smaller upset distances than needed for roughly flashed surfaces will be satisfactory for most metals. Some heat-resistant alloys may require upset distances as large as 1 to 1.25 times the section thickness. Satisfactory welds are made in aluminum with upset distances about 50 percent greater than those employed with steels of similar thicknesses.

Upset Current

As discussed under post-heating above, in some cases the weld zone may tend to cool too rapidly after flashing is terminated. This may result in inadequate upset or cold cracking of the upset metal. The joint temperature can be maintained during upset by resistance heating with current supplied by the welding transformer. The current magnitude is commonly controlled electronically. Normally, upset current would be terminated at the end of upset. If the flash is to be mechanically trimmed immediately after welding, upset current may be maintained for an additional period to achieve the desired temperature for trimming.

15.6 WELD QUALITY

Effect of Welding Variables

Weld quality is significantly affected by the specific welding variables selected for the application. Table 15.3 indicates the effects of several variables on quality when they are excessive or insufficient in magnitude. Each variable is considered individually, although more than one can produce the same result. Common defects found in flash welds are discussed below.

TABLE 15.3 Effect of variables on flash and upset weld quality

	Voltage	Rate	Time	Current	Distance or Force
Excessive	Deep craters are formed that cause voids and oxide inclusions in the weld; cast metal in weld	Tendency to freeze	Metal too plastic to upset properly		Tendency to upset too much plastic metal: flow lines bent perpendicular to base metal
Insufficient	Tendency to freeze; metal not plastic enough for proper upset	Intermittent flashing, which makes it difficult to develop sufficient heat in the metal for proper upset	Not plastic enough for proper upset: cracks in upset	Longitudinal cracking through weld area; inclusions and voids not properly forced out of the weld	Failure to force molten metal and oxides from the weld; voids

Base Metal Structure

Metallurgical discontinuities that often originate from conditions present in the base metal can usually be minimized by specifying necessary qualities in the materials selected. The inherent fibrous structure of wrought mill products may cause anisotropic mechanical behavior. An out-turned fibrous structure at the weld line often results in some decrease in mechanical properties as compared to the base metal, particularly in ductility. The decrease in ductility caused by flash welding is normally insignificant, except in two cases:

- The base material may be extremely non-homogeneous. Examples are severely banded steels, alloys with excessive stringer-type inclusions, and mill products with seams and cold shuts produced during the fabrication process.
- The upset distance may be excessive.

When the upset distance is excessive, the fibrous structure may be completely reoriented transverse to the original structure.

Oxides

Another source of metallurgical discontinuities is the entrapment of oxides at the weld interface. Such defects are rare since proper upset should expel any oxides formed during the flashing operation.

Flat Spots

Flat spots are metallurgical discontinuities that are usually limited to ferrous alloys. Their exact cause is not certain. They appear on a fractured surface through the weld interface in the form of smooth, irregularly shaped areas. There is excellent correlation between the location of flat spots and localized regions of carbon segregation in steels. In many cases, the cooling rates associated with flash welds are rapid enough to produce brittle, high-carbon martensite at areas on the flashing interface where the carbon content happens to be greater than the nominal composition of the alloy. Micro-hardness tests and metallographic examination have confirmed the presence of high-carbon martensite in the region surrounding a "flat spot" in almost every case, even in plain carbon steels. Furthermore, steels with banded microstructures appeared significantly more susceptible to this type of defect than unbanded steels.

Die Burns

Burns are discontinuities produced by local overheating of the base metal at the interface between the clamping die and the part surface. They can usually be avoided by keeping the parts clean and mating them properly with the dies.

Voids

Voids are usually the result of either insufficient upset or excessive flashing voltage. Deep craters produced on the faying surfaces by excessive flashing voltage may not be completely eliminated during upset. Such discontinuities are usually discovered when the welding procedure is being qualified.

Cracking

The type of discontinuity known as a crack may be internal or external. It may be related to the metallurgical characteristics of the metal. Alloys that exhibit low ductility over some elevated temperature range may be susceptible to internal hot cracking. Such alloys, known as "hot-short" alloys, are somewhat difficult to flash weld, but usually can be successfully welded with the proper conditions. Cold cracking may occur in hardenable steels. It can usually be eliminated by welding with conditions that moderate the weld cooling rate, coupled with post-welding heat treatment as soon as possible after welding.

Mechanical Discontinuities

Mechanical discontinuities include misalignment of the faying surfaces prior to welding, and non-uniform upset during welding. These discontinuities are easily detected by visual inspection. Misalignment of the parts is corrected by adjustment of the clamping dies and fixtures. Non-uniform upset may be caused by part misalignment, insufficient clamping force, or excessive die opening at the start of upset. The latter can be corrected by decreasing the initial die opening and then adjusting the welding schedule, if necessary.

15.7 FLASH WELDING APPLICATIONS

Base Metals

Many ferrous and non-ferrous alloys can be flash welded; typical metals are carbon and low-alloy steels, stainless steels, aluminum alloys, nickel alloys, and copper alloys. Titanium alloys can be flash welded, but an inert gas shield to displace air from around the joint is necessary to minimize embrittlement. Dissimilar metals may be flash welded if their upsetting characteristics are similar. Some dissimilarity can be overcome with a difference in the initial extensions between the clamping dies, adjustment of flashing distance, and selection of welding variables. Typical examples are welding of aluminum to copper or a nickel alloy to steel.

Typical Products

1. The automotive industry uses wheel rims produced from flash welded rings that are formed from flat cold rolled steel stock.

2. The electrical industry uses motor and generator frames produced by flash welding plate and bar stock previously rolled into cylindrical form. Cylindrical transformer cases, circular flanges, and seals for power transformer cases are other examples.
3. The aerospace industry uses flash welds in the manufacture of landing gear struts, control assemblies, hollow propeller blades, and rings for jet engines and rocket casings.
4. The petroleum industry uses oil drilling pipe with fittings attached by flash welding.
5. Several major railroads use flash welding to join relatively high-carbon steel track. In many cases, welding is done in the field using welding machines and portable generating equipment mounted on railroad cars.
6. Miter joints are sometimes used in the production of rectangular frames for windows, doors, and other architectural trim. These products are commonly made of plain carbon and stainless steels, aluminum alloys, brasses, and bronzes. Usually, the service loads are limited, but appearance requirements of the finished joints are stringent.
7. Most commonly welded metals are low-carbon steel, low-alloy steel, tool steel, stainless steel, copper alloys, aluminum alloys, nickel alloys, molybdenum alloys, magnesium alloys and titanium alloys.
8. Thick pipes, ends of band saws, frames, aircraft landing gears are produced by flash welding.
9. Flash welding is also used for butt welding of matching sections, chain links, railway lines, window frames, aero-engine rings, car wheel rims, metal strip in rolling mills, etc.

15.8 ADVANTAGES OF THE PROCESS

Butt joints between parts with similar cross section can be made by friction welding and upset welding, as well as by flash welding. The major difference between friction welding and upset and flash welding is that the heat for friction welding is developed by rubbing friction between the faying surfaces, rather than from electrical resistance. Upset welding is similar to flash welding except that no flashing action occurs. Listed below are some important advantages of flash welding:

1. Cross-sectioned shapes other than circular can be flash welded: for example, angles, H-sections, and rectangles. Rotation of parts is not required (flexible cross section).
2. Parts of similar cross sections can be welded with their axes aligned or at an angle to each other, within limits (flexible positioning for similar cross section).

3. The molten metal film on the faying surfaces and its ejection during upset act to remove impurities from the interface.
4. Preparation of the faying surfaces is not critical, except for large parts that may require a bevel to initiate flashing.
5. Rings of various cross sections can be welded.
6. The heat affected zones of flash welds are much narrower than those of upset welds.

15.9 LIMITATIONS OF THE PROCESS

1. The high single-phase power demand produces unbalance on three-phase primary power lines.
2. The molten metal particles ejected during flashing present a fire hazard, may injure the operator, and may damage shafts and bearings. The operator should wear face and eye protection, and a barrier or shield should be used to block flying sparks.
3. Removal of flash and upset metal is generally necessary and may require special equipment.
4. Alignment of workpieces with small cross sections is sometimes difficult.
5. The parts to be joined must have almost identical cross sections.
6. Cast iron, lead and its alloys cannot be welded by this process.
7. Heavy shapes cannot be welded.
8. Not suitable for lap joint.
9. There is a chance of fire hazard in this process.

15.10 INTRODUCTION: UPSET WELDING

Upset welding is a resistance welding process that produces coalescence over the entire area of the faying surfaces, or progressively along a butt joint, by the heat obtained from the resistance to the flow of welding current through the area where those surfaces are in contact. Pressure is used to complete the weld.

15.11 PRINCIPLES OF OPERATION (FIGURE 15.7)

Basically, upset welding is used to join the metal process end-to-end. In this process, the metal piece usually bars or rods of the same cross section are held in suitable clamp or vice. Projecting length of the piece between the clamps is adjusted according to their cross section. A longer projection will offer more resistance. After adjustment current is switched on and the faying surfaces are heated up to fusion temperature, then pressure is applied.

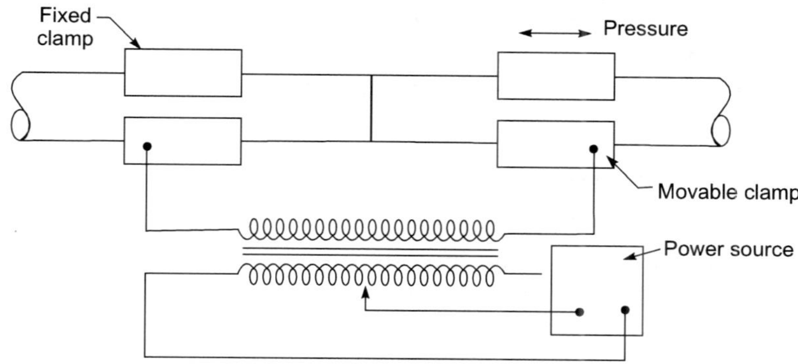

Figure 15.7 Upset welding

Sequence of Operations

The essential operational steps to produce an upset welded butt joint are as follows:

1. Load the machine with the parts aligned end-to-end
2. Clamp the parts securely
3. Apply a welding force
4. Initiate the welding current
5. Apply an upset force
6. Cut off the welding current
7. Release the upset force
8. Unclamp the weldment
9. Return the movable platen and unload the weldment

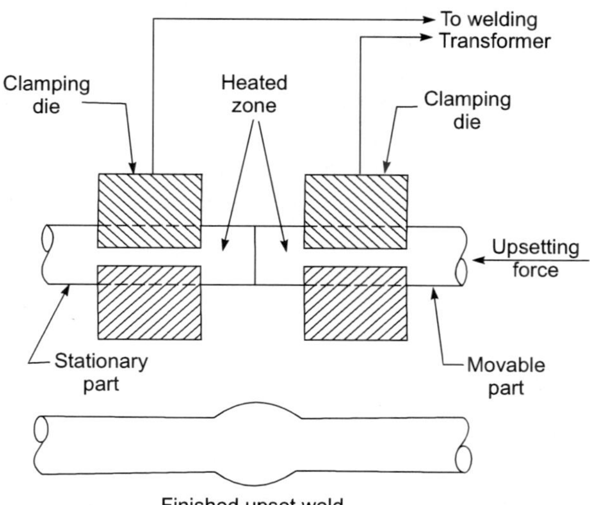

Figure 15.8 Upset welding of bars, rods and pipes

The general arrangement for upset welding is shown in Figure 15.8. One clamping die is stationary and the other is movable to accomplish upset. Upset force is applied through the moveable clamping die or a mechanical backup, or both.

15.12 JOINT PREPARATION

For uniform heating, the faying surfaces should be flat, comparatively smooth, and perpendicular to the direction of the upsetting force. Prior to welding, they should be cleaned to remove any dirt, oil, oxidation, or other materials that may impede welding. The contact resistance between the faying surfaces is a function of the smoothness and cleanliness of the surfaces and the contact pressure. This resistance varies inversely with the contact pressure, provided the other factors are constant. As the temperature at the joint increases, the contact resistance changes, but it finally becomes zero when the weld is formed. Upset welding differs from flash welding in that no flashing takes place at any time during the welding cycle. Generally, force and current are maintained throughout the welding cycle. The force is kept low at first to promote high initial contact resistance between the two parts. It is increased to a higher value to upset the joint when the welding temperature is reached. After the prescribed upset is accomplished, the welding current is turned off and the force is removed.

15.13 EQUIPMENT (FIGURE 15.9)

Equipment for upset welding is generally designed to weld a particular family of alloys, such as steels, within a size range based on cross-sectional area. The mechanical capacity and electrical characteristics of the machine are matched to that application. Special designs may be required for certain aluminum alloys to provide close control of upset force. Electric current for heating is provided by a resistance welding transformer. It converts line power to low voltage, high current power. No-load secondary voltages range from about 0.4 to 8 V. Secondary current is controlled by a transformer tap switch or by electronic phase shift. Basically, an upset welding machine has two platens, one of which is stationary and the other movable. The clamping dies are mounted on these platens. The clamps operate either in straight line motion or through an arc about an axis, depending upon the application. Force for upset butt welding is produced generally by a mechanical, pneumatic, or hydraulic system.

15.14 HEAT BALANCE

The upset process is generally used to join together two pieces of the same alloy and same cross-sectional geometry. In this case, heat balance should be uniform across the joint. If the parts to be welded are similar in composition and cross section but of unequal mass, the part of larger mass should project from the clamping die somewhat farther than the other part. With dissimilar metals, the one with higher electrical conductivity should extend farther from the clamp than the other. In upset welding large parts do not make good contact with each other, it is sometimes advantageous to interrupt the welding current periodically to allow the heat to distribute evenly into the parts.

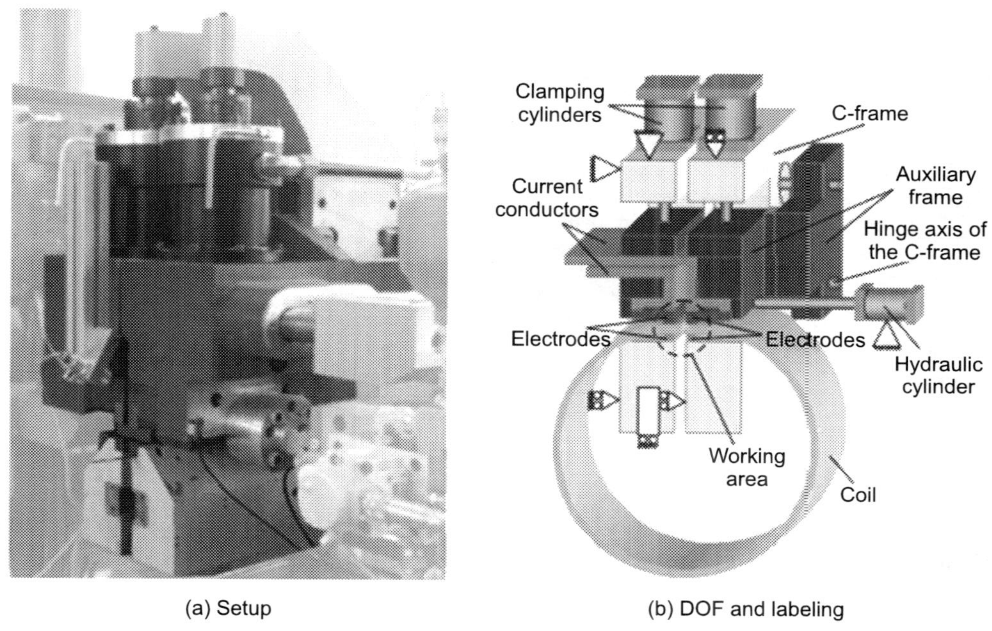

Figure 15.9 Equipment

15.15 PROCESS VARIATIONS

Upset welding has two variations:

1. Joining two sections of the same cross section end-to-end (butt joint).
2. Continuous welding of butt joint seams in roll formed products, such as pipe and tubing.

The first variation can also be accomplished by flash welding and friction welding. The second variation is also done with high frequency welding.

Butt Weld Cycle (Figure 15.10)

In the butt welding cycle, the parts are brought together first, a current is applied and most of the heating occurs just by the I^2R heating. There should be no flashing in the upset butt process. And because when we start heating, the parts begin to heat and expand, it tends to push the platen back a little until it reaches a temperature where the parts get hot enough for general yielding and the force of the platen begins to push the parts back again. When the part finally gets hot enough, full upset is applied where a rapid closure occurs.

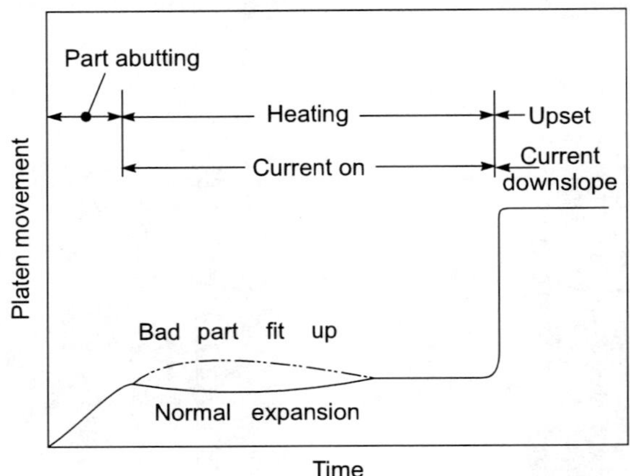

Figure 15.10 Schematic of butt weld cycle

15.16 METALS WELDED

A wide variety of metals in the form of wire, bar, strip, and tubing can be joined end-to-end by upset welding.

These include:

1. Carbon steels
2. Stainless steels
3. Aluminum alloys
4. Brass
5. Nickel alloys
6. Electrical resistance alloys

15.17 COMPARISON BETWEEN FLASH WELDING AND UPSET BUTT WELDING

1. No arching or flashing takes place in upset butt welding as in the case of flash welding.
2. Less current is consumed in flash welding but welding time is more.
3. Heat application in flash welding precedes pressure application whereas in upset flash welding, constant pressure is applied during heat processes particularly to eliminate flashing only.
4. Dissimilar metal can weld efficiently.
5. Flash welding can be used to weld the high-alloy steel.

15.18 APPLICATIONS

Upset welding is used in wire mills and in the manufacture of products made from wire. In wire mill applications, the process is used to join wire coils to each other to facilitate continuous processing. The process is also used to fabricate a wide variety of products from bar, strip, and tubing.

Upset welding can be used in the manufacturing of circular objects that have ends of the same size. This includes wire rings, tubular rings, burner rings, wheel rims. It can be used to weld large structures compared to other resistance welding techniques.

15.19 WELD QUALITY

Butt joints made by upset welding are as strong as un-welded base metal itself. With proper procedures, welds made in wires are difficult to locate after they have passed through a subsequent drawing process. In many instances, the welds are then considered part of the continuous wire. Upset welds may be evaluated by tension testing. The tensile properties are compared to those of the base metal. Metallographic and dye penetrant inspection techniques are also used. A common method for evaluating a butt weld in wire is a bend test. A welded sample is clamped in a vise with the weld interface located one wire diameter from the vise jaws. The sample then is bent back and forth until it breaks into two. If the fracture is through the weld interface and shows complete fusion, or if it occurs outside the weld, the weld quality is considered satisfactory.

15.20 INSPECTION AND TESTING

Upset welds can be inspected and tested in the same manner as flash welds. In general, the quality requirements for upset welds are not as stringent as those specified for flash welds. The process normally cannot produce welds with the consistency available with flash welding.

15.21 ADVANTAGES OF UPSET WELDING

1. It is suitable for joining many alloys that are difficult to weld using fusion processes.
2. Since there is no melting, the weld metal retains many of the characteristics of the base metal.
3. Resulting welds have a hot worked structure and thereby, have higher strength than fusion welding in the same mate.
4. Compositional gradients are not introduced, second-phase materials are minimally disrupted.

Review Questions

1. Define is flash welding and upset welding.
2. Explain with a neat sketch the process of flash welding.
3. What are the welding procedures followed in flash welding?
4. What are the process variables used in flash welding?
5. What is upset pressure in upset welding?
6. What are the advantages of flash welding?
7. What are the disadvantages of flash welding?
8. What are the advantages of upset welding?
9. What are the disadvantages of upset welding?
10. What are the applications of upset and flash welding?

16

CHAPTER

High-Frequency Welding

16.0 INTRODUCTION

High-frequency welding includes those processes in which coalescence of metals is produced by the heat generated from the electrical resistance of the work to the high-frequency current, usually with the application of an upsetting force to produce a forged weld.

There are two processes that utilize high-frequency current to produce the heat for welding: *high-frequency resistance welding (HFRW)* and *high-frequency induction welding (HFIW)*, sometimes called *induction resistance welding*. Heating of the work in the weld area and the resulting weld are essentially identical with both processes. With HFRW, the current is conducted into the work through electrical contacts that physically touch the work. With HFIW, the current is induced in the work by coupling with an induction coil. There is no physical electrical contact with the work.

With low frequency (50 to 360 Hz), direct current or "square wave" resistance welding, much higher currents are required to heat the metal and large electrical contacts must be placed very close to the desired weld area. The voltage drop across the weld is very low, and the current flows along the path of least resistance from one electrode to the other. With high-frequency welding, by contrast, the current is concentrated at the surface of the part. The location of this concentrated current path in the part can be controlled by the relative position of the surfaces to be welded and the location of the electrical contacts or induction coil. Heating to welding temperature can be accomplished with a much lower current than with low frequency or direct current resistance welding.

Although the welding process depends upon the heat generated by the resistance of the metal to high frequency, other factors must also be considered for successful high-frequency welding. Because of the concentrated high-frequency heats, only a small volume of metal just where the weld is to take place melts. Therefore, this process is extremely energy efficient and can be done at the high speeds. Maximum speeds are normally limited by mechanical considerations of materials handling, forming and cutting. Minimum speeds are limited by material properties and weld quality requirements.

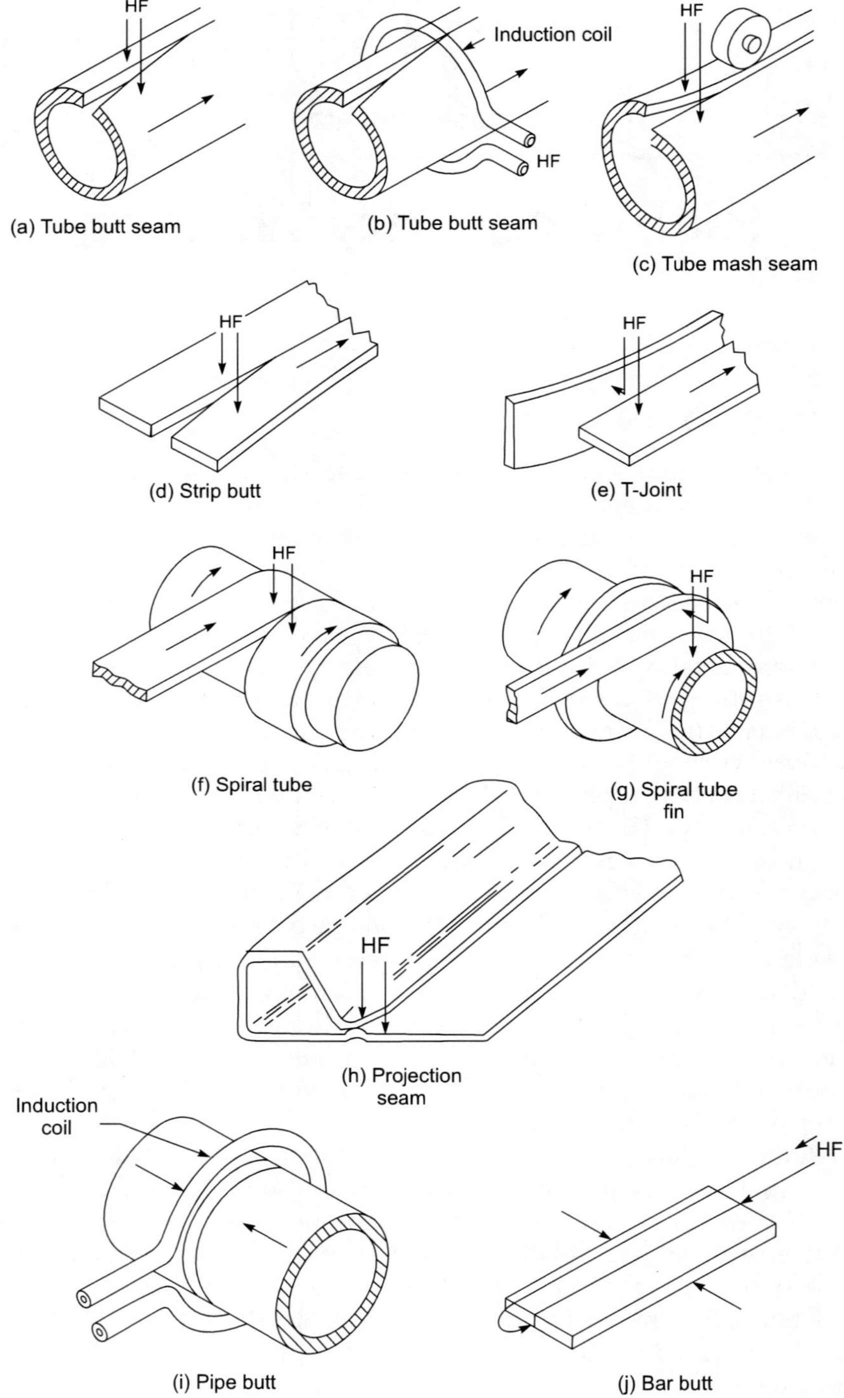

Figure 16.1 Basic high-frequency welding applications

The fit of the surfaces to be joined and the manner in which they are brought together is important if high-quality joints are to be produced. Flux is not usually used but can be introduced to the weld area in an inert gas stream. Inert gas shielding of the welding area is generally needed only for joining reactive metals, such as titanium and also for certain stainless steel products.

Typical high-frequency welding applications are shown in Figure 16.1.

16.1 HISTORY

In the late 1940s and early 1950s, emphasis was on the development of procedures and equipment for high-frequency induction butt end welding of pipes and tubes. Successful welding operations were conducted using 10 kHz motor-generator equipped with induction coils that could be opened for removing the finished work. The first mobile installation for joining pipe in the field was utilized by a utility company in the streets of New York City. The unit produced welded butt joints in pipe up to 12 inches (305 mm) in diameter with wall thickness up to 5/16 inches (8 mm). With this process, pipe ends were pressed together and induction heated to forge welding temperature in approximately 60 seconds.

In 1949, the first high-frequency induction welding system for continuous "forged welding" of the longitudinal seam in small steel tubes was developed. It used a 10 kHz motor-generator with split return induction coil over the seam similar to those now used for continuous normalizing of the welded seam in pipes. This unit was successful and operated for many years, but it had the disadvantage of heating a large portion of the tube. The unit was eventually replaced by a 400 kHz HFRW system.

In 1952, Thomas Crawford made tests on continuous induction welding of the longitudinal seam of various metal tubes including some with cables inside. The tests were made at a frequency of 400 kHz and employed an induction coil surrounding the tube. Induction welding of tube and pipe has become very successful for smaller diameters up to about 6 inches (150 mm). It is now becoming more popular for larger diameters, up to 20 inches (500 mm) in some installations.

In 1952, W. C. Rudd and Robert Stanton invented a 400 kHz process for welding a large variety of continuous joints. The Rudd and Stanton process introduced the welding current directly into the work by means of sliding contacts. The sliding contacts permitted the production of butt, lap, and tee joints in pipe, strip, and structural products. Successful tests were made in 1954 on butt welding the longitudinal seam in aluminum tubing. This led to the first commercial installation in 1955, for the welding of aluminum irrigation tubing, and the mill operated for merely twenty years.

Subsequently, over 3000 installations of various forms of the system had been installed around the world. A large variety of metals had been welded. High-frequency welding is an automated process, and is not adaptable to manual welding.

16.2 FUNDAMENTALS OF THE PROCESS

High-frequency current in metal conductors tends to flow at the surface of the material at relatively shallow depth. This is commonly called the *skin effect*. The effect of frequency on the depth of penetration for several metals is shown in Figure 16.2. For example, the depth of current penetration in steel at 800°C is about 0.8 mm at 450 kHz and nearly 5.5 mm at 10 kHz. At room temperature, the penetration in steel is about 0.05 and 0.25 mm at 450 and 10 kHz, respectively.

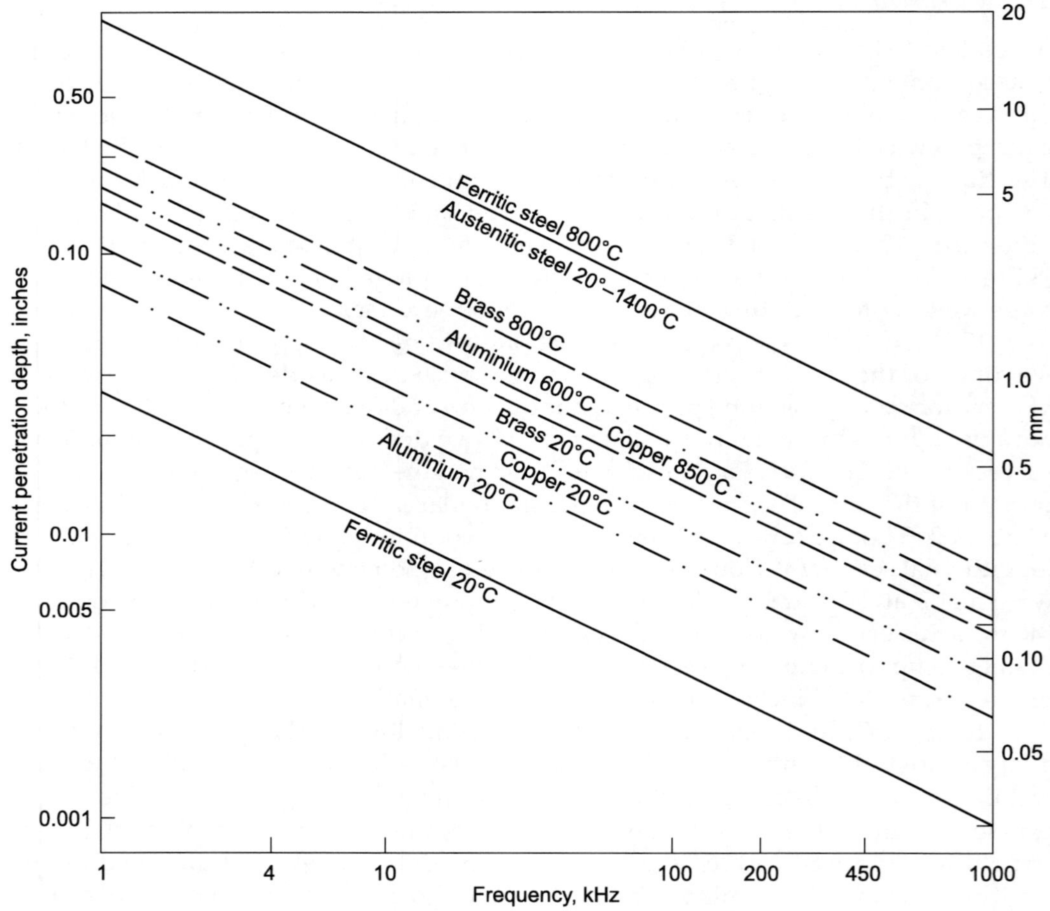

Figure 16.2 Effect of frequency on depth of current penetration into various metals at selected temperatures

The high-frequency current path at the surface of the workpiece is controlled by the nearness of its own return flow path. This phenomenon, called *proximity effect*, is illustrated in Figure 16.3.

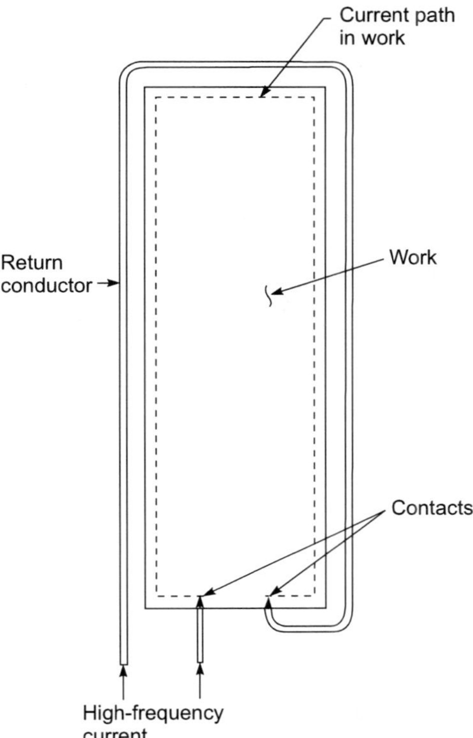

Figure 16.3 Restriction of the flow path of high-frequency current by the proximity effect of the return conductor

Both the skin effect and proximity effect become more pronounced with increasing frequency. Therefore, the effective resistance of the current path in the work increases as the frequency increases because as frequency increases, the current is confined to a shallower and narrower path. This current concentration is advantageous because extremely high heating rates and high temperatures can be achieved in a localized area. Moreover, these high temperature concentrations can be positioned where they are needed at the surfaces to be welded.

Current patterns (high-temperature concentrations) in the workpiece at frequencies of 60 Hz and 10,000 Hz are illustrated in Figure 16.4. In Figure 16.4(a), the return conductor is positioned parallel and close to the plate. The 60 Hz current flowing in the steel plate travels in opposite phase to the current in the adjacent proximity conductor.

In this case, the size and shape of the proximity conductor have a negligible effect on the distribution of the current in the steel plate. As a result, the current flows uniformly throughout the cross section of the plate.

When a 10,000 Hz current is applied to the same system, as shown in Figure 16.4(b), the current in the work is confined to a relatively narrow band immediately beneath the proximity conductor. This narrow band is the path of lowest inductive reactance to the current in the plate. The shape and magnetic surroundings of the proximity conductor have considerable effect on the distribution of the current in the work, but have no effect on the depth of current penetration. Two round proximity conductors at different distances from the work are illustrated in Figure 16.4(c). The closer proximity conductor develops more confined current path. A rectangular proximity conductor with the narrow edge at the same distance from the work as the close round conductor exhibits a broader current distribution in the workpiece. If a magnetic core is placed around the proximity conductor, the current is further confined and heating takes place directly beneath the proximity conductor as shown. If the two conductors with currents flowing in opposite directions are sheets placed edge to edge in a plane with a small gap between them, the proximity effect will cause the two adjacent edges to heat. The skin effect will confine the current to a shallow depth at those edges. This situation occurs during the butt welding of the longitudinal seam of tube and pipe using either the HFIW or HFRW process.

Almost all high-frequency welding techniques employ some force to bring the heated metals into intimate contact during coalescence. During the application of force, an upset or flash occurs in the weld area. In many cases, the flash is removed after welding.

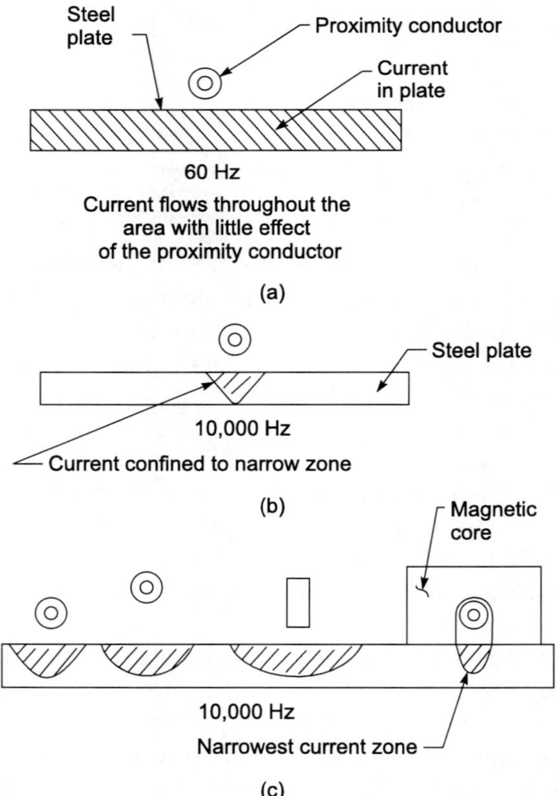

Figure 16.4 Current depth and distribution adjacent to various proximity conductors

16.3 EQUIPMENT

Power Sources

Except for few special applications, such as tube end and strip butt welding, vacuum tube oscillators with output power ranging from 50 to 1200 kW at frequencies ranging from 200 to 500 kHz are used for high-frequency welding. A basic circuit for a typical high-frequency oscillator is shown in Figure 16.5.

Units can be manufactured for special line voltages, but typical input voltages are either 460 V, 60 Hz, 3-phase, or 380 V, 50 Hz, 3-phase. After the incoming circuit breaker and contactor, there is a 3-phase SCR (thyristor) voltage regulator. The regulator is designed both to maintain a preset voltage when the line voltage varies and also to enable the preset voltage to be controlled either automatically or by the welder operator. In automatic operation, control is dependent upon variables such as the weld temperature or the mill speed.

The plate transformer converts this controlled voltage to a high voltage which is then rectified to provide the direct current required for the oscillator circuit. The filter choke and filter capacitor reduce the ripple in the DC to an acceptable level, typically less than one percent. The oscillator circuit converts the direct current to high-frequency alternating current for the output transformer. The output transformer converts the high voltage-low current power to the low voltage-high current power required for welding.

Vacuum tube oscillators inherently have high output impedance (high voltage-low current) and must be fed into high impedance loads. The inductors and the workpiece contact circuits in high-frequency welding are low impedance (low voltage-high current) loads. An impedance matching output transformer is required to transfer energy efficiently from the oscillator to the work. Power is transferred most efficiently from the high-frequency generator to the work when the impedance of the work circuit matches the impedance of the generator. For induction welding, variable impedance matching transformers are often used in which the primary winding can be moved relative to the secondary winding to match the relatively wide impedance range typical of induction applications.

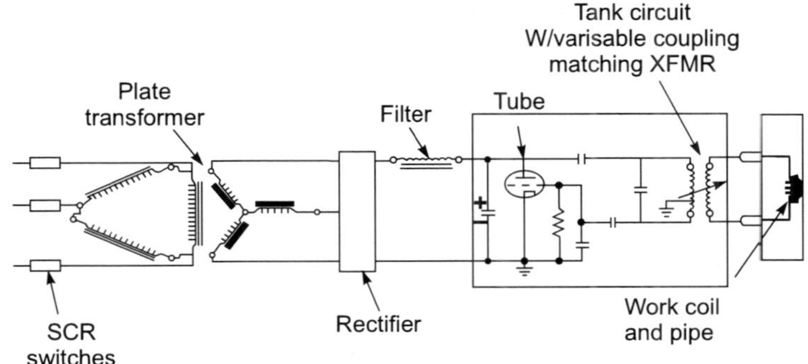

Figure 16.5 Schematic circuit for a typical high-frequency oscillator

The secondary winding of the impedance matching transformer is in series with the induction coil or the contacts and workpiece. This forms the low voltage-high current welding system. The connecting leads should have the lowest possible impedance to obtain high efficiency and to minimize the voltage drop in the leads. This may be achieved by using short, wide leads made of flat copper plate separated by approximately 1.6 mm of insulation. Power losses in poorly designed leads or incorrectly matched transformers can seriously degrade the performance of a high-frequency welding system.

16.4 PROCESS VARIATIONS

High-Frequency Induction Welding

High-frequency induction welding (HFIW) is generally used to weld continuous seam tubes and pipes. The tube is formed from metal strip in a continuous roll-forming mill and enters the weld area with the edges to be welded slightly separated. In the weld area, they converge in a "vee" until they touch the weld point. An induction coil, typically made of copper tubing, or copper sheet with attached cooling tubes, encircles the tube slightly ahead of the weld point. A current is induced by this induction coil which flows both around the tube immediately underneath the coil and along the tube edges between the induction coil and the weld point. This is illustrated in Figure 16.6.

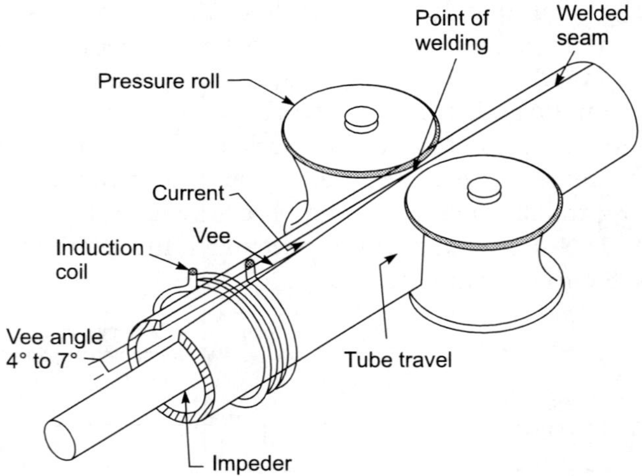

Figure 16.6 Joining a tube seam by high-frequency induction welding

The high-frequency current follows in a localized path down one side of the vee consisting of the two edges being welded and back along the other due to the skin and proximity effects. The edges are resistance heated to a shallow depth. Welding speed and power level are adjusted so that the two edges are at welding temperature when they reach the weld point. At that point, pressure rolls forge the hot edges together and upset them to produce a weld. Hot metal containing impurities from the faying surfaces of the joint is squeezed out in both directions. The upset metal is normally trimmed off flush with the base metal. In the case of very thin wall tubes where the wall thickness is of the order of two times the depth of current penetration or less, a magnetic core called an *impeder* is required inside the tube. An impeder limits the current flowing around the inside circumference of the tube, which detracts from the current available to heat the metal at the vee and reduces the efficiency of the process.

The impeder is normally made of a ferrite material, and it must be cooled to prevent its temperature rising to the curie point where it becomes non-magnetic. As the pipe or tube gets larger, the losses around the outside circumference of the tube become larger relative to the heating in the vee, making the process progressively less efficient. Therefore, induction welding is more efficient for smaller sizes of tube and pipe, and high-frequency resistance welding becomes more competitive for larger sizes.

Butt Welding of Hollow Pieces

HFIW of individual pieces can be done only when the induced current can circulate in a closed circuit path. A typical application is the welding of butt joints between the sections of pipe or tubing. A narrow induction coil is placed around the joint. High-frequency current in the coil induces a circulating current concentrated in the area of the butted pipe joint which is heated rapidly. When the metal reaches the welding temperature, it is upset to produce a forge weld. The placement of the coil in this application is shown in Figure 16.7. This process is used for welding high-pressure boiler tubes with diameters of about 25 to 75 mm and with wall thicknesses up to about 10 mm. It has also been used for joining pipes up to about 300 mm in diameter in the field. Welding times range from 10 to 60 seconds per joint.

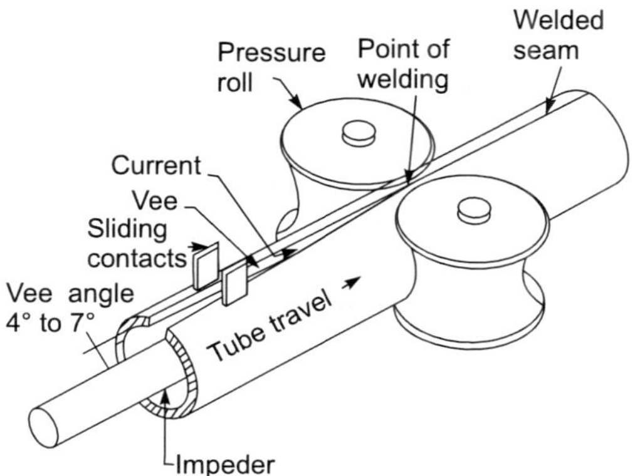

Figure 16.7 Joining a tube seam by high-frequency resistance welding

Finite Length Welding

Techniques are available for butt welding the ends of two strips. This is done by passing a high-frequency current through the joint area. The current is introduced at each end of the joint by small contacts and is confined to the area of the joint by a proximity conductor. Generally, a magnetic core is used to assist in narrowing the current path, as shown in Figure 16.8. By selecting proper frequency, the depth of penetration of the current can be adjusted to heat the joint through its thickness. When the joint reaches the welding temperature, forging force is applied, and the hot metal is upset. Joints are made in this fashion at rates as high as 1000 joints per hour. The use of a proximity conductor in the HFRW of strip butt joints is illustrated in Figure 16.8.

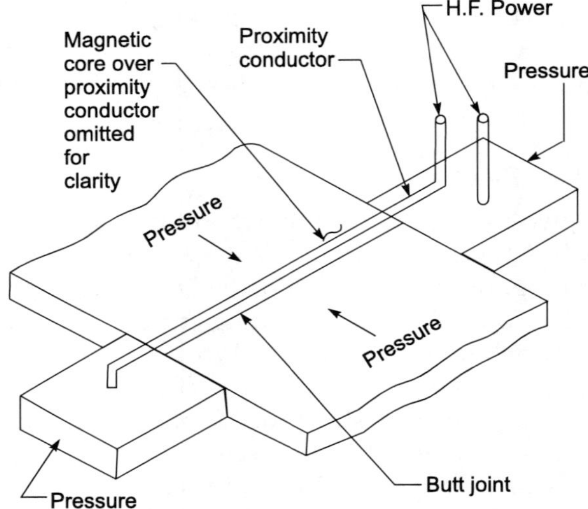

Figure 16.8 Joining strips using high-frequency welding

16.5 APPLICATIONS

Almost all engineering metals and alloys can and have been joined by high-frequency welding. The exceptions are metals that cannot be hot formed, are unstable at elevated temperatures, or have unsatisfactory properties that cannot be restored by mechanical or thermal post-weld treatments. Such a material would be cast iron. Reactive metals can be protected with an inert atmosphere. Inert shielding may be unnecessary because the weld cycle time is very short. High-conductivity metals such as pure copper are satisfactorily welded, and dissimilar metals can be readily joined. However, in dissimilar metals, the weld temperature is limited to the lower melting point of the two metals.

Metals with large non-metallic inclusions, with large grain size, and with damaged or contaminated faying surfaces can be difficult or impossible to weld satisfactorily.

Uniformity of dimensions, strength, and electromagnetic and thermophysical properties are desirable for consistent high quality welding.

16.6 ADVANTAGES AND LIMITATIONS

High-frequency welding processes offer several advantages over low frequency and direct current resistance welding processes. One characteristic of the high-frequency processes is that they can produce welds with very narrow heat affected zones. The high-frequency welding current tends to flow only near the surface of the metal because of the "skin effect" and along a narrow controlled path because of the "proximity effect". The heat for welding, therefore, is developed in a small volume of metal along the surfaces to be joined. A narrow heat affected zone is generally desirable because it tends to give a stronger welded joint than with the wider zone produced by many other welding processes. With some alloys, the narrow heat affected zone and absence of cast structure may eliminate the need for post-weld heat treatment to improve the metallurgical characteristics of the welded joint.

The shallow and narrow current flow path results in extremely high rates and therefore, high welding speeds and less power consumption. A major advantage of the continuous high-frequency welding process is its ability to weld at very high speeds. For instance, a 1/2 inch (12 mm) wall steel pipe with a diameter ranging from **8** to 48 inches (200 to 1200 mm) can be HFRW welded at speeds over 100 fpm (30 m/s) with a 600 kW output welder. Smaller tube diameters such as 1 to 2 inches (25 to 50 mm) with light walls in the range of 0.25 to 0.065 inche (0.6 to 1.7 mm) can be induction welded at speeds ranging from 200 to 800 fpm (60 to 240 m/min.), using welders with 100 kW to 400 kW output.

High-frequency welding can also be used to weld very thin wall tubes. Wall thicknesses down to less than 0.005 inch (0.13 mm) are presently welded on continuous production mills. The process is equally adaptable to large diameter pipes with wall thicknesses up to 1.0 inch (25 mm).

The process is adaptable to many metals including low-carbon and alloy steels, ferritic and austenitic stainless steels, and many aluminum, copper, titanium, and nickel alloys.

Because the time at welding temperature is very short and the heat is localized, oxidation and discoloration as well as distortion of the part are minimal. Materials that would normally be damaged from prolonged exposure to heat can be welded with high-frequency welding processes. For example, electrical cable sheathing with the cable inside is welded with HFIW. High-frequency welding power sources have a balanced 3-phase input power system. Using conventional vacuum tube welding power sources, as much as 60% of the energy is converted into useful heat in the work. Solid state high-frequency power sources, which have even higher efficiencies, are now becoming available. As with any process, there are also limitations. The equipment operates in the radio frequency range, and therefore, special care must be taken in its installation, operation, and maintenance to avoid radiation interference in the plant's vicinity. As a general rule, the minimum speed in carbon steel is about 25 fpm (0.125 m/s). For products which are required in small quantities, the process may be uneconomical unless the technical advantages justify the application. Because the process utilizes localized heating in the joint area, proper fit-up is important. Equipment is usually incorporated into the mill or line operation and must be fully automated. The process is limited to the use of coil, flat, or tubular stock with a constant joint symmetry throughout the length of the part. Any disruption in the current path or change in the shape of the vee can cause significant problems. Also, special precautions must be taken to protect the operators and plant personnel from the hazards of high-frequency current.

Review Questions

1. What is high frequency welding?
2. Explain with a neat sketch the process of (high frequency) HF welding.
3. What is skin effect and proximity effect in HF welding?
4. What are the process variables used in HF welding?
5. What is the purpose of oscillator in HF welding?
6. What are the advantages of HF welding?
7. What are the disadvantages of HF welding?
8. What are the advantages of HF welding?
9. What are the process variations of HF welding?
10. How butt welding of hollow tubes is done using HF welding?

17 CHAPTER

Percussion Welding

17.0 INTRODUCTION

Percussion welding is a joining process that produces coalescence with an arc resulting from a rapid discharge of electrical energy. Pressure is applied percussively during or immediately following the electrical discharge. In general, the term "percussion welding" is used in the electronics industry for joining wires, contacts, leads, and similar items to a flat surface.

In percussion welding, the two parts are initially separated by a small projection on one part, or one part is moved towards the other. At the proper time, an arc is initiated between them. This arc heats the faying surfaces of both parts to welding temperature. Then, an impact force drives the parts together to produce a welded joint. There are basically two variations of the percussion process: capacitor discharge and magnetic force. Although the steps may differ in certain applications because of process variations, the essential sequence of events in making a percussion weld is as follows:

1. Load and clamp the parts into the machine.
2. Apply a low force on the parts or release the driving system.
3. Establish an arc between the faying surfaces with high voltage to ionize the gas between the parts or with high current to melt and vaporize a projection on one part.
4. Move the parts together percussively with an applied force to extinguish the arc and consummate a weld.
5. Turn off the current.
6. Release the force.
7. Unclamp the welded assembly.
8. Unload the machine.

Percussion welding is similar to capacitor discharge stud welding. The differences between the two processes lie in the applications and the type of power source. Percussion welding may be used to join equal cross sections of wires, rods, and tubes. Welding current is supplied by a capacitor storage bank in these applications. The process may also be used to weld wires or contacts to large flat areas with power from a capacitor bank or a transformer.

17.1 PRINCIPLES OF OPERATION

In percussion welding, two pieces are welded by a high intensity short duration arc followed by rapid or percussive impacting of the workpieces. The molten surfaces are then squeezed together by collision and some of the metal is forced out to the side of the joint. The current density is very high, and this melts a thin layer of metal on the faying surfaces in a few milliseconds (Figure 17.1).

There are two process variations that differ in the type of power supply, method of arc initiation, and work drive motion.

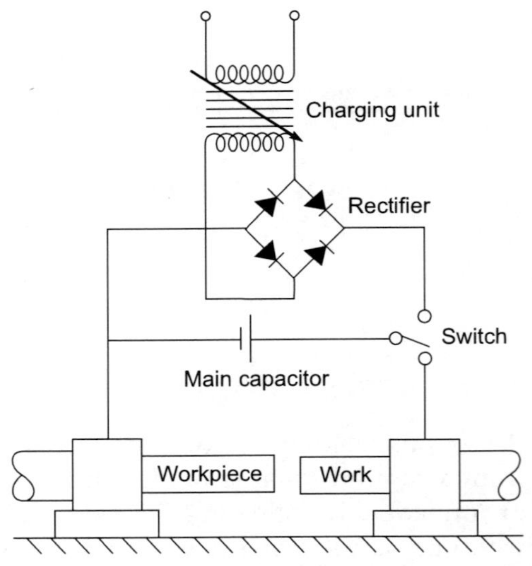

Figure 17.1 Percussion welding

17.2 TYPES OF PERCUSSION WELDING

17.2.1 Capacitor Discharge Percussion Welding (Figure 17.2)

With the capacitor discharge method, power is furnished by a capacitor storage bank. The arc is initiated by the voltage across the terminals of the capacitor bank (charging voltage) or a superimposed high voltage pulse. Motion may be imparted to the movable part by mechanical or pneumatic means.

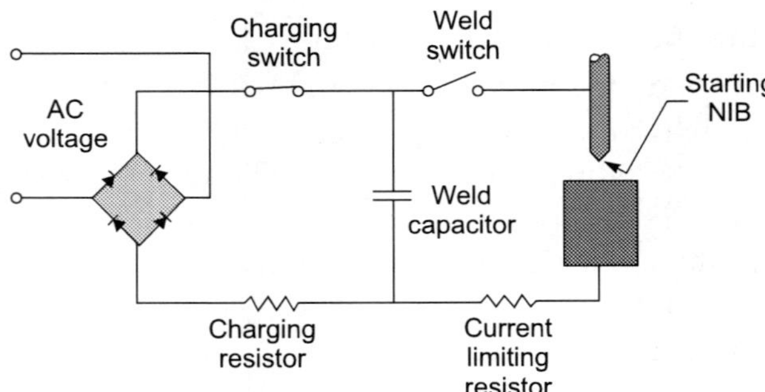

Figure 17.2 Capacitor discharge percussion welding

A typical circuit is shown in Figure 17.2, The capacitor(s) is charged by direct current from a rectifier or generator. This welding energy is stored at 50 to 300 VDC and later discharged through either mechanical or electrical switching methods into the primary of a high-current welding transformer. The controls for this weld power supply are:

1. Voltage applied – An adjustment to voltage stored in the capacitors.
2. Amount of capacitance – An adjustment of the number of capacitors.
3. Weld current – An adjustment of the taps on the welding transformer to vary the voltage on the secondary winding.

17.2.2 Magnetic Force Percussion Welding

For magnetic force welding, power is supplied by a welding transformer. The arc is initiated by vaporizing a small projection on one part with high current from the transformer. The vaporized metal provides an arc path. The percussive force is applied to the joint by an electromagnet that is synchronized with the welding current. Magnetic force percussion welds arc is made in less than one half cycle of 60 Hz. Consequently, the timing between the initiation of the arc and the application of magnetic force is critical.

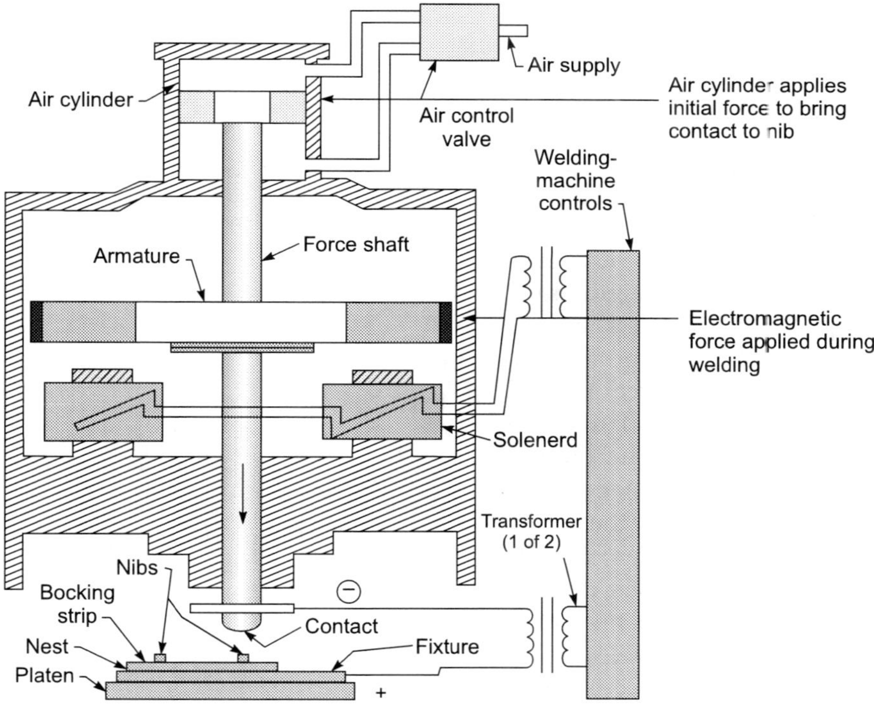

Figure 17.3 Magnetic force percussion welding

Welding Machines

Magnetic force welding machines use a low-voltage power source (20 to 35 V) from a transformer, a projection-type arc starter, and an electromagnetic system to produce the weld force. A unit generally consists of a modified press-type resistance welding machine with specially designed transformer, controls, and tooling.

Magnetic force percussion welding machines usually have an independent power source for the electromagnet so that the force magnitude and time of application can be varied to coincide with the initiation of the welding current. This is accomplished by using two separate transformers, one for welding power and one for the electromagnet power. The acceleration of the force member can be controlled by adjusting the magnitude of the electromagnet current, thereby providing a control for the duration of arc time.

Since welding is done during one-half cycle of 50 Hz, current is unidirectional. In some cases, the polarity of the two workpieces may have some effect on weld quality. In general, the same conditions with respect to polarity that prevail in direct current (DC) arc welding also are in effect in percussion welding. The current is always passing through the transformer in the same direction and the core may become partially saturated. Consequently, the electrical controls should provide a low-amplitude one half-cycle pulse in the opposite direction to remove the remaining flux from the transformer and electromagnet. This can be done during the loading time.

17.3 JOINT DESIGN

For welding two flat surfaces together, a projection similar to that for resistance welding must be formed on one piece, as shown in Figure 17.4. Its diameter and height must be developed for each application. The diameter must be large enough to support the initial force applied to the parts but too small to carry the welding current. The height determines the gap between the faying surfaces and thus, the initial arc voltage. When large area contacts are welded, two projections may be required. The surfaces to be joined must be flat and parallel during welding so that arcing occurs over the entire area. Areas that are not melted will probably not weld when impacted together.

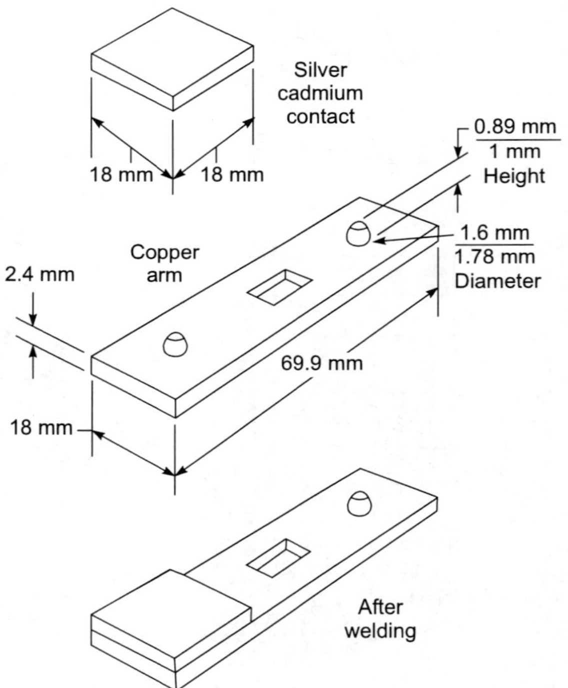

Figure 17.4 Typical design of a magnetic force, percussion-welded contact assembly

Voltage and Current

It is necessary to establish and maintain the desired magnitude of voltage and current for the required weld area. These are determined by the projection desired, the capacity of the welding transformer, and the impedance of the secondary circuit. The transformer should have low impedance with secondary voltages higher than those commonly used in resistance welding.

Arc Time

Arc time can be considered as the time beginning with the explosion of the projection and ending when the two parts come together and the arc is extinguished. The timing between the initiation of the arc and the application of magnetic force is critical.

The arc time is a function of:

1. Magnitude of magnetic force
2. Timing of the magnetic force with relation to the welding current
3. Inertia or mass of the moving parts in the force
4. Height of the projection
5. Magnitude of the welding current and the diameter of the projection

Acceleration of the movable head is directly proportional to the magnetic force applied and inversely proportional to the mass. The acceleration of the movable head with the two-transformer system can be controlled by adjusting the magnitude of the force current, which thereby provides a duration control for arc time, within limits.

17.4 EFFECT ON METALS WELDED

Heat Effects

A percussion weld is made in an extremely short time. It may take milliseconds when using magnetic force welding. Because of this short time, the heat affected zones of percussion welds are shallow, usually less than 0.25 mm. There is little oxidation of the mating surfaces and a minimum of alloying between dissimilar metals. Since the heat affected depth is so small, heat treated metals may be welded without softening them. The heat input is so concentrated and of such short duration that heat-sensitive components near the weld area are not affected by the welding cycle. Heat balance between parts is usually not a factor of concern. Since percussion welding is essentially a DC process, polarity of the most parts involved may be important in some cases, as in arc welding.

Metal Loss

The metal loss that occurs during a percussion weld is not as great as in arc stud welding. The loss varies with the area of the weld and the type of welding machine. Metal loss can generally be ignored for parts to be joined by capacitor discharge percussion welding. However, it should be considered in magnetic force percussion welding.

Flash

Flash is the metal that is expelled at high velocity from the weld interface during percussion welding. It can damage adjacent tooling and may affect accuracy of assembly. Any flash attached to the weld joint should be removed so that it does not cause a problem in service.

Weld Quality

The quality of percussion welds can be determined by metallographic examination and mechanical tests. Metal metallographic examination will show the weld interface and the widths of the heat affected zones. In the case of dissimilar metals, it may reveal the degree of alloying at the interface. Micro-hardness tests on a metallographic section may indicate the effect of welding on the base metal. Welded joints may be tested in tension, bending, or shear, depending upon the joint design. The effect of vibration may be important in some applications. The test method should be designed to qualify the welding procedures and weld joint properties for the intended applications.

17.5 SAFETY

Mechanical

The welding machine should be equipped with appropriate safety devices to prevent injury to the operator's hand or other parts of the body. Initiating devices, such as push buttons or foot switches, should be arranged and guarded to prevent them from being actuated inadvertently. Machine guards, fixtures, or operating controls should prevent the hands of the operator from entering between the work holding clamps or the parts to be welded. Dual hand controls, latches, presence-sensing devices, or any similar device may be employed to prevent operation in an unsafe manner.

Electrical

All doors and access panels on machines and controls should be kept locked or interlocked to prevent access by unauthorized personnel. When the equipment utilizes capacitors for energy storage, the interlocks should interrupt the power and discharge all the capacitors through a suitable resistive load when the panel door is open. A manually operated switch or other positive device should also be provided in addition to the mechanical interlock or contacts. Use of this device will assure complete discharge of the capacitors, a lock out procedure should be followed prior to working with the electrical or hydraulic systems.

Safety Considerations for Personnel

Flash guards of suitable fire-resistant material should be provided to protect the operator from sparks and fires. In addition, personal eye protection with suitable shaded lenses should be worn by the operator. When the welding operations produce high noise levels, operating personnel should be provided with ear protection devices. Metal fumes produced during welding operations should be moved by local ventilating systems.

17.6 APPLICATIONS

1. **The magnetic force method** is primarily used for joining electrical contacts to contactor arms. Combinations include copper to copper, silver-tungsten to copper, silver oxide to copper, and silver-cadmium oxide to brass. Areas from 26 to 820 mm^2 are welded. Some metal loss occurs at the weld interface and in most instances, some flash must be removed from the periphery of the weld.
2. **The capacitor discharge method** is usually employed to produce the following type of joints:
   - Butt joints between wires or rods
   - Lead wire ends to flat conductors or terminals
   - Contacts to relay arms

   The wire is usually made of copper and may be solid or stranded, bare or tinned. The rods are usually copper, brass, or nickel-silver. Other alloys such as steel, alumel, chromal, aluminum, and tantalum may be welded to themselves or to other materials. The method is also applicable to reactive, refractory, and dissimilar metal welds, because the short weld time limits contamination of the reactive metals and the formation of low-strength intermetallic zones in the joints.
3. **Industrial Uses:** Companies using percussion welding are mainly those in the electrical contact or component field. Large contact assemblies for relays and contactors are usually made in magnetic force percussion welding machines. Such machines can be automated for high production. Hand-held capacitor discharge equipment may be used to weld wires to pins, this is particularly applicable to aerospace equipment that is subject to shock and vibration. The process is also used to weld electronic components to terminals (Figure 17.5).

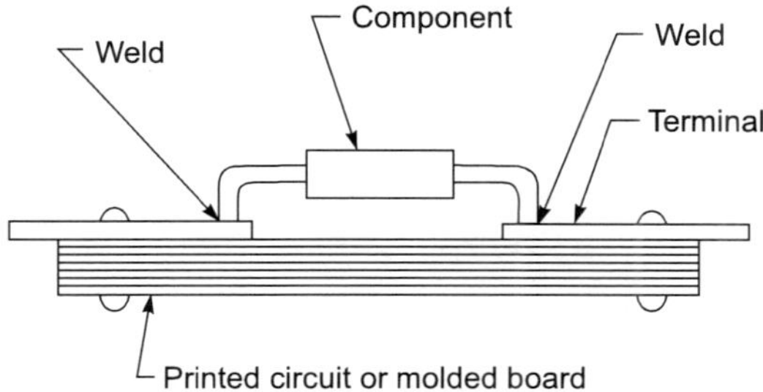

Figure 17.5 Application of percussion welding

This process is applicable for welding of stellite tips to tools, silver contact tips to copper, copper to aluminum. It is also used in telephone industries. This process is used for welding fine wire leads to filaments in lamps.

17.7 ADVANTAGES OF PERCUSSION WELDING

1. The extreme brevity of the arc in both versions of percussion welding limits melting to a very thin layer on the faying surfaces. Consequently, there is very little upset or flash on the periphery of the welded joint (but enough to remove impurities from the joint). Heat-treated or cold worked metals can be welded without annealing them.
2. Filler metal is not used and there is no cast metal at the weld interface.
3. A percussion welded joint usually possesses higher strength and conductivity than a brazed joint. Unlike brazing, no special flux or atmosphere is required.
4. A particular advantage of the capacitor discharge method is that the capacitor charging rate is easily controlled and is also low compared to the discharge rate. The line power factor is better than with a single-phase AC machine. Both these factors give good operating efficiency and low power line demand.
5. Percussion welding can tolerate a slight amount of contamination on the faying surfaces because expulsion of the thin molten layer tends to carry any contaminants out of the joint.

17.8 LIMITATIONS OF PERCUSSION WELDING

1. The percussion welding process is limited to butt joints between two similar sections and to flat pads or contacts joined to flat surfaces. In addition, the total area that can be joined is limited since control of the arc path between two large surfaces is difficult. Joints between two similar sections can usually be accomplished more economically by other processes.
2. Percussion welding is usually confined to the joining of dissimilar metals not normally considered weldable by other processes, and to the production of joints where avoidance of upset is imperative.
3. Another limitation of this process is that two separate pieces must be joined. It cannot be used to weld a ring from one piece.

Review Questions

1. What is percussion welding?
2. Explain with a neat sketch the process of percussion welding.
3. How does a percussion welding machine work?

4. Explain magnetic force percussion welding.
5. Explain capacitive assisted percussion welding.
6. What are the applications of percussion welding?
7. What are the advantages of percussion welding?
8. What are the limitations of percussion welding?
9. What are the heat effects and metal loss in percussion welding.
10. Explain the different types of percussion welding used in industry?

SECTION C

Solid State Welding Processes

18
CHAPTER

Friction Welding

18.0 INTRODUCTION

Friction welding (FRW) is a solid-state welding process which produces the weld by the heat obtained from mechanically induced sliding motion between two rubbing surfaces. The work parts are held together under pressure and one part is rotated with respect to the other to generate frictional heat at the junction. When a suitable high temperature is reached rotational motion is ceased and additional pressure is applied and joint is produced. Filler metal, flux, and shielding gas are not required in this process. The weld produced is characterized by a narrow heat affected zone, presence of plastically deformed material around the weld (flash), and absence of a fusion zone.

Technically, as no melting of the part occurs, friction welding is not actually a welding process in a traditional way, but a forging technique. But, due to its similarities with traditional welding, the term frictional welding is commonly used. Friction welding is used with metals and thermoplastics in a wide variety of aviation and automotive applications.

18.1 METHODS OF OPERATION (FIGURE 18.1)

1. There are two variations of friction welding process. In the original process, one part is held stationary and the other part is rotated by a motor which maintains a constant rotational speed.

   The two parts are brought in contact under pressure for a specified period of time with a specific pressure. Rotational power is disengaged from the rotating piece and pressure is increased. When the rotating piece stops, the weld is completed. This process can be controlled when speed, pressure and time are closely regulated.
2. The other variation is called inertia welding. Here, a flywheel is revolved by a motor until a preset speed is reached. It, in turn, rotates one of the pieces to be welded. The motor is disengaged from the flywheel and the other part to be

welded is brought in contact under pressure with the rotating piece. During the predetermined time, the rotational speed of the part is reduced, the flywheel is brought to an immediate stop and additional pressure is provided to complete the weld.

Both methods utilize frictional heat and produce welds of similar quality, however slightly better control is claimed with the original process.

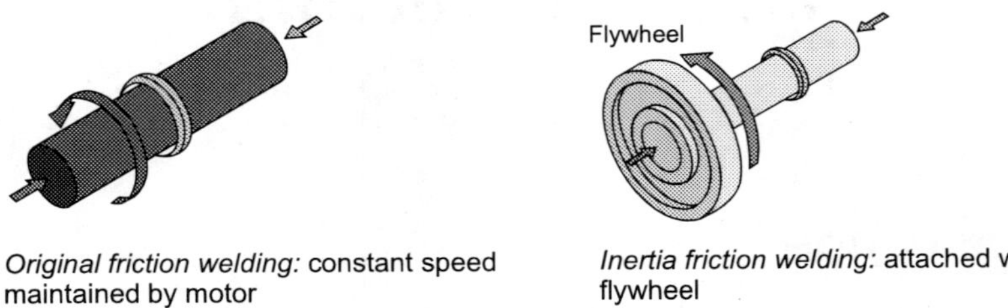

Figure 18.1 Methods of operation

18.2 WORKING PRINCIPLE

In friction welding, the heat required to produce the joint is generated by friction heating at the interface. The components to be joined are first prepared to have smooth, square cut surfaces. One piece is held stationary while the other is rotated against it at high speed. A low contact pressure may be applied initially to permit cleaning of the surfaces by burnishing. This pressure is then increased and the contacting friction quickly generates enough heat to raise the abutting surfaces to the welding temperature.

As soon as this temperature is reached, rotation is stopped and the pressure is maintained or increased to complete the weld. The softened material is squeezed out to form a flash. A forged structure is formed in the joint. If desired, the flash can be removed by subsequent machining action. Friction welding has been used to join steel bars up to 100 mm in diameter and tubes with outer diameter up to 100 mm.

18.3 BASIC STEPS IN FRICTION WELDING (FIGURE 18.2)

1. One workpiece is rotated and the other is held stationary.
2. When the appropriate rotational speed is reached, the two workpieces are brought together and an axial force is applied.
3. Rubbing at the interface heats the workpiece locally and upsetting starts.
4. Finally, rotation of one of the workpieces stops and upsetting is completed.

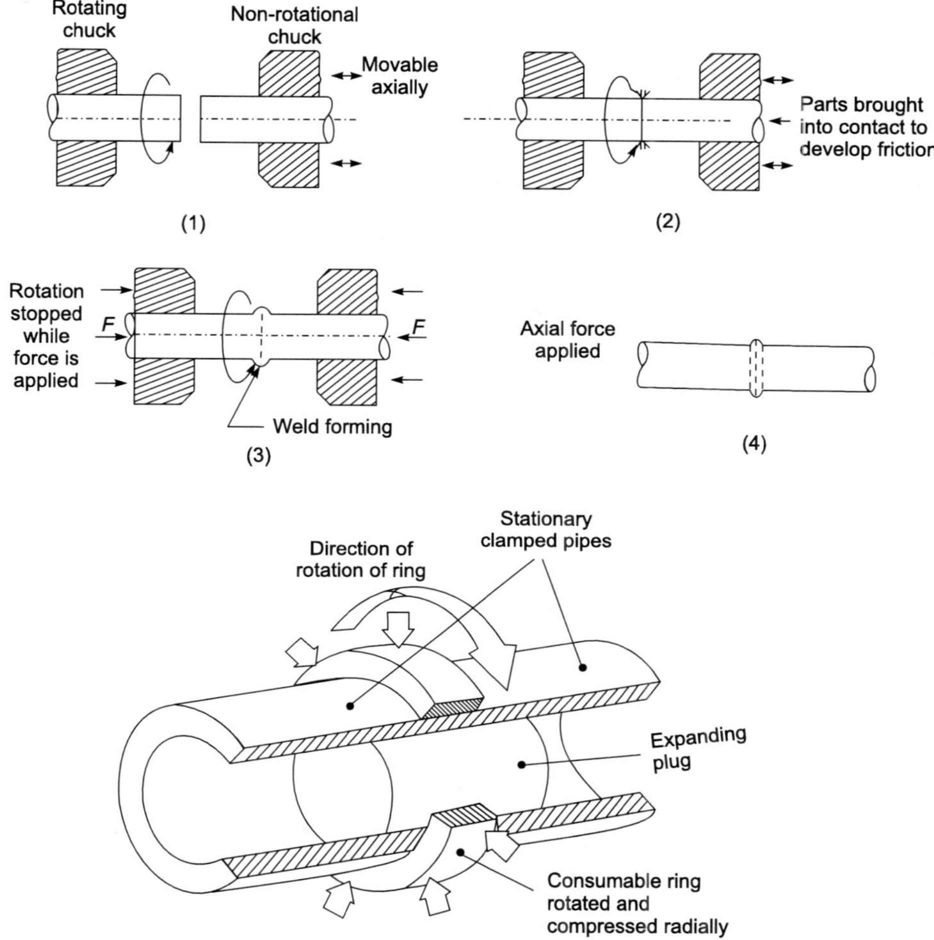

Figure 18.2 Basic steps in friction welding

18.4 THEORETICAL CONSIDERATIONS

The fundamental consideration of the process is based on the law that friction force is proportional to the applied normal load, L, thus,

$$F = \mu L \tag{18.1}$$

where, μ is the coefficient of friction which increases with increase in load and is also dependent on the speed. According to Vill, frictional force can be expressed by the following equation,

$$F = \alpha A + \beta L \tag{18.2}$$

where, A is the area of contact and α and β are constants. For high values of pressure, the first term is very small and thus = βL, where, β is nearly equal to μ so that the basic law of friction holds good.

From the start of the operation till the weld is completed after applying brakes, the frictional force varies. To study its effects on different phases of the process, it is convenient to do so by analysing the time-torque relationship shown in Figure 18.3. The initial peak in the torque curve is due to dry friction but soon after follows the second phase of the process, wherein seizure and rupture takes place at the high points of the contact. The average temperature during the second phase is only 100 to 200°C. The quick rise and the fluctuating nature of the curve is due to the change from marginal or boundary layer friction with μ = 0.1 to 0.2 to pure friction with $\mu > 0.3$. The torque increases as the points of seizure increase and finally molten metal may appear at these points of contact and act as a lubricant and the average temperature of the interface may rise to 900 to 1100°C.

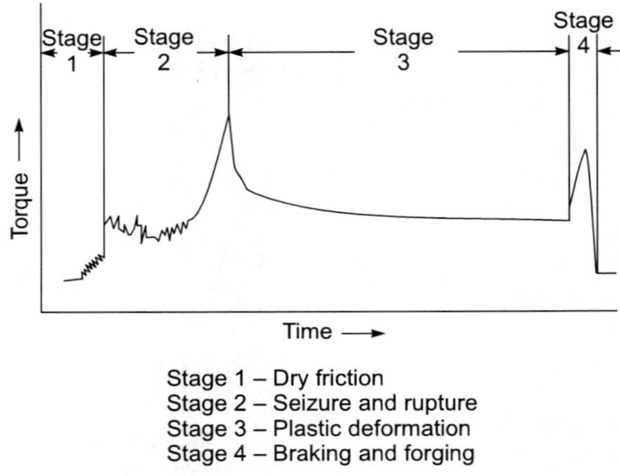

Figure 18.3 Time-torque relation for continuous drive friction welding process

Only 13% of the total heat is produced during the first two stages ($T_1 + T_2$), while the rest is produced during the third stage (T_3). Increased speed, instead of reducing the duration of the process, increases it as is evident from Figure 18.4. This is because enhanced speed results in decreased heating intensity. Heat energy generated per unit area of the faying surfaces is given by the following expression,

$$H = 2\frac{PK}{nR}10^2\,(\text{W/mm}^2) \tag{18.3}$$

where,

H = Heat generated, W/mm^2
P = Applied pressure, N/mm^2
R = Work radius, mm
N = Rpm
K = A constant = 8×10^7 mm^2/min^2 for low-carbon steel

The second stage may cover nearly 30 to 70% of the total time; however, this phase is non-productive so it is intended to reduce its time period to increase productivity. This is usually done by increasing the clamping pressure to a maximum possible value.

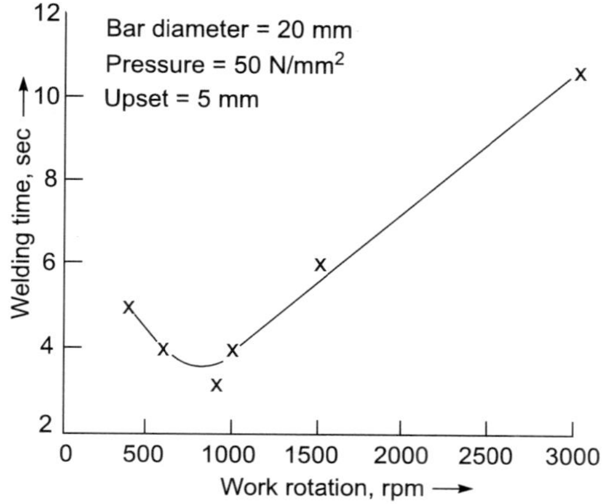

Figure 18.4 Effect of rotational speed on duration of welding

The power required is proportional to the axial pressure and the duration of the third stage is inversely proportional to the axial pressure. Thus, for optimal results, the axial load should be kept low during the initial phase and increased gradually or may be applied in two stages.

The maximum temperature attained is controlled by the applied axial load. When viscosity or strength of the plastic metal is low, the metal gets thrown out by the centrifugal force under a low axial load, for example, in the welding of copper.

In friction welding of dissimilar metal combinations, such as stainless steel to carbon steel, the plane of maximum temperature can move away from the interface; with high speed, it moves into the stainless steel so that one half of the flash is bimetallic. In this case, decreasing the rotational speed provides the desired result and at a certain speed, the interface again becomes the plane of maximum temperature and thus, a plane of maximum shear strain rate.

The applied pressure is perhaps the most important single factor as it controls the temperature and determines the torque required. The heat input rate is proportional to the product of torque and the rotational speed. Rotational speed has to be such that certain minimum or threshold power is exceeded. If the power is above the threshold, the process is self-regulating. If too much power is applied, the width of the sheared zone of metal increases. If the power applied is just above the threshold, it will take a long time for the required temperature to be attained and the heat affected zone will be wide. The most important variable is the unit pressure applied during rotation and the recommended values for some of the metals are given in Table 18.1.

TABLE 18.1 Friction pressure for different metals

Metal	Friction Pressure (N/mm^2)
Mild steel	5500
Medium carbon steel	7000
Nickel-chromium steel	11000
Aluminum alloys	8000
Nimonic alloys (80/20 Ni/Cr)	14000

The sliding velocity varies from zero at the centre of the workpiece to a maximum at the peripheral surface, and the radius at two-thirds the diameter of the workpiece is used for calculations. Longer heating times result in more material for forging and for optimal results, there should be adequate heated material available for forging when the rotation is stopped. Excessive initial pressure results in excessive squeezing of heated metal, leaving only relatively cold metal to be forged when the forging pressure is applied.

18.5 PROCESS CHARACTERISTICS

One of the two workpieces is rotated at a constant speed throughout the operation except when the brakes are applied at the final stage; the process is therefore, often referred to as *continuous drive friction welding*.

The workpieces rub together under pressure for a predetermined heating time or until a preset axial shortening takes place. The drive is then disengaged and the rotation of work is stopped by applying brakes. The axial pressure is maintained or increased, to forge the meta, till the weld cools. Figure 18.5 shows how the process parameters change during welding when the force at the end is increased to forge the joint. Welds in mild steel can be made by merely maintaining the pressure constant.

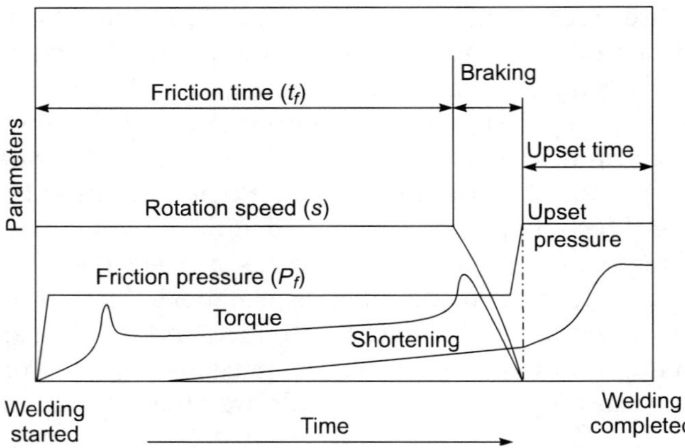

Figure 18.5 Variations of process parameters with time for continuous drive friction welding

With the decrease in speed, the thickness of the hot plasticized band increases and the torque drops to zero as the rotation stops.

Bonding mechanism in friction welding of dissimilar metals is more complex. Due to mechanical mixing and diffusion, some alloying possibly occurs in a very narrow band at the interface. The properties of this narrow band may have considerable influence on the overall joint performance. Mechanical mixing and interlocking may also help in bonding. Due to these complexities, prediction of weldability of dissimilar metals is very difficult and it needs to be established for a particular application by a series of tests, designed specifically for the purpose.

18.6 FRICTION WELDING MACHINES AND EQUIPMENT (FIGURE 18.6)

The major components of a friction welding machine, as shown in Figure 18.6, include:

1. Driven head
2. Clamping arrangements
3. Rotating and upsetting mechanisms
4. Controls
5. Braking mechanism

One of the workpieces to be welded is held firmly in self-centering head and the other is held in a centering chuck that is mounted on a rotatable spindle, which is driven by a motor usually through a variable speed drive.

Rotating chucks must be well-balanced, have high strength and provide good gripping power. Collet chucks meet these requirements well and are therefore, most frequently used.

The gripping mechanism of the chucks must be rigid and resist the applied thrust. Serrated gripping jaws are recommended for maximum clamping reliability.

Attempts to use lathe for friction welding were not much successful because it lacked the rigidity of the structure and efficient gripping. A lathe is not designed for endurance need for the friction welding nor is it provided with quick disengagements of the process. Also, to ease the braking problem, a low moment of inertia is required in the rotating parts.

Friction welding machines are required to control accurately three variables, viz., axial thrust, speed of rotation, and degree of upset. Plain carbon and low-alloy steels require a forge pressure of 15 to 30 N/mm^2, while tungsten steels require pressure in the range of 225 to 400 N/mm^2. These latter values are comparable to the pressure used in flash butt welding. When the rate of application of pressure by a hydraulic system is found to be low, it is replaced by a pneumatic system.

The duration of welding on a 25 mm diameter bar should be from 5 to 7 seconds. This can be achieved by a surface speed of 75 to 600 m/min, which corresponds to about 1000 rpm. Higher speeds may give better impact strength and are, therefore, desirable for hollow sections and high hot strength alloys.

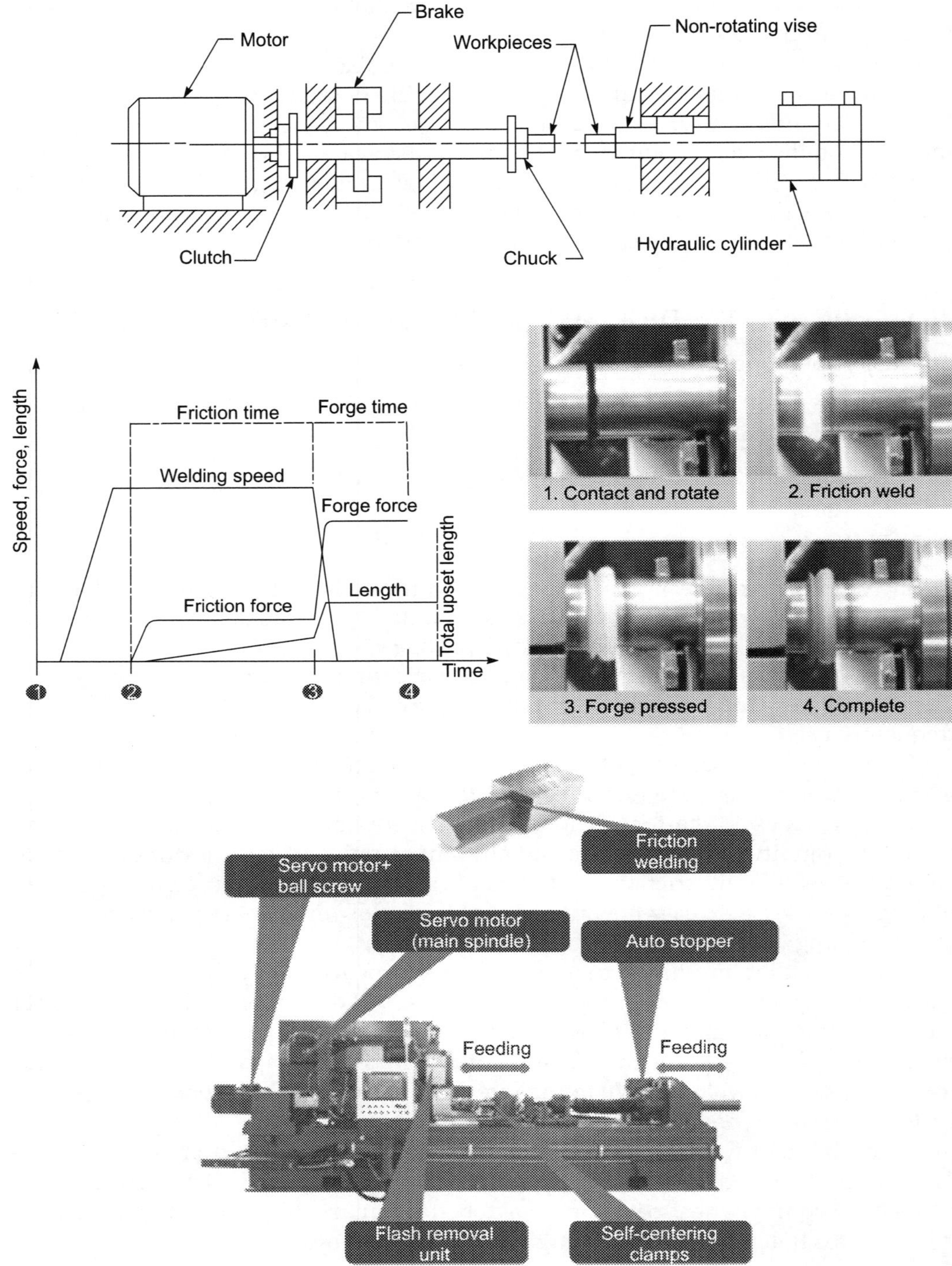

Figure 18.6 A friction welding machine

Control of the welding operation may be by the time or by the extent of upset. The latter method is applied by limit switches arranged to increase the pressure to cause upset after a certain amount of shortening has occurred. Long times tend to allow the heat to spread back to regions behind the interface and therefore, result in heavy upsets, which are costly to remove after the operation is over. The accuracy of upset is expected to be within 0.1 mm.

Controls by time sequence are satisfactory for where constant surface condition cannot be assured and welds are of secondary importance. When time control is used, high rotational speeds are preferred. The speed for mild steel is selected on the basis of stock diameter and is given by the expression,

$$nd = (1.2 \text{ to } 6.0) \times 10^4 \tag{18.4}$$

where, n is the rpm and d is the stock diameter in mm.

The smaller values of the constant refer to welding with high rates of input.

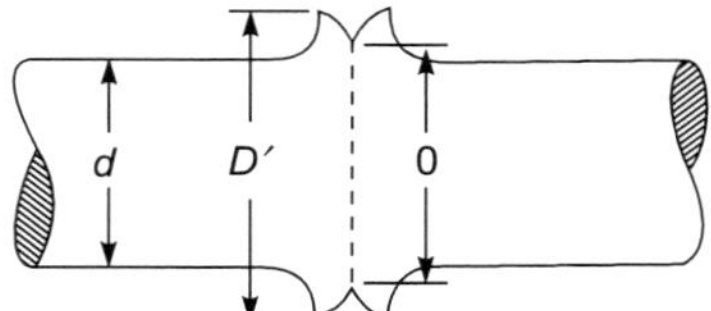

Figure 18.7 Configuration of a friction welded bar

A typical section of a friction weld, between solid bars, with upset is shown in Figure 18.7. The maximum extent of upset is known as the *apparent upset*, while the biggest diameter decides the extent of *real upset*. With reference to the above figure, these values can be expressed by the following relationships:

$$\text{Apparent upset, } U_A = \frac{D' - d}{d} \times 100\% \tag{18.5}$$

$$\text{Real upset, } U_R = \frac{D - d}{d} \times 100\% \tag{18.6}$$

First, braking is provided to stop the rotation quickly at the end of specified heating time or after a designed amount of axial shortening of the weldment. This provides desired control of overall weldment length and broadens the acceptable range of welding variables for critical applications. With small diameters, where the speed is high, very fast braking is required and this can be achieved by electric clutches, motor braking or friction braking. Also, arrangement can be made to release the originally stationary specimen towards the end of the heating cycle.

There are essentially two types of machines:

(i) Low power machines with power ratings of order 12 W/mm^2

(ii) High power machines with 35 to 115 W/mm^2 of the material welded.

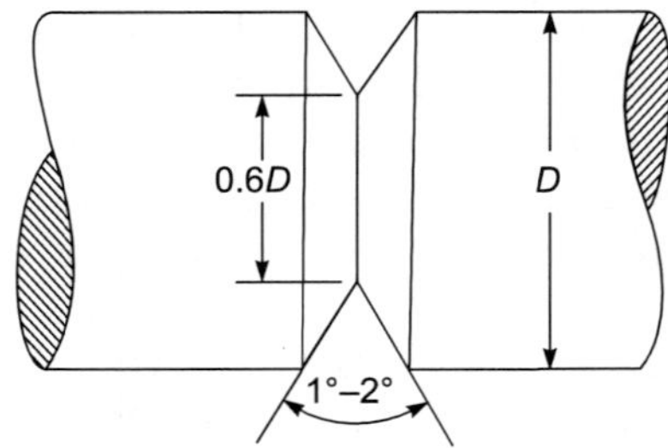

Figure 18.8 Use of bevelled bars to increase machine capacity

If adequate pressure is available, it is possible to increase the capacity of a machine by using a bevelling technique, as shown in Figure 18.8.

18.7 WELDING VARIABLES

Three major variables in continuous drive friction welding are:

(i) Rotational speed
(ii) Axil pressure
(iii) Heating time

Rotational Speed: Rotational speed provides the necessary relative velocity at the faying surfaces. Its magnitude depends upon the metal being welded and for steels, the tangential velocity for both solid and tubular workpieces, should be in the range of 75 to 110 m/min. Tangential speeds lower than 75 m/min result in excessive torque with consequential clamping problems, non-uniform upset and tearing of metal at the joint. Friction welding machines for production purposes, handling 50 to 100 mm diameter workpieces, usually operate at speeds varying between 90 and 200 m/min.

High rotational speeds are useful for welding but axial pressure and heating time must be carefully controlled to avoid overheating of the weld zone, particularly for welding quench hardenable steels to control cooling rate and possible cracking.

In dissimilar metal welds, low rotational speeds can minimise the formation of brittle intermetallic compounds; however in general, for controlling weld quality, rotational speed is not considered a critical parameter.

Axial Pressure: The applied axial pressure controls the temperature gradient in the weld zone, the power required for the machine, and the axial shortening of the workpiece. The specific pressure depends upon the metal being welded and the joint configuration. It can be used to compensate for heat loss to a large body as in the case of tube-to-tubeplate welds.

Applied pressure must be high enough during the heating phase so as to keep the faying surfaces in close contact to avoid oxidation. Joint properties may often be improved if the applied pressure is increased at the end of the heating phase.

For making sound welds in mild steels, the heating pressure used is usually 30 to 60 N/mm^2, while the forging pressure may lie in the range of 75 to 150 N/mm^2, and the commonly used values are 55 to 135 N/mm^2. However, higher forging pressures are required for high hot strength alloys like, stainless steels and nickel based alloys. If a preheating effect is required, then the axial pressure of 20 N/mm^2 is applied initially for a short period, which is then raised to the normal heating pressure.

Heating Time: Heating time is controlled depending upon whether a fixed preset time is allowed for heating or the extent of the axial upset is to be within the specified limits.

Excessive time limits productivity and results in wastage of material; while insufficient time may result in uneven heating as well as entrapped oxide and unbonded areas at the interface. The duration of welding for a 25 mm diameter bar, as already mentioned, should be from 5 to 7 seconds at a rotational speed of 1000 rpm.

18.8 WELD PROPERTIES

One of the attractive features of friction welding is the metallurgical quality of the welds; the rapidity of heat generation produces an almost negligible heat affected zone. Due to good control of temperature and as the plastic metal is subjected to hot working during the heating phase, and to cold working during the forging phase, this results in a weld with extremely fine grain structure.

Metallurgical examination shows no evidence of melting as the measured temperatures or steels are usually in the range of 1260 to 1330°C. However, rapidity of friction welding leads to high cooling rates which result in higher hardness of weld zone. Friction welds in hardenable steels, therefore, most often have to be annealed after welding.

Many times, friction welding is employed to join dissimilar metals with a view to economising in the use of expensive alloy steels and high temperature alloys. It has been shown that 18/8 (Cr/Ni) stainless steel is actually work hardened from 200 to 250 VHN on the stainless steel side due to the diffusion of carbon in it.

In a weld between 18/8 stainless steel and 20% Cr-Moo steel, the hardness of Cr-Mo steel is increased from 175 to 405 VHN but it can be reduced to 250 VHN by annealing. The increased hardness of stainless steel, however, remains unaffected by annealing.

Satisfactory welds between aluminum and stainless steel can be produced without the formation of brittle intermetallic compound layer. However, welds between aluminum and mild steel and aluminum and copper may result in the formation of intermetallic compounds at the interface, which may be reduced in the latter case by increasing the forging pressure to about 200 N/mm^2.

18.9 APPLICATIONS

Friction welding is frequently used for applications where one of the components to be joined has axial symmetry. Compared with flash welding, friction welding has the advantage of cleanliness and balanced steady load on the mains; also it can be installed along with other machine tools and can be easily automated for high rate production.

Almost any metal then can be hot forged and material used in dry bearing application can also be friction welded. However, some metals may require post-weld heat treatment to remove the effect of quench hardening at the weld interface. Free machining alloys are difficult to friction weld as they often result in welds with planes of weakness in weld zone due to redistribution of inclusions. Such welds often exhibit lower values of strength, ductility and notch toughness.

Friction welding can be employed to weld solid material from 5 to 100 mm diameter or the equivalent areas in tubes and pipes. Due to solid state joining, friction welds have high quality performance both in similar and dissimilar combinations.

Carbon steels up to 1.1% C can be easily welded with the exception of free cutting variety. Ni-Cr steels up to 18% Ni and 8% Cr and dissimilar steels can be easily welded; the range may include welding of 18/8 stainless steel to 2.25% Cr-Mo steel.

Welding of steel, due to its lower conductivity and greater plastic range, is comparatively much easier than the welding of non-ferrous metals and their combinations.

The largest single user of friction welding is the automobile industry for large production of components like axle casings for cars and heavy vehicles; the joint produced is shown in Figure 18.9. Another important use is the manufacture of twist drills in which tungsten high speed steel ends are welded to carbon steel shanks.

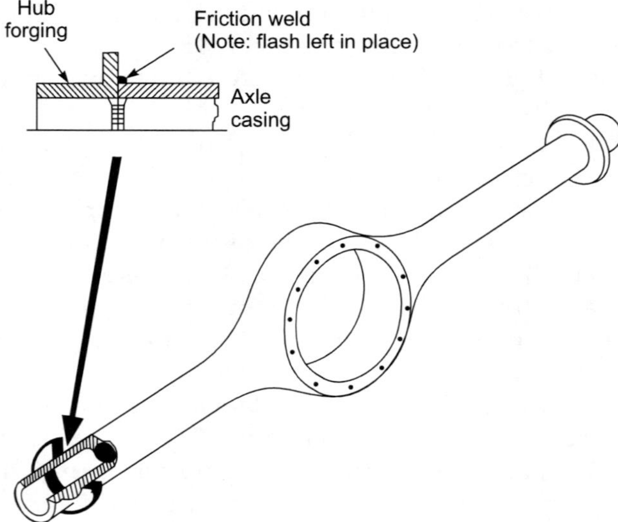

Figure 18.9 A typical friction welded hub end forging to the axle casing for a heavy vehicle

One of the most important applications of friction welding is welding of studs to plates of any thickness; another application of the process is the production of marine engine valves, the valves so produced are as good as or superior to those produced by forging. Impeller used in the turbo charger of a diesel engine may be made by friction welding of a carbon steel shaft to an investment cast heat-resisting austenitic steel.

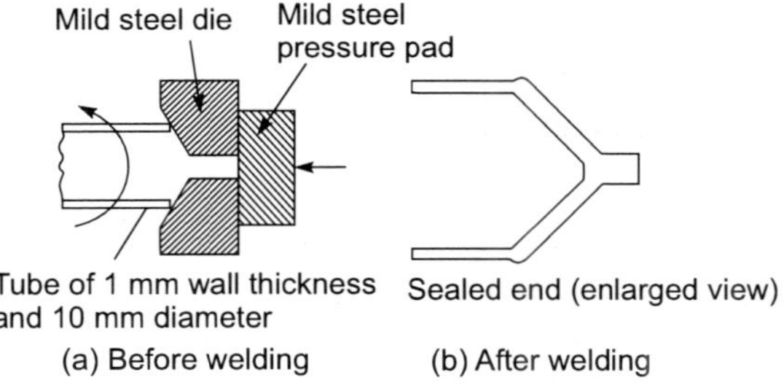

Figure 18.10 A setup for friction welded end sealing of tubes

End sealing of tubes, as shown in Figure 18.10, can also be achieved by friction welding. Dissimilar welded combinations may include joining of stainless steel to zirconium. Figure 18.11 shows the edge preparation for joining stainless steel tube to a zirconium bar. For successful joining of dissimilar metal combinations, it is essential to use high rotational speeds (more than 3200 rpm), so as to reduce the thickness of the intermetallic zone to the minimum.

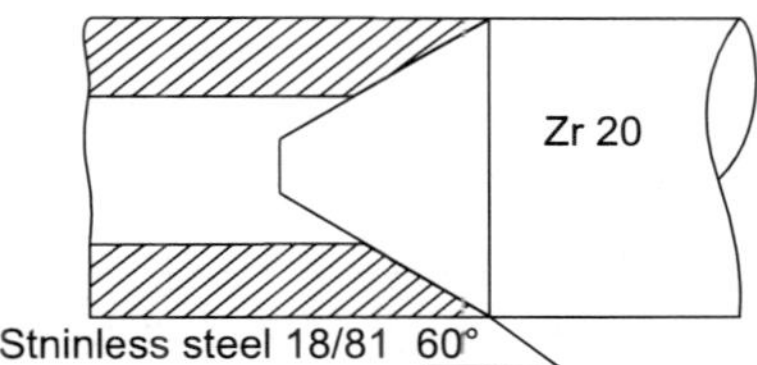

Figure 18.11 Edge preparation for friction welding of stainless steel tube to zirconium bar

18.10 ADVANTAGES AND DISADVANTAGES

Friction welding, like any other welding process, has its specific advantages and disadvantages.

Advantages

1. No filler metal is needed. Operation is quite simple.
2. Flux and shielding gas are not required.
3. The process is environmentally clean; no arcs, sparks, smoke or fumes are generated by clean parts.
4. Surface cleanliness is not as significant, compared with other welding processes, since friction welding tends to disrupt and displace surface films.
5. There are narrow heat affected zones.

6. Friction welding is suitable for welding most engineering materials and is well suited for joining many dissimilar metal combinations.
7. In most cases, the weld strength is as strong as or stronger than the weaker of the two materials being joined.
8. Operators are not required to have manual welding skills.
9. The process is easily automated for mass production. Welding takes place in a few seconds once parameter is set.
10. Welds are made rapidly compared to other welding processes.
11. Plant requirements (space, power, special foundations, etc.) are minimal.
12. Cost effective method of joining highly dissimilar materials and geometries

    Materials that can be welded include aluminum and its alloys, alloy steels, low-carbon steel, carbon steel, stainless steel, tantalum, stainless steel to carbon, steel, alloy steel to carbon steel, aluminum to stainless steel, copper to aluminum, etc.

Disadvantages

1. In general, one workpiece must have an axis of symmetry and be capable of being rotated about that axis.
2. Preparation and alignment of the workpieces may be critical for developing uniform rubbing and heating, particularly with diameters greater than 50 mm.
3. Capital equipment and tooling costs are high.
4. Dry bearing and non-forgeable materials cannot be also friction welded.
5. If both parts are longer than 1 m, special methods are used.
6. Free-machining alloys are difficult to weld.
7. All types of metal cannot be joined by this process.

Worked Examples

Example 1: Two steel bars each of diameter 10 mm are coaxially friction welded end-to-end at an axial pressure of 200 MPa and rotational speed of 400 rpm. The coefficient of friction between the mating faces of the rotating bars is 0.50. The torque is assumed to act at 3/4th radius of the rotating bar. What is the power consumed in kW at the interface of welding? **(GATE 2010)**

Solution: Force (F) = Pressure × area × coefficient of friction

$$\text{Force } (F) = 200 \times \frac{\pi}{4} \times 10^2 \times 0.5 = 7854 \text{ N}$$

$$\text{Torque} = F \times \frac{3}{4} \times \text{radius}$$

$$\text{Torque} = 7854 \times \frac{3}{4} \times 5 \times 10^{-3} = 29.45 \text{ Nm}$$

$$\text{Power} = \frac{2\pi NT}{60000} = \frac{2\pi \times 4000 \times 29.45}{60000} = 12.33 \text{ kW}$$

Example 2: Find the power rating of a machine to friction weld a high-tensile steel (0.2% C, 1% Cr, 0.4% Ni) having a forging temperature of 900°C and a strength at that temperature of 125 N/mm^2. With a rotational speed of 3000 rpm, an upset gather of 2.8 mm in 10 mm diameter material is allowed. Assume that the coefficient of friction, $\mu = 1$, and that under continuous shear, the shear strength of the material is equal to the compressive strength and that the torque acts at two-thirds the radius of the workpiece.

Solution: $L = 125 \text{ N/mm}^2$ $\quad N = 3000$ rpm

$\mu = 1$ $\quad d = 10$ mm

Frictional force/area, $F = \mu L$

$= 1 \times 125 = 125 \text{ N/mm}^2$

Torque, T = force × area × torque arm

$$= F \times A \times \frac{2}{3} r$$

$$= F \times \left(\frac{\pi}{4} d^2\right) \times \left(\frac{2}{3} r\right)$$

$$= 125 \times \left(\frac{\pi}{4} \times 10^2\right) \times \left(\frac{2}{3} \times 5\right)$$

= 32738 N-mm

= 32.738 N-m

$$\text{Power of machine, } P = \frac{\text{Torque} \times \dfrac{2\pi n}{60}}{1000} \text{ kW}$$

$$= \frac{32.738 \times \dfrac{2\pi \times 3000}{60}}{1000} \text{ kW}$$

= 10.3 kW

Example 3: The energy applied in friction welding is given by the formula $E = IS^2/C$, where, I is the moment of inertia of the flywheel, S is the spindle speed in rpm, and C is a constant of proportionality (5873, when the moment of inertia is given in lb-ft^2). For a spindle speed of 600 rpm and an operation in which a steel tube (3.5 inches OD, 0.25 inch wall thickness) is welded to a flat frame, what is the required moment of inertia of the flywheel if all of the energy is used to heat the weld zone (approximated as the material 0.25 inch deep and directly below the tube)? Assume that 1.4 ft-lbm is needed to melt the electrode.

Solution: The moment of inertia of flywheel can be calculated as

$$E = \frac{IS^2}{C}$$

$$1.4 = \frac{I600^2}{5873}$$

Solving, we get $I = 0.0228$ lb-ft^2

Example 4: Two pipes of inner diameter 100 mm and outer diameter 110 mm, each are joined by flash butt welding using 30 V power supply. At the interference, 1 mm of material melts from each pipe which has a resistance of 42.4 W. If the unit melt energy is 64.4 MJ/m^3. What is the time required for welding.

Solution: Given inner diameter = 100 mm, outer diameter = 110 mm, voltage = 30 volts

Resistance = 42.4 ohms. Unit energy to melt material is $U_m = 64.4$ MJ/m^3

Each pipe melts 1 mm of material, so thickness of material melt

$$t = 2 \times 1 = 2 \text{ mm}$$

Melting energy in the entire volume can be written as

$$H = \text{Area} \times \text{thickness} \times U_m = \frac{\pi(d_o^2 - d_i^2) \times t \times U_m}{4}$$

$$H = \frac{\pi(110^2 - 100^2) \times 2 \times 10^{-3} \times 64.4 \times 10^6}{4} = 212.32 \text{ J}$$

The amount of heat generated at the contact area of the element to be welded is

$$H = I^2Rt = \frac{V^2}{R}t$$

$$212.32 = \frac{30^2}{42.4}t$$

Solving we get t = 10 seconds

Review Questions

1. What is friction welding?
2. Explain with a neat sketch the process of friction welding.
3. How does a friction weld machine work?
4. Explain the basic steps used in friction welding.
5. Explain theoretical considerations in friction welding.
6. What are the applications of friction welding?
7. What are the advantages of friction welding?
8. What are the limitations of friction welding?
9. What are the welding parameters used in friction welding?

19

CHAPTER

Friction Stir Welding

19.0 INTRODUCTION

Friction stir welding is a solid-state welding process invented in 1990. Since in this process welding is carried without melting the parent metal, high quality weld can be obtained. Certain welding defects like solidification cracking, porosity, oxidation can be eliminated using friction stir welding process. In this process, a cylindrical shouldered tool with a profiled pin is inserted into the joint line between the two pieces being joined. As the tool rotates, frictional heat will be generated between the wear resistant pin and the two workpieces which are connected in butt joint and clamped on to a backing bar. The heat generated will not melt the metal, instead it softens and is joined together as the pin traverses the entire joint.

The friction heat generated as the tool moves plasticizes the material. Here, it consolidates and cools, forming a solid state weld. As there is no melting of the material, the same fine grained structure can be maintained in the weld joint. As there is no filler material added in this process, the weld is nominally identical in composition to the base material. Although no compositional changes occur, the stirring action and the associated heat result in significant difference between the weld nugget and the surrounding material. The original grain structure will be radically changed, often resulting in a fine grain size, generally assumed to be a consequence of recrystallisation process. The primary particle fragmentation may also occur. Friction stir welding has several advantages over arc welding. The lower energy inputs of friction stir welding are expected to limit distortion and residual stress in the welded structures. Moreover, as it is a friction stir welding, the problems of porosity, slag inclusion, hydrogen cracking, etc. can be eliminated. Summary of friction based welding is shown in Figure 19.1.

A schematic diagram of friction stir welding operation is shown in Figure 19.2. The plates are placed on a backup plate and the plate edges are clamped rigidly to prevent lateral movement during the welding operation.

Welding is carried out with a specially designed cylindrical tool, rotating at several hundred rpm. The tool consists of a nib extending from the shoulder that is slightly downward in the joint line. The nib may have diameter one-third of the cylindrical tool and typically has a length slightly less than the thickness of the plate being welded. The nib is forced into the plates at the joint until shoulder contacts the plate surface. The vertical penetration of the nib into the plates is halted when the shoulder makes contact with the plate surface. A downward forging pressure from the shoulder helps to prevent the expulsion of softened material, in addition to providing supplementary frictional heating.

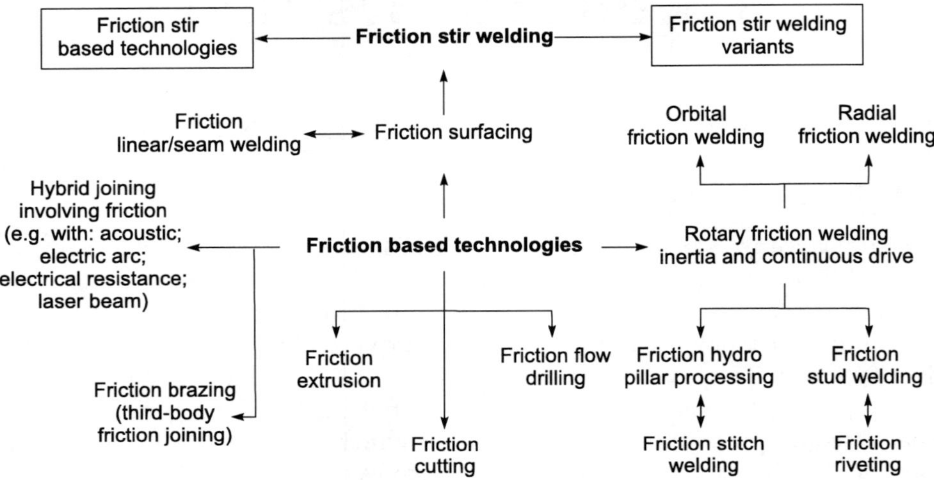

Figure 19.1 Friction-based welding technologies

The surface friction of the shoulder with the plates creates additional heat and plasticizes the cylindrical metal column around the inserted pin and the immediate material under the shoulder. As the tool is moved forward, material is forced to flow from the leading edge to the trailing edge of the tool, and the material that flows around the tool undergoes extreme levels of plastic deformation (Figure 19.2).

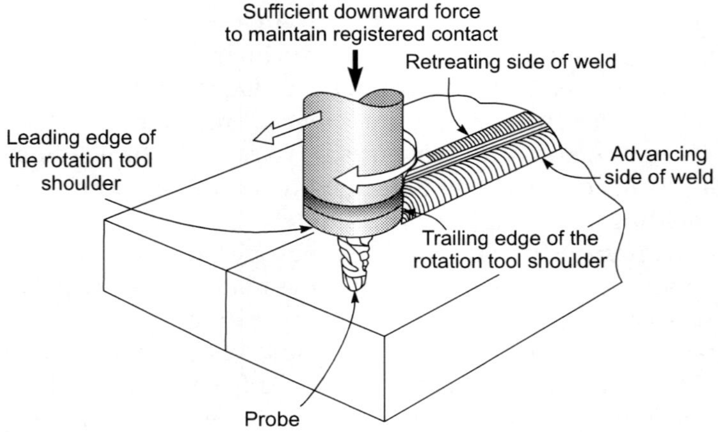

Figure 19.2 Schematic diagram of the friction stir welding process

As a result, weld thus generated usually leads to thinning down of the parent metal by about 3 to 6% of the original thickness. The stirring action due to the rotating tool, plasticizes the metal within a narrow zone while transporting metal from the leading face of the pin to the trailing edges. As the tool passes, the weld cools, thereby joining the two plates together. Since the maximum temperature created by friction stir welding process is usually less than 80% of the melting temperature of the welding material, friction stir welding can avoid welding defects caused by traditional fusion welding.

A significant benefit of friction stir welding is that it has a few variables that must be controlled. In a fusion weld, for instance, technicians have to closely monitor the voltage, amperage, wire feed, travel speed, shield gas, and arc gap.

In friction stir welding, there are only three variables: tool rotation speed, travel speed, and pressure, which are to be controlled. Important process parameters include the tool rpm, travel speed, tool dimensions and the downward force on the tool. Heating is caused by rubbing of the tool faces against the workpiece and by visco-plastic dissipation of mechanical energy. There is a unique combination of shear and normal forces in FSW. An understanding of the contribution of these two forces is important for developing a friction stir welding process.

19.1 TOOL GEOMETRY

Friction stir welding (FSW) is partially implemented for joining low melting temperature materials like aluminum alloys. The main hurdle in FSW is the tool geometry as the joint quality depends on this. Optimum tool design for producing the desired joint quality as well as enable higher welding speed and longer tool life is the main criteria. Applications in high melting point alloys, such as mild steel, stainless steel, titanium, remain limited due to problems in terms of tool life and required force.

The welding tool forms the heart of friction stir welding process. In the early days of the friction stir welding process, the tools used to be of simple geometry (Figure 19.3). Complex FSW tools, such as the Whorl™ and MX Triflute™ tools, have been developed by TWI. They provided improved welding process as well as extended the thickness range for welding.

The shape of the bottom of the tool shoulder affects material flow around the tool nib and contributes to preventing the escape of plasticized material. They can be flat or concave, smooth or grooved, with concentric or spiral grooves. A concave shoulder has the advantage, when compared with a flat bottom, of directing material flow to the center close the tool nib. The tool nib may have a diameter one-third of the cylindrical tool and typically has a length slightly less than the thickness of the workpiece.

The effect of skew motion of the FSW tool on the weld quality has been investigated by TWI. In this, an orbital motion of the tool is achieved by inclining the longitudinal axis of the tool with respect to the retaining machine spindle. This process, referred to as the skew-stir technique, increases the ratio of the "swept" volume to the physical volume of the tool. This is usually referred to as the dynamic to static volume ratio.

It has been observed that this parameter has a significant effect towards achieving the required level of both, weld quality and process efficiency.

Figure 19.3 Typical friction stir welding tools

19.2 TOOL SHOULDER MATERIAL AND BACKING MATERIAL

The tool shoulder material plays an important role in the heat generation process of friction stir welding. Compared to tool steel, the shoulder made from zirconia engineering ceramic seems to generate approximately 30 to 70% more heat depending on the welding parameters.

The welding efficiency is affected by the heat loss through tool as well as steel backing bars. Using a tool material having low thermal conductivity along with suitable non-conducting backing bar, this heat loss can be substantially reduced. This will enable increased welding speed. Thus, a combination of low thermal conductivity tool material like SS 660 with zirconia coated tool shoulder and zirconia backing bar can significantly improve process efficiency through increase in welding speed. If full advantage is to be taken regarding welding efficiency, an increase in welding speed necessitates a similar increase in rotational speed to get sound welds without line porosity and weld defects.

19.3 HEAT GENERATION (FIGURE 19.4)

Heat generated in a friction stir welding process increases with increasing tool rotation speed and tool downward force. Travel speed influences the heat input per unit weld length, which results in affecting metal flow around the tool nib. Other relevant parameters are the time of indentation of the tool, shoulder radius and shoulder angle of the FSW tool.

The time of indentation of the tool is the period between the instant the tool contacts the workpiece and the instant the tool begins moving along the joint. During this period, generated heat spreads in the vicinity of the tool nib, softening the adjacent material and stabilizing material flow around the tool nib. If this period is too short, defects can appear in the initial part of the weld. Time can range usually from 5 to 30 seconds. The tool shoulder angle allows a gradual increase in the pressure on the top surface of the plates being welded and helps to direct the material flow. Commonly used tool angle is up to 3°.

In estimating the heat generation, two extreme conditions and an intermediate condition are assumed. In the first estimation, a sticking interface condition ($d = 1$) is assumed and in the second estimation, a pure sliding ($d = 0$) interface described by a Coulomb friction condition is assumed. In case of the sticking condition, the shearing is assumed to occur in a layer very close to the interface and in the sliding condition, the shear is assumed to take place at the contact interface. These two types of estimations are distinguished by the assumptions under which the shear stress contact is introduced.

The third estimation is used in the case where the partial sliding/sticking condition is assumed.

During the friction stir welding process, heat is generated at or close to the contact surfaces, which have complex geometries depending on the tool geometry. For analytical estimation, a simplified tool geometry with a conical pin and concave shoulder surface is assumed.

The concave shoulder surface is characterized by the cone angle α, which is zero in case of a flat shoulder. The tool geometry and the components of heat generated at various surfaces are shown in Figure 19.4. Q_1 is the heat generated under the tool shoulder, Q_2 at the tool pin side and Q_3 at the tool pin tip, hence the total heat generation is given by,

$$Q_{total} = Q_1 + Q_2 + Q_3$$

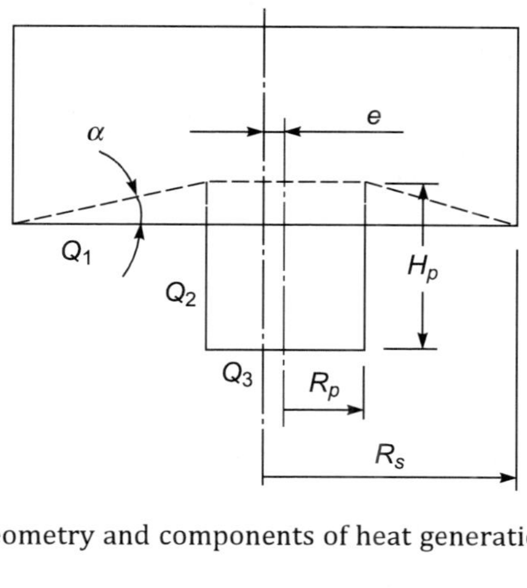

Figure 19.4 Tool geometry and components of heat generation at various surfaces

Heat generation due to friction of various surfaces is given by equation (19.1).

$$dQ = \omega dM = \omega r dF = \omega r \tau_{\text{contact}} dA \tag{19.1}$$

where,

M = Torque (Nm)
F = Force (N)
A = Area (m^2)
ω = Tool angular rotation speed (rad s^{-1})
r = Position along tool radius (m)
τ_{contact} = Shear stress (Pa), which is a product of frictional coefficient m and normal pressure

In this expression, only the rotation power is considered to contribute to heat generation. The effect due to the translatory movement of the tool is neglected. The various tool surfaces in contact with the plates will be either horizontal, vertical or conical with respect to the rotation axis of the tool, as shown schematically in Figure 19.5.

The following subscripts are used to characterize the orientation of the surfaces:

H = Horizontal (perpendicular to the rotation axis, circular surface).
V = Vertical (parallel to the rotation axis, cylindrical surface).
C = Conical (tilted with respect to rotation axis, conical surface).

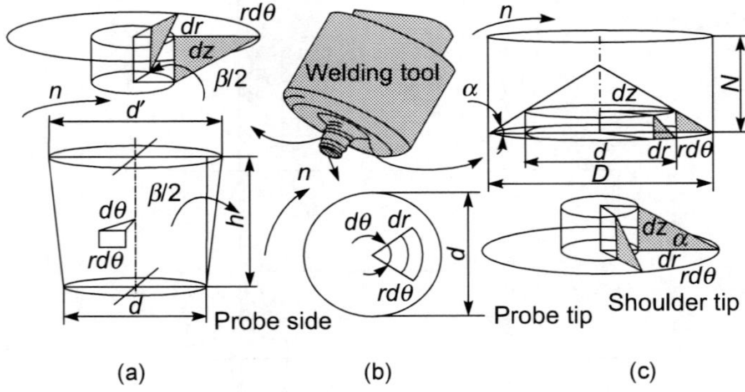

Figure 19.5 Schematic representation of various surface orientations (a) Vertical, (b) Horizontal, (c) Conical

Heat Generation from Horizontal Surface

Heat generation due to friction of the horizontal circular surface, shown in Figure 19.5(b), is obtained from the following considerations. Assuming that the horizontal surface is subjected to uniform shear stress (τ_{contact}) one obtains, considering an infinitesimal area,

$$dA_H = r d\theta dr$$

The force and the torque acting on it are given by,

$$dF_H = \tau_{\text{contact}}\, dA_H$$
$$dM_H = r dF_H$$

Therefore, from equation (19.1), heat generation from this segment will be given by as shown in equation (19.2).

$$dQ_H = \omega r dF_H = \omega r^2 \tau_{\text{contact}}\, d\theta dr \tag{19.2}$$

Heat Generation from Vertical Surface

Similarly, the heat generation due to the friction of the vertical surface, shown in Figure 19.5(a) will be given by as shown in equation (19.3),

$$dQ_v = \omega r dF_v = \omega r^2 \tau_{\text{contact}} d\theta dz \tag{19.3}$$

Heat Generation from Conical Surface

The forces generated on the conical surface as shown in Figure 19.5(c) are split into its vertical and horizontal components, given by equation (19.4)

$$dF_c = dF_H + dF_v \tag{19.4}$$

The projection of the tilted segment area is given by,

$$dz = \tan\alpha\, dr$$

Therefore,

$$dA_v = r d\theta dz = r d\theta \tan\alpha\, dr \text{ and } dA_H = r d\theta dr \tag{19.5}$$

From equations (19.4) and (19.5), one obtains,

$$dF_c = \tau_{\text{contact}} dA_H + \tau_{\text{contact}}\, dA_v = \tau_{\text{contact}} r d\theta dr\,(1 + \tan\alpha) \tag{19.6}$$

Therefore, heat generated at the conical surface is given by,

$$dQ_c = \omega r dF_c = \omega r^2 \tau_{\text{contact}}\, d\theta dr\,(1 + \tan\alpha) \tag{19.7}$$

Heat Generation from the Shoulder

Considering a concave shoulder surface and integrating equation (19.7) over the shoulder surface from $R_{\text{pin}}(R_p)$ to R_{shoulder} (R_s), heat generated due to the shoulder friction is given by,

$$Q_1 = \int_0^{2\pi}\int_{R_p}^{R_s} \omega\pi\tau_{\text{contact}}\, r^2(1+\tan\alpha)\,dr d\theta = \frac{2}{3}\pi\tau_{\text{contact}}\,\omega(R_s^3 - R_p^3)(1+\tan\alpha) \tag{19.8}$$

Heat Generation from the Tool Nib

The tool nib geometry is taken as a vertical cylindrical surface with a flat tip with radius R_p and nib height H_p. Heat is generated from the friction of nib side surface (Q_2) and the tip flat surface (Q_3). Integrating equation (19.3) over the nib side surface gives

$$Q_2 = \int_0^{2\pi} \int_0^{H_p} \omega\tau_{\text{contact}} R_p^2 dz d\theta = 2\pi\tau_{\text{constant}}\, \omega R_p^2 H_p \tag{19.9}$$

Integrating equation (19.2) over the pin tip surface yields

$$Q_3 = \int_0^{2\pi} \int_0^{R_p} \omega\tau_{\text{contact}} r^2 dr d\theta = \frac{2}{3}\pi\tau_{\text{contact}}\, \omega R_p^3 \tag{19.10}$$

Total Heat Generation

Total heat generated due to the friction of the cylindrical nib and concave shoulder of a FSW tool is given by

$$Q_{\text{total}} = Q_1 + Q_2 + Q_3$$

$$= \frac{2}{3}\pi\tau_{\text{contact}}\, \omega(R_s^3 - R_p^3)(1 + \tan\alpha) + R_p^3 + 3R_p^2 H_p \tag{19.11}$$

whereas in case of a flat shoulder, the heat generation expression simplifies to

$$Q_{\text{total}} = \frac{2}{3}\pi\tau_{\text{contact}}\, \omega(R_s^3 + 3R_p^2 H_p) \tag{19.12}$$

Heat Transfer into Tool and Workpiece

Heat generated due to the friction flows into the workpiece as well as the tool. The amount of heat conducted into the workpiece determines the quality of the weld, residual stress and distortion of the workpiece. The amount of the heat that flows to the tool dictates the life of the tool and the capability of the tool for the joining process. For instance, insufficient heat from the friction could lead to breakage of the tool nib, since the material is not soft enough.

Majority of the heat generated from the friction, depending on the thermal conductivity of the tool material, about 95% is transferred to the workpiece and only 5% flows into the tool. The maximum temperature created by friction stir welding process ranges from 80 to 90% of the melting temperature of the material being welded.

Heat Transfer in Tool

Friction stir welding normally starts from the beginning of the joint and finishes at the other end of the joint. In the central part of the plate along the longitudinal direction, the heat transfer process in the tool attains approximately a steady state. As shown in Figure 19.6, the heat flow in the tool and the machine head involves Q_3, Q_4 and q_1, where, Q_3 is the part of the heat getting conducted in the tool, q_1 is the heat lost from the surface of the tool to the environment through convection and Q_4 is transferred to the machine head in which the tool is mounted. Therefore,

$$Q_3 + Q_4 + q_1 \tag{a}$$

Heat Transfer in Workpiece (Figure 19.6)

Considering energy balance at any instant during the friction stir welding is given as,

$$Q_1 = Q_2 + q_2 + Q \tag{b}$$

where, Q_1 is the heat flux coming from the friction between the tool and the workpiece, Q_2 is the heat conducted from the bottom surface of the workpiece to the backing plate on the machine, q_2 is the heat lost from the surface of the workpiece to the environment through convection, and Q is the increase in heat content in the workpiece. The heat transfer process in the entire workpiece remains transient in nature.

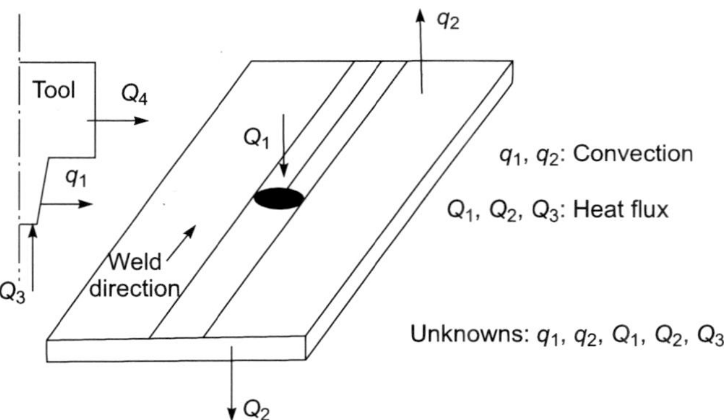

Figure 19.6 Heat transfer in tool and workpiece in friction stir welding

19.4 BASIC FSW METALLURGY

Although the friction stir welding process is now used commercially in a wide range of applications, this is despite the fact that for many alloys, the metallurgy of the welded joint is still not fully understood. Various researchers have tried to study the material flow during FSW.

From these studies and other experiments, three effective zones with differing material flow and associated temperature changes have been identified. These zones were initially classified by Threadgill, and subsequently revised when it became apparent that the microstructural changes in aluminum alloys were not replicated in most other metallic materials. The three effective zones consist of differing levels of material flow and associated temperature changes, and are described in Table 19.1.

TABLE 19.1 Typical micro-structural zones found in friction stir welds

Region	Material Flow	Temperature
Weld nugget	High	High
Thermo-mechanically affected zone (TMAZ)	Low	Medium
Heat-affected zone	None	Medium

19.5 WELD NUGGET

A typical micrograph through an aluminum alloy friction stir weld in case of 5083 aluminum alloy is shown in Figure 19.7. The center of the weld, commonly referred to as the 'nugget zone', consists of a very fine grained, less than 4 μm, microstructure as seen in Figure 3.11. The majority of the grain boundaries within the nugget zone is believed to form through dynamic recrystallisation during the stirring process. The width of the nugget zone depends on the combination of tool design, welding parameters, and alloy composition. The mechanical properties of the weld joint is much higher than the properties of the parent metals. This is attributed to the grain refinement that takes place in the weld nugget as seen in Figure 19.8.

19.6 THERMO-MECHANICALLY AFFECTED ZONE (TMAZ)

Outside the nugget zone, aluminum alloy friction stir welds contain two other distinct microstructural zones (Figure 19.7). In the region immediately surrounding the nugget zone, the parent alloy grain structure becomes both heated and deformed, although to a lesser extent than for the nugget zone (this zone is referred to as the TMAZ, Thermo-Mechanically Affected Zone).

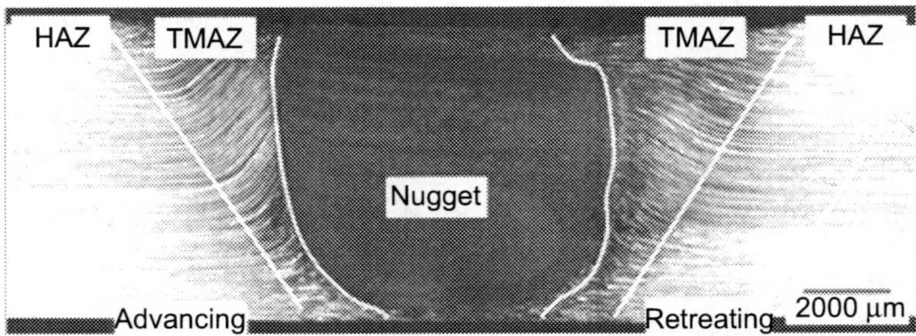

Figure 19.7 Typical macrostructure of aluminum alloy FSW

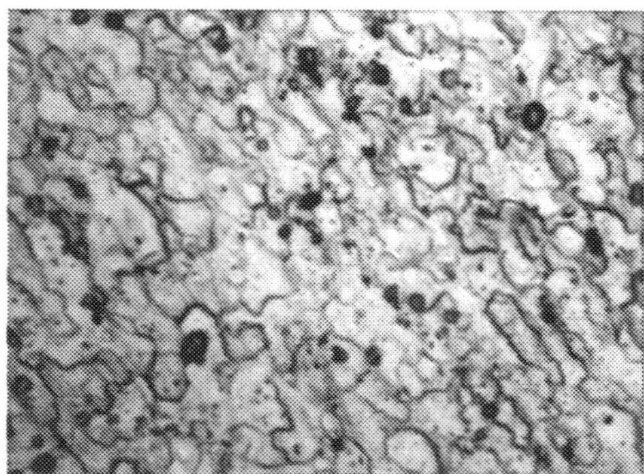

Figure 19.8 Fine-grained microstructure at the weld nugget

This often leads to a region of partially recrystallised grains in which many of the fibrous grains, which are normally aligned in the rolling direction, are rotated. This can be very dangerous as the newly aligned high angle grain boundaries can become susceptible to stress corrosion cracking.

In other materials, the whole of the TMAZ can be recrystallised, with no apparent distinction between that region and the nugget zone. This is a feature of materials such as pure titanium, copper, and austenitic stainless steels, which have no thermally induced phase transformation, which would induce transformation in the absence of strain. The thermally induced phase transformation that occurs in ferritic steels can make interpretation of the microstructure, and precise definition of the TMAZ IHAZ boundary, difficult.

19.7 HEAT AFFECTED ZONE (HAZ)

Surrounding the TMAZ is the heat affected zone. The heat affected zone of friction stir welded 5083 aluminum alloy is shown in Figure 19.9. The stirring effect and the temperature attained at the HAZ leads to grain refinement as observed in Figure 19.9.

Figure 19.9 HAZ of friction stir welded 5083 Al alloy

19.8 DEFECTS AND THEIR DETECTION

The types of defect resulting from the solidification process associated with conventional fusion welding methods are not expected in a solid-state joining technique. However, defects can occur in friction stir welds if any of the process variables which control the quality of the weld deviate from the optimum. The process variables are:

- Tool rotation speed
- Weld travel speed
- Downward force
- Tool geometry

The possible defects that may occur due to improper welding parameters are:

- Lack of penetration
- Kissing bonds
- Wormholes
- Root toe defects
- Improper joint strength

The rate of heat input decreases with increasing travel speed, which reduces softening of the material in the vicinity of the tool nib, making plastic flow more difficult. High-travel speeds therefore, may cause defects, such as cavities. For low tool rotation speed, low downward force and high-travel speed external defects may form for welds in some aluminum alloys. The increase of the downward force moves defects to the interior of the weld.

The ratio of tool rotation speed vs travel speed is sometimes used to distinguish between hot welds having high ratio, and cold welds with low ratio. Hot welds are less sensitive to defect formation but may exhibit more significant changes in microstructure and mechanical properties than cold welds in aluminum alloys.

Too fast a welding speed or insufficient downward force applied to the weld can lead to the formation of voids. Whereas either too short a nib depth or tool plunge can cause joint line defects at the weld root. Joint line remnants can take the form of dispersed oxides if pre-weld cleaning and plate edge preparation are not done properly. Inadequate nib depth or tool plunge can be one of the causes of kissing bonds. Kissing bonds, also referred to as zero-volume disbonds, present a particular problem, as compressive forces can produce closure of the surfaces, hiding the disbond.

The FSW process involves extensive stirring of material and thus, introduces the possibility of defects into the weld region. Detecting these kinds of flaws will probably require a high resolution NDT technique. Ultrasonic testing offers the best possibility of detecting a kissing bond due to the technique's high resolution and high sensitivity. Dye penetration techniques using fluorescent dyes can also be used to identify kissing bonds in welds produced under non-optimum conditions. Pores and cavities in the weld can be detected by conventional X-ray images and ultrasonic scans.

Because of the very nature of FSW process, any defect present can occur at a variety of angles and orientations. This presents a challenge to the usual weld inspection techniques, which generally deal with directionally specific defects. Consequently, detection of FSW defects has proved to be difficult using conventional techniques. One technique used with success for defect detection in FSW of thin aluminum sheets is phased array ultrasonics. This method typically uses a phased array of 128 elements, pulsed with differing delays, and generating an ultrasonic beam which can be swept through a range of angles to detect misoriented (non-axial or transverse) defects. Improved detection can be achieved by incorporating a 2-D matrix transducer to facilitate lateral, rather than simple linear scanning.

Multi-sensor systems for on-line weld inspection with real-time results display have been tried for non-destructive testing of FSW joints. The techniques involved for such real-time NDT are:

- Charged coupled device (CCD) cameras for computer-aided visual inspection.
- Multi-coil eddy current probes for the detection of surface defects as an alternative to dye penetrants.
- Ultrasonic phased arrays for inner discontinuities.

With such NDT techniques, on-line process monitoring is possible. This coupled with feedback provided to the welding device, is desirable, but until it can guarantee defect-free welds, some form of conventional NDT is necessary.

19.9 ADVANTAGES AND DISADVANTAGES OF THE FSW PROCESS

In general, FSW is claimed to produce stronger, lighter and more efficient welds than other joining processes. It has the ability to join dissimilar materials, and materials previously thought to be unweldable, or not recommended for fusion welding. An obvious advantage of a solid-state welding process is the absence of gravitation effects, thereby facilitating its use in all positions, namely, horizontal, vertical, overhead, and orbital. Specific advantages claimed for the process include:

- Absence of melting, resulting in less weld contamination.
- No consumables.
- Minimal edge preparation required.
- Very controllable process, with high productivity.
- High integrity welds with low shrinkage and little porosity.
- Relatively simple process, with low running costs.
- Fine grained microstructure in the weld region, leading to improved mechanical properties.

- Low distortion and residual stress levels, compared to arc welding processes.
- Environment-friendly; relatively quiet process, with absence of fumes and spatter.
- No gas shielding required.
- Able to join wrought to cast products.
- There is no weld pool, it can operate in all positions.

The current limitations of the FSW process are as follows.

- Welding speeds slower than those of some fusion welding processes.
- Backing bar required.
- Keyhole at the end of each weld.
- Postweld machining required.
- Large down forces required with heavy duty clamping necessary to hold the plates together.
- Less flexible when compared to manual arc welding processes.

Review Questions

1. What is friction stir welding?
2. Explain with a neat sketch the process of friction stir welding.
3. How does a friction stir weld machine work?
4. Explain the basic steps used in friction stir welding.
5. Explain thermal considerations in friction stir welding.
6. What are the applications of friction stir welding?
7. What are the advantages of friction stir welding?
8. What are the limitations of friction stir welding?
9. What are the welding parameters used in friction stir welding?
10. Explain the heat affected zone in FSW.

Section D

Gas Welding and Brazing Processes

Oxyfuel Welding

20.0 INTRODUCTION

Gas welding is a non-pressure, fusion welding process in which the required heat is supplied by a high-temperature flame obtained by the mixtures of two gases. The common mixtures of gases are oxygen-acetylene, oxygen-hydrogen and air-acetylene.

The main advantage of gas welding is that its temperature is controllable. This is necessary for delicate works. It can weld most of the common materials. The equipment is inexpensive and versatile. Also, flame can be used for cutting, metal spraying, braze welding, brazing and soldering.

Oxyfuel welding (OFW) is a group of welding processes which join metals by heating with a fuel gas flame or flares with or without the application of pressure and with or without the use of filler metal. Oxyfuel welding includes any welding operation that makes use of a fuel gas combined with oxygen as a heating medium. The process involves melting of the base metal and a filler metal, if used, by means of the flame produced at the tip of a welding torch.

In OFW, oxygen and fuel gas are mixed in proper proportions in a mixing chamber which may be part of the welding tip assembly. Molten metal from the plate edges and filler metal, if used, intermix in a common molten pool. Upon cooling, they coalesce to form a continuous piece.

There are three major processes in this group: oxyacetylene welding, oxyhydrogen welding and pressure gas welding. Of these three processes, oxyacetylene welding is the most popular and versatile process.

20.1 OXYACETYLENE WELDING PROCESS

Here, heat is obtained by a flame got by the combustion of oxygen and acetylene. As a rule, no pressure is required. Oxygen is stored in a cylinder at about 130 to 150 bars. Acetylene is commercially available in cylinders. But in many cases, it can be generated by dissolving CaC_2 in water. Acetylene is stored in combination with acetone, since this gas cannot be stored safely above 1 bar. The acetylene cylinders are filled with a porous filter, saturated with acetone in which acetylene gas can be compressed. Stored this way, it can be kept at a pressure of 13 to 15 bars safely. The porous materials used are asbestos, charcoal, silk fibre, etc.

Oxyacetylene welding (gas welding) is an autogenous process, in which two parts of the same or different metals are joined by causing the edges to melt by an oxyacetylene flame and unite while molten without the aid of hammering or compression. Filler metal is added by inserting it into the molten puddle of the base metal. The puddle then solidifies, making the weld bead. The oxyacetylene flame is made by mixing oxygen and acetylene gases in a special welding torch or blowpipe, producing, when burned, heat of 6300°C, which is more than twice the melting temperature of common metals. This flame, while being of intense heat is of very small size.

The use of oxyacetylene welding dates back to the middle 1800's, where a mixture of oxygen and hydrogen was used to produce a hot flame and that was used in the making of jewellery. It wasn't until the late 1890's when gas welding developed into the process that we know today. An oxyacetylene outfit is portable, less expensive and more versatile than an electric welding setup. By using the proper tips, rods and fluxes, almost any metal can be welded, heated or cut using the oxyacetylene process.

20.2 OPERATING PRINCIPLE

In oxyacetylene welding, the oxygen is supplied from steel cylinders and the acetylene either from cylinder or from an acetylene generator, which can be of medium-pressure or low-pressure type. With cylinder gas, the pressure is reduced to 0.06 N/mm^2 or under, according to the work by means of a pressure reducing valve and the acetylene is passed to the blowpipe, where it is mixed with oxygen in approximately equal proportions, and finally passed into the nozzle or tip to be burnt.

Burning of oxygen and acetylene is accomplished in five steps. First, the cylinder pressure is released to a working-line pressure by means of special regulators. The gases are transported, by hoses usually, to the torch body. These gases are mixed in the mixing chamber, and then this mixture is ejected into the atmosphere. All that is needed now to supply combustion is for this mixture to be ignited. When the spark of ignition is applied, the triangle of combustion is then complete. The three requirements for oxyacetylene flame are an oxygen source, an acetylene source and a proper ignition temperature. To control combustion, one needs only to adjust or remove one of these three elements.

Oxyacetylene Flame

Combustion of oxygen (O) and acetylene (C_2H_2) to support maximum combustion has been found to be

$$2C_2H_2 + 5O_2$$

That is, two volumes of acetylene combined with five volumes of oxygen for the complete burning of the two gases. Upon the complete burning or combustion, the residue is

$$4CO_2 + 2H_2O$$

or four volumes of carbon dioxide and two volumes of water vapor, however, this burn does not happen one time. Normally, it occurs in two stages. First, equal volumes of oxygen and acetylene are ejected from the torch tip to burn in the atmosphere ($2C_2H_2 + 2O_2$). When these equal amounts of the gases go through combustion, the inner cone is created and is readily visible at the end of the torch tip. The inner cone is referred to as the primary combustion zone on the primary stage of combustion. This first combustion has a residue of four volumes of carbon monoxide and two volumes of hydrogen. Since both these residual gases are capable of supporting combustion, when they are liberated, they ignite and burn at lower temperature other than the primary burn of acetylene and oxygen are consumed. The four volumes of carbon monoxide combine with two volumes of oxygen from the atmosphere and burn in the secondary combustion zone or outer envelope, which is also called the secondary combustion. This combustion can be expressed chemically as $4CO_2 + 2O_2$ and liberates four volumes of carbon dioxide ($4CO_2$).

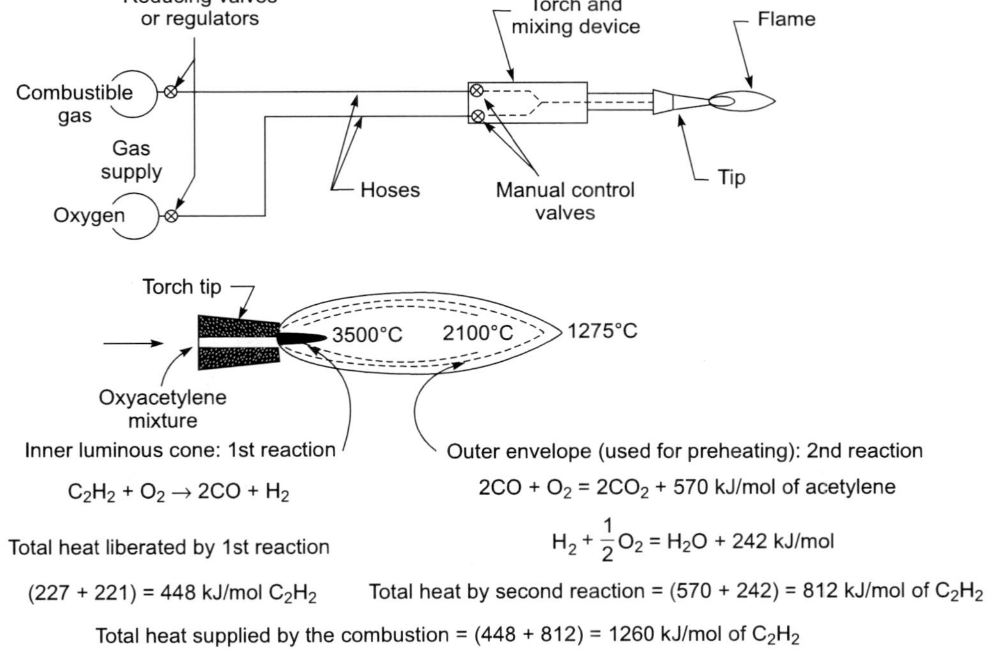

Figure 20.1 Chemical reactions at inner and outer envelopes of flame

The last volume of oxygen combines in the secondary combustion zone with the remaining two hydrogen volumes that resulted from the original combustion release. These two gases react and liberate two volumes of water vapor ($2H_2O$).

The complete combustion of oxyacetylene flame can be represented by the following reactions:

$$2C_2H_2 + 5O_2 = 4CO_2 + 2H_2O$$
$$2H_2O = 4CO + O_2$$

The approximate range that can be attained by the oxyacetylene flame is from 5000 to 6300°F at the inner cone, around 3800°F at the extreme end of the secondary combustion envelope. These temperatures can be varied to some extent by changing the mixture of gases or by causing an improper balance in the volumes of oxygen and acetylene ejected from the tip of the torch. This imbalance is usually accomplished by adjusting the needle valves on the blowpipe or the torch body. Figure 20.1 shows the various temperature zones and chemical reactions that occur in an oxyacetylene flame.

20.3 EQUIPMENT

The basic oxyacetylene outfit consists of a cylinder of oxygen, cylinder of acetylene, oxygen and acetylene regulators, hoses to carry the gases, valves, torch and torch tips as shown in Figure 20.2. The gases are generally stored at high pressure in steel cylinders which are seamless and heat treated properly to give good strength and toughness. Cylinders are designed for specific gases and are not generally interchangeable. It is important that only appropriate fitting should be used for delivery of compressed gas. The oxygen cylinder is equipped with a high pressure valve, valve protector cap and safety plug. Acetylene cylinders are packed with porous filler, which is saturated with acetone. They are also fitted with high pressure valve, valve protector cap and fusible plugs.

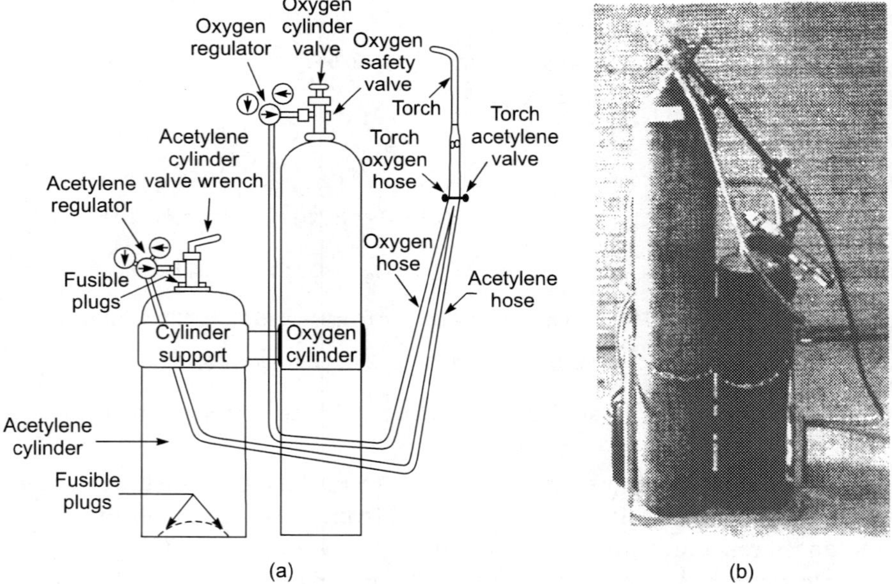

Figure 20.2 Welding equipment

Oxygen Cylinder

The oxygen required for both high- and low-pressure systems in solid drawn steel cylinders is at a pressure of 17.1 N/mm^2. The cylinders are rated according to the amount of gas they contain, 8.5 m^3 being the usual size but very large cylinders containing 800 m^3 are available. The volume of oxygen contained in the cylinder is approximately proportional to the pressure; hence for every 10 liters of this oxygen use, the pressure drops about 0.02 N/mm^2. This enables us to tell how much oxygen remains in the cylinder. The oxygen cylinder is provided with a valve threaded right hand and is painted black. On to this valve, which contains a screw type tap, the pressure regulator and pressure gage are screwed. The regulator adjusts the pressure that is required at the blowpipe. Since grease and oil can catch fire spontaneously when in contact with pure oxygen under pressure, they must never be used on any account upon any part of the apparatus. Leakages of oxygen can be detected by the application of a soap solution, the leak is indicated by the soap bubbles. Never test leakages with a naked flame.

Large consumers find it more economical to purchase oxygen in liquid form, which is delivered and stored at –297.3°C. Liquid oxygen (LOX) is generally supplied in large storage tanks in 160 litre containers. The major drawback of these large containers, however, is that the average evaporation loss is 1.5 percent per 24 hour storage. Therefore, they are economical only for those consumers who use large amounts of oxygen daily.

Acetylene Cylinder

Acetylene gas, a hydrocarbon with the chemical formula C_2H_2, is not found in a free state in nature. It is generally produced by the carbide-to-water method, which allows water to react upon calcium carbide in a carbide-to-water generator. The reaction between water and carbide is instantaneous; carbon in the carbide combines with the hydrogen in water, forming acetylene gas. Thus, chemical reaction is

$$CaC_2 + 2H_2O \rightarrow C_2H_2 + Ca(OH)_2$$

Acetylene gas under pressure becomes very unstable and in free state will explode before reaching a pressure of 2 bars. This instability places special requirements on the storage of acetylene. A storage cylinder is filled with a mixture of calcium silicate, a material that is 92% porous. The cylinder is then filled with acetone, which is the solvent agent of acetylene gas and which has an absorptive capacity up to 35 volumes of acetylene per volume of acetone per atmosphere of pressure. This enables about 420 volumes of acetylene to be compressed at 250 psi. Under these conditions, the gas is present in the form which is to be used. Acetylene comes out of the acetone solution at a slow constant rate as the pressure on the cylinder is released. The rate, however, depends on the temperature of the gas (Figure 20.3).

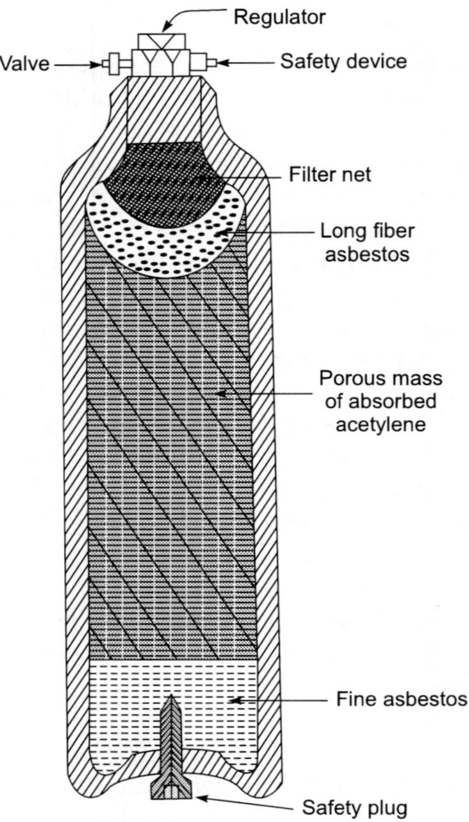

Figure 20.3 Section through acetylene cylinder

The pressure of acetylene is usually about 1.5 N/mm^2, a typical capacity being 5700 liters. The gas leaves the cylinder through a valve after passing through a filter pad. The valve has a screw tap fitted and is screwed left-hand. The cylinder is painted maroon and a regulator (also screwed left-hand) is necessary to reduce the pressure to 0.013-0.05 N/mm^2 as required at the blowpipe.

The amount of dissolved acetylene in a cylinder cannot be determined with any accuracy from the pressure gage reading, since it is in the dissolved condition. The most accurate way to determine the quantity of gas in a cylinder is to weigh it, and subtract this weight from the weight of the full cylinder, which is usually stamped on the label attached to the cylinder. Volume of gas remaining in the cylinder is calculated by remembering that 1 liter of acetylene weighs 1.1 gm.

Acetylene is highly inflammable and no naked lights should be held near a leaking cylinder, valve or tube. Leaks can be detected by smell or by soap bubbles. If any part of the acetylene apparatus catches fire, immediately shut the acetylene valve on the cylinder. The cylinder should be stored and used in an upright position.

Safety Precautions of Acetylene Cylinder

Following safety tips are needed for careful use of acetylene cylinders:

1. Acetylene is highly flammable and so necked flame should not be brought close to the cylinder.
2. Never use the hammer for opening and closing of spindle valve.
3. Use the cap on valve when cylinder is not in use.
4. When under pressure of 203 kPa and above, acetylene is unstable and a sight shock can cause it to explode, even in the absence of the oxygen or air. By dissolving purified and dried acetylene in liquid acetone, it can be used to store about 7.79 m^3 of acetylene under the pressure of 1.7 MPa.
5. Never transfer gas from one cylinder to another.
6. Acetylene cylinder should never be used as supporting element.
7. Keep the acetylene cylinder always in upward position so as to keep the acetone from escaping during use.
8. Keep the acetylene cylinder away from sunlight and heat.
9. Ragging and rolling of cylinder should be avoided.
10. Not more than 20% of each cylinder's gas capacity must be drawn off per hour; otherwise acetone used to dissolve the acetylene in the cylinder may be drawn out with the gas. In the high-pressure acetylene system, the gas is stored in cylinders having an average gas capacity of 3398 to 6230 liters. These cylinders may be used singly by individual welders, in which case pressure of up to 105 kg/cm^2 may be used at the blowpipe. Alternatively as a centralised manifold system feeding multiple welding points, in which case, for safety reasons, the line pressure is not allowed to exceed 063 kg/cm^2 (Figure 20.4).

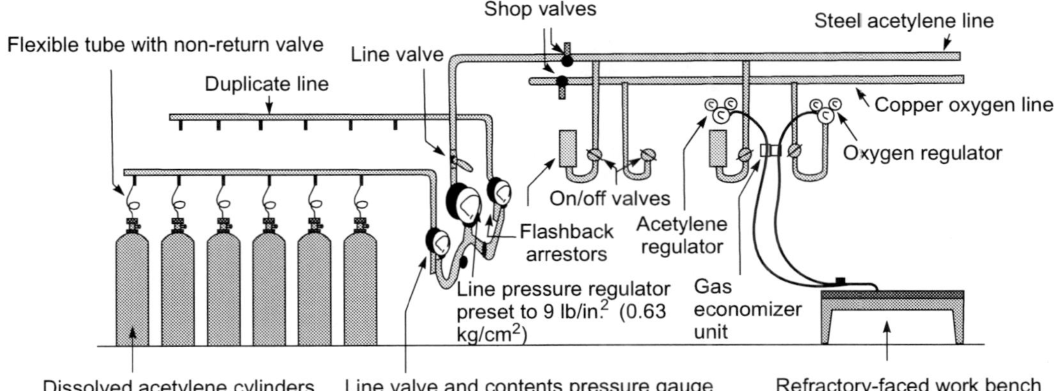

Figure 20.4 Manifold system on high pressure acetylene supply line

The number of acetylene cylinders required on a manifold is related to the total amount of acetylene required at any one time. Manifolded supplies may be used for both acetylene and oxygen cylinders, although where the oxygen consumption is high, it is more practical to use an oxygen generator/evaporator plant, oxygen being delivered and stored in the more compact liquid form.

Acetylene Gas Generator

If larger quantities of acetylene gas are being consumed, it is much cheaper to generate the gas at the place of use with the help of acetylene gas generators. Acetylene gas is generated by two methods; carbide to water method and water to carbide method. Carbide to water type generator is preferred over water to carbide types as the former uses complete reaction of water and calcium carbide and is more efficient. A typical calcium carbide to water type acetylene generator is shown in Figure 20.5. Small size of calcium carbide particles is stored in hopper at top of the generator where calcium carbide is fed down to the water by a feed mechanism and acetylene is produced as a result of action of water on calcium carbide. At the bottom of the generator, calcium hydroxide (sludge) is collected.

Low pressure generator can deliver gas pressures less than 0.1 bar. With this kind of generator, only the injector type of blowpipes can be used. Low pressure generator is portable and produces gas up to 15 LPM (litres per minute).

Medium pressure regulator is considered stationary and it can produce acetylene up to 3000 LPM. This type of generator is commonly used.

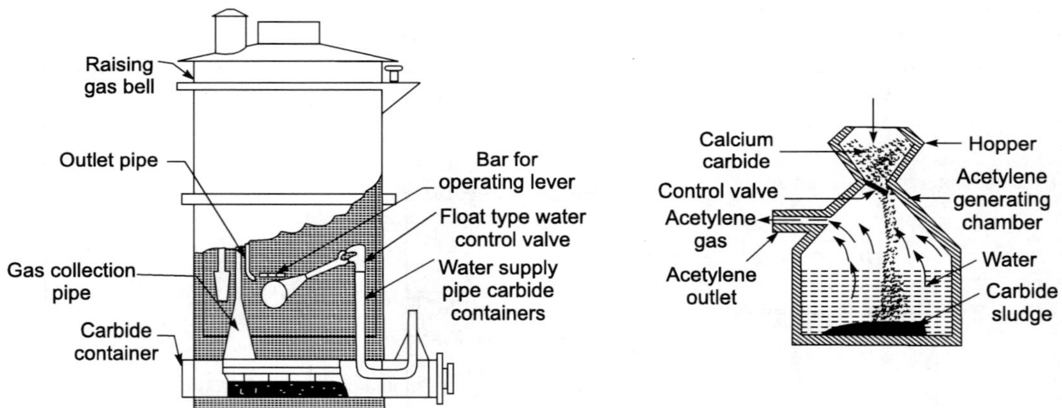

Figure 20.5 Acetylene generator

Disadvantages of Acetylene Generator

1. Greater safety precautions are required.
2. Labour is required to charge carbide and clean the sludge.

3. Gas obtained is not pure.
4. There is tendency to pressure fluctuations with resultant unsteady flame, especially in the case of low pressure type generator.

Backfire or Popping

Backfire is a small explosion of the flame followed by either extinguishing of flame or continued burning of gases. It is basically due to pre-ignition of gases. The causes for backfire are as follows:

1. When gases come out too slowly from torch under low pressure due to small tip used. Pre-ignition or burning of gases may take place within torch.
2. If tip of the torch become overheated.
3. Due to deposition of hot metal particles inside the torch tip in such a way, torch becomes overheated. This results in pre-ignition of gases within the torch.

Flashback

Flashback is quite different from backfire phenomenon. A flashback occurs when the flame disappears from the tip and travels back inside the hose. Flame makes a hissing noise if there is a flashback. To prevent flashback, both gases should be shut off immediately to avoid combustion within the torch. Torch should be allowed to cool before restart. Flashback may also be due to a clogged barrel or mixer passage and excessive pressure of oxygen. Sometimes, organic oxides in oxygen hose may also be responsible for flashback.

Regulators

They reduce the pressure from the cylinder and supply the gas to the gas hose at useable pressure and maintain at a constant value regardless of the variation of the pressure at the source. Regulators are made of various ranges of inlet and outlet pressures. They can be adjusted within the working pressure range by turning and adjusting screw or a handle. Most of the regulators are equipped with two pressure gages; one on inlet side and other on the outlet.

Basically, regulators work by admitting the high pressure through a valve which is operated by a flexible diaphragm. By turning the regulator, adjusting the knob or screw in or out, causes a spring in the regulator to operate the diaphragm which opens or closes a valve in the regulator. This, in turn, regulates the outer pressure and flow. By turning the adjusting knob in, one can increase the flow and pressure and vice versa. Most regulators have two gages. One shows the inlet pressure from the cylinder (the high-pressure gage) and the other (low-pressure gage) shows the working pressure being supplied from the regulator. There are regulators that are made for heavy duty or rough service but are not equipped with gages (referred to as gage less) and have a scale in the regulator body that is used to make pressure adjustments.

There are two general types of regulators; single-stage and two-stage regulators.

Single-Stage Pressure Regulator: In this type, the gas enters the regulator at the base from a finely sintered metal filter. The cylinder pressure is indicated on the first pressure gage. The gas then enters the body of the regulator (*R*) through the aperture (*A*), which is controlled by valve (*V*), as shown in Figure 20.6.

The pressure inside the regulator rises until it is sufficient to overcome the pressure of the spring (*S*) which loads the diaphragm (*D*). The diaphragm is, therefore, pushed back and the valve (*V*) to which it is attached, closes the aperture (*A*) and prevents any more gas from entering the regulator.

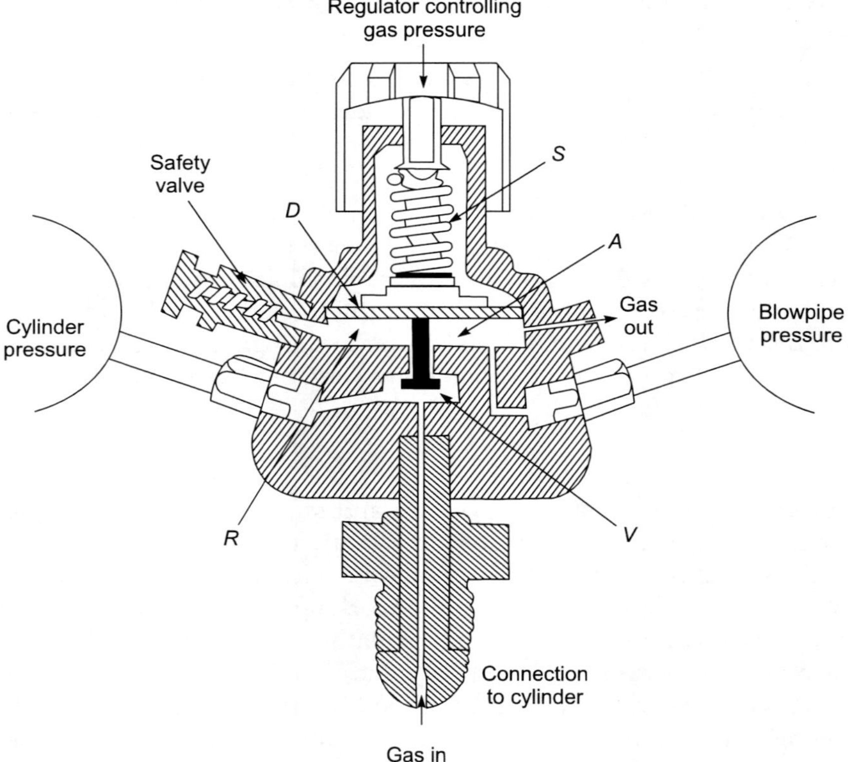

Figure 20.6 Single-stage regulator

The outlet side is fitted with a pressure gage which indicates the working pressure or the blowpipe. Upon gas being drawn off from the outlet side, the pressure inside the regulator body fails, the diaphragm is pushed back by the spring and the valve opens – letting more gas in from the cylinder. Single-stage regulators are suitable for general welding purpose with maximum outlet pressure.

Two-Stage Pressure Regulator: A double-stage or two-stage regulator consists of two sets of diaphragm, valve and spring, as shown in Figure 20.7. The first stage is adjustable. It reduces the pressure of the cylinder, when the gas enters from the first to the second stage, the pressure is adjustable (according to the thickness of the metal). The two-stage regulator reduces the pressure in two stages and gives a much more stable output pressure than the single-stage regulator. The first stage, which is preset, reduces the pressure from the cylinder, then the gas at that pressure passes into the second stage – from which it emerges at a pressure set by the pressure adjusting control screw attached to the diaphragm. A high flow safety relief valve is fitted to avoid abnormal pressures which, if created, are quickly vented to the atmosphere, the valve reseating itself when pressure returns to normal. Regulator bodies are of brass forgings fitted with safety valve set to relieve the excess gas pressures. To identify the regulator, the oxy regulator (usually right hand threaded) is painted black and acetylene regulator (usually left hand threaded) is painted maroon.

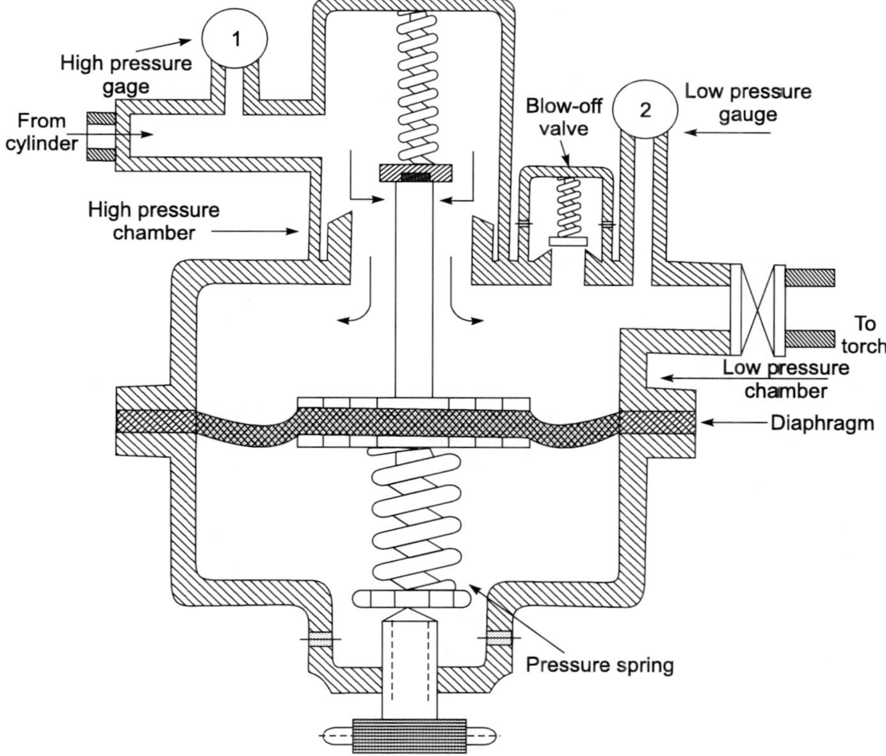

Figure 20.7 Two-stage pressure regulator

Comparison between single and two-stage regulator is given in Table 20.1. Table 20.2 compares oxygen and acetylene regulators.

TABLE 20.1 Comparison between single- and two-stage regulators

	Single-Stage Regulator	Two-Stage Regulator
1.	Reduction of pressure from the cylinder pressure to the welding pressure takes place in single stage.	Reduction of pressure takes place in two stages.
2.	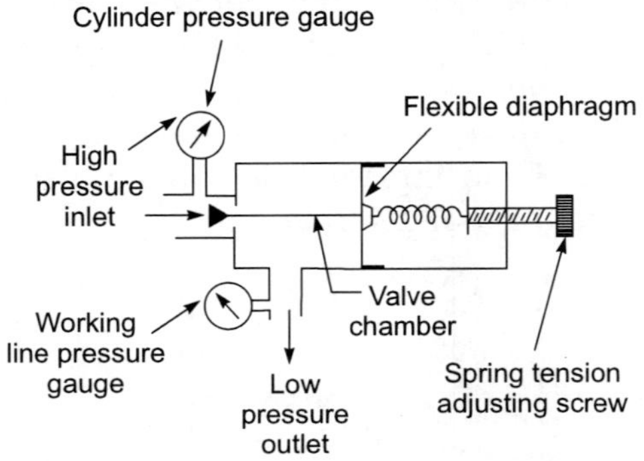 Cylinder pressure gauge Flexible diaphragm High pressure inlet Working line pressure gauge Valve chamber Low pressure outlet Spring tension adjusting screw	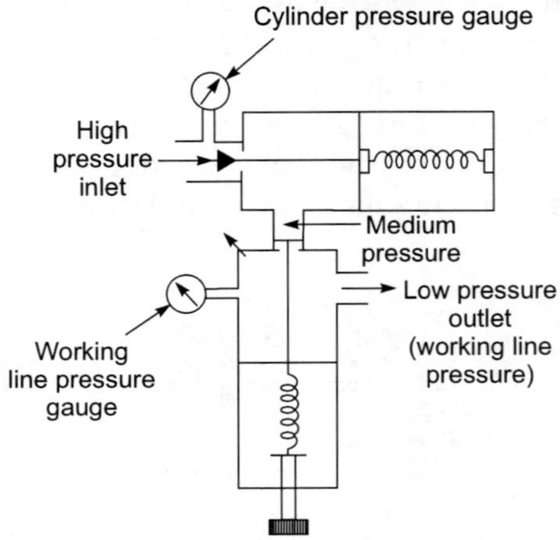 Cylinder pressure gauge High pressure inlet Medium pressure Low pressure outlet (working line pressure) Working line pressure gauge
	Pressure in the hoses is controlled by applying pressure to the spring through an adjusting screw. The spring applies pressure to a flexible rubber diaphragm, which is connected to the (high pressure) valve. The gas from the cylinder flows past the open valve, enters the valve chamber and builds up pressure on the diaphragm. When the pressure inside the valve chamber is slightly greater than the spring tension supporting the diaphragm, the valve closes the high pressure inlet, and stops the flow of any more gas into the valve chamber. The gas thus trapped inside the valve chamber is allowed to flow out through the low pressure outlet into the welding hose, which in turn, relieves the chamber pressure, opens the valve again and the process repeats.	The principle of pressure reduction in a two-stage regulator is exactly the same as in a single-stage regulator. But the pressure is reduced in two stages instead of one, using two diaphragms and two control valves, so that the pressure reduction ratio is less abrupt. The gas pressure is automatically reduced by the first, preset stage, the second stage of pressure reduction being controlled by the welder by way of the adjustable screw and spring valve.
3.	A single-stage regulator is all that actually is needed for both oxygen regulation and acetylene regulation for oxyacetylene welding.	A two-stage regulator is used with cylinders and manifolds.

Contd.

4.	Single-stage regulators may be classified as follows, depending upon the method used to close the seat mechanism and thereby control the gas flow: (a) Direct or nozzle type (b) Inverse or stem type In (a), the seat is forced to close against the gas flow (away from the nozzle) by the incoming high pressure gas. In (b), the incoming high pressure gas forces the seat towards the nozzle in the direction of the gas flow.	A two-stage regulator may consist of: (a) Two inverse type of stages. (b) Two direct type stages. (c) First direct type stage and second inverse type stage. Two-stage regulators of mixed stage type are the most common.
5.	A single-stage regulator tends to freeze in cold weather. This is because a sudden expansion and resulting drop in initial pressure causes rapid cooling of the gas involved. The moisture present in the gas, thus, results in the formation of ice on or near the regulator nozzle, which causes irregular seating of the seat on the nozzle and therefore, substantial pressure fluctuations.	In the first stage, the regulator tends to freeze in cold weather. This is because a sudden expansion and resulting drop in initial pressure causes rapid cooling of the gas involved. However, the second stage is not affected much and there is no noticeable pressure differences.

TABLE 20.2 Comparison between oxygen and acetylene regulators

	Acetylene regulator	Oxygen regulator
1.	The cylinder and hose connections have left-handed threads on the acetylene regulator.	There are right-hand threads in this case.
2.	Acetylene connection nuts have chamfers or grooves cut in them.	Nuts are plain, i.e., with no chamfer or grooves.
3.	Color band on acetylene regulator is maroon or red.	It is either blue or black on the oxygen regulator.
	Grooved nut	Plan nut
4.	The inlet or high-pressure gage on the regulator reads up to 100 bars on the oxygen regulator and upto 8 bars on the acetylene regulator.	
5.	The outlet or low-pressure gage on the regulator reads up to 4.8 bars on the oxygen regulator and up to 1 bar on the acetylene regulator.	

Welding Hoses

Flexible hoses permit the gas cylinders and regulators to be kept at a safe distance from the working area. Two separate hoses for oxygen and acetylene will be required. The cylinders, regulators and torch are usually connected together by a double line rubber hose, which comes in black, red and green colors. The green hose is usually used to transport oxygen; the red hose is usually the fuel-gas hose and the black hose is usually used to transport other industrially available welding gases. The use of colors is not standardized, but the collared hoses are many times used in this fashion. The hose is manufactured in various diameters. The more popular diameters are 3/16, 1/4, 5/16, 3/8 and 1/2 inches. These industrial hoses are usually measured by inside diameter and come in standard lengths of 25, 50, and 75 ft.

As a safety precaution, all the oxygen and compressed air hoses are right-hand threaded, and all fuel carrying hoses are left-hand threaded. This standard arrangement prevents oxidants from combining with fuel, which could result in an explosion.

Certain precautions should be taken when using reinforced rubber hoses: only one gas should be used in the hose; the hose should never be patched or repaired; and hot metal should never touch the hose.

Check Valves

For combustion to occur, fuel and oxygen have to be mixed. This should only happen in the torch mixer or the torch tip. Sometimes, due to improper operation, fuel and oxygen could feed back into the hoses and cause combination in the hoses or regulators, which is undesirable. Check valves when installed between the hoses and torch prevent this back flow as they close if a reverse flow starts. Check valves should be used with all torches.

Flashback Arrestors

A flashback, which is a rapid high pressure flame in the hose can occur if there are no check valves or the check valves fail to operate due to improper installation. Once a flashback starts, check valves cannot stop it, but a flashback arrestor can. The arrestor connects the same as the check valve, in the hose at the torch or regulator and contains a trap that is spring loaded that cuts off the gas flow in the event of a flashback. Both check valves and flashback arrestors are very important safety devices and should be included on every oxyacetylene outfit.

Welding Torches

Welding torch controls the operating characteristics of the welding flame and enables manipulation of the flame during welding. The two gases are mixed in desired volumes in it and are burnt at the end of the tip. The torch assembly consists of the handle, oxygen and fuel gas valves and a mixing chamber. Welding tips or a cutting attachment can be used with the handle, allowing it to be used for welding, heating and cutting operations. Oxygen and fuel gases flow inside the handle which are blend in the mixing chamber or tip. It is at the tip that the mixed gases are ignited. There are two basic mixer types, the equal or medium pressure type (also known as balance or positive pressure type) and the injector type. The equal pressure type is the most common and is used with fuel gas pressures above 1 psi. Oxygen and fuel gas enter the torch at almost equal pressures. The injector type is used when fuel gas pressures are less than 1 psi. In this type, oxygen at high pressure pulls the fuel gas into the mixing chamber.

Welding torch or blowpipes are also classified as high-pressure and low-pressure welding blowpipe or torch. The high-pressure blowpipe is a mixing device to supply equal volumes of oxygen and acetylene to the nozzle and is filled with regulating valves to vary the pressure of the gases as required. The blowpipe is made of bronze or brass. The blowpipe is divided in three parts – shank, neck and nozzle.

High-Pressure Blowpipe: A selection of shank is supplied with each blowpipe provided with two controllers. The two gases mix in the mixing chamber. The mixed gases are then supplied to the nozzle made of copper, which has good resistance to heat. A high-pressure blowpipe cannot be used in a low-pressure system. Figure 20.8 shows the high-pressure blowpipe.

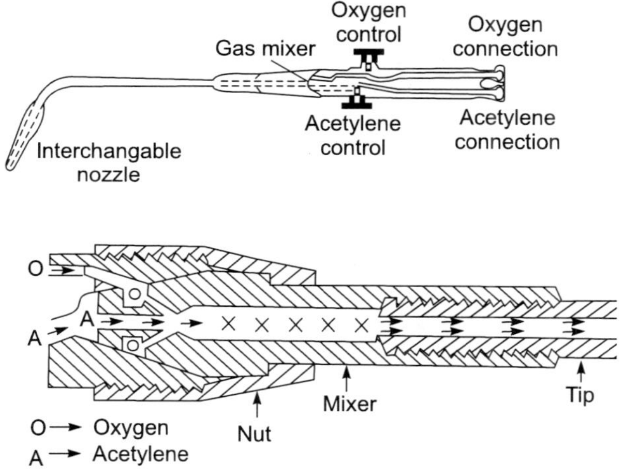

Figure 20.8 High-pressure blowpipe

Low-Pressure Blowpipe: It is a device to produce and adjust a suitable oxyacetylene flame in which the fuel gas is generated at low pressure. The low-pressure blowpipe has an injector nozzle inside its body through which the high-pressure oxygen streams out. This oxygen draws the low-pressure acetylene into the mixing chamber and gives it necessary velocity to produce a steady flame. The injector also helps prevent back firing. If too much metal is removed from the nozzle, it tends to backfire. The low-pressure blowpipe is more expensive than high-pressure blowpipe as it can be used on high-pressure system if required. Figure 20.9 shows a low-pressure blowpipe.

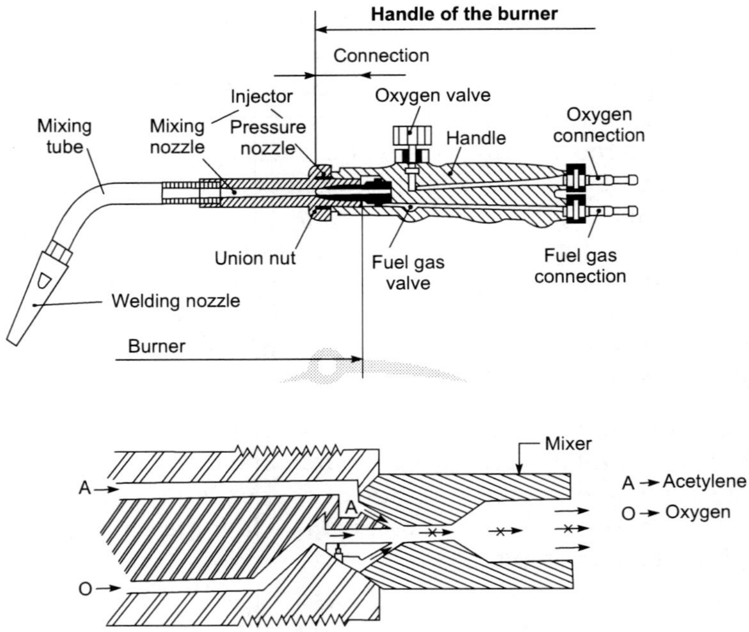

Figure 20.9 Low-pressure blowpipe

To eliminate the difficulties, an injection is provided in the blowpipe to supply acetylene gas in equal volume with oxygen. Through the narrow hole of the centre of the injector flows oxygen at high velocity around the injector, it creates a vacuum. The low-pressure acetylene is drawn along with the oxygen. This high-pressure mixture then passes through the nozzle at a high velocity. A modern system uses a combined welding blowpipe and metal cutting torch. In this, the shank is arranged so that a full range of nozzles or a cutting head can be fitted. This type of blowpipe is cheaper than the ordinary set of cutting and welding.

Welding Tips

The welding tip is mounted on the end of the torch handle and through it, the oxygen and fuel gas mixture feed the flame. Tips are available in a variety of shapes and sizes to fit any welding job and are identified by a number. The larger the number, the larger the hole in the tip and the thicker the metal that can be welded. Welding tips have one hole and cutting tips have a centrally located hole with a number of small holes located around it in circular pattern. Oxygen comes from the centre hole with the preheat flame coming from the holes around it. Many factors determine the size of tip to be used, but mainly the thickness of the metal to be welded determines which tip size to use.

Goggles: It is essential to protect eyes from harmful heat and ultraviolet and infrared rays. In addition, spark lighter, apron, gloves, etc., are also used.

Welding Rods: The welding rods usually have the same composition as the part being welded. The size of the welding rod depends on the type of the weld being made, shape and thickness of the members to be joined. The welding rod material should flow thoroughly and freely and unite readily with the base metal to produce clean welds. A small welding rod may be subjected to the danger of burning and a large rod may not melt through and may enter the joint in the plastic state rather than in molten condition. Welding rods are available in various sizes and in standard length.

Fluxes: Fluxes are used to remove the impurities and oxide present on the surfaces of the metal to be joined and to obtain a satisfactory bond. The fluxes react with the oxides and form fusible slags at the welding temperature which floats to the top of the molten pool and does not interfere with the deposition of filler metal. It also protects the molten pool from atmospheric oxygen. The oxides will have a higher melting point than the metal itself. These oxides do not flow from the welding zone; remain on the surface, become entrapped in the solidifying metal and interfere with the addition of filler metal which is prevented from fluxes. The fluxes are available in powder forms, paste forms or thick slurry or in the form of a coating in the welding rod.

No flux is used for gas welding of steels. Fluxes are employed in gas welding of cast iron, stainless steel and most non-ferrous metals other than lead and zinc.

20.4 OPERATING PARAMETERS

Important parameters which affect the oxyacetylene welding process are:

(i) Flame adjustment
(ii) Welding techniques
(iii) Weld movement
(iv) Weld position

Flame Adjustment

Oxygen and acetylene can be injected from the torch tip in three possible gas mixtures by adjusting the needle valve on the torch: excess of acetylene, equal mixture, or excess of oxygen. Based on it, three types of oxyacetylene flames can be produced:

1. The carburizing flame (reducing)
2. The balanced flame (neutral)
3. The oxidizing flame

The carburizing or reducing flame has excess of acetylene and it is characterized by three stages of combustion instead of two as in other two types of flames. The extra combustion stage, called the intermediate feather, can be adjusted by the amount of acetylene imbalance induced at the torch body needle valve. The length of the intermediate feather is usually measured by eye, in terms of the inner-core length. A 2X carburizing flame, then, would be a feather approximately two times as long as the inner core.

The carburizing flame does not completely consume the available carbon; therefore, its burning temperature is much lower and the leftover carbon is forced into the metal. This action is characterized by the molten weld puddle appearing to boil. After this carbon-rich weld solidifies, the welding will have a pitted surface; these pits can extend through the entire weld bead. The bead will also be hardened and extremely brittle because of the excessive amount of carbon that was injected into the molten base metal. This excess of carbon, however, is an ideal condition when welding high-carbon steel.

Upon further adjustment of the torch needle valves, the intermediate feather can be drawn into the inner cone. Oxygen is usually increased in order to maintain the minimal amount of acetylene flow. Increase of oxygen flow causes the intermediate feather to recede into the inner cone. The instant the feather disappears into the cone, oxygen and acetylene mixture produces balanced or neutral flame. This two-stage neutral flame, which should have a symmetrical inner cone and should make a hissing sound, is the most used flame in both, the welding and the flame cutting of metals. It has little effect upon the base metal properties. The inner cone of the neutral flame is not hot enough to melt most of the commercial metals, and the secondary combustion envelope is an excellent cleanser and the protector of ferrous metals.

Further opening of the oxygen needle valve on the torch body shortens the inner cone approximately two-tenths of its original length. The inner cone then loses its symmetric shape and the flame gives off a loud roar. These three actions are the characteristics of oxidizing flame, which is also the hottest flame that can be produced by any oxygen-fuel source. The oxidizing flame injects oxygen into the molten metal of the weld puddle, causing the metal to oxidize or burn too quickly, as demonstrated by the bright sparks that are thrown off the weld puddle.

Also, the excess of oxygen causes the weld bead and the surrounding area to have a scummy or dirty appearance. However, a slightly oxidizing flame is helpful when welding most copper-base metals, zinc-base metals, and a few types of ferrous metals, such as manganese steel and cast iron. The oxidizing atmosphere creates a base metal oxide that protects the base metal. For example, in welding brass, the zinc alloyed in the copper has a tendency to separate and fume away. The formation of a covering copper oxide prevents zinc from dissipating. Figure 20.10 below shows different oxyacetylene flames.

Summary of different flames generated at the torch tip.

Neutral Flame: When one part of acetylene is mixed with one part of oxygen and burnt, resulting flame will be neutral flame. This flame consists of two envelopes; outer and inner cones. Temperature of inner core is around 3200°C. **Ratio of acetylene: oxygen = 1:1**. There is no chemical change in molten metal and therefore, no oxidation of molten metal. The neutral flame is clearly indicated by well-defined white inner core at the torch tip. This is surrounded by the outer envelope which is slightly luminous. The neutral flame is shown in Figure 20.10. Chemical reactions in neutral flame are shown in Figure 20.11. Normally used for welding mild steel, cast iron, copper, stainless steel and aluminum.	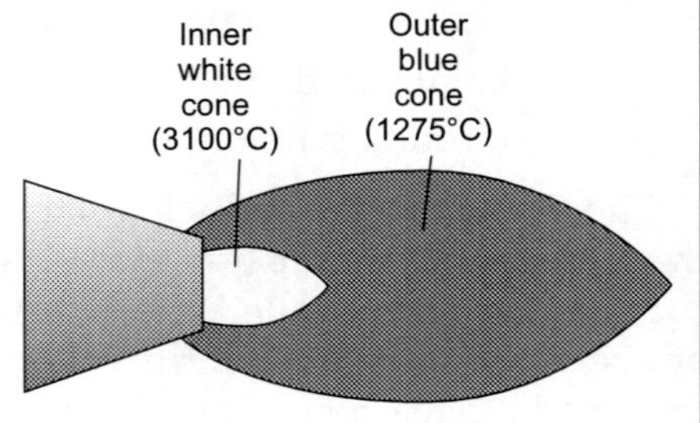 **Figure 20.10** Neutral flame

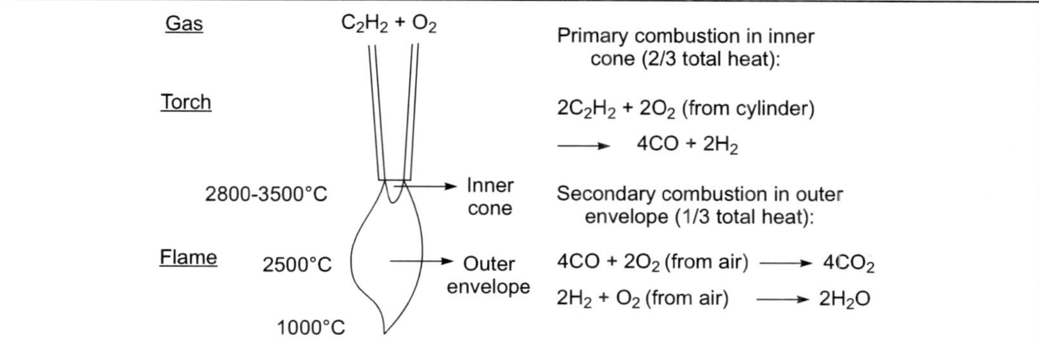

Figure 20.11 Chemical reactions & temperature distribution in a neutral flame.

Oxidising Flame (**Figure 20.12**): The ratio of **oxygen to acetylene is 1.5:1** in oxidizing flame. Temperature of inner core is around 3500°C. It is hotter than neutral flame. It is used for copper-base metal, zinc-base metal and bronze-base metal. This flame is similar to neutral flame except that the inner luminous core is much shorter. This flame is used for welding brass, bronze but is undesirable in other applications as it has a strong tendency to oxidize the metal being welded. An oxidizing flame is often used as the preheating flame for gas cutting applications. In the welding of brass and bronze, it is desirable to form a tough oxide film over the molten metal to prevent vaporisation of zinc and oxidation of the molten metal.

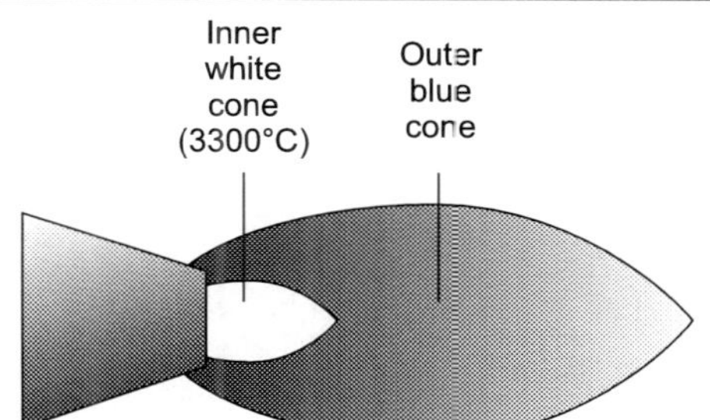

Figure 20.12 Oxidizing flame

Carburizing Flame or Reducing Flame (Figure 20.13): Acetylene is more in carburising flame compared to oxygen. Temperature of the carburizing flame is 3150°C, lowest of all flames. A reducing flame is used for annealing, to soften steel sheets and welding monel metal, nickel and other non-ferrous alloys. When excess acetylene is used, there will be three zones instead of two. There will be intermediate core of whitish color between the luminous core and the outer envelop, the length of which can be determined by the amount of excess acetylene. This is used for welding monel metal, nickel, certain alloy steels and materials like stellite and calamnay.

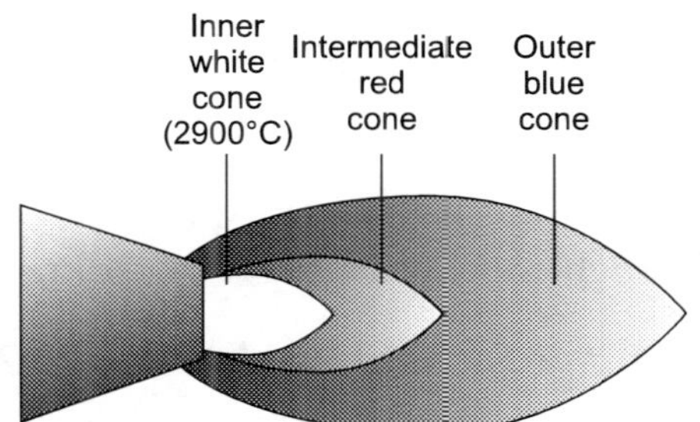

Figure 20.13 Carburizing flame

Table 20.3 gives the list of ferrous and non-ferrous materials that can be welded by oxyacetylene welding processes.

TABLE 20.3 List of Materials that can be welded by oxyacetylene processes

Base Metal	Filler-Metal Type	Flame Type	Flux Type
Aluminums	(a)	Slightly reducing	Aluminum
Brasses	Navy brass	Slightly oxidizing	Borax
Bronzes	Copper tin	Slightly oxidizing	Borax
Copper	Copper	Neutral	(b)
Copper-nickel	Copper-nickel	Reducing	(b)
Inconel	(a)	Slightly reducing	Fluoride
Iron, cast	Cast iron	Neutral	Borax
Iron, wrought	Steel	Neutral	(b)
Lead	Lead	Slightly reducing	(b)
Monel	(a)	Slightly reducing	Monel
Nickel	Nickel	Slightly reducing	(b)
Nickel-silver	Nickel-silver	Reducing	(b)
Low-alloy steel	Steel	Slightly reducing	(b)
High-carbon steel	Steel	Reducing	(b)
Low-carbon steel	Steel	Neutral	(b)
Medium-carbon steel	Steel	Slightly reducing	(b)
Stainless steel	(a)	Slightly reducing	Stainless steel

(a) Match base metal
(b) No flux required

20.5 GAS WELDING TECHNIQUES

Techniques for welding mild steel are as follows:

1. Leftward or forward welding
2. Rightward or backward welding

Leftward or Forward Welding (Figure 20.14)

The leftward method of welding is also known as forward welding. It is the oldest and most widely established method for the butt-welding of steel plates. It is the most satisfactory method, both from the point of view of the economy of the joint and the resultant mechanical properties of the weld. In this process, welding is performed from right hand end and welding torch is held in right hand while filler rod held in the left hand. Torch making angle of 60° to 70° with workpieces. Filler rod is held at about 30° to the workpiece. Leftward technique is usually used on relatively thin metals, i.e., having thickness less than 5 mm. When workpiece thickness is over 3 mm bevel, the plate edges to produce a V-joint so that good root fusion may be achieved. The included angle of V-joint is 80° to 90°. This large volume weld is uneconomical in terms of time, weld metal deposited, and quantity of gases used. It may also distort weldment. Long welding time also leads to overheating of the weld area and thus, the weld metal may have coarse grain. When welding materials over 6.5 mm thick, it is difficult to attain even penetration at the bottom of the V and therefore, the quality of the weld decreases as plate thickness increases. The leftward technique requires careful manipulation to guard against excessive melting of the base metal and filler metal. The influence of the base metal on the properties of the weld metal can be very deep. Another disadvantage of leftward technique is that the look of the joint edges is interrupted and it is necessary to remove the end of the rod (which slows down the welding process). This itself results in the oxides formed on the tip of the rod being deposited into the weld pool when the weld recommences. Parameters are shown in Table 20.4 and joints details in Figure 20.15.

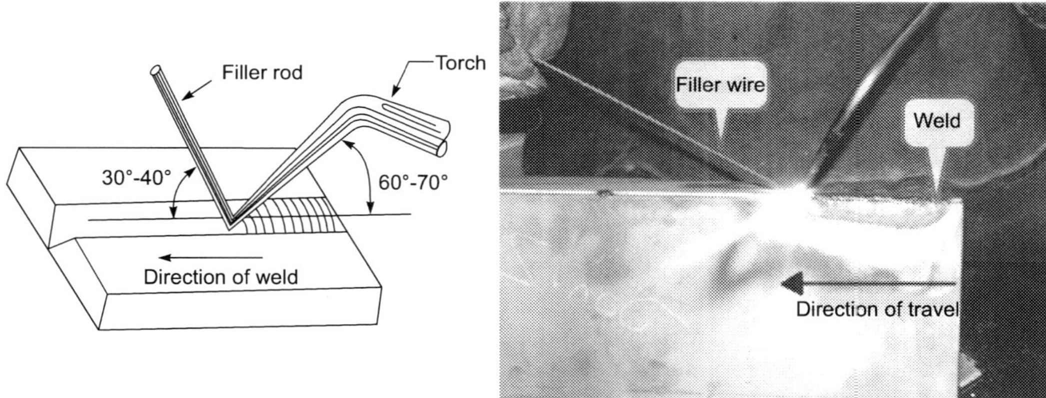

Figure 20.14 Leftward or forward welding

TABLE 20.4 Leftward or forward welding parameters

Metal Thickness in mm	C.C.M.S. Filler Rod Diameter in mm	Blowpipe Nozzle Size	Edge Preparation	Root Gap in mm	Flux to be Used
0.8	1.6	1	Flange	Nil	For gas welding of mild steel, no flux is required.
1.6 to 2.0	1.6	3	Square	2	
2.5	2	5	Square	2	
3.15	2.5	7	Square	3	
4.0	3,15	7	80° Vee	3	
5.0	3.15	13	80° Vee	3	

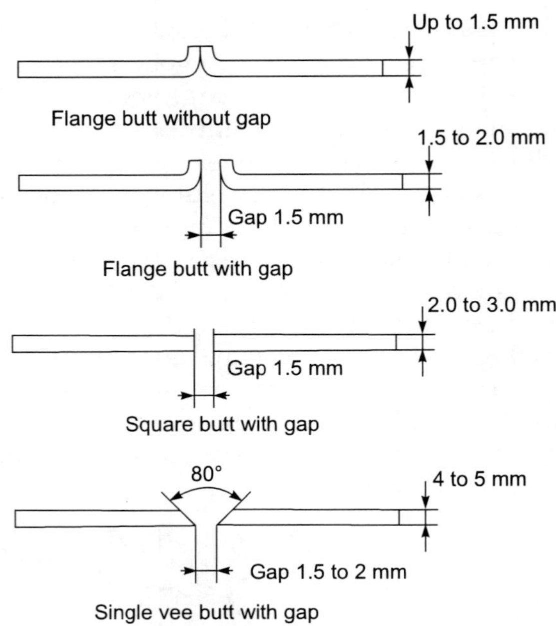

Figure 20.15 Joints for leftward or forward welding

Rightward or Backward Welding (Figure 20.16)

The welder holds the welding torch in the right hand and the filler rod in the left. Welding begins at the left hand end of the joint and precedes towards the right, hence the name rightward technique. The direction of welding is opposite to that when employing the leftward technique. The torch flame in rightward technique is directed towards the completed weld and the filler rod remains between the flame and the completed weld section.

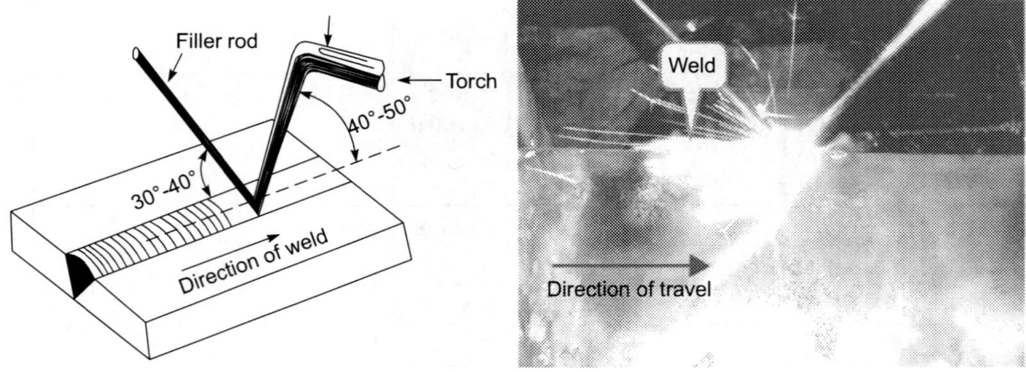

Figure 20.16 Joints for rightward or backward welding

Since the flame is constantly directed towards the V ahead of the weld puddle, no side-wise motion of the welding torch is necessary. As a result, a narrower V groove can be utilized than in leftward welding. This provides a greater control and reduced welding cost.

During welding, the filler rod can be moved in circles (within the puddle) or semi-circles (back and forth around the puddle). In rightward welding, the weld puddle is less fluid and this results in a slightly different appearance of the weld surface. The ripples are heavier and spaced further apart. The rightward techniques are used on heavier or thicker (above 5 mm) base metals. Weld with penetration of approximately 12 mm can be achieved in a single pass. Parameters are given in Table 20.4 and joint details in Figure 20.17.

TABLE 20.5 Leftward or forward welding parameters

Metal Thickness in mm	C.C.M.S. Filler Rod Diameter in mm	Blowpipe Nozzle Size	Edge Preparation	Root Gap in mm	Flux to be Used
5	3.15	10	Square	2.5	For gas welding of mild steel, no flux is required.
6.3	4.0	13	Square	3.0	
8	5.0	18	60° Vee	3.0	
10 to 16	6.3	18	60° Vee	4.0	
Above 16	6.3	25	60° double Vee	3.0	

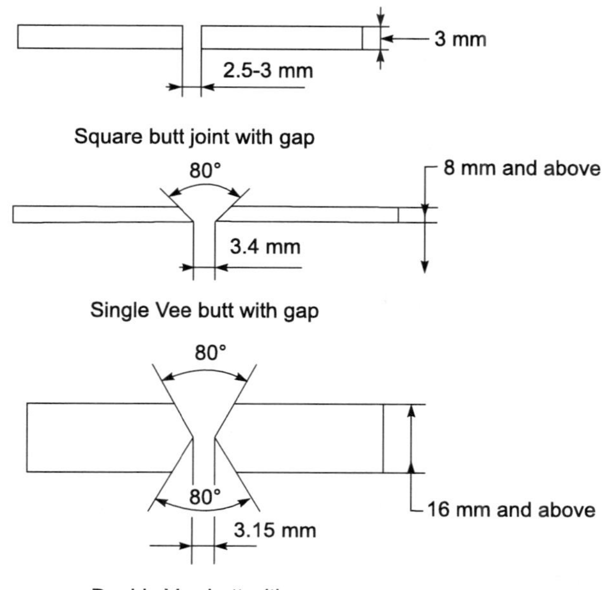

Figure 20.17 Joints for backward or rightward welding

Welding rate in relation to plate thickness is given in Table 20.6.

TABLE 20.6 Welding rate for left and right ward method

Welding Rate, m/h	Plate Thickness	
	Leftward Method, mm	Rightward Method, mm
6.1 to 7.6	0.8, 2.4	–
7.6 to 9.1	1.6	–
5.5 to 6.1	3.2	–
4.6 to 5.5	4.0	–
3.7 to 4.6	4.8	4.8
3.0 to 3.7	–	6.4
2.1 to 2.4	–	7.9
1.8 to 2.1	–	9.5
1.4 to 1.5	–	12.7
1.1 to 1.3	–	15.9
0.9 to 1.0	–	19.1
0.6 to 0.7	–	25.4

Advantages of Rightward Welding Over Leftward Welding

1. Consumption of filler rod is less. Welds are made faster.
2. Less chance for expansion or contraction.
3. Flame anneals the weld portion; therefore the mechanical properties are superior.
4. Better control of weld as the molten pool can be viewed clearly.
5. Less risk of excessive oxidation as the reducing flame covers the weld.
6. For welding of bigger thickness, where bevelling of plate edge becomes necessary, included angle can be 60° only.

7. Less quantity of deposited metal as compared to left-hand welding.
8. Better quality of welding as compared to left-hand techniques.

20.6 ADVANTAGES OF GAS WELDING

1. It is one of the most versatile methods of welding. It can be used for preheating, postheating, welding, braze welding, and torch brazing and it can be converted for oxygen cutting.
2. Control a welder can exercise over the rate of heat input, the temperature of weld zone and the oxidizing or reducing potential of the welding atmosphere. If the rate of the heat input from the flame is properly coordinated with the speed of welding, it is possible to control the size and viscosity of the weld puddle.
3. Weld bead size and shape, and weld puddle viscosity are also controlled in the welding processes because the filler metal is added independently of the welding heat source.
4. Oxyfuel welding is ideally suited to the welding of thin sheet, tubes, and small diameter pipe. It is also used for repair welding. Thick section welds, except for repair work is not economical.
5. Rate of heat generation is very less as compared to arc welding processes.
6. As the sources of heat and filler metal are separate, the welder has control over filler metal deposition rates. Heat can be applied preferentially to the base metal or the filler metal.
7. Welding equipment can be transported from one place to another easily and the cost of the equipment is low.
8. HAZ is very narrow in gas welding.
9. Maintenance cost of gas welding equipment is less.
10. There is no need of electrical power supply.
11. Equipment cost is not so high as compared to other welding processes.
12. With proper technique, practically all metals can be welded.

20.7 LIMITATIONS OF GAS WELDING

1. Heavy sections cannot be joined efficiently.
2. Proper penetration cannot be achieved.
3. All types of metals cannot be joined using gas welding.
4. Flux used in certain welding processes produces fumes that are harmful to eyes, nose, throat and lungs.

5. More safety is recommended.
6. Acetylene and oxygen gases are rather expensive.
7. Prolonged heating of the joint results in large heat affected zone.
8. Proper storage is needed for acetylene cylinder.
9. Refractory metals, such as ceramics, tungsten, etc., cannot be welded.

20.8 APPLICATIONS OF GAS WELDING

1. The oxyacetylene torch can be applied to three basic types of welding: fusion welding, braze welding and brazing.
2. The oxyacetylene fusion welding process can be used to weld various ferrous metals such as cast iron, various varieties of steels, stainless steel and non-ferrous metals such as aluminum, copper, nickel, magnesium, lead and other alloys.
3. Neutral flame is generally employed for fusion welding of ferrous metals, in which the secondary combustion of the flame protects the molten puddle sufficiently.
4. In fusion welding of cast iron and chromium-nickel steel castings, a flux is used to help remove oxides from the molten puddle.
5. The oxyacetylene fusion welding process is applicable for a wide range of ferrous metals and is used more for repairs than as the production welding procedure.
6. In fusion of welding aluminum, care has to be taken for three basic things:
   (a) Aluminum does not give any indication of change in temperature as the ferrous material does. Its color does not change until all at once it suddenly reaches its melting point and collapses. Therefore, welder has to use some method, such as carpenter's blue chalk, rubbing a pine stick, striking the cold metal with a hammer, or using commercially available temperature measuring devices to determine the temperature of aluminum.
   (b) Aluminum has extreme hot shortness, i.e., it is weak at elevated temperatures. It is highly thermally conductive, which means heat travels through it quickly and spreads rapidly through the mass of the aluminum. Therefore, welding jig has to be used, which acts as source to absorb some of the heat of the base metal. It also prevents the metal from buckling and from possible distortion during welding. Preheating also solves the problems to a great extent.
   (c) There is a refractory oxide skin to the surface of all aluminum which has thrice the melting point of the base metal. It prevents the base metal and the filler metal from fusing properly. This oxide skin is removed by the use of fluxes consisting of sodium or potassium chloride which penetrates the oxide in metal and seperates the oxide film from the base metal.

7. The welding technique used for fusion welding of copper is same as used for the welding of steel, the only difference is the high fluidity of the molten copper. Therefore, the tip size of the oxyacetylene torch should be one or two sizes larger than the ones used for the same thickness of the steel.
8. During the welding of nickel and its alloys, pure acetylene gas should be used, as it is highly susceptible to impurities on the metal surface, which leads to embrittlement, porosity, or centre bead cracking. Also, it should be ensured that oxidizing flame is not used as it would burn out the oxidizing elements to the base metal and can cause embrittlement in the weld zone.
9. Welding of magnesium and its alloys is generally done in the similar manner as of aluminum.

20.9 OXYHYDROGEN WELDING

The temperature achieved is quite low (2400°C) compared to that in oxyacetylene welding (3200°C). As such, this is suited for welding thin sheets, low-melting alloys and in some brazing works. Flame adjustments are quite different in oxyhydrogen welding because there is no distinguishing color to guide the gas proportions. An advantage of this process is that no oxide is formed on the surface of the weld if the reducing atmosphere is used.

With the pressure in a full cylinder at least 17 bars and full oxygen cylinder at 150 bars, a way is needed to lower these cylinder pressures to desired working pressures for use in the torch. This is accomplished by using an adjustable pressure regulator. The regulator also maintains a steady working pressure drops from use.

20.10 AIR ACETYLENE WELDING

It is generally used for lead welding, low-temperature brazing and soldering operations. The torch used here is similar to Bunsen burner in constructions. This process uses the lowest temperature amongst the gas welding processes. Acetylene is supplied to the torch from the cylinder and air is drawn to the torch from the atmosphere. Air inlet to the torch can be adjusted to control the volume of air to the torch.

20.11 OXYFLAME CUTTING (FIGURE 20.18)

Oxyflame cutting is an efficient method of cutting steel products. It can only be successfully employed on materials that have a lower ignition temperature than their melting point, e.g., carbon steels or low alloy steels. Materials such as aluminum and stainless steel, cannot be successfully cut because their oxide layer melts at a higher temperature than their melting point.

Principles of the Process

The process of oxyflame cutting makes use of the burning reaction between heated iron and oxygen. When iron is heated to above 815 °C (called the ignition temperature), it readily combines with oxygen. The resulting reaction produces iron oxide. This reaction (combustion) also gives off extra heat, which keeps the process of oxidation going. The important point to note is that the reaction occurs at a lower temperature than the melting point of steel (approximately 1500°C). The molten iron oxide together with some free iron, which runs off as molten slag, is removed by the kinetic energy of an introduced jet of oxygen, thus exposing more preheated iron and iron oxide. The properties of the commonly used fuel gases is given in Table 20.7.

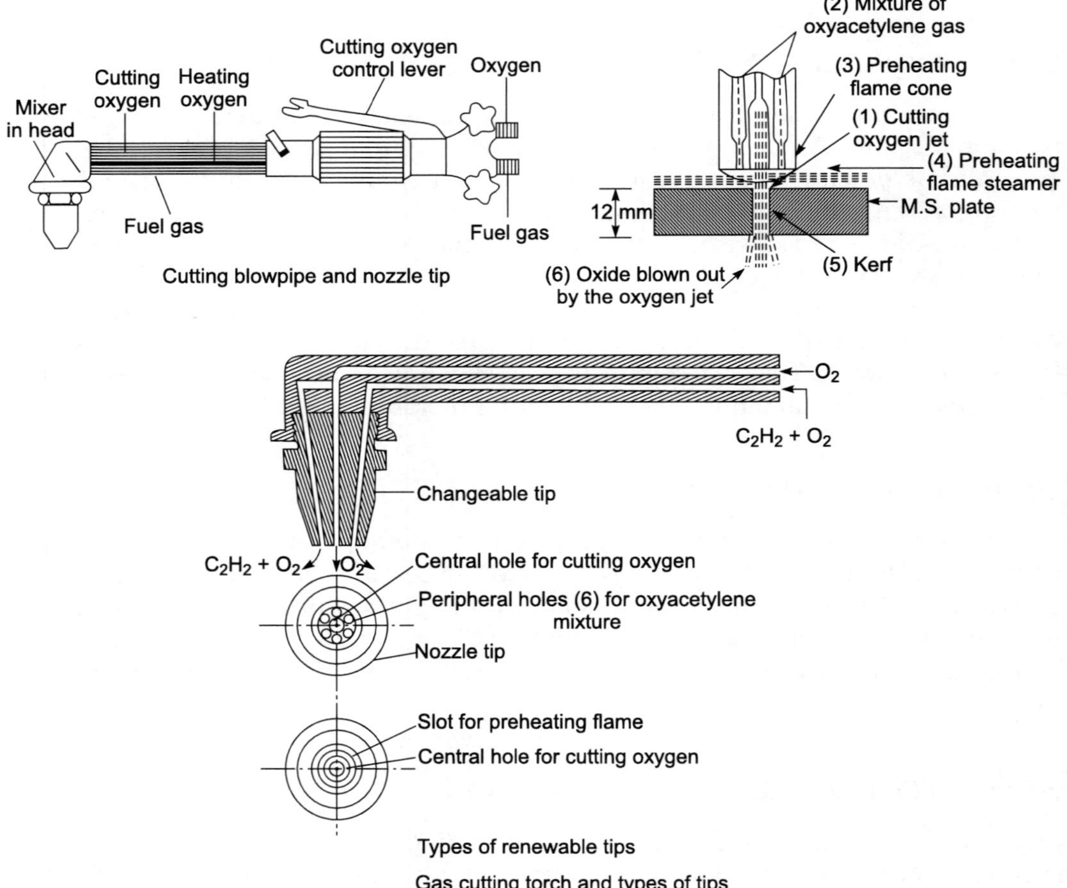

Figure 20.18 Oxyflame cutting

TABLE 20.7 Characteristics of various gases

Characteristics	Oxygen	Acetylene	Fuel Gases		
			Fuel Gases (Propane)	Natural Gas (Methane)	Hydrogen
Chemical term	O_2	C_2H_2	C_2H_3	CH_4	H_2
Production	Air liquidation fract. distillation	Carbide gasification chemical-processes	Natural oil distillation	Natural location	Chlor-alkali-electroysis, electrolysis of water
Use	Oxy-gas welding, steel production, chem. industry, medicine, rocketry	Oxy-gas welding chem, industry synthetic material, fertilizer	Fuel gas, household, industry heating, oxy-gas welding	Household, industry heating, oxy-gas welding	Welding of aluminum and load, flame-cutting of higher thickness, furnace brazing
Purity/ commercial	> 99.5%	> 99.0%	> 99.8%		> 99.75
Characteristics	Not toxic, tasteless odourless, colorless, not combustible, combustible aid	Not toxic, colorless, in higher doses anaesthetic, not odourless	Not toxic, odourless, high volume weight, rubber and synthetic solvent	Not toxic, odourless, colorless	Not toxic, odourless, tasteless, colorless, combustible
Impurities	N_2, Ar, H_2O	Ar, H_2O	H_2, N_2, O_2, CH_4		O_2, H_2O
Volume weight [kg/m^3] 15°C/1 bar Air: 1,209 [kg/m^3]	1.337	1.095	1.874	0.671	0.084
Boiling temperature°C	–183	–84	–43	–162	–253
Thermal value H_2 [kJ/m^3]	–	57120	93000	≈ 36000	10750
Ignition range in air % gas in O_2	–	2.4 + 80 2.4 + 93	2 + 10 2 + 55	4 + 17 4 + 60	4 + 76 4 + 95

Contd.

Min. ignition temper. with air °C with O_2	–	335 295	510 490	640 690	510 450
Ignition velocity air m/s O_2	–	1.3 13.5	0.3 0.7	0.4 3.3	2.7 8.9
Max. temperature of flame with air °C with O_2	–	2325 3200	1925 2850	1920 2750	2045 2650
Power of flame kJ/cm^2/s	–	≈45	≈ 11.0	≈ 13.0	≈ 14.0

Oxidation

Oxyflame cutting relies on the process of oxidation. Most elements have the property of combining or uniting with one another to form new substances which have chemical and physical properties entirely different from those of the two substances which entered into the combination. This new substance is called a compound and the process is known as a chemical combination or chemical reaction, e.g., iron will combine with sulphur to form iron sulphide. Oxidation is a chemical reaction in which oxygen combines with another element to form an oxide, e.g., iron + oxygen = iron oxide. When steel (an alloy of iron) is exposed to the air (nitrogen and oxygen) at room temperature and moisture is present, a film of oxides (rust) will form on the surface. Heat accelerates this process and when steel is heated to about 815°C (or above) and is not protected from the atmosphere, rapid oxidation of the surface of the metal occurs, i.e., oxygen combines with the iron, an element in the steel, to form oxides. In welding, this oxidation results in the weld becoming porous, hard and brittle. Therefore, during welding, care has to be taken to prevent oxygen coming into contact with the metal. Ignition ranges of several fuel gases with mixture of air and oxygen is given in Figure 20.19.

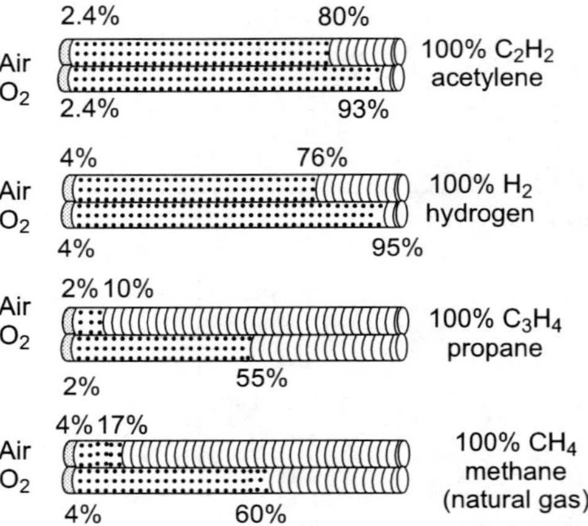

Figure 20.19 Ignition ranges of several fuel gases

Equipment

Oxyflame cutting equipment, in its simplest form, requires a gas supply consisting of cylinders of oxygen and acetylene, regulators for both gases, flashback arrestors, correct couplings and hoses, a blowpipe, a cutting blowpipe attachment and cutting nozzles.

Adjusting pressures when a cutting attachment is fitted to a multi-purpose blowpipe, the blowpipe oxygen valve should always be kept fully open. Oxygen flow is then controlled via the preheat oxygen valve, and the cutting lever of the cutting attachment. Acetylene should be set to a 100 kPa maximum with the acetylene blowpipe valve open approximately one quarter of a turn. Oxygen should be set to the appropriate pressure (Table 20.8), with the cutting lever fully depressed. Use your pressure chart as a guide. Incorrect pressures are wasteful, cause flashbacks, and can be dangerous.

TABLE 20.8 Pressures for cutting

Thickness of Plate (mm)	Size of Nozzle	Oxygen Pressure (kPa)	Acetylene Pressure (kPa)
3	8	100	100
6	8	180	100
12	12	200	100
20	12	235	100
25	15	180	100
40	15	300	100
50	15	350	100

Application and Limitation

Oxyacetylene can only cut low- to medium-carbon steel and wrought iron. High-carbon steels cannot be cut because melting point is very close to the temperature of the flame and so the slag from the cutting action does not eject as sparks, but rather mixes with the clean melt near the cut. This keeps the oxygen from reaching the clean metal and burning it. In the case of cast iron, graphite between the grains and the shape of the grains themselves interfere with the cutting action of torch.

Safety in Handling Acetylene Gas

The gas employed in gas welding may form explosive mixture if incorrectly handled and burning of gasses must be controlled within the recommended limits for the safety of the welder and the surrounding property. The following precautions on safety rules should be taken to make the welding safer:

1. Only approved apparatus is to be used (torches, valves, regulators, etc.).
2. Acetylene should not be generated greater than 1 bar pressure.
3. Cylinder should be as per standard specifications and should be provided with all the accessories. They should be kept away from sources of heat and should be stored in well ventilated dry location. They should not be kept in cupboards and lockers.
4. They should be protected from excessive rise in temperature.
5. Valve protection caps should be placed when the cylinder is not in use.
6. Oxygen cylinders should not be placed near to fuel gas cylinders.
7. A platform or a cradle should be used when transporting cylinders by crane.
8. Cylinders should not be drawn or struck with each other, slings or magnets should not be used to carry. They should not be lifted using valve protection caps. They can be rolled on one edge.
9. Cylinder should not be placed in places where they become part of the electric circuit.
10. Safety devices in cylinders should not be tampered with.
11. Fuel gas should never be used without reducing the pressure.
12. Oxygen cylinder and accessories should be kept away from oil or grease substances, as these might ignite violently in presence of oxygen under pressure.
13. Regulators and valves should be used only for the gas and at pressures for which they are intended.
14. The hoses should be of good quality rubber and as per the standards.
15. Hoses should not be rimmed and should not be bent to control the pressure of the gas.

Worked Examples

Example 1: During gas welding of copper sheets a neutral flame was used consuming 8 liters of acetylene but the weld was found to be defective. So it was rewelded using oxidizing flame. What is the likely consumption of oxygen in liters? **(GATE 1993)**

Solution: In oxidizing flame, ratio of oxygen to acetylene is 1.15 to 1.5. Oxygen consumption for 8 liters of acetylene is 9.2 to 12 liters.

Example 2: In oxyacetylene, arc, and laser beam cutting, the processes basically involve melting of the workpiece. If an 80 mm diameter hole is to be cut from a 250 mm diameter, 12 mm thick plate, plot the mean temperature rise in the plate as a function of kerf. Assume that one half of the energy goes into the plate and one half goes into the blank. *(Manufacturing Process:* **Kalpakjian***)*

Solution: The volume melted is given by

$$V = \pi Dth = \pi \times 80\ (\text{mm}) \times 12\ (\text{mm}) \times t = 30106t$$

where, t is the kerf width in mm.

The energy input,

$$E = \frac{\text{Specific energy required to melt} \times \text{volume melted}}{2}$$

Energy is divided by 2, because half of energy goes into the blank,

$$E = \frac{u \times 30106t}{2} = 1508ut$$

The volume of blank,

$$V = \frac{\pi d^2 h}{4} = \frac{\pi(250^2 - 80^2) \times 12}{4} = 5.29 \times 10^{-4}\ \text{m}^3$$

The temperature rise in the blank,

$$\Delta T = \frac{E}{\rho V C_p}$$

$$\Delta T = \frac{1508ut}{\rho \times 5.29 \times 10^{-4} \times C_p} = 2.86 \times 10^6 \times (t)\left(\frac{u}{\rho C_p}\right)$$

It can be seen that temperature rise is a function of t and it is plotted for some material as given in Figure 20.20.

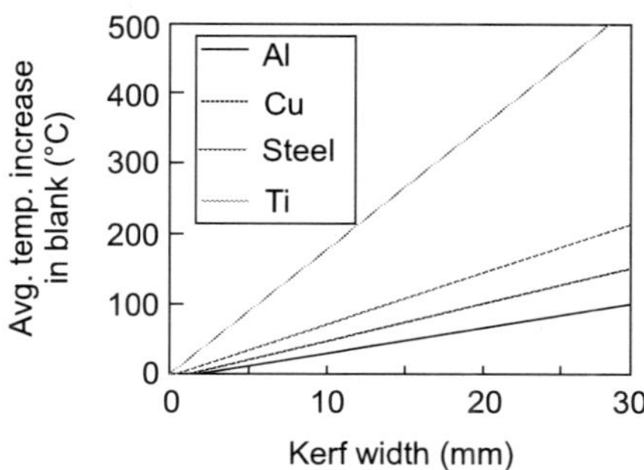

Figure 20.20

Example 3: An oxyacetylene torch supplies 0.3 m^3 of acetylene per hour and an equal volume rate of oxygen for an oxyacetylene welding (OAW) operation on 4.5

mm steel. Heat generated by combustion is transferred to the work surface with an efficiency of 0.25 and 75% of heat from the flame is concentrated in a circular area on the work surface that is 9 mm in diameter. Find (a) rate of heat liberated during combustion, (b) rate of heat transferred to the work surface, and (c) average power density in circular area.

Solution: Refer to Figure 20.21.

(a) The rate of heat generated by the torch is the product of the volume rate of acetylene times the heat of combustion

$$HR = \frac{(0.3\ \text{m}^3/\text{hour})(55 \times 10^6\ \text{J/m}^3)}{3600} = 4583\ \text{joules/second}$$

(b) With a heat transfer efficiency $= f_1 = 0.25$ and the rate of heat received at the work surface,

$$HR \times f_1 = 4583 \frac{\text{Joules}}{\text{second}} \times 0.25 = 1146\ \text{joules/second}$$

(c) The area of the circle in which 75 % of the heat flame is concentrated is

$$A = \frac{\pi \times 9^2}{4} = 63.6\ \text{mm}^3$$

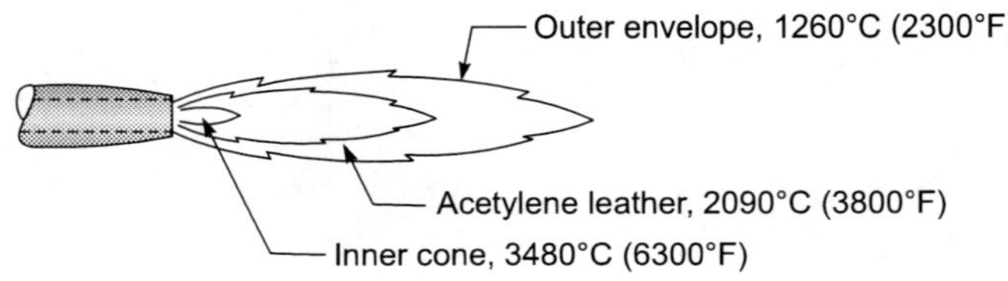

Figure 20.21

The power density in the circle is found by dividing the available heat by the area of the circle

$$\text{Power density, } PD = \frac{0.75 \times 1146}{63.6} = 13.5\ \text{W/mm}^2$$

Example 4: Suppose in above Example 3, the fuel used in the welding operation is MAPP (Methylacetylene-propadiene, developed by Dow Chemicals, USA) (C_3H_4) instead of acetylene, and the proportion of heat concentrated in the 9 mm circle is 60% instead of 75 %. Compute (a) rate of heat liberated during combustion, (b) rate of heat transferred to the work surface, and (c) average power density in the circular area.

Solution:

(a) The rate of heat generated by the torch is the product of the volume rate of MAPP times the heat of combustion,

$$HR = \frac{(0.3\ \text{m}^3/\text{hour})(97 \times 10^6\ \text{J/m}^3)}{3600} = 7642\ \text{joules/second}$$

(b) With a heat transfer efficiency, $f_1 = 0.25$ and the rate of heat received at the work surface,

$$HR \times f_1 = 7642 \frac{\text{Joules}}{\text{Second}} \times 0.25 = 1910\ \text{joules/second}$$

(c) The area of the circle in which 75 % of the heat flame is concentrated is

$$A = \frac{\pi \times 9^2}{4} = 63.6\ \text{mm}^3$$

The power density in the circle is found by dividing the available heat by the area of the circle,

$$PD = \frac{0.60 \times 1910}{63.6} = 18.5\ \text{W/mm}^2$$

MAPP is the fuel that competes most closely with acetylene in burning temperature and heating value is methyacetylene–propadiene. It is the fuel developed by DOW chemical company and sold under trade name MAPP. It can be stored under pressure as liquid, thus avoiding the storage problems associated with acetylene.

Example 5: During gas welding of copper sheets a neutral flame was used consuming 8 litres of acetylene. But the weld was found to be defective. So, it was rewelded using an oxidizing flame. The likely consumption of oxygen in litres will be ____________. **(GATE 1993)**

(a) 6 (b) 8 (c) 10 (d) 16

Solution: Oxidizing flame, $O_2 : C_2H_2 = 1.5 : 1.15$

For 8 L of acetylene

$$O_2 = \frac{1.5}{1.15} \times 8 = 10.4 \cong 10\ \text{L}$$

so correct answer is C.

Review Questions

1. Explain the principle of operation of oxyacetylene welding process.
2. What equipment are used in oxyacetylene welding process?
3. Explain the chemistry involved in the oxyacetylene flame.
4. What are the safety precautions to be followed for acetylene cylinder?
5. What are the disadvantages of acetylene generator?
6. Compare single- and two-stage regulators.
7. What are the applications of gas welding?
8. What are the advantages of gas welding?
9. What are the limitations of gas welding?
10. Explain gas cutting operation with a neat sketch.

21

CHAPTER

Brazing, Braze Welding and Soldering

21.0 INTRODUCTION

Brazing is a group of welding processes which produce coalescence of materials by heating to a suitable temperature and using a filler metal having liquidus above 449°C and below the solidus of the base metal. The filler metal is distributed between the closely fitted surfaces of the joint by capillary attraction. Brazing is different from soldering. Soldering employs a filler metal having a liquidus below 449°C.

Brazing and soldering both are solid/liquid processes, and primarily involve three steps (Figure 21.1).

1. Heating of plates to be joined using suitable heat source.
2. Placing and melting of solder or brazing material, followed by heating to molten state.
3. Filling of molten filler metal between the faying surfaces of the components to be joined by capillary action.

Brazing must meet each of three criteria:

(a) The parts must be joined without melting the base metals.
(b) The filler metal must have a liquidus temperature above 449°C.
(c) The filler metal must wet the base metal surfaces and be drawn onto or held in the joint by capillary attraction.

Brazing is not the same as braze welding, which uses a brazing filler metal that is melted and deposited in fillers and grooves exactly at the points it is to be used. Brazing filler metal is also distributed by capillary action. Limited base metal fusion may occur in braze welding.

To achieve a good joint using any of the various brazing processes, the parts must be properly cleaned and protected by either flux or the atmosphere during heating to prevent excessive oxidation. The parts must provide a capillary for the filler metal when properly aligned, and the heating process should provide proper brazing temperatures and heat distribution.

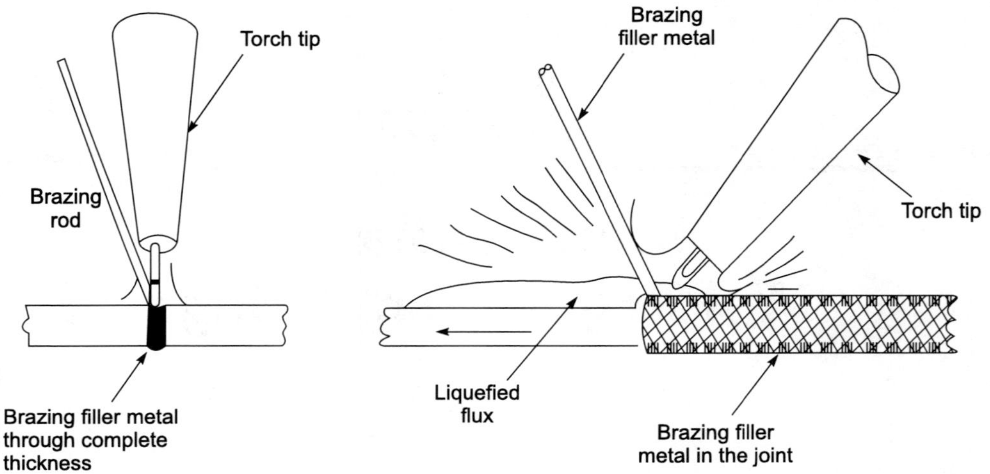

Figure 21.1 Brazing steps

21.1 BRAZING PROCESS

1. Capillary flow is the most important physical principle which ensures good brazements, providing to molten filler metal both adjoining surfaces.
2. The joint must also be properly spaced to permit efficient capillary action and resulting coalescence. More specifically, capillarity is a result of surface tension between base metal(s), filler metal, flux or atmosphere, and the contact angle between base and filler metals. In practice, brazing filler metal flow characteristics are also influenced by considerations involving fluidity, viscosity, vapor pressure, gravity, and by the effects of any metallurgical reactions between the filler and base metals.
3. The brazed joint, in general, is one of a relatively large area and very small thickness. In the simplest application of the process, the surfaces to be joined are cleaned to remove contaminants and oxide. Next, they are coated with a flux or a material capable of dissolving solid metal oxides present and preventing new oxidation. The joint area is then heated until the flux melts and cleans the base metals, which are protected against further oxidation by the liquid flux layer.
4. Brazing filler metal is then melted at some point on the surface of the joint area. Capillary attraction is much higher between the base and filler metals than that between the base metal and flux. Therefore, the flux is removed by the filler metal. The joint, upon cooling to room temperature, will be filled with solid filled metal. The solid flux will be found on the joint surface.
5. High fluidity is a desirable characteristic of brazing filler metal because capillary attraction may be insufficient to cause a viscous filler metal to run into the tight fitting joints.

6. Brazing is sometimes done with an active gas, such as hydrogen, or in an inert gas or vacuum. Atmosphere brazing eliminates the necessity for post-cleaning and ensures absence of corrosive mineral flux residue. Carbon steels, stainless steels, and superalloy components are widely processed in the atmosphere of reacted gases, dry hydrogen, dissociated ammonia, argon, and vacuum. Large vacuum furnace is used to braze zirconium, titanium, stainless steels, and the refractory metals. With good processing procedures, aluminum alloys can also be vacuum furnace brazed with excellent results.
7. Brazing is preferred for making high-strength metallurgical bonds, and preserving base metal properties because it is economical.

Steps in Brazing Operation

1. Surfaces to be joined are thoroughly cleaned with carbon tetrachloride or trichloroethylene to remove oil, greece, etc. Oxides and scratches can be removed by pickling treatment, followed by hot water rinsing.
2. The joint is coated with flux.
3. The workpiece is then assembled and supported.
4. The fluxed parts are preheated and brought to the proper brazing temperature prior to the application of brazing alloys.
5. Brazing filler metal is added and allowed to solidify.
6. The torch is cleaned usually by hot water or steam to remove the residual fluxes.

21.2 TYPES OF BRAZING PROCESSES

Generally, brazing processes are specified according to the heating methods (sources) of industrial significance. Whatever the process used, the filler metal has a melting point above 840°F (450°C) but below the base metal and distributed in the joint by capillary attraction. The brazing processes are:

(a) Torch brazing
(b) Furnace brazing
(c) Induction brazing
(d) Resistance brazing
(e) Dip brazing
(f) Infrared brazing

21.2.1 Torch Brazing (Figure 21.2)

Torch brazing is performed by heating with a gas torch of a proper composition, and appropriate flux. This depends on the temperature and heat amount required. The fuel gas (acetylene, propane, city gas, etc.) may be burned with air, compressed air, or oxygen.

Brazing filler metal may be preplaced at the joint in the form of rings, washers, strips, slugs, or powder or it may be fed from handled filler metal in wire or rod form. In any case, proper cleaning and fluxing are essential.

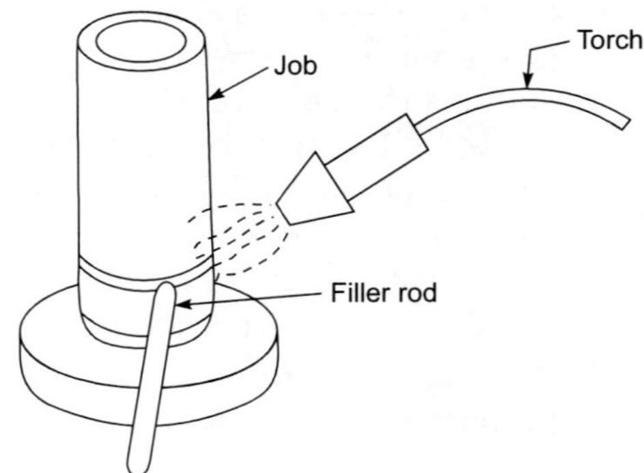

Figure 21.2 Torch brazing

For manual torch brazing, the torch may be equipped with a single tip, either single or multiple flames. Manual torch brazing is particularly useful on assemblies involving sections of unequal mass. Welding machine operations can be set up where the production rate allows, using one or several torches equipped with single or multiple flame tips. The machine may be designed to move either the work or torches, or both.

21.2.2 Furnace Brazing (Figures 21.3 and 21.4)

Furnace brazing is used extensively where the parts to be brazed can be assembled with the brazing filler material in the form of wire, foil, fillings, slug, powder, paste or tape. This process is particularly applicable for high production brazing. Fluxing is employed except when an atmosphere is specifically introduced in the furnace to perform the same function. Most of the high production brazing is done in a reducing gas atmosphere, such as hydrogen and combusted gases that are either exothermic (formed with heat evolution) or endothermic (formed with heat absorption). Pure inert gases, such as argon or helium, are used to obtain special atmospheric properties.

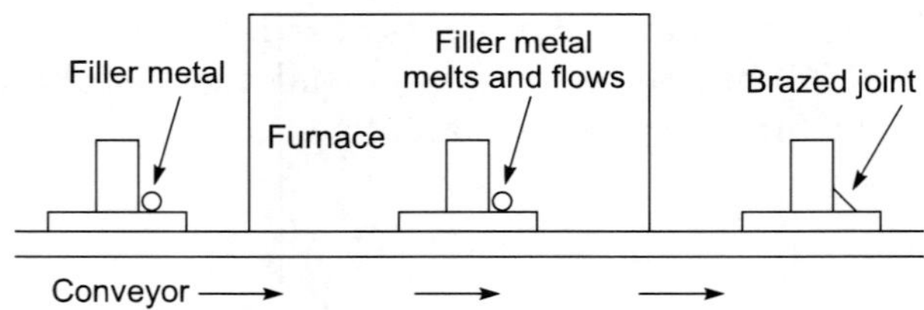

Figure 21.3 Furnace brazing process

A large volume of furnace brazing is performed in vacuum, which prevents oxidation and often eliminates the need for flux. Vaccum brazing is widely used in aerospace and nuclear fields, where reactive metals are joined or where entrapped fluxes would be intolerable. If the vaccum is maintained by continuous pumping, it will remove volatile constituents liberated during brazing. The types of furnaces generally used are either batch or contiguous. These furnaces are usually heated by electrical resistance elements, gas or oil, and should have automatic time and temperature controls. Cooling is sometimes accomplished by cooling chambers, which either are placed over the hot retort or are an integral part of the furnace design. Forced atmosphere injection is another method of cooling. Parts may be placed in the furnace singly, in batches, or on a continuous conveyor.

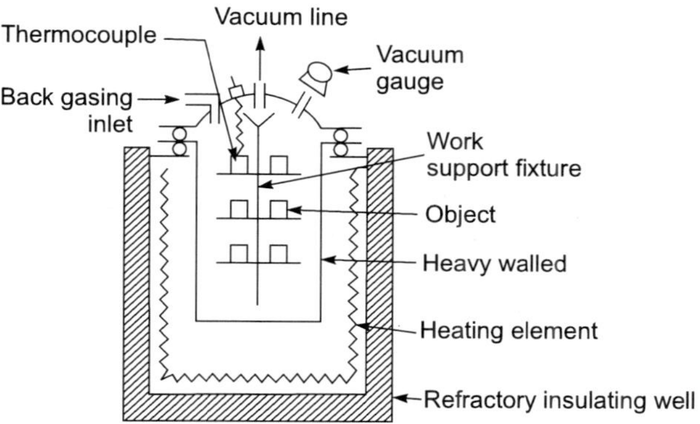

Figure 21.4 Furnace brazing details

Vacuum brazing is relatively an economical method of accurately controlled brazing atmosphere. Vacuum provides the surface cleanliness needed for good wetting and flow of filler materials without the use of fluxes. Base metals containing chromium and silicon can be easily vacuum brazed where a very pure, low dew point atmosphere gas would otherwise be required.

Advantages

1. Excellent control of brazing temperature.
2. Ideal for mass production.
3. No skilled labour is required.
4. There is uniform distribution of the brazing alloys in the joint.
5. Since the heating is uniform and not localised, there is no residual stresses induced in the joint.
6. Bright surfaces can be obtained because the parts will be free from scale because of controlled atmosphere.
7. The conveyor type furnace allows high production rate.

Limitations

1. Localised heating of the work is not possible.
2. Suitable fixtures are desired.

21.2.3 Induction Brazing (Figure 21.5)

Induction brazing involves heating the components to be brazed to the brazing temperature using induction heating. This is based on the following physical principle: if a metal object, which is to be heated, is exposed to an electromagnetic alternating field, an electric current is induced therein. The flow of current subsequently heats the metal.

In this process, the heat necessary to braze metals is obtained from a high frequency electric current through the motor-generator, resonant sparkplug, and vaccum tube oscillator. It is induced or produced without electric or magnetic contact in the parts (metals). The parts are placed in or near a water cooled coil carrying the alternate current. They do not form any part of the electrical circuit. The brazing filler metal normally is preplaced. Careful design of the joint and coil is necessary to assure that surfaces of all the metals of the joint reach the brazing temperature at the same time. Flux is employed except when an atmosphere is specially introduced to perform the same function.

The equipment consists of tongs or clamps with the electrodes attached at the end of each arm.

The tongs should preferably be water cooled to avoid overheating. The arms are current carrying conductors attached by leads to transformer. Direct current may be used but is comparatively expensive. Resistance welding machines are also used. The electrodes may be carbon, graphite, refractory materials, or copper alloys according to the required conductivity.

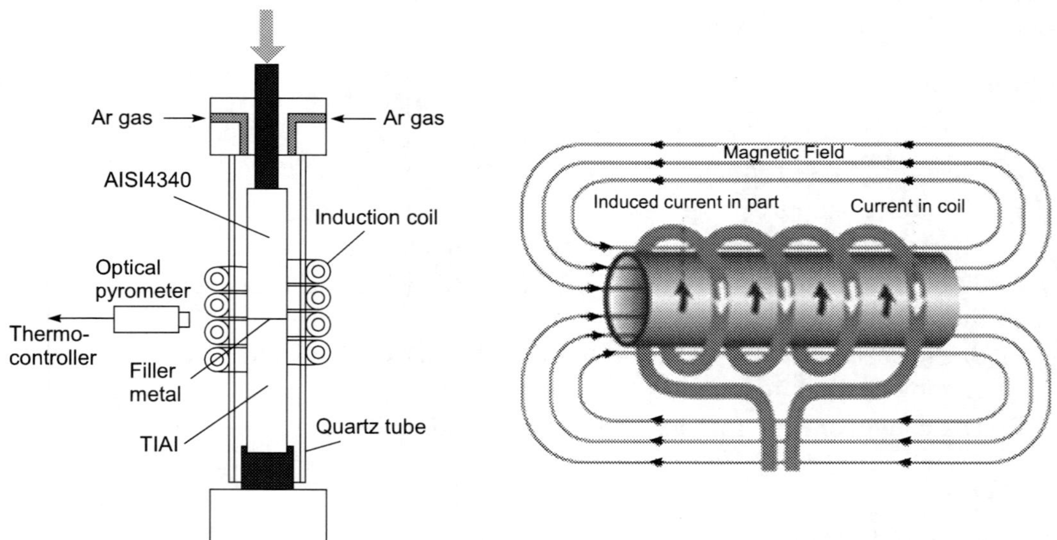

Figure 21.5 Induction brazing

Advantages

1. Rapid heating, therefore high production rate.
2. The tooling is not very expensive and uniform results are obtained.
3. Heating can be confined to a localised area. Close control of temperature can be achieved.
4. High efficiency.
5. Low maintenance.

Limitations

1. High initial cost, therefore suited only for mass production.
2. There is likely to be some oxidation.
3. There is limitation to the size of the parts.
4. The assembly should preferably have regular shape.

Applications of Induction Brazing

1. Brazing of cemented carbides (e.g., drills, circular saw blades, paper knives, milling cutters) in the tooling industry.
2. Brazing of metallic eyeglass frames, in particular made from titanium in the eyeglasses manufacturing industry.
3. Brazing of contact materials for all types of switches in the electrical industry.
4. Brazing of thin-walled tubes within the automotive industry.
5. Brazing of compressor connectors for refrigeration and air-conditioning technology.
6. Brazing of pipe connections for refrigeration and air-conditioning technology.

21.2.4 Resistance Brazing (Figure 21.6)

The heat necessary for resistance brazing is obtained from the resistance to the flow of an electric current through the electrodes and the joint to be brazed. The parts comprising the joint form a part of the electric circuit. The brazing filler metal in some convenient form is preplaced or face fed. Fluxing is done with due attention to the conductivity of the fluxes. Flux is employed except when an atmosphere is specifically introduced to perform the same function. The parts to be brazed are placed between the two electrodes, and proper pressure and current are applied. The pressure has to be maintained until the joint has solidified. In some cases, both electrodes are placed on the same side of the joint with suitable backing to maintain the required pressure.

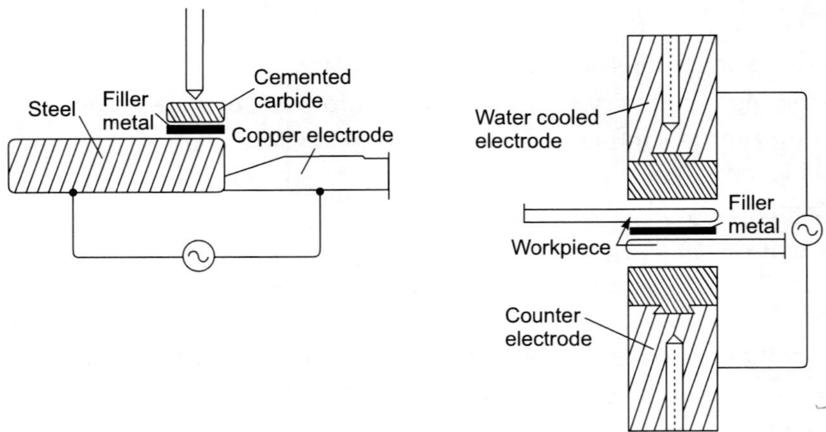

Figure 21.6 Furnace brazing details

Advantages

1. Rapid heating.
2. Localised heating.
3. Moderate production rate.
4. Reduced fire hazards.
5. Very effective for brazing long seams.

Limitations

1. Requires individual heating of parts.
2. There is likely to be some distortions.
3. Size of the assembly that can be brazed is limited.

21.2.5 DIP BRAZING (FIGURE 21.7)

There are two methods of dip brazing:

1. Chemical bath dip brazing
2. Molten metal bath dip brazing

In **chemical bath dip brazing,** the brazing filler metal, in suitable form, is preplaced and the assembly is immersed in a bath of molten salt. The salt bath furnishes the heat necessary for brazing and usually provides necessary protection from oxidation; if not, a suitable flux should be used. The salt bath is contained in a metal or other suitable pot, also called the furnace, which is heated from the outside through the wall of the pot, by means of electrical resistance units placed in the bath, or by the I^2R loss in the bath itself.

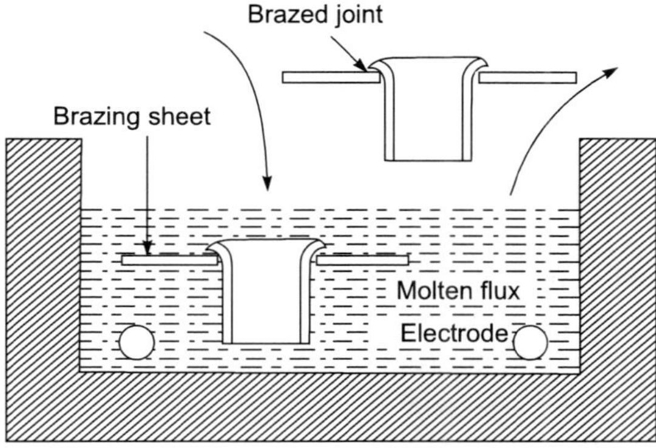

Figure 21.7 Dip brazing

Advantages

1. Rapid heating.
2. Easy control of temperature.
3. Protection from oxidation.
4. Flux is not required for most of the operation.

Limitations

1. Undesirable carryover of salt.
2. Cleaning of work pieces material is needed to remove salt.
3. Chance of some oxidation during cooling.
4. Danger of explosion from wet work.
5. Fixtures are desired.

In **molten metal bath dip brazing,** the parts are immersed in a bath of a molten brazing filler metal contained in a suitable pot. The parts must be cleaned and fluxed if necessary. A cover of flux should be maintained over the molten bath to protect it from oxidation. This method is largely confined to brazing small parts, such as wires or narrow strips of metal. The end of wires or parts must be held firmly together when they are removed from the bath until the brazing filler metal has fully solidified.

Advantages

1. Rapid localized heating.
2. Accurate temperature control.

Limitations

1. Suitable fixtures are desired.
2. There is a danger of explosion; wet parts are dipped in the bath.
3. The duration of dipping is critical.

21.2.6 Infrared Brazing (Figure 21.8)

Infrared heat is radiant heat obtained below the red rays in the spectrum. While with very 'black' source, there is sane visible light, the principal heating is done by radiation. Heat source lamps, capable of delivering up to 5000 W of radiant energy, are commercially available. Reflectors are placed to concentrate the heat so the lamps need not follow the contour of the part.

Assemblies to be brazed are supported in a position that enables the energy to impinge on the part. In some applications, only the assembly itself is enclosed. There are, however, applications where the assembly and the lamps are placed in a bell jar or retort that can be evacuated, or in which an inert gas atmosphere can be maintained. The assembly is then heated to a controlled temperature, as indicated by thermocouple. The part is moved to the cooling plates after brazing.

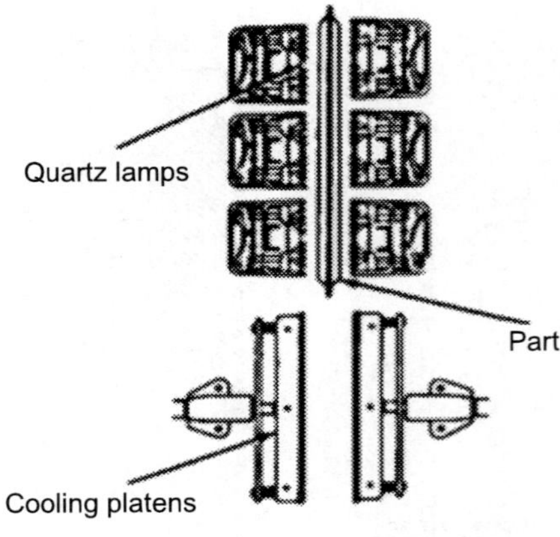

Figure 21.8 Infrared brazing

21.3 BRAZING FLUXES

Main functions of brazing fluxes are:

1. To prevent oxidation of the base metal.
2. To dissolve oxides that are formed during heating.
3. To lower the surfaces tension of the molten brazing metal, thus promoting its flow into the joint.

Surfaces should be cleaned before the flux is applied. Because the melting point and chemical composition of various oxides are different, no one flux will be satisfactory for all uses. Therefore, flux must be selected not only for the filler metal used but also for the metal to be joined and for the brazing method to be employed.

Brazing fluxes are generally classified as:

1. **High-temperature fluxes** are intended for use with copper, brass brazing alloys at a temperature of 1500°F. The main ingredients are borax and boric acid.
2. **General-purpose low-temperature fluxes** usually melt at temperature below 1500°F and are used with silver brazing alloys and copper phosphorus alloys. There may be combinations of boric acid, sodium tetraborate, calcined borax, strantalum chloride, barium chloride, Zn chloride, potassium chloride, potassium and sodium phosphates.
3. **Special-purpose fluxes** depend upon the application. Fluxes are available in the form of paste, dry powder or saturated solution.

21.4 BRAZING FILLER METALS

For satisfactory use in brazing applications, brazing filler metals must possess the following properties:

1. The ability to form brazed joints possessing suitable mechanical and physical properties for intended application.
2. A melting point or melting range compatible with the base metals being joined and sufficient fluidity at brazing temperature to flow and distribute into properly prepared joints by capillary action.
3. A composition of sufficient homogeneity and stability to minimize separation of constitutes (liquation) under the brazing conditions to be encountered.
4. The ability to wet the surfaces of the base metals being joined and form a strong, sound bond.
5. Depending on the requirements, ability to produce or avoid base metal-filler metal interactions.

TABLE 21.1 Comparison among welding, soldering and brazing

S. No.	Welding	Soldering	Brazing
1.	Welding joints are strongest joints used to bear the load. Strength of the welding portion of the joint is usually more than the strength of base metal.	Soldering joints are weakest joints out of the three. Not meant to bear the load. Use to make electrical contacts generally.	Brazing joints are weaker than welding joints but stronger than soldering joints. They can be used to bear the load up to some extent.
2.	Temperature required is 3800°C in welding joints.	Temperature requirement is up to 450°C in soldering joints.	Temperature may go to 600°C in brazing joints.
3.	To join, workpieces need to be heated till their melting point.	Heating of the workpieces is not required.	Workpieces are heated below their melting point.
4.	Mechanical properties of base metal may change at the joint due to heating and cooling.	No change in mechanical properties after joining.	May change in mechanical properties of joint but it is almost negligible.
5.	Heat cost is involved and high skill level is required.	Cost involved and skill requirements are very low.	Cost involved and skill required are in between the other two.

Contd.

6	Heat treatment is generally required to eliminate undesirable effects of welding.	No heat treatment is required.	No heat treatment is required after brazing.
7	No preheating of workpiece is required before welding as it is carried out at high temperature.	Preheating of workpieces before soldering is good for making good quality joint.	Preheating is desirable to make strong joints as brazing is carried out at relatively low temperature.

21.5 ADVANTAGES AND LIMITATIONS OF BRAZING

Advantages

The uses of brazing are quite numerous. It is possible to produce at low cost, a ductile and high-strength joint which is resistant to fatigue or corrosion. Steel, copper, brass, nickel, bronze, monel CI, Al, and Mg can be brazed to themselves. Dissimilar metals and parts of these sections can be joined easily. The other advantages are:

1. The joint requires no further finishing.
2. Assemblies can be brazed in stress-free conditions.
3. Materials of different thicknesses can be joined easily.
4. There is no flash or weld spatter.
5. Process can be easily mechanized.
6. It is economical for complex assemblies as it is a simple way to join large joint areas.
7. Metallurgical properties of the base metal are not seriously affected.
8. Dissimilar metals can be joined.
9. Wrapping is eliminated because of excellent stress and heat distribution.
10. The process is quite fast.

Disadvantages

1. All the brazing alloys lose their strength at elevated temperature.
2. Color mismatch between the filler and the part being brazed and lower joint strength than that of the part material.
3. If the joint is exposed to corrosive media, the filler metal used must have the required resistance. Base metal erosion may occur. Also, the formation of brittle intermetallics or other phases can make the resulting joint too brittle to be acceptable.
4. Since the brazing processes use a molten metal to flow between the materials to be joined, there is the possibility of liquid metal interaction, which is unfavorable.
5. Highly skilled technicians are required.

21.6 APPLICATIONS OF BRAZING

Three major industries using brazing are the electrical industry, the utensil manufacturing industry, and the maintenance industry.

In the maintenance field, brazing is widely used for repairing cast iron components.

21.7 COMPARISON BETWEEN BRAZING AND WELDING

Welding and brazing are two types of processes that are used for joining different metal parts. These two methods are also used to fix or join broken parts or to fill the gaps in metals. Though these two processes have the same use, they have different mechanisms. They have differences in the use of temperature and the melting of filler and base metals.

Brazing is similar to soldering. For brazing, the parts of the two metals that are to be joined should be free from oxides. In brazing, the metals that have to be joined together are not heated to their melting points, but only the filler metal is heated just above the melting point.

A difference that can be seen between brazing and welding is in the temperatures. In welding, high temperatures are needed. But in brazing, the temperature is a bit lower than that of the temperature used in welding.

1. Welding and brazing are two types of processes that are used for joining different metal parts, fix broken parts, and fill the gaps in metals.
2. Welding is a process by which the base metal as well as the filler metal is melted, and each forms a molten material or a weld pool. This weld pool solidifies to make a strong joint.
3. Brazing is similar to soldering. The filler metal is melted in-between the parts that have to be joined. The wetting that is formed in between the joints gets solidified and gives the joint more strength.

The decision to weld or to braze comes down to the specifics of the application. Here are the factors to consider:

Size of the Assembly: If you want to join large assemblies, welding is more appropriate. The tendency of larger assemblies to disperse heat can be a stumbling block for brazing, making it hard to reach the temperatures required for the filler metal to flow. In contrast, the concentrated heat of welding and its ability to trace a joint allow it to excel in joining big assemblies.

Thickness: Both methods are sound choices when metal sections are on the thick side, around 0.5 inches or more. Brazing has the edge, however, when it comes to thin sections. The high heat of welding can warp or burn through such sections. In contrast, brazing can help you avoid distortion.

Shape of Joint: While both welding and brazing can create spot joints, the localized heat of welding offers the advantage of speed and low cost. However, when it comes to linear joints, the manual tracing required by welding makes it the less convenient choice. Additionally, brazing can easily draw the filler metal into straight, curved or irregular joint configurations.

Types of Materials: Brazing soundly beats welding when joining dissimilar metals. As long as the filler material is metallurgically compatible with both base metals and melts at a lower temperature, brazing can create strong joints with barely any alteration of the base metals' properties. In contrast, base materials are melted during welding, so joining two dissimilar metals using this method can entail complex and costly techniques.

Production Volume: In manual jobs, the factors—size, thickness, shape of joint and materials—guide our decisions, but when part volumes are in the hundreds or thousands, production techniques and cost are the most important considerations. While both brazing and welding can be automated, brazing allows for more degrees of automation. With welding, we are usually forced to choose between two extremes—either weld manually, one by one, or use pricey, cutting-edge equipment for large runs. In contrast, brazing can support medium-sized runs.

Appearance: With a neat and discreet strip, brazed joints are generally more attractive than welded ones, which have non-uniform beads. Finishing operations are not required after brazing.

21.8 BRAZE WELDING

Braze welding is a method of welding whereby a groove, fillet, plug or slot weld is made using a non-ferrous filler metal having a melting point below that of the base metal but above 427°C. The filler metal is not distributed in the joint by capillary action.

Braze welding differs from brazing, in that capillary attraction is not used to distribute the filler metal and the metal is deposited by gravity. This process is mainly applied for such fabrications like metal furniture, bicycles, automobiles, refrigerators and household appliances. But its main use is for repair and maintenance work, particularly for ferrous castings. The filler used for braze welding is many a time referred to as bronze welding rod. Cu-Si alloys and Ni-Ag alloys may also be used as filler metal. The fluxes should be capable of removing the oxide film from the base metal. The fluxes for braze welding of cast iron or cast steel consist of iron oxide or manganese dioxide.

Since the bond strength of braze welded joint depends on bond area, the bond area may be increased by beveling the edges of the joint. Edges can be prepared by grinding on machining, except for cast iron for which chipping is preferred. The surface to be braze welded should be thoroughly clean as in soldering and brazing.

Oxyacetylene flame is normally used as a source of heat for braze welding. The bond strength depends on proper heating of base metal and proper deposition of filler metal. Overheating results in a porous deposit. To avoid overheating, the torch tip should be held at a small angle to the work surface.

Carbon arc and inert gas heated arc methods may also be used for braze welding when Cu-Si alloy is used as the filler. Though preheating is not essential, it is applied to the casting to ensure success. A preheat of about 900°F is generally used. Though no postheating is necessary, cooling should be recovered by wrapping the pieces with asbestos.

21.9 COMPARISON BETWEEN BRAZING AND BRAZE WELDING (TABLE 21.2)

In braze welding, the surfaces to be joined are heated to temperature corresponding to the melting point of the filler rod used. The filler rod, when molten, adheres to the surface by virtue of surface tension or it tins the surfaces and forms a sound joint on solidification.

TABLE 21.2 Comparison between braze welding and brazing

Braze Welding	Brazing
Filler material is not distributed by capillary action.	Filler material is distributed by capillary action.
Filler alloy is deposited directly at the point where it is desired.	Filler alloy is fed to one or a few points in the assembly and it is drawn into the rest of the joint by capillary attraction.
The melting point of the filler metal is below the parent metal.	The melting point of the filler metal is above the parent metal.

21.10 BRAZE WELDING FILLER METALS

Braze welding is used primarily on steel and cast iron. It is also used to join copper, nickel and nickel based alloys. Table 21.3 shows the composition of filler materials. Up to 1% Sn improves strength and corrosion resistance. Filler rod with 10% Ni increases weld metal strength, gives a whiter deposit and is thus, used for braze welding when closer color match is important. At room temperature, the joint strength usually ranges from 2800 to 4200 kg/cm^2, depending upon the filler material used. The strength of the joint drops off very quickly at temperature above 2600°C.

TABLE 21.3 Composition of filler metals

Sr. No.	% Composition*					Liquidus Temperature, °C
	Cu	Zn	Sn	Fe	Ni	
1.	60	39	1	—	—	900
2.	60	38	1	1	—	890
3.	60	37.5	1	1	0.5	890
4.	50	40	—	—	10	935

* Small additions of Sn, Fe, Mn and Si if made:
(a) tend to increase strength,
(b) improve flow characteristics, and
(c) reduce volatilization of zinc.

21.11 BRAZE WELDING FLUXES

The purposes of using braze welding fluxes are as follows:

1. Remove oxide from the edges of the metal, giving a chemically clean surface on to which the filler metal will flow.
2. Protect the heated edges of base metal from the oxygen in the atompshere.
3. Float oxide and impurities introduced into the molten pool to the surface, where they do not harm.
4. To clean the base metals and help tinning operations.
5. To suppress the formation of oxides due to fumes from the filler metal.

The flux may be applied by the following ways:

1. By dipping the heated end of the filler rod in the flux.
2. By brushing flux on the joint before braze welding.
3. By using flux coated filler rod.
4. By introducing the flux through the oxyfuel gas flame.

21.12 BRAZE WELDING PROCEDURE

The braze welding procedure consists of the following steps:

1. Joint Preparation

The joint preparation is similar to gas welding process. Square grooves are used for the plate thickness up to 2 mm and less. Since the bond strength of a braze welded joint depends on the bond area, the latter is increased for strength purposes by beveling the edges of the joint. For example, for thickness over 2 mm, single or double V grooves are prepared with 90° to 120° included angle to provide large bond areas between the base metal and the filler metal. Fillet, plug and slot welds present naturally open faces. The joint may be prepared by grinding or chipping or machining. Chipping is preferred for preparing the edges of cast iron work materials.

2. Cleaning of Surfaces

Joint surfaces should be thoroughly cleaned for getting maximum bond strength. All oxides, dirt, grease, oil and other foreign matter, that inhibit bonding, are removed from the surfaces to be braze welded. Both mechanical and chemical methods (solution cleaning) can be used to clean surfaces prior to braze welding. Mechanical methods include grinding, cleaning with wire brush, emery paper, steel wool, etc. For cleaning any of the cast iron parts before braze welding, the use of salt water is the best. However, if a salt bath is not available, from the ground cast iron surfaces, the graphitic smear can be removed by quickly heating the surface until it is dull red and then wire brushing it after it cools. The oil film from ground cast iron surface can be removed by painting the surface with chemically pure HCl. After 15 minutes, the surface must be scrubbed with a wire brush and cold clean water.

3. Aligning of Components and Clamping

The parts to be braze welded must be correctly aligned and clamped in position in a fixture. Tack weld may be used as an added precaution.

4. Preheating of the Parts

The parts to be braze welded are heated from 425 to 480°C. Preheating is done to prevent cracking from thermally induced stresses in large cast iron parts. Preheating also reduces the amount of heat and time and hence, the cost required to braze weld the parts. The extent of preheating depends on the size of the casting or assembly; larger pieces require more extensive preheating than small workpieces. Preheating may be local, depending upon the size of the casting. After brazing, cast iron parts should be allowed to cool slowly to minimize residual stresses and cracking.

5. Flame Adjustment

A slightly oxidizing flame is used for braze welding of cast iron. For carbon steels, the oxyacetylene flame is used with neutral flame. To avoid overheating, the torch tip should be held at a small angle to the work surface.

21.13 METHOD OF APPLYING FLUX AND FILLER ROD (FIGURE 21.9)

Heat the bronze rod, dip it into the flux and apply it to the surface. Flux may also be sprinkled on thick joints during heating with an oxyfuel gas torch. In heating the rod, do not apply the inner core of the flame directly to the rod. Heat the base metal using a slightly oxidizing flame until the metal begins to turn red. Melt a little bronze rod onto the surface and allow it to spread along the entire seam. The flow of this thin film of bronze is called tinning operation. Tinning is necessary for satisfactory braze welding. After tinning, start depositing the proper size beads over the seam. Use a slightly circular torch motion and ruin the beads as in regular fusion welding with a filler rod. In case of grooved workpieces, use several passes to fill the Vee. When braze welding is in a vertical position, first build up a slight shell at the bottom which then acts as support for the further metal. To ensure an even rate of cooling, heat sufficient area around the joint. Remove the excess flux and other particles using steel wool or a stiff wire brush.

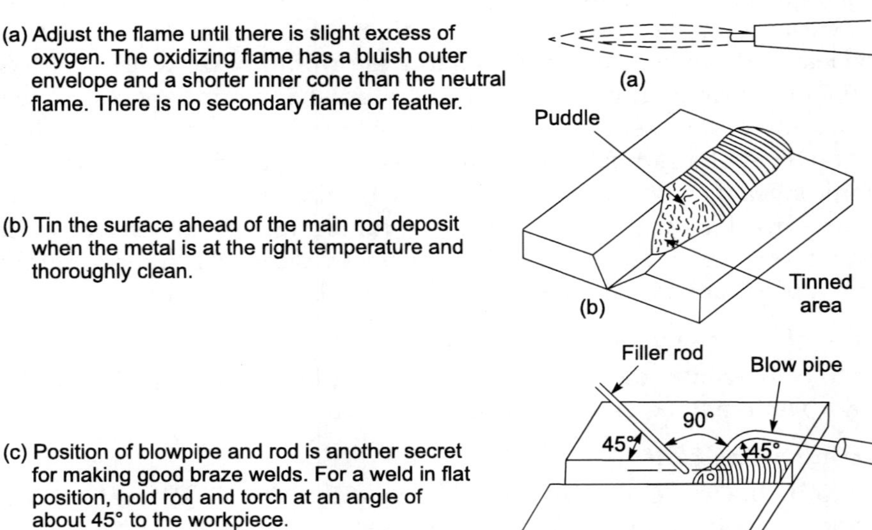

Figure 21.9 Braze welding procedure

21.14 METALLURGY OF BRAZE WELDING

Studies of bond between copper alloys and cast iron show three sources of force to bind braze welded joints. Wetting is the first action needed to accomplish bonding. The atomic force between metals at their interfaces takes place due to atomic diffusion between the brazing filler metal and the base metal in a narrow zone at the interface. Then, alloying happens due to diffusion of the metals in a narrow zone at the interface. In other words, constituents of filler metal (Cu, Zn, Sn, etc.) diffuse into the base metal (e.g., cast iron) and correspondingly base metal constituents diffuse into the filler metal in a narrow zone at the interface between the copper brazing alloy and the base metal. The third is intergrannular penetration. In this case, the brazing alloy may slightly penetrate the grain boundaries of the base metal and contribute to bond strength.

Advantages of Braze Welding

1. Requires less preheat.
2. Greater welding speed can be used.

3. Less likely to crack metals like cast iron.
4. Little or no finishing is required.
5. Joints are extremely smooth.
6. No flash or weld spatter.
7. Low temperatures are employed.
8. Since less heat is required to braze weld the component, the distortion produced is also less as compared to that in conventional welding.
9. Braze welding has lower fuel consumption and higher welding speeds.
10. Dissimilar metals (copper to steel), not amenable to fusion welding, may be joined.
11. The weld deposit is relatively soft and ductile, providing machinability and low residual stresses.
12. Joints with good strength and generous fillets are produced.
13. Brittle metals, like gray cast iron, can be joined using these processes.
14. The equipment is simple and suitable for on-site repair work.

21.15 COMPARISON BETWEEN BRAZING AND BRAZE WELDING

Braze welding is essentially a gas welding technique while brazing represents a whole group of welding processes. They mainly differ in: (1) joint design, (2) joint preparation, and (3) method of filler metal distribution.

As braze welding is mainly a gas welding technique, the joint designs are similar to those in gas welding. The edges are often beveled, but in brazing the joints are fitted closely together. The filler distribution in brazing is by capillary attraction while in braze welding, the filler metal is fused where it is deposited. However, they have one common feature, that is, both use non-ferrous filler metal that melts above 800°F but below the melting point of the base metal.

21.16 APPLICATIONS OF BRAZE WELDING

Braze welding is often used as low-temperature substitute for gas welding or a low-cost substitute for brazing. Braze welding may be employed for such fabrications as metal furniture, bicycles, automobiles, refrigerators and household appliances. Braze welding is mainly used for repair and maintenance work, particularly where ferrous castings and steel machine parts are involved.

21.17 SOLDERING

It is a method of joining two or more metal sheets by means of a fusible alloy or metal, called solder, applied in the molten state. The melting point of the filler metal is below 425°C. Generally, alloys of lead and tin having a melting range of 150 to 380°C, are mainly used for soldering.

Soldering is a group of welding processes which produce coalescence of materials by heating them to a suitable temperature and by using a filler metal having a liquidus, not exceeding 450°C and below the solidus of the base materials. The filler metal is distributed between the closely fitted surfaces of the joint by capillary attraction.

The metals mostly joined by soldering are iron, copper, nickel, lead, tin, zinc and many of their alloys. Aluminum can be soldered by special means.

The strength of the soldered joint depends on the surface alloying and mechanical bonding between the component parts. By increasing the tin content of the tin-lead solders, the wetting properties and the joint strength are improved. The soft solders may also contain antimony, bismuth, zinc, cadmium, silver, etc. The addition of Ag helps in higher temperature service.

21.18 BASIC STEPS IN SOLDERING

Soldering requires that the surface of the metal must be clean. A flux is generally used to get rid of surface oxides and to promote wetting and obtain intimate contact between the solder and the base metal. All dirt, oil and grease should be removed before the flux is applied.

(a) Joint Fit-Up

Clearances between the parts being joined should be such that the solder can be drawn into the space between them by capillarity, but not so large that the solder cannot fill the gap. A clearance of 0.15 mm is suitable for most work except when precoated metals are used, in which case a clearance of 0.025 mm or less is advisable.

(b) Precleaning

A clean, oxide-free surface is imperative to ensure uniform quality and a sound soldered joint. If all grease, oil, dirt, and oxides have been removed from the base metal before soldering, there is a much better chance of obtaining a sound joint because only then, uniform capillary attraction can be obtained.

(c) Application of Flux

Soldering fluxes are of three types: corrosive, mild and non-corrosive. Zinc chloride is the most efficient but it is corrosive. Ammonium chloride is other popular corrosive flux. These are quick acting and help to produce efficient joints rapidly.

The flux that is applied to the surfaces to be soldered, should have the following characteristics:

1. It should be fluid and effective in removing oxides and other non-metallic materials that might be present at soldering temperatures.
2. It should be a barrier to reoxidation of the metal surface that has been previously cleaned.
3. It should permit displacement by the solder.
4. It should promote wetting of the surface by the solder.

(d) Application of Heat

After the flux is applied, the next step in a soldering operation generally, is the application of heat. A number of different heating methods are used.

(e) Applying the Solder

Soldering takes place in two steps: wetting the metal surfaces and then filling the gap between them with solder. The two steps can be carried out separately or together, depending upon the conditions dictated by the application. In general, each step is better done separately because the conditions can be more easily controlled. It is frequently desirable to precoat the base metal, especially if it is difficult to solder, with solder or solderable metals.

(f) Cooling the Joint

After the surfaces of the joint have been wetted and the space between them is filled with solder, the next step is to cool the joint to room temperature. Proper jigging, assembly, or controlled cooling may be employed to prevent excessive deformation of the joint or failure of the joint during solidification of the solder. The solder should be cooled and solidified as rapidly as possible, commensurate with the requirements of the assembly and the solder used, as slow cooling may cause excessive alloying, resulting in embrittlement. Fast cooling from too high a temperature, may cause warpage and may also cause small fractures in the solder.

Cooling may normally be achieved either by conducting the heat away to the main mass of the assembly or by accelerating it with a water spray or air blast. The cooling method should be varied to suit each individual job.

(g) Flux Residue Treatment

After the soldered joint is completed, there is a flux residue which may or may not be removed, depending upon its degree of corrosiveness. Non-corrosive fluxes, which generally have a rosin base, do not require removal unless appearance is a prime factor or the joint area is to be painted or otherwise coated. On the other hand, corrosive fluxes, such as those having a zinc chloride or other corrosive base, leave a fused residue which, if not removed, would most likely cause corrosion. The self-neutralizing fluxes vary from slightly to highly corrosive and should be treated accordingly as to the removal of the residue.

21.19 SOLDERS AND PHASE DIAGRAMS

A better understanding of the nature of solders, and how to select one for a specific application, can be obtained by examining the melting characteristics of metals and alloys. Pure metals transform from the solid to liquid state at one temperature. The melting of alloys is more complicated because they may melt over a temperature range. Any alloy system can best be studied by examining the phase diagram, which shows the melting characteristics in relation to chemical composition.

The tin-lead phase diagram is shown in Figure 21.10. The terms used are defined as follows:

The solidus temperature is the highest temperature at which a metal or solder is completely solid (curve *ACEDB* of Figure 21.10). The liquidus temperature is the lowest temperature at which a metal or solder is completely liquid (*AEB* of Figure 21.10). Melting point and flow point are terms which have been in common use, but they have not always been applied with the same meaning. For this reason, the terms solidus temperature and liquidus temperature, which can be more clearly defined, will be used. Eutectic compositions are those specific solder compositions that melt at one temperature and not over a range. In this respect, eutectic solders behave like pure metals. In any binary system having a eutectic means that composition where two descending liquidus curves meet. Thus, the eutectic composition (point *E* in Figure 21.10) has a lower liquidus temperature than its neighboring compositions.

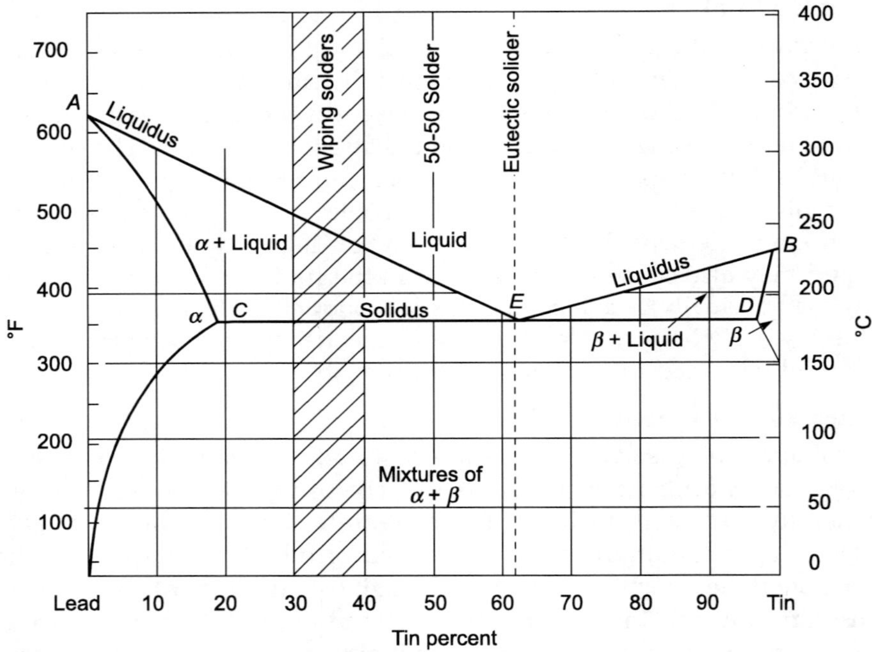

Figure 21.10 Phase diagram for tin-lead alloy system

As shown in Figure 21.10, 100% lead has a melting point of 327°C (point *A*), whereas 100% tin has a melting point of 232°C (point *B*). It observed that the tin-lead solders containing from 19.5% tin (point C) to 97.5% tin (point *D*) have the same solidus temperature, i.e., 183°C. Figure shows that the eutectic composition is approximately 63% tin and 37% lead (point *E*). When this composition melts, it becomes completely liquid at 183°C. Any composition other than the eutectic composition does not become completely liquid until a higher temperature is reached. For example: 50% tin-50% lead solder has a solidus temperature of 183°C and liquidus temperature of 216°C. This combination of tin and lead begins to melt at 183°C and becomes completely liquid at 216°C. At temperatures between the solidus and liquidus lines, the solder is partially melted. The region between the solidus (*ACEDB*) and liquidus (*AEB*) lines is called the melting range.

21.20 TYPES OF SOLDERS

Solders are available in many forms like ingots, cast bars, cakes, wire solders, triangular solders, pellets, etc. The selection of the solder depends on:

1. Cost of material involved.
2. Cost of heating the parts to the required temperature.
3. Strength and ductility required in the final joint.
4. Required resistance to corrosion.
5. Operating temperature of the soldered joint.
6. Color of the joint.
7. Wetting or spreading characteristic.
8. Solidus temperature.
9. Liquidus temperature.

Tin-Lead Solders

The tin-lead alloys are the most widely used solders and are used for joining most metals. Joint clearances of 0.1 to 0.15 mm are optimum, but variations are allowable in specific instances. Capillary attraction, as a force to fill gaps with solder, does not function with clearances greater than 0.25 mm. All cleaning and soldering processes may be used with the tin-lead solders. Fluxes of all types are used with these solders. The selection depends on the type of metals to be joined. The treatment of the flux residues is dictated by the flux used. These solders have good corrosion resistance to most of the common media. Some characteristics of the tin-lead solders are shown in Table 21.4.

The 2A and 5A solders have relatively high solidus temperatures with a short melting range.

The wetting and flow characteristics are poorer than those of the higher tin content solders, which necessitate extra care in surface preparation. The high lead containing solders, used for some automotive radiators, have better strength properties at 150°C than tin-lead solders containing more tin. The high soldering temperature limits the use of organic base fluxes, such as rosin or those of the intermediate type. The 5A solder is particularly adaptable to torch, dip, induction, or oven soldering. The low tin solders are used for sealing precoated containers, coating and joining metals, and for moderately elevated temperature uses. The 10B, 15B, and 20B solders have lower liquidus and solidus temperatures but wider melting ranges than the 5A solder. The wetting and flow characteristics are also better. Joint clearances are the same as already described. Extreme care must be taken to avoid movement of the solder joint during solidification to prevent hot tearing. Fluxes of all types and all heating methods are applicable. These solders are used for fabricating automobile radiators and coating and joining of metals where service temperatures are low enough to permit their use.

TABLE 21.4 Some characteristics of the tin-lead solders

ASTM Alloy Grade	Fed. Spec. QQ-S-571	Tin % Desired	Lead % Nominal	Antimony %			Silver % Desired	Melting Range			
				Minimum	Desired	Maximum		Solidus		Liquidus	
								°C	°F	°C	°F
70A		70	30	–	–	0.12	–	183	361	192	378
70B	Sn 70	70	30	0.20	–	0.50	–				
63A		63	37	–	–	0.12	–	183	361	183	361
63B	Sn 63	63	37	0.20	–	0.50	–				
	Sn 62	62	36	0.20		0.50	2				
60A		60	40	–	–	0.12	–	183	361	190	374
60B	Sn 60	60	40	0.20	–	0.50	–				
50A		50	50	–	–	0.12	–	183	361	216	421
50B	Sn 50	50	50	0.20	–	0.50	–				
45A		45	55	–	–	0.12	–	183	361	227	441
45B		45	55	0.20	–	0.50	–				
40A		40	60	–	–	0.12	–	183	361	238	460
40B	Sn 40	40	60	0.20	–	0.50	–				
40C		40	58	1.8	2.0	2.4	–	185	365	231	448
35A		35	65	–	–	0.25	–	183	361	247	477
35B	Pb 35	35	65	0.20	–	0.50	–				
35C	Sn 35	35	63.2	1.6	1.8	2.0	–	185	365	243	470
30A		30	70	–	–	0.25	–	183	361	255	491
30B	Pb 30	30	70	0.20	–	0.50	–				
30C	Sn 30	30	68.4	1.4	1.6	1.8	–	185	364	250	482
25A		25	75	–	–	0.25	–	183	361	266	511
25B		25	75	0.20	–	0.50	–				
25C		25	73.7	1.1	1.3	1.5	–	184	364	263	504
20B	Pb20	20	80	0.20	–	0.50	–	183	361	277	531
20C	Sn 20	20	79	0.80	1.0	1.20	–	184	363	270	517

The 25A and 30A solders have lower liquidus temperatures than all previously mentioned alloys but have the same solidus temperature as the 20B solder. Therefore, the melting range is narrower. All standard cleaning, fluxing, and soldering techniques can be used with these solders. Machine and torch soldering are widely used. Many automobile radiators and cans are made with solders of this type.

The 35A, 40A, 45A, and 50A solders have liquidus temperatures low enough to be easily worked. The solidus temperature is the same as that for 20B to 30A solders. The melting ranges, therefore, are relatively narrow. Solders of this group have the best combination of wetting properties, strength, and economy and, as such, are widely used. These tin-lead solders are the general-purpose solders and are used for wiping and sweating solders. Soldering automobile radiator cores, electrical connections, roofing seams, and heating units are but a few of the typical applications of these solders. The 40% tin-60% lead solder has become a popular general-purpose solder and is used extensively in sheet metal work. It is used as a rosin cored wire for radio and television applications.

The 60A and 63A solders are generally referred to as fine solders and are used wherever temperature requirements are critical. These solders are most commonly used for wave and dip soldering of electronic assemblies. All methods of cleaning, fluxing, and heating may be used with these solders.

The 70A solder is a special-purpose solder used where high tin content is necessary. All soldering techniques are applicable.

Tin-Antimony-Lead Solders

Antimony may be present as an impurity in solder, or deliberately added. The solders for which 0.2 to 0.5% antimony is the specified range are generally classed as B solders, i.e., 20B, 30B, 40B, etc. Federal Specification QQ-S-571 requires the presence of antimony to prevent the possible phase change from beta tin to alpha tin (called tin pest), with the accompanying change in volume and drastic loss of solder strength. These solders may normally be used, except in special cases, for the same applications as the A solders. Antimony may be added to a tin-lead solder as a substitute for some of the tin. The addition of antimony up to 6% of the tin content increases the mechanical properties of the solder but with slight impairment of the soldering characteristics.

Tin-Antimony Solder

The tin-antimony solder shown in Table 21.5 has excellent soldering and strength characteristics. It is useful for applications where moderately elevated temperature is a factor. It has a higher electrical conductivity than the tin-lead solders and is also recommended in applications such as in food handling vessels, where lead contamination must be avoided. This solder may be easily applied with rosin fluxes.

TABLE 21.5 Tin-antimony solder

ASTM Alloy Grade	Fed. Spec. QQ-S-571	Composition (Weight %)		Temperature			
		Tin	Antimony	Solidus		Liquidus	
				°C	°F	°C	°F
95TA	Sb5	95	5	233	452	240	464

Tin-Silver Solders

The characteristics of the tin-silver solder, listed in Table 21.6, are similar to those of the tin-antimony solder. The tin-silver solders are usually used for fine instrument work and some specialty tube joining because the cost is prohibitive for general-purpose soldering. They are easy to apply with rosin flux.

TABLE 21.6 Tin-silver solder

ASTM Alloy Grade	Fed. Spec. QQ-S-571	Composition (Weight %)		Temperature			
		Tin	Silver	Solidus		Liquidus	
				°C	°F	°C	°F
96.5TS	Sn 96	96.5	3.5	221	430	221	430

Tin-Zinc Solders

A large number of tin-zinc solders, some of which are listed in Table 21.7, have come into use for joining aluminum. Galvanic corrosion of soldered joints in aluminum is minimized if the metals in the joint are close to each other in the electrochemical series. Thus, alloys containing 70 to 80% tin, balance zinc, have been highly recommended. These alloys have liquidus temperatures between 260 and 310°C. In recent years, the tendency has been to add 1 to 2% aluminum or to raise the zinc content to as high as 40%. These solders are more corrosion resistant but they have higher liquidus temperatures and are therefore, a little more difficult to apply.

TABLE 21.7 Tin-zinc solder

Composition (Weight %)		Temperature			
		Solidus		Liquidus	
Sn	Zn	°C	°F	°C	°F
91	9	199	390	199	390
80	20	199	390	270	518
70	30	199	390	311	592
60	40	199	390	341	645
30	70	199	390	376	708

Lead-Silver Solders

Lead-silver solders, listed in Table 21.8, have solidus temperatures high enough to make them useful where strength at moderately elevated temperatures is required. Although pure lead melts at 327°C (621°F), a comparatively high temperature for solder, it is not used because lead normally does not wet steel, cast iron, or copper and its alloys. The addition of silver to lead results in alloys which will more readily wet steel and copper. Flow characteristics, however, are very poor. The lead-silver solders are susceptible to humid atmospheric corrosion in storage and may become unusable as solders. The addition of 1% tin to a lead-silver solder increases its wetting and flow characteristics and, in addition, reduces its susceptibility to humid atmospheric corrosion. The addition of tin to a lead-silver solder containing more than 1.75% silver, causes the segregation of tin-silver intermetallic crystals. Therefore, silver content is generally limited to 1.5% when tin is to be added.

TABLE 21.8 Lead-silver solder

ASTM Alloy Grade	Fed. Spec. QQ-S-571	Composition (Weight %)			Temperature			
		Pb	Ag	Sn	Solidus		Liquidus	
					°C	°F	°C	°F
2.5 S	Ag 2.5	97.5	2.5	–	304	579	304	579
5.5 S	Ag 5.5	94.5	5.5	–	304	579	365	689
1.5 S	Ag 1.5	97.5	1.5	1.0	309	588	309	588

Cadmium-Silver Solder

The 95% cadmium-5% silver solder has melting characteristics as shown in Table 21.9. The primary use of this solder is in applications where service temperatures are higher than permissible with lower melting solders. At room temperature, butt joints in copper can be made to produce tensile strengths of 170 MPa. At 220°C a tensile strength of 18 MPa can be obtained. Joining of aluminum to itself or dissimilar metals is possible with cadmium-silver solder. However, as is generally true in joining aluminum with dissimilar metals, electrolytic corrosion must be considered. Improper use of this solder may lead to health hazards.

TABLE 21.9 Cadmium-silver solder

Composition (Weight %)		Temperature			
		Solidus		Liquidus	
Cd	Ag	°C	°F	°C	°F
95	5	338	640	393	740

Cadmium-Zinc Solders

The cadmium-zinc solders, listed in Table 21.10, are useful for soldering aluminum. These solders develop joints with intermediate strength and corrosion resistance when used with the proper flux. The 40% cadmium-60% zinc solder has found considerable use in the spot soldering of aluminum lamp bases. Improper use of these solders may lead to health hazards.

TABLE 21.10 Cadmium-zinc solder

Composition (Weight %)		Temperature			
		Solidus		Liquidus	
Cd	Zn	°C	°F	°C	°F
82.5	17.5	265	509	265	509
40	60	265	509	335	635
10	90	265	509	399	750

Zinc-Aluminum Solder

The zinc-base solder, shown in Table 21.11, is specifically for use on aluminum and develops joints with high strength and good corrosion resistance. The solidus temperature of the solder is high, which limits its use to applications where soldering temperatures in excess of 370°C can be tolerated. This solder is extensively used in ultrasonic soldering of aluminum heat exchanger return bends.

TABLE 21.11 Zinc-aluminum solder

Composition (Weight %)		Temperature			
		Solidus		Liquidus	
Zn	Al	°C	°F	°C	°F
95	5	382	720	382	720

Fusible Alloys

Bismuth containing solders, the so-called fusible alloys, are useful for soldering operations where a soldering temperature below 183°C (lower than that available with the tin-lead solders) is required. The melting characteristics and compositions of a representative group of fusible alloys are shown in Table 21.12. Fusible alloys have the following applications:

1. Soldering heat treated surfaces where higher soldering temperatures would result in a softening of the part.

2. Soldering joints where adjacent material is temperature-sensitive and would deteriorate if a higher soldering temperature is necessary.

TABLE 21.12 Typical fusible alloys

Composition (Weight %)					Solidus		Liquidus		Yield Temperature	
Sn	Bi	Pb	Cd	Others	°C	°F	°C	°F	°C	°F
13.3	50.5	26.7	10	–	70	158	70	158	70	158
15.5	52.5	32.0	–	–	90	203	90	203	90	203
14.5	48.0	28.5	–	Antimony 9.0	103	217	227	440	116	241
–	55.5	44.5	–	–	124	255	124	255	124	255
43	57.0	–	–	–	138	281	138	281	138	281

3. Step soldering operations where a low soldering temperature is necessary in order to avoid destroying a nearby joint that has been made with a higher melting temperature solder.
4. On temperature sensing devices where the failure of a soldered joint is required at a relatively low temperature, which is known as the yield temperature.

Indium Solders

Indium solders possess certain properties which make them valuable for some special applications. Their usefulness for any particular application should be checked with the supplier. The melting characteristics and compositions of a representative group of indium solders are shown in Table 21.13. The standard 97.5% lead-2.5% silver solder does not wet most metals well. Adding 1 to 2% indium to this solder improves its wetting properties. Thus, a higher melting solder can be used without precoating the parts to be soldered. The lead-silver-indium solders are especially applicable where alkaline corrosion is a problem.

Solders containing tin, lead, and indium having more than 25% indium also show good resistance to corrosion by alkaline solutions. However, they start melting at a much lower temperature and have a wider pasty range than the lead-silver-indium solder. A 50% indium-50% tin solder adheres to glass readily and may be used for glass-to-metal and glass-to-glass soldering. The low vapor pressure of this alloy makes it useful for seals in vacuum systems.

Indium solders generally do not require special techniques during use. The low melting indium solders containing bismuth do require the use of acid fluxes or precoating. All the heating methods, fluxes and techniques which are used with the common tin-lead solders are applicable with indium solders.

TABLE 21.13 Indium solders

Composition (Weight %)					Temperature			
Tin	Indium	Bismuth	Lead	Cadmium	Solidus		Liquidus	
					°C	°F	°C	°F
8.3	19.1	44.7	22.6	5.3	47	117	47	117
12	21	49	18	—	58	136	58	136
12.8	4.0	48.0	25.6	9.6	61	142	65	149
50	50	—	—	—	117	243	127	260
48	52	—	—	—	117	243	117	243

21.21 FLUXES

A soldering flux is a liquid, solid, or gaseous material which, when heated, is capable of promoting or accelerating the wetting of metals by solder. The purpose of a soldering flux is to remove and exclude oxides and other impurities from the joint being soldered. Anything that interferes with the attainment of uniform contact between the surface of the base metal and the molten solder will prevent the formation of a sound joint. An efficient flux removes films and oxides from the metal and solder and prevents reoxidation of the surfaces when heated. It is designed to lower the surface tension of the molten solder so that the solder will flow readily and adhere to the metal. The flux should be readily displaced from the metal by the molten solder.

Theories of Flux Action (Figure 21.11)

The flux removes the oxide film from the base metal and solder by reacting with and loosening the film and floating it off into the main body of the flux. Because of the refractory nature of many oxide films, it has been suggested that the flux wets, coagulates, and suspends the oxide, which has been loosened by a penetrating and reducing action. The molten flux then forms a protective blanket over the bare metal, which prevents the film from reforming.

Molten solder displaces the flux and reacts with the base metal to form an intermolecular bond. The solder layer builds up in thickness and when the heat is removed, it solidifies.

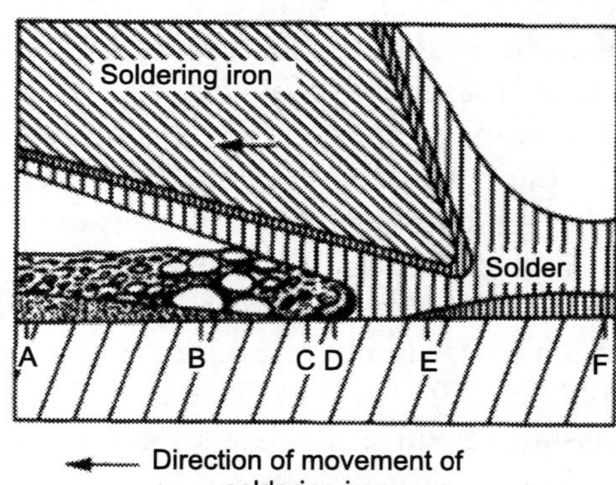

Figure 21.11 Theories of flux action

21.22 TYPES OF FLUX

Fluxes are frequently classified on the basis of their residues. They are divided in three main groups: corrosive, intermediate, and non-corrosive fluxes. The mildest flux that will perform satisfactorily in a specific application should always be selected.

Corrosive Fluxes

The corrosive fluxes, consisting of inorganic acids and salts, are used to best advantage where conditions require rapid and highly activated fluxing. They can be applied as solutions, pastes, or as dry salts and function equally well with all heating methods, since they do not char or burn. Corrosive fluxes can be formulated to be stable over various temperature ranges. They are more versatile in this respect than the less corrosive fluxes. Corrosive fluxes are almost always required when the higher melting temperature solders are used.

Corrosive fluxes can be formulated to penetrate the most tenacious of the oxide films. Commercial fluxes are available for specific applications in the form of dilute and concentrated solutions, as pastes, or as fillers for acid core solder wire. The corrosive fluxes have one distinct disadvantage: the residue remains chemically active after soldering. The residue, if not removed, may cause severe corrosion at the joint. Adjoining areas may also be attacked by residues from the spray or flux vapors. For this reason, corrosive fluxes are not used to solder closed containers, such as thermostats or bellows, nor to solder assembled electrical equipment.

Intermediate Fluxes

The intermediate fluxes, as a class, are weaker fluxes than the inorganic salt types. They consist mainly of mild organic acids and bases and certain of their derivatives, such as the hydrohalides. These fluxes are active at soldering temperatures, but the period of activity is short because of their susceptibility to thermal decomposition. Their tendency to volatilize, char, or burn when heated limits their use with torch or flame heating. They are useful, however, in quick spot soldering operations and when properly used, their residue is relatively inert and easily removed with water.

Non-corrosive Fluxes

All non-corrosive fluxes have rosin as a common ingredient. Rosin has unique physical and chemical properties which make it ideal as a flux. It melts at 127°C and remains active in the molten state up to 315°C. The active constituent of rosin (abietic acid) is inert in the solid state, active when molten; and returns to an inactive state when cooled. Thus, it is widely used in the electrical and electronics industries because the flux residue is non-corrosive and non-conductive.

Paste Fluxes

It is sometimes convenient to have the flux in paste form. Paste fluxes can be more easily localized at the joint and have the advantage of not draining off the surface or spreading to other parts of the work where the flux would be harmful. The paste-forming ingredients may be water, petroleum jelly, tallow, or lanolin, with glycerine or other moisture-retaining substances. If the pastes contain inorganic salts, such as zinc or ammonium chloride, they are classed as corrosive fluxes.

21.23 SELECTING THE FLUX

The following factors influence the choice of flux:

1. The assembly being soldered
2. Accessibility of the part for cleaning after soldering
3. Solderability of the base metals
4. Rate of soldering required
5. Heating method

Soldering of complicated electrical equipment requires the choice of a non-corrosive flux, since corrosive residues cannot be tolerated and post-cleaning is virtually impossible. Corrosive fluxes can be used when the parts can be thoroughly washed after soldering as, for example, in the assembly of radiators.

Where a small degree of corrosion can be tolerated and removal of the flux residue is impractical, the intermediate fluxes, properly used, are sufficiently active and fast for soldering the more difficult-to-solder metals. The more corrosive fluxes are often demanded, however, because of increased speed in soldering. Although the base metal is a big factor in flux selection, the converse is also sometimes the case. Thus, in the electrical industry, difficult-to solder metals are precoated with metals such as silver, tin, cadmium and copper to permit the use of rosin fluxes. Methods of heating may govern the choice of flux and conversely, the flux may determine the heating method. The inorganic salt-type fluxes can be effectively used with any heating method, since they do not char or decompose readily. The intermediate fluxes and rosin-base fluxes, however, are sensitive to the heating method. Since they are of essentially organic origin, they decompose and become ineffective when overheated. Torch or flame heating methods, unless carefully controlled, are not generally recommended for intermediate and rosin-base fluxes.

Joint Design

In general, solders have lower strength properties than the materials to which they are joined. Structurally, loaded joints must therefore be carefully evaluated so that they will be capable of sustaining the applied stresses for an adequate lifetime. Long-term joint properties are more important than short-term tensile or shear tests in the determination of an appropriate joint cross-sectional area once the design has been selected. Bulk solder alloy properties must not be utilized in design, since they are not reflected in actual joint strengths.

Electrical conductivity of soldered connections is also an important factor. Solder resistivity values vary widely and must be considered in selecting a suitable joint design or connection.

The soldered joint should relate to the maximum expected capacity of the electrical circuit and be designed to prevent localized heating or resistance changes that would influence overall circuit performance. Some typical joints are shown in Figure 21.12.

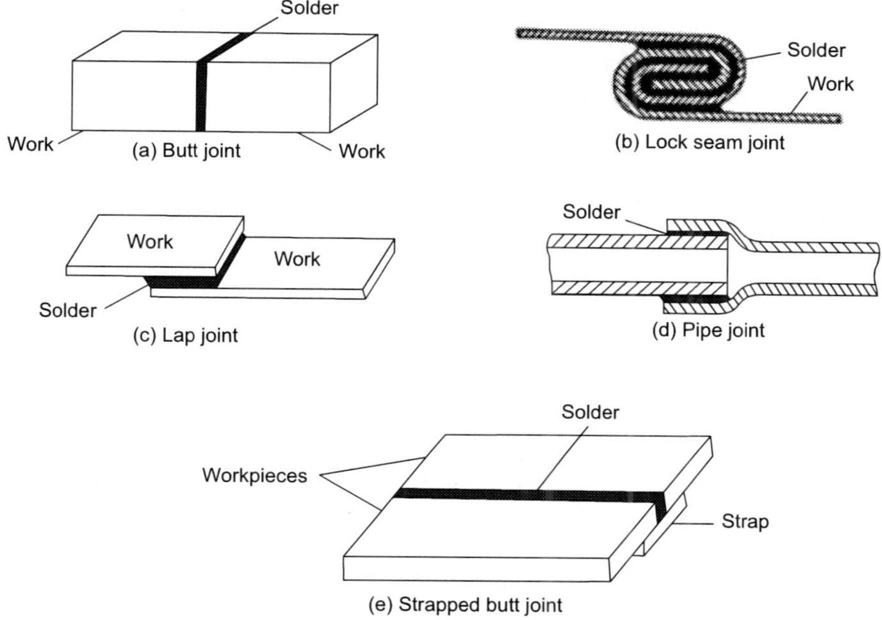

Figure 21.12 Joint designs frequently used in soldering

Precleaning and Surface Preparation

Proper surface preparation is essential for successful soldering. The more frequent preleaning methods are degreasing, acid cleaning, mechanical abrasion, and etching.

Degreasing

Organic films, such as oils and greases, are frequently encountered on the surface of metals to be soldered. Such oils and greases must be removed because they prevent wetting action by the flux and solder. Degreasing may be accomplished by immersion of the parts in a liquid or suspension of the parts in vapors of a suitable solvent.

Acid Cleaning

The purpose of pickling or acid cleaning is to remove rust, scale, and oxides or sulfides from the metal to provide a chemically clean surface for soldering. The inorganic acidshydrochloric, sulfuric, orthphosphoric, nitric, and hydrofluoric, singly and mixed, all fulfill this function, although hydrochloric and sulfuric acid are most widely used.

Mechanical Preparation

Various abrasive techniques are frequently employed to clean metallic surfaces before soldering. They are effective and economical methods but have one definite limitation:

particles of the abrasive may become embedded in the surface being cleaned. These abrasive materials (sand, grit, ceramic, steel wool, etc.) are generally not solderable. Although the surface may appear clean, if sufficient abrasive particles to significantly reduce the anchorage area have been embedded in the surface, the result is reduced solderability. A simple solderability test should be performed following abrasive cleaning. An etch treatment following abrasive cleaning may be required to remove sufficient surface material to eliminate the embedded abrasive.

Etching

The removal of a small amount of material from the surface to be soldered is a common cleaning and repair technique. A non-plated copper surface particularly lends itself to this technique. Copper etchants such as ferric chloride, copper chloride, and ammonium persulfate are used.

Precoating

The coating of the base metal surfaces with a more solderable metal or alloy prior to the soldering operation is sometimes desirable. Coatings of tin, copper, silver, cadmium, iron, nickel, and alloys of tin-lead, tin-zinc, and tin-copper are used for this purpose. The advantages of precoating are twofold: soldering is more rapid and uniform, and strong acidic fluxes are avoided at the assembly. The precoating of metals which have tenacious oxide films, such as aluminum, aluminum bronzes, highly alloyed steels, and cast iron, is almost mandatory. Precoating of steel, brass, and copper, although not absolutely essential, is of great value in some applications.

21.24 EQUIPMENT, PROCESSES, AND PROCEDURES

Soldering Irons

The soldering iron (Figure 21.13) should provide constant heat to parts being soldered, ensuring that the parts are joined using minimal contact time, thereby safeguarding the components in close proximity and areas adjacent to the soldering connection are not adversely affected by heat absorption.

Figure 21.13 Soldering gun

Soldering Iron Tips

Materials

The properties required for soldering iron tips are:

1. High thermal conductivity to ensure that heat transfer is rapid and efficient.
2. Ease of tinning to ensure a liquid metallic path through which the heat of the tip surface may be readily transmitted to the work.
3. Low oxidation to ensure good heat transfer from tip to work and to prevent the tip from freezing in the soldering iron.
4. Resistance to corrosion from soldering fluxes if acid core, acid paste, and water soluble fluxes are used.
5. Resistance to erosion by the molten solder. Four basic types of tips can be used:
   - Copper
   - Iron plated with coated shank
   - Iron plated with stainless steel shank
   - Calorized

Processes

Torch Soldering

Torch soldering is commonly used for automotive body work, plumbing and structural joints, and in locations where electricity is not readily available. Torch selection and gas mixture is governed by the size, mass, and configuration of the assembly to be soldered. The flame from a torch will heat large masses of material rapidly but is likely to cause burning or carbonization of the flux. This is less likely to occur when flux core solders with chloride fluxes are used. One way to prevent carbonization or decomposition is to preheat the assembly (without causing excessive oxidation) before applying the solder and flux. The elevated temperature of the flame from a torch can cause damage to heat-sensitive components or to areas adjacent to the soldered connection.

Dip Soldering

Dip soldering is useful and cost effective because an entire unit of any number of joints can be soldered merely by dipping the pre-fluxed part in a bath of molten solder. It is necessary to use jigs or fixtures to contain the unit and keep the proper clearance at the joint until the solder solidifies. A preliminary treatment of the unit such as degreasing, cleaning, and fluxing is also required before dip soldering. Care should be taken when immersing the parts in the pot. The molten bath of solder supplies both the heat and solder necessary to complete the joint. The solder pot should be large enough so that at a given rate of production, the units being dipped will not appreciably lower the temperature of the solder bath. Pots of adequate size allow the use of lower operating temperatures while still supplying sufficient heat for soldering.

Spray Gun Soldering

Two types of guns are used to spray solder. The first uses propane with oxygen or natural gas with air to heat and spray a continuously fed solid solder wire of approximately 3 mm diameter. With ordinary procedures, about 90% of the solder wire is melted by the flame of the gun, and contact with the workpiece is made by the solder in a semiliquid form. The workpiece then supplies the balance of the heat required to melt and flow the solder. The solder is then wiped automatically or manually. Adjustments can be made within the spray gun to deposit completely liquified solder or a series of fine drops. The second type of spray gun has a small, electrically heated cone into which solid solder wire approximately of 6 mm in diameter, is fed. Through an orifice in the small end of the cone, the molten solder is directed into a compressed air stream which transmits the solder a distance of 25 to 75 mm.

Induction Heating

Induction heating generally, is applicable for soldering operations having the following requirements:

1. Large-scale production
2. Localized application of heat
3. Minimum oxidation of surface adjacent to the joint

Resistance Heating

In resistance heating, the work to be soldered is connected either between a ground and a movable electrode or between two movable electrodes to complete an electrical circuit. The heat is applied to the joint both by the electrical resistance of the metal being soldered and by conductance from the electrode, which is usually carbon. Resistance soldering equipment consists of a heavy-duty variable transformer, which converts the normal line voltage to a lower voltage with correspondingly increased amperage. A wide variety of accessories can be attached to the transformer. In one method of resistance soldering, the work is attached to a ground lead by either an alligator clip or C-clamp. The single movable electrode used in conjunction with the ground attachment is carbon mounted on a non-conducting handle. A variation may be achieved by fixing the electrode in position and bringing the grounded work to be soldered into contact with the electrode while simultaneously applying the solder. Another method uses a two-circuit soldering tool consisting of two carbon electrodes mounted on a non-conductive handle, eliminating the necessity of a ground lead.

Oven Soldering

Ovens have long been used successfully for high-production soldering. Although conveyorized setups normally result in higher productivity than batch type operations, both are commonly used; in either case, large production runs are needed to justify the cost of the furnaces and required tooling.

Focused Infrared Soldering

Optical soldering systems are available which are based on focusing infrared light (radiant energy) on the joint by means of a lens. Lamps ranging from 45 to 1500 watts, can be used for different application requirements. The devices can be programmed through a silicon controlled power supply with an internal timer.

Hot Gas Soldering

The principle is to use a fine jet of inert gas, heated to above the liquidus of the solder. The gas acts as a heat transfer medium and as a blanket to reduce access of air around the joint.

Wave Soldering

A liquid wave is generated by circulating molten solder by a pump in an appropriately designed soldering machine. The prime functions of the wave are to serve as a heat source and heat transfer medium and to supply solder to the joint area. A properly functioning solder wave, as a consequence of its geometry, thermodynamics, and fluid mechanics characteristics, will contribute to the wetting of the metal surfaces, promote through-hole penetration, and ensure formation of reliable solder joints and fillets. A wave soldering production line includes fluxing, preheating, and soldering stations and a means of conveyance of the assembly. In-line cleaning and drying can also be included in the operation.

Methods of Flux Application

The method used for wave fluxing is the application of flux using the liquid wave principle to form a wave of flux, which touches the workpiece while the assembly passes through it. By this method, the flux coats the areas to be soldered.

Foam Fluxing: The flux foam is generated from liquid flux by means of a porous medium immersed in the flux. Low-pressure air is forced through the pores of the diffuser and generates fine bubbles of foam. These are guided to the surface by a nozzle to form a foam head or wave through which the assembly is passed.

Brush Fluxing: A rotating brush partly immersed in flux is used as a means to transfer the flux to the workpiece.

Spray Fluxing: Flux is applied to the workpiece by means of jets or spray nozzles. One method of spray fluxing employs a drum with fine stainless screen partially immersed in flux and rotated in it. The flux wets the screen, and air jets inside the drum blow off the flux as minute droplets in the direction of the assembly. The amount of flux transferred in unit time is controlled by the rotational speed of the drum and the air pressure.

Preheating

The essential function of preheating is ·the evaporation of the flux solvent. Proper preheating will also promote wetting and reduced thermal shock. A preheat temperature of 75 to 80°C is usually employed for evaporation of fluxes, a preheat temperature somewhere above the boiling point of water may be necessary. Fluxes with other solvents may require extended preheat times. Printed wiring boards, when heavily loaded with connector parts, may require higher preheat temperatures.

Soldering Station

The essential feature of the soldering machine is the generation of a wave of molten solder.

Modern systems are capable of pumping wave widths (or lengths) from 50 to 600 mm, and wave heights up to 20 mm. They have relatively large solder capacities to maintain soldering temperature and provide satisfactory flow patterns. An automatic solder feed mechanism is used on high production units to maintain a constant solder level without affecting the thermal balance of the pot.

Conveyance

Conveyors move parts through the soldering station and are frequently designed to be integrated with component assembly, fluxing, and preheating and cleaning operations to form one continuous production line. Conveyors are designed to provide a smooth, vibration-free movement of the printed circuit assembly at fixed or adjustable slopes, ranging from horizontal to 8 degrees and speeds to 6 m/min (20 ft/min). There are basically two types of conveyor. One is a chain conveyor, which requires the use of board holding carriers to secure the workpiece or pallets. The other is an adjustable width finger-type conveyance for use when a large variety of sizes printed wiring boards are to be wave soldered. The fingers are usually made of titanium to resist flux, high temperatures, and prevent solder adherence. Multitrack systems are a variation of the finger type, which permit soldering printed circuit assemblies of two different sizes simultaneously.

Flux Removal

Adequate cleaning is particularly important in printed circuit applications. The techniques for flux removal can be divided into two basic approaches: batch-type cleaning, in which the operation is separated in time and space from soldering, and in-line cleaning, where the cleaning positions follow immediately after the soldering position, forming one continuous system.

Batch cleaning includes the use of dishwasher-type cleaners, ultrasonic dip tanks, and vapor degreasers. These methods are generally used for small parts and low volume processing systems as part of a hybrid process. **In-line cleaning,** particularly where a production volume exists, has become the generally accepted method. Cleaning stations utilize liquid waves, immersion tanks, forced sprays, rotating brushes, ultrasonic tanks, vapor immersion, and combinations of the above with the appropriate solvent for the flux to be removed. Drying stations following cleaning employ air knives, infrared, and air blasts. The latest development for high production cleaning is the use of biodegradable water detergent solutions in combination with multiple stage in-line spray cleaning systems.

Aluminum Soldering

Aluminum soldering presents the following problems:

1. Aluminum has a tough oxide layer, which is difficult to remove and which forms easily on exposure to air also. Aloxide is chemically very inert. Thus, a strong flux is required to clean the surface. Only special fluxes, designed for aluminum soldering, are effectively used.
2. Under regular conditions soldering to aluminum with tin-lead solders cause a galvano-potential beyond the tolerable range of fast deterioration of the joints resulting under humidity condition. It is found that there will be an electromotive potential of 1.53 volts between aluminum and tin-lead solder. This problem can be overcome by using a Zn-Mg-Al solder, which could have the small potential as Al and thus, be suitable.
3. Aluminum has an extremely fast rate of heat transfer. Therefore, heating of aluminum is often a problem when localized heating is required.
4. It has a large coefficient of thermal expansion. Therefore, if the heating is not localized there is likely to be heat distortion. The melting point of aluminum is close to that of soldering alloys, therefore heating should be properly controlled such that aluminum is not melted in the joint area.

The chemical fluxes which are effective on aluminum contain organic fluorides together with a heat-stable salt. Sometimes, also contains metal fluorides, inorganic chlorides and ammonia compounds.

Reaction fluxes containing zinc chloride and sometimes tin chloride in combination with other solids can also be used as fluxes. At the soldering temperature, these metal chlorides are reduced by aluminum to form volatile aluminum chloride and either metallic zinc or tin.

TABLE 21.14 Comparison between soldering and brazing

Soldering	Brazing
Melting point of solder/filler rod is lower than 427°C.	Melting point of filler rod is greater than 427°C but lower than the melting point of parent metal.
Soldered joints may be corrosive.	Brazed joints are corrosion resistant.
Soldered joints are weaker than brazed joints and may weaken at high temperature.	Brazing joints are stronger and withstand high service temperature.
Equipment cost is low.	Equipment cost is high.
The most common type of filler metal used is 60:40, tin-lead alloy.	Brass alloys are commonly used in brazing.

21.25 ADVANTAGES OF SOLDERING

1. Low investment cost.
2. Properties of parent metal are not affected due to low operating temperature.
3. Good and effective sealing in fabrication as compared to other process, like riveting, spot welding and bolted joints.
4. Process is simple. Dissimilar materials can be joined.
5. Thin wall parts can be joined easily.
6. Lower energy than brazing or fusion welding.
7. Variety of heating methods are available.
8. Good electrical and thermal conductivity in joint.
9. Rework and repair can be carried out easily.

21.26 DISADVANTAGES OF SOLDERING

1. Large sections cannot be joined.
2. Fluxes may contain toxic components.
3. Soldering joints cannot be used for high-temperature applications. Weakening or melting of joint in elevated temperature is possible.
4. Joint strength is low unless, reinforced by mechanical means.
5. Careful removal of the flux residuals is required in order to prevent corrosion.

Review Questions

1. Explain the principle of operation of brazing.
2. What are the equipment used in brazing.
3. Explain the chemistry involved in brazing and soldering.
4. What are the safety precautions to be followed for soldering?
5. What are the disadvantages of brazing?
6. Compare brazing and braze welding.
7. What are the applications of soldering?
8. What are the advantages of brazing?
9. What are the limitations of soldering?
10. Compare soldering, brazing and braze welding.

Section E

Advanced Welding Processes

22
CHAPTER

Ultrasonic Welding

22.0 INTRODUCTION

Ultrasonic welding is a solid-state welding process which produces a joint by the local application of high frequency vibrating energy as the work parts are held together under pressure. Welding occurs when the ultrasonic tip or electrode, the energy coupling device, is clamped against the workpieces and is made to oscillate in a plane parallel to the weld interface.

In ultrasonic welding, coalescence is produced between a wide range of similar and dissimilar combinations of metals by local application of high frequency vibratory energy in a short cycle and without melting the materials being joined.

22.1 OPERATION (FIGURES 22.1 AND 22.2)

Ultrasonic welding uses combination of pressure and high frequency vibration motions to form solid state bond. Material to be welded is clamped between the transducer and backup piece. Then ultrasonic vibrations are induced into the materials and the combined clamping pressure and oscillating forces introduce dynamic stresses in the

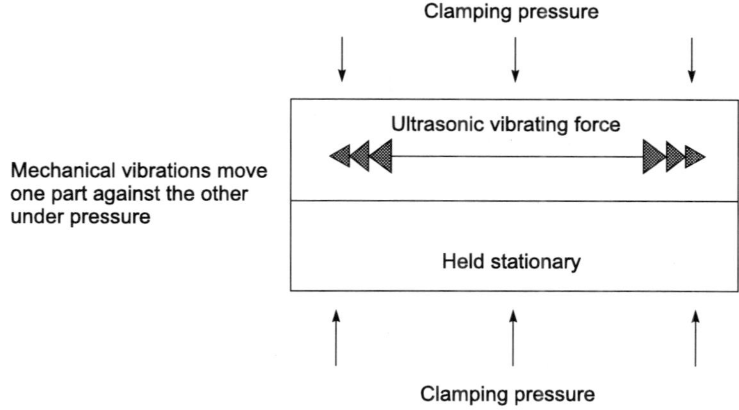

Figure 22.1 Ultrasonic welding force and pressure distributions

base metal. As a result, minute deformations will be produced and cause a moderate temperature rise at the weld zone. Due to internal friction, materials are heated up to nearly 35% of their melting temperature at the contact point. This coupled with clamping pressure provides for coalescence across the interface to produce weld. Ultrasonic energy also aids in cleaning the weld area by breaking up oxide films and removes it. The vibratory energy comes from transducer coupled to the work by various types of tooling, which can range from tips similar to resistance welding tips to resistance roll welding electrode wheels.

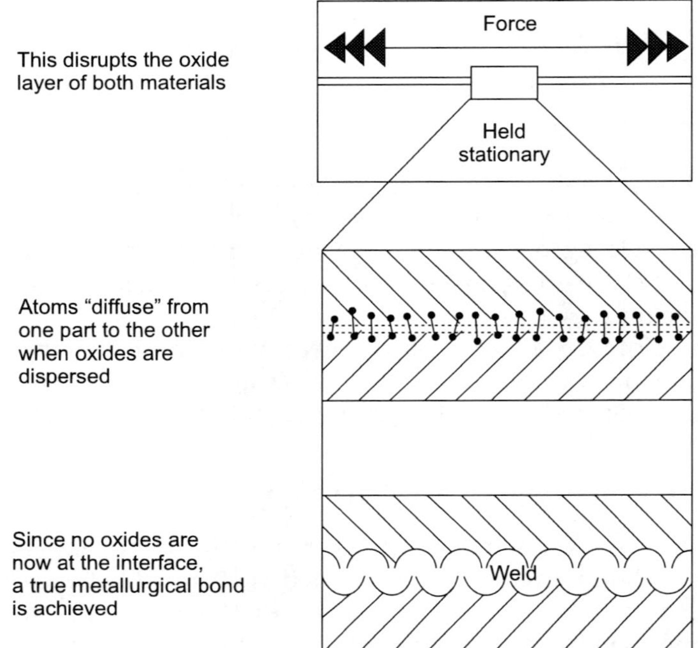

Figure 22.2 Ultrasonic welding mechanism

22.2 DESCRIPTION OF PROCESS (FIGURE 22.3)

One of the newest and most interesting fields of joining technology is ultrasonic welding. In this process, high frequency vibrations are combined with pressure to join two materials together quickly and securely, without producing significant amounts of heat. These factors give it many advantages over traditional heat-based welding techniques. These include the ability to weld metals of significantly dissimilar melting points; metals that normally form brittle alloys at the weld junction; and welds that are in close proximity to heat-sensitive components, such as electronics or plastic components. Finally, ultrasonic welds are made without consumables, such as solder or filler that would ordinarily be used for the connection and with far less energy usage than traditional joining techniques. There are some restrictions on the types of joints that can be made with ultrasonic welding. One of these is that it is restricted primarily to non-ferrous metals and plastics. Another is that at least one of the parts must be relatively light, as it would take a tremendous amount of energy to vibrate a heavy part at the necessary frequency. This restriction, unfortunately, limits the process to small components and wires.

Ultrasonic welding is a solid-state welding process, used since 1950s. It uses the high frequency vibrations to join the materials. Workpieces are clamped together under a moderate static force normal to their interface and oscillating shear stresses of ultrasonic frequency is applied parallel to the plane of interface for about a second or even less.

First, a power supply is used to convert a low frequency electrical energy into high frequency electrical impulses. High frequency electrical impulses are then converted into mechanical vibration using suitable transducer. A coupling system transmits and applies high frequency mechanical vibrations to the weld interface under a clamping force.

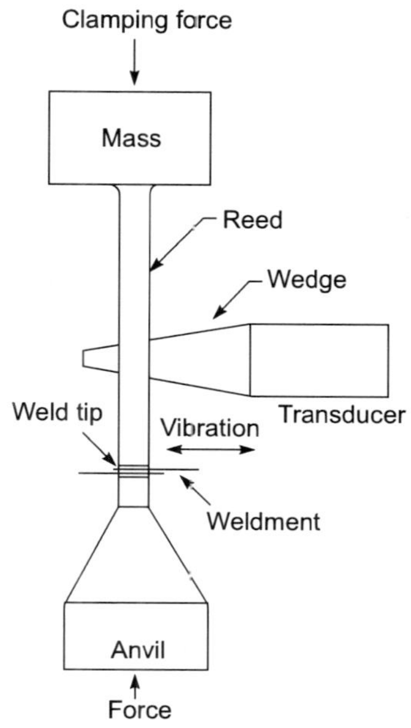

Figure 22.3 Ultrasonic welding processes

The clamping forces produce a uniform and symmetrical stress that radiates into the material from the clamping point. As the lateral welding force is applied, the stress distribution in the material shifts to one side of the clamp point. This represents half of a vibration cycle. During the second half of the vibration cycle, lateral forces reverse direction and stress distribution within the workpiece shifts to the opposite side of the clamping point. This rapid change in subsurface stresses occurs tens of thousands of times per second and results in the dispersion of surface oxides and films, the smoothing of surface irregularities through plastic deformation and ultimately, inter-atomic bonding through atomic diffusion.

The welding interface experiences maximum temperature, that is, only 35 to 50% of the melting point of the material. Because there is no melting at the weld interface, metals may be joined without the use of fluxes or shievering gases and without the chance of producing undesirable porosity or gas.

Because of these capabilities, the use of USW has been expanding rapidly within the electronics and electric motor industries for wire and sheet welding. Now it is also used in aerospace and automotive industries for more critical structures.

22.3 DESCRIPTION OF EQUIPMENT

A schematic layout of USM is shown in Figure 22.4. The function of an ultrasonic welding machine is to generate and deliver high frequency mechanical motion to a clamped workpiece. This is accomplished through the four subsystems of a USW machine. These are:

1. Power supply
2. Transducer
3. Coupling system
4. Clamping system

Power Supply

The ultrasonic welding power supply converts 50 or 60 Hz electrical signal to a much high frequency signal. This is accomplished through the use of a frequency generator and amplifier. The frequency generators used for various welding applications are capable of operating at frequencies from 10 to 75 kHz. Typically, the higher frequencies are used for lower power, light duty jobs.

The second stage of the power supply consists of amplifier circuits that amplify this high frequency signal to a usable level for the particular job. This may range from only a few watts for light duty jobs, such as wire welding to power level as high as 16,000 W for sheet and plate welding.

Transducer

The function of the transducer is to convert the electrical energy from the power supply into the mechanical energy that will perform welding. There are two types of transducers used in USW, depending upon the application. They are:

- Magnetostrictive transducers
- Piezoelectric transducers

Magnetostrictive transducer: The property of the magnetostriction is used in this transducer. Magnetostrictive materials change length under the influence of varying magnetic flux density. In this transducer, a heavy gage wire is wrapped around a laminated stack of nickel sheets to form an electromagnet.

Every time a pulse of current is passed through the wire; a magnetic response is induced within the stack. This results in a momentary change of about 0.05 mm in the overall length of the stack. The magnetostrictive transducer is generally used for application requiring lower powers. It can be characterised as low efficiency device, although it is very rugged and long lived.

Piezoelectric transducer: Piezoelectic materials change length slightly under the influence of current directly passing through it. These transducers are made of a stack of ceramic disks, such as lead zirconate titanate.

Piezoelectric transducers are used for high power applications. However, because the ceramic material is brittle, it is far more fragile than the magnetostrictive transducer. Both types of transducers require either air or water cooling to remove waste heat generated from their oscillations.

Coupling System

The function of coupling system is to mechanically transmit the high frequency vibrations from the transducer to the workpiece. There are two methods for transmitting energy to the weld; the wedge-reed system and the lateral drive system.

The wedge-reed system, which is applicable only to spot welding, consists of a transducer that drives a wedge-shaped coupler attached to a vertical reed (as in Figure 22.3). When the transducer vibrates the reed, nodes and antinodes are produced along the length of the reed. The length of the reed is selected so that at a particular operating frequency, a node (which is a point of maximum displacement) will occur at the welding tip. If an application requires very high power, several transducers can be linked to the reed at the proper nodal points.

The reed and coupler must be made of a material which has good fatigue resistance. Titanium, monel or stainless steel are used for this purpose. The clamping force is provided by tip as well as anvil. The function of anvil is to hold the bottom piece of metal stationary against the lateral forces of the vibrations.

Wedge-reed systems are able to provide very high clamping force but of a low amplitude; meaning that this type of system is capable of high power applications such as large area bonding, heavy section welding and stranded wire welding.

Lateral drive system consists of a horizontally mounted transducer attached to the coupling horn and tip as shown in Figure 22.4. This produces a tip motion that is also parallel to the weld interface. The clamping force is generated from the pivoting motion of the ultrasonic assembly. This system can be used to either low power spot welds or continuous seam welding.

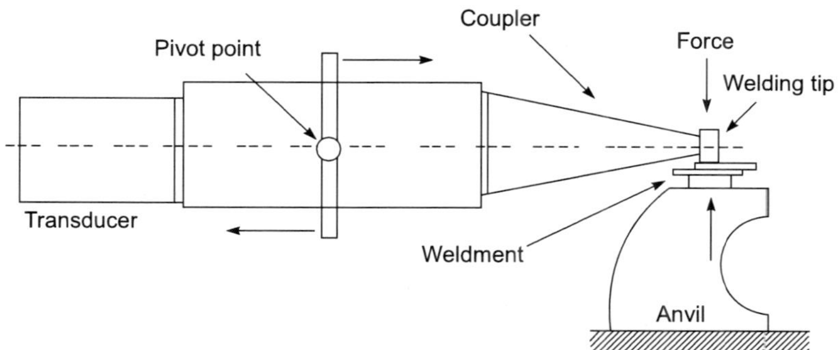

Figure 22.4 Lateral drive USW systems

In the lateral drive system, energy from the transducer first travels through a tapered coupler or horn. The horn is designed with a decreasing cross-sectional area, which results in an increased amplitude at its output end. In effect, the horn acts as a mechanical step-up transformer with respect to the weld tip motion.

Energy from the horn is coupled into the welding tip made of materials, such as tool steel or nickel alloys. The lateral drive system is not only capable of performing spot welds, but with proper welding tip, it can produce high speed continuous lap seam welds. This is accomplished through the use of a rotating roller tip.

Because of the low clamping force and high mechanical amplitude inherent in lateral drive system, this method is most often limited to low thickness applications.

Clamping System

Clamping systems provide the force being applied through either the welding tip or the workpiece itself as well as providing the static force necessary to hold the workpiece together during the weld cycle. The amount of clamping force required for a particular job increases when welding softer material or as the weld contact area increases. Simple spring system is used where the power is low. Pneumatic system is used for medium power and hydraulic systems are used for the highest powers.

22.4 PROCESS PARAMETERS

The establishment of a set of process parameters is based on workpiece material thickness, ambient material temperature, surface condition and material properties. Therefore, if the alloy, surface finish or weld geometry are changed, the parameters will also have to be adjusted. The parameters that are adjustable in USW are the power, clamping force, welding time and frequency.

Power

For ultrasonic spot weld, amount of power can range from a few watts for wire welding, to as much as 16,000 W for heavy selection welding. Seam welds are performed with power levels that range from 100 to 500 W. As power is increased, welding thickness ability is also increased and cycle time is decreased.

Clamping Force

The amount of clamping force used for ultrasonic welds can vary from less than 100 grams to thousands of kilograms. Clamping force must be increased as the contact area of the weld zone increases. For example, light forces would be used for wire welding applications and larger forces would be used for plate welding. Insufficient force can result in tip sticking, too much force needlessly increases power requirement.

Welding Time

Ultrasonic spot welding is usually performed in welding process of 0.05 to 1 second each. Extended welding times are an indication that not enough power is being used. Insufficient power and subsequently longer cycles, increase the possibility of tip sticking, increase heating of the workpiece and increase the chances of marring the part surface at the welding tip location. Therefore, it is desirable to use the highest possible power levels to minimize the welding cycle time.

The parameter of time is controlled by traverse rate in ultrasonic seam welding. The transverse rate determines the amount of energy delivered per unit length of the weld and thus, determines the weld penetration. Aluminum foil 0.025 mm thick can be welded at a rate of 4.5 m/min and 0.152 mm thick aluminum at 0.46 m/min.

Welding Pressure

It is the pressure between the anvil and the sonotrode applied on the parts to be welded. The welding pressure should be constant for the process. It is usually developed pneumatically and totals up to 400 to 1500 N.

Trigger Point

The point on time at which ultrasonic welding starts is called trigger point. It is usually a function of the clamping and compaction of the parts or function of stroke length. The ultrasonic is automatically triggered for the welding time as soon as the sonotrode work surface touches the workpiece and certain clamping force has been built up. Whether pressure should be initiated while sonotrode is oscillating or before the oscillation starts depends on the application.

The adjusting element is a pressure regulated valve for the activation of the air cylinders.

The trigger point has to be chosen individually and is determined by welding experiments. It is one element of the parameter welding time.

Amplitude

Amplitude is the oscillation length of sonotrode. It should be kept constant throughout the process. Range of amplitude is defined by a mechanical booster and adjustments are made with electrical selective switch. Amplitude can be adjusted quickly and very precisely electrically, instead of time consuming mounting and dismounting of boosters with different amplitude transmission ratios. It is usually between 15 and 45 microns for a half wave. Individual magnitude of the amplitude is determined by the shape and construction material of the welded parts. In addition, there is a correlation between welding time, welding energy and welding pressure. Proper magnitude will be determined during test welding.

Welding Tools

It is the decisive factor for determining quality of the weld. If the surface on the work surface of the sonotrode begins to wear, it may be judicious to change one or several parameters until a reconditioning or exchange of tool becomes necessary.

Ultrasonic Frequency

Frequency remains constant in the process since it depends on the generators working frequency. It is decisive for the construction of machine and complete transducer system. Most equipment work on the frequency in the order of 20 kHz, but in small scale applications, equipment with 35 and 40 kHz is used.

22.5 PROCESS CAPABILITY

Because of the non-thermal nature of ultrasonic welding process, a wide range of metals and metal combinations can be successfully welded.

Figure 22.5 illustrates the combinations of metals that are applicable to USW. Blank areas on the chart indicate combinations that either have not been successfully welded or that have not yet been tried.

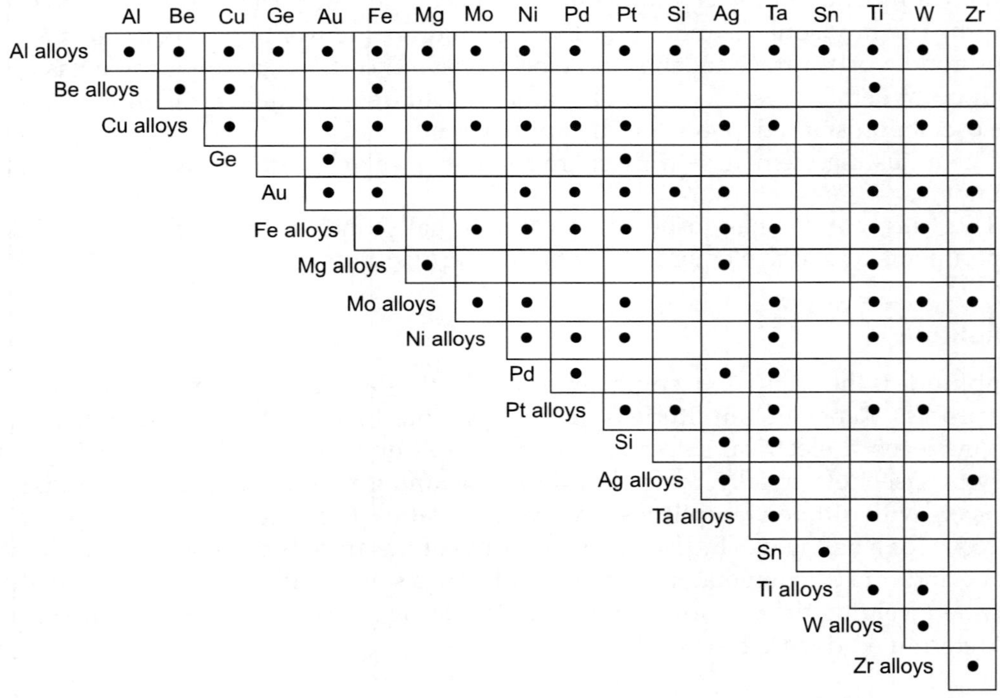

	Al	Be	Cu	Ge	Au	Fe	Mg	Mo	Ni	Pd	Pt	Si	Ag	Ta	Sn	Ti	W	Zr
Al alloys	•	•	•	•	•	•	•	•	•	•	•	•	•	•	•	•	•	•
Be alloys		•	•			•										•		
Cu alloys			•		•	•	•	•	•	•	•		•	•		•	•	•
Ge					•						•							
Au					•	•			•	•	•	•	•			•	•	•
Fe alloys						•		•	•	•	•		•	•		•	•	•
Mg alloys							•						•			•		
Mo alloys								•	•		•			•		•	•	•
Ni alloys									•	•	•			•		•	•	
Pd										•			•	•				
Pt alloys											•	•		•		•	•	
Si													•	•				
Ag alloys													•	•				•
Ta alloys														•		•	•	
Sn															•			
Ti alloys																•	•	
W alloys																	•	
Zr alloys																		•

Figure 22.5 Metal combinations that can be welded using USW

The materials to weld by USW are aluminum, copper, gold, silver, platinum, their alloys and steels. Refractory and some high strength metals such as molybdenum, tungsten, zirconium, tantalum, columbium, titanium can also be welded by USW.

The welding thickness is limited by the thickness of the top-joint member. Multiple layers of foil can be spot or seam welded together, depending upon the thickness of the layers to be welded and the power of the machine. As many as 20 layers of foil have been successfully welded in production application.

Aluminum sheet can be welded in thickness upto 2.54 mm and refractories are limited to maximum thickness of 0.38 to 1.02 mm. Wires from 0.025 to 0.76 mm in diameter are within the capabilities of USW while the minimum thickness of sheet metal that can be welded reliably is 0.005 mm.

When producing multiple spot welds on a single workpiece, possibility of fracturing the previously made welds exists if the workpiece is in resonance with the operating frequency of the machine. All workpieces have natural resonant frequencies, and these frequencies are mostly different from those of the operating system. If by chance, resonance problems occur, they can be eliminated by modifying the workpiece dimensions, changing the clamping location or method or adding temporary weights to various locations on the workpiece. The surface of an ultrasonically welded part has a rough surface at the points where the welding tip and anvil contact the part.

The ultrasonic welds exhibit greater ductility in comparison to resistance welding because no melting has occurred. Furthermore, ultrasonic welding offers cost and quantity advantages because of simplified surface preparation requirement, lower energy consumption and the ability to successfully weld workpieces that are grossly different in composition, properties or thickness.

22.6 APPLICATIONS OF USW

1. This is traditionally used for electronic, electrical and defense industries.

   In electronic appliances like switches, sensors and data storage keys are fabricated using ultrasonic welding.
2. Ultrasonic welding is also used to make medical parts like filters, catheters, medical garment and masks.
3. It is used for welding wires and containers for microcircuits and for attaching wires and contacts to motor and switches.
4. It is used for hermetic sealing of volatile substances and in fabricating nuclear fuel elements.
5. It is used for continuous seam welding to assemble components of corrugated heat exchangers.
6. Packing applications like blister packs, pouches, tubes, storage containers and carton spouts can be fabricated using ultrasonic welding.
7. Ultrasonic welding is used both to connect anodes and cathodes, and to interconnect cells of lithium ion batteries.
8. Threaded inserts, grub screws or other parts can be ultrasonically embedded in thermoplastics (Figure 22.6).

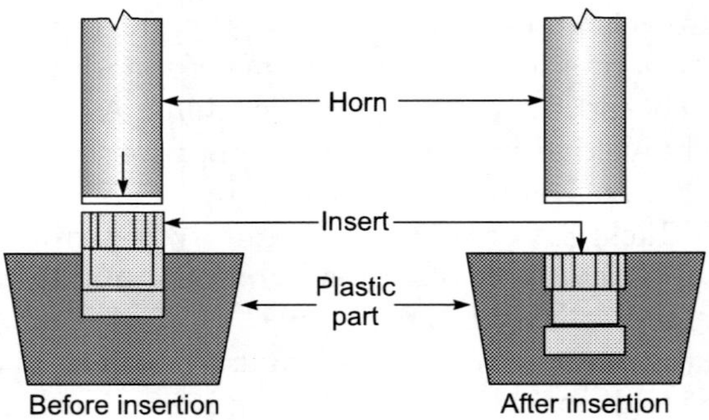

Figure 22.6 USM of thermoplastic

9. They can be used for melting and forming of a plastic stud to retain or lock another material, often dissimilar, in place (Figure 22.7).

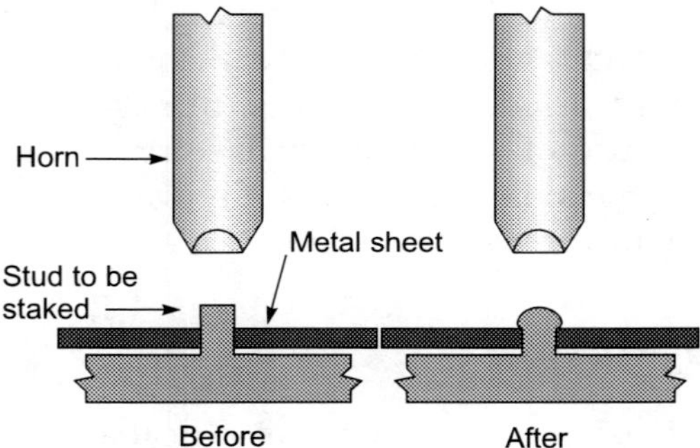

Figure 22.7 USM of thermoplastic

10. Used for joining two thermoplastic parts with no preformed hole or energy director. This technique is particularly suited to large parts or parts with complicated geometry and thermoformed or blow-molded parts without a joint (Figure 22.8).

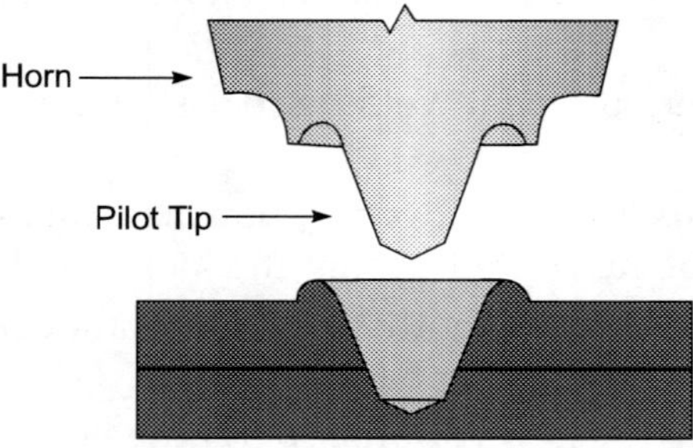

Figure 22.8 USM of thermoplastic parts with no performed hole

11. Ultrasonic welding is used in the automotive industry to fabricate headlamp parts, dashboards, buttons and switches, fuel filter, fluid vessels, seat-belt locks, electronic key fobs, lamp assemblies and air ducts.

22.7 ADVANTAGES OF USW

1. Surface preparation and surface cleanliness is not critical.
2. No defects are produced from arc, gases and filler metals.
3. Welding of glass is possible.
4. Surface deformation produced by USW is minimum.
5. Very thin metals and plastics can be welded.
6. Dissimilar metals having vastly different melting points can be joined.
7. Thin and thick sections can be joined together by USW.
8. Due to low temperature requirement, the characteristics of the materials are not altered and are continued through weld zone.
9. It is easy to weld heat conductivity and high conductivity materials, such as gold, silver, copper, aluminum, etc.
10. It can be applied to a variety of combinations of material welding.
11. The power consumption is less.
12. Very low deformation of the workpieces surfaces.
13. High quality weld is obtained.
14. The process may be integrated into automated production lines.
15. Moderate operator skill level is enough.

22.8 DISADVANTAGES OF USW

1. Capital investment in ultrasonic machines is very high compared to resistance welding.
2. Only small and thin parts may be welded. The process is restricted to join thin materials, such as sheet foil and wire. The maximum thickness of 3 mm for softer materials (such as aluminum) and 1 mm for harder materials.
3. Materials, being welded tend to stick to the weld tip and anvil.
4. Parts may get damaged due to resonance problem.
5. Workpieces and equipment components may fatigue at the reciprocating loads provided by ultrasonic vibration.
6. Bonding mechanism of ultrasonic welding is quite complex and welding tip is subjected to cracking and sticking.

Review Questions

1. Explain the principle of operation of USW.
2. What are the equipment used in USW?
3. Explain the mechanism involved in the USW.
4. What are the process capabilities of USW?
5. What are the process parameters used in USW?
6. Compare different transducers used in USW.
7. What are the applications of USW?
8. What are the advantages of USW?
9. What are the limitations of USW?
10. Explain the role of coupling systems in USW.

23 CHAPTER

Plasma Arc Welding

23.0 INTRODUCTION

Plasma arc welding (PAW) is an arc welding process that produces coalescence of metals by heating them with a constricted arc between an electrode and the workpiece (transferred arc) or between the electrode and the constricting nozzle (non-transferred arc). Shielding is generally obtained from the hot, ionized gas issuing from the torch. This plasma gas is usually supplemented by an auxiliary source of shielding gas. Shielding gas may be a single inert gas or a mixture of inert gases. Pressure is not applied, and filler metal may or may not be added.

Plasma arc welding, like gas tungsten arc welding (GTAW), uses a non-consumable electrode. The PAW torch has a nozzle that creates a gas chamber surrounding the electrode. The arc heats the gas fed into the chamber to a temperature where it becomes ionized and conducts electricity. This ionized gas is defined as *plasma.* Plasma issues from the nozzle orifice at a temperature of about 30,000°F (16,700°C).

Plasma arc welding can be applied to join most metals in all positions. It provides better directional control of the arc and smaller heat affected zones as compared to GTAW. The narrow, constricted arc pattern is also more tolerant to variations in torch standoff distance.

The major disadvantage of plasma arc welding is the relatively high expense of the equipment. Also, compared to GTAW, there are more process control variables, requiring more complex welding procedures and more extensive operator training.

23.1 HISTORY OF PLASMA ARC

One of the earliest plasma arc systems was a gas vortex stabilized device, introduced by Schonherr in 1909. In this unit, gas was blown tangentially into a tube through which an arc was struck. The centrifugal force of the gas stabilized the arc along the axis of the tube by creating a low pressure axial core. Arcs up to several meters in length were produced, and the system proved useful for arc studies.

Gerdien and Lotz, built a water vortex arc stabilizing device in 1922. In this device, water injected tangentially into the center of a tube was swirled around the inner surface and ejected at the ends. When an arc struck between carbon electrodes was passed through the tube, the water concentrated the arc along its axis. The Gerdien and Lotz invention had no practical metal working applications because of the rapid consumption of its carbon electrodes and the presence of water vapor in the plasma jets.

While working on the arc melting of refractory metals in 1953, Gage observed the similarity in appearance between a long electric arc and an ordinary gas flame. Efforts to control the heat intensity and velocity of the arc led to the development of the modern plasma arc torch.

The first practical plasma arc metal working tool was a cutting torch introduced in 1955. This device was similar to a gas tungsten arc welding torch in that it used a tungsten electrode and a "plasma" gas. However, the electrode was recessed in the torch, and the arc was constricted by passing it through an orifice in the torch nozzle. The usual circuitry for gas tungsten arc welding was supplemented in the plasma arc cutting torch with a pilot arc circuit for arc initiation.

Commercial equipment for plasma arc surfacing emerged in 1961, and plasma arc welding was introduced in 1963.

23.2 MODE OF OPERATION

Plasma arc welding is basically an extension of the gas tungsten arc welding (GTAW) process. However, it has a much higher arc energy density and higher plasma gas velocity by virtue of the arc plasma being forced through a constricting nozzle, as shown in Figure 23.1.

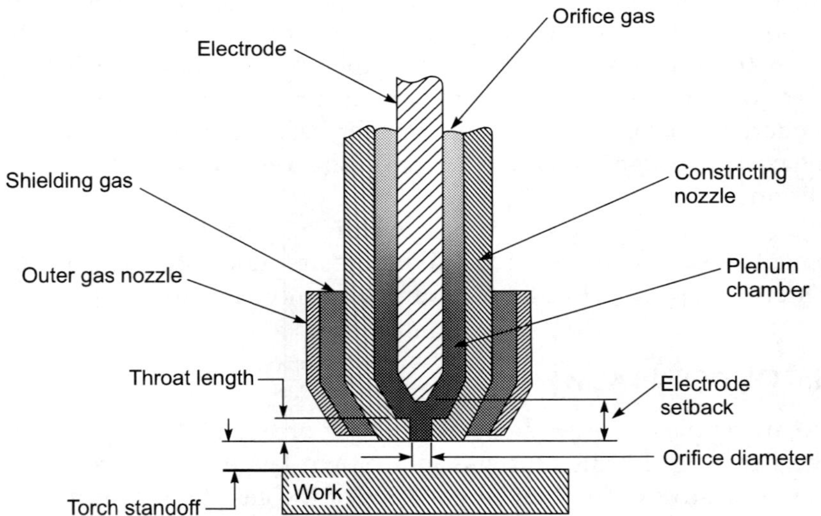

Figure 23.1 Plasma arc welding

The orifice gas is the gas directed through the torch to surround the electrode. It becomes ionized in the arc to form the plasma, and issues from the orifice in the torch nozzle as the plasma jet. For most operations, auxiliary shielding gas is provided through an outer gas cup, similar to gas tungsten arc welding. The purpose of the auxiliary shielding gas is to blanket the area of arc plasma impingement on the workpiece to avoid contamination of the weld pool.

The arc constricting nozzle, through which the arc plasma passes, has two main dimensions: orifice diameter and throat length. The orifice may be cylindrical or have a converging or diverging taper.

The distance that the electrode is recessed within the torch is the electrode setback. The dimension from the outer face of the torch nozzle to the workpiece is known as the *torch standoff distance*.

The plenum or plenum chamber is the space between the inside wall of the constricting nozzle and the electrode. The orifice gas is directed into this chamber and then through the orifice towards the work. A tangential vector may be imparted to the gas flow to form a swirl through the orifice.

The basic arrangement of both GTAW and PAW systems is shown in Figure 23.2. The electrode in the GTAW torch extends beyond the end of the shielding gas nozzle, the gas tungsten arc is not constricted and assumes an approximately conical shape, producing a relatively wide heat pattern on the workpiece. For a given welding current, the area of impingement of the cone-shaped arc on the workpiece varies with the electrode-to-work distance. Thus, a small change in arc length produces a relatively large change in heat input per unit area.

By contrast, the electrode in the plasma arc torch is recessed within the constricting nozzle. The arc is collimated and focused by the constricting nozzle on a relatively small area of the workpiece. Because the shape of the arc is essentially cylindrical, there is very little change in the area of contact on the workpiece as torch standoff varies. Thus, the PAW process is less sensitive to variations in torch-to-work distance than the GTAW process.

Since the electrode of the plasma arc torch is recessed inside the arc-constricting nozzle, it is not possible for the electrode to touch the workpiece. This feature greatly reduces the possibility of contaminating the weld with electrode metal.

As the orifice gas passes through the plenum chamber of the plasma arc torch, it is heated by the arc, expands and exits through the constricting orifice at high velocity. Since too powerful a gas jet can cause turbulence in the weld puddle, orifice gas flow rates are generally held within 0.25 to 5 L/min. The orifice gas alone is not generally adequate to the weld pool from atmospheric contamination. Therefore, auxiliary shielding gas must be provided through an outer gas nozzle. Typical shielding gas flow rates are in the range of 10 to 30 L/min.

23.3 PURPOSE OF ARC CONSTRUCTION

Several improvements in performance over open arc operation (GTAW) can be obtained by passing the plasma arc through a small orifice. The most noticeable improvement is the directional stability of the plasma jet. A conventional gas tungsten arc is attracted to the nearest work connection back to the power supply, and it is deflected by low strength magnetic fields. On the other hand, as a plasma jet is comparatively stiff, it tends to go in the direction in which it is pointed, and it is less affected by magnetic fields.

High current densities and high energy concentration can be produced by arc constriction. The higher current densities result in higher temperatures in the plasma arc.

The higher temperatures and electrical changes brought about by constricting an arc are compared in Figure 23.2. The left side of this figure represents a normal unconstricted tungsten arc operating at 200 A, DCEN, in argon at a flow rate of 19 L/min. The right side illustrates an arc, with the same current and gas flow that is constricted by passing it through a 4.8 mm diameter orifice. Under these conditions, the constricted arc shows a 100 percent increase in arc power and a 30 percent increase in temperature over the open arc. The spectroscopic methods used to measure the temperature of arcs are based on the analysis and interpretation of emission.

The increased temperature of the constricted arc is not its chief advantage, since the temperature in the gas tungsten arc far exceeds the melting points of the metals generally welded by the process.

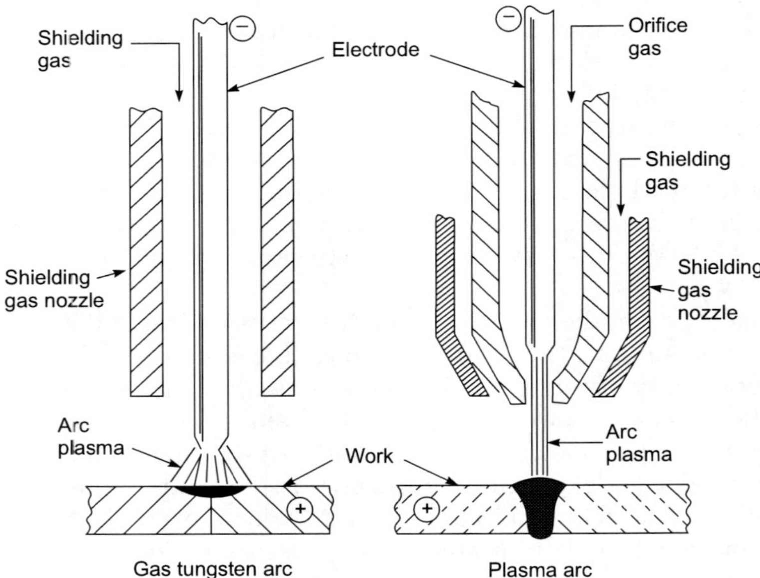

Figure 23.2 Comparison of gas tungsten arc and plasma arc welding process

The main advantages of the plasma arc are its directional stability and focusing effect brought about by arc constriction, and its relative insensitivity to variations in torch standoff distance.

The plasma arc efficiently uses the arc energy. The degree of arc collimation, arc force, energy density on the workpiece, and other characteristics are primarily functions of the following:

1. Plasma current
2. Orifice diameter and shape
3. Type of orifice gas
4. Flow rate of orifice gas
5. Type of shield gas

The fundamental differences in the plasma arc metal working processes arise from the relationship of these five factors. They can be adjusted to provide very high or very low thermal energies. For example, the high energy concentration and high jet velocity necessary for plasma arc cutting dictate a high arc current, a small diameter orifice, high orifice gas flow rate, and a gas having high thermal conductivity. For welding, on the other hand, a low plasma jet velocity is necessary to prevent weld metal expulsion from the workpiece. This calls for larger orifices, considerably lower gas flow rates and lower arc currents.

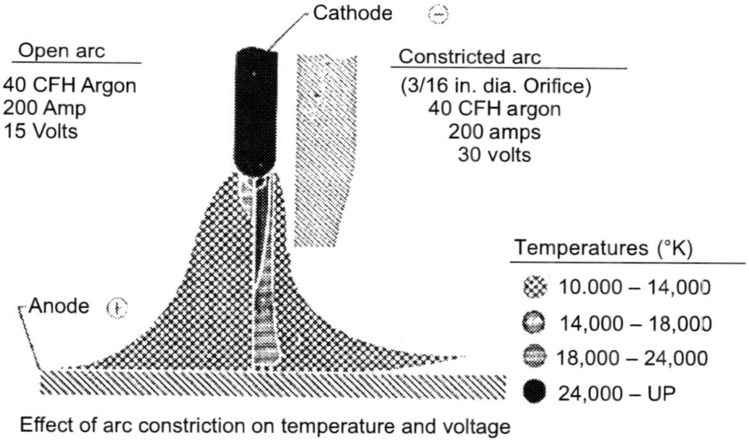

Figure 23.3 Effect of arc constriction on temperature and voltage

Constricted arc is much more effective than an open arc for heating the gas to be used for a particular operation. When the gas passes directly through a constricted arc, it is exposed to higher energy concentrations than when it passes around a conventional gas tungsten arc, as shown in Figure 23.3.

23.4 ARC MODE (FIGURE 23.4)

Two arc modes are used in plasma arc welding: transferred arc and non-transferred arc. Figure 23.5 illustrates the two modes. With a transferred arc, the arc "transfers" from the electrode to the workpiece. The workpiece is part of the electrical circuit, and heat is obtained from the anode spot on the workpiece as well as from the plasma jet.

With a non-transferred arc, the arc is established and maintained between the electrode and the constricting orifice. The arc plasma is forced through the orifice by the plasma gas. The workpiece is not in the arc circuit. Useful heat is obtained from the plasma jet only.

Transferred arcs have the advantage of greater energy transfer to the work, and this is the mode generally used for welding. Non-transferred arcs are useful for cutting and joining non-conductive workpieces or for applications where relatively low energy concentration is desirable.

If there is insufficient orifice gas flow or excessive arc current for a given nozzle geometry, or if the nozzle is touched to the work, the nozzle may be damaged by a phenomenon known as double arcing. In double arcing, the metallic torch nozzle forms part of the current path from the electrode back to the power supply. In essence, two arcs arc formed, as shown in Figure 23.4. The first arc is from the electrode to the nozzle, and the second is from the nozzle to the work. Heat generated at the cathode and anode spots, formed where the two arcs attach to the nozzle, invariably causes damage to this part.

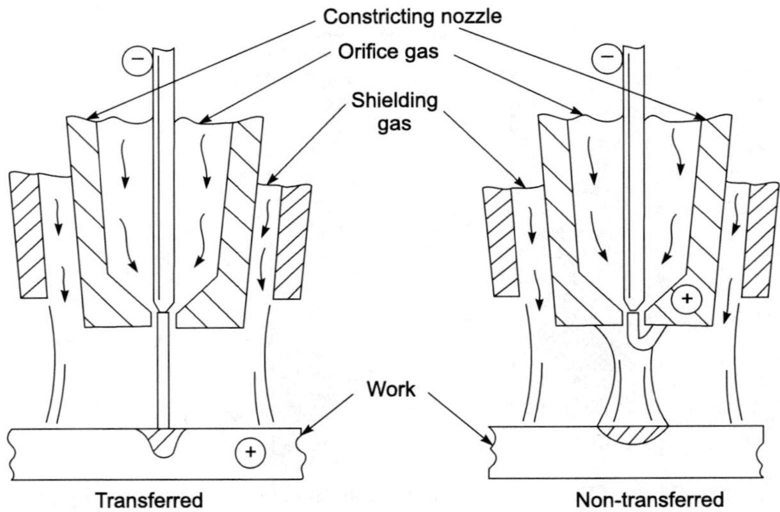

Figure 23.4 Transferred and non-transferred plasma arc modes

23.5 TYPES OF WELDING CURRENT

Direct current, electrode negative (DCEN) power is used for most plasma arc welding applications. A pure tungsten or thoriated tungsten electrode and a transferred arc are used for most applications. The current range for DCEN plasma welding is from approximately 0.1 to 500 amp. Current pulsation is often used. Steel alloys, stainless steels, nickel alloys, and titanium are commonly welded. Direct current, electrode positive (DCEP) is used to a limited extent for welding aluminum. Excessive electrode heating is the primary limitation using electrode positive. Its maximum current is usually less than 100 amp.

Sine wave alternating current with continuous high frequency stabilization can be used to weld aluminum and magnesium alloys. The current range is generally limited between 10 and 100 A. Higher amperages generally cannot be used due to excessive electrode deterioration during the electrode positive current cycle. The primary reason for using alternating current when welding magnesium and aluminum alloys is oxide removal. During the electrode positive half-cycle of alternating current, positive ions are released from the electrode which bombard the oxides on the surface of the workpiece. This bombardment, called cathodic etching, removes the oxides and exposes clean aluminum alloy to be welded. Square wave AC has largely replaced sine wave AC for welding aluminum and magnesium alloys.

Square wave alternating current, with unbalanced electrode positive and negative current half-cycles (variable polarity plasma arc) is highly efficient for welding magnesium and aluminum alloys. High frequency stabilization is not required. Unbalanced square wave retains the desired property of cathodic etching for oxide removal, with most of the energy anodic for maximum heat transfer.

23.6 ARC LENGTH (FIGURE 23.5)

The columnar nature of the constricted arc makes the plasma arc process less sensitive to variations in arc length than the gas tungsten arc process. Since the unconstricted gas tungsten arc has a conical shape, the area of heat input to the workpiece varies as the square of the arc length. A small change in arc length, therefore, causes a relatively large change in the unit area heat transfer rate. With the essentially cylindrical plasma jet, however, as the arc length is varied within normal limits, the area of heat input and the intensity of the arc are virtually constant.

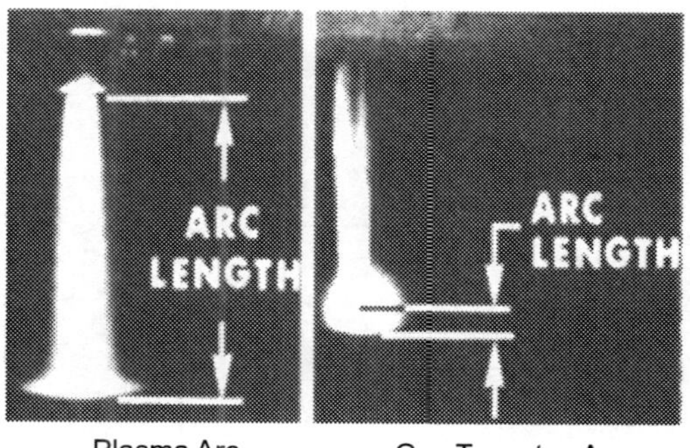

Figure 23.5 Comparisons of plasma arc and gas tungsten arc lengths commonly used for welding very thin metal sections at 10 A

The collimated plasma jet also permits the use of a much longer torch-to-work distance (torch standoff) than is possible with the GTAW process, and reduces the operator's skill required to manipulate the torch. Typical arc lengths used to weld thin gage material at approximately 10 amperes are shown in Figure 23.6. The plasma arc is approximately 1/4 inch (6.4 mm) long as compared to the 0.06 inch (1.5 mm) long gas tungsten arc.

23.7 PROCESS TECHNIQUES (FIGURE 23.6)

Process of work material cleaning and filler metal addition is same as TIG welding. Temperature experienced in a tungsten arc used for TIG welding is of the order of 11,000°C. The plasma arc welding process is well accepted in the fabrication, aerospace, and nuclear industries. It offers process fabrication latitude and economy, while maintaining quality and weld joint reliability. All metals weldable with the GTAW process can be satisfactorily welded with the PAW process; therefore, a few exceptions are required in the establishment of weldment acceptance specifications.

The advantage of the low-current plasma arc process, operating in the range of 0.1 to 50 A, is the very stable and controllable arc available at low currents for welding thin materials. Through the use of a pilot arc, transferred arc starting is more reliable than GTAW starting at these current levels. Because of the collimated arc, edge joints result in uniform bead contours when welded manually or automatically. Other applications include turbine blades, seal edges, bellows, pacemakers, and diaphragms. For these applications, plasma arc welding has often been the economical choice over laser welding.

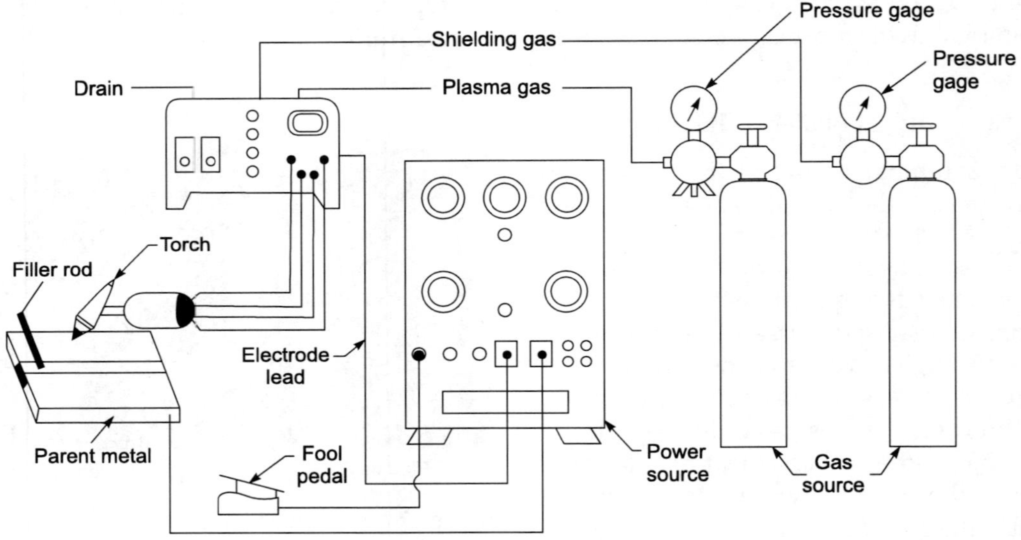

Figure 23.6 PAW process

High-current welding procedures, in the range of 50 to 400 A, often use the melt-in mode, which produces a weld similar to that obtained with conventional gas tungsten arc welding. The melt-in mode is generally preferred over the gas tungsten arc process in mechanized applications for consistent control of weld quality. Again, due to arc stability and stiffness, arc penetration into the weld joint is more controlled. It can also reduce welding time. Applications include welding lamination stacks, tube mill welding, cladding, and cover passes on keyhole welding.

23.8 ADVANTAGES

The low-current and high-current (melt-in) modes have the following advantages over gas tungsten arc welding:

1. Energy concentration is greater, hence
   (a) Welding speeds are higher in some applications.
   (b) Lower current is needed to produce a given weld and results in less shrinkage. Distortion may be reduced by as much as 50 percent.
   (c) Penetration can be controlled by varying welding variables.
2. Arc stability is improved.
3. Arc column has greater directional stability.
4. Narrower beads (higher depth-to-width ratio) for a given penetration, resulting in less distortion.
5. Need for fixturing is less for some applications.
6. Where the addition of filler metal is desirable, this operation is much easier since torch standoff distance is generous and the electrode cannot touch the filler or puddle. This also results in less downtime for tungsten repointing and eliminates tungsten contamination of the weld.
7. Reasonable variations in torch standoff distance have little effect on bead width or heat concentration at the work; this makes out-of-position welding much easier.

23.9 LIMITATIONS

Some of the limitations associated with low-current and high-current (melt-in) plasma arc welding include:

1. Due to the narrow constricted arc, the process has little tolerance for joint misalignment.
2. Manual plasma welding torches are generally more difficult to manipulate than a comparable GTAW torch.
3. For consistent weld quality, the constricting nozzle must be well maintained and regularly inspected for signs of deterioration.
4. Consumption of inert gas is high.
5. Noise level is high.

23.10 KEYHOLE WELDING TECHNIQUE

In plasma arc welding of certain ranges of metal thicknesses, appropriate combinations of plasma gas flow, arc current, and weld travel speed produce a relatively small weld pool with a hole penetrating completely through the base metal. It is called a *keyhole* and is shown in Figure 23.7. The keyhole technique is generally performed in the downhand position on material thicknesses ranging from 1.6 to 9.5 mm. However, using appropriate welding conditions on certain metal thicknesses, keyhole welding can be done in any position. The plasma arc process is the only gas shielded welding process typically operated with this unusual characteristic.

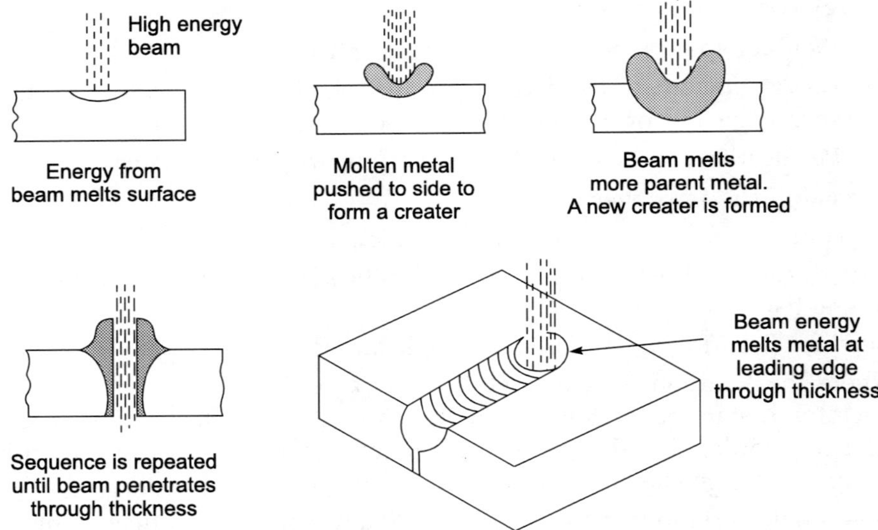

Figure 23.7 Pictorial representation of the keyhole in plasma arc welding

In the keyhole operation, molten metal is displaced to the top bead surface by the plasma stream (as it penetrates the weld joint) to form the keyhole. As the plasma arc torch is moved along the weld joint, metal melted at the front side of the keyhole by the arc flows around the plasma stream to the rear where the weld pool progressively solidifies. The principal advantage of keyhole welding is making welds in a single pass.

An open keyhole also forms an escape path through the thin molten lining for impurities to flow to the surface and gases to be expelled before solidification. The maximum weld pool volume and the resultant underbead root surface profile are largely determined by the force balance between the surface tension of the molten weld metal, the plasma arc current, and the velocity of the ionized gas exiting the orifice.

The high-current keyhole technique of welding operates just below the conditions that would actually cut rather than weld. For cutting, slightly higher orifice gas velocity blows the molten metal away. In welding, the lower gas velocity lets surface tension hold the molten metal in the joint. Consequently, orifice gas flow rates for welding are critical and must be closely controlled. Flow no more than 0.12 L/min is recommended. This rate is quite small.

A summary of the advantages of plasma welding with the keyhole technique over GTAW include:

1. The plasma gas flushing through the open keyhole helps to remove gases that would, under other circumstances, be trapped as porosity in molten metal.
2. The symmetrical fusion zone of the keyhole weld reduces the tendency for transverse distortion.

3. Greater joint penetration permits a reduction in the number of passes required for a given joint. Many welds can be completed in a single pass.
4. Square butt joints are generally used, thereby reducing joint preparation and machining costs.

Limitations of the plasma welding keyhole technique include:

1. Welding procedures involve more process variables and can have narrow operating windows.
2. Increased operator's skill is required for manual operation, particularly on thicker materials.
3. Plasma welding is not a all position welding process.
4. The plasma torch must be well maintained for consistent operation.

Comparison between the plasma arc welding and TIG welding is given in Table 23.1.

TABLE 23.1 Comparison between plasma and TIG welding

Plasma Arc Welding	TIG Welding
1. Two gases (inert) are used, one for plasma gas and other for shielding purpose.	1. Only one gas used, which forms plasma as well as shields the arc and molten weld pool.
2. It uses a constricted arc.	2. It uses a non-constricted arc.
3. Electrodes remain within nozzle and therefore, chance of tungsten inclusion and electrode contamination is zero.	3. Change of tungsten inclusion and electrode contamination if carelessly welded.
4. Very high temperature of the order of about 11,000°C can be achieved.	4. Temperature of the order about 4000°C can be achieved.
5. Deeper penetration can be achieved.	5. Penetration obtain by TIG is not so deeper.
6. Input heat distribution in the job is improved.	6. Input heat distribution is not so improved in TIG as PAW.
7. No more filler metal required.	7. More filer metal required.
8. Faster metal deposition rate can be achieved.	8. Metal deposition rate is not so faster.
9. Inert gas consumption is high.	9. Inert gas consumption is low.
10. Welding equipment cost is high.	10. Welding equipment cost is not so high.
11. Cutting of hard and brittle material is possible.	11. Cutting of hard and brittle material is not possible.

23.11 EQUIPMENT

The basic equipment for plasma arc welding is shown in Figure 23.8. Plasma arc welding is done with both manual and mechanized equipment.

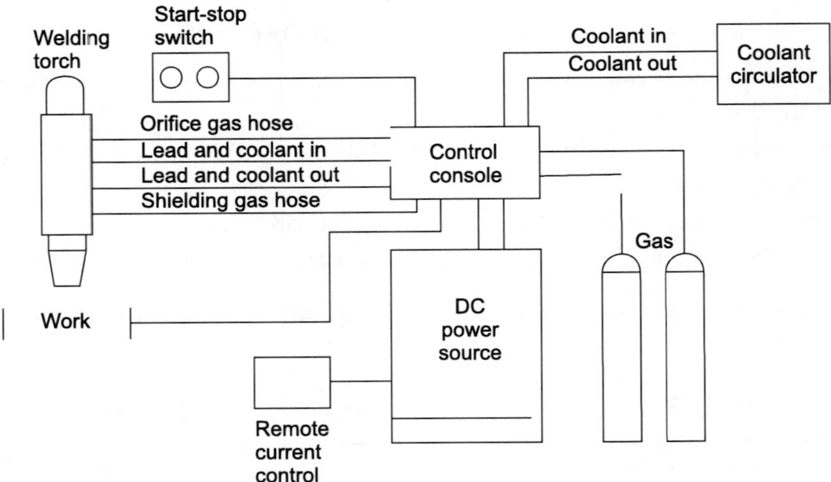

Figure 23.8 Typical equipment for plasma arc welding

A complete system for manual plasma arc welding consists of a torch, control console, power source, orifice and shielding gas supplies, source of torch coolant, and accessories such as an on-off switch, gas flow timers, and remote current control. Equipment is presently available for operation in the current range of 0.1 to 225 A, DCEN.

Mechanized equipment must be used to achieve the high welding speeds and deep penetration advantages associated with high-current plasma arc welding. A typical mechanized installation consists of a power source, control unit, machine welding torch, torch stand or travel carriage, coolant source, high frequency power generator, and supplies of shielding gases. Accessory units, such as an arc voltage control and filler wire feed system, may be used as required. Machine welding torches are available for welding with currents up to 500 A, DCEN.

23.12 ARC INITIATION

The plasma arc cannot be initiated with the normal techniques used with gas tungsten arc welding. Since the electrode is recessed in the constricting nozzle, it cannot be touch-started against the workpiece. It is first necessary to ignite a low-current pilot arc between the electrode and the constricting nozzle. Pilot arc power is normally provided either by a separate power supply within the control console or from the welding power supply itself. The pilot arc is generally initiated through the use of high-frequency AC power or by a high-voltage DC pulse superimposed on the welding circuit. These methods help to break down the arc gap and ionize the orifice gas so that it may conduct the pilot arc current.

The basic circuitry for a plasma arc welding system with a high frequency generator is shown in Figure 23.9. The constricting nozzle is connected to the power source positive terminal through a current limiting resistor. A low current pilot arc is initiated between the electrode and the nozzle by the high frequency generator. The electrical circuit is completed through the resistor. The ionized gas from the pilot arc forms a low resistance path between the electrode and work. When the power source is energized, the main arc is initiated between the electrode and the work. The pilot arc is used only to assist in starting the main arc. After the main arc starts, the pilot arc may be extinguished.

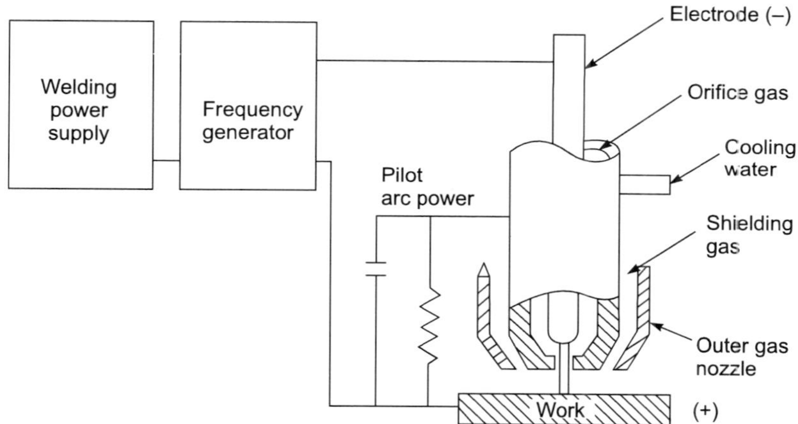

Figure 23.9 Plasma arc welding system with high frequency pilot arc initiation

23.13 POWER SUPPLY – NON PULSED

Steady-current DC power sources for plasma arc welding come in various amperage capacities from 0.1 A up to several hundred amperes. Conventional-type power sources with a drooping volt-amp characteristic are used for DCEN plasma arc welding. These are typically the same types of power sources that are used for GTAW and are available in amperage ranges from 0.1 A up to 400 A with 60 percent through 100 percent duty cycles. Rectifier type units are preferred over motor generator units because of electrical output characteristics.

A rectifier with an open circuit voltage in the range of 65 to 80 volts is satisfactory for plasma arc welding with argon or with argon-hydrogen gas mixtures containing up to 7 percent hydrogen. However, if helium or argon-hydrogen gas mixture containing more than 7 percent hydrogen is used, additional open circuit voltage is required for reliable arc ignition. This may be obtained by connecting two power sources in series. If erratic arc ignition is experienced, another approach is to strike the arc in pure argon and then switch over to the desired argon-hydrogen mixture or to helium for the welding operation. Constant current power sources are available with several options, such as a programmed upslope of current, a programmed taper or decay of weld current, and a programmed downslope of weld current. These special features of the power source are used for various applications, primarily automatic welding.

23.14 POWER SUPPLY – PULSED CURRENT

For some plasma arc welding applications, it is essential that pulsed current be used. Pulsed current power sources similar to those used with gas tungsten arc welding are used for plasma arc welding. A pulsed current power source is a conventional drooping volt-amp characteristic power source which has the capability of pulsing to a high level referred to as a peak current. Pulsed current power sources used for plasma arc welding have variable pulse frequencies and pulse width ratios. Figure 23.10 gives definition of terms for pulsed current.

Transistorized, inverter, and SCR-type power sources are available with pulsed current capabilities built into them. For conventional steady-current power sources, add-on packages are available which provide pulsed current within a limited range of pulse frequencies. Upslope and downslope of weld current can be obtained on pulsed current power sources.

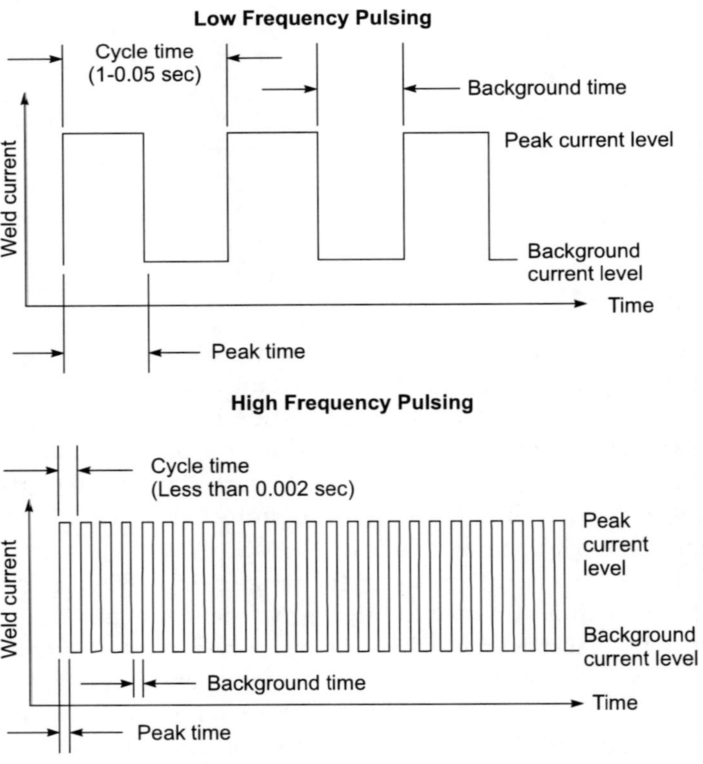

Figure 23.10 Pulsed current terminology

23.15 VARIABLE POLARITY PLASMA ARC WELDING

Keyhole plasma arc welding of aluminum has been done using square wave AC with variable polarity (VPPA). The variable polarity waveform is shown in Figure 23.11. This type of waveform, where the duration and magnitude of the DCEN and DCEP current excursions can be controlled separately, is obtainable using solid-state technology.

Cleanliness of the aluminum workpiece surface is of the utmost importance in avoiding porosity in the weld. Usual cleaning procedures involve mild alkaline solutions or vapor degreasing. Welding should occur soon after cleaning. With the variable polarity process, oxide removal before welding is not required for most aluminum alloys. However, the 5000 series alloys with their extremely tenacious surface oxides require removal of the surface oxides by scraping prior to welding.

Although a DC pilot arc is maintained during VPPA to stabilize the welding arc, continuous high frequency is not used.

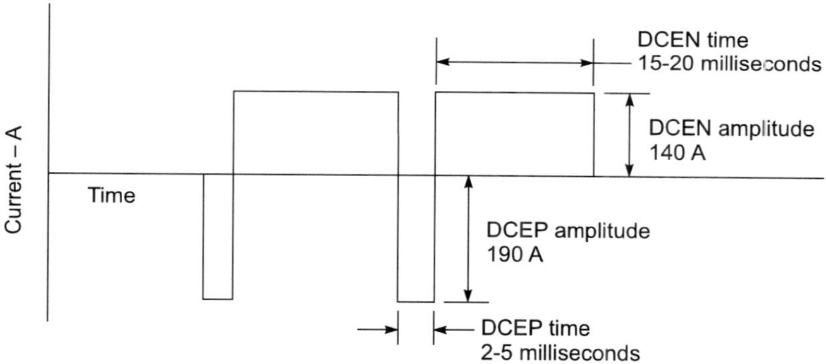

Figure 23.11 Typical variable polarity current waveform

23.16 DURATION OF CURRENT EXCURSIONS

The most important variable keyhole plasma arc welding of aluminum was found to be DCEN and DCEP duration times. The proper ratio of DCEN and DCEP times was determined empirically. Best results were obtained with DCEN current flowing from 15 to 20 milliseconds and DCEP current flowing from 2 to 5 milliseconds. Refer to Figure 23.11. With DCEP durations shorter than 2 milliseconds, the weld was porous. When the DCEP time exceeded 6 milliseconds, tungsten deterioration and double-arcing tendencies became apparent.

Note that the DCEP current amplitude, shown in Figure 23.11, is greater than the DCEN current amplitude. The added DCEP current provides an additional spike of cleaning action to break up surface oxides on the workpiece, yet produces minimal heat input on the electrode and torch orifice. Appropriate cleaning of the weld face and root can be accomplished by increasing the DCEP current (up to 80 A).

23.17 APPLICATION

Typical keyhole VPPA welding conditions for 0.25 in thick aluminum in the flat, horizontal and overhead positions are shown in Table 23.2.

Table 23.2 VPPA welding conditions for 0.250 inch thick aluminum in the flat, horizontal and overhead positions

Position	Flat		Horizontal		Overhead	
Material thickness, inch (mm)	1/4	(6.4)	1/4	(6.4)	1/4	(6.4)
Type aluminum	2219		3003		1100	
Filler metal diameter, inch (mm)	1/16	(1.6)	1/16	(1.6)	1/16	(1.6)
Filler metal grade	2319		4043		4043	
DCEN welding current, A	140		140		170	
DCEN welding time, ms	19		19		19	
Additional DCEP current, A	50		60		80	
DCEP current time, ms	3		4		4	
Plasma gas cfh (L/min) (start)	Ar 2	(0.9)	Ar 2.5	(1.2)	Ar 2.5	(1.2)
Plasma gas cfh (L/min) (run)	Ar 5	(2.4)	Ar 4.5	(2.1)	Ar 5	(2.4)
Shielding gas flow cfh (L/min)	Ar 30	(14)	Ar 40	(19)	Ar 45	(21)
Electrode size, in. (mm)	1/8	(3.2)	1 /8	(3.2)	1 /8	(3.2)
Travel speed, ipm (mm/s)	8	(3.4)	8	(3.4)	7.5	(3.2)

The process has been used to make single pass, full penetration welds on space shuttle fuel tanks, hulls for hydrofoil boats, and aluminum tanks.

23.18 PLASMA CONTROL CONSOLE

The plasma console is a device that controls the main plasma arc functions. A typical plasma console contains controls for setting plasma gas flow, shielding gas flow, provide a junction box for gas and water hoses, may or may not provide a high frequency circuit for the starting of the pilot arc, and may have a small power source that will provide the current to the pilot arc. Other features that might be provided are a high-low option so that plasma gas flow rate can be easily switched between melt-in mode and keyhole mode, and an arc pressure gage that would measure the plasma gas back-pressure at the orifice.

The plasma console might have built-in capability for programmed upslope and downslope of the plasma gas to start and close a keyhole, and the plasma console might also have the water circulator built into the same package. Figure 23.12 illustrates a plasma control console.

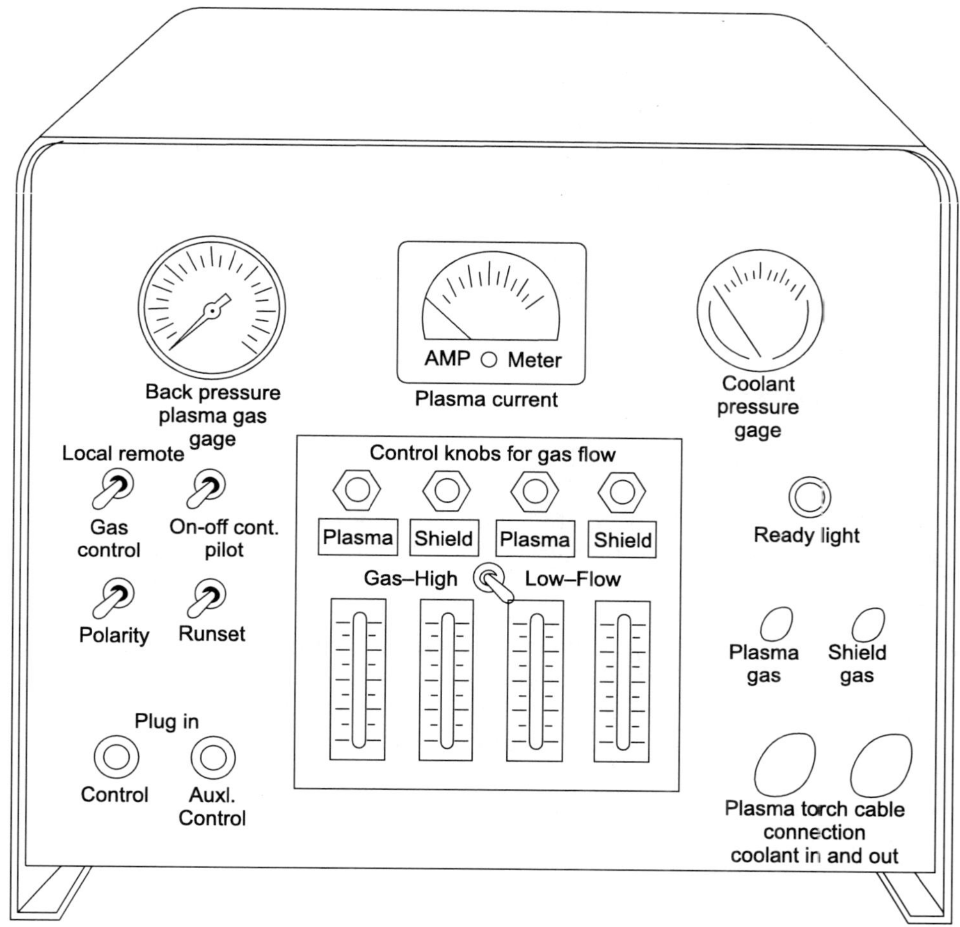

Figure 23.12 Plasma welding control console

23.19 WELDING TORCHES

Plasma arc welding torches are more complex than those used for gas tungsten arc welding. A series of passages are necessary to supply the torch with orifice gas, shielding gas, and liquid coolant to chill the constricting nozzle assembly.

In most instances, two dual-function cables provide both electrical energy and the circulating coolant. One cable supplies current for the pilot arc, while the other supplies the welding current. Two additional hoses provide the orifice (plasma) and shielding gases. Cooling water is necessary to dissipate the heat generated in the constricting nozzle by the pilot arc and main welding arc. The electrode holder assembly for a plasma arc welding torch is made of copper, and most are designed to center the electrode automatically within the central section of the torch nozzle. Any misalignment of the electrode in this section may cause melting of the copper nozzle near the orifice, possible weld contamination, and undercutting.

The orifice gas supplied to the torch has a low flow rate and thus, does not provide sufficient gas to protect the molten weld pool from atmospheric contamination. In addition, turbulence is caused by the high velocity of the plasma stream while keyhole welding further reduces effectiveness of plasma gas coverage. Necessary shielding gas is supplied through the shielding gas nozzle that surrounds the orifice tip portion of the torch. In some applications, additional shielding gas trailers are required for further protection.

23.20 MANUAL TORCHES

A cross section of a typical torch design for manual plasma arc welding is shown in Figure 23.13. The torch generally is lightweight and has a handle, some device for securing the tungsten electrode in position and conducting current to it, separate passages for orifice and shielding gases, a water cooled constricting copper nozzle with inlet and outlet passages, and a shielding gas nozzle (usually made of a ceramic material).

Manual plasma arc welding torches are available with 70 degree and 90 degree head angles. They are available for operation on DCEN at currents up to 225 A, and also for operation on DCEP up to about 70 A. DCEP is used to a limited extent with tungsten or water cooled copper electrodes for welding aluminum.

Controls for welding current and gas are usually separate from the torch and are operated either by foot control or automatically. Holders to mount the torch for mechanized application are available.

23.21 MACHINE TORCHES (FIGURE 23.13)

Torches for automated plasma arc welding are similar to manual torches, except they are designed with straight-line or offset configurations.

Mechanized plasma arc welding torches are available for operation on either DCEN, DCEP, or square wave AC, with current ratings generally ranging from 50 to 500 A. DCEN is used with a tungsten electrode for most welding applications, with an optional mode of pulsed current welding. In the pulsing mode, the current is allowed to fluctuate between two set currents, a higher and a lower level of amperage, which allows the molten weld puddle to solidify at the lower level. The pulsing mode can help alleviate problems of distortion by reducing the total heat input along the weld joint. Also available is variable polarity (square wave AC) plasma arc welding for welding aluminum.

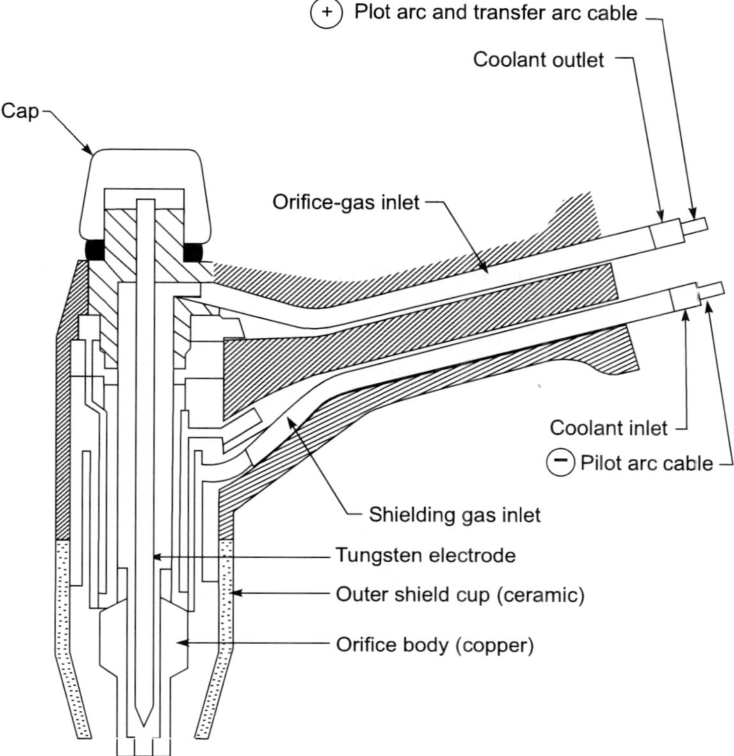

Figure 23.13 Typical manual torch head design

23.22 ARC CONSTRICTING NOZZLES

Nozzles in a wide variety have been designed for use in PAW. These include single orifice and multiple orifice nozzles, with holes arranged in circles, rows, and other geometric patterns. Single orifice nozzles are the most widely used. Among the multiple port nozzles, the design most widely used has the center orifice bracketed by two smaller auxiliary gas orifices, all with a common centerline. These two types are shown in Figure 23.14.

With a single orifice nozzle, the arc and all of the plasma gas pass through the single orifice. With the multiport nozzle, the arc and some of the orifice gas pass through the larger center orifice, while the remainder of the gas moves through the smaller auxiliary orifices.

Multiport nozzles can be used to an advantage on several types of weld joints. When the multiport nozzle is aligned to place the common centerline of all three ports perpendicular to the weld groove, the plasma jet is concentrated on the joint by these adjacent gas streams. The results are a narrowed weld bead and higher welding speeds.

Each given orifice size and orifice gas flow rate has a maximum current rating. For example, an orifice with 2.1 mm diameter might be rated at 75 A with 0.9 L/min argon flow rate. If the flow rate of the orifice gas dropped below 0.9 L/min, the maximum current rating of the orifice would also decrease.

The electrode in the plasma arc torch is recessed in the arc constricting nozzle. As the arc passes through the nozzle, it is collimated so that the arc heat is concentrated onto a relatively small area of the workpiece. The increased heat concentration, along with the characteristically more forceful plasma stream, produces a narrower weld fusion zone while increasing weld penetration and preventing arc wander.

During normal operation, the arc column within the torch nozzle is surrounded by a layer of non-ionized gas. This layer of relatively cool non-conductive gas at the nozzle wall provides thermal and electrical insulation that protects the inner surface of the nozzle. The most commonly used nozzle material is copper. Copper nozzles are water cooled, and can be used to constrict an arc plasma and can withstand temperature up to 16600°C. If the protective layer of gas is disturbed, as when there is insufficient orifice gas flow, or if there is excessive arc current for a particular nozzle geometry, double arcing may occur and damage the nozzle.

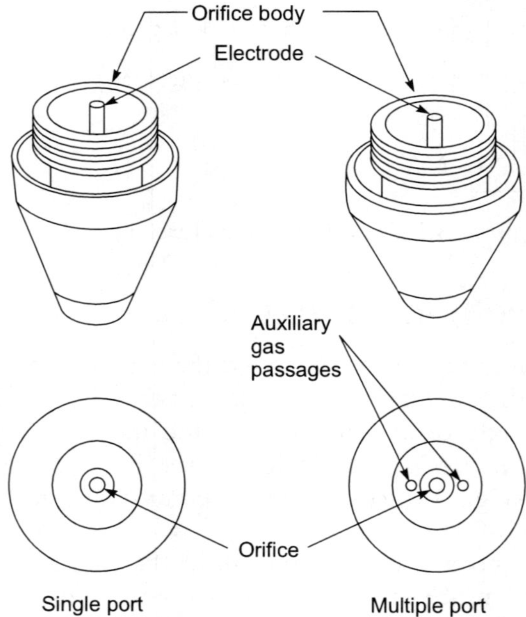

Figure 23.14 Single and multiple port constricting nozzles

23.23 ACCESSORY EQUIPMENT

Wire Feeders

As with the GTAW process, conventional filler wire feed systems can be used with the PAW process. The filler metal is added to the leading edge of the weld pool or the keyhole at a predetermined feed rate. A wire feed system may alleviate occurrences of the undercut or underfill when welding thicker materials.

Hot wire feed systems may also be used and should be fed into the trailing edge of the weld pool. Initiation and termination of wire feed may be controlled and programmed with automatic welding equipment.

Arc Voltage Control

Since the plasma arc process is relatively insensitive to arc length variations, arc voltage control equipment is not necessary for many applications. Arc voltage control can nevertheless be used when plasma arc welding is uneven or contoured joint geometries. The control must be deactivated or "locked-out" when the current or plasma gas flow rate is sloped during weld starts or crater filling because changing these variables will also cause a change in the arc voltage.

Positioning Equipment

Positioning equipment for PAW is similar to that used for GTAW, depending on the application, either the workpiece may be manipulated or the torch motion can be controlled. Workpiece manipulation generally involves a rotary positioner with the capability of tilt control. Moving the torch while the workpiece remains stationary, requires a carriage on tracks or a side beam carriage for following linear joints. Combining the movement of the torch and workpiece as a system requires computer programming for coordinating the operations.

23.24 MATERIALS

The plasma arc welding process can be used to join all metals weldable by the GTAW process. Most material thicknesses from 0.3 to 6.4 mm can be welded in one pass with a transferred arc. All metals except aluminum and magnesium and their alloys are welded with DCEN. Square wave AC is used to effectively remove refractory oxides when welding aluminum and magnesium. AC welding reduces the current capacity of the electrode unless the power source is capable of minimizing the duration of the positive electrode cycle. One-pass keyhole welds can be made in aluminum alloys up to 12.7 mm thick.

Metallurgical effects of the heat from the plasma and gas tungsten arc welding processes are similar, except the smaller diameter plasma arc will usually melt less base metal, resulting in narrower and deeper penetration. Preheat, postheat, and gas shielding procedures are similar for both processes. Each base material has its requirements that maximize weld quality.

23.25 CONSUMABLES

Filler Metal

Filler metals used to weld the work base materials are the same as those used with the GTAW and GMAW processes. They are added in rod form for manual welding or wire form for mechanized welding. Table 23.3 lists the AWS (American Welding Society) specifications for appropriate filler metals.

TABLE 23.3 AWS specifications for filler metals used for plasma arc welding

AWS Specification	Filler Metals
A5.7	Copper and copper alloy welding rods
A5.9	Corrosion-resistant chromium and chromium-nickel steel bare electrodes
A5.10	Aluminum and its alloys welding rods and bare electrodes
A5.14	Nickel and its alloys bare welding rods and electrodes
A5.16	Titanium and its alloys bare welding rods and electrodes
A5.18	Mild steel electrodes for gas metal arc welding
A5.19	Magnesium alloys welding rods and bare electrodes
A5.24	Zirconium and its alloys bare welding rods and electrodes

Electrodes

The electrode is the same as used for gas tungsten arc welding. Pure tungsten rods and tungsten with small additions of thoria, zirconia, or ceria may be used for DCEN welding. Electrodes are made to ANSI (American National Standards Institute)/AWS A5.12, specification for tungsten arc welding electrodes. Pure tungsten electrodes are generally selected for AC welding.

Direct current electrode positive welding is not widely practiced with the plasma arc welding process because of its intense electrode heating and reduced current capacity. The arc end of the electrode is ground to a cone with an angle of 20 to 60 degrees, as specified by the torch maker.

It is essential to have a smooth concentric shape. The electrode collet must hold the electrode in the exact center of the nozzle orifice. A gage is usually specified to set the electrode's axial position. If these precautions are not observed, the consistency of the welds will suffer, and excessive deterioration of the constricting nozzle will occur.

For square wave AC welding, the electrode is usually prepared with a balled or flat end. The recommended shapes help prevent electrode overheating and provide greater current carrying capability.

Gases

The choice of gases to be used for plasma arc welding depends on the metal to be welded. For many PAW applications, the shielding gas is often the same as the orifice gas. Typical gases used to weld various metals are shown in Table 23.4.

TABLE 23.4 Gas selection guide for high current plasma arc welding*

Metal		Thickness		Welding Technique	
		in.	mm	Keyhole	Melt-in
Carbon steel	under	1/8	3.2	Ar	Ar
(aluminum killed)	over	1/8	3.2	Ar	75% He-25% Ar
Low alloy steel	under	1/8	3.2	Ar	Ar
	over	1/8	3.2	Ar	75% He-25% Ar
Stainless steel	under	1/8	3.2	Ar, 92.5% Ar-7.5% H_2	Ar
	over	1/8	3.2	Ar, 95% Ar-5% H_2	75% He-25% Ar
Copper	under	3/32	2.8	Ar	75% He-25% Ar
	over	3/32	2.8	Not recommended[b]	He
Nickel alloys	under	1/8	3.2	Ar, 92.5% Ar-7.5% H_2	Ar
	over	1/8	3.2	Ar, 95% Ar-5% H_2	75% Re-25% Ar
Reactive metals	under	1/4	1/4	Ar	Ar
	over	1/4	6.4	Ar-He (50 to 75% He)	75% He-25% Ar

* Gas selections are for both orifice and shielding gases.

b. The underbead will not form correctly. The technique can be used for copper-zinc alloys only.

The orifice gas must be inert with respect to the tungsten electrode to avoid rapid deterioration of the electrode. The shielding gases are generally inert. Active shielding gas can be used if it does not adversely affect the weld properties.

Argon is the preferred orifice gas for low current plasma arc welding because its low ionization potential assures a dependable pilot arc and reliable arc starting. Since the pilot arc is used only to maintain ionization in the plenum chamber, pilot arc current is not critical; it can remain fixed for a wide variety of operating conditions. The recommended orifice gas flow rates are usually less than 1 ft^3/h (0.5 L/min) and the pilot arc current may be fixed at 5 A.

Typical shielding gases for low current welding are given in Table 23.5. Argon is used for welding carbon steel, high-strength steels, and reactive metals, such as titanium, tantalum, and zirconium alloys.

Argon-hydrogen mixtures are often used as the orifice and shielding gases for making keyhole welds in stainless steel, nickel base alloys, and copper-nickel alloys. Permissible hydrogen percentages vary, from the 5 percent used on 1/4 inch (6.4 mm) thick stainless steel to the 15 percent used for the highest welding speeds on 0.150 inch (3.8 mm) and thinner wall stainless tubing in tube mills. In general, the thinner the workpiece, the higher the permissible percentage of hydrogen in the gas mixture, up to 15 percent maximum. However, when argon-hydrogen mixtures are used as an orifice gas, the rating of the orifice diameter for a given welding current is usually reduced because of the higher arc temperature.

TABLE 23.5 Shielding gas selection guide for low current plasma arc welding*

Metal	Thickness		Welding Technique	
	in.	mm	Keyhole	Melt-in
Aluminum .... under	1/16	1.6	Not recommended	Ar, He
over	1/16	1.6	He	He
Carbon steel .... under	1/16	1.6	Not recommended	Ar, 25% He-75% Ar
(aluminum killed) over	1/16	1.6	Ar, 75% He-25% Ar	Ar, 75% He-25% Ar
Low alloy steel .... under	1/16	1.6	Not recommended	Ar, He, Ar-H_2 (1-5% H_2)
over	1/16	1.6	75% He-25% Ar, Ar-H_2 (1-5% H_2)	Ar, He, Ar-H_2 (1-5% H_2)
Stainless steel	All		75% He-25% Ar, Ar-H_2 (1-5% H_2)	Ar, He, Ar-H_2 (1-5% H_2)
Copper .... under	1/16	1.6	Not recommended	25% He-75% Ar, 75% He-25% Ar, He
over	1/16	1.6	75% He-25% Ar, He	He
Nickel alloys	All		Ar, 75% He-25% Ar, Ar-H_2 (1-5% H_2)	Ar, He, Ar-H_2 (1-5% H_2)
Reactive metals .... under	1/16	1.6	Ar, 75% He-25% Ar, He	Ar
over	1/16	1.6	Ar, 75% He-25% Ar, He	Ar, 75% He-25% Ar

* Gas selections are for shielding gas only. Argon is the orifice gas in all cases.

Additions of hydrogen to argon, produce a hotter arc and more efficient heat transfer to the workpiece. In this way, higher welding speeds are obtained with the given arc current. The amount of hydrogen that can be used in the mixture is limited because excessive hydrogen additions tend to cause porosity or cracking in the weld bead. With the plasma keyhole technique, a given metal thickness can be welded with higher percentages of hydrogen than are possible with the gas tungsten arc welding process. The ability to use higher percentages of hydrogen without inducing porosity may be associated with the keyhole effect and the different solidification pattern it produces.

Helium additions to argon, produce a hotter arc for a given arc current. The mixture must contain at least 40 percent helium before a significant change in heat can be detected; mixtures containing over 75 percent helium behave in the same way as pure helium. Argon-helium mixtures containing between 50 and 75 percent helium are generally used for making keyhole welds in heavier titanium and aluminum sections, and for filler passes on all metals when additional heat and wider heat pattern are desirable.

The shielding gas provided through the gas shielding nozzle and around the nozzle can be argon, an argon-hydrogen mixture, or an argon-helium mixture, depending on the welding application. Shielding gas flow rates are usually in the range of 10 to 15 L/min for low current applications; for high current welding, flow rates 15 to 30 L/min are used.

Use of helium as an orifice gas increases the heat load on the torch nozzle and reduces its service life and current capacity. Because of the lower mass of helium, it is difficult, at reasonable flow rates, to obtain a keyhole condition with this gas. Therefore, helium is used only for making melt-in welds.

Since the shielding gas does not contact the tungsten electrode, reactive gases such as CO_2, can sometimes be used. Flow rates for CO_2 are in the range of 10 to 15 L/min. Plasma welding of lamination stacks often uses 75% Ar-25% CO_2 as the shielding gas.

When the gas flow and current must be varied during the weld, or at the start and end of a keyhole weld, an electronic programmable gas control system is used.

Table 23.6 gives the summary of plasma welding gases.

Table 23.6 Summary of plasma welding gases

Gas	Material	Benefits	Laminations
Argon plasma/ Argon shield	All including reactive metals (Ti)	Inert	Lowest heat input
Argon plasma/ Argon shield	All particularly high conductivity materials	Inert, higher heat input	May reduce arc pressure and limit keyhole technique
Argon plasma/ Argon 5% Hydrogen shield	Austenitic steels	Improves constriction and heat transfer. Removes oxide	Cannot be used with hydrogen-sensitive materials, such as ferritic alloys steels, copper-aluminum

Back Purge and Trailing Gases

When welding metals such as titanium, zirconium, or tantalum, it is essential to shield the hot metals from atmospheric contamination until they have cooled below the reaction point. Auxiliary shielding provided by backup and trailing shields are necessary. A trailing shield device can be attached to the rear of a plasma torch. The trailing shield provides increased gas coverage and time for cooling.

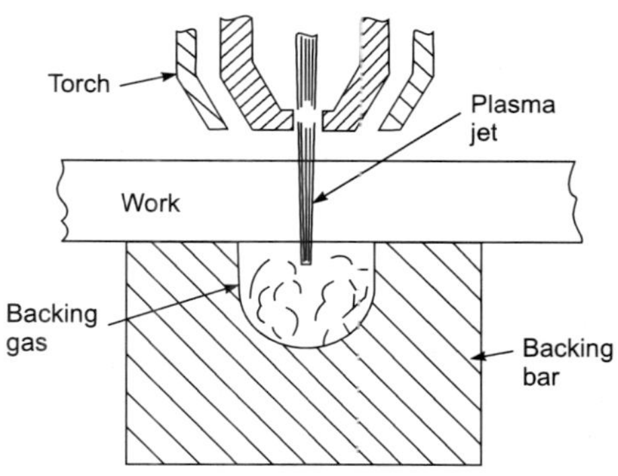

Figure 23.15 Typical backing bar for keyhole plasma arc welding

PAW of highly reactive materials can be done in a welding chamber or a glove box similar to GTAW. For such metals as stainless steel, where the root of the weld is exposed to the atmosphere, a back purge of a non-reactive gas such as argon, helium, or nitrogen is used (Figure 23.15).

23.26 FIT-UP AND FIXTURING

Fit-up and fixturing for PAW should be the same as used for GTAW. Joint edges for butt fusion welds should be in intimate contact and clamp down pressure should be applied to the weld joint during welding. Burrs, notches and gaps along the weld joint can cause separation of the weld joint because of melt back and should be eliminated.

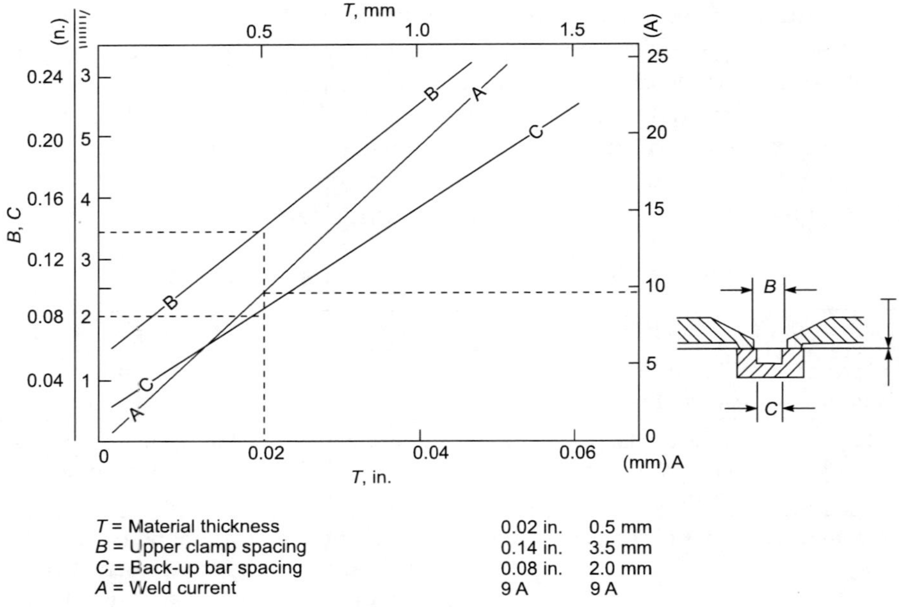

Figure 23.16 Graph showing welding data for low current butt welding of stainless steel (on the right, the jig geometry is schematically represented)

Fit-up, fixturing, and cleanliness are critical when welding very thin materials, such as 0.004, in stainless steel. Joint fit-up must be precise and consideration should be given to chill bars to remove heat from the joint.

Figure 23.16 shows a graph that gives a general idea of fixturing and clamping for welding of 304 stainless steel.

Weld Backing

Weld backing is used to locate parts, take heat away from the weld, and provide gas backup to shield the underside of the weld area.

Chill clamping and gas backup also help reduce the heat affected zone of the weld, The type of material being welded will largely determine if gas backup is needed. Metals such as titanium, zirconium, and niobium, need inert gas backup. Some weld joints may not allow gas backup because of design constraints. Gas backup inside of parts can also help cool and protect items such as electronic components that might be inside a part during welding.

Torch Positions

PAW can be done in all positions similar to GTAW welding. Users must remember that the plasma arc is very directional and when welding, it is more difficult to keep the arc exactly on the weld joint. The arc does not bend to follow the centerline of the joint. Side-to-side variations of the torch must be extremely limited.

Fortunately, in plasma arc welding the variation of stand off distance (distance between work and torch) is not so critical when compared to GTAW welding.

Filler metal can be used with plasma welding in the same manner as with GTAW welding.

23.27 WELDING PROCEDURES

Welding procedures for a variety of materials are listed in Tables 23.7 to 23.10.

TABLE 23.7 Typical plasma arc welding conditions for butt joints in stainless steel

Thickness		Travel Speed		Current (DCSP)	Arc Voltage	Nozzle Type[a]	Gas Flow[b]				Remarks[d]
							Orifice[c]		Shield[c]		
in.	mm	in./min	mm/s	A	V		ft³/H	L/min	ft³/H	L/min	
0.092	2.4	24	10	115	30	111 M	6	3	35	17	Keyhole, square-groove weld
0.125	3.2	30	13	145	32	111 M	10	5	35	17	Keyhole, square-groove weld
0.187	4.8	16	7	165	36	136 M	13	6	45	21	Keyhole, square-groove weld
0.250	6.4	14	6	240	38	136 M	18	8	50	24	Keyhole, square-groove weld

a. Nozzle type: number designates orifice diameter in thousandths of an inch; "M" designates design.

b. Gas underbead shielding is required for all welds.

c. Gas used: 95% Ar-5% H.

d. Torch standoff: 3/16 in. (4.8 mm).

Manual Welding

Usually, manual welding with plasma is best done in the low current range from 0.1 A to 50.0 A. When manually welding above 50.0 A, the GTAW process is often easier for the operator to manipulate and more economical. Manual PAW is often used for welding wire mesh when each wire has to be joined or edge fusion welded. Because standoff distance in PAW is not so critical as in GTAW, PAW is superior for this application. Usually, the standoff distance can be raised or lowered as much as 0.035 inch (0.9 mm) without affecting the weld quality.

Mechanized Melt-in Welding

Mechanized melt-in PAW is very popular, especially for welding small intricate components, such as medical equipment, lighting components, instrumentation, batteries, wires, and bellows.

In many applications, plasma is used with microprocessor controls which control initial current, upslope, pulsing, downslope, and final current.

Because high frequency is used only to start the pilot arc, there is no high frequency burst when the arc is transferred for welding. This allows plasma to be used around electronic test equipment, robots, microprocessor controls, and programmable controls, without the need to isolate or shield components that would need protection in automated systems.

TABLE 23.8 Typical plasma arc welding conditions for butt joints in carbon and low alloy steels

Metal	Thickness		Travel Speed		Current (DCSP) A	Arc Voltage V	Nozzle Type[a]	Gas Flow[b]				Remarks[d]
								Orifice[c]		Shield[c]		
	in.	mm	in./min	mm/s				ft³/H	L/min	ft³/H	L/min	
Mild steel	0.125	3.2	12	5	185	28	111 M	13	6	60	28	Keyhole, square-groove weld
4130 steel	0.170	4.3	10	4	200	29	136 M	12	6	60	28	Keyhole, square-groove weld, 1.2 mm (3/64 in.) diam. filler wire added at 30 in./min (13 mm/s)
D6AC steel	0.250	6.4	14	6	275	33	136 M	15	7	60	28	Keyhole, square-groove weld, 600°F (315°C) preheat

[a] Nozzle type: number designates orifice diameter in thousandths of an inch; "M" designates design.
[b] Gas underbead shielding is required for all welds.
[c] Gas used: argon.
[d] Torch standoff: 3/64 in. (1.2 mm) for all welds.

Mechanized Keyhole Welding

It is recommended that plasma keyhole welding should be done in an automated mode. Keyhole welding requires accurate control of travel speed, plasma gas flow, and wire feed speed. The development of good mass flow controllers has led to more accurate control of the plasma gas during welding.

TABLE 23.9 Typical plasma arc welding conditions for butt joints in titanium

Thickness		Travel Speed		Current (DCSP)	Arc Voltage	Nozzle Type[a]	Gas Flow[b]				Remarks[c]
							Orifice		Shield		
in.	mm	in./min	mm/s	A	V		ft³/H	L/min	ft³/H	L/min	
0.125	3.2	20	8.5	185	21	111 M	8[d]	3.8	60[d]	28	Keyhole, square-groove weld
0.187	4.8	13	5.5	175	25	136 M	18[d]	9	60[d]	28	Keyhole, square-groove weld
0.390	9.9	10	4.2	225	38	136 M	32[e]	15	60[e]	28	Keyhole, square-groove weld
0.500	12.7	10	4.2	270	36	136 M	27[f]	13	60[f]	28	Keyhole, square-groove weld

[a.] Nozzle type: number designates orifice diameter in thousandths of an inch; "M" designates design.
[b.] Gas underbead shielding is required for all welds.
[c.] Torch standoff: 3/16 in. (4.8 mm).
[d.] Gas used: argon.
[e.] Gas used: 75% He-25% Ar.
[f.] Gas used: 50% He-50% Ar.

Table 23.10 Typical plasma arc welding conditions for welding stainless steels-low amperage

Thickness		Type of Weld	Travel Speed		Current (DCSP) A	Orifice Dia.		Gas Flow Orifice[a,b,c]		Torch Standoff		Electrode Diam.		Remarks
in.	mm		in./min	mm /s		in.	mm	d/h	L/min	in.	mm	in.	mm	
0.030	0.76	Square-groove weld, butt joint	5.0	2	11	0.030	0.76	0.6	0.3	1/4	6.4	0.040	1.0	Mechanized
0.060	1.5	Square-groove weld, butt joint	5.5	2	28	0.047	1.2	0.8	0.4	1/4	6.4	0.060	1.5	Mechanized
0.030	0.76	Fillet weld, tee joint	–	–	8	0.030	0.76	0.6	0.3	1/4	6.4	0.040	1.0	Manual, filler metal[d]
0.060	1.5	Fillet weld, tee joint	–	–	22	0.047	1.2	0.8	0.4	1/4	6.4	0.060	1.5	Manual, filler metal[d]
0.030	0.76	Fillet weld, lap joint	–	–	9	0.030	0.76	0.3	0.6	3/8	9.5	0.040	1.0	Manual, filler metal[d]
0.060	1.5	Fillet weld, lap joint	–	–	22	0.047	1.2	0.8	0.4	3/8	9.5	0.060	1.5	Manual, filler metal[e]

[a.] Orifice gas: argon.
[b.] Shielding gas: 95% Ar-5% H at 20 ft 3/h (10 L/min).
[c.] Gas underbead shielding: argon at 10 ft 3/h (5 L/min).
[d.] Filler wire: 0.045 inch (1.1 mm) diameter 310 stainless steel.
[e.] Filler wire: 0.055 inch (1.4 mm) diameter 310 stainless steel.

23.28 WELD QUALITY CONTROL

Plasma arc weld discontinuities include both surface and subsurface types as listed in Table 23.11. Surface discontinuities such as reinforcement, underfill, undercut, and mismatch, which are associated with weld bead contour and joint alignment, are easily detected visually or dimensionally. Lack of penetration is also detected visually through the absence of a root bead. Weld cracks, open to the surface arc, are usually detected using dye penetrants. Finally, surface contamination, which results from insufficient shielding gas coverage, is usually revealed by severe discoloration of the weld bead or adjacent heat affected zones.

Subsurface weld discontinuities are generally more prevalent in manual PAW than in mechanized PAW. In either case, subsurface discontinuities are detected primarily using radiography or ultrasonic testing. Porosity is the most common subsurface discontinuity.

Tunneling is a severe void which runs along the axis of the joint. This discontinuity may result from a combination of torch misalignment and improper welding variables, particularly travel speed.

Lack of fusion discontinuities are most prevalent in repair areas, either single or multipass. The discontinuities result from insufficient heat input to obtain complete fusion.

Subsurface contamination in PAW can result when copper from the torch nozzle is expelled into the weld. This condition usually results when the torch nozzle gets too close to the weld, overheats, and copper melts into the puddle. The resulting contamination, which may be detrimental, may be undetectable welding by conventional NDT procedures. The best way to avoid copper contamination is by properly training the operator and by developing good torch manipulation techniques.

TABLE 23.11 Plasma arc weld discontinuities

Surface Discontinuities	Internal Discontinuities
Reinforcement	Porosity
Underfill	Tunneling (voids)
Undercut	Lack of fusion
Mismatch	Contamination
Lack of penetration	Cracks
Cracks	
Contamination	

23.29 SAFETY RECOMMENDATIONS

When welding with a transferred arc at currents up to 5 A, spectacles with side shields or other types of eye protection with a No. 6 filter lens are recommended. Although face protection is not normally required for this current range, its use depends on personal preference. When welding with transferred arc currents between 5 and 15 A, a full face plastic shield is recommended in addition to eye protection with a No. 6 filter lens. At current levels over 15 A, a standard welder's helmet with the proper shade of filter plate for the current being used is required.

When a pilot arc is operated continuously, normal precautions should be used for protection against arc flash and heat burns. Suitable clothing must be worn to protect exposed skin from arc radiation. Welding power should be turned off before electrodes are adjusted or replaced. Adequate eye protection should be used when observation of a high frequency discharge is required to center the electrode.

Accessory equipment, such as wire feeders, arc voltage controls, and oscillators, should be properly grounded. If they are not grounded, insulation breakdown might cause these units to become electrically "hot" with respect to ground.

Adequate ventilation should be used, particularly when welding metals with high copper, lead, zinc, or beryllium contents.

23.30 ADVANTAGES OF PLASMA WELDING

1. Less sensitivity to changes in arc length.
2. Recessed electrode reduces the possibility of tungsten inclusions in the weld and can substantially increase the period between electrode dressing resulting in increased life.
3. Weld in a single-pass upto 6 mm plates in square butt position and 10 mm plates in only two passes.
4. Keyhole mode of welding gives smaller heat affected zone resulting in reduced strength loss at the joint for heat treated metals, promotes less grain growth, which gives better ductility.
5. Higher welding speeds and thus reduced weld time, results in less embrittlement by carbides and complex intermetallic compounds for stainless steel and super-alloys.
6. Less residual stress and distortion.
7. Less filler metal required in keyhole mode significantly reduces porosity.

23.31 APPLICATIONS

(a) *Micro plasma welding:* Traditionally used for welding thin sheets (0.1 mm) and wire and mesh sections. The needle-like stiff arc minimizes arc wander and distortion.

(b) *Medium current welding:* Alternative to conventional TIG. The advantages are deeper penetration (from higher plasma gas flow) and greater tolerance to surface contamination including coatings. The major disadvantage lies in the bulkiness of the torch, making manual welding more difficult.

(c) *Keyhole welding:* Deep penetration and high welding speeds. Can penetrate plate thicknesses up to 10 mm but when welding using a single pass, it is more usual to limit the thickness to 6 mm. The normal method is to use the keyhole mode with filler to ensure smooth weld bead profile. For thickness up to 15 mm a V joint preparation is used with a 6 mm root face. This technique is only suitable for mechanized welding. It can be used for positional welding, usually with current pulsing. It is normally applied in high speed welding of thicker sheet material (over 3 mm) in the flat position.

Review Questions

1. Explain the principle of operation of PAW.
2. Explain the equipment used in PAW.
3. What is the difference between transferred arc and non-transferred arc?
4. Compare gas tungsten arc and plasma arc welding processes.
5. Discuss the effect of arc constriction on temperature and voltage.
6. Compare TIG and PAW.
7. Discuss the weld quality produced by PAW.
8. What are the important applications of PAW?
9. What are the advantages of PAW?
10. What are the disadvantages of PAW?

24
CHAPTER

Electron Beam Welding

24.0 INTRODUCTION

In 1958, a German engineer, Steigerwald, designed prototype electron beam equipment for welding application. Later on, this technology spread in other industrial applications.

Electron beam welding (EBW) is a fusion joining process that produces coalescence of materials with heat obtained by impinging a beam composed primarily of high energy electrons onto the joint to be welded: transformation of mechanical energy into heat energy.

The heart of the EBW process is the electron beam gun/column assembly. Electrons are generated by heating a negatively charged emitting material to its thermionic emission temperature range, causing electrons to boil of this emitter or cathode and be attracted to positively charged anode.

In this method, the joint to be welded is heated by bombarding it with a dense stream of high velocity electron. The kinetic energy of electron is changed to heat energy due to impact. The electron beam is produced in high vacuum environment by a electron gun usually consisting of tungsten or tantalum cathode, a grid and anode. The electrons emitted from the cathode are gathered, accelerated to high velocity and shaped into the beam by electrical feeds. The beam is focused using electromagnetic focusing controlled magnetic lenses.

Electron beams have high resolution and long depth of field due to their shorter wavelength. Other features of beam include their high power densities, ability to catalyze many chemical reactions, controllability and compatibility with high vacuum.

Electron beam welding is an advanced joining operation. Unlike arc, it utilizes the kinetic energy of the electrons to generate heat and thereby fusing the base metal. Using this fusion welding technique, metals, alloys and even ceramics are welded. Electron beam welding can be used for cutting and soldering.

24.1 POWER DENSITY OF HEAT SOURCE

Consider directing a 1.5 kW hair drier very closely to a 304 stainless steel sheet 1.6 mm thick. Obviously, the power spreads out over an area of roughly 50 mm diameter, and the sheet just heats up gradually but will not melt. With GTAW at 1.5 kW, however, the arc concentrates on a small area of about 6 mm diameter and can easily produce a weld pool. This example clearly demonstrates the importance of the power density of the heat source in welding.

The heat sources for the gas, arc, and high-energy beam welding processes are a gas flame, an electric arc, and a high-energy beam, respectively. The power density increases from a gas flame to an electric arc and a high-energy beam. As shown in Figure 24.1, as the power density of the heat source increases, the heat input to the workpiece that is required for welding decreases. The portion of the workpiece material exposed to a gas flame heats up so slowly that, before any melting occurs, a large amount of heat is already conducted away into the bulk of the workpiece. Excessive heating can cause damage to the workpiece, including weakening and distortion. On the contrary, the same material exposed to a sharply focused electron or laser beam can melt or even vaporize to form a deep *keyhole* instantaneously, and before much heat is conducted away into the bulk of the workpiece, welding is completed.

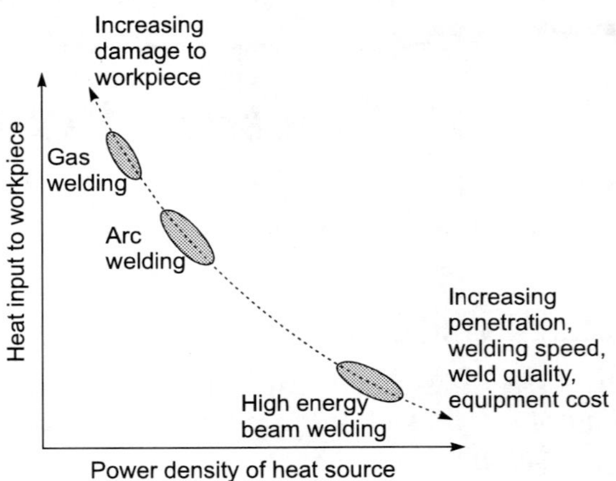

Figure 24.1 Variation of heat input to the workpiece with power density of the heat source

Therefore, the advantages of increasing the power density of the heat source are deeper weld penetration, higher welding speeds, and better weld quality with less damage to the workpiece, as indicated in Figure 24.1. Figure 24.2 shows that the weld strength (of aluminum alloys) increases as the heat input per unit length of the weld per unit thickness of the workpiece decreases. Figure 24.3 (a) shows that angular distortion is much smaller in EBW than in GTAW (2). Unfortunately, as shown in Figure 24.3 (b), the costs of laser and electron beam welding machines are very high.

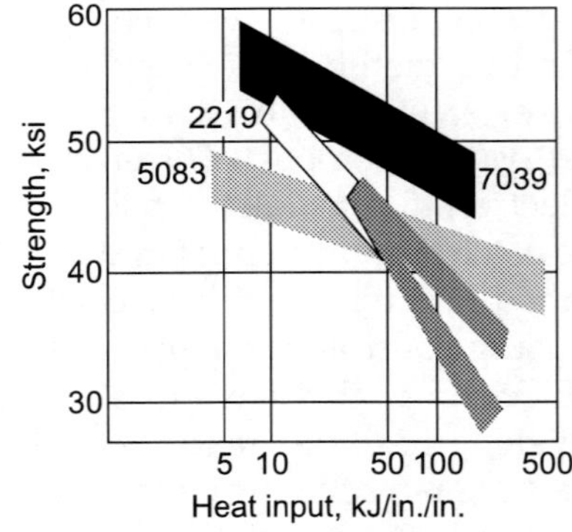

Figure 24.2 Variation of weld strength with heat input per unit length of weld per unit thickness of workpiece

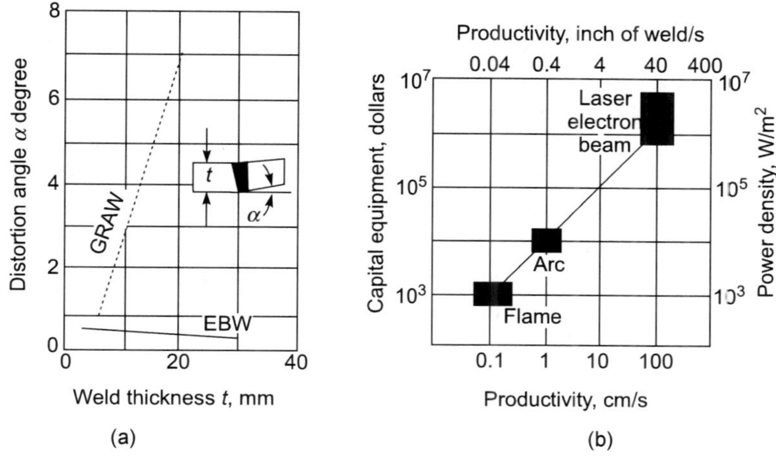

Figure 24.3 Comparison between welding processes: (a) Angular distortion; (b) Capital equipment cost

24.2 CLASSIFICATION OF ELECTRON BEAM WELDING

EBW machines can be classified as:

1. Based on accelerating voltage
   (i) High voltage EBW (150 kV)
   (ii) Low voltage EBW (60 kV)
2. Based on vacuum
   (i) High vacuum
   (ii) Fine vacuum
   (iii) Atmospheric
3. Based on the machine configuration
   (i) Conveyor-type EBW
   (ii) Clock configuration EBW
   (iii) Local vacuum machine
   (iv) Mobile vacuum machine
   (v) Micro and fine welding machine

24.3 OPERATION

The principle of operation of EBW is: fusion occurs in a specific confined area of the base metal by bombarding with high velocity electrons, which are accelerated to about half the speed of light. As the electrons strike the workpiece, their energy is converted into heat, instantly vaporizing the metal under temperatures near 25,000°C. The heat penetration is so deep that it can weld much thicker workpieces than by other welding processes. The depth-to-width ratio of the weld bead is as much as

15:1. Since the electron beam is tightly focused, the total heat input is much lower than other welding processes. As a result, the effect of welding on the surrounding material is minimal, heat affected zone is small. Rapid cooling and less distortion are advantages in EBW but it leads to cracking in high carbon steels.

The operation is performed in vacuum to prevent reduction of electron velocities. If vacuum is not used, electrons would strike the small particles of the atmosphere, reducing their velocity and decreasing their heating ability. Most vacuum chambers house workpiece, cathode, focusing device, remainder of gun, and electron beam gun.

In this process, electrons are emitted from the heated filament called cathode, and then focused into a beam which is directed at welding point. The electron is accelerated between the workpiece and cathode filament by applying large potential difference. The cathode is placed in the electron beam gun which emits electrons, which are converged by a series of focusing lenses, which converge on one another to form the beam. The electrons are accelerated by tremendous potential difference between cathode and anode. Depending on the voltage, the speed can vary from 30,000 to 1,20,000 miles/sec.

Principle of Electron Emission

The release of electrons from the cathode is governed by the photoelectric effect, explained by Albert Einstein. The cathode is a metal which releases electrons from its outer orbits of its atomic structure. The release of electrons is achieved either by supplying energy in the form of radiation or by electrical resistance heating. If these photons strike with enough velocity, electrons from the metal will be drawn with great velocity. The intensity of radiation and number of electrons emitted are directly proportional, and is the prime factor controlling the intensity of the electron beam. The ease with which a metal will give up its electrons is called the thermionic function of the metal. The intensity of electronic beam is called the emission current.

When the electron beam strikes the welding point, the kinetic energy of the high speed electrons is converted into heat. The heat generated melts and fuses the metal. Because of the impact of electrons the workpiece can melt and even boil. Due to high local temperatures, ceramics can be even cut or welded.

First, the stream of electrons penetrates a thick layer of metal to a depth of about 0.001 inch. As the stream of electrons moves deeper into the metal, the electrons are scattered, slowed down, stopped by atoms in crystalline structure due to collision, resulting in the heating of a pear-shaped volume of metal. Some heated metal vaporizes and a minute hole is produced by vaporization and similarly, in different layers, until complete penetration of the base plate. The penetration is accomplished in a fraction of a second, depending on the energy released, type, and thickness of material. If the electron beam is allowed to transverse the metal, same sequence is repeated along the weld seam. The metal that is melted is redeposited in the hole, called key holing technique, and makes it possible for EBW to be used as cutting or as a soldering device by increasing the voltage. Increasing voltage prevents solidification.

24.4 EQUIPMENT

The schematic arrangement of EBW is shown in Figure 24.4. It consists of three main components: beam generation, beam manipulation and forming and working chamber.

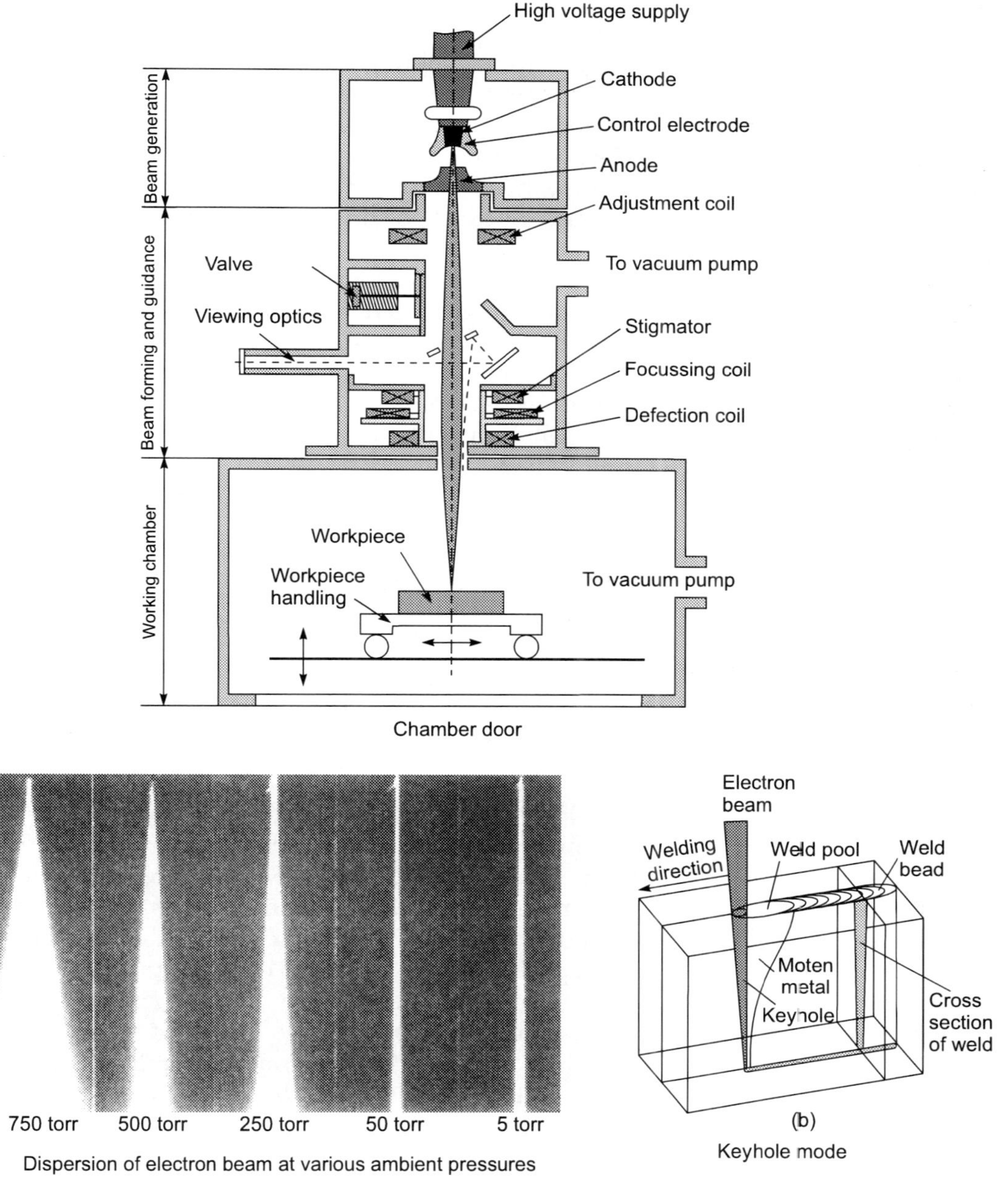

Figure 24.4 EBM setup details

A tungsten cathode, heated under vacuum, emits electrons by thermal emission. Heating of the tungsten cathode may be carried out directly – by filament current – or indirectly – as, for example, by coiled filaments. The electrons are accelerated by high voltage between the cathode and the anode.

A modulating electrode, called grid cup (also called Wehnelt cylinder), is positioned between anode and cathode, regulates the electron flow.

The electron beam which diverges after passing through anode, however, obtains the power density which is necessary for welding only after having passed the adjacent alignment and the focusing system. One or several electromagnetic focusing lenses bundle the beam onto the workpiece inside the vacuum chamber. A deflection coil assists in maintaining the electron beam oscillating motion.

An additional stigmator coil may help to correct aberrations of the lenses. A viewing optic or a video system allows the exact positioning of the electron beam onto the weld groove. The core piece of the electron beam welding machine is the electron beam gun, where the electron beam is generated under high vacuum.

The tightly focussed electron beam diverges rapidly under atmospheric pressure, caused by scattering and ionization development with air. As it would, here, lose power density and efficiency, the welding process is, as a rule, carried out under medium or high vacuum. The necessary vacuum is generated in separate vacuum pumps for working chamber and beam gun.

Major equipment required in EBW are:

- **Power Source**

  Obtain power from ordinary line voltage of 220 to 440 V, 3 phase, 60 Hz alternating current. The EBW pulls lower current (usually less than 1 A) but high voltages (60 to 200 kV), and hence classified as high voltage and low voltage machines. High voltage draws a current of 40 mA and provides a depth-to-width ratio of 25 : 1, whereas low voltage gives 12 : 1. The power of a beam is an indicator of its ability to do work and it determines the power density, generally about 40 to 4000 kW/cm^2 or 100 to 10,000 kW/cm^2.

- **Electron Beam Gun**

  There are two types of electron beam guns:

  Work accelerated gun: They produce high energy electrons by utilizing the difference in potential between cathode and the work. They are low voltage machines operating at current ranging from 250 to 1000 mA. The gun is mounted inside a vacuum chamber and is capable of revolving around the workpiece. As the material melts and vaporizes, it leaves a film over the gun making it necessary to clean low voltage units frequently. The voltage power and the current of the beam and surface of the work can also be varied by changing the distance between anode and cathode. The penetration depth is less due to low (about half) operating voltages but its advantages are low cost, easier maintenance, simpler design, and easier operation.

Self-accelerated guns: In this type, the cathode and anode are enclosed within the gun. The gun is similar to that of work accelerated gun except that the anode is placed inside the gun which attracts negatively charged particles. The anode performs the same function as the positively charged workpiece when the work accelerated gun is used. They are available in the range of 1,50,000 to 3,00,000 V.

The main components of electron gun are given below:

The Electron Emitter: The source of the electron beam is an emitter cathode, which is usually a metal with low thermionic function. Generally, they are made of tungsten or tantalum which is heated to nearly 2600 K by electrical resistance. They emit about 6e18 electrons per sec per square centimeter. The filament can also be heated using an outside radiation. An AC supply of about 3 V can be tied across the filament to produce the electricity needed for heating. The anode is a perforated plate with high positive potential which attracts negatively charged electrons, producing a current flow through the perforated anode.

The gun produces and accelerates the electrons using a tungsten emitter that emits electrons when heated. The electrons are then attracted to the anode inside the tool, where they are collected and focused using electro-magnetically controlled magnetic lenses. These components are housed in electron beam gun column in which a hard vacuum is maintained (about 0.00001 torr).

The Optics System: Radiations like X-ray, infrared ray, UV ray are produced from the electron gun during operation. The welder must be protected from these, therefore safe optical systems are used to afford protection. It includes a lens and a mirror system that focuses the light rays for visual monitoring and for alignment of the work table to the beam.

Beaming & Focusing Devices: The electron beam produced by the emitter should produce maximum power and hence, it should be focused to a fine point of concentration. It is accomplished by electromagnetic or electrostatic distortion of the beam.

In electromagnetic distortion, electromagnetic lens is used to focus the beam, which includes a coil encased in iron with a gap in the bore. The electron beam is rotated and refracted to converge. The focal length of the beam can be controlled by DC current and the beam can be positioned into axis direction by using deflection plates. Beam movement depends on DC voltage.

In electrostatic distortion, a lens disc is placed closer to the filament and the electrons passing through the holes are affected by the electric field of force that surrounds the hole and thus, causes it to converge. By moving the lens disc closer or farther, the diameter of the beam can be controlled.

Vacuum Chamber: The vacuum chamber performs the function of reducing the atmosphere to vacuum of about 5e-5(m) of mercury, where normal atmosphere is 760 mm of mercury. To avoid the reduction of velocities of the electrons from the cathode due to collision with particles in the air, the setup is placed in a vacuum chamber. The vacuum is about 0.0000001 of atmospheric pressure. When this vacuum is achieved, the chamber is less than 1 ppm of atmosphere. To achieve this, 2 pumps are used, roughing and diffusion pumps. The roughing pump reduces pressure to 0.1 mm of Hg and using diffusion pump further, the required pressure is reached. The props and system required to reach this vacuum are both expensive and complex.

- **Work Table & Fixtures**

  The electron beam is concentrated to a fine point, automatic fixtures are necessary to make sure welding, cutting, piercing are precisely done. The operation can be done in a straight line around rectangles or circles or in zig-zag paths or any combination of these.

24.5 PROCESS PARAMETERS OF ELECTRON BEAM MACHINING

Main process parameters are:

1. Welding current/beam current
2. Welding voltage/accelerating voltage
3. Groove area/beam spot size
4. Energy input/beam power
5. Welding time
6. Beam offset

Increasing the accelerating voltage or beam current increases the depth of penetration. The product of these two variables is equal to the overall beam power, which in turn, determines the amount of material melted for a given welding transverse rate. Maintaining all other variable constants an increase in welding traverse rate decreases the depth of penetration, energy input per unit length, and the width of weld bead.

The diameter of the electron beam spot on the workpiece is influenced by the gun design, beam current, and focus current and standoff distance. Most electron beam systems are capable of focusing the beam to diameter as narrow as 0.25 mm. However, beam alignment is difficult with narrow beam. Larger spot size upto 1.27 mm are used when shallow, wide weld nuggets are desired. Figure 24.5 shows the effect of beam offset (BO).

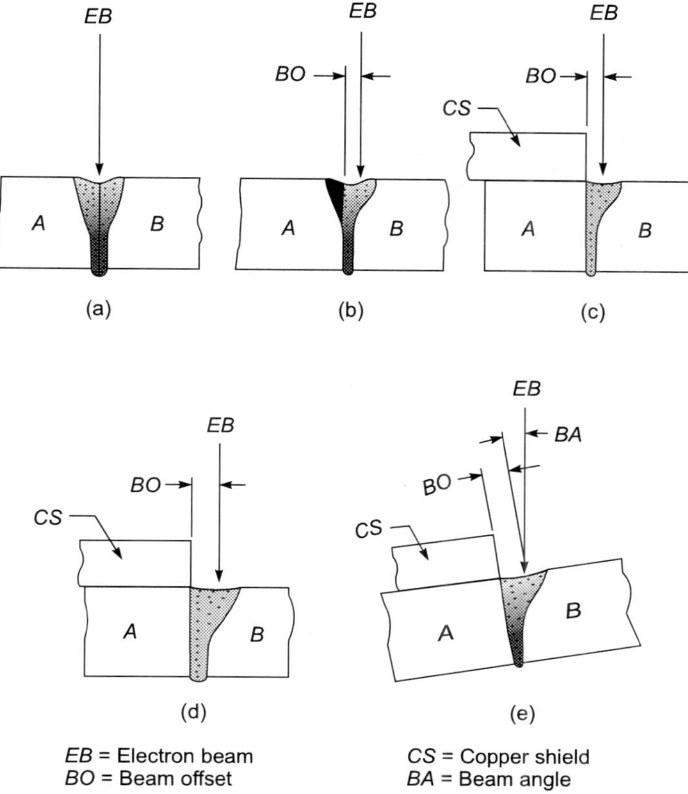

Figure 24.5 Process parameters of EBM

The level of vacuum in which welding is performed, also affects the performance of the process. As pressure increases, welding penetration decreases and width of the fusion zone increases (Figure 24.6).

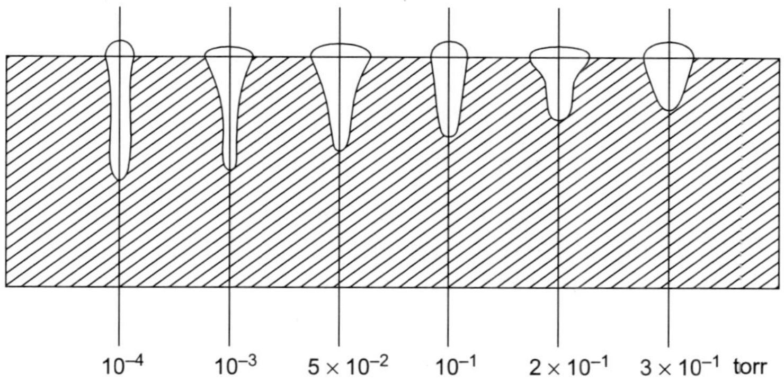

Figure 24.6 Influence of level of vacuum on penetration

Typical characteristics of EBM are given below:

Welding current	0.27 A
Welding voltage	1,50,000 V
Number of passes	1
Groove area	896 mm^2
Filler material	No
Melting efficiency	7.7 kg/hour
Energy input	64×10^3 kJ
Welding time	27 min

24.6 OPERATING PARAMETERS

The important parameters which characterize the electron beam welding are:

I. Welding Methods

According to different welding environments, EBW is classified into three, they are:

1. **Hard vacuum EBW:** First method, the vacuum maintained in the chamber is extremely high (0.00001 torr). Up to 15 cm thickness can be welded. Stand-off distance can be as great as 0.7 m. While it is the most efficient of the 3 modes, the disadvantages are that the evacuation for such high vacuum is difficult and the high cost of entire machine.
2. **Soft vacuum EBW:** Vacuum pressure under 0.01 torr comes under it. It allows large welding chambers, and easy evacuation but the stand-off distance is reduced to half and material thickness to 5 cm.
3. **Non-vacuum or out-of-vacuum EBW:** It is performed in atmospheric pressure. The stand-off distance diminishes to 4 cm and material thickness is about 5 cm. However, it allows any size of the work to be welded since the size of the chamber is no longer a factor.

II. Heat Generation

The amount of heat generated at welding is determined by power density.

- Power density is increased by increasing the voltage and decreasing the diameter of the hole. With higher, there are fewer electrons for the same amount of power and less scattering of the bombarding electron, results in smaller beam diameter.
- Voltages vary from 10 to 150 kV and spot sizes vary from 0.001 to 0.250 inch. To produce a power density of 25,000,000 W/inch^2, power levels that are more adequate in order to vaporize the metal.
- Low voltages are required for thin beams but penetration is decreased.

III. Electron Beam Intensity

The intensity of the beam is controlled by several factors:

- Speed of the beam across the workpiece. Others include, diameter of the beam, velocity of electrons, and beam current.
- Varying any of these parameters, optimum density and size of the beam can be found for particular materials that are to be fused.

IV. Weld Bead Penetration

Weld bead penetration depends on the amount of heat, which in turn, depends on the number and speed of electrons, diameter of the beam, distance between anode and cathode, and travel speed.

- Greater the beam current, more will be the heat and penetration, and vice versa for travel speed. While the travel speed decreases heat input and reduces penetration.
- Diameter of beam varied by focusing system focusing below the surface increases penetration and decreases above the surface.
- Potential difference is high, which increases velocity, thereby increasing kinetic energy and hence, the heat and penetration.
- Penetration is proportional to voltage and amperage. Penetration is inversely proportional to the welding speed. The greater the welding speed, less penetration is accomplished by the electron beam.
- Penetration is affected by thermal conductivity. Weld penetration increases when thermal conductivity of the metal decreases.

24.7 PROCESS CAPABILITIES OF EBW

1. EBM can perform single-pass, deep penetration welds while maintaining low thermal distortion. Plates of 0.1 to 300 mm can be welded using EBW.
2. It can perform multiple tier welding, i.e., it can simultaneously weld two or more layers of materials that are separated by distances in the order of few millimeter.
3. This process has the ability to weld thick workpieces with large gaps.
4. EBW can be used to weld dissimilar materials in nuclear industries.
5. It can be used to weld stainless steels, low-carbon steels, copper and its alloys, nickel based alloys, aluminum alloys, refractory materials such as tantalum, titanium, niobium, zirconium, etc.
6. Possible welded joints using EBW are shown in Figure 24.7. To make exact and uniform joints, one should have good knowledge about beam geometry, effect of welding parameters, interaction between the beam and work, etc.

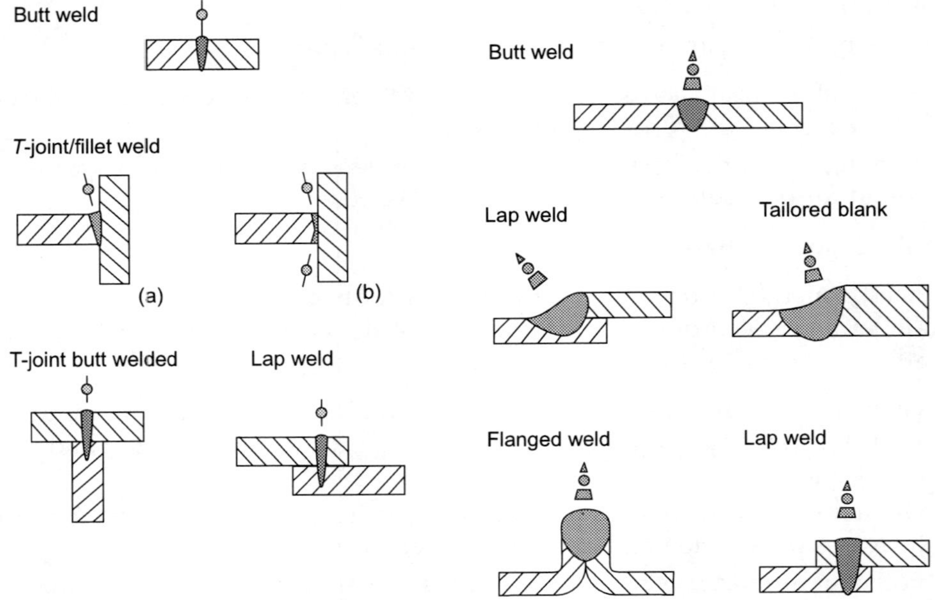

Figure 24.7 Typical joint designs for EBW

24.8 MATHEMATICS OF ELECTRON BEAM

Free electrons move under the effect of electric or magnetic fields and may be accelerated greatly. Even with relatively small potential difference, electron velocity can be that of the speed of light.

Consider an electron that moves in electric and magnetic field. The electron is subjected to Lorentz force of

$$F_L = \vec{F}_e + \vec{F}_m \tag{1}$$

F_L = Lorentz force

$\vec{F}_e$ = Force due to electric field

$\vec{F}_m$ = Force due to magnetic field

Also,

$$\vec{F}_e = \text{Force due to electric field} = -eE_s$$

$$\vec{F}_m = \text{Force due to magnetic field} = -e\mu V_s H$$

where,

e = Charge of the electron,

E_s = Electrostatic field

V_s = Speed of the electron,

μ = Magenetic permeability

H = Strength of magnetic field

The magnetic field is always at right angles to the direction of electron movement. Hence it does no work, but can only change the direction of electron movement. Thus, the kinetic energy of the electron is not affected by the magnetic field having no component acting in the direction of the electron movement.

The sum of the kinetic and potential energies of electrons in motion in a given electric field is constant, when

$$\frac{1}{2}m_e V_s^2 - eE_s = C_o \tag{2}$$

At the terminal conditions, $V_s = 0, E_s = 0, C_o = 0$ thus,

$$\frac{1}{2}m_e V_s^2 = eE_s$$

From which the speed of the electron is given by

$$V_s = \sqrt{\frac{2eE_s}{m_e}} \frac{m}{s} \tag{3}$$

where,

e = Charge of the electron in columbs
E_s = Electrostatic field in volts
V_s = Speed of the electron in m/s

The static mass of electron, $m_e = 9.1 \times 10^{-28}$ gm and $e = 1.6 \times 10^{-19}$ columbs

$$V_s = \sqrt{\frac{2eE_s}{m_e}} = \sqrt{\frac{2 \times 1.6 \times 10^{19} \times E_s}{9.1 \times 10^{-28}}} = 600\sqrt{E_s} \tag{4}$$

Thus, the variation of the speed of the electron is a function of the potential difference of the electric field. The order of this velocity can be calculated with velocity of the light, when

$$\frac{V_s}{C_L} = \frac{600\sqrt{E_s}}{930 \times 10^3} \tag{5}$$

where, C_L is the velocity of light in km/sec.

If the applied voltage lies in the range of 10^6 to 10^8 volts, the order of electron velocity would lie in the range of the velocity of the light, thus imparting a very high amount of kinetic energy to the electrons.

Electrons moving in the electric field can accumulate considerable energy when they reach anode surface. A rapid change in momentum takes place through the collisions with the atoms of the workpiece metal when the kinetic energy of the electrons is transferred to the atoms of the work material under bombardment. The various technological processes like micro-welding, melting and cutting and making thin films can be carried out by controlling the concentrated thermal energy obtained through electron bombardment.

24.9 MECHANICS OF ELECTRON BEAM

Electrons are the smallest stable elementary particles with a mass of 9.1 × 10^{-28} gm with a negative charge of 1.6 × 10^{-19} columbs. If the initial velocity is negligible, then the electron velocity at the striking is given by

$$V_s = \sqrt{\frac{2eE_s}{m_e}} = \sqrt{\frac{2\times1.6\times10^{19}\times E_s}{9.1\times10^{-28}}} = 600\sqrt{E_s}\,\frac{\text{km}}{\text{sec}}$$

Power of the electron beam is given by

$$P_b = E_s e n_e l_b$$

$$n_e l_b = \text{Number of electrons per second}$$

$$n_e = \text{Number of electrons per second per amp}$$

$$n_e = 6.3\times10^{-1}\,\frac{\text{electron}}{\text{sec}}\,\text{amp}$$

$$en_e = 1.6\times10^{-19}\times6.3\times10^{-1}\,\frac{\text{electron}}{\text{sec}}\,\text{amp} = 1$$

Therefore power is given by,

$$P_b = E_s l_b \qquad (6)$$

$$I_b = \text{Beam current in amp}$$

The pressure from the electron beam and reaction of vapors from the evaporating material causes the stable condition of cone penetration and movement of the material from its leading wall in the zone of crystallization.

The force of the electron beam on the molten material is given by

$$F_b = n_e m_e V_s \qquad (7)$$

Combining (3) and (7)

$$F_b = \frac{I_b}{e} m_e \sqrt{\frac{2eE_s}{m_e}}$$

$$F_b = \frac{I_b\sqrt{E_s}}{e}\sqrt{\frac{2m_e}{e}}\ \text{newton} \qquad (8)$$

In CGS unit,

$$F_b = \frac{I_b\sqrt{E_s}}{e}\sqrt{\frac{2m_e}{e}}$$

$$= \frac{I_b\sqrt{E_s}}{e}\sqrt{\frac{2\times 9.1\times 10^{-2}}{1.6\times 10^{-19}\times 10^{3}}}\times 10^{5} = 0.34 I_b\sqrt{E_s}\ \text{dynes}$$

If I_b is expressed in current per cm^2, equation (8) would give the pressure due to electron bombardment per cm^2.

The force on the fused metal arising out of the evaporation of metal from the surface can be determined by equating the velocity of atoms due to thermal energy acquired by the work material due to electron bombardment with the mean velocity of atoms of metal evaporating from the surface of the material.

According to the classical kinetic theory of heat, each atom of a crystalline solid acquires energy, on an average of $k\theta$,

$$\frac{1}{2}m_a V_a^2 = k\theta$$

m_a = Mass of one atom of the workpiece, kg
V_a = Thermal velocity acquired, m/sec
θ = Temperature raised through electron bombardment, k
k = Boltszmann constant = 1.38×10^{-23} J/k/atom

Thus, the thermal velocity,

$$V_a = \sqrt{\frac{2k\theta}{m_a}} \quad (9)$$

The mean velocity of evaporation is evaluated from the total force of reaction, when

$$F_n = n_a m_a V_e \quad (10)$$

n_a = Number of electrons in the beam
m_a = Mass of electrons in the state of rest, kg
V_e = Mean velocity of evaporation, m/s

From equation (10)

$$V_e = \frac{F_n}{n_a m_a} \quad (11)$$

Equating (9) and (11), we get

$$\frac{F_n}{n_a m_a} = \sqrt{\frac{2k\theta}{m_a}}$$

From which we get

$$F_n = n_a\sqrt{2k\theta m_a} \tag{12}$$

Calculations indicate that F_n determined from equation (12) is considerably greater than F_b calculated from equation 8. Thus, the pressure caused by evaporating reaction is much greater than the force caused by direct bombardment of electrons.

Thus, if the beam is moved, the metal on the leading wall of the fused zone will be subjected to simultaneous action of electron beam pressure and evaporating metal reaction, while the rear wall of the fused zone does not undergo this pressure.

24.10 APPLICATIONS OF ELECTRON BEAM WELDING (EBW)

1. EBW is used extensively in automobile, aircraft, space, electrical, electronics industries, and nuclear power plants.
2. EBW can be used to weld almost all steels, high-strength aluminum alloys, nimonics, magnesium alloys, copper alloys, titanium, tungsten, gold and other material combinations, like uranium with aluminum, magnesium, beryllium, zirconium and stainless steel.
3. In nuclear industry, EBW is used for sealing uranium fuel elements into cans of aluminum, magnesium, beryllium, zirconium and stainless steel.
4. In aerospace industry, EBW is used to weld titanium fuel tanks for the space rockets. Nimonic alloys used in gas turbines are welded using EBW.
5. In automobile industry, gears are joined using this process. Automatic transmission elements are welded using EBW.
6. Repair, modifications of complex and costly aircraft and rocket components are welded by EBW.
7. In electronics industries, EBW is used to join dissimilar materials. Tungsten and molybdenum rings are joined using this process.
8. EBW is used for welding of automobile catalytic converters. Four layers of 1.27 mm thick stainless steel plates are welded using EBW.
9. It is used for welding of gas turbine engine components.
10. EBW is used to weld bimetal strips. Bimetal strips are commonly used in electronics industries.
11. EBW is used for fabrication of large coupling hubs. Clutch fingers on large disks are welded using EBW.
12. EBW is used in welding of medical implants.
13. It can be used nearly in all areas where TIG welding is used, commonly used to weld stainless steel, superalloys and reactive and refractory metals.
14. Beryllium welding can be performed. With other welding processes, beryllium would vaporize rather than melting. Various tests with electron beam reduce the problem and sheets up to 1/16 inch thick can be welded with a single pass.

15. Dissimilar metal combinations can be welded.
16. During welding of plain carbon steel in vacuum, deoxidizer must be used to prevent porosity due to the formation of gases.
17. Several previously unweldable metals are welded with EBW. Aluminum has been welded with high voltage and various butt positions. Even though porosity was found in all cases, strength of the material was not affected. High strength steels are also welded.
18. The use of EBW process is limited to edge, butt, spot, or fillet welds.

24.11 ADVANTAGES OF ELECTRON BEAM WELDING

1. Both thin and thick plates can be welded using EBW. Thickness ranges from 0.1 mm to 300 mm. Electron beam weldment exhibits a high depth-to-width ratio. This feature allows for single-pass welding of thick joints.
2. Can be used to make very narrow seams. Aspect ratio up to 50:1 can be achieved.
3. Total heat input to weld zone is less. This gives less weld distortion.
4. Can be used to weld preprocessed parts.
5. Welding speed is high and hence, high production rates.
6. No shield gas is used.
7. Directly converts electrical energy into beam output mechanical energy. Thus, the process is extremely efficient.
8. Many materials can be welded using EBW. This is also used to weld dissimilar materials.
9. Gives good gap bridging.
10. The heat generation rate is 5000 times higher than conventional welding. Beam travel speed is high compared to conduction rate. The heat input per unit length for a given depth of penetration can be much lower than with arc welding. The resulting narrow weld zone results in low distortion and fewer deleterious thermal effects.
11. The amount of the melt zone is small compared to conventional welding processes and it is easily controllable. Superior control over penetration and other weld dimensions are proportional.
12. Instantaneous melting of parts is possible in EBW. As a result, high depth to width ratio can be achieved in EBW.
13. It can be used to weld highly conductive materials.
14. Improves joint strength by 15% to 25% as compared to arc welding process.
15. Gives a narrow heat affected zone (HAZ). Geometric shapes and dimensions are highly stable, particularly when it is used as a finish operation.

16. A high purity environment is used so it eliminates oxide and tungsten inclusions and removes impurities. The weld metal has a fine crystalline structure.
17. Being an automated process, the welded joint quality is consistent.
18. The weld can be made deeper, narrower and less tapered than any other process.
19. Refractory and reactive metals can be welded.
20. Dissimilar metals and metals with high thermal conductivity, such as copper, can be welded.
21. The ability to project the beam over distance of several centimeters in vacuum often allows weld to be made in otherwise inaccessible locations.
22. Rapid travel speeds are possible because of the high melting rates associated with this concentrated heat source. This reduces welding time and increases productivity and energy efficiency.
23. Reasonably square butt joints in both thick and relatively thin plates can be welded in one pass without filler metal addition.
24. Hermetic closures can be welded with high or medium vacuum modes of operation while retaining a vacuum inside the component.
25. The beam of electrons can be magnetically deflected to produce various shapes of welds and magnetically oscillated to improve weld quality or increase penetration.
26. The forced beam of electrons has a relatively long depth of focus, which will accommodate a broad range of work distances.
27. Full penetration, single-pass welds with nearly parallel sides, and exhibiting nearly symmetrical shrinkage, can be produced.

24.12 DISADVANTAGES OF ELECTRON BEAM WELDING

1. Capital costs are substantially higher than those of arc welding equipment, depending on the volume of parts to be produced. However, the final per piece part costs attainable with EBW can be highly competitive.
2. Preparation for welds with high depth-to-width ratio requires precision machining of the joint edges, exacting joint alignment, and good fit up.
3. Electron beam gets deflected easily by magnetic field. While welding ferrous alloy materials, parts must be demagnetized before welding to avoid unwanted beam deflection.
4. High cooling rates of parts make it hard and microcracks may develop.
5. X-ray shielding is required.
6. Large workpieces cannot be welded due to limitation on size of vacuum chamber.
7. Special jigs are required on the chamber work table.
8. There is a time loss for evacuating the vacuum chamber of EBW.

9. The rapid solidification rates achieved can cause cracking in highly constrained low ferrite stainless steel.
10. For high and medium vacuum welding, work chamber size must be large enough to accommodate the assembly operation. The time needed to evacuate the chamber will influence production costs.
11. Partial penetration welds with high depth-to-width ratios are susceptible to root voids and porosity.
12. Because the electron beam is deflected by magnetic fields, non-magnetic or properly degaussed metals should be used for tooling and fixturing close to the beam path.
13. With the non-vacuum mode of electron beam welding, the restriction on work distance from the bottom of the electron beam gun column to the work will limit the product design in areas directly adjacent to the weld joint.
14. With all modes of EBW, radiation shielding must be maintained to ensure that there is no exposure of personnel to the X-radiation generated by EBW.
15. Adequate ventilation is required with non-vacuum EBW, to ensure proper removal of ozone and other harmful gases formed during this mode of EBW.

24.13 COMPARISON BETWEEN EBW AND LBW (LASER BEAM WELDING)

LBW	EBW
Deep penetration in many materials is possible but not in metals that reflect the laser light of specific wavelength.	Deep penetration is possible in all materials.
EBW	LBW (absorbed) LBW (refected)
Cab be narrow in key hole mode.	Very narrow welds are possible.
High energy density/low linear heat input.	High energy density/low linear heat input.
Can operate in air, inert gas and vacuum.	Mostly performed in vacuum.
Usually require tight fitting joints.	Also requires tight fitting joints.
Very inefficient electrically (12%).	Very efficient electrically (99%).
No X-rays are produced.	Generates X-ray radiations.
Equipment cost is high.	Equipment cost is high.

Worked Examples

Example 1: The voltage in an EBW operation is 45 kV. The beam current is 60 milliamp. The electron beam is focused on a circular area that is 0.25 mm in diameter. The heat transfer factor is 0.87. Calculate the average power density in the area in watt/mm^2.

Solution: Power density, $PD = \dfrac{f_1 \times EI}{A} = \dfrac{0.87 \times 45000 \times 60 \times 10^{-3}}{A} = \dfrac{2349 \text{ watts}}{A}$

$$\text{Area, } A = \frac{\pi \times 0.25^2}{4} = 0.0491 \text{ mm}^2$$

$$\text{Power density, } PD = \frac{2349}{0.0491} = 47853 \text{ watts/mm}^2$$

Example 2: An electron beam welding operation is to be accomplished to butt weld two sheet metal parts that are 3.0 mm thick. The unit melting energy = 5.0 J/mm^3. The weld joint is to be 0.35 mm wide, so that the cross section of the fused metal is 0.35 mm by 3.0 mm. If accelerating voltage = 25 kV, beam current = 30 milliamp, heat transfer factor $f_1 = 0.85$, and melting factor $f_1 = 0.85$, determine the travel speed at which this weld can be made along the seam.

Solution: Available heat for welding, $R_{HW} = f_1 f_2 EI = U_m A_w v$

$$\text{Travel velocity, } v = \frac{f_1 f_2 EI}{U_m A_w}$$

A_w = cross-sectional area of weld seam $= 0.35 \times 3 = 1.05 \text{ mm}^2$

$$\text{Travel velocity, } v = \frac{f_1 f_2 EI}{U_m A_w} = \frac{0.85 \times 0.75 \times 25000 \times 30 \times 10^{-3}}{5 \times 1.05} = 91.05 \text{ mm/s}$$

Example 3: Calculate the velocity of electron impingement in electric field having strength of 2.5 × 10^5 volts. If the beam current is 2 × 10^{-5} amp, calculate the power of the electron beam.

Solution: $V_s = \sqrt{\dfrac{2eE_s}{m_e}} = 600\sqrt{E_s}\,\dfrac{\text{km}}{\text{sec}} = 600\sqrt{2.5 \times 10^6} = 3 \times 10^5 \dfrac{\text{km}}{\text{sec}}$

$$P_b = E_s I_b = 2.5 \times 10^5 \times 2 \times 10^{-5} = 5 \text{ watts}$$

Example 4: Calculate the electron beam pressure, if the current density is 2 × 10^{-3} amp/cm^2 and the voltage is 4 × 10^6 volts.

Solution: $F_b = \frac{I_b\sqrt{E_s}}{e}\sqrt{\frac{2\times 9.1\times 10^{-28}}{1.6\times 10^{-19}\times 10^3}}\times 10^5 = 0.34 I_b\sqrt{E_s}$ dynes

If I_b is expressed in current per cm^2, equation (8) would give the pressure due to electron bombardment per cm^2

$$F_b = 0.34 I_b\sqrt{E_s}\text{ dynes} = 0.34\times 2\times 10^{-3}\times\sqrt{4\times 10^6} = 1.36\text{ dynes / cm}^2$$

Example 5: Estimate the thermal velocity acquired by an electron of the work material due to electron bombardment, if the vaporization temperature of the work material concerned is 3327°C.

Solution: m_a = Mass of one atom of the workpiece = 9.1 × 10^{-28} gm
V_a = Thermal velocity acquired, m/sec
θ = 3327 + 273 = 3600 K
k = Boltszmann constant = 1.38 × 10^{-2} J/k/atom

Thus the thermal velocity is given by,

$$V_a = \sqrt{\frac{2k\theta}{m_a}}$$

$$V_a = \sqrt{\frac{2\times 1.38\times 10^{-2}\times 3600}{9.1\times 10^{-28}}} = 1.04\times 10^5\text{ m / sec}$$

Example 6: Calculate the thermal velocity acquired by an electron of the work material due to electron bombardment, if the vaporization temperature of the work material is 3500°C and mass of the one atom of workpiece is 9.1×10^{-28} gm.

Solution: m_a = Mass of one atom of the workpiece = 9.1 × 10^{-28} gm
V_a = Thermal velocity acquired, m/sec
θ = 3500 + 273 k = 3773 K
k = Boltszmann constant = 1.38 × 10^{-23} J/k/atom

Thus, the thermal velocity is given by,

$$V_a = \sqrt{\frac{2k\theta}{m_a}}$$

$$V_a = \sqrt{\frac{2\times 1.38\times 10^{-23}\times 3773}{9.1\times 10^{-28}}} = 0.106\times 10^5\text{ m / sec}$$

24.14 RECENT DEVELOPMENT IN EBW

Although the basic process has remained unchanged for many years, some significant advances have been made in the maximum power available, the operating techniques and the equipment.

(a) EBW Beam Power

Whilst for many applications, beam power of up to 25 kW, is quite adequate, there has been attempt to extend the weldable thickness range, particularly for out of vacuum applications and systems with output powers up to 200 kW have been built.

(b) Chamber Loading Systems

Chamber loading time may be considerably reduced, particularly for large volume of small components, by using an integrated loading system incorporating vacuum seals and progressive pressure reduction. The shuttle transfer system developed and patented by Wentgate Dynaweld is shown in Figure 24.8 and involves the use of custom designed workpiece carriers, into which the component to be welded is loaded. The carrier or shuttle which is fed into the welding chamber via a feed tube is equipped with rings which seal the enclosure and allow pre-evacuation.

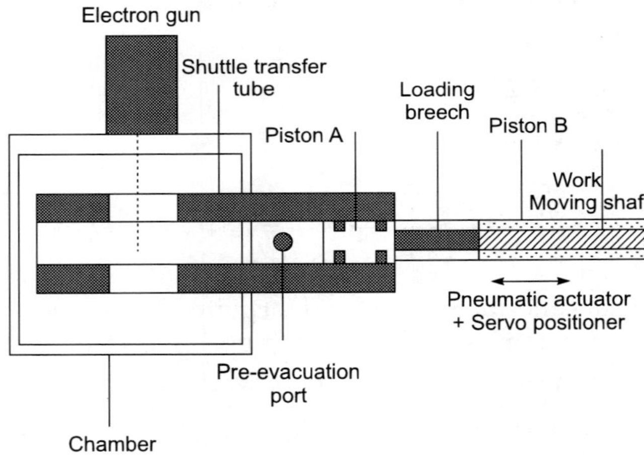

Figure 24.8 Shuttle system for rapid loading of small components

(c) Vapor Shields and Beam Traps

The collection of metal vapor inside the electron gun can cause unstable operation, and mechanical shields are often incorporated to exclude vapor from the gun area. A further improvement in this area is the development of the magnetic trap, developed by the welding institute in UK (Figure 24.9). This system has proved very effective for preventing gun discharges when welding aluminum alloys.

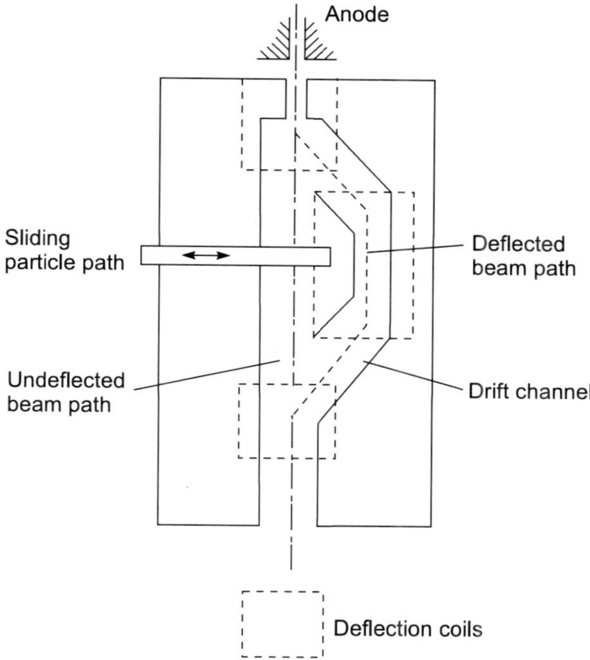

Figure 24.9 The three-bend magnetic trap

(d) Computer Control

The incorporation of CNC and computer control is common in current EBW systems. The computer can control workpiece positioning, operating cycle and operating parameters. These systems simplify parameter selection and may often be linked to integrated manufacturing cells. By using telecommunication data links, it is also possible to perform remote setting and diagnostics.

(e) Non-vacuum EBW

Non-vacuum EBW is more feasible as beam powers have increased. Production systems are in operation, welding up to 6 mm thick aluminum and the development of systems for use in robotic installations is in progress.

(f) Power Sources for EBW

The solid state power source technology may be applied to EBW power sources to provide improved control and consistency. Invertor designs that offer high electrical efficiency are particularly suitable for these applications.

Comparison of high energy welding processes is given in Table 24.1.

Table 24.1 Comparison of high energy welding processes

Feature	Plasma Keyhole	Laser	Electron Beam
Energy density (W m^{-2})	3×10^{10}	3×10^{11}	10^{13}
Thickness range (mm)	Up to 12	Up to 20	Up to 200
Travel speed	Higher than GTAW	High	Very high
Materials weldable	All, but aluminum needs DEEP or AC	Most, but difficult on reflective surface (e.g., Cu, Ag)	Most, but not low vapor pressure (e.g., Cd, Zn and coated metal)
System requirements	Simple; as for automated GTAW	Highly automated plus screening safety interlocks	Usually vacuum system, X-ray screens and high level of automation
Capital cost	Slightly more than GTAW	10 to 50 times GTAW	10 to 50 times GTAW
Running cost	Low	High	Very high
Cost of weld	Depends on application		

Review Questions

1. Explain the principle of operation of EBW.
2. Explain the equipment used in EBW.
3. What is the function of vacuum system in EBW?
4. Discuss the mechanics of electron beam.
5. Discuss the recent developments in EBW.
6. Compare EBW and PAW.
7. Discuss the weld quality produced by EBW.
8. What are the important applications of EBW?
9. What are the advantages of EBW?
10. What are the disadvantages of EBW?

25

CHAPTER

Laser Beam Welding

25.0 INTRODUCTION

Laser beam welding is also a radiant energy welding process in which coalescence is generated by the heat obtained from application of a concentrated coherent light beam falling on the surface that is to be joined. Laser beam welding, like electron beam welding, can weld any material. Heat-affected zone is a narrow zone due to very high energy intensity and high speed of LBW can also be achieved.

LASER is an acronym for Light Amplification by Stimulated Emission of Radiation. Laser is an electromagnetic radiation. In laser beam machining, work material is melted and evaporated by means of an intense, monochromatic beam of light called the laser. The heat produced in the small area where the laser beam strikes can melt almost any of the known material. LBW is used in machining difficult to machine material. The light energy emitted by laser has several characteristics.

1. **Spectral Purity:** The laser is highly monochromatic and permits the use of simple lenses for focusing beams. Expensive color corrected lens are not required.
2. **Directivity:** The laser is highly collimated with less divergence angle.
3. **High Focused Power Density:** As a result of its small beam divergence, all of the laser beam energy can be collimated with simple optics and focussed onto a small area. Power density can be increased by proper focusing. The power density of the beam and interaction with the workpiece determines whether the laser beam can be used to perform cutting, welding, marking and heat treatment.

25.1 LASING ACTION AND POPULATION INVERSION

The operation of laser depends on the naturally occurring transitions between the energy levels of an atom. Planck assumed that energy levels of atoms are equally spaced. Einstein put forward his viewpoint that light was really photons and could be regarded as particles with energy. Further, Bohr had pointed out that any system of atoms has energy levels but they are not necessarily equally spaced like Planck's oscillator.

Let the horizontal lines shown in the Figure 25.1 indicate the allowed values of energy. When the electron is free (when its energy is positive), it can have any energy moving at any speed. But bound energies are not arbitrary but discrete levels of allowed values such as E_0, E_1, E_2 ... as shown in Figure 25.1. If an atom is initially at any of the excited states, it does not remain in that state. Sooner or later, it drops to the lower energy state and radiates the energy in the form of light. The frequency of emitted light is determined from the principle of conservation of energy between the difference of energy levels and quantum mechanics. The transition from energy E_2 to energy E_0 is related by

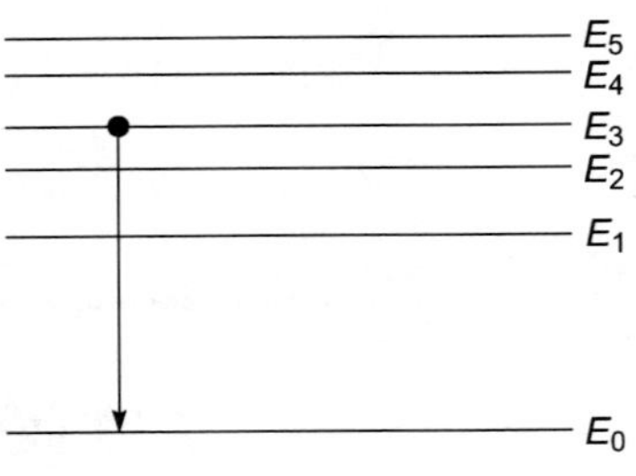

Figure 25.1

$$h\omega = E_2 - E_0$$

h = Planck's constant

ω = Frequency of liberated light

E_2, E_0 are energy levels between which transition occurs.

Einstein introduced the idea of interaction of radiation with matter. Consider any two discrete energy levels of atom, p^{th} and q^{th} level, as shown in Figure 25.2. Einstein proposed that when such an atom has light of right frequency acting on it, it can absorb that photon of light and can make a transition from state q to state p.

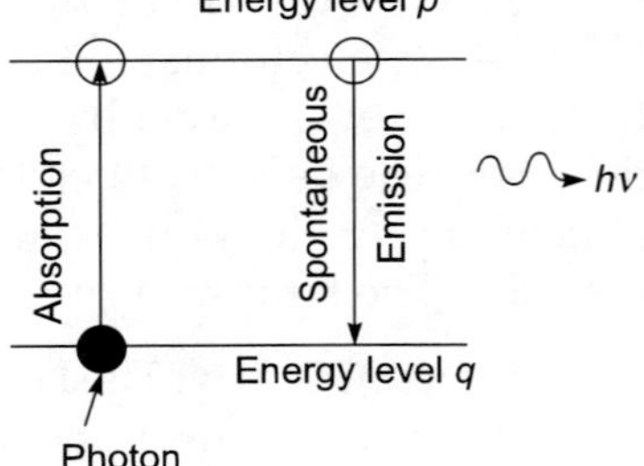

Figure 25.2

The probability of this phenomenon occurring per second depends on:

(i) Energy levels of two states between which transition takes place

(ii) Intensity of the light acting on the atom

Absorption: Phenomenon of movement of atoms from lower to higher energy level is known as absorption (Figure 25.3).

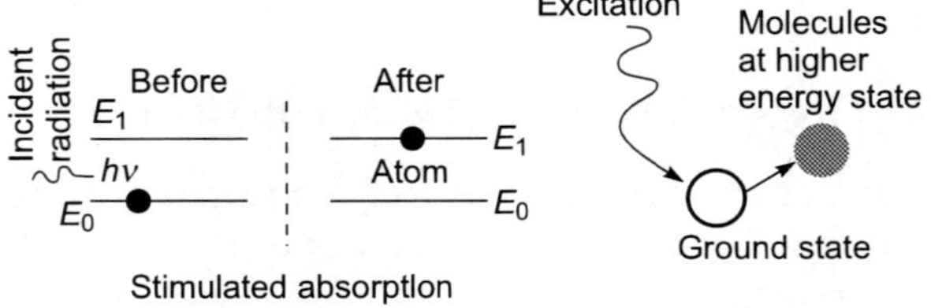

Figure 25.3

Emission: Movement of atom from the higher energy level (p) back to lower energy level (q) is called emission. Einstein proposed that there are two processes by which such emission can occur.

1. Spontaneous Emission: In which no light is present at the higher energy level and atom in an excited state could fall to the lower energy level by emitting a photon and the process is independent of light intensity (Figure 25.4).

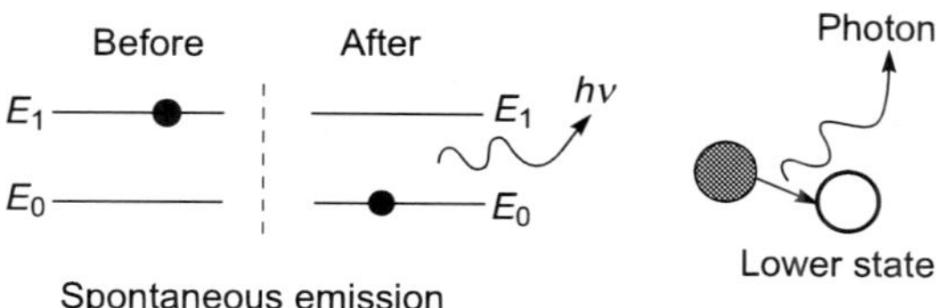

Figure 25.4

2. Stimulated or Induced Emission: In which the emission is influenced by the presence of light. If the light of right frequency acts on an atom, an increased rate of photons will be emitted and such emission would be proportional to the intensity of light (Figure 25.5).

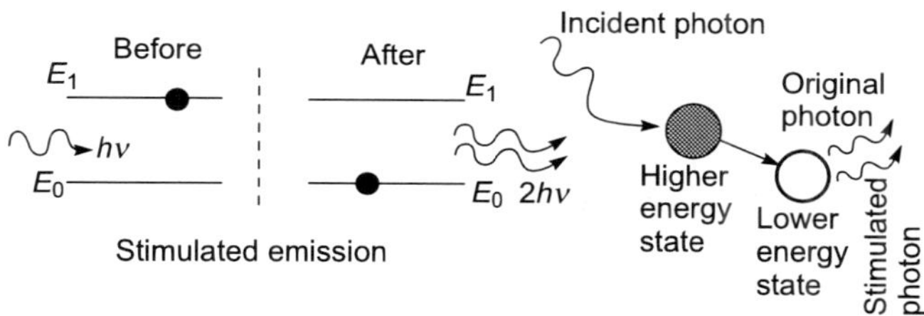

Figure 25.5

If the absorption rate for a given energy level is known, we can mathematically calculate the spontaneous emission rate and the stimulated emission rate or any combination.

In the presence of light of appropriate frequency, stimulated emission will occur in the upper energy level when the atoms begin to emit and a chain reaction will occur by causing more to emit and the whole avalanche would dump down together. This is called lasing action. Stimulated emission is the basis of the laser operation.

Stimulated emission is in phase with the incident wave which is amplified but is also attenuated by absorption. From Maxwell-Boltzmann relation, it is known that the number of atoms in q^{th} level is greater than the number of atoms in p^{th} level, when energy $E_p > E_q$. Under this circumstance, with the shining of the light, the process of absorption will predominate. But if the number of atoms at p^{th} level is made greater than the number of atoms in q^{th} level, then population inversion occurs. Population inversion is required for stimulated emission to predominate. Comparison between spontaneous and stimulated emission is given in Table 25.1.

TABLE 25.1 Comparison between spontaneous and stimulated emissions

Spontaneous Emission	Stimulated Emission
1. The spontaneous emission was postulated by Bohr.	1. The stimulated emission was postulated by Einstein.
2. Additional photons are not required in spontaneous emission.	2. Additional photons are required in stimulated emission.
3. One photon is emitted in spontaneous emission.	3. Two photons are emitted in stimulated emission.
4. The emitted radiation is poly-monochromatic.	4. The emitted radiation is monochromatic.
5. The emitted radiation is incoherent.	5. The emitted radiation is coherent.
6. The emitted radiation is less intense.	6. The emitted radiation is highly intense.
7. The emitted radiation has less directionality. Example: Light from sodium or mercury lamp.	7. The emitted radiation has high directionality. Example: Light from laser source.

25.2 METHODS TO ACHIEVE POPULATION INVERSION

Different methods of pumping are shown in Figure 25.6.

Population inversion can be achieved by:

1. Optical pumping (example: ruby laser)
2. Direct electron excitation (example: argon laser)
3. Inelastic atom to atom collision (example: helium neon laser)

(a) Optical Pumping

As shown in Figure 25.6(a), an external light source is used to produce a high population of some particular energy in the laser medium by selective optical absorption. Optical excitation is used in solid-state lasers, such as ruby laser.

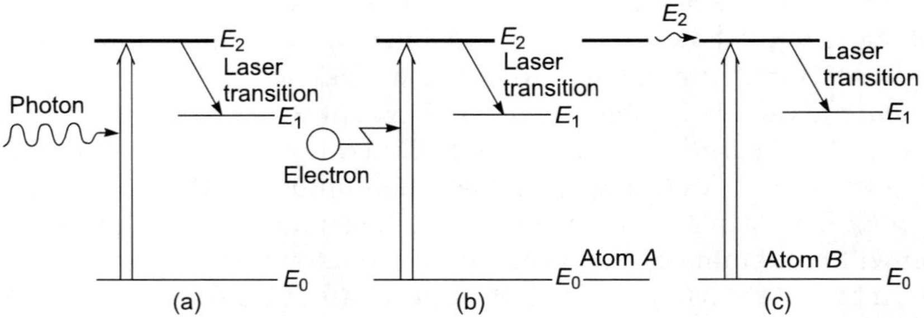

Figure 25.6

(b) Direct Electron Excitation (Argon Laser)

In gaseous ion (example, argon) laser, direct electron excitation in a gaseous discharge is used to produce the population inversion. The laser medium itself carries the discharge current. For the appropriate conditions of pressure and current, the electrons in the discharge may directly excite the active atoms to produce higher populations in certain levels, compared to lower ones.

(c) Inelastic Atom-Atom Collisions (Helium Neon Laser)

A combination of gases is used. Say A and B are the gases and both have some excited states A^* and B^*. The transfer of excitation may occur between the two atoms such that

$$A^* + B \rightarrow A + B^*$$

If the excited state of one of the atoms, say A^* is metastable, then the presence of gas B will serve as an outlet for the excitation. As a result, the excited level of atom B may become more highly populated than some lower level to which the atom B can decay by radiation. This condition arises with the helium neon laser. Excited helium atoms excite a neon atom and laser transition occurs.

Figure 25.7 schematically shows working of a laser. There is a gas in a cylindrical glass vessel. This gas is called the lasing medium. One end of the glass is blocked with a 100% reflective mirror and the other end is having a partially reflective mirror. Population inversion can be carried out by exciting the gas atoms or molecules by pumping them with flash lamps. Then stimulated emission would initiate lasing action. Stimulated emission of photons could be in all directions. Most of the stimulated photons, not along the longitudinal direction, would be lost and generate waste heat. The photons in the longitudinal direction would form coherent, highly directional, intense laser beam.

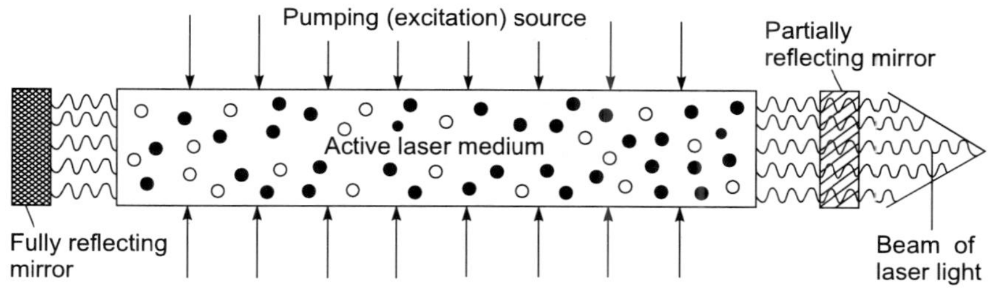

Figure 25.7

25.3 TYPES OF LASERS

Depending on the lasing medium, lasers can be classified as:

1. Solid-state lasers
2. Gas lasers

Solid-state lasers can be further classified as

1. Ruby laser
2. Nd: glass laser
3. Nd: YAG laser

Gas lasers can be further classified as

1. Helium neon lasers
2. Argon lasers
3. Carbon dioxide lasers.

Summary of all the lasers is shown in the Table 25.2.

Solid-State Lasers

The laser materials used in this process are ruby, neodymium glass (Nd-glass) and Neodymium–yttrium aluminum garnet (Nd-YAG). Of these, Nd-glass and Nd-YAG are widely used in machining applications. These laser materials are fabricated into rods and their ends finished to high optical tolerances. The method used to inject energy into the material is by generating a intense light flux which can be absorbed by these laser materials and then converted into a collimated laser beam. The light flux of high intensity is provided by xenon filled flash lamps. The life of this flash lamp is around 10,000 to 1,00,000 pulses. Nd-glass laser operates only in a pulsed mode, with a pulse duration of a few billionths of a second to a maximum of 1 millionth of a second. Nd-YAG lasers can be operated either pulsed or in a continuous wave to produce continuous power output.

Working of Solid State Nd-YAG Laser

Solid-state lasers consist of the following components:

- A solid, crystalline lasing medium that can be pumped to a higher state of energy.
- A pumping system, to pump energy into the lasing medium (usually an optical flash tube).
- A resonator (usually a pair of mirrors mounted at each end of the laser), to bounce stimulated light through the lasing medium.

The flash tube stimulates the lasing medium to higher energy level. When atoms, ions and molecules go back to their ground state, they emit a photon of light in a random direction. These spontaneous photon emissions interact with excited atoms, ions, and molecules and give laser light.

TABLE 25.2 Commercial lasers available for machining, welding and trimming

Laser Type	Wavelength (mm)	Mode of Operation	Power (W)	Pulse, Pulses per Second	Length of Time	Application	Comments
Argon	0.4880 0.5145	Pulse	20 peak: 0.005 average	60	50 μs	Scribing thin films	Power low
Ruby	0.6943	Pulse	2×10^5 peak	Low (5 to 10)	0.2-7 ms	Large material removal in one pulse, drilling diamond dies, spot welding	Often uneconomical
Nd-glass	1.06	Pulse	2×10^6 peak	Low (0.2)	0.5-10 ms	Large material removal in one pulse	Often uneconomical
Nd-YAG	1.06	Continuous	1000	—	—	Welding	Compact, economical at low powers
Nd-YAG	1.06	Repetitively Q-switched	3×10^5 peak; 30 average	1-24,000	50-250 ns	Resistor trimming, electronics	Compact and economical
Nd-YAG	1.06	Pulse	400	300	0.5-7 ms	Spot weld, drill	
CO_2	10.6	Continuous	15,000	—	—	Cutting organic materials, oxygen-assisted metals	Very bulky at high powers
CO_2	10.6	Repetitively Q-switched	75,000 peak 1.5 average	400	50-200 ns	Resistor trimming	Bulky but economical
CO_2	10.6	Pulse	100 average	100	100 μs and up	Welding, hole production, cutting	Bulky but economical
KrF (excimer)	0.248	Pulse	Up to several kW	100-1000	15-45 ns	Micromachining, industrial materials processing, and laser annealing	Short wavelength, high energy, and high average power
XeCI (excimer)	0.308	Pulse	200	300	40-50 ns		

Figure 25.8 shows an electrical circuit for operation of a solid state laser. The flash tube is operated in pulsed mode by charging and discharging of the capacitor. Thus, the pulse on time is decided by the resistance on the flash tube side and pulse off time is decided by the charging resistance. There is also a high voltage switching supply for ignition of pulses.

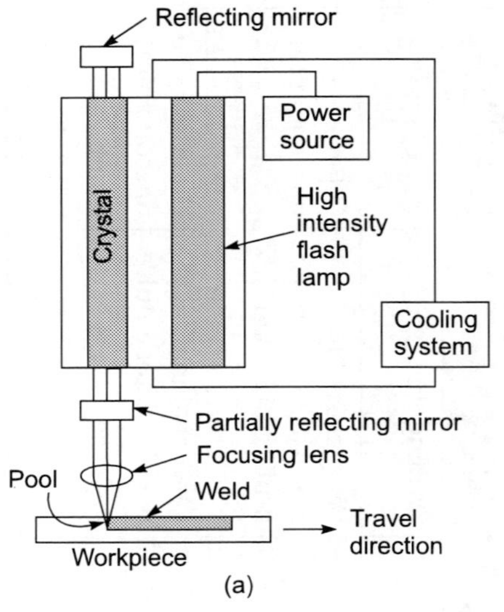

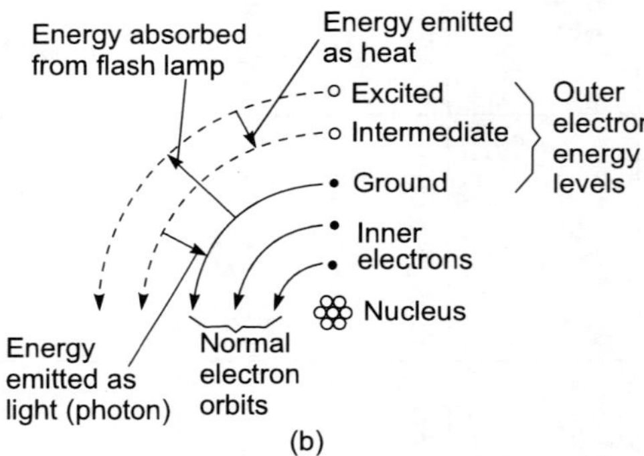

Figure 25.8 Solid state laser

CO_2 Gas Laser

CO_2 laser system uses three gases; namely, nitrogen, carbon dioxide and helium, flowing through a glass discharge tube. Nitrogen functions as an intermediary between electrical energy and vibrational energy of CO_2 molecules. Thus, nitrogen helps to sustain the gas plasma. Helium cools the gas mixture so that it may be re-excited again. Construction of CO_2 laser is shown in Figure 25.9. When high voltage is applied at the two ends, gas plasma is formed. The energy of this discharge leads to population inversion and lasing action. At the two ends of the laser, there are two mirrors. One is 100% reflector and other is partial reflector. The 100 % reflector redirects the photons inside the gas tube and partial reflector allows a part of the laser beam to be redirected to output port, where it is used for material processing. Laser tubes are cooled externally.

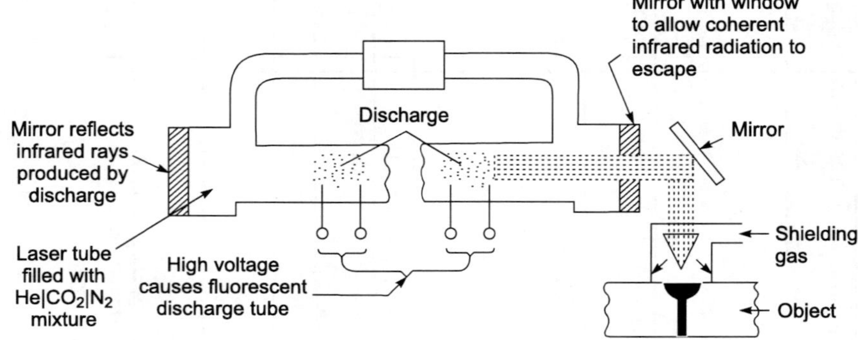

Figure 25.9 CO_2 laser setup

The power of CO_2 laser is typically around 100 watts per metre of tube length. To get high power laser, a rather long tube is required, which is not practical from manufacturing point of view. Therefore, for optimal use of floor space, high powered CO_2 lasers are made of folded design.

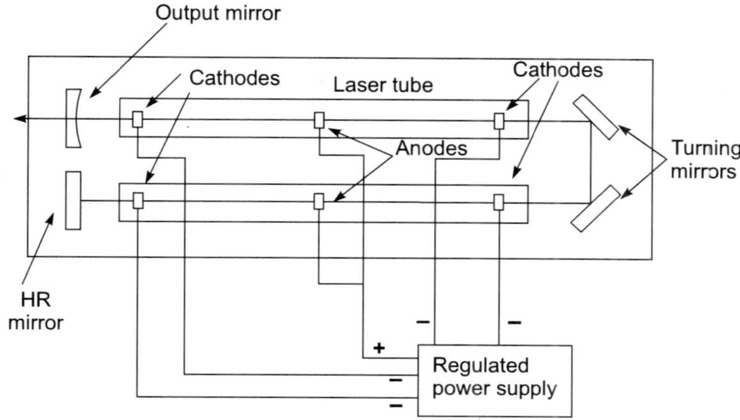

Figure 25.10 CO_2 laser setup: folded type

Figure 25.10 shows a folded axial flow laser. In folded laser, there are a few 100% reflective turning mirrors for maneuvering the laser beam from gas supply as well as high voltage supply, as shown in Figure 25.10.

Figure 25.11 (a) shows LBW in the keyholing mode. Figure 25.11(b) shows a weld in a 13 mm thick A633 steel made with a 15 kW CO_2 laser at 20 mm/s.

The advantages of CO_2 lasers over solid lasers are:

- Relatively high and continuous output.

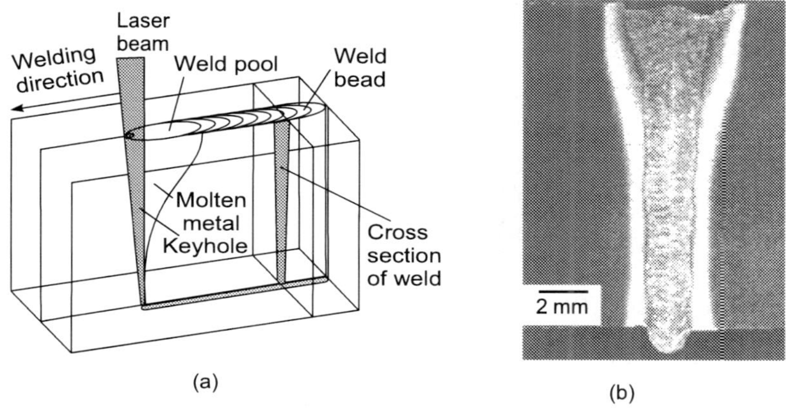

Figure 25.11 (a) Keyhole mode (b) Welding of A633 using CO_2 welding

- High conversion efficiency. Efficiency of ruby laser is around 1% whereas for YAG laser, it is around 10%.
- CO_2 laser has a few maintenance problems unlike the solid lasers which require frequent replacement of flash lamps.
- Divergence is small and are available in many configurations.
- They are compact.
- High absorption of their output wavelength by many materials.
- Higher power levels.

Commonly used Gases in Laser Welding

The following gases are generally used for gas lasers as lasing medium:

- Helium-neon
- Argon
- Carbon dioxide

Helium-Neon	Argon	Carbon dioxide
The most common and inexpensive gas laser, the helium-neon laser is usually constructed to operate in the red at 632.8 nm. It can also be constructed to produce laser action in the green at 543.5 nm and in the infrared at 1523 nm.	The argon laser can be operated as a continuous gas laser at about 25 different wavelengths in the visible between 408.9 and 686.1 nm, but is best known for its most efficient transitions in the green at 488 nm and 514.5 nm. Operating at much higher powers than the helium-neon gas laser, it is not uncommon to achieve 30 to 100 watts of continuous power using several transitions. This output is produced in a hot plasma and takes extremely high power, typically 9 to 12 kW, so these are large and expensive devices.	The carbon dioxide gas laser is capable of continuous output powers above 10 kilowatts. It is also capable of extremely high power pulse operation. It exhibits laser action at several infrared frequencies but none in the visible. The CO_2 laser is the most efficient laser, capable of operating at more than 30% efficiency.

25.4 BEAM PARAMETERS

Process criteria are generally influenced by the beam parameters as enumerated below:

Lasing Material	Ruby	Nd-YAG	Nd-Glass	CO_2
Type	Solid state	Solid state	Solid state	Gas
Composition	0.03-0.07% Nd in Al_2O_3	1% Nd doped with yittrium, aluminum and garnet	2-6% Nd in glass	CO_2 + He + N_2 (typically 3:8:4)
Wavelength	0.69 micron	1.066 micron	1.064 micron	10.6 micron
Efficiency	1%	2%	2%	10-15%
Beam mode	Pulsed	Pulsed	Pulsed	Pulsed
Spot size	0.015 mm	0.015 mm	0.025 mm	0.075 mm
Beam output	10-100 W	10-1000 W	10-100 W	0.1-10 kW
Peak power	200 kW	400 kW	200 kW	100 kW
Beam divergence	5-7 mRad	1-5 mRad	5-7 mRad	0.1-10 mRad
Pulse rate	1-10 pps	1-300 pps	1-3 pps	Continuous wave

Laser Welding: Laser can be used to weld dissimilar materials. There are two types:

1. Conduction laser welding
2. Penetration laser welding

Conduction Laser Welding	Penetration Laser Welding
Low intensity lasers are used	High intensity lasers are used
Amount of material removed depends on conduction of laser into workpiece	High density laser vaporizes and melts the workpiece and weld nugget is formed

Laser is also used for spot welding and seam welding.

25.5 ADVANTAGES OF LBW

1. Can be performed inside plastic housing/transparent glass.
2. Large variety of material can be joined by this process.
3. Easily weld dissimilar materials.
4. Low thermal input.
5. Laser can be transmitted over long distance with minimum loss of power.

6. Small heat affected zone, and neglible thermal damage or effect on adjacent regions.
7. Easy control of beam configuration and size of exposed area.
8. Vaporization rate of metals can be controlled.
9. Heat treated materials can be welded and properties will not get affected.

25.6 DISADVANTAGES OF LBW

1. Overall efficiency is low.
2. Effective safety procedure is required.
3. Rapid cooling rate may cause cracking in certain metals.
4. High initial cost of equipment and consumables.
5. Optical surface can be damaged.
6. Maintenance cost is quite high.

25.7 APPLICATIONS OF LASER BEAM WELDING

1. Cutting and welding is possible.
2. Small electronic components and integrated circuitry in electronic industries.
3. Welding of nuclear plant components.
4. High melting point of metal can be welded.
5. Large dissimilar metals such as Cu, Ni, Al, SS, Tn, Zr, etc. can be easily welded by this process.

Lasers in Arc Welding

As shown in Figure 25.12, laser assisted gas metal arc welding (LAGMAW) has demonstrated significantly greater penetration than conventional GMAW. In addition to direct heating, the laser beam acts to focus the arc by heating its path through the arc. This increases ionization and hence, the conductivity of the arc along the beam path and helps focus the arc energy along the path. It has been suggested that combining the arc power with a 5 kW CO_2 laser, LAGMAW has the potential to achieve weld penetration in mild steel equivalent to that of a 20 to 25 kW laser. A lower power CO (not CO_2) laser of 7 W and 1 mm diameter can initiate, guide, and focus an Ar–1% CO gas-tungsten arc.

Advantages and Disadvantages Over EBW

Like EBW, LBW can produce deep and narrow welds at high welding speeds, with a narrow heat affected zone and little distortion of the workpiece.

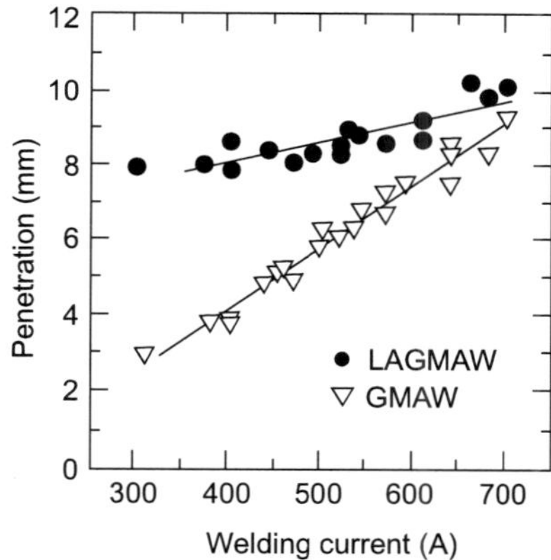

Figure 25.12 Weld penetration in GMAW and laser assisted GMAW using CO_2 laser at 5.7 kW. Reprinted from Hyatt et al. Courtesy of American Welding Society.

It can be used for welding dissimilar metals or parts varying greatly in mass and size. Unlike EBW, vacuum and X-ray shielding are not required in LBW. However, the very high reflectivity of a laser beam by the metal surface is a major drawback, as already mentioned. Like EBW, the equipment cost is very high, and precise joint fit-up and alignment are required.

Worked Example

Example 1: In a laser beam welding process, what is the quantity of heat per unit time (J/sec) that is transferred to the material if the heat is concentrated in a circle with a diameter of 0.2 mm? Assume the power density provided is 9000 W/mm^2.

Solution: Power density is 9000 W/mm^2 for laser beam welding

$$P = PD \times A = 9000 \times \pi \times 0.2^2/4 = 283 \text{ W} = 283 \text{ J/sec}$$

Review Questions

1. Explain the principle of operation of LBW.
2. Explain the equipment used in LBW.
3. What is spontaneous emission?
4. Discuss the mechanics of population inversion.
5. Discuss the different types of lasers used in LBW.
6. Compare EBW and LBW.
7. Discuss the weld quality produced by LBW.
8. What are the important applications of LBW?
9. What are the advantages of LBW?
10. What are the disadvantages of LBW?

26
CHAPTER

Underwater Welding

26.0 INTRODUCTION

Underwater welding, as the name implies, is the welding of components in the water. The fact that electric arc could operate was known for over 100 years. But the first underwater welding was successfully carried out in 1930. In this chapter, we will cover the most common types of underwater welding and throw light on the most recent advancements in the field.

Underwater welding is a vital tool for fabrication underwater. It is mainly used to repair ships, offshore oil platforms and pipelines. Welding underwater can either take place in an environment directly exposed to water (wet welding) or in a dry enclosure at the hydrostatic pressure of the sea, created near the area to be welded (dry welding).

26.1 COMPARISON BETWEEN UNDERWATER AND NORMAL AIR WELDING

1. Electrodes are painted for water-proofing.
2. Electrode core wire is usually the same as air welding but in the case of welding of high strength steel inside water, using wet welding technique, a core wire of stainless steel or special steel is used.
3. The flux coating in common use is that of rutile type. Iron oxide covering, which is not very common in air welding, has been found to be more advantageous.
4. In air welding, a gap is maintained between the electrode and base plate. For maintaining an arc in water, it is necessary to keep the electrode in contact with the plate. A slight pressure is also maintained. Cooling action of water on flux coating and water proof paint results in the formation of a barrel at the end of electrode. Arc burns inside this barrel space (Figure 26.1).

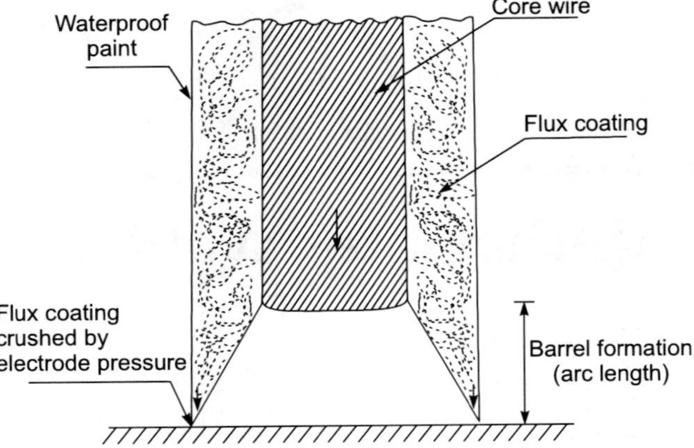

Figure 26.1 Barrel formation

5. Underwater arc is surrounded by a bubble of stream and gases. The pressure on the arc equals the atmospheric pressure plus the pressure of the water column above the arc, as shown in Figure 26.2. The pressure around the arc, thus increases with depth. This affects arc behavior and equilibrium of chemical reactions which affect weld chemistry. Carbon, silicon and manganese content of the weld metal increases with depth with corresponding change in properties.

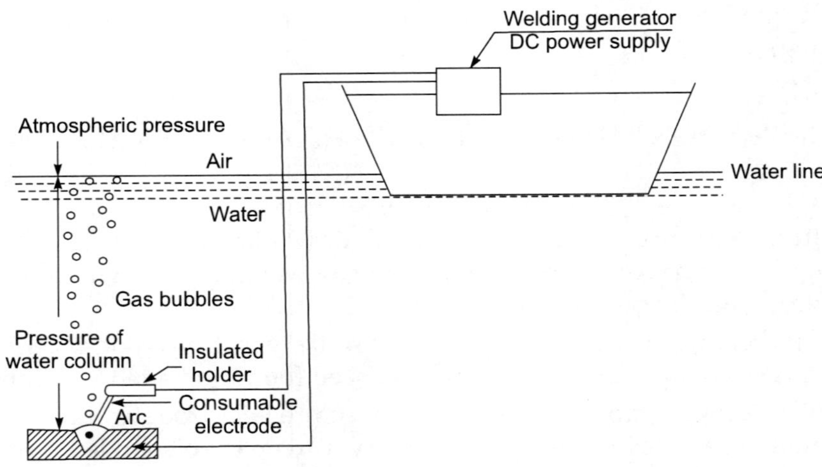

Figure 26.2 Pressure in underwater welding

6. Cooling rates in air welding could be controlled by change in arc energy input. There is far less scope for doing this, as the voltage and current during underwater welding have a close range.
7. Hydrogen and oxygen levels are normal in air welding while weld metal and heat affected zones, hydrogen and oxygen levels are in excess of those in air

welding. This is due to increased amount of hydrogen and oxygen in arc bubble.

8. Electrode holders are insulated.

26.2 TYPES OF UNDERWATER WELDING

1. Wet welding
2. Dry welding
3. Local cavity welding
4. Portable dry spot welding

26.2.1 Wet Welding (Figure 26.3)

- In wet welding, manual metal arc welding (MNAW) with direct current electrode negative (DCEN) is used.
- The direct current rating should be between 300 and 400 A.
- The welding circuit includes a positive type switch, usually a knife switch, operated on the surface and commanded by the welder.
- The knife switch in the electrode circuit is capable of breaking the full welding current and is used for safety reasons. The power should be connected to the electrode holder only during welding.

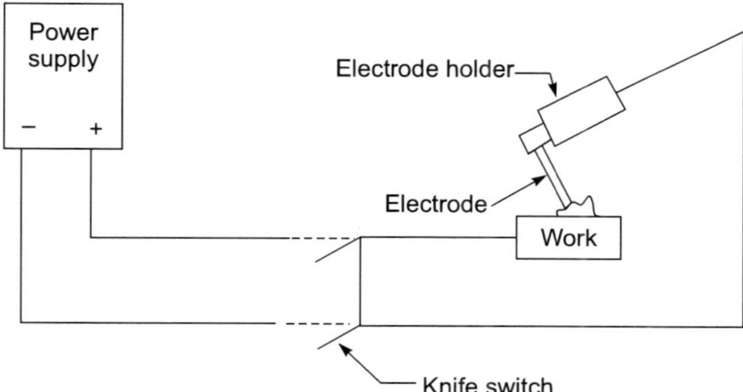

Figure 26.3 Knife switch arrangement in wet underwater welding

Following things are to be taken care of while doing underwater wet welding:

(a) When DCEP (direct current electrode positive) is used, electrolysis will take place and cause rapid deterioration of any metallic component in the holder.

(b) AC current cannot be used because of safety reasons and difficulty of maintaining an arc underwater.

(c) The electrode holder should have extra insulation and a twist-type head for gripping the electrode.

(d) Electrode must be waterproof and all connections must be thoroughly insulated. If insulation is weak, seawater will come in contact with the metal conductor and part of the current will leak. In addition, there will be rapid deterioration of the copper cable at the point of leak.

The poor quality of welds made in the wet is due to the problem of heat transfer, welder visibility and the presence of hydrogen in the arc atmosphere. When the base metal and the arc area are surrounded by water, there is no temperature of heat buildup of the base metal at the weld. This creates high temperature gradient or quenching effect, which reduces the ductility of the weld metal. The arc area is composed of a high concentration of water vapor. The arc atmosphere of hydrogen and oxygen of the water vapor is absorbed in the molten weld metal and contributes to porosity and hydrogen cracking. In addition, welders working underwater are restricted in their efforts to see and manipulate the welding arc: under ideal conditions, the welds produced in the wet with covered electrodes are marginal. They may be used for short periods but should be replaced as quickly as possible. Improvement in underwater welding electrodes helps in the wet weld quality.

Efforts have been made to produce a bubble of a gas in which the weld can be made. This technique has not been able to ensure the good quality welds made with covered electrodes in the wet.

The general arrangements for underwater in the wet welding are shown in Figure 26.4. The power source for underwater should always be a DC machine rated at 300 or 400 A. Generator welding machines are often employed for underwater welding in

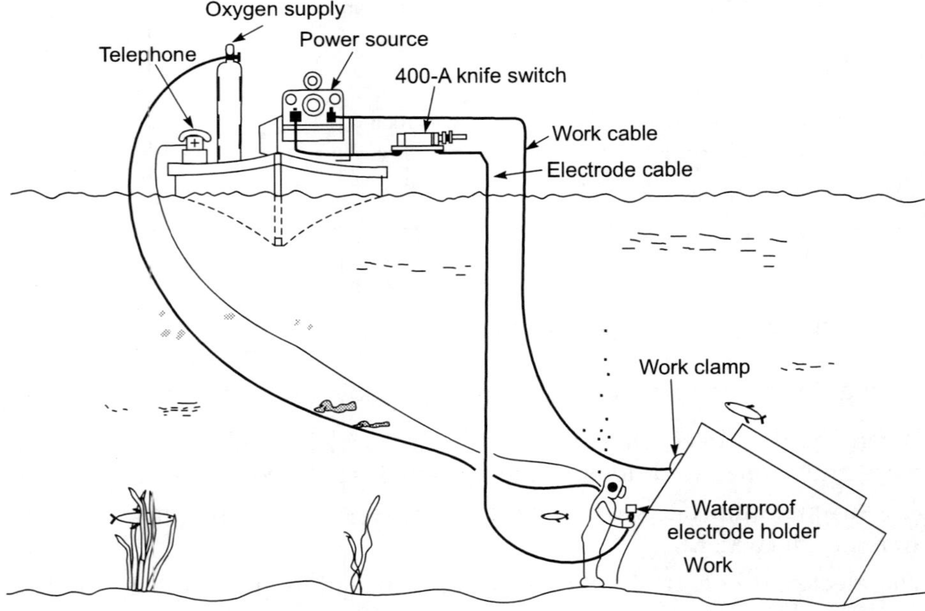

Figure 26.4 Arrangement for underwater welding in the wet

the wet. The frame of welding machine must be connected to the ship. The welding circuit must include a positive switch, usually a knife switch, that is operated on the surface upon the command of the welder. The knife switch in the electrode circuit must be capable of breaking full welding current. It is required for safety reasons. The welding power should be connected to the electrode holder only while the welder is welding. Direct current with electrode negative (straight polarity) is used. Special electrode holders with insulation against water are employed.

The underwater electrode holder accommodates two sizes of electrodes, normally 4.8 and 4 mm. The electrode types are normally used to meet E6013 classification and must be waterproofed. This is done by wrapping them with waterproof tape or by dipping them into sodium silicate mix or other waterproofing material. Electrodes for underwater welding are available commercially.

The welding lead and work lead should be at least 2/0 size, and the insulation must be perfect. If the total length of the lead exceeds 100 m, they should be paralleled. With paralleled leads to the electrode holder, the last 1 m should be single cable. All connections must be thoroughly insulated so that water cannot come in contact with the metal parts. If the insulation breaks, the seat water will contact the copper conductor and part of the current will leak and will not be available at the arc. In addition, there will be rapid deterioration of copper cable at the break. The workpiece lead should be connected to the work being welded within 1 m of the point of welding. Welding in the wet is shown in Figure 26.5.

Figure 26.5 Welding in the wet

A special underwater cutting torch uses the oxygen arc cutting processes with a tubular steel covered electrode. The torch is fully insulated and uses a twist type collect for gripping the electrode. It includes an oxygen valve and connections for attaching the welding lead and an oxygen hose. It is equipped to handle up to 7.9 mm tubular electrodes. In this process, the arc is stuck in the normal fashion and oxygen is fed through the centre hole of the electrode to provide the cutting action. The normal electrical connections are employed.

The need to produce high quality weld underwater has increased as oil and gas are found in deep water. Most offshore exploration, drilling and production until recently, were in water ranging from 10 to 16 m deep. When a pipeline needs to be repaired, it is raised to the surface. Repaired and lowered back to ocean floor. Exploration, drilling and production are moving into deeper water up to 305 m depth. Modifications and work must be done on the ocean floor. More pipelines are damaged and there is necessity for making tie-ins of pipelines on the ocean floor. The repairs and tie-ins must be high quality welds to prohibit the possibility of leaks or oil spills. This type of work is now done in depth of 61 to 182 m.

Bead Manipulation in Wet Welding Processes

Deposition of stringer beads (Figure 26.6) is used for enhancing the weld bead strength. The advantages of stringer beads are:

1. Easy control over speed control
2. Uniform bead surface
3. Good arc stability
4. Reduced risk of slag inclusion
5. Reduced chances of undercutting
6. Consistent and satisfactory penetrations
7. Ease of welding in low visibility conditions

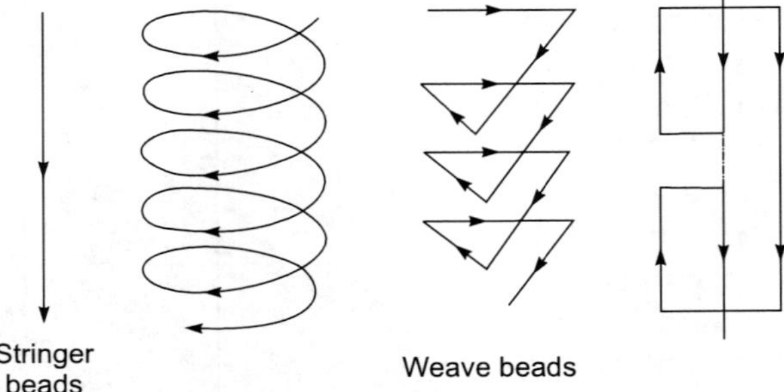

Figure 26.6 Types of bead manipulation

The following precautions are to be taken for good weld beads:

1. The joints should be well fitted.
2. Should be free from rust, oil, paint, etc.
3. No abrupt changes in weld contours.
4. The ends of the short beads or tacks should be thoroughly cleaned and hammered to give a smooth surface.
5. The bead or layer deposited should be cleaned of slag, spatter or globules before superimposing additional runs. Van der Willingen described the use of a special wrapped heavy coated iron powder electrode, which gives high deposition rate and excellent torch weld characteristics.

Advantages of Wet Welding

- The low cost and versatility of wet welding makes this method highly desirable.
- It is faster than dry welding.
- Portions of offshore structure can be reached by the welder that could not be welded using other methods.
- No enclosures are required and no time is lost in building it.
- Equipment needed for mobilization of a wet welded job is minimal.

Disadvantage of Wet Welding

- Rapid quenching, which increases the tensile strength but decreases the ductility and impact strength of the weldment and increases porosity and hardness.
- Hydrogen embrittlement, cracks and microscopic fissures are produced as with increase in depth, hydrogen content increases and dissolves in the heat affected zone. These cracks can take the shape of a larger one and can result into failure of the structure.
- Poor visibility underwater, thus welder cannot weld properly sometimes.
- Wet welding is not suitable for mild steel.
- Deep water welding is difficult.
- Electrode coating should be protected by layer of wax, varnish or cellulose.

26.2.2 Dry Welding Process

The development of welding in the dry environment makes it possible to produce high quality weld joints that meet X-ray code requirement. A number of welding processes are used for welding in the dry: the shielded metal arc, the gas tungsten arc, the plasma arc, the gas metal arc, and the flux cored arc welding process. The shielded metal arc welding is rarely used for welding in dry environment, because of larger amount of smoke and fumes produced. The gas tungsten arc welding is used to produce welds that meet quality requirement. The gas tungsten is relatively slow, but it is acceptable since the welding operation is a small part of the total repair operation. Efforts are ongoing to develop plasma arc welding more fully for deep water operations. Successful application of gas metal arc welding in the dry environment has been made to depths as great as 51 m.

There are two types of dry underwater welding. One involves a welding chamber or habitat and is known as hyperbaric welding. The habitat or welding chamber provides the welder with all the necessary equipment for welding and related work in dry environment. The weld chamber is made so that it can be sealed to the part to be welded. Since most work is on pipes, arrangements are made to seal the habitat to the pipe. The bottom of the chamber is exposed to open water and is covered by grating. The pressure in the chamber is equal to the water pressure at the operating depth. Figure 26.7 shows a welding habitat.

Life support equipment involves a two-way telephone and video camera for continuous observation. Life support atmosphere for the welder, power for tools and for welding, and the gas supply for the welding atmosphere. Habitats usually include an atmospheric conditioning system, used to both cool and filter the atmosphere in the chamber. Filtering is important since metal vapor is released during welding. Air conditioning is employed since heat is generated by welding. For gas metal arc welding, the welding power source is normally on the surface. Welding cables are lowered to the habitat to provide power for the arc. The electrode wire, wire feeder, and the control unit are located in the habitat. The wire feeder must be protected from the high pressure and high humidity conditions in the chamber. The electrode wire must also be protected from the humidity.

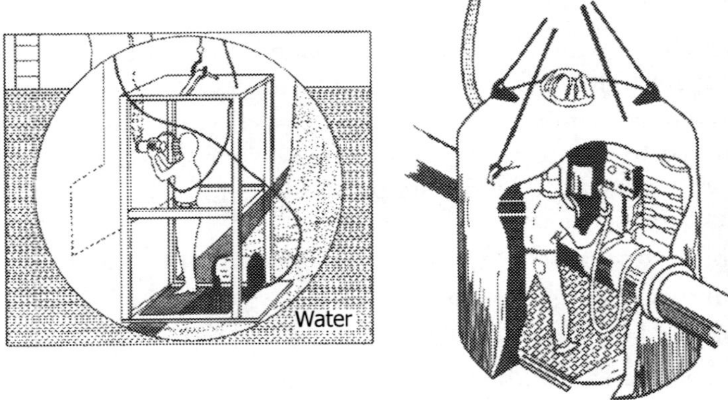

Figure 26.7 Underwater pressure habitat

The gas for breathing and for welding is designed for use at high pressures involved. The pressure in the habitat increases 1 atmosphere or 1 bar for 10 m depth. The water pressure must be equalized by the pressure of atmosphere within the habitat. This high pressure creates problems for the crew and for welding.

With the shielded arc metal arc welding, the problem involves the removal and filtering of the atmosphere in habitat. With gas tungsten arc welding, very little smoke and fumes are generated, but the inert gas used for welding disrupts the breathing atmosphere. For saturated diving work, the breathing atmosphere of the welder is based on the working depth. Premixed gas is used and the oxygen content is based on the depth.

Welding with gas metal arc in a habitat presents specific problems. In shallow depths, the welding in the habitat is essentially the same as welding on the surface. When welding at the depth of 35 m, the pressure is about 4 atmospheres. The weld metal quality is essentially the same as the surface made welds. However, the welding voltage increases and welds bead penetration increases. As the depth is increased, the atmosphere pressure increases and the arc becomes more constricted, and this leads to increased arc voltage and increased penetration and higher burn-off rates. This makes the weld pool difficult to handle. At depths beyond 35 m, the weld pool becomes increasingly difficult to control and an increased amount of smoke is generated. Special welding power sources are required.

Advantages of Dry Welding

- Welder's safety is ensured as welding performed in a chamber is immune to ocean currents and marine animals and is well illuminated and has its own environment control system.
- Good quality welds are produced as hydrogen level is much lower and sea water is not present to quench the weld.
- Surface monitoring can be done. Joint preparation, pipe alignment, NDT inspection, etc., are monitored visually.
- Non-destructive test is also facilitated by dry habitat environment.

Disadvantage of Dry Welding

- Large quantities of complex equipment and support equipment are required. Also, the chamber in which the welding is done is very complex.
- Cost of habitat welding is very high and increases with depth because at greater depths, the arc constricts and consequently, higher voltages are required which increases the cost.
- Same chamber cannot be used for another job, if it is a different one.

26.3 LOCAL CAVITY WELDING

The process, developed by the Japanese, makes use of special torch where the gas is fed into the torch to create a dry zone in the region surrounded by water. A water jet acts as a shutoff wall to the exterior water. The obtained cavity is stable and durable to the external and internal perturbations and to torch travel. The stability of the arc and bead formations is excellent. Figure 26.8 shows local chamber welding.

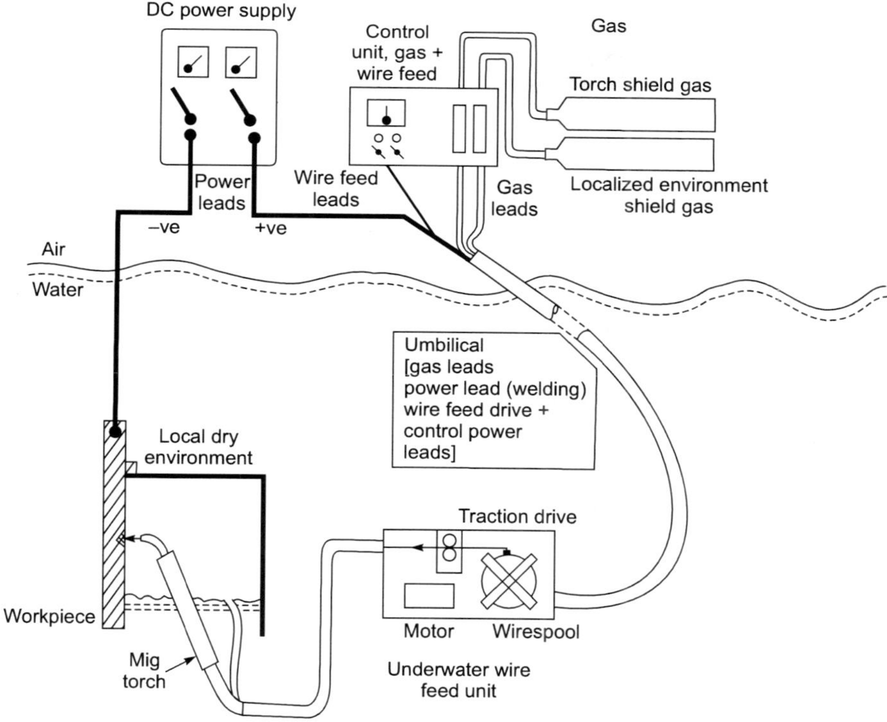

Figure 26.8 Local cavity welding

26.4 PORTABLE DRY SPOT WELDING

In this method, weldment is enclosed in dry environment, usually transparent plexiglass box and welder is submerged in water. Weld properties are similar to air welds. No heavy equipment is needed. Gas and wire feeding is quite difficult and therefore, MIG is used. Figure 26.9 shows portable dry spot welding process.

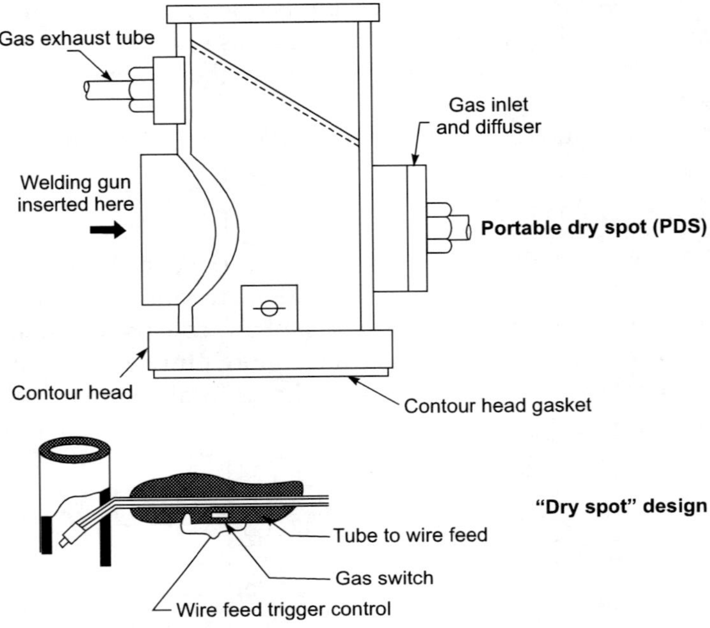

Figure 26.9 Portable dry spot welding processes

26.5 WET-DRY WELDING

Improvements have been made to provide more flexibility for gas metal arc welding when welding underwater. The dry hyperbaric chambers or habitats are extremely expensive and must be designed for specific applications. It is now possible to take gas metal arc welding outside the habitat by the use of special nozzles or chambers that surround the torch. In using this apparatus, the welder is in the wet or in water, but the nozzle of the welding gun and material to be welded is in the dry atmosphere. The gun and nozzle are in this small chamber, but the wire feeder and electrode wire supply are in another water-tight pressurized enclosure. The pressure of the shielding gas coming through the system is more than the pressure of the water at the operating level. The gas flows through the wire feeder enclosed through the electrode conduit to the torch, where it provides the shielding atmosphere for welding. It also provides the pressure to evacuate the water to provide a dry atmosphere. The dry gas environment chambers are relatively inexpensive, small and light in weight. They are provided with flexible seals to be used against the part being welded. They can be handled or made with clamps for quick attachment to the part to be welded. The gun in hand is manipulated inside the small chamber in the same way as on the surface. The chambers are made of transparent material or have a sufficient number of windows so that the welder can see inside to properly manipulate and direct the welding arc. This technique can be used for welding with gas metal arc up to 35 m below the surface. High quality welds that meet code requirements can be produced.

26.6 OPERATING PARAMETERS OF UNDERWATER WELDING

The main factors influencing underwater welding are:

- rapid cooling rate provided by the surrounding water quench,
- influence of pressure, increasing with depth,
- limiting carbon equivalent permissible,
- concern for hydrogen cracking, and
- presence of porosity and its influence on fatigue life.

Electric Arc

- The process most commonly used for underwater welding is SMAW although, FCAW (flux coated arc welding) has also been employed. Wet underwater welding has been carried out as a building or repair technique at depths of 100 m and down to 200 m under sea level and over.
- As already discussed, underwater welding by the electric arc is performed by either the wet environment or in a protective chamber to be constructed around the weld zone and evacuated of water by compressed helium gas, to create dry conditions.
- The cost of hyperbaric underwater welding is much higher than that of wet welding. Assembling and dismantling can be very expensive in deep sea environment.

Underwater arc is exposed to two basic mechanisms of compression and constriction. Underwater arc is surrounded by a bubble. Hydrogen content (about 93%) of the arc bubble atmosphere, together with water surrounding, compresses the arc and at the same time, it has a severe cooling effect on the arc column compared to normal air welding. This causes the arc constriction. This compression and constriction of arc column results in a higher current density in underwater arc. Further, in straight polarity welding, the limited geometrical dimensions of the electrode end prevent the free expansion of the cathode spot with increase in welding current. The arc is thus, constricted. This apparently explains the fact that the volt-ampere characteristic curves of an underwater arc are concave or rising. Due to these compressive forces, the increase in cross-sectional area of the arc lags behind the given increase in the welding current, thereby raising the current density of field intensity. Thus, to maintain same arc conditions, the current should be increased by 10% atmosphere (10 meters of water) of additional pressure. These higher current densities produce high arc temperatures. Temperature of arc column at different currents and depths is given in Table 26.1.

TABLE 26.1 Temperature of arc column at different currents and depths

Welding Condition			Temperature of Arc Column, K	
Depth, m	Current, amp	Effective Dia. of Arc Column, cm	Thin Wire Electrodes	Stick Electrodes
10	100	0.202	8400*	9300
10	200	0.205	9200*	10200
10	300	0.210	9750	10700
10	400	0.260	10150	11100
10	500	0.317	10650	11500
20	300	—	10000	11000
40	300	—	10300	11300
60	300	—	10400	11500
80	300	—	10600	11700
100	300	—	10800	11800

Note: Calculations based on assumption that arc column is a cylinder of arc length 2 mm. Stick electrode air-arc temperature is 6000°K.

Arc Shape: Basic shape of arc column is cylindrical for metal arc welding and truncated cone with its base on work for thin CO_2 welding. Metal transfer characteristics for the two types of welding processes are given in Table 26.2.

TABLE 26.2 Rates of metallurgical reactions in various methods of underwater welding

Characteristics of Metal Transfer	Thin Wire Without CO_2		Thin Wire with CO_2	EPS-52 Covered Electrode
	Salt Water	Fresh Water	Fresh Water	
Drop transfer* per second	12	16	23	44
Lifetime of drop,** second	0.1700	0.1305	0.0575	0.0254
Average weight of one drop, gm	0.1670	0.0804	0.1100	0.1100
Volume of one drop in mm^3	21.4	10.3	14.1	14.1
Coefficient of reactivity of the process, Cn	21.8	16.77	7.37	3.26
Arc voltage, volt	39	40	39	39 (S.P.)
Arc current, amp	240	250	240	240

* Drop transfer throughout.

** Lifetime of drop has largest apparent effect.

Arc Atmosphere

A peculiar feature of underwater welding is an arc bubble which is maintained around the arc. The size of the bubble fluctuates between small bubble, barely covering the arc column and a large bubble of 10 to 15 mm diameter, that eventually breaks away from the weld puddle and floats to the surface, leaving behind a nucleus bubble with a diameter of 6 to 9 mm. This phenomenon of bubble growth and its breakaway occurs at an approximate rate of 15 times per second at 150 mm of water depth. Gases generated per second for E6013, E6027, E7024 are 40 cc, 50 cc and 60 cc, respectively. The gas bubble consists of 62 to 82% hydrogen, 11 to 24% carbon monoxide, 4 to 6% carbon dioxide and remaining 3% is nitrogen and metallic and mineral salt vapors.

Arc Stability

During underwater welding, the arc-voltage and current values fluctuate. A stability factor for comparing arc performance was defined as maximum current divided by minimum current. The arc is considered to be stable for values of this factor near one. For values much higher than one, the arc is considered unstable. One cause of these fluctuations is the variation in voltage due to changes in arc length during metal transfer. Another cause of the fluctuation is collapsing of thick flux covering, occurring every 0.3 second or less during the arc welding. Different electrodes produce different levels of stability. It has been found that E7024 is more stable than E6027. While E6013 was found comparatively unstable because of its coating being thinner than the other two. Arc has been found to be more stable in salt water than in fresh water. This is due to ease of ionization of seawater, but there is more current leakage in seawater (up to about 65 to 110 amp, at an open circuit voltage of 83 to 89 volts).

Metal Transfer

Normally, the metal transfers in droplets (globular). Occasionally, a large drop forms and short circuits the arc. Drop transfer frequency is 80 to 100 drops per second for the coated electrodes. Drop transfer frequency depends on the type of electrode, in addition to other factors. Underwater arc is constricted and produces a higher core temperature of 9000 K to 11,000 K at 10 m depth, as compared to 5000 to 6000 K for air welding. This increased temperature causes fast melting rate for plates as well as electrodes. The weld puddle, which otherwise has been uncontrollable, solidifies rapidly due to quenching effect of water.

Water Quenching

In normal dry conditions and the time for cooling from 800 to 500°C is generally between 8 and 16 sec.

- In wet conditions, this time reduces to about 1 to 6 sec.
- This rapid quenching produces significant amount of heat affected zone.
- Forms martensite in carbon steels and has significant consequences on microstructure and properties of welds.

Depth Influence

- Due to pressure increase with increasing depth, there is decrease of silicon and manganese and an increase in carbon in the weld pool.
- Arc is less stable with increasing depth because of higher ionization potential of hydrogen, making it more difficult to sustain the arc.
- Increase in H_2 underwater, increases the danger of hydrogen cracking.

Hydrogen Cracking

To limit hydrogen cracking underwater, certain measures can be attempted:

- Designing the weld for minimum stress, using improved fit up.
- Reduced weld deposits or using temper beads to partially relieve the stresses of a previous weld.
- Using consumable electrodes like austenitic stainless or nickel electrodes, that can hold a higher concentration of hydrogen in molten weld pools.
- Introducing carbonates in the flux coatings, which deliver alternate gases in the weld plasma.
- Using specific elements in the weld pools, which getter the hydrogen by forming hydrides.

Porosity

Another concern for wet water welding is metal pore formation, which increases with depth. It can be minimized by altering the following parameters:

- A high current with DCEN
- A short arc
- A fast travel speed
- A low current with DCEP

Polarity

Electrode negative polarity produces less undercuts and spatter, better bead shape, more regular welds and minimum corrosion damage to the electrode holder. Polarity makes little difference to weld appearance or visibility. Barrel length was, however, more than with electrode positive. Electrode positive or negative polarity can be used for underwater welding.

Salinity of Sea Water

Electrical conductivity of water is found to increase with salinity. For bottom sea water, it is approximately 0.03 mohs per cubic centimeter. It is easy to initiate and maintain arc in saline water, penetration is increased. In MIG welding, salinity is reported to increase droplet size, reduce the number of drops per unit time and power consumption. Salinity improves bead shape of underwater welds.

Carbon Equivalent

For wet repair welding, the carbon equivalent should be limited to max. 0.40.

$$CE = \%C + \%Mn/6 + \%(Cr+Mo+V)/5 + \%(Ni + Cu)/15$$

Hardness of 400 vickers measured under a load of 10 kg can be found in the weld. When the carbon equivalent is about 0.4%, the major changes in microstructure and in mechanical properties occur in the first 50 m. At greater depths, the weld metal composition remains fairly constant.

If necessary, compositional modifications of consumables for wet underwater welding must be studied to provide for suitable properties.

Table 26.3 shows the summary of likely effects of underwater welding conditions on potential weldability.

TABLE 26.3 Summary of likely effects of underwater welding conditions on potential weldability

Aspect of Weldability	Wet Welding	Local Chamber Welding	Habitat Welding
1. Hydrogen cracking	Very high increased risk of cracking	Probably some increased risk, particularly at great depths	Probably some increased risk, particularly at great depths
2. Solidification cracking	Some increased risk with depth	Some increased risk with depth	Some increased risk with depth
3. Lamellar tearing	Possible increased risk, particularly at depth	Possible increased risk, particularly at depth	Possible increased risks, particularly at depth
4. HAZ toughness	Probable deterioration	Little effect anticipated, except possible slight deterioration immediately after welding	No effect anticipated, except possible slight deterioration immediately after welding
5. Weld metal toughness	Deterioration	Possible effect at depth, dependent on composition	Possible effect at depth, dependent on composition
6. Stress corrosion	Increased risk	No effect	No effect
7. Fatigue	Possible deterioration in life	Possible deterioration in life	Possible deterioration in life

26.7 ELECTRODES USED

Various electrodes are used for underwater welding. Some of them are given below.

(a) **Cellulosic:** These electrodes give a harsh digging arc, resulting in high penetration. In underwater welding, the currents used are high to maintain the arc. This has been found to aggregate the situation and produce more undercuts and convex beads. The results are not good even with reverse polarity. E6010 has been found to spatter violently, gives irregular beads and produces clouds of black smoke, while E6011 gives almost no spatter and produces continuous beads. It means that the presence of substances which ionize easily, improves the electrode performance.

(b) **Rutile:** Rutile electrodes have been found superior to cellulosic and inferior to acidic. Plain rutile electrodes are inferior to iron powder type. Lightly coated rutile electrodes, E6013, have been recommended by US navy in their manual on underwater cutting in 1953.

(c) **Oxidizing:** Electrodes give satisfactory welds but the welds are inferior in strength and ductility as compared to acid and rutile electrodes. A comparison of various electrodes is given in Table 26.4.

Iron Powder: These electrodes give high deposition rate and excellent drag welding characteristics. These electrodes give more stable arc and fine droplets transfer with occasional short circuits. They are superior to rutile electrodes. Heavily coated rutile E7024 and iron oxide E6027, give higher heat inputs than basic and rutile. For E6013, better coating has to be designed to eliminate chipping of the outside coating during welding. Arc elongation effect is more serious in E7024 and E6027 and therefore, the discrepancy between the machine current setting and the actual measured value is 15 to 25 amp.

From the above discussion, it is clear that none of the existing electrodes for air welding can be used directly for underwater welding and special welding electrodes have to be developed to avoid the difficulties encountered in the use of the existing air welding electrodes.

26.8 REQUIRED CHARACTERISTICS OF ELECTRODES FOR UNDERWATER WELDING

Flux covering for underwater welding should have some special characteristics in addition to usual characteristics required in air welding. Because of arc constriction effect, the current density of underwater arc column is more and therefore, deeper penetration is obtained in underwater welding. The arc should therefore, have soft behavior. Purely cellulosic electrodes are unsuitable for underwater welding as their arc is harsh and have digging tendency. Arc should have high stability to counter the extinguishing effect of water. Because of poor visibility conditions, the coating should give easily removable slag to assist in multipass welding. Coating should be non-conducting and non-hygroscopic by applying suitable insulating and waterproof paint on flux coating.

TABLE 26.4 Strength characteristics of various coated electrodes used in underwater welding

S. No.	Investigator	Type of Electrodes	Water Proofing Coating	Yield Strength, N/mm^2	Ultimate Tensile Strength, N/mm^2	Percentage Reduction in Area	Impact Strength, Joules
1	Berthet and Kermshon	(i) Acid	Vinyl lacquer	460.6	490	8.5	40-28
		(ii) Rutile	-do-	416.5	436.1	5.5	33.6–27.4
		(ii) Oxide	-do-	372.4	436.1	17.5	34.4–33.6
2	Hibshman and Jensen	(i) Oxide coated	–	279.3	387.1	–	–
		(ii) Organic coating	–	343.0	377.0	–	–
3	Silva & Hazlett	(i) Rutile, E-6013	–	470.4	558.6	14.3	23°C : 13.6 0°C : 9.6 –15°C:8.0
		(ii) Heavily coated rutile, E-7023	–	470.4	588.0	16.0	23°C:19.2, 0°C:12.8 –18°C:9.3
		(iii) Iron oxide, E-6027	–	509.6	646.8	13.1	23°C:24.45 0°C:10.64 –18°C:8.32
4	Grubbs	Multipass stick rutile E-6013	–	509.6	588–656.6	6.10	70°F, 32.8 30°F, 29.92 0°F, 23.2 –30°F, 13.6
5	Madatov	Iron power	–	372.4–490	–	–	–
6	Meloney	Rutile, E-6013	–	–	539–705.6	14–19.3	68°F, 37.28 32°F, 20.48 –60°F, 9.92

Soft Arc Behavior

Rutile and iron coating give soft arc. Iron oxide coated electrodes give better strength and ductility than plain rutile ones in flat and horizontal position. For multipass, all position welding, these rods fail because the solidified flux on bead surface is difficult to remove for subsequent pass to be made. Rutile electrodes are therefore preferred.

High Arc Stability

Because of the extinguishing effect of cold water surrounding the arc, the problems of arc stability in water deserve special attention. Compounds having low ionization potential (salts of potassium and cesium) or compounds that promote electron emission, tend to stabilize the arc in shielded metal arc welding. By manipulating electrode coating composition, an arc with better stability can be obtained. With a very stable arc, weaving of the weld beads may also be possible. This will permit larger heat inputs to the weld per unit length, larger bead size and less hardening.

Non-conducting and Non-hygroscopic Coating

Ordinary coatings, which are invariably porous, absorb water when immersed in water. The moist coating gives porous welds and permits current leakage. To protect the electrode from these two effects, waterproofing non-conducting paints are used. Shellac, paraffin wax, ucilon or celluloid dissolved in acetone are used.

26.9 DEVELOPMENTS IN UNDERWATER WELDING

Underwater Friction Welding Developments: The application of friction welding as a repair procedure to be conducted underwater, was attempted in England, by The Welding Institute.

One of the most important advantages of friction welding is the absence of a liquid phase with the complete elimination of the influence of gases (hydrogen and nitrogen) and the fact that the metal composition is not altered, with consequent uniformity of results independent of depth.

Therefore, a procedure demonstrated at a certain depth need not be requalified at another one, which is a requirement for all other welding underwater processes.

26.10 PRECAUTIONS TO BE TAKEN FOR UNDERWATER WELDING

- Adequate electric insulation for the welder.
- Shutting off the electricity supply immediately after the arc is extinguished.
- Limiting the open circuit voltage of welding sets.
- Hydrogen and oxygen pocket build-up must be avoided as they are potentially explosive.
- Standby divers, emergency air supply and decompression chambers, to avoid nitrogen narcosis following rapid surfacing after saturation diving.
- Inspection postwelding is difficult. Hence, there is a risk that welding defects may remain undetected.

26.11 APPLICATIONS

- Used to repair ships, offshore platforms, and pipelines.
- Steel is the most common metal welded by this technique.
- Research into dry welding at depths up to 1000 m are ongoing.
- For other underwater applications where high strength is necessary.

26.12 DEVELOPMENT IN UNDERWATER WELDING (FIGURE 26.10)

Wet welding has been used as an underwater welding technique for a long time and is still in use. With recent acceleration in building offshore structures, underwater welding has assumed increasing importance. This has led to the development of alternative welding methods like friction welding, explosive welding and stud welding.

Figure 26.10 Development in underwater welding

Wet MNAW is still used for underwater repairs, but the qualities of wet welds are poor and are prone to hydrogen cracking. Dry hyperbaric welds are better than wet welds. Present trends are toward automation, e.g., THOR-1 (TIG Hyperbaric Orbital Robot is developed, where divers perform pipefitting, install the track and orbital head on the pipe and rest of the process is automated).

Development of diver-less welding is an even greater challenge, calling for major developments like pipe preparation and aligning, automatic electrode and wire reel changing functions. This is in testing stage in deep waters. Explosive and friction welding are also to be tested in deep waters.

Review Questions

1. Explain the principle of operation of UWW.
2. Explain the equipment used in UWW.
3. What is dry welding?
4. Discuss the advantages of dry welding.
5. Discuss the types of UWW.
6. Compare conventional welding with UWW.
7. Discuss the recent developments in UWW.
8. What are the important applications of UWW?
9. What are the advantages of UWW?
10. What are the disadvantages of UWW?

SECTION F

Allied Welding Processes

27

CHAPTER

Explosive Welding

27.0 INTRODUCTION

Explosion welding (EXW) is defined as a solid-state process that produces a weld by high-velocity impact of the workpieces as a result of the controlled explosive detonation. Although, many in the explosion welding industry use the word 'bonding' to describe a welded joint, 'welding' is the preferred term for joining metals using explosives because an atomic bond occurs and is predominant at the interface. Explosive welding is well known for its capability to directly join a wide variety of both similar and dissimilar combinations of metals that cannot be joined by any other technique and which are not metallurgically compatible with each other, like stainless steel, titanium, titanium copper, aluminum. Metallic glasses are cladded successfully using this process.

27.1 FUNDAMENTALS OF EXPLOSION WELDING (FIGURE 27.1)

Explosive welding uses a controlled explosive detonation to force two metals together at high pressure. The resultant composite system is joined with a durable, metallurgical bond. Explosive welding under high velocity impact, was first recognized by Garl in 1944. Explosive welding was first recognized as a possibility in 1957 in the United States, when it was observed by Philip Chuck that metal sheets being explosively formed, occasionally stuck to the metal dies. The process has been developed fully with large applications in the manufacturing industry. It has been found to be possible to weld together combinations of metals, which is impossible by other means.

Explosives detonated on the surface of a metal, generate a high pressure pulse which propels the metal at a very high speed. When this metal collides with another metal surface at an angle, welding occurs. As the surfaces of the metals collide, a jet is produced at the collision interface which is necessary for welding as this jet cleans the metal and allows for pure metal surfaces to join under extremely high pressure.

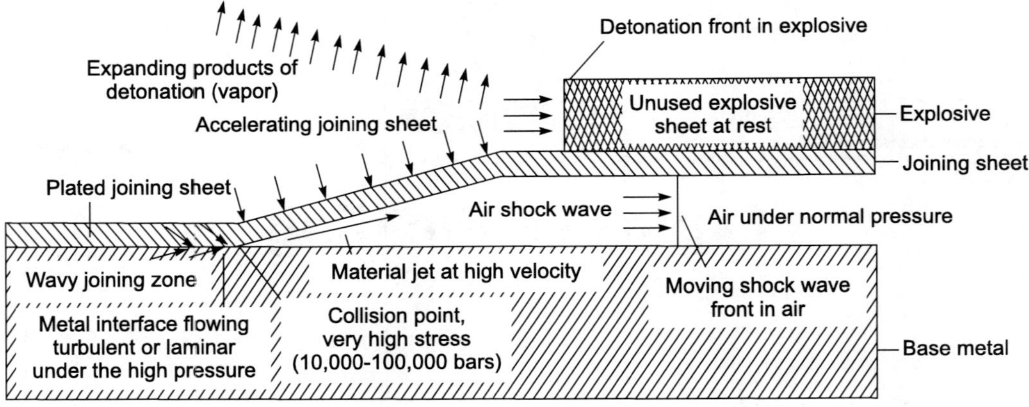

Figure 27.1 Explosive welding

27.2 PRINCIPLE OF OPERATION

The nature of the interface between the impacting components depends upon the velocity with which they strike against each other. A flat interface is formed if collision velocity is below the critical value for a particular combination of materials being welded. Such welds are not considered good because small variations, in collosion conditions can result in lack of bonding and thus, an unacceptable weld.

Welds made with collision velocities above the critical values have wavy interface, as shown in Figure 27.2, with the amplitude of waves varying between 0.1 and 4 mm and wavelength from 0.25 to 5 mm, depending on welding conditions.

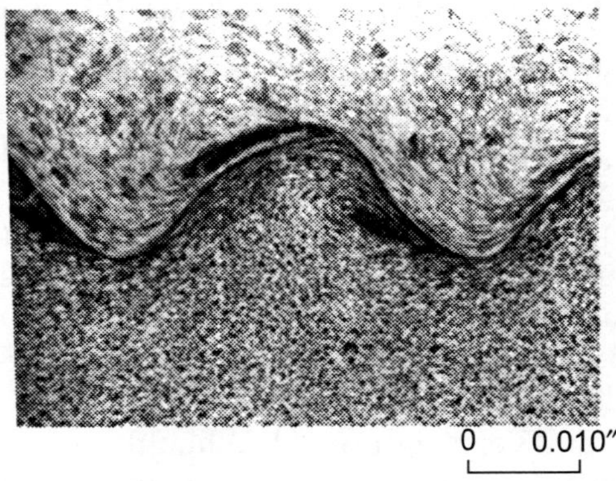

Figure 27.2 Explosive weld with wavy interface

Welds with such an interface have better mechanical properties than those with flat interfaces. In such welds, a phenomenon known as surface jetting is also observed so that a small jet of metal is formed from the metals of two impacting components, as shown in Figure 27.3. Such a jet is freely expelled at the edge of the joint, however, if it is trapped, it results in rippling effect.

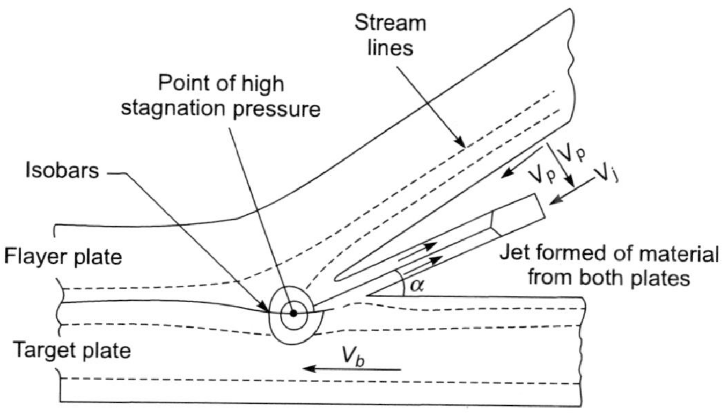

Figure 27.3 Jet formation in explosion welding

In explosive welding set up shown in Figure 27.4, the impact velocity becomes plate velocity V_p and it must be high enough for the impact pressure to exceed the yield stress of the material by a considerable margin. The collision point velocity, V_{cp}, that is the velocity at which collision point moves along the surface being joined, must also be less than the velocity of sound in the two materials. The relationship between the different velocities is shown in the vector diagram of Figure 27.5, where, V is the impact velocity, V_j is jet velocity, V_b is base plate velocity and α is the angle of incidence which becomes actual standoff angle γ as shown in Figure 27.6.

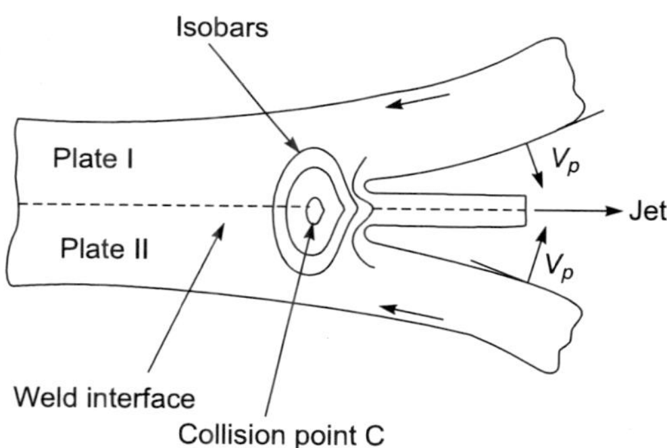

Figure 27.4 Jet formation in explosion welding under symmetrical angular standoff conditions

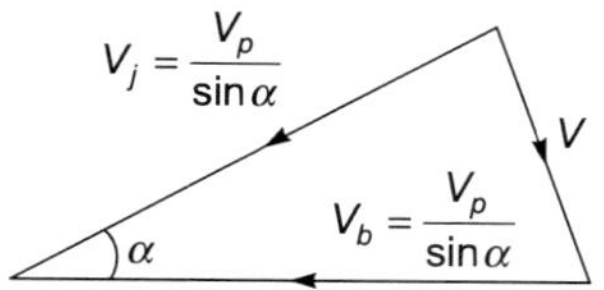

For velocities and angle refer to Figure 27.5

V – Impact velocity

V_j – Jet velocity

V_b – Target plate velocity

α – Angle of incidence

Figure 27.5 Velocity diagram for explosion welding

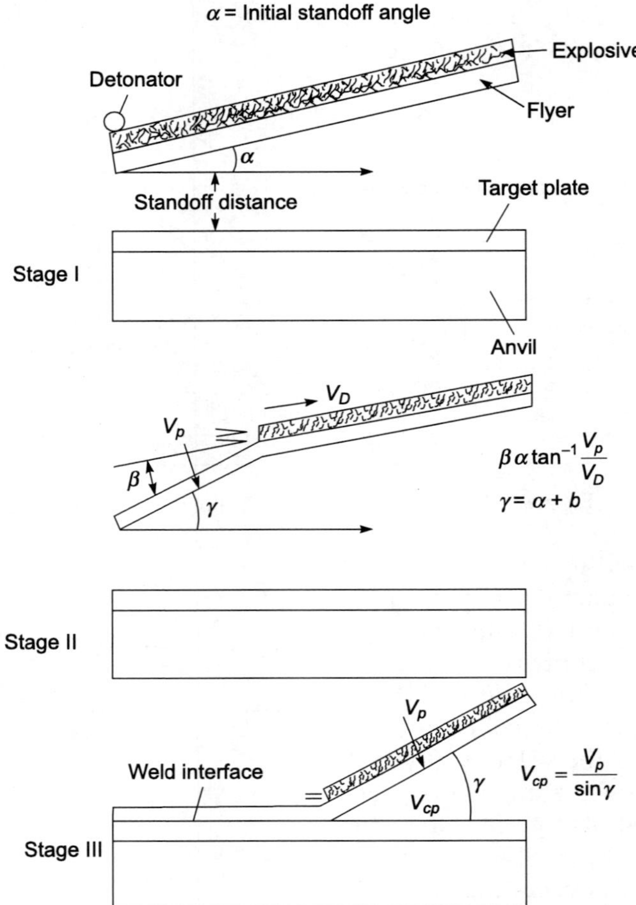

Figure 27.6 Different stages in explosion welding

The explosive welds are made either of the two setups shown in Figure 27.7. The welds are best made with parallel configurations of components in which only one plate is accelerated. In such a setup, the detonation velocity of the explosive must be less than the velocity of sound in the material to be joined, in order to satisfy the condition that the collision point velocity, V_{cp}, must be subsonic. It is, however, difficult to fulfill this condition with most of the explosives as evident from Table 27.1. The detonation velocity of the explosive must be less than about 120% of the sonic velocity, V_s, of the material being welded.

TABLE 27.1 Properties of some well known explosives

Explosive	Detonation Velocity
RDX (Cyclotrimethylene trinitramine, $C_3H_6N_6O_6$)	8100-8200 m/sec
PETN (Pentaerythritol tetranitrate, $C_5H_8N_{12}O_4$)	8190-8300 m/sec
TNT (Trinitrotoluene, $C_7H_5N_3O_6$)	6600-6700 m/sec
Tetryl (Trinitrophenylmethylinitramine, $C_7H_5O_8N_5$)	7800-7900 m/sec
Lead azide (N_6Pb)	5010 m/sec
Ammonium nitrate (NH_4NO_3)	2655 m/sec
Du point	7100 m/sec
Metabel	7000 m/sec

$$V_s = \sqrt{\frac{k}{\rho}} \quad \text{and} \quad k = \frac{E}{1-2\mu}$$

where,

k = Adiabatic bulk modules, dynes/cm^2
ρ = Density of material, grams/cm^2
E = Youngs modulus
μ = Poisson´s ratio

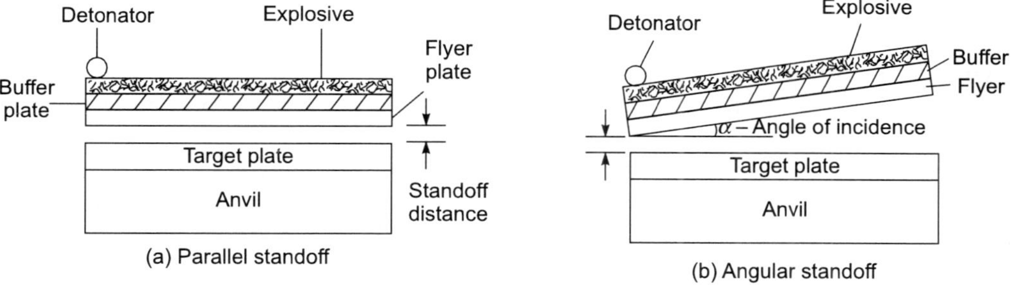

Figure 27.7 Components arrangement for parallel standoff method and angular standoff method

If the sonic velocity of the explosive is greater than 120%, the sonic velocity of the material with higher sonic velocity, a shock wave develops. This results in an extremely steep rise to the maximum pressure (maximum pressure is equal to the denotation pressure of the explosives). In such a case, the material just in front of the shock wave experiences no pressure, while the material just behind the shock wave is compressed to peak pressure and density. The shock wave travels through the material at a supersonic velocity and creates significant plastic deformation locally and results in considerable hardening, known as shock hardening.

The second type of detonation is when the detonation velocity is between 100% and 120% of the sonic velocity of the material being welded. This results in a detached shock wave that travels slightly ahead of the detonation.

When the detonation velocity is less than the sonic velocity of the metal, the pressure generated by the expanding gases, and which is imparted to the metal, moves faster than the detonation. Though no shock wave is produced but the rising pressure reaches its peak value.

In detached shock wave and no shock wave cases, pressure generated is ahead of the collision point of metal plates,. If a sufficiently large pressure is generated, it will cause the metal just ahead of the collision point to flow as a jet into the space between the plates. This high velocity jet effaces the material that removes the unwanted oxides and other unwanted surface films. At the collision point, the newly cleaned metal surfaces impact at high pressure, typically between 0.5 to 6 GPa. Also, a significant amount of heat is generated upon detonation of the explosive, however as detonation is completed within a few hundred microseconds, so a very little part it flows into the metal. Thus, no bulk diffusion takes place and a weld with only localized melting is produced.

It is, therefore, better to use the angular setup in which the velocity of the collision point is a function of the plate velocity and the initial standoff angle. The relationship is given by

$$V_{cp} = \frac{V_p}{\sin\left[\alpha + \tan^{-1}\left(\frac{V_p}{V_D}\right)\right]}$$

The plate velocity (V_p) is related to the mass of the plate and the explosive as well as the impulse (per unit mass) of the explosive. Knowing these parameters, V_p can be calculated.

In the angular setup, the wavelength of the ripples is directly related to the collision point velocity, while the shape of the ripples depends upon the plated velocity. Crested waves are most often produced with high plate velocity. The impact conditions for parallel plate setup are related by the following relation:

$$V_p = 2V_{cp}\sin\left(\frac{\gamma}{2}\right)$$

where, V_{cp} is the impact or collision point velocity, which is equal to the detonation velocity, V_D, of the explosives and γ is referred to as the dynamic bend angle. It is the angle created between the flyer and the target plates at the impact point, while V_p is the plate collision velocity at the point of impact.

Typically, the detonation velocity ranges from 1200 to 3800 m/s, depending upon the metal to be welded. The standoff distance, which is an independent variable like V_D, is selected to achieve a specific dynamic bend angle and velocity of the impact.

The dynamic bend angle is a dependent variable that is controlled by the detonation velocity, V_D and standoff distance. Typical values for γ are between 2 and 24°. This results in a plate collision velocity at the impact point of about 200 to 500 m/sec.

An important aspect of explosive welding is the flow pattern in the region of the collision point. Under subsonic condition metal behaves as non-viscous compressible fluid. Due to jet formation, oxide films and absorbed gases are completely removed from the weld. However, when the jet becomes unstable, the gases and oxide films may get entrapped. Entrapment can either result in continuous molten layer of 1/2 to 250 micrometer thickness or in the formation of a ripped interface, which often has localized fused zones on the forward side of the crest.

For explosive welding, it is required to impart subsonic velocity to the flyer plate. This has to be done with an explosive which has constant detonation velocity of 6000 m/s. The weight of the explosives required for a specific welding job is determined by trial and error method and there appears to be linear relationship between the ratio of $\dfrac{\text{Weight of explosive}}{\text{Weight of flyer plate}}$ and the flyer velocity V_p. A ratio of 0.5 gives a plate velocity of 900 m/sec for Du point sheet explosives EL506 using a thin layer of rubber as buffer. For successful explosives welding, it is required that the velocities of two plates must be similar and this necessitates that the angle of inclination between them should be small, as shown in Figure 27.8. With low angle, the impact velocity required to produce waves at the interface becomes greater.

When the explosive welding is carried out in normal atmospheric pressure, gas between the plates provides the cushioning effect, which not only necessitates higher minimum velocity but may also lead to inconsistent results. For welding aluminum in vacuum of about 1 mm of Hg, the collision velocity should be about 150 to 300 m/sec with an included angle of 1 to 2°. In order to accelerate the plates being welded to this velocity, the standoff distance should be equal to 1/4 to 1/2 times the plate thickness, as marked in Figure 27.8.

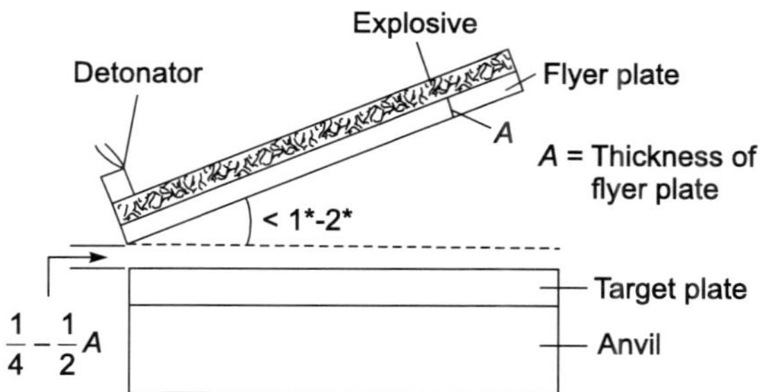

Figure 27.8 Angular standoff method with small angle of inclination to keep the velocities of two plates nearly equal during explosion welding

The standoff distance is held by use of a shim. There are many types of shims which are designed to be consumed as the jet so as not to adversely affect the weld.

If the effective angle attained by the flyer plate is too small, the velocity will be highly supersonic and no waves will be formed at the interface. Ideally, the detonation velocity of the explosion should be subsonic. However, it is rarely possible in practice as detonation velocities exceed 5500 m/s, while the velocity of sound in steel, which is among the highest among metal, is only 5200 m/sec as shown in Table 27.2.

TABLE 27.2 Velocity of sound in different metals

Material	Velocity of Sound (m/s)
Lead	2100
Copper	4200
Steels	4800-5200
Aluminum	5500

No special surface cleaning treatment is required for explosive welding, however greases, if present in the surface, must be removed. Dirt or oxide, if present, will get accumulated near the crest of the ripples and may lead to reduce the strength of the joint.

The pressure corresponding to a plate velocity of 120 m/sec on copper is 2400 MPa and for a velocity of 220 m/s on aluminum is 6200 MPa. These pressures are adequate to force metal through cracks in the oxide film and to weld it. It is also reported that even when the surfaces of 18/8 stainless steel and mild steel were covered with adherent layer of black oxide, they will be welded satisfactorily with the desired rippled interface.

27.3 EXPLOSIVE SELECTION

For a particular welding application, the appropriate explosive and amount per unit area should be chosen based on parameters like:

(a) Required detonation velocity
(b) Uniformity of detonation (so that constant collision velocity is achieved from start to finish of the weld)

An important process parameter in explosive welding is the amount of explosive used for a particular explosive, collision velocity of flyer plate is a function of the amount of explosive charge. As the explosive detonates, part of the chemical energy dissipates in air and the rest is transferred to the flyer plate as kinetic energy.

Excessive loading should be avoided to prevent interfacial melting (excessive kinetic energy of flyer plate may heat up the interface to very high temperatures), as these solidified melt pockets reduce the bond strength.

27.4 OPERATING PARAMETERS

Explosive welding requires two metal surfaces to collide at high relative velocity and pressure at proper orientation for maximum plastic interaction. The major techniques employed for this can be classified into:

(a) Contact Technique

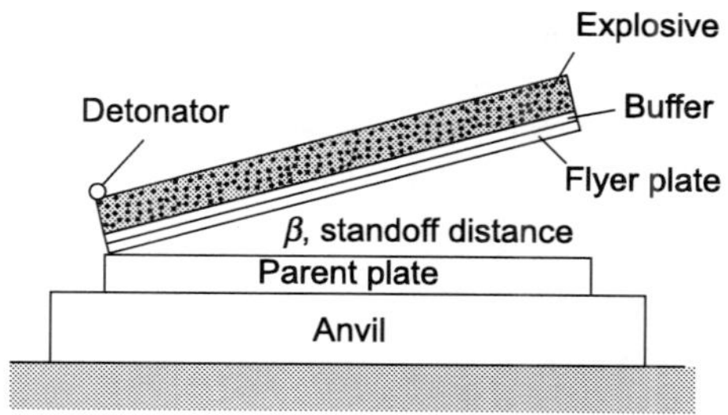

Figure 27.9 Contact technique

In this configuration, the base component and flyer plate are kept at an angle to each other with one end touching the base component as shown in Figure 27.9.

(b) Impact Technique (Figure 27.10)

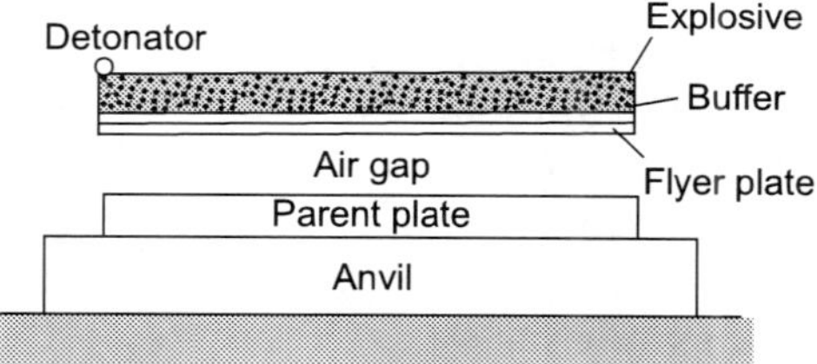

Figure 27.10 Impact technique

In this configuration, the base component and flyer plate are kept parallel to each other as shown in Figure 27.10.

The major process variables in explosive welding are:

1. **Impact Velocity:** It depends on the ratio of weight of explosive to the weight of flyer plate and on the contact angle. There is a minimum velocity for each material, below which welding does not take place. This minimum velocity is determined by the magnitude at which projectile material becomes sufficiently plastic on impact to form divided jet.
2. **Standoff Distance:** Increasing standoff distance increases the angle of approach, resulting in increased size of the wave which reaches a maximum and then decreases with further increase in standoff distance. In parallel setup, a standoff

distance between 0.5 and 2 times the flyer plate is normally used; the lesser standoff distance used with explosives having high detonation velocity.

3. **Angle of Approach:** For successful explosive, this is required to be between 5° and 25°. With parallel setup, this angle can only develop if there is a proper standoff distance.

27.5 SELECTION OF BONDING PARAMETERS

The preceding discussion introduced the primary variables used to predict explosive bonding parameters. The following four conditions are used to define a window of acceptable bonding parameters:

- Critical angle for jetting
- Critical flow transition velocity
- Minimum flyer plate impact for jetting, based on analysis of the stress required to cause flow
- Maximum flyer plate impact for the limitation of weld defects, based on jet entrapment

The critical angle for jetting is used to define the upper boundary for collision point velocities. The lower limit in collision velocity is determined by the transition from laminar to turbulent flow, where turbulent flow is preferred. The other two limiting conditions are specified by using the physical and mechanical properties of the material to determine the maximum amount of energy required to either produce hydrodynamic flow (that is, overcome a critical shear stress) or avoid extensive melting. Once the correct set of explosive parameters has been determined by utilizing either predictive models or an empirical approach, the production of explosively welded metal is relatively a simple task.

27.6 APPLICATIONS

Applications include welding and the bimetallic cladding of metals. The most important use of explosion welding is the manufacturing of clad plate. The cladding of dissimilar metals by explosion welding takes advantage of the best properties of both metals; one metal may provide protection from a harsh environment while the other may provide economic benefits. Explosion welds are often used for metal transition joints, either for electrical or mechanical purposes. Explosion welding can also be used as a joining process to increase mechanical strength between dissimilar metals that are not weldable by fusion processes. Metals that typically are combined include various grades of steel, stainless steel, nickel alloys, aluminum, copper alloys, titanium, zirconium and tantalum.

- Joining of tube-to-tube sheets for the manufacture of heat exchangers.
- As a repair tool for repairing leaking tube-to-tube sheets.
- Tube plugging.
- Remote joining in hazardous environments.
- Joining of dissimilar metals – aluminum to steel, titanium alloys to Cr-Ni steel, Cu to stainless steel, tungsten to steel, etc.
- Attaching cooling fins.
- Applications in pressure vessels, ship building industry, cryogenic industry, etc.
- Cladding of base metals with thinner alloys (most widely used).
- Joining of pipes in a socket joint (new).

27.7 ADVANTAGES

- Can bond many dissimilar, normally unweldable metals.
- Minimum fixturings/jigs.
- Simplicity of the process.
- Extremely large surfaces can be bonded.
- Wide range of thickness can be explosively cladded together.
- No effect on parent properties.
- Small quantity of explosive is used.

27.8 LIMITATIONS

- Metals must have high enough impact resistance and ductility. Not suitable for brittle materials.
- Noise and blast can require operator protection, vacuum chambers, buried in sand/water.
- Use of explosives in industrial area is restricted due to noise and ground vibrations caused by explosions.
- Geometries welded must be simple – flat, cylindrical, conical.
- Storage of explosives is a problem.
- If the thickness of material is more than 60 mm, then large amount of explosives is required.
- Berrillium, glass, ceramics, tungsten, boron, cannot be joined by explosive welding.

27.9 CLADDING

Cladding improves the corrosion resistance and heat transfer characteristics of base metal. Clad metal plates consist of thin layer of an expensive, corrosion-resistant metal, like titanium or stainless steel, which is metallurgically combined with a less expensive, less corrosion-resistant, thicker base metal, typically carbon steel. They are used in the construction of heavy, corrosion-resistant pressure vessels and heat exchangers for upstream oil and gas, oil refinery and similar industries (Figure 27.11).

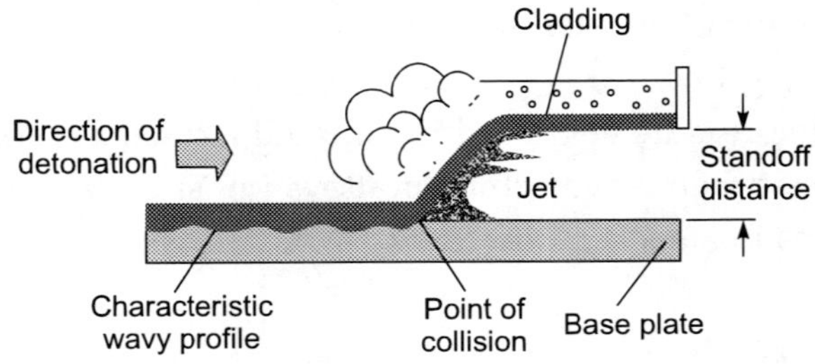

Figure 27.11 Cladding

Various technologies were developed to produce materials with properties similar to a solid metal, but at lower cost. The most popular of these are explosive welding, rollbonding and weld overlaying. There are limitations on the product that each can produce. For an end user, the choice depends on the type of application and thickness required.

27.10 EXPLOSIVE WELD CLADDING

- It retains the properties of original metals, like corrosion resistance and mechanical properties, unlike rollbonding and weld overlaying.
- Explosive welding is the only process that can bond titanium, zirconium, and tantalum metals to a base metal.
- Alloy chemistry is constant over the entire thickness of the product.
- The two metals do not come apart even under the most extreme circumstances when fabricated properly.
- It is suitable to produce products with cladding thickness of 0.04 to 2 inches and base thickness of 0.25 to 40 inches.

27.11 EXPLOSIVE WELDED CLADDING PROCESS (FIGURE 27.12)

- The materials are first inspected, mating surfaces are ground, and the metal plates are assembled for cladding.

- A sheet of cladder is placed over a parallel plate of backer material and the cladder material is covered with a layer of specifically formulated explosive.
- A stand-off distance is maintained.
- The explosion is initiated on one side of the cladder material which travels across it, forcing it down on the backer. The explosion occurs in one thousandth of a second. The collision causes a thin layer of mating surfaces to be spelled away in a jet, removing oxides and surface contaminants immediately ahead of the collision.
- The extreme pressures force the two metal components together, creating a metallurgical bond between them. This produces a strong, ductile, continuous metallurgical weld over the clad surface.
- Resulting clad plates are flattened and cut and inspected to assure conformance with internationally accepted product specifications.

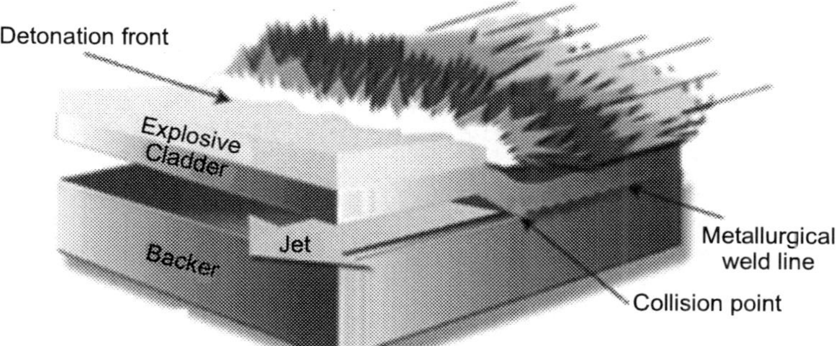

Figure 27.12 Cladding explosion cladding process

The cladding materials are typically titanium, stainless steel, aluminum, copper alloys, nickel alloys, tantalum, and zirconium. The base metals are typically carbon steel, alloy steel, stainless steel and aluminum.

Worked Examples

Example 1: Using a weight ratio of $\dfrac{\text{Weight of explosive}}{\text{Weight of flyer plate}} = 0.3$, the flyer plate velocity comes to 540 m/sec. Find the flyer plate to target plate included angle (α) so that the collision point velocity (V_{cp}) is kept subsonic (<5000 m/sec) for welding steel plates having a detonation velocity of 7100 m/sec.

Solution: $V_{cp} \leq 5000\dfrac{\text{m}}{\text{sec}}$

$$V_p = 540 \frac{\text{m}}{\text{sec}}$$

$$V_D = 7100 \frac{\text{m}}{\text{sec}}$$

α to be found out

Now using the equation,

$$V_{cp} = \frac{V_p}{\sin\left[\alpha + \tan^{-1}\left(\frac{V_p}{V_D}\right)\right]}$$

$$5000 = \frac{540}{\sin\left[\alpha + \tan^{-1}\left(\frac{5400}{7100}\right)\right]}$$

$$= \frac{540}{\sin\left[\alpha + \tan^{-1}(0.0761)\right]}$$

$$= \frac{540}{\sin\left[\alpha + 4.35\right]}$$

$$\sin\left[\alpha + 4.35\right] = \frac{540}{5000}$$

$$[\alpha + 4.35] = 6.20°$$

$$\alpha = 6.20° - 4.35° = 1.85°$$

Example 2: Select a suitable explosive from the three given in Table below for explosion welding of aluminum plates with 2° included angle, if the flyer plate velocity is to be 900 m/sec. Velocity of sound in aluminum is 5500 m/sec.

S. No.	Explosive	Detonation Velocity
1.	TNT	6700 m/sec
2.	RDX	8200 m/sec
3.	Du point	7100 m/sec

$$V_{cp} \leq 5500 \frac{\text{m}}{\text{sec}}$$

$$V_p = 900 \frac{\text{m}}{\text{sec}}$$

$$\alpha = 2°$$

For TNT

$$V_{cp} = \frac{V_p}{\sin\left[\alpha + \tan^{-1}\left(\frac{V_p}{V_D}\right)\right]}$$

$$= \frac{900}{\sin\left[\alpha + \tan^{-1}\left(\frac{900}{6700}\right)\right]} = 5368 \text{ m/sec}$$

For RDX

$$V_{cp} = \frac{900}{\sin\left[\alpha + \tan^{-1}\left(\frac{900}{8200}\right)\right]} = 6265 \text{ m/sec}$$

For Du pont

$$V_{cp} = \frac{900}{\sin\left[\alpha + \tan^{-1}\left(\frac{900}{7100}\right)\right]} = 5615 \text{ m/sec}$$

As V_{cp} has to be less than 5500 m/sec, therefore only TNT is a suitable explosive,

$$5000 = \frac{540}{\sin\left[\alpha + \tan^{-1}(0.0761)\right]}$$

$$= \frac{540}{\sin[\alpha + 4.35]}$$

$$\sin[\alpha + 4.35] = \frac{540}{5000}$$

$$[\alpha + 4.35] = 6.20°$$

$$\alpha = 6.20° - 4.35° = 1.85°$$

Review Questions

1. Explain the principle of operation of explosive welding.
2. Explain the equipment used in explosive welding.
3. What are the different stages in explosion welding?
4. Discuss the jet formation in explosive welding.
5. Discuss some properties of explosives.
6. What is explosive cladding?
7. Discuss the angular and parallel standoff explosive welding.
8. What are the important applications of explosive welding?
9. What are the advantages of explosive welding?
10. What are the disadvantages of explosive welding?

28 CHAPTER

Metal Flame Spraying/Metallizing

28.0 INTRODUCTION

Dr. Ulrich Schoop developed the basic principles of metal spraying. In 1910, he designed a nozzle that allowed an annular flame to heat powder particles as they propelled onto a surface. His further work led to the utilizaton of wire instead of powder and the equipment came to be known as wire metallizing gun. Metallizing or metal spraying terms denoted the application of metallic coating to a base metal. Later on, when it became possible to deposit materials by this process, a more general term known as flame spraying, was coined.

28.1 DEFINITIONS

- **Metallizing or Metal:** Spraying is the process of applying molten metal to a base metal surface in the form of a fine spray that bonds to the base metal and forms a metal coating.
- **Flame Spraying:** It defines the deposition of materials, both metallic and non-metallic, by the use of a combustion flame.
- **Thermal Spraying:** It is a process that deposits finely divided particles of metals, intermetallics, metallic oxides or plastics in the molten or semi-molten condition to form an adherent coating on suitable substrate. It covers all deposititition processes by which surfacing materials are made molten or semi-molten by a heat source and propelled onto the base metal surface by rapidly moving gas stream (metallizing, flame spraying, electric arc spraying, plasma spraying and detonation spraying).

28.2 SURFACE PREPARATION

Since the bond between base metal and deposited metal is purely mechanical, proper surface preparation of base metal is required for proper mechanical interlocking of sprayed and base metals. This is done in the following steps:

(a) Base metal is cleaned of all foreign substances like oxides, oil, water, etc. Oxide and other coatings may be removed mechanically by machining, grinding, sanding, etc. Oil and grease may be removed chemically.

(b) Base metal is roughened by forming keys or openings for sprayed metal to penetrate them, resulting in a bonding action. The following methods are employed for this purpose:

1. *Rough threading (for shafts)*
2. *Grit blasting (for large surfaces):* Medium grade clean sharp sand, steel grit and aluminum oxides are used as abrasive materials.
3. *Elecric bonding:* It is used when surface is too hard to be roughened by grit blasting or machining. It is used to deposit a bonding coat of nickel alloy on the surface. In this process, a nickel or nickel alloy electrode is used to fuse small irregular particles to the surface which is to be metallized. It is not suitable for thin deposits; it may produce slight distortion. It also lowers fatigue resistance of materials and hence, not recommended for highly stressed or cyclically loaded components.
4. *Studding:* Implies placing studs (a machine screw or a rivet) on the base metal surface to increase bonding qualities of sprayed metal. It is not done on large surfaces as it consumes a lot of time.
5. *Sanding or grinding:* Rotary hand grinders are used to prepare hard alloy surfaces that cannot be effectively undercut or rough threaded.
6. *Rough machining or grooving:* Grooving involves cutting a series of parallel grooves, using a rounded grooving tool. The lands between the grooves may then be rolled over with a knurling tool.
7. *Spray bonding (molybdenum and exothermically reactive materials):* Since almost any oxide-free metal adheres to molybdenum, a layer of it is uniformly sprayed onto the base metal to produce a surface rough enough to receive a metallizing spray.

28.3 CONCEPT OF FLAME SPRAYING (METALLIZING) (FIGURE 28.1)

The material to be sprayed is taken in powder or wire form, is fed into the oxyacetylene flame, there it melts by the heat of the (neutral) flame, gets atomized and blown by a stream of compressed air onto the base metal surface, where it spreads around and interlocks with projections, imbed in pits, and freezes quickly upon contact with the base metal surface. Separate particles overlap and intertwine with one another to form a coherent structure. The sprayed material is generally applied in layers less than 0.25 mm thick.

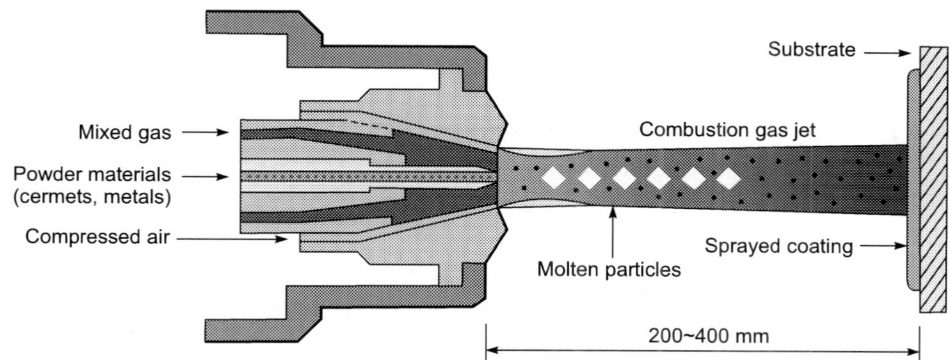

Figure 28.1 Flame spraying (metallizing) process

28.4 CLASSIFICATION OF THERMAL SPRAYING PROCESSES

A. Those using heat source as:
   (i) Combustion flame
   (ii) Electric arc
   (iii) Plasma arc
   (iv) Detonation gun

B. Those using surfacing material in the form of:
   (i) Powder
   (ii) Wire or rod

28.4.1 Wire Spray Process (Figure 28.2)

This is probably the most commonly used metallizing process employed. It is a fast and low-cost method of applying a wide variety of metal coatings, such as carbon steel, stainless steel, brass and bronze, over a grit blasted or nickel-nickel aluminide undercoated substrate. Such coatings are often used for shaft build-up and general restoration of worn or mismatched parts.

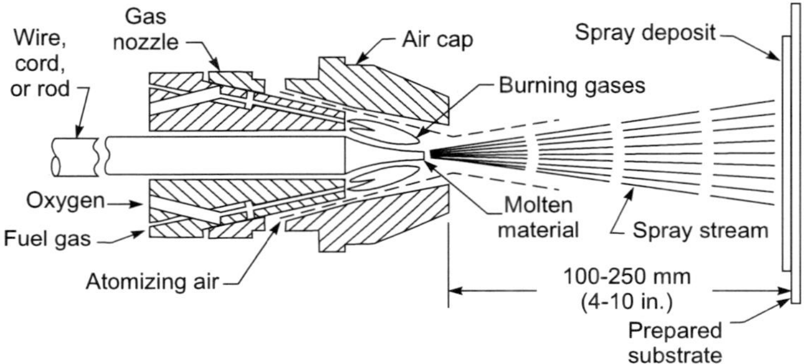

Figure 28.2 Wire spray process

A wire metallizing system consists of the following equipment and supplies:

1. A wire-type spray gun; resembles an oxyfuel gas torch with a hole for the wire through the centre of its tip. The gun also has an arrangement to drive a pair of knurled rolls to feed the wire into the flame. An air cap encloses the tip of the gun and directs a blast of compressed air to pick up the molten metal and project it, thoroughly atomized, onto the base metal.
2. A wire reel and straightener.
3. An oxygen cylinder and pressure regulator.
4. An acetylene (fuel gas) cylinder and pressure regulator.
5. Flowmeters for acetylene, oxygen and compressed air.
6. A dry compressed air supply, an air filter and an air pressure control.
   The compressed air:
   (i) forms an air envelope around the outside of the flame and the liquid metal.
   (ii) atomizes the molten wire metal that has been melted by oxyacetylene flame.
   (iii) cools the substrate, thus maintaining its temperature below 205°.
   (iv) accelerates the molten metal particles on its path to the base metal and thus, deposits them from there.
7. Acetylene, oxygen and air hoses.
8. Flame arresters; a wire metal spraying (or metallizing) system.

28.4.2 Powder Spray Process (Figure 28.3)

In powder spray process, metal powder (less than 200 mesh size) is carried by compressed air from powder container to the spray gun, where it feeds through a centre orifice in the nozzle and from there, into the oxyacetylene flame. As it enters the flame, the powder melts and the resulting molten globules are blown against the job surface in the same manner as in a wire metal spray gun. An advantage of the powdered process is its ability to apply a larger variety of metal, ceramic and cermet (ceramic/metallic mixture) materials.

As compared to wire metallizing equipment, powder spraying is simpler and lighter in weight. One problem experienced in powder spray process is the difficulty of maintaining a steady flow of powder. Moreover, deposition efficiency is low.

A powder spray system consists of the following equipment and supplies:

1. A powder spray gun
2. An oxygen cylinder, pressure regulator and flowmeter
3. An acetylene cylinder, pressure regulator and flowmeter
4. A dry compressed air supply, its pressure control and flowmeter
5. A powder container
6. Oxygen, acetylene and air hoses

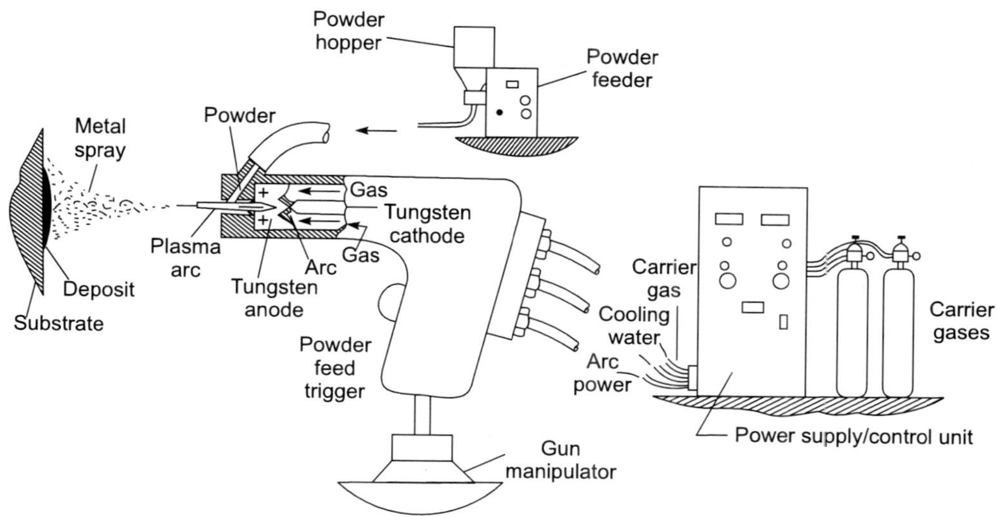

Figure 28.3 Powder spray process

28.4.3 Electric Arc Spraying (Figure 28.4)

In some cases, the adhesion of a flame sprayed coating is unsatisfactory because it is impossible to attain a sufficiently high temperature in the particles on impact. This problem has been overcome by electric arc spraying, where maximum temperatures of 4000°C have been achieved in the arc.

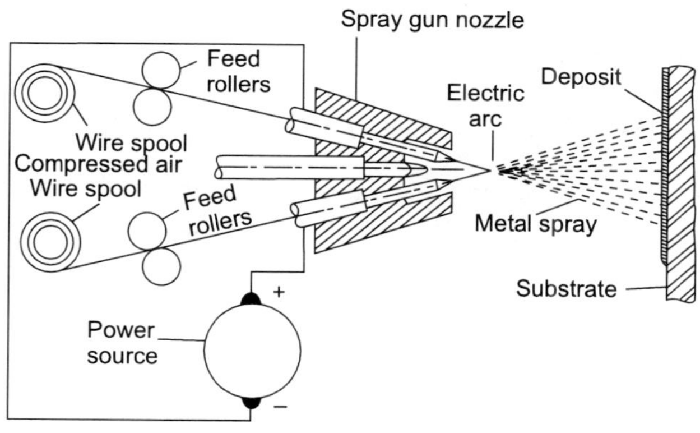

Figure 28.4 Electric arc spraying

Concept

An electric arc is maintained between two wires of the metal, which is required to be deposited on the workpiece surface. The metal melted by the arc heat is then propelled or sprayed onto the workpiece surface by using an air jet directed across the arc and at the workpiece surface to be sprayed.

Equipment

The electric arc spraying system consists of the following equipment:

1. Wire reels
2. Switchgear and control console to adjust current, voltage, air pressure and rate of wire feed as required
3. Air supply input to the system
4. Spray gun. It consists of the following:
   (i) The case which houses the wire guides and the wire feed parts and gearing, the main electrical conducting parts, the power controls and the atomizing air passage.
   (ii) The wire feed air motor.
   (iii) The electrical wire guides and arc shield.
   (iv) The atomizing air nozzle.
5. Wire feed air motor
6. Power supply (for the two metal wires)
7. A motor generator or a solid state power source (50 to 650 A) for the arc spray system. The motor generator set has either constant voltage or constant current regulation, depending upon application
8. Wire feed rolls energized by air motor
9. Atomizing air nozzle to produce atomizing air jet
10. Spray wire of the metal that is to be deposited on the workpiece
11. DC arc power input line

Process parameters summary is given in Table 28.1.

TABLE 28.1 Process parameters summary

Process	Coating Material Form	Heat Source	Flame Temp°C	Gas Velocity, m/s	Porosity, %	Coating Adhesion, MPa
Plasma spray	Powder	Plasma flame	12,000-16,000	500-600	2-5	40-70
Wire arc spray	Wire	Electric arc	5000-6000	<300	5-10	28-41
Wire flame	Wire	Oxyfuel combustion	3000	<300	5-10	14-21
HVOF	Powder	Oxy-gas fuel combustion	3200	1200	1-2	>70

Arc Spraying Operation (Figure 28.5)

The surface to be sprayed must be clean and rough, although considerably less roughness is required for electric arc spraying. Sand and aluminum oxide should not be employed for preparing the base metal surface.

The two metal wires are advanced by the feed mechanism under full electric current and they touch at a low contact pressure and a small point-like contact surface. Because of the high density of the current and the extreme heat of the contact surfaces, the arc strikes immediately.

The arc creates a molten condition at the ends of the electrode wires and the molten particles are swept off at high velocities (about 100 particles/second) by a gas (e.g., nitrogen) or compressed air jet. The vibrations of the wire ends as they make and break contact assist the spraying action. Thus, the molten particle stream is impinged upon the workpiece surfaces being sprayed, resulting in the deposition of a high quality coating.

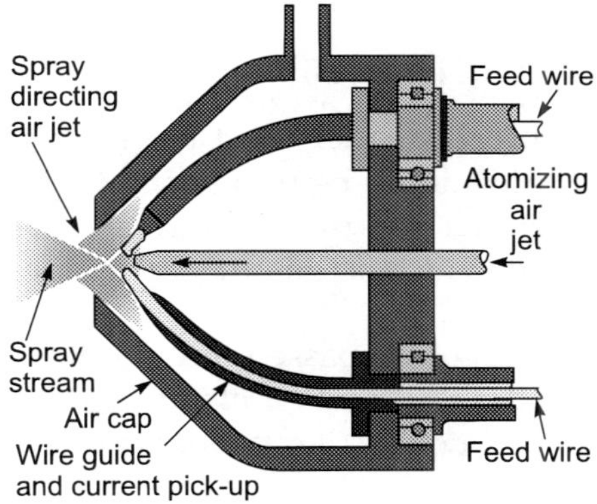

Figure 28.5 Electric arc spray process

Advantages

(i) Due to the high rate of deposition of metal in electric arc spraying, it is approximately 70% cheaper than flame spraying.

(ii) No combustion gases are required, thus eliminating all problems and dangers associated with these gases.

(iii) Arc spraying process can be interrupted and resumed at any time by simply disengaging and reengaging the wire feed mechanism.

(iv) Arc spraying provides high level of thermal efficiency.

(v) In this process, two wire ends are resistance heated between the short circuit and contact points with 40 to 50% of energy produced in the electric arc.

(vi) Arc spraying requires less surface preparation than flame spraying.

(vii) Arc sprayed particles and the coating layer is oxidized as much as 50 to 70% less than flame sprayed particles, resulting in higher resistance to tensile, compressive and bending stresses.

(viii) Arc can be struck between two wires of different materials, such as stainless steel and aluminum. In this situation, these two materials will intermingle and be deposited on the base metal, resulting in a long wearing surface with a desirable bearing surface.

Disadvantages

(i) Considerable quantities of carbon are burnt out of steel wires and a loss of Mn and Si can be expected because a protective gas is not employed.
(ii) The arc spraying velocity is lower than that of flame spraying.
(iii) Coatings do not normally resist high temperatures or great surface pressures because of their limited adhesion, neither can they resist stressing.
(iv) Adequate resistance to wear can be achieved, but the performance cannot be compared to a surface built up by welding.
(v) The process creates fumes, dust and odours around the work.

28.4.4 Plasma Arc Spraying (Figure 28.6)

All the dust, oil and other foreign matter must be removed from the surface to be sprayed. After that, the surface should be roughened by (steel) grit blasting or acid etching. In plasma arc spraying, a non-transferred, direct current, high intensity arc is struck between the gun body which acts as an (Cu) anode and a thoriated tungsten cathode (electrode).

An inert/gas, for example, argon, is passed through this arc, where it is heated to the plasma state (20,000 to 30,000°F) and accelerated to supersonic speeds. Argon, heated by arc, escapes out of the nozzle (i.e., copper anode) as a hot, brightly shining flame having a length of about 3 to 7 cm.

Powdered metal is metered into the plasma stream, which reaches molten stage much more quickly than in flame spraying. (The powdered material melts in the air and not in the gun because of the high temperatures of the plasma arc). The powdered metal is then carried to the workpiece surface in a molten state by the high velocity of the plasma jet.

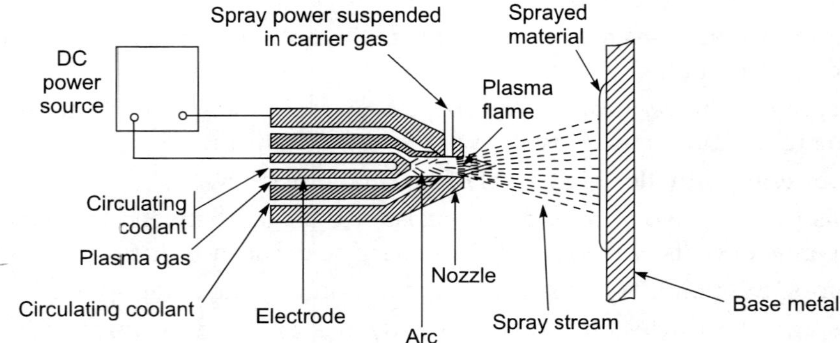

Figure 28.6 Plasma arc spray process

Equipment Required

Plasma Spray Gun

In a plasma spray gun, an arc is contained within a water cooled tube into which gas is injected. The arc path is between a tungsten cathode in a water cooled holder and a hollow water cooled copper anode.

The gas picks up energy from the arc and issues from the device in a configuration resembling an open welding flame. Spray guns have a handling capacity of about 40 kW. DC is used at 50 to 500 volts and 80 to 800 amperes. Several nozzle configurations can be used to accommodate different plasma gases and to spray different types of powders.

Power Supply

Selenium rectifier or silicon rectifier sets of 100% duty cycle–constant current type are used as (DC) power units.

Gas Supply

In a plasma spray system, gases are used for the following purposes:

- To form plasma (such as, nitrogen and argon).
- As auxiliary gas, mixed in small quantities (5 to 10%) with the gas forming plasma (such as hydrogen and helium).
- As a powder carrying gas. The powder carrier gas and the plasma generating gas have a common source.

Powder Metal Supply

The plasma torches are usually of a type that permit only powder materials to be used. The grain size of the powder is 40 to 50 mesh to 5 micron size. Powder feed mechanisms are of two types:

(i) An aspirator-type feeder that is composed of powder hopper, which contains a gas nozzle that aspirates powder adjacent to the jet into the discharge line.

(ii) A mechanical metering type feeder that utilizes the metering action of a screw to deliver precise, measurable amounts of powder to a mixing chamber, where the powder is introduced into the carrier gas stream.

Control Console. It controls

(i) Plasma gas flow rate

(ii) Arc current

(iii) Starting and stopping functions

(iv) Powder feed unit.

Sprayable Materials: Plasma spraying can be used to spray metals, ceramics (oxides and carbides), cermets, and composites, as listed in Table 28.2.

TABLE 28.2 Materials commonly employed for plasma spraying

Metals	Oxides	Carbides	Cermets	Composites
Aluminum	Alumina	Chromium carbide	Alumina	Metal/graphite
Chromium	Chromium oxide	Titanium carbide	Nickel	Powder
Copper	Magnesia	Tungsten carbide	Alumina	Metal/molybdenum
Molybdenum	Titania		Ni-aluminide	Disulphide
Nickel	Zirconia		Magnesia-nickel	Powder
Nickel-chromium	Magnesium zirconate		Zirconia-nickel	
Alloys			Zirconia-nickel	
Tantalum			Aluminde	
Tungsten			Tungsten carbide/cobalt	

Metals vary from soft metals, such as aluminum and zinc, for corrosion-resistance applications to cobalt-base hard facing material for wear resistance applications.

The most popular ceramic coatings are aluminum oxide and chromium oxide or mixtures of chromia and silica. These are used mainly for wear resistance applications. Ceramics such as yttria-stabilized zirconia, magnesium zirconate, and calcia-stabilized zirconia are used for thermal barrier coatings on engine components and the like. Aluminum oxide and magnesia/alumina are often used for electrical insulation applications.

The most popular cermet consumables for plasma spraying are tungsten carbide/cobalt for wear resistance applications. The composite consumables such as metal/graphite powders and metal/molybdenum disulphide powders are generally used for special applications.

Advantages

(i) Almost any base material or substrate can be coated with the plasma spray process.
(ii) High bond strength is obtained.
(iii) The coating is stronger and denser.
(iv) Heat distortion problems are minimum.

Disadvantages

(i) There is a high noise level, ranging around 100 dB; this may damage the inner ear.
(ii) There is intense ultraviolet and infrared radiations, which may cause sunburn, making it essential to provide protection to eyes.

Applications

(i) Plasma spraying is mainly used for producing wear-resistant, temperature-resistant, and heat insulating coatings.

(ii) Making rocket parts, such as nozzles.

28.4.5 Detonation Gun Spraying (Figure 28.7)

It is a metal spraying process which makes use of a specially designed gun, into the chamber of which are put metered amounts of oxygen and acetylene, along with powder particles of coating material suspended in nitrogen.

The high kinetic energy of each particle is converted to additional heat upon impact with the base metal surface, thereby producing a bond that is both metallurgical and mechanical in nature.

Unlike metallizing and plasma torches which operate continuously, the detonation gun fires 4 to 8 times a second, thereby forming a laminar coating on the workpiece. Successive detonations can build upcoatings to a maximum thickness of 0.75 mm. The gun-to-workpiece distance is 50 to 100 mm.

Auxiliary cooling is employed to keep workpiece temperature between 65 and 150°C. Because of the extreme noise caused by the firing of gun, the process has been completely automated and the gun is usually located within concrete sound absorbing walls and is aimed and fired by remote control.

As compared to plasma or metallized coatings, coatings made by detonation process are more costly but they have much greater wear resistance.

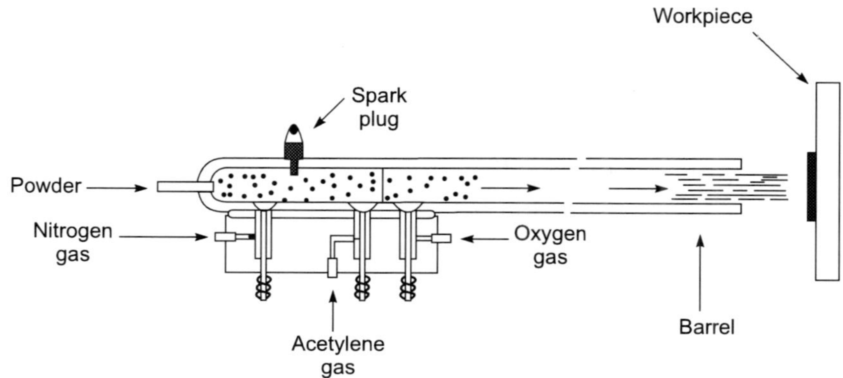

Figure 28.7 Detonation gun spray process

Applications

A variety of coatings can be done including tungsten carbide, aluminum oxide, and chromium carbide.

Only metallic surfaces can be sprayed as the non-metallic surfaces can get eroded due to the high velocity.

Advantages

- Major advantage is the porosity and oil retention capacity of the sprayed metal.
- Metal of required properties can be coated only where it is required.
- Parts that are sprayed remain cool during the process, thus avoiding warpage.
- Zinc and aluminum can be coated to any practical desired thickness for corrosion resistance purposes.

Disadvantages

- Cannot be used to add strength as all coated metal will be brittle. Thus, it is used only where strength addition is not required.

Applications

- To obtain corrosion resistance by spraying Zn and Al.
- Build up worn out surfaces of machine components.
- To mechanically correct a component that is not up to specifications.
- Increase hardness and wear resistance of a surface.
- Apply thin coating of noble metal to specific areas where conventional plating would not be viable.
- Obtain reflective surfaces (spraying of Al to make mirrors).
- Obtain conductive surface on a pure conductor.
- Provide decorative.
- Effects on architectural works.

Review Questions

1. Explain the principle of operation of thermal spraying.
2. Explain the equipment used in thermal spraying.
3. What are the differences between flame and thermal methods?
4. Discuss the surface preparations used in thermal spraying.
5. Describe classification of thermal spraying.
6. What are the materials commonly used in plasma spraying?
7. Discuss the detonation gun spraying method.
8. What are the important applications of thermal spraying?
9. What are the advantages of thermal spraying?
10. What are the disadvantages of thermal spraying?

SECTION G

Miscellaneous Topics in Welding

29

CHAPTER

Welded Design and Fabrications

29.0 INTRODUCTION

The process of joining two or more pieces of metals by fusion or by fusing the edges of plates with or without pressure and filler material is called welding. In welded fabrication, welds are used to join different parts. Such junctions of parts are called weld joints and are defined as the locations where two or more members are to be joined. The loads in welded structures are transferred from one member to another through welds placed in such joints, thus here we try to understand the basic weld design principles and also the weld symbols used on a drawing, specifying where the welds are located.

29.1 TYPES OF JOINTS

Welding is done to join two different parts. Such junctions of parts are called 'weld joints'. Various members such as plates, sheets, pipes, angular cross sections, etc. are welded. It is the placement of these members that help defining the types of joints. The various types of joints are given in Table 29.1.

TABLE 29.1 Types of joints

S. No.	Type of Joint	Description	Figure
1.	Butt joint	Joint between two members lying in the same plane.	
2.	Tee joint	Joint between two members located at approximately right angles to each other in the form of alphabet *T*.	

Contd.

3.	Lap joint	Joint between two overlapping members.	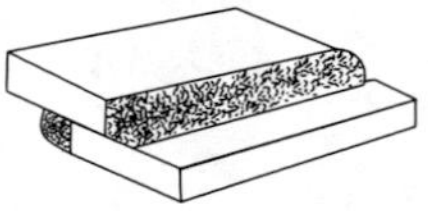
4.	Corner joint	Joint between two members located at the corners at right angles to each other.	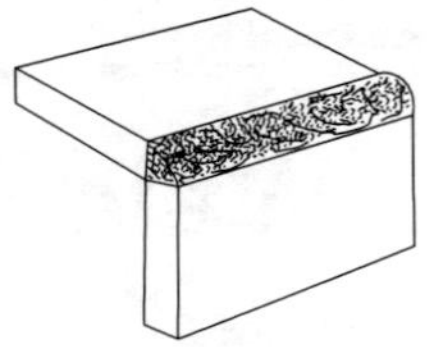
5.	Edge joint	Joint between the edges of two or more parallel members.	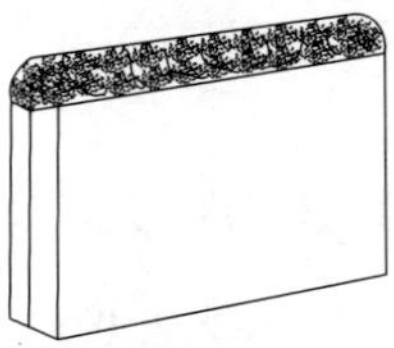

29.2 TYPES OF WELDS

When two members to be joined are placed in any joint configuration as mentioned in the table above, depending on location and geometry of the welding to be done and the shape and dimensions of the surfaces to be joined, weld can be classified into eight distinct types, as shown in Table 29.2.

TABLE 29.2 Types of welds

S. No.	Type of Weld	Description	Figure
1.	Fillet weld	A weld of approximately triangular cross section, joining two surfaces nearly at right angle to each other.	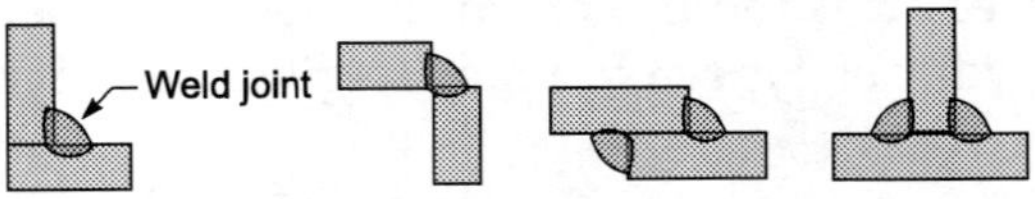 Weld joint Fillet weld on corner joint Fillet weld on lap joint Fillet weld on *T* joint
2.	Groove weld	The weld consisting of an opening between two part surfaces, which provides space to contain weld metal. Groove welds are used on all joints, except lap joints.	(a) Square groove weld (b) Single bevel groove weld (c) Single V groove weld 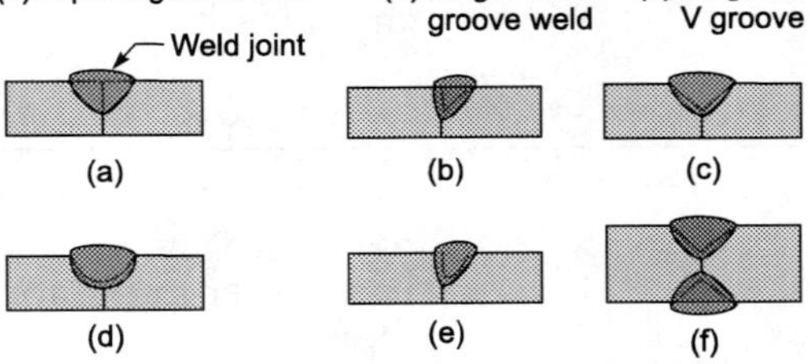 Weld joint (a) (b) (c) (d) (e) (f) (d) Single U groove weld (e) Single J groove weld (f) Double V groove weld for thicker sections

Contd.

3.	Back or backing weld	A back weld is where a second weld is created on the back side of the joint after the primary groove weld is completed. Conversely, a backing weld is a weld that the welder performs first.	Back weld; Ist; 2nd; Backing weld; 2nd; Ist
4.	Flange weld	Thin sheet metals are often flanged to produce edge flange or corner flange welds. These welds are typically made without adding filler metal because the flange melts and provides all the filler needed.	Flange weld; Two sheet metal parts
5.	Plug or slot weld	A weld between two members made by welding within a slot in one of the members. Slot welds are deposited in elongated holes for thicker plates.	Plug weld; Hole in top part; Slot in top part for slot weld
6.	Spot or projection weld	Spot welds are deposited during resistance welding process, where the weld at the interface of the members is joined.	Spot welds; Partial cut-away view; Two sheet metal parts; Cut-away view showing fused (welded) section
7.	Seam weld	Seam welding is a variation of resistance spot welding. The only difference is that it is formed by a series of overlapping spot welds.	Cut-away view to show fused (welded) seam; Seam weld; Overlapping section; Sheet metal part

Contd.

8.	Surface weld	A weld which consists of one or more stringer beads or weave beads deposited on an unbroken surface in order to establish desired dimensions or properties.	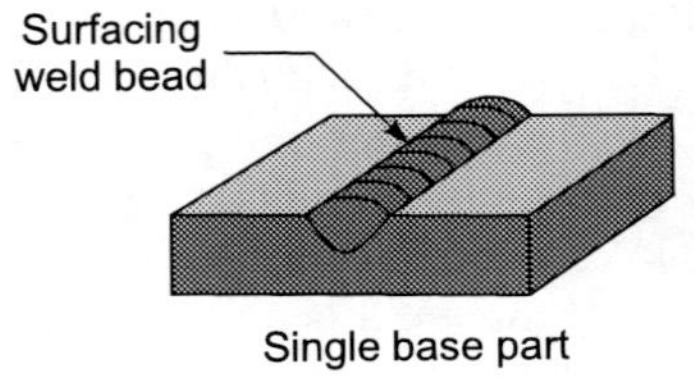 Surfacing weld bead Single base part

29.3 WELDING & WELD SYMBOLS

Special symbols are used to specify where welds are to be located, the type of joint to be used, as well as the size and amount of weld metal to be deposited in the joint. These symbols have been standardized by the American Welding Society (AWS). Welding symbols provide the means of placing complete welding information on drawings.

Basic Weld Symbols

A weld symbol indicates the required type of weld, and basic weld symbols are shown in Table 29.3.

TABLE 29.3 Weld symbols

S. No.	Designation	Illustration	Symbol
1.	Butt weld between plates with raised edges (the raised edges being melted completely)		
2.	Square butt weld		
3.	Single V butt weld		
4.	Single bevel butt weld		
5.	Single V butt weld with broad root face		
6.	Single bevel butt weld with broad root face		
7.	Single U butt weld (parallel or sloping edges)		
8.	Single U butt weld		

Contd.

9.	Backing run; back or backing weld		
10.	Fillet weld		
11.	Plug weld; plug or slot weld		
12.	Spot weld		
13.	Seam weld		

Basic Welding Symbol (Figure 29.1)

This is a method of representing the weld symbol in drawings. The assembled "welding symbol" consists of the following eight elements:

- **Reference Line:** A horizontal line with an arrow inclined to it at one end.
- **Arrow:** It points to a spot on the drawing where the welding is to be done.
- **Basic Weld Symbols:** Like, fillet, butt joints, etc.
- **Dimensions and Other Data:** Dimensions of a weld are shown on the same side of the reference line as the weld symbol. The size of the weld is shown to the left of the weld symbol and the length of the weld is placed in the right.
- **Supplementary Symbols:** They complement the basic symbols and provide additional requirements or instructions to the welder, based on the shape of the weld surface.

Flush	Convex	Concave

- **Finish Symbols:** Finish symbols are used to indicate the method of finishing. ("C" = chipping, "G" = grinding, "H" = hammering, "M" = machining, etc.)
- **Tail:** When a specification process, test or other reference is needed to clarify a welding symbol, it is put in tail of the welding symbol.
- **Specification, Process, or Other Reference:** These are a set of letter designations, used to notify specification, type of welding process, etc. Detailed welding designations are given in Table 29.4.

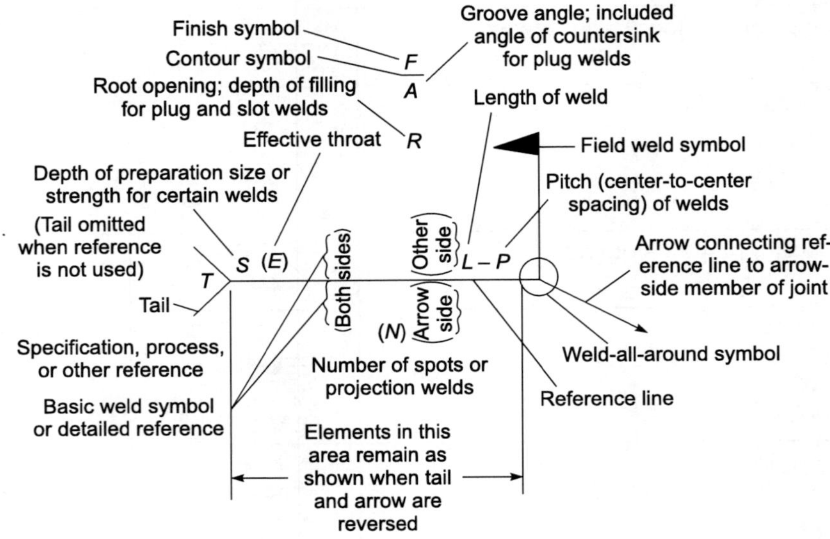

Figure 29.1 Welding symbols

TABLE 29.4 Welding processes and letter designations

Group	Welding Process	Letter Designation
Arc welding	Carbon arc	CAW
	Flux cored arc	FCAW
	Gas metal arc	GMAW
	Gas tungsten arc	GTAW
	Plasma arc	PAW
	Shielded metal arc	SMAW
	Stud arc	SW
	Submerged arc	SAW
Brazing	Diffusion brazing	DFB
	Dip brazing	DB
	Furnace brazing	FB
	Induction brazing	IB
	Infrared brazing	IRB
	Resistance brazing	RB
	Torch brazing	TB
Oxyfuel gas welding	Oxyacetylene welding	OAW
	Oxyhydrogen welding	OHW
	Pressure gas welding	PGW

Contd.

Resistance welding	Flash welding	FW
	High frequency resistance	HFRW
	Percussion welding	PEW
	Resistance projection welding	RPW
	Resistance seam welding	RSEW
	Resistance spot welding	RSW
	Upset welding	UW
Solid state welding	Cold welding	CW
	Diffusion welding	DFW
	Explosion welding	EXW
	Forge welding	FOW
	Friction welding	FRW
	Hot pressure welding	HPW
	Roll welding	ROW
	Ultrasonic welding	USW
Soldering	Dip soldering	DS
	Furnace soldering	FS
	Induction soldering	IS
	Infrared soldering	IRS
	Iron soldering	INS
	Resistance soldering	RS
	Torch soldering	TS
	Wave soldering	WS
Other welding processes	Electron beam	EBW
	Electroslag	ESW
	Induction	IW
	Laser beam	LBW

29.4 PRINCIPLES OF WELD JOINT DESIGN

The major factors that influence the selection of joint design are:

1. Load requirements
2. Ease of welding
3. Cost

General Design Principles

1. The design should satisfy load, fatigue, creep, bending, and torsion requirements.

2. Unduly high factor of safety increases the cost. Therefore, it must be avoided.
3. Joint should be selected in such a way that the amount of weld metal is minimized. This reduces distortion and residual stresses.
4. Maximum number of parts should be joined with one-pass weld (Figure 29.2).

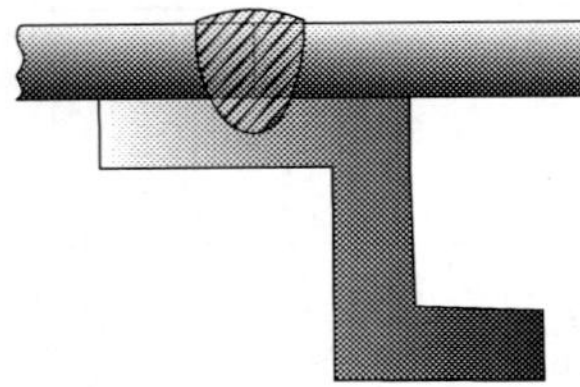

Figure 29.2 Three parts joined with one-pass weld

5. Standard rolled sections, plates, and bar should be used for economy and availability.
6. Proper planning is required to reduce the number of parts, which will reduce the assembly and amount of welding.
7. Casting or forging can be used in place of a complex weldment to simplify design.
8. Fillet welds are always designed on the basis of shear strength on the throat. Thus, flat 45° is more economical than concave welds (Figure 29.3).

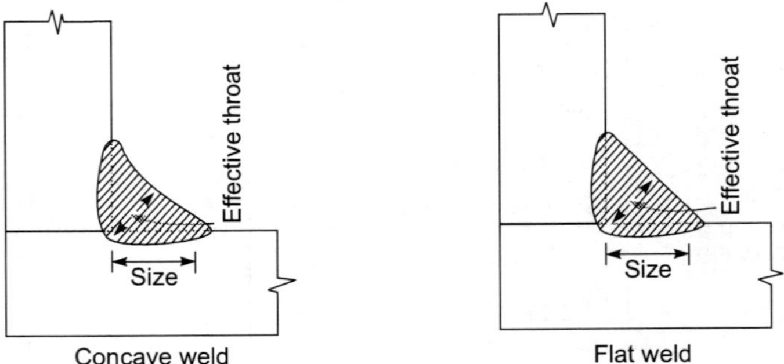

Figure 29.3 Fillet weld designs

9. Create a corner by bending or forming rather than by welding the two pieces together.
10. Bend flanges on plates rather than welding flanges on it.
11. Instead of using expensive alloy components, inexpensive component with surface weld can be used to provide wear resistance.
12. Easy access of welding spots results in good welds.
13. The weld size should be correct. The leg size of a fillet weld should be three-fourths of the plate thickness, if full strength is required. If rigidity is the main requirement, 50% of the full strength of the weld is sufficient.

14. Where possible, use a square groove joint together with a welding process capable of deep joint penetration or welding within the groove, for example electroslag, electrogas or narrow gap welding.
15. For corner joints in steel, the depth of the groove should be such that it intersects the full thickness of 50 mm and over.
16. Double fillet weld is better than groove weld for T-joint when fatigue is not of primary concern.

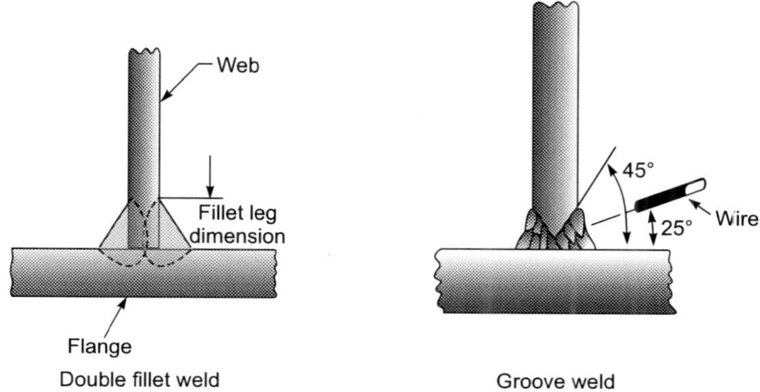

Figure 29.4 Double fillet weld versus groove weld

17. The size of the weld should always be designed with reference to the size of the thinner member and it should be placed in the section of least thickness.
18. Weldments should be designed so that the position in which the welds are made shall have the following order of preference:
    (a) Fillet welds: flat, horizontal, vertical, overhead.
    (b) Groove welds: flat, vertical, horizontal, overhead.
19. In dynamic loading situations, the joint profile is much more significant and it is important to avoid sudden change in sections that may act as stress raiser.
20. Welds placed in transverse shear are 30% stronger than those placed in longitudinal shear.
21. Where design permits, fillet welds are used in preference to groove welds due to economy.
22. A joint of the type that requires minimum amount of weld metal should be selected. Smallest practical groove angle and root opening should be chosen. This reduces distortion and residual stresses as well as the need and cost of postweld stress relieving and straightening (Figure 29.5).

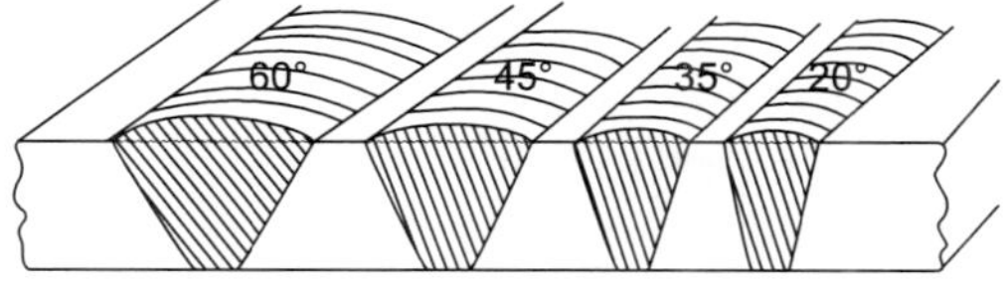

Figure 29.5 Possibilities of making a butt weld with different groove angles

23. On thick plates, double U or V groove welds should be used to reduce the amount of weld metal and control distortion (Figure 29.6).

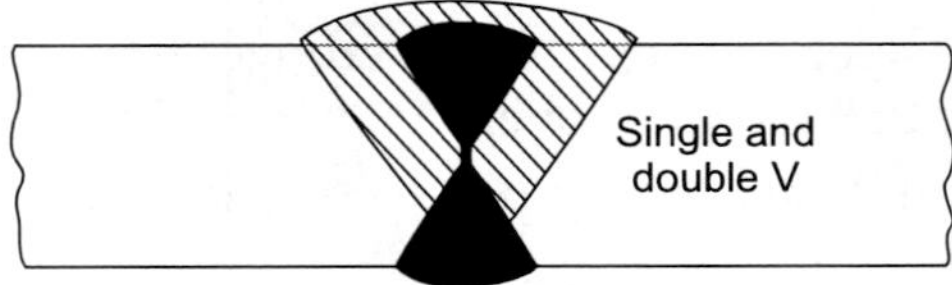

Figure 29.6 Schematic representation of weld metal required for single and double V groove welded butt joints

24. Maximum use of a welded joint should be made. Several parts can be joined with one-pass weld (Figure 29.7).

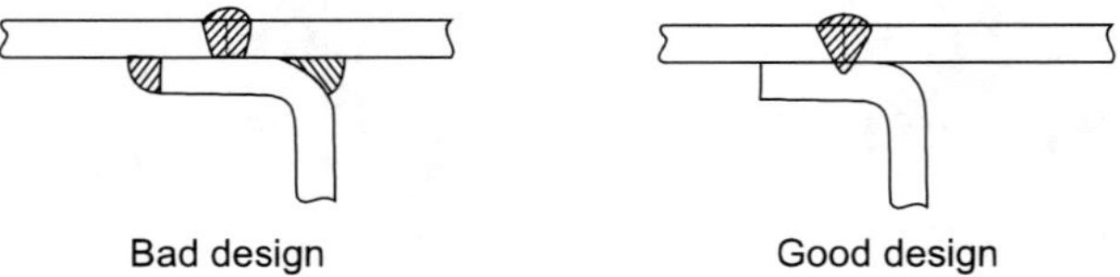

Figure 29.7 Maximum use of welded joint

25. Fillet welds are always designed on the basis of shear strength at the throat. Flat 45° fillet welds are more economical than fillet welds, as the strength still depends on the actual throat distance (Figure 29.8).

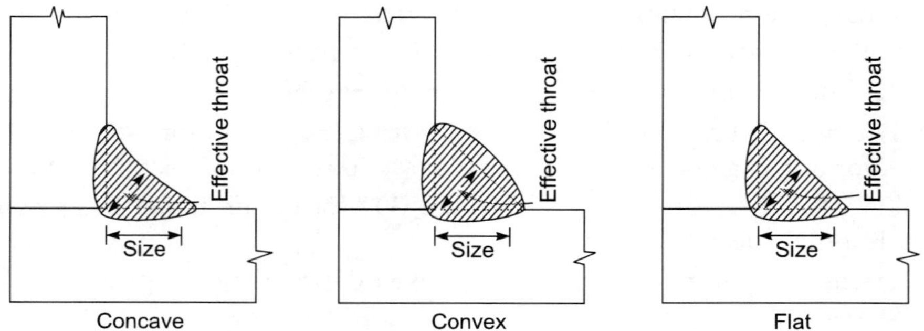

Figure 29.8 Design of fillet welds

26. Standard rolled sections, plates, and bar should be used for economy and availability.
27. Plan the design to minimize the number of pieces, which will reduce assembly time and the amount of welding.
28. Create a corner by bending or forming rather than by welding the two pieces together.
29. Bend flanges on plates rather than welding flanges on it.

30. Casting or forging can be used in place of a complex weldment to simplify design and reduce manufacturing costs.
31. Instead of using expensive alloy components, inexpensive component with a surface weld can be used to provide wear resistance.
32. Joints can be made better, faster and more economically if it is easily accessible.
33. The weld size should be correct. The leg size of a fillet weld should be three-fourths of the plate thickness if full strength is required. If rigidity is the main requirement, 50% of the full strength of the weld is sufficient (Figure 29.9).

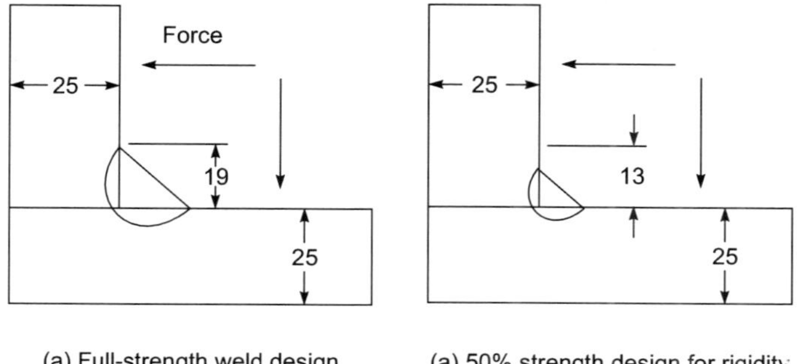

Figure 29.9 Correct weld size

34. Where possible, use a square groove joint together with a welding process capable of deep joint penetration or welding within the groove, for example, electroslag, electrogas or narrow gap welding.
35. For corner joints in steel, the depth of the groove should be such that it intersects the full thickness of the material to minimize lamellar tearing, especially in thickness of 50 mm and over (Figure 29.10).

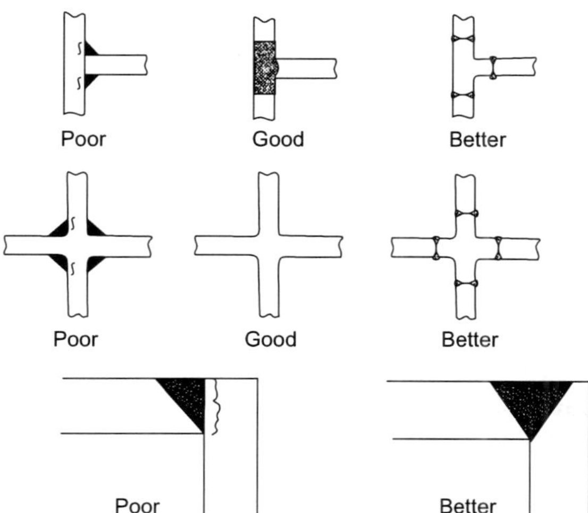

Figure 29.10 Edge preparations to avoid lamellar tearing in a corner joint

36. Double fillet weld is better than groove weld for T joint when fatigue is not of primary concern (Figure 29.10).
37. The size of the weld should always be designed with reference to the size of the thinner member and it should be placed in the section of least thickness.
38. Weldments should be designed so that the position in which the welds are made shall have the following order of preference:
    (a) Fillet welds: flat, horizontal, vertical, overhead.
    (b) Groove welds: flat, vertical, horizontal, overhead.
39. In dynamic loading situations the joint profile is much more significant and it is important to avoid sudden change in sections that may act as stress raiser. This is particularly important in joints subjected to fatigue loading.
40. Welds placed in transverse shear are 30% stronger than those placed in longitudinal shear.
41. Where the design permits, fillet welds are used in preference to groove welds for economy. A combination of fillet and groove welds can also be designed where the required size is more than 16 mm, to provide the effective throat.

29.5 WELDED FABRICATIONS

Due to immense technical advancements in the field of welding technology in terms of better power supplies, better consumables, etc., it is possible to make high quality welded joints which have better functional properties.

The direct impact of this is seen in the fact that most of the heavy structures that were employing riveted joints have switched over to welding. Structures made using these joints have also gained the confidence of the fabricators.

Welded joints are used in:

- Boilers and pressure vessels
- Ships
- Pipelines
- Bridges
- Offshore structures, etc.

Welding of Pressure Vessels

Fired and unfired pressure vessels are used in a large number in various industries, like chemicals and fertilisers, power plants, cryogenic applications, etc. Currently, welding is widely used to fabricate them. Due to the critical nature of the application of these structures, some special considerations have to be made while welding them.

In general, the fabrication of pressure vessels may be divided into the following three steps:

1. Material selection
2. Welding process selection and fabrication
3. Preheat and postweld heat treatment

Material Selection

The material selected should satisfy the following criteria:

1. It should have a high resistance to corrosion.
2. It should withstand and perform suitably well at the operating temperature range for which it is intended.
3. The formability and the weldability must be within the acceptable limits.

Welding Process Selection and Fabrication

The next step is to fabricate the selected material into the required shape. This involves mechanical operations like rolling, press working, assembly welding, postweld heat treatment and inspection. The welding process that is selected must suit the need, i.e., appropriate process has to be selected for different requirements like welding of the shell, welding of the end plates, nozzle and tube-to-tube plate welding.

The following welding processes are used for the same.

Electroslag Welding

Advantages

- Most economical when used for plates of thickness more than 50 mm.
- The desired thickness can be obtained in a single pass, using one or more electrodes.

Disadvantages

- Cumbersome when used for circumferential welding as the shells would have to be placed vertically.
- Results in coarse grain structures, which would have to be refined by postweld heat treatment.

Submerged Arc Welding

Advantages

- More economical than electroslag welding for wall thicknesses up to 45 mm.
- More cost effective and simpler than electroslag welding when used for circumferential joints.

Gas Metal Arc Welding/Flux Cored Arc Welding

Advantages

- Finds its use for pressure vessels of small diameter where internal access is a major problem. This type of welding is used to form the base of sufficient thickness for subsequent submerged arc welding.

Disadvantage

- Cannot be used for welding of thick walled pressure vessels.

Shielded Metal Arc Welding

Advantages

- Can be used in welding on-site vessels, where there is a restriction of space rendering it is impossible to use automatic processes.
- Widely used in welding of nozzles and tubes to the pressure vessels.

Apart from the processes mentioned above, many other methods are also employed, e.g., automatic TIG process is widely used to weld tube-to-tube plate in case of fabricating heat exchangers in the petrochemical industries. The same is done by explosive welding for pressure vessels.

Preheat and Postweld Heat Treatment of Pressure Vessels

Preheating is heating the base material to a specific temperature before the welding process begins.

The temperature to which the material is to be preheated, depends on the following factors:

- Composition
- Thickness
- Welding process
- Arc energy

Preheating is generally not continued during welding as heat from the weld source ensures that the temperature does not fall below the required temperature. The temperature up to which preheat is to be carried out varies from material to material. But, as a general rule, no welding should be carried out when the temperature of the parent metal, within 150 mm of the joint, is less than 5°C.

Preheat is done for the following reasons:

- To heat the material above the temperature at which fracture may occur during fabrication.
- It also leads to reduction in shrinkage stresses in the material.

- It may also be done to slow the rate of cooling to ensure higher diffusion of hydrogen from the materials so as to make them more resistant to cracking (due to hydrogen embrittlement). The slower cooling rate ensures that the material would remain at higher temperatures for longer duration and this is what increases the diffusion process.

Postweld heat treatment (PWHT), as the name suggests, is heating after the welding process to relieve the material of any stresses during the process. It is employed in pressure vessels to avoid the problem of brittle fracture, especially in the case of thick walled vessels.

PWHT can be avoided for thin walled vessels of mild steel with a maximum wall thickness of 50 mm. PWHT must be applied to the complete vessel as a whole by making use of a large furnace. In some cases, it may not be possible to do so. In such cases, we go for locally treating the weld by making use of electrically-heated pads arranged around the periphery of the weld. But heating this way may not fetch as good results as seen in heating the entire vessel as a whole, making the material more susceptible to stress corrosion cracking, due to residual stresses.

Selection of temperature and the treatment time depends on the composition of the material as well as its thickness.

PWHT may also be used to reduce the hardness values in the heat affected zone (HAZ), particularly in the case of alloy steels and also, to reduce the effect of welding on the mechanical properties of the parent metal. It is also employed to improve the properties of the weld material.

29.6 PIPE WELDING

Pipe welding is the simplest and easiest way of joining sections of pipes.

Its main advantages are:

- It eliminates the need for using complicated joint designs using special equipment.
- Welded pipes have reduced flow restrictions when compared to other mechanical connections.
- Installation cost is less.

Pipe welding is a recognizable profession today and though it is similar to other welding operations, welders develop some specific skills unique to the field. Due to the fact that most pipelines carry hazardous materials, welders are required to clear certain tests for getting certified.

Pipe Welding Positions (Figure 29.11)

The four positions used in pipe welding are as follows:

- Horizontal rolled position (1G)

- Horizontal fixed position (5G)
- Pipe inclined position (6G)
- Vertical position (2G)

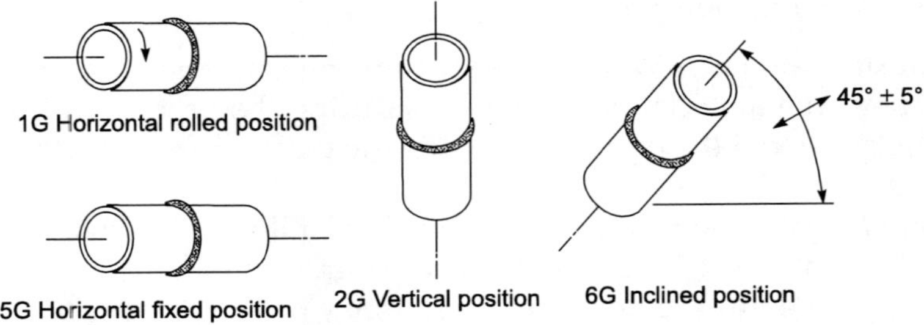

Figure 29.11 Weld positions in pipes

Pipe Welding Procedures (Figures 29.12 and 29.13)

Some standard procedures are followed in welding of pipes. In general, the following should be kept in mind when pipe joints are welded:

- If welding cannot be done in a single pass, we go for multiple passes with each layer having a maximum thickness of 3.2 mm
- Layers should be deposited with either a weaving or oscillating motion.
- Each layer should be thoroughly cleaned before depositing the next to ensure that no slag is trapped between the layers.
- For wall thicknesses up to 18 mm, either the single V or single U type of butt joint may be used. But, for greater wall thicknesses, only the single U can be used.

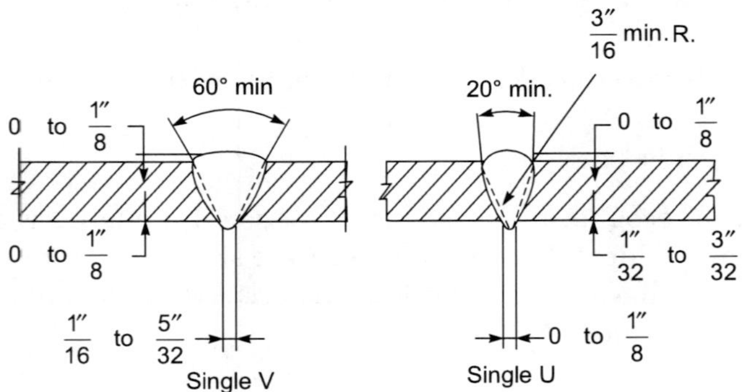

Figure 29.12 Butt joints

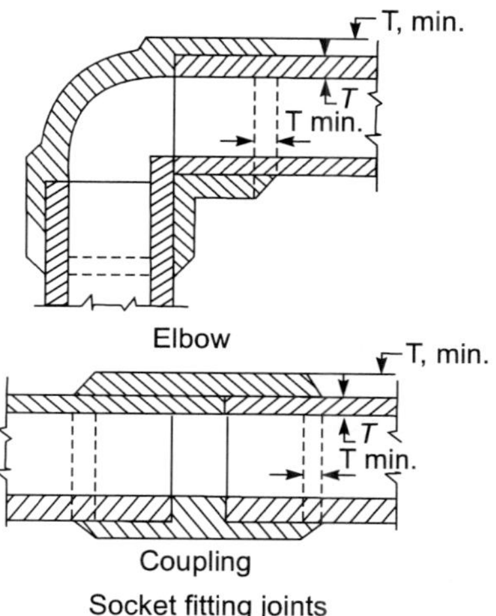

Socket fitting joints

P-42. Fillet welded slip-on flanges

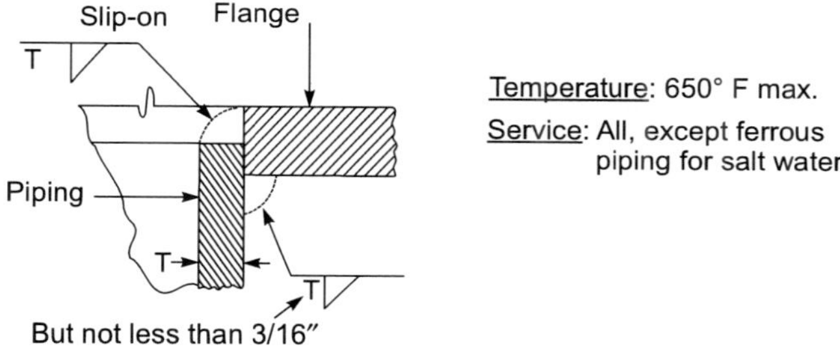

P-43. Fillet and bevel welded slip-on flange

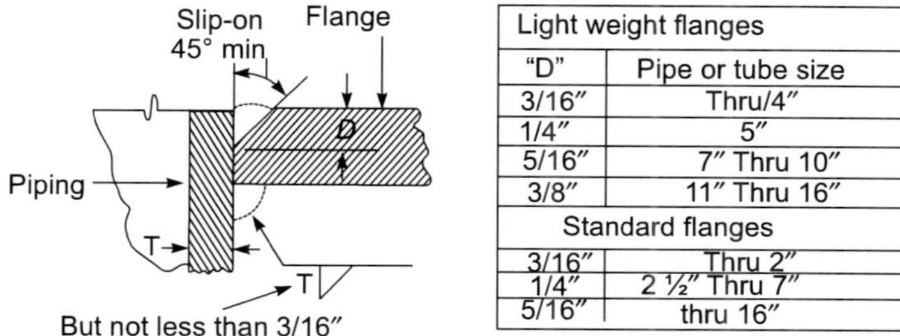

Light weight flanges	
"D"	Pipe or tube size
3/16″	Thru/4″
1/4″	5″
5/16″	7″ Thru 10″
3/8″	11″ Thru 16″
Standard flanges	
3/16″	Thru 2″
1/4″	2 ½″ Thru 7″
5/16″	thru 16″

Temperature: 650° F max.

Service: All, including salt water

Figure 29.13 Flange connections

Fillet welds are used to weld slip-on and threaded flanges to pipes. It can be given on both sides and also in combination with a bevel weld on one side, depending on the strength and the type of service required.

Before welding, the pipe and the flange should be aligned with each other and this is done with the help of a flange square (Figure 29.14).

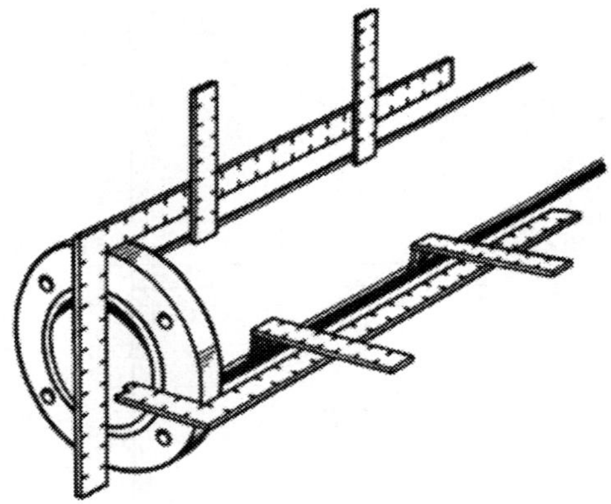

Figure 29.14 Flange alignment

Fillet welds may also be used in welding screw or socket couplings to pipes. For obtaining tightness and making the joint leak proof, we may make use of a seal weld. But it does not, however, contribute to the strength of the joint.

Joint Preparation and Fit-up

Before welding the joint, it should be made sure that the surfaces to be joined are free of all loose scale, slag, rust, paint, oil and other foreign matters. Also, it has to be made sure that the surfaces are smooth and uniform.

The bevels should be accurately cut by grinding, machining or using a gas cutting torch. While in the field, the welding operator usually makes the cut using a gas torch. While bevelling, it should be made sure that as little metal as possible is removed to obtain complete fusion and penetration.

Proper bevelling reduces the amount of filler metal and hence, helps in saving time and lowering the cost of operation. It also leads to reduction in the strain on the weld.

After preparation of the joint, we have to align the sections that are to be joined before welding them. Only a deviation up to 20 percent of the thickness of the pipe is allowed.

The following methods are employed to ensure proper alignment:

- **Using Clamps/Jigs**

  For a small diameter pipe, a piece of angle iron may be used. But for sections of larger diameter, a section of channel or I-beam is used.

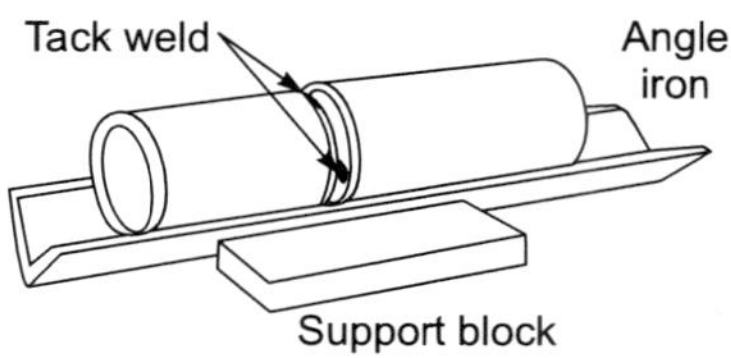

Figure 29.15 Angle iron jig

- **Tack Welding**
  - Tack welds are used to hold the sections in place temporarily.
  - Number of tack welds required depends on the diameter of the pipe.
  - For a 1/2 inch diameter pipe, two welds are placed directly opposite to one another.
  - For a standard pipe, 4 tack welds are required.
  - The welds must be of the same quality as the final weld.
  - The tack weld surface must be cleaned and should have good fusion.
  - The length should not exceed twice the pipe thickness and the depth should be less than two-thirds of the pipe thickness.
- **Spacers/Chill Rings/Backing Rings**

  They are accurately machined metal pieces which conform to the dimensions of the designed joint. They serve the following purposes:
  - Maintain the specified root opening.
  - Provide convenient location for tack welding.
  - Aid the pipe alignment.
  - Prevent spatter and formation of slag or icicles.

Electrode Selection

Before welding, one has to select an electrode best suited for the job. The selected electrode must ensure that proper penetration and fusion take place without any defects, such as undercutting and slag inclusions. Also, the electrode has to be supplied with current in the range specified by its manufacturer.

Review Questions

1. Discuss the basic types of welded joints.
2. Discuss various types of welds.

3. Explain how weld symbols are designated.
4. How is welding process designated using letter designation?
5. Discuss the major factors that influence the selection of a joint design.
6. State general design principles.
7. Discuss about welding process selection and fabrication.
8. What are the advantages of pipe welding?
9. Explain all pipe welding positions.
10. Explain joint preparation and fit-up for pipe welding operation.

Thermal Considerations in Welding

30.0 INTRODUCTION

During fusion welding, thermal cycles produced by the moving heat source cause physical state changes, metallurgical phase transformation, and thermal stress and metal movement.

The finished product may contain physical discontinuities that are due to excessively rapid solidification, or adverse microstructures that are due to inappropriate cooling, or residual stress and distortion that are due to the existence of incompatible plastic strains.

In order to analyse these problems, we need a welding heat flow model using mathematical tools for the thermal data analysis, design iterations, or the systematic investigation of the thermal characteristics of any welding parameter.

30.1 HEAT FLOW IN WELDING

For controlling metallurgical events in welding, the thermal conditions in and near the weld metal must be established.

Important factors to be considered are:

- The distribution of maximum temperatures in the weld heat affected zone.
- Time and temperature.
- Cooling rates in the weld and in the heat affected zone.
- The solidification rate of the weld metal.

Heat Source Efficiency

The heat source efficiency is defined as:

$$\eta = \frac{\text{Energy transferred to the workpiece}}{\text{Energy generated by heat source}}$$

Only a portion of the power provided by the heat source is transferred to the workpiece and the remaining portion is lost to the surroundings. Consequently, $\eta < 1$.

In arc welding process, efficiency can be expressed as:

$$\eta = Qt / EIt$$

where,

Q = The rate of heat transfer from the heat source to the workpiece.
t = The welding time.
E = Constant voltage of arc welding.
I = Constant current of arc voltage.

If the heat source efficiency η is known, the heat transfer rate to the workpiece, Q, can be easily determined from the above equation.

Heat Source Efficiency of Various Welding Processes (Figure 30.1)

Laser Beam Welding: The heat source efficiency is very low because of the high reflectivity of metal surface. This can be improved by surface modification as: roughening, oxidizing, coating.

Plasma Arc Welding: The heat source efficiency is much higher than LBW because reflectivity is not a problem.

Gas Tungsten Arc Welding: In AC GTAW, electrons bombard the workpiece only during the electrode-negative half cycle, and the arc efficiency is thus, lower.

Gas Metal Arc Welding: Heat transfer to the electrode can be transferred back to the workpiece through metal droplets, thus improving the arc efficiency.

Submerged Arc Welding: The arc is covered with a thermally insulating blanket of molten slag and granular flux, thus reducing heat losses to the surroundings and improving the arc efficiency.

Electron Beam Welding: The keyhole in EBW acts like a "blackbody", trapping the energy from the electron beam. So, the efficiency of the electron beam is very high.

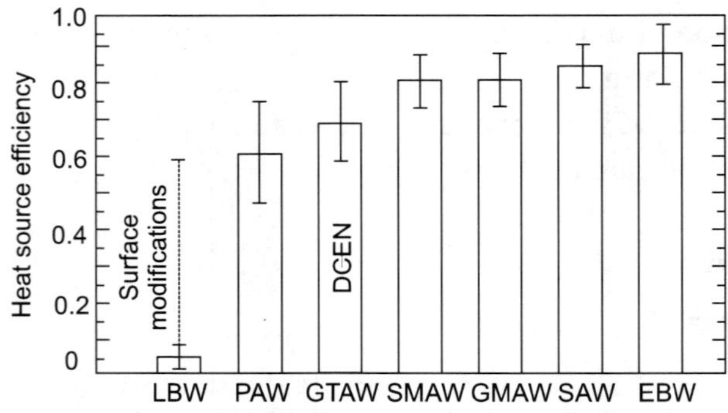

Figure 30.1 Heat source efficiencies in several welding processes

The amount of heat, input to the weld at its rate determines the geometry of the weld bead deposited and the width of the heat affected zone. It also affects the microstructure of the weld and heat affected zone, which in turn, affects the mechanical properties of the joints. In the following paragraphs, we shall discuss the factors like determination of heat input to the weld, maximum heat input rate, in fusion welding of plates, and resistance welding of thin sheets.

The discussion will also include the heat flow in welding peak temperatures reached adjacent to the weld and in the HAZ, estimation of the width of HAZ and the effect of preheat of this width. Determination of cooling rates has also been included in the discussion as it affects the weld microstructure and consequently, the mechanical properties of the welds. The following sections provide practical working equations for consumable electrode welding applications and other weld processes. The following quantities can be estimated using the heat flow equations:

1. Peak temperatures
2. Width of HAZ
3. Cooling rates
4. Solidification rates

Before going into the details of the above equations, let us first concentrate on the heat input to the weld.

30.2 HEAT INPUT TO THE WELD

The heat input, Q in watts, in the case of arc welding is given by:

$$Q = VI \text{ joules/sec} \tag{1}$$

For the melting of the weld at the joint, the exact amount of heat that enters the joint can be calculated (for an electrode moving at a speed of S_w mm/s) using the following relation:

$$H = \frac{Q}{S_w} \text{ joules/mm} \tag{2}$$

But the actual heat utilized by the joint depends upon how effectively this heat is transferred from electrode tip to the joint. Hence, heat transfer efficiency factor (f_1) enters the calculations of net heat available at the joint.

$$H_{\text{net}} = \frac{f_1 VI}{S_w} \text{ joules/min} \tag{3}$$

All of this net heat is not used for melting since part of it is conducted away to the base plate. The heat actually used for melting (H_m) can be obtained by another efficiency factor, f_2.

$$H_m = \frac{f_1 f_2 VI}{S_w} \quad (4)$$

where, $f_2 = \dfrac{\text{Heat required to melt the joint}}{\text{Net heat supplied}}$

Example 1: Calculate the melting efficiency in the case of arc welding of steel with a current of 200 A at 20 V. The travel speed is 5 mm/s, and the cross-sectional area of the joint is 20 mm^2. Heat required to melt steel may be taken as 10 J/mm^3 and heat transfer efficiency is 0.85.

Solution: Volume of base metal melted = 20 × 5 = 100 mm^3/s

Heat required for melting = 100 × 10 = 1000

$$f_2 = \frac{1000}{f_1 VI} = \frac{1000}{0.85 \times 20 \times 200} = 0.2941 = 29.41\%$$

30.3 RELATION BETWEEN WELD CROSS SECTION AND ENERGY INPUT

There is a simple but important relationship between the weld metal cross section, A_w, and energy input:

$$A_w = \frac{f_2 H_{net}}{Q} = \frac{f_1 f_2 H}{Q}$$

where, $H_{net} = \dfrac{f_1 EI}{v}$ J/s

Q = Heat required for melting in joules/mm^3

E = Voltage supplied in volts

I = Current consumed in amp

v = Welding speed in mm/s

$A_w = (A_m + A_r)$ in mm^2

$A_w = A_m$, if no filler metal is added

$$A_w = \frac{f_1 f_2 EI}{vQ}$$

where, $A_w = (A_m + A_r)$

f_1 = Heat transfer efficiency from electrode to plate

f_2 = Melting efficiency $= \dfrac{\text{Heat required to melt the joint}}{\text{Net heat supplied}}$

H_{net} = Net heat available at the weld joint (J/s)

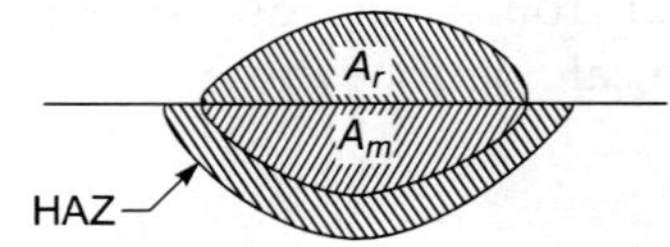

Heat Source	f_1
MMA/GMA	0.8-0.66
SAW	0.9-0.99
GTAW	0.21-0.48

Example 2: An arc weld pass is made on steel under the following conditions:

$E = 20$ V $\qquad I = 200$ A $\qquad v = 5$ mm/s

$f_1 = 0.9$ $\qquad f_2 = 0.3$ $\qquad Q = 10$ J/mm^3

Estimate the cross-sectional area of the weld pass.

Solution: $A_w = \dfrac{(0.9)(0.3)(20)(200)}{(5)(10)} = 21.6 \text{ mm}^2$

30.4 HEAT INPUT RATE

In many situations, we are interested in determining the minimum heat input rate in watts required to form a weld of a given width 'w' in a 'V' groove as shown in a Figure 30.2. It can be calculated for two-dimensional heat source or a three-dimensional heat source using equations (5) and (6), respectively.

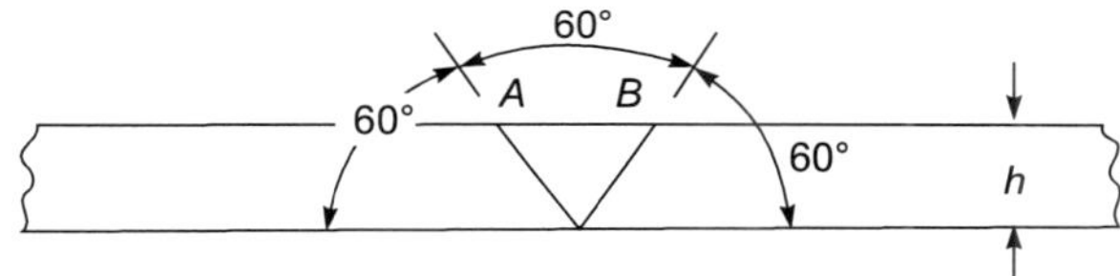

Figure 30.2 Plate geometry for calculating the heat input rate

The following symbols are used in these equations:

α = thermal diffusivity of the work in (m^2/s)

K = thermal conductivity of work material (W/m-°C)

K_{steel} = 43.6 W/m -°C; $\alpha_{steel} = 1.2 \times 10^{-5}$ m^2/s

ρ = density and C = specific heat

θ_m = M.P. of metal

W = weld width (m)

h = plate thickness (m)

v = welding speed (m/s)

θ_m = M.P. of steel = 1530°C

θ_0 = room temp = 30°C

ρ_c = 0.0044

For two-dimensional heat source

$$Q = 8K\theta_m h\left(\frac{1}{5} + \frac{vw}{4\alpha}\right) \tag{5}$$

and, for three-dimensional heat source

$$Q = \frac{5}{4}\,\pi\omega m\mathrm{K}\theta_m\left(\frac{2}{5} + \frac{vw}{4\alpha}\right) \tag{6}$$

It can be observed from these equations that "$v\omega/\alpha$" is the most important parameter. Theoretical results fail to accommodate many practical difficulties, e.g.,

1. Inhomogeneous conducting medium (liquid pool + solid)
2. Absorption and rejection of the latent heat at the forward and rear edges, respectively, of the weld pool.

Still the two equations provide a good estimate.

In arc welding with short circuit transfer, the heat input is given by

$$Q = CVI \tag{7}$$

where, V = arc voltage

I = arc current

C = fraction of total time for which the arc is on.

In a butt welding process using arc welding, the arc power was found to be 2.5 kVA.

The process is used to weld two plates of steel 3 mm thick, with 60° V-edge preparation angle.

Example 3: Determine the maximum possible welding speed. The metal transfer is short circuit type. type and the arc is on for 85% of the total time given.

Solution: The rate of heat input is given as

$$Q = CVI$$

$$= 0.85 \times 2.5 \times 10^3 w = 2.12 \times 10^3 w$$

The minimum weld width to be maintained

$$w = AB = 2\sqrt{3}\ \text{mm} = 2\sqrt{3} \times 10^{-3}\ \text{m}$$

$$\theta_m = (1530 - 30) = 1500°\text{C} \qquad h = 3 \times 10^{-3}\ \text{m}$$

As in the welding of thin plates, the source of heat can be approximated as a line source. Thus, using equation (1)

$$Q = 8K\theta_m h\left(\frac{1}{5} + \frac{vw}{4\alpha}\right)$$

$$2.12 \times 10^3 = 8 \times 43.6 \times 1500 \times 3\left(\frac{1}{5} + \frac{vw}{4\alpha}\right)10^{-3}$$

$$\left(0.2 + \frac{vw}{4\alpha}\right) = 1.35$$

$$v = \frac{1.15 \times 4\alpha}{w}$$

$$w_{\min} = 2\sqrt{3} \times 10^{-3}\ \text{m}$$

$$v = \frac{1.15 \times 4 \times 1.2 \times 10^{-5}}{2\sqrt{3} \times 10^{-3}}$$

$$= 0.016 \frac{\text{m}}{\text{sec}} = 0.95 \text{ m/min}$$

30.5 HEAT FLOW EQUATIONS

An important parameter that needs to be calculated is the peak temperature point in the material during welding. The cooling rate from this peak temperature determines the metallurgical transformations likely to take place in the HAZ.

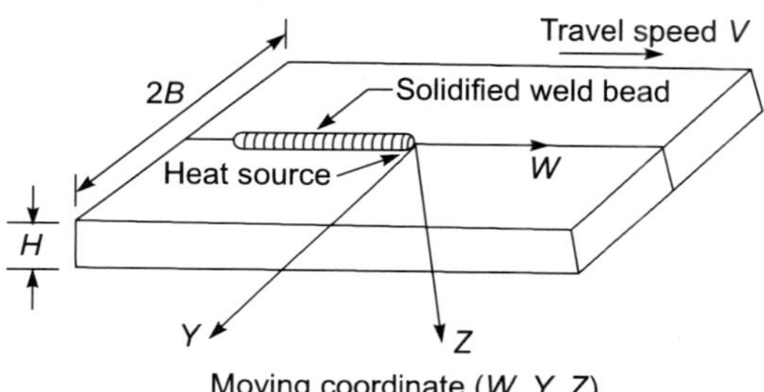

Figure 30.3 Heat flow

Peak temperatures can be calculated using the following equation

$$\frac{1}{(T_p - T_0)} = \frac{\sqrt{2\pi e}\rho_c t y}{H_{\text{net}}} + \frac{1}{T_m - T_0} \qquad (8)$$

where, e = base of natural logarithm = 2.7828128

Thus, $\sqrt{2\pi e} = \sqrt{2 \times 3.14 \times 2.7828128} = 4.13$

$\rho_c = 0.0044$

Peak Temperature (T_p)

Peak Temperature Equation: For a single pass full penetration butt weld in sheet or plate, peak temperature in the base metal adjacent to the weld T_p in HAZ region

$$\frac{1}{(T_p - T_0)} = \frac{4.13\rho_c t y}{H_{\text{net}}} + \frac{1}{T_m - T_0} \qquad (9)$$

is given by equation (9) where,

T_p = The peak or max. temp.°C, at a distance Ymm from the weld fusion boundary (this eq. does not apply for temps. within the weld metal)

t = Plate thickness

T_0 = Initial plate temperature°C

T_m = Melting temperature of base metal

ρ_c = 0.0044

Uses of this equation

1. Determining peak temperature in specific locations in HAZ.
2. Estimating width of HAZ.
3. Determining effect of preheat on width of HAZ.

Example 4: A single full penetration weld pass is made on steel using the following parameters:

$E = 20$ V, $I = 200$ A, $v = 5$ mm/s, $T_o = 25$°C, $T_m = 1510$°C

$\rho_c = 0.0044$ J/mm^3.°C, $t = 5$ mm, $f_1 = 0.9$ $H_{net} = 720$ J /mm

Calculate the peak temperatures at distances of 1.5 and 3.0 mm from the weld fusion boundary.

Solution:

(i) At $Y = 1.5$ mm

$$\frac{1}{T_p - 25} = \frac{4.13(0.0044)\,5(L5)}{720} + \frac{1}{1510 - 25}$$

$$T_p = 1184°C$$

Note that at $Y = 0$, $T_p = T_m$

(ii) At $Y = 3.0$ mm

$$\frac{1}{T_p - 25} = \frac{4.13(0.0044)\,5(3)}{720} + \frac{1}{1510 - 25}$$

$$T_p = 976°C$$

30.6 WIDTH OF HEAT ZONE

For this calculation, the outer extremity of the HAZ must be clearly identified with a specific peak temperature. For example, for most carbon or alloy steels, there is a distinct etching boundary (as observed on polished and etched weld cross section), corresponding to a peak temperature of 730°C.

Now the problem reduces to the determination of the distance (y_z) at which T_p = 730°C.

$$\frac{1}{730-25}=\frac{4.13(0.0044)5(y_z)}{720}+\frac{1}{1510-25}$$

$$y_z = 5.9 \text{ mm}$$

Thus, a region 5.9 mm wide, adjacent to the fusion boundary, will be structurally changed, i.e., it may be affected by the heat of welding.

If the steel plate is preheated to 200°C, its effect is to widen the HAZ width. This plate was tempered at 430°C. Any temp. above this 430°C, modifies its property. Now T_p becomes 430°

$$\frac{1}{430-200}=\frac{4.13(0.0044)5(y_z)}{720}+\frac{1}{1510-200}$$

$$y_z = 28.4 \text{ mm}$$

Without preheat, this width would be

$$\frac{1}{430-25}=\frac{4.13(0.0044)5(y_z)}{720}+\frac{1}{1510-25}=14.2 \text{ mm}$$

Thus, preheating has doubled the width of HAZ.

Finally, if the net energy input is increased 50% to × 1.5 × 720 = 1080 J/mm

$$\frac{1}{430-25}=\frac{4.13(0.0044)5(y_z)}{1080}+\frac{1}{1510-25}$$

$$y_z = 21.3 \text{ mm}$$

The weld width is also increased by 50%.

30.7 COOLING RATES

Calculation and comparison of cooling rates require careful specification of conditions, because it varies with position and time. Most useful method is to determine the cooling rate on the center line of the weld at the instant the metal is passing through a particular temperature of interest, T_c. At temperatures well below melting, the cooling rate in the weld and its immediate HAZ are substantially independent of position.

In carbon and low-alloy steels, the temperature of interest is best taken near the pearlite "nose" temperature on the TTT (time-temperature transformation) diagrams. The exact temperature is not critical but should be the same for all calculations and comparisons. A value of $T_c = 550$, which is quite satisfactory for most steels.

For thick plates requiring several passes (more than six) to complete the joint the cooling rate (for the first pass or each pass) R is given by:

$$R=\frac{2\pi K\left(T_c T_c\right)^2}{H_{\text{net}}}$$

where, R = cooling rate at a point on the weld centerline, °C/s at just that moment when point is cooling past T_c.

K = Thermal conductivity of the metal J/mm-s°C
T_c = Temperature at which cooling rate is calculated
T_o = Initial plate temperature, °C

The cooling rate is maximum at the weld centereline. The above equation gives this maximum cooling rate. At fusion boundary, it is only a few percent lower. Thus, this equation applies to the entire weld and the HAZ. If the plates are thin, requiring fewer than four passes, then

$$R = 2\pi K\rho_c \left(\frac{t}{H_{\text{net}}}\right)^2 (T_c - T_0)^3 \tag{10}$$

where, t = Thickness of base metal mm
ρ = Density of metal, g/mm^3
C = Specific heat of base metal, J/g °C

In thick plates, the heat flow is three dimensional. Equation (10) applies to small bead-on-plate welds on thin plates.

Relative plate thickness factor, T, is defined as follows to distinguish between thick and thin plates.

$$\tau = h\sqrt{\frac{\rho_c (T_c - T_0)}{H_{\text{net}}}}$$

$T \leq 0.75$ thin plate equation is valid
$T \geq 0.75$ thick plate equation is valid.

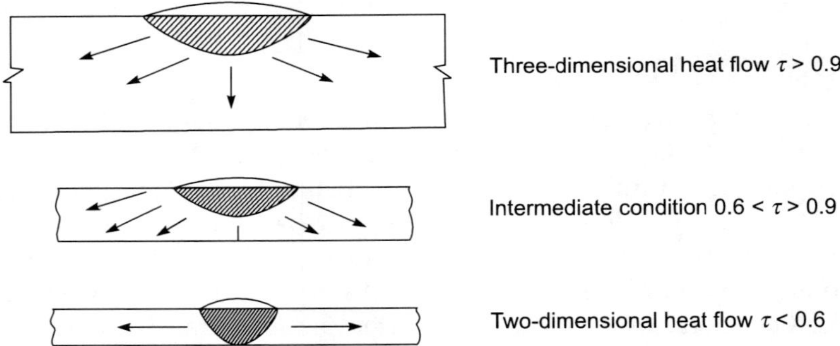

Figure 30.4 Relative plate thickness factor (τ) for cooling rate calculations

Example 5: Find the best welding speed for welding 6 mm steel plates with an ambient temperature of 30°C with the welding transformer set at 25 V and current passing is 300 A. The arc efficiency is 0.9 and possible travel speeds are 6 to 9 mm/s. The limiting cooling rate for satisfactory performance is 6°C/s at a temperature of 550°C.

Solution: Given T_o = 30°C, T_c = 550°C, K = 0.028 J/mm-s-°C, R = 6°C/s, V = 25 V, I = 300 A, h = 6 mm, f_1 = 0.9, ρ_c = 0.0044 J/mm³°C.

Assume a travel speed of 9 mm/s

$$\text{Heat input, } H_{\text{net}} = \frac{f_1 VI}{v} = \frac{0.9 \times 25 \times 300}{9} = 750 \text{ J/mm}$$

To check whether it is a thick or thin plate

$$\tau = h\sqrt{\frac{\rho_c (T_c - T_0)}{H_{\text{net}}}} = 6\sqrt{\frac{0.0044(550 - 30)}{750}} = 0.3314$$

Being less than 0.6, it is a thin plate. Cooling rate will be calculated by using the thin plate equation

$$R = 2\pi K \rho_c \left(\frac{h}{H_{\text{net}}}\right)^2 (T_c - T_0)^3$$

$$= 2\pi \times 0.028 \times 0.0044\left(\frac{6}{750}\right)^2 (550 - 30)^3 = 6.9659\text{°C/s}$$

This value is higher than the critical cooling rate required, we may reduce the travel speed to 8 mm/s and recalculate the cooling rate.

This cooling rate is higher than the limiting cooling rate of 6°C/s (given) at a temperature of 550°C. We, therefore, reduce the travel speed to 8 m/s and recalculate.

$$v = 8 \text{ mm/s}$$

$$\text{Heat input, } \quad H_{\text{net}} = \frac{0.9 \times 25 \times 300}{8} = 843.75 \text{ J/mm}$$

To check whether it is a thick or thin plate:

$$\tau = h\sqrt{\frac{\rho_c (T_c - T_0)}{H_{\text{net}}}} = 6\sqrt{\frac{0.0044(550 - 30)}{843.75}} = 0.312$$

This being less than 0.6, it is a thin plate. Using thin plate equation for cooling rate

$$R = 2\pi K \rho_c \left(\frac{h}{H_{\text{net}}}\right)^2 (T_c - T_0)^3$$

$$= 2\pi \times 0.028 \times 0.0044\left(\frac{6}{843.75}\right)^2 (550 - 30)^3 = 5.504\text{°C/s}$$

This is a satisfactory cooling rate, the welding speed can be finalized at 8 mm/s.

These equations could also be used to calculate the preheat temperature required to avoid martensitic transformation in the weld zone.

30.8 CONTACT-RESISTANCE HEAT SOURCE (FIGURE 30.5)

The electrical resistance could be used as a source of heat. It could be:

(a) contact resistance of interfaces, or
(b) resistance of molten flux and slag.

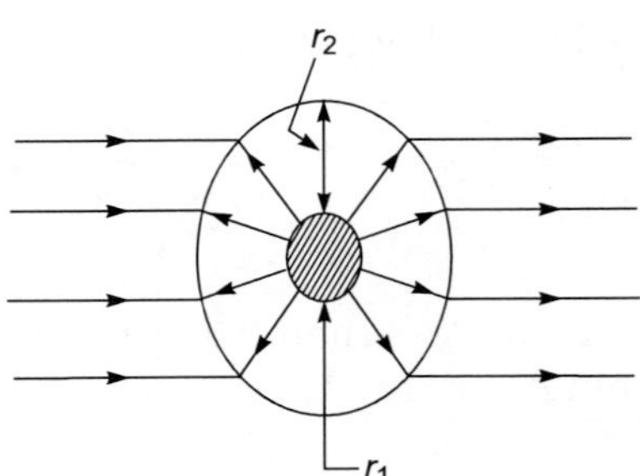

Figure 30.5

Resistance of each hemispherical constriction,

$$R = \rho(r_2 - r_1)/s$$

where, ρ = Resistivity of material
$(r_2 - r_1)$ = Length of current path
S = Geometric mean area of the two hemispheres of radii r_1 and r_2, respectively.

$$= \sqrt{(2\pi r_2^{\,2})(2\pi r_1^{\,2})} = 2\pi r_1 r_2$$

Now, $$R = \frac{\rho(r_2 - r_1)}{2\pi r_1 r_2} = \frac{\rho}{2\pi r_1} \text{ as } r_2 >> r_1$$

Total constriction resistance, R_c of n such spheres/unit area

$$R_c = \frac{1}{n}\frac{\rho}{\pi r_1} = \frac{\rho}{n\pi r_1}$$

This approximation does not cause an error of more than 15%

Thus, $$R_c = 0.85\rho/n\pi r_1$$

Heat generation rate by this contact resistance with an applied voltage of V is $Q = V^2/R_c$ per unit area.

However, after a very short time ($\approx$ 0.001 sec) the contact resistance drops to 1/10th of original value due to softening of material and increase in temperature.

Example 6: In resistance welding process, applied voltage = 5 V, bridges formed, $n = 25/\text{cm}^2$, bridge radius, r_1 = 0.1 mm = 0.01 cm and resistivity of material is

$\rho = 2 \times 10^{-5}$ ohm-cm. Calculate the contact resistance.

Solution: $$R_c = \frac{0.85\,\rho}{n\pi r_1} = \frac{0.85 \times 2 \times 10^{-5}}{25 \times 3.14 \times 0.01}$$

$$= 0.00022 \text{ ohm-cm}^2$$

Rate of heat generated/unit area,

$$Q = \frac{V^2}{R_c} = \frac{5^2}{0.00022}$$

$$= 1.136 \times 105 \text{ W/cm}^2$$

Example 7: Two different pairs of sheets of the same material are to be spot welded. In one pair, there are 25 bridges/cm^2 and the average radius of each bridge is 0.1 mm. The other pair of sheets contains 50 bridges/cm^2 with the same average radius of each bridge. Determine the ratio of the voltages to be applied in these two cases to generate the same rate of heating/unit area.

(Exercise problem in *Manufacturing Process*: Ghosh and Mallik)

Solution: The rate of heat generated by contact resistance with an applied voltage V is

$$\frac{V^2}{R_c}$$

$$R_c = \frac{0.85\,\rho}{n\pi r_1}$$

ρ = Resistivity of the material
V = Applied voltage
R_c = Constriction resistance
n = Number of bridges/cm^2
r = Radius of bridge (average)

Case 1. Rate of heat generated/unit area

$$= \frac{V_1^2}{R_{c_1}} = \frac{V_1^2 \times 25 \times \pi \times r}{0.85\rho}$$

Case 2. Rate of heat generated/unit area

$$= \frac{V_2^2}{R_{c_2}} = \frac{V_2^2 \times 50 \times \pi \times r}{0.85\rho}$$

For equal heat to be generated

$$\frac{V_1^2 \times 25 \times \pi \times r}{0.85\rho} = \frac{V_2^2 \times 50 \times \pi \times r}{0.85\rho}$$

$$\left(\frac{V_1}{V_2}\right)^2 = 2 \frac{V_1}{V_2} = 1.414$$

Example 8: The voltage-arc length characteristic of a DC arc is given by: $V = (20 + 4l)$ volts, where l is the arc length in mm. During a welding operation, it is expected that the arc length will vary between 4 and 6 mm. It is desired that the welding current is limited to the range 450 to 550 A. Assuming a linear power source characteristic, determine the open circuit voltage, and short circuit current of the power source.

(Exercise problem in *Manufacturing Process*: Ghosh and Mallik)

Solution: DC arc voltage, $V = 20 + 4l$

Arc length varies between 4 and 6 mm.

It is desired that welding current should be between 450 and 550 A (difference 100 A)

Assume a linear power source characteristic.

Find open circuit voltage and short circuit current voltage variation range:

$$\left.\begin{aligned} V = 20 + 4 \times 4 = 36 \text{ V} \\ 20 + 4 \times 6 = 44 \text{ V } 8 \text{ V} \end{aligned}\right\} 8 \text{ V}$$

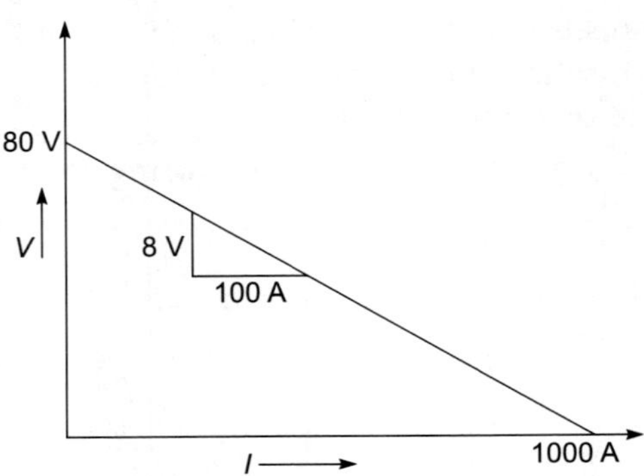

Figure 30.6 Voltage-current relationship

Current range (450 to 550) $\cong$ 100 A

$$\text{Slope} = \frac{8}{100} = 0.08$$

$$V = C - mI = C - \frac{8}{100}I$$

$$36 = C - \frac{8}{100} \times 550$$

$$C = 80$$

Thus V = 80 − 0.08I

$$V = C - 0.08I$$

$$V = 80 - 0.08I$$

When $V = 0$

$$I = \frac{80}{0.8} = 1000 \text{ A}$$

Short circuit current = 1000 A

Open circuit voltage = 80 V

Example 9: During an experimental investigation, the arc voltage has been found to be related with arc length as $V = (22 + 41)$ volts. The power source characteristics are as follows:

$$\left(\frac{V}{V_0}\right)^2 + 2\left(\frac{I}{I_0}\right) = 1$$

where, V_o = open circuit voltage and I_0 = open circuit current. In one of the observations, V_o = 90 volts and I_0 = 1000 amp. What will be the values of welding currents for arc lengths of 3 mm and 5 mm with corresponding arc voltage of 30 volts and 40 volts?

(Exercise problem in *Manufacturing Process*: Ghosh and Mallik)

Solution: Using the data given

$$\left(\frac{30}{90}\right)^2 + 2\left(\frac{I}{1000}\right) = 1$$

$$I_1 = \frac{8}{9 \times 2} \times 1000 = 444.44 \text{ amp}$$

$$\left(\frac{40}{90}\right)^2 + 2\left(\frac{I}{1000}\right) = 1$$

$$I_2 = \left(1 - \frac{16}{81}\right) \times \frac{1}{2} \times 1000 = 400.61 \text{ amp}$$

The values of welding currents are 444.44 amp and 400.61 amp, corresponding to arc voltages of 30 and 40 volts, respectively.

Review Questions

1. What are the important quantities that can be estimated using heat flow equation?
2. Write an expression for heat input in weld.
3. Derive the relation between weld cross section and energy input.
4. What are the equations used to calculate heat input rate for 2D and 3D problems?
5. Discuss the practical applications of heat flow equations.
6. Discuss about the width of HAZ.
7. How are cooling rates calculated in welding?
8. What is contact resistance in resistance welding?
9. How residual stresses occur in welds?
10. Briefly explain stress relieving treatment of welds.

31

CHAPTER

Basic Welding Metallurgy

31.0 INTRODUCTION

The study of welding metallurgy requires consideration of the following metallurgical phenomena that play important role in fusion welding. Both parent as well as electrode melt and then resolidify. This causes redistribution of micro constituents and the alloying elements in weld metal zone. The unmelted parent material is subjected to a complex heat treatment in the form of temperature gradient from the melting range to room temperature. The changes in structure that result due to welding depend upon:

- Welding process used
- Filler rod composition
- The conditions under which weld is made
- The chemical composition of base material
- Temperature and phase changes, which induce volume changes, resulting in residual stresses

31.1 SOLIDIFICATION

Solidification of liquid metal commences at specific points called nuclei, which grow into grain or clusters of crystals and the grains with differing orientations meet at grain boundaries. Thus, the starting point for a grain is nucleus. It may commence from the wall of container or from foreign solid particles suspended in liquids. These are examples of heterogeneous nucleation. In cases where such nucleation is not present, nuclei must be provided by liquid itself. So, nucleation or commencement of solidification process requires super cooling of liquid metal. In arc welding, the molten weld pool is surrounded by solid metal, which provides an ideal nucleation site. Thus, no homogeneous nucleation and super cooling is negligible. As the nucleation grows, some of them are pinched off, causing fewer but larger grains in the weld metal, which has a marked effect on its grain properties.

31.1.1 Gas Metal Reactions

The molten weld metal produced during arc welding must be protected from oxygen and nitrogen because their ingress into weld makes it brittle and porous. These gases ionize and become highly reactive in the electric arc and thus, become harmful. Protection of arc and weld pool is necessary in arc welding processes. This can be done by giving flux covering in manual arc welding processes. Deoxidizing materials, such as manganese and aluminum, prevent oxygen from reacting with carbon and prevents porosity. They can be provided with flux and filler wire. They react preferentially with oxygen, thus:

$$Si + 2O \rightarrow SiO_2$$

$$2Al + 3O \rightarrow Al_2O_3$$

$$Mn + O \rightarrow MnO$$

Such oxides may rise to surface as slag or remain entrapped in the from of complex oxides. But they have very little effect on mechanical properties due to being extremely fine and widely dispersed. Water vapor is another harmful gas to be avoided. It dissociates in arc to form oxygen and hydrogen. Oxygen makes weld brittle and porous. Hydrogen in excess causes weld metal embrittlement. It also forms cracks in HAZ of welded joints of high-carbon steels.

31.1.2 Slag Metal Reaction

Fluxes used in manual metal arc welding are converted to liquid slag in the arc and play an important role in the flux shielding process. Primary objective of a slag is to remove the deoxidation products or the non-metallic from the molten weld metal. The quantity depends on the welding consumables used, which, in most cases are silicates of aluminum, manganese, iron, etc. They rise to the surface of weld pool and become slag, however a minute proportion of slag remains entrapped in weld and depends on the quantity of non-metallic generated in weld pool. If arc environment is strongly oxidizing, it calls for more deoxidation and gives rise to more non-metallic. With reducing conditions, these impurities are provided in smaller amounts and thus, the weld metal is much cleaner. Another important function of the welding slags is to control the amounts of sulphur and phosphorus in weld metal. Basic slags produced from lime like fluorspar type fluxes are capable of lowering the content of these harmful elements in weld metal. This increases ductility and toughness of material. Acidic fluxes, because of their chemical nature, are unable to perform such useful functions. Slag metal reactions and the non-metallic give rise to weld defect known as hot cracking. It occurs during the last stages of weld metal solidification when interdendritic liquid has much lower freezing point than bulk dendrite. This small liquid region is subjected to shrinkage stresses which occur due to solidification processes and it develops interdendritic cracks. They are known as hot cracks because they occur at temperatures near bulk solidification temperatures. They are likely to occur whenever the compositional variations in the weld metal results in the production of low melting interdendritic liquid.

31.1.3 Solid Metal Reactions

Welds are subjected to rapid cooling rates and unless the carbon content is kept very low, the products are likely to be martensite and other hard constituents rather than an aggregate of ferrite and carbide. If hydrogen is also present in the arc atmosphere, it can promote cracking in weld metal. The cracking defects occur after the weld has cooled to low temperatures (even below 150°C), so it is known as cold cracking. Desired mechanical properties must be obtained in the welded condition, since heat treatment is ruled out. In some instances, heat treatment is given to postweld metal to relieve residual stresses and softens hardened heat affected zone. The tensile and other properties of weld metal whose carbon content is low, are obtained by solid solution and fine particle strengthening. Most weld metals naturally develop a fine solidification structure. The dendrites are large and yet the fine dendrite branching between major dendrite columns produces a fine microstructure, which is encouraged by the strengthening mechanisms. Three distinct zones are present in welded joints (Figure 31.1).

- Weld metal zone – Grain growth region
- Heat affected zone – Grain refined region, transition region
- Unaffected base metal or parent metal

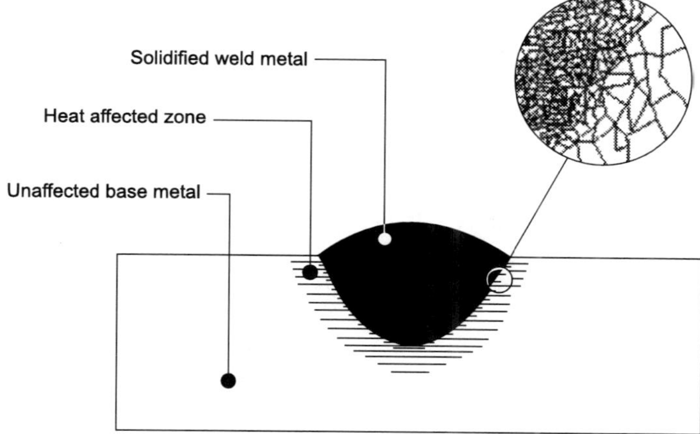

Figure 31.1 Various zones in welded joint

31.2 WELD METAL ZONE

Formed when weld metal solidifies from the molten state and is a mixture of parent metal and electrode. Its microstructure reflects the cooling rate. Table 31.1 shows the cooling rate and microstructure type.

TABLE 31.1 Cooling rates for different microstructures

Microstructure Type	Cooling Rate
Martensite	Fast
Fine pearlite	Medium
Coarse pearlite	Slow

Depending upon the composition and solidification rates, weld solidifies in cellular growth mode and dendritic growth mode. Both modes cause segregation of alloying elements and thus, weld metal is less homogeneous and therefore, has different properties from parent metal.

31.3 HEAT AFFECTED ZONE (HAZ)

It is composed of parent metal that did not melt but was heated at a high temperature for a sufficient period for grain growth to occur. The properties and microstructure have been altered by the heat of welding. The HAZ is subjected to complex thermal cycles (sudden heating followed by rapid cooling) and consists of graded structures ringing the weld bead. HAZ is the weakest of the welds and most welding failures originate in heat affected zone. HAZ is often defined by response of welded joints to hardness or etching effect tests. The HAZ in low-carbon steel, welded in one run with coated electrodes or by submerged arc process, contains three metallurgically distinguished regions:

- Grain growth region
- Grain refined region
- Transition region

Grain Growth Region

Grain growth region is immediately adjacent to the weld metal zone (fusion boundary). In this zone, parent metal has been heated to a temperature well above the upper critical (A_3) temperature (above 1150°C). This results in grain growth or coarsening of the structure.

The maximum grain size and the extent of this grain growth region increase as the cooling rate decreases, and are greatest for electroslag welds.

Grain Refined Region

Adjacent to the grain growth region is the grain refined zone. The refined zone indicates that in this region, the parent metal has been heated to just above the A_3 (950 to 1150°C) temperature, where grain refinement is completed and the finest grain structure exists.

Transition Zone

In the transition zone, a temperature range exists between the A_1 and A_3 (750 to 950°C) transformation temperatures, where partial allotropic recrystallization takes place. The structure of the transition zone of the low-carbon steel in which the ferrite grains have not been altered but the pearlite region is made much finer.

This change is produced by heating to the critical range which transforms the pearlite into austenite and by subsequent cooling, reforming the pearlite.

31.4 UNAFFECTED PARENT METAL

Outside the heat affected zone is the parent metal that was not heated sufficiently to change its microstructure. The typical grain structure of the parent metal remains same as it was before welding.

31.5 MICROSTRUCTURAL PRODUCTS IN WELDMENTS

In welding, the weld metal is quickly melted and then it resolidifies under the equivalent of chill casting conditions. Due to thermal variations, different microstructural products are developed in the weld metal and HAZ. The mechanical performance of the weldment depends upon the microstructures obtained in the weld metal and HAZ. Because steels are the most welded material therefore, most of the microstructural products discussed in this section relate to steels. Microstructural phases observed in Fe-C diagram (under equilibrium conditions) are:

1. **Austenite:** Austenite is the solid solution of carbon and/or other alloying elements (e.g., Mn, Ni, etc.) in gamma iron (FCC).
2. **Ferrite:** Ferrite is BCC iron phase with very limited solubility for carbon. Ferrite is soft and ductile.
3. **Cementite:** Cementite or iron carbide (Fe_3C) contains 6.67% carbon by weight. It is a very hard and brittle interstitial compound of low tensile strength but high compressive strength.
4. **Pearlite:** Pearlite consists of alternate lamellae of ferrite and somewhere between ferrite and cementite.

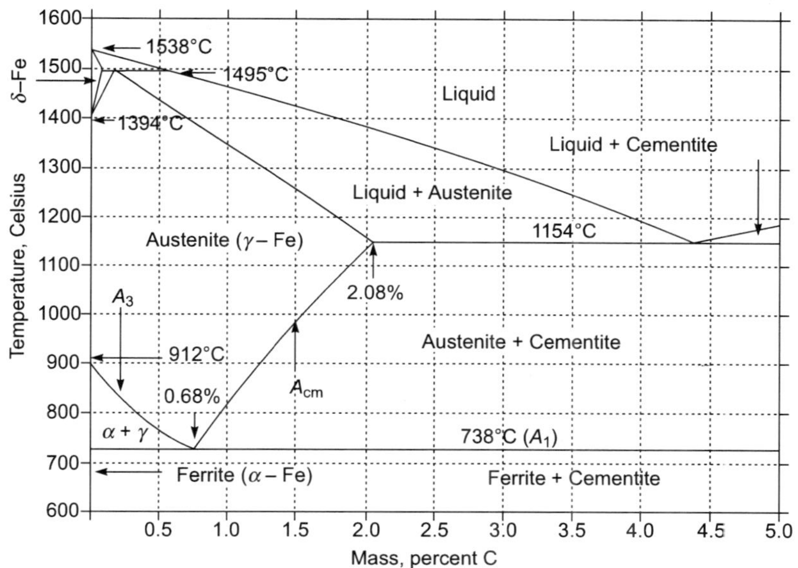

Figure 31.2 Iron-carbon diagram

Most fusion welding processes subject parent metals to high rates of heating and cooling. Fe-C diagram does not explain the decomposition of austenite under non-equilibrium conditions or conditions involving faster rates of cooling than equilibrium cooling as it is observed in welds. Several structures (e.g., martensite) not appearing on the equilibrium diagram may be found in microstructures of welds in steels.

On the other hand, TTT (time temperature transformation) diagram and CCT (continuous cooling transformation) are more practical diagrams. They show what structures can be expected after various rates of cooling and under continuous cooling conditions, as what actually happens in welding.

Different microstructural phases encountered in fusion welding of steel can be grouped into three types:

1. Primary microstructures
2. Secondary microstructures
3. Tertiary microstructures

31.5.1 Primary Microstructures in Weldments

Primary microstructure is the one obtained by direct solidification of the molten metal, usually results in a cast structure with columnar grains. The length axis of these grains is nearly parallel to the direction of the temperature gradient during solidification. Two major primary microstructural products encountered in carbon steel welds are delta-ferrite and austenite as already discussed.

31.5.2 Secondary Microstructures in Weldments

Secondary microstructure is the one obtained by transformation of solid phase austenite into different solid phases like: pearlite, bainite, martensite, grain boundary ferrite, ferrite side plates, lath ferrite acicular ferrite, polygonal ferrite and retained austenite.

Grain Boundary Ferrite: Grain boundary ferrite (GBF) is often known as proeutectoid ferrite. During their transformation, grain boundary ferrites are formed from the original austenite, which appear on the grain boundary of the prior austenite grains.

Ferrite Side Plates: Ferrite side plates, which are also known as Widmanstatten ferrite or ferrite with aligned M-A-C (martensite, austenite, cementite), appear as long needles with length to width ratios of at least 20:1. This microstructural constituent grows from grain boundary ferrite into the austenite as packers of parallel plates. Ferrite side plates, like the grain boundary ferrites, are also detrimental to toughness, due to the possible presence of precipitated carbide, retained austenite and martensite along the ferrite plates.

Lath Ferrite: It is an intragranular austenite transformation product resembling lower bainite. It is found amongst acicular ferrite or side plate structures. Unlike ferrite side plate, lath ferrite is not attached to grain boundary ferrite. Lath ferrite is also detrimental to toughness on the similar considerations as the ferrite side plates.

Polygonal Ferrite: Polygonal ferrite occurs in the form of coarse ferrite islands inside the prior austenite grain. Polygonal ferrite like grain boundary ferrite, is detrimental to toughness because of its coarse grain structure.

Acicular Ferrite: Acicular ferrite is formed in the interior of the original austenite grains by direct nucleation from the inclusions, resulting in randomly oriented short ferrite needles with a basket weave pattern. This interlocking nature together with its fine grain size provides maximum resistance to crack propagation, hence increasing the toughness and strength of the weldment.

Bainite: Bainite is the product of intermediate transformation with fast or very fast cooling rates. While fast cooling rate is said to give coarse or upper bainite, very fast cooling rates result in fine or lower bainite structures. Upper bainite microstructure consists of ferrite containing carbides aligned along ferrite axis whereas, lower bainite microstructure may be likened to a ferrite containing small carbide grains, aligned parallel to each other and at 60° to the ferrite axis.

Martensite: If the steel is quenched from austenite condition, the excess carbon will be trapped within the lattice to form the supersaturated solid solution of carbon in iron called martensite. It has a fine needle-like structure, which is very strong and hard but is very brittle. Martensite crystallizes in the tetragonal system, with a body centered unit cell very close to a cube.

Retained Austenite: During quenching of steel, the transformation from austenite to martensite may not take place completely. Thus, some austenite may be retained at room temperature, known as retained austenite. The retained austenite may be transformed to martensite by the tempering process. Presence of martensite and retained austenite adversely affect the toughness of the ferritic weld metal.

31.5.3 Tertiary Microstructures in Weldments

Tertiary microstructure is encountered only in multirun welds where, due to successive runs the previous runs, are heated to austenite region and thus, result in products of different fineness affecting the mechanical strength and toughness due to the normalizing effect of heating by successive runs.

31.6 DETAILED DISCUSSION ON VARIOUS ZONES IN A WELDMENT

When filler metal is added to a weld, its composition is modified by mechanical mixing with material melted from the base metal. This region, which constitutes the majority of the weld, is a composition intermediate between that of the base metal and filler metal and depends upon the degree of dilution with base metal. This zone is known as the composite zone.

Close to the fusion line, a small boundary layer exists at the fusion line where the melting front has just touched. This layer is called the unmixed zone, since it has essentially no dilution. The unmixed zone is located in the boundary layer, but is sometimes found well inside the fusion zone. The unmixed zone has been found to be subjected to corrosion attack.

31.6.1 Micro Segregation

During solidification of an alloy, the morphology of the emerging solid is determined by thermal gradient *G* and growth rate *R*, besides the composition (solute content *C* and partition coefficient *k*). Partitioning of solute into the liquid during solidification leads to a region ahead of the solid-liquid interface, which has cooled below its solidification temperature by virtue of solute diffusion. This is illustrated for a binary alloy of *A* and *B* in Figure 31.3. This region of constitutional supercoiling makes the moving *SL* interface unstable and the growth front develops into elongated cells or dendrites, depending on the solidification parameters and composition.

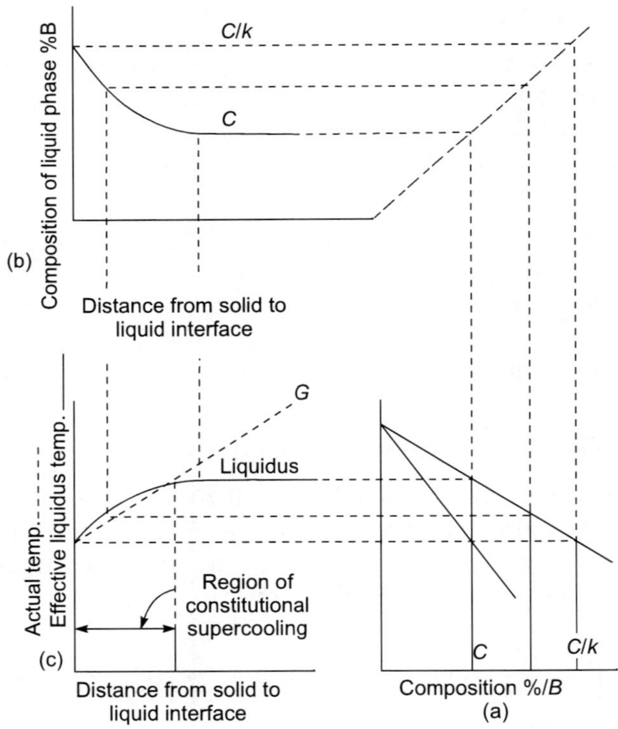

Figure 31.3 Schematic illustration of constitutional supercooling in a binary alloy: (a) binary phase diagram showing portion of alloy, (b) variation of liquid phase composition with distance from the S-L interface, and (c) variation of effective liquidus temperature showing constitutionally supercooled zone

This is illustrated in Figure 31.4, where the growth morphologies have been mapped as a function of C_o/k and G/R. It is observed that for higher solute contents or partitioning tendency ($k < 1$), the structures are more dendrites. More refined microstructures are obtained at high growth rates and thermal gradients associated with high energy density process, such as EBW or LBW, while increasing the heat input increases segregation and dendrite arm spacing.

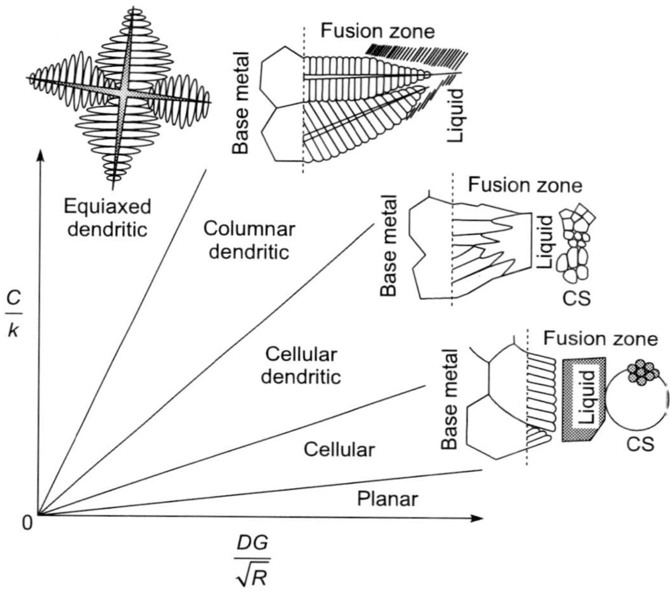

Figure 31.4 Solidification growth morphology of alloys as a function of compositional and thermal parameters

Due to solute partitioning between the solid and the remaining liquid, which takes place according to equilibrium phase diagram, the cell/dendrite cores have a different composition from the intercellular region. This micro segregation, which can be quite severe during welding of alloys because of the high solidification rates, can lead to deviations from the design composition in localized regions, with implications on the strength and corrosion behavior of the weld. For estimating micro segregation, a volume element at the tip of a columnar dendrite/secondary arm is usually considered. The solute profiles that can result, depend upon the extent of convective mixing in the weld pool and diffusion in the solid state, which equalizes the composition behind the *S-L* interface, as given in Figure 31.5. In complete equilibrium solidification, a liquid of composition, C_0, emerges as a solid of the same composition. When diffusion in the solid is neglected and only diffusion in the liquid is taken into account, profile *b* is obtained, while profile *c* results for complex mixing in the liquid and no diffusion in the solid. Actual weld solidification falls somewhere between the two extremes, represented by *b* and *c*, as the diffusion in the solid is not negligible and the profile approximately follows that of *d*, which is closer to the profile *c* than that of *b*.

Segregation of solute and impurity elements during welding is undesirable as it introduces homogeneity in mechanical properties and decreases corrosion resistance of the weld metal. The other important deleterious effect of segregation is the enhanced tendency for cracking and micro fissuring. The segregation of impurity elements such as *S* and *P* is particularly strong during welding, since they exhibit poor solubility in the solid and have low partition coefficients (*k* between 0.01 and 0.1). This strong segregation leaves the inter-crystalline regions in the last solidifying pockets highly enriched in solute. This leads to the formation of low melting eutectics, which may solidify as much as a few hundred degrees below the solidus temperature of the alloy.

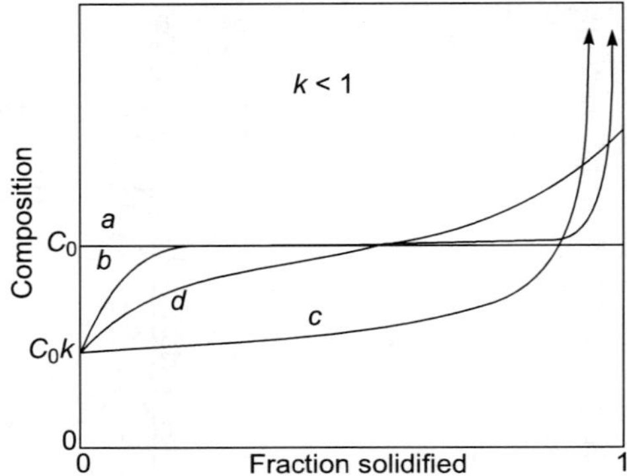

Figure 31.5 Schematic diagram of compositional variation in a volume element solidified under (a) equilibrium conditions, (b) diffusion in the liquid considered, (c) full convective mixing in the liquid considered, and (d) actual case, where full conventional mixing in the liquid and diffusion in the solid influence composition CD is the initial composition of the alloy and *k*, the partition coefficient

31.6.2 Hot Cracking

Under the action of shrinkage stresses in a restrained weld, the already solidified part can crack when wetted by low melting eutectic liquid. The ratio of interfacial energies of the solid and liquid phases is an important determinant of hot cracking during welding. The effect of this ratio on wetting angle is given by:

$$\frac{\gamma_{SL}}{\gamma_{SS}} = \frac{1}{\left(2\cos\frac{\theta}{2}\right)}$$

where, γ_{SL} and γ_{SS} are the interfacial energies of the *S-L* and *S-S* boundaries, respectively and θ is the wetting angle. When wetting angle is *z*, wetting of *SS* boundary is enhanced and cracking is promoted. The variation of wetting angle with change in the interfacial energy ratio is shown in Figure 31.6. The wetting angle has been found to be sensitive for segregating highly surface active elements, such as *S*, *O* and *Se*.

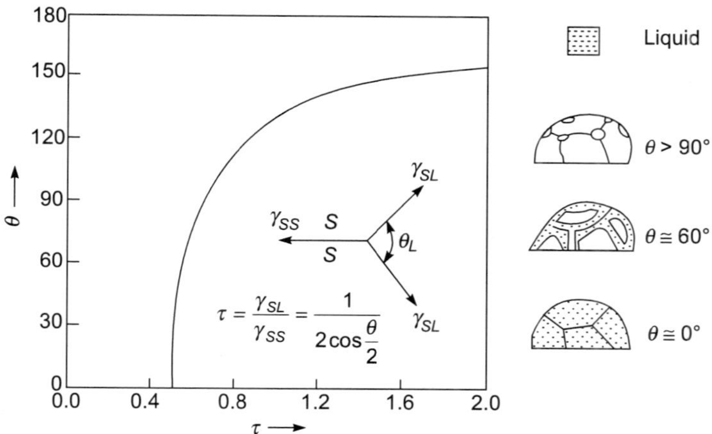

Figure 31.6 Variation of dihedral angle θ with interfacial energy ratio r

The conditions conducive to hot cracking in a particular alloy system are not easy to predict, since the characteristics of the liquid films and their interaction behavior with the solid are not readily available parameters. This places emphasis on actual weld ability testing for establishing the composition limits and welding conditions to avoid cracking.

31.6.3 Partially Melted Zone: Liquation Cracking

Completely surrounding the moving weld pool is a region where the peak temperatures experience fall between the liquidus and the effective solidus. This zone is shown in Figure 31.7. The segregation invariably present in commercial alloys ensures that the effective solidus is always below the equilibrium solidus for the nominal alloy composition. In addition, interactions between precipitates, inclusions and the matrix cause a depression of the solidus. At the high temperatures near the fusion line, rapid grain boundary migration would occur as a result of natural grain growth. However, the liquated regions, since they have a composition similar to that of the matrix, readily wet the intercepting boundaries and form molten films which arrest further grain boundary migration. During the migration process, the grain boundaries are said to "sweep" along with them impurity elements, such as *S*, *P*, *Si*, etc., which are very

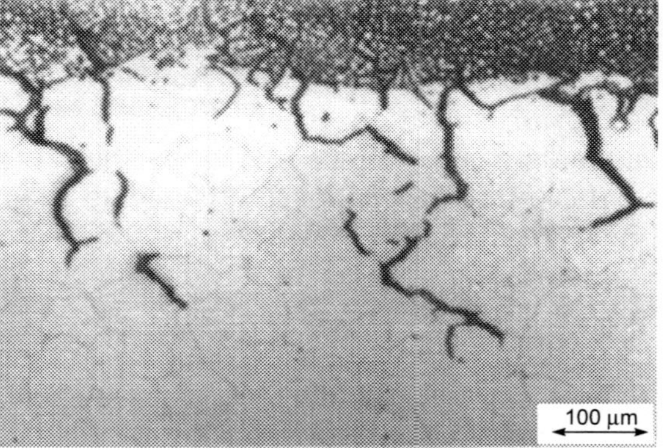

Figure 31.7 Liquation cracking in the partially melted zone of the HAZ in nitrogen-bearing 316L stainless steel

effective in depressing the melting point of the grain boundary. Under these conditions, any restraining force would lead to cracking, as shown in Figure 31.7, which illustrates liquation cracking in a nitrogen-bearing type 316L stainless steel. The thickened grain boundaries along the partially melted zone, are clearly visible. However, cracking extends quite deeply into the HAZ.

31.6.4 Weld Pool Convection-Weld Penetration

Convection is responsible for the shape of the weld puddle and ensures uniform distribution of heat and solutes within the melt zone. Convection currents are set up in the weld pool by (i) surface-tension induced flows, (ii) electromagnetic forces created by the arc current and magnetic fields, (iii) buoyancy forces created by density differences between hot and cold metal, and (iv) aerodynamic drag forces by gas flow on the surface of weld pool. Of these forces surface tension, tension forces are dominant in giving shape to the penetration profile during welding. The origin of surface tension driven forces is illustrated in Figure 31.8. The variation of surface tension with temperature is a strong function of the concentration of surface-active elements in the weld pool. When the coefficient of surface tension is negative, as in Figure 31.8 (a), the centre of the weld pool directly under the arc experiences a lower surface tension by virtue of its higher temperature. As the cooler peripheral regions have greater surface tension, the surface flow of liquid sets up an outward current. Such a movement, which produces weak convection currents, results in shallow weld puddles. Poor and inconsistent weld penetration occurs when there is insufficient concentration of surface-active elements. The inconsistency in penetration profile occurs in the absence of a dominant factor among the various convective forces. When element, such as *S* or *O*, is present above the critical concentration, a positive coefficient of surface tension prevails (Figure 31.8 (b)) and the direction of convection is reversed. Now the surface flow is inwards the centre of weld puddle. The hot liquid under the arc is pushed down, which results in a digging action, increasing the depth-width ratio and providing consistent weld profiles.

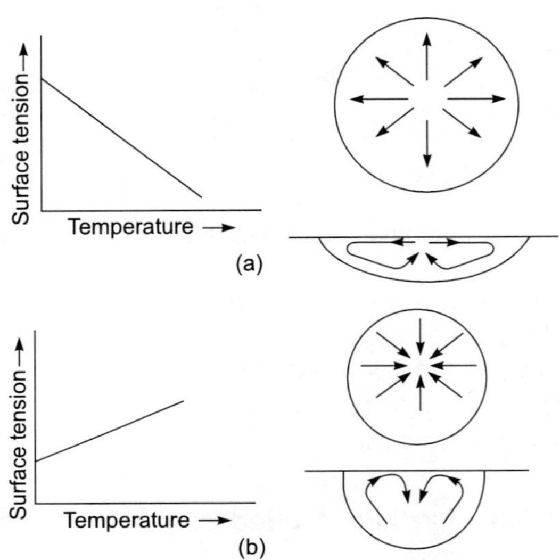

Figure 31.8 Different convective flow patterns produced by different temperature coefficients of surface tension (a) positive, (b) negative

31.6.5 Heat-Affected Zone Microstructure

The relevant portion of the iron-carbon phase diagram along with a schematic sketch of a weld and HAZ is shown in Figure 31.9. During heating part of the weld thermal cycle, the peak temperature to which base metal is heated varies with distance from the fusion line, as shown in the figure. If the peak temperature exceeds the lower critical temperature A_1, ferrite (stable at room temperature with a BCC crystal structure) transforms into austenite (high temperature phase with FCC crystal structure). In the region of the HAZ, where the temperature exceeds the upper critical temperature A_3, this transformation will be complete and an austenitic microstructure is formed. At temperatures higher than 1300°C, extensive growth of the austenite grains takes place. The region of the HAZ where extensive grain growth of the weld metal takes place is referred to as coarse grained HAZ (CGHAZ). The region of the HAZ next to it, where peak temperature is in the range of 900 to 1200°C and austenite grain size remains small, is called fine grained HAZ (FGHAZ). The region of the HAZ heated to the temperature range in between the two critical temperatures A_1 and A_3, is often referred to as inter-critical HAZ (ICHAZ).

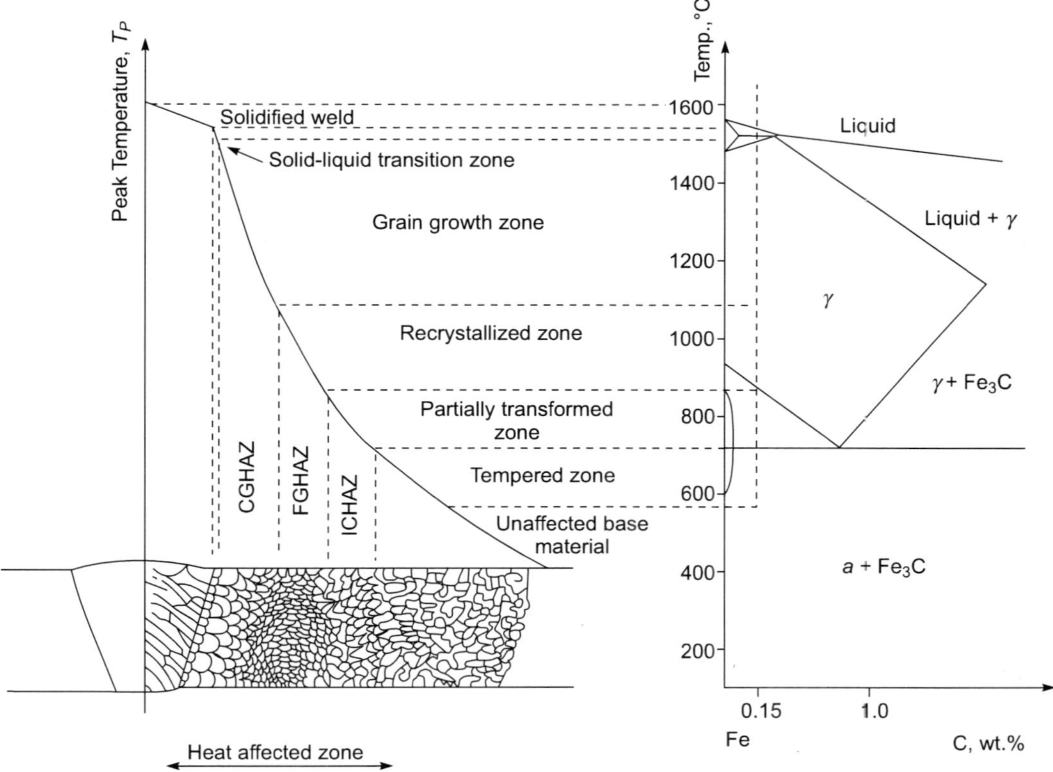

Figure 31.9 Schematic diagram of various subzones of the heat affected zone approximately corresponding to the alloy of 0.15 wt C indicated on Fe-C equilibrium diagram

During the cooling cycle, the austenite formed at high temperature transforms back to ferrite or other metastable phases and this transformation is best represented with the help of a continuous cooling temperature (CCT) diagram. A typical CCT diagram is shown in Figure 31.10. The lines shown in the figure are the boundaries which define the start and finish of transformation of austenite into various phases like, ferrite, pearlite (a mixture of ferrite and carbides), bainite or martensite. An actual cooling curve can be inserted into this diagram and the phases formed in that region of the weld to which the cooling curve corresponds can be obtained. It can be seen that for low cooling rate, as experienced in high heat input welding, i.e., submerged arc welding, electroslag welding, etc., or in welds produced with sufficient preheat, austenite transforms predominantly into a mixture of ferrite and carbides. If the cooling rate is very high, this transformation is prevented and low temperature transformation, bainite or martensite are formed.

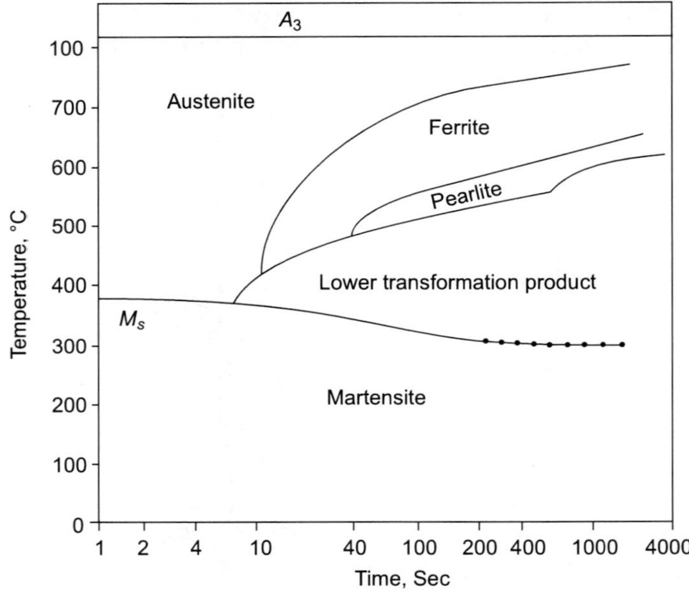

Figure 31.10 Example of a CCT diagram for a medium strength C-Mn steel

Zone of the weld, the phases formed from transformation may also vary with distance. Further, CCT curves defining start and finish of ferrite transformation shift toward the right with increase in alloy content and prior austenite grain size. As both grain size and cooling rate are maximum for CGHAZ, the probability of this region transforming into bainite or martensite is very high. This explains why CGH has maximum hardness and poor toughness properties compared to the rest of the HAZ.

As indicated above, the effect of alloying elements on the CCT diagram is to shift the ferrite transformation curves towards the right. Thus, for a given cooling rate, the HAZ of an alloy steel contains higher volume fraction of bainite than that of its steel with low *c* content and without any significant alloying elements. This property of the steel to produce a hard microstructure when cooled from austenite phase is often referred to as hardenability and it increases with alloy content and austenite grain size. Hence, in high-carbon or high-alloyed steel welds, HAZ microstructure is fully martensitic irrespective of the weld heat input or the preheat temperature used to reduce the cooling rate. For example, in the case of SMAW, HAZ microstructure of mild steel is a mixture of carbides and ferrite, while that of 2.25cr-lMo steel is fully bainitic and that of 9Cr-lMo steel is fully martensitic even though welding is carried out without any preheating in the case of the first and with preheating in the other two.

Review Questions

1. What factors in welding cause changes in the structure of the material?
2. Explain the process of solidification.
3. Discuss solid metal and slag metal reactions.
4. What is heat affected zone?
5. Discuss different microstructural phases in the fusion welding of steel.
6. Explain primary microstructures in weldments.
7. Explain secondary microstructures in weldments.
8. Describe tertiary microstructures in weldments.
9. What is hot cracking?
10. Explain liquation cracking.

32
CHAPTER

Residual Stress and Distortion

32.0 INTRODUCTION

Residual stresses are self-balancing, internal system of stresses arising from non-uniform mechanical of plastic flow. These stresses exist in a body when all mechanical or thermal strains are removed.

During welding, a weldment experiences complex temperature changes that cause transient thermal stresses and non-uniform distribution of elastic strains that are produced in the weld and in the regions near to it. Due to this, both distortion and residual stresses result in the welded part. These stresses are elastic in nature and exist in a body without any external force. Hence, they are self-equilibrating type, i.e., the summation of the forces and moments about any point in the cross section of the body is equal to zero.

The most obvious manifestation of the mechanism producing residual stress is distortion, which cannot be considered in isolation from residual stress. In simple terms, high conditions of restraint during welding limit distortion but impose large local strains at the welded joint, giving rise to higher residual stress, which greatly increases the risk of failure. Conversely, low restraint does not inhibit the local contractions, which result in larger displacements or distortion and lower levels of residual stress.

32.1 BASIC MECHANISM (FIGURE 32.1)

Residual stresses in welded components arise directly from the localized shrinkage associated with the cooling down of the hot weld metal which is restrained, either internally or externally, by the cold sections of the fabrication surrounding the weldment. But, this does not accurately describe the true mechanism, which in reality, is a highly transient condition, being modified continually, depending on the nature of the joint, the number of passes involved, and to some extent, the welding parameters themselves. At any moment, any one of the influencing parameters could be different, making study of the mechanism an extremely complex exercise.

The cycle starts at room temperature, *O*, the initial thermal expansion is restrained by the surrounding cold material and thus, generating plastic compressive stresses until the yield strength reaches at position *A*.

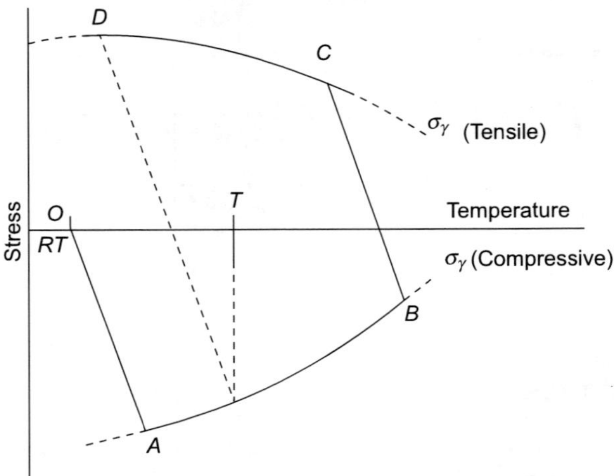

Figure 32.1 Basic mechanism of residual stress

With further increase in temperature, plastic strains occur following the material's characteristic compressive yielding curve. On cooling the response is once again elastic, but opposite to the heating cycle, until the tensile yield strength is achieved. Cooling back to room temperature then follows the material's characteristic tensile yield strength curve, thus generating yield magnitude residual stresses at room temperature. Also, the magnitude of residual stress is a function of the temperature difference experienced. In the case considered, any temperature rise exceeding *T* will always generate yield magnitude elastic residual stresses, whereas at lower temperatures, the shrinkage strains are not sufficient to reach the tensile yield limit and will result in less than yield magnitude elastic residual stress. The critical temperature difference is a function of material yield strength, which for typical mild steel is of the order of 175°C, with reference to room temperature. The peak magnitude of the longitudinal residual stress is usually of the order of yield stress level. The magnitude of the transverse residual stress is much lower than the magnitude of the longitudinal stress component. Often, a ratio of 1:4 to 1:3 is reported for the magnitude of both residual stress components.

32.2 TYPES OF RESIDUAL STRESSES (FIGURE 32.2)

Residual stresses are commonly classified into two groups as either macro or micro. Macro residual stresses, or residual stresses of the first kind, are those of engineering nature and which are measured over a gage length that encompasses several grains. Micro residual stresses, or residual stresses of the second kind, relate to stress systems set up by microstructural inhomogeneities, which can either be confined within a single grain or a particular set of grains of the same preferred orientation.

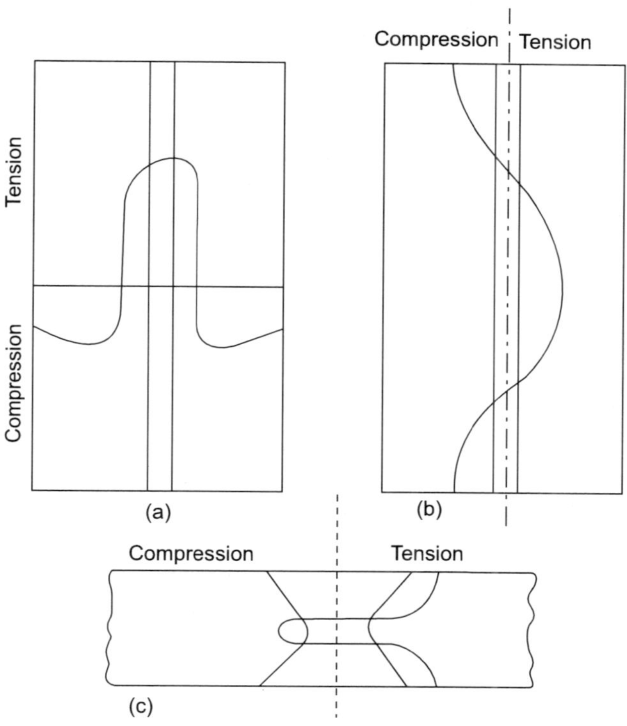

Figure 32.2 Typical residual stress distributions in plates without fixed ends (a) longitudinal, (b) transverse, and (c) through thickness

Though doubts do exist, over which of these mechanisms is to be considered important to stress corrosion cracking (brittle fracture or fatigue initiation) it is likely that both types can contribute depending on the situation. However, it is suggested that for all mechanisms, it is the first kind that is of fundamental interest. The two measurement techniques used in this study, namely, strain relaxation technique and X-ray diffraction technique, measure macro residual stresses.

32.3 FACTORS INFLUENCING RESIDUAL STRESSES

Among the various factors influencing residual stress magnitude and distribution, the material properties, specimen dimension, heat input, welding process and welding sequence are considered to be the most important ones.

32.3.1 Material Properties

The magnitude of residual stresses is affected by:

1. The temperature distribution in the weldment characterized by thermal conductivity and diffusivity.
2. The thermal expansion characteristics of the material.
3. The mechanical properties of the material at elevated temperature.

Residual stresses as high as the yield stress of the weld metal and the base metal are produced on carbon steel weldments. Residual stresses near the weld zone in aluminum joints are as high as the yield stress of the metal. In the case of high strength steels, the residual stresses in the welds are often considerably lower than the yield stress values of the parent material.

32.3.2 Specimen Dimension

Reports indicate that the specimen length in the case of low-carbon steel, butt welds should be at least longer than 457 mm to produce high tensile stresses in the longitudinal direction. Longitudinal residual stresses become uniform in the central region for welds longer than 457 mm. Specimen length seem to have little effect on the transverse residual stresses. The effect of specimen width is negligible as long as it is several times the width of the residual stress zone. As far as the effect of the plate thickness is concerned, it is found that when weldments are made in plates thicker than 25 mm, residual stresses in the thickness direction can become significant. However, lack of a simple and reliable measuring technique appears to limit the studies conducted so far.

32.3.3 Welding Processes

It is generally believed that similar residual stresses are produced in welds made by various arc processes, including the shielded metal arc, submerged arc, gas metal arc and gas tungsten arc processes.

32.3.4 Welding Sequence

As far as residual stresses along the weld are concerned, the effect of welding sequence is minor. High tensile longitudinal stresses are found in all welds tested. Block welding sequences generally produce less shrinkage, less strain energy, and less reaction than multilayer sequences.

32.3.5 Different Sources of Residual Stresses

Welding residual stresses arise not only because of the variation in shrinkage of differently heated areas but also as a result of a surface quenching effect and transformation of austenite.

(a) Residual Stresses Owing to the Shrinkage Process of the Seam and HAZ (Figure 32.3)

An important source of residual stress is the difference in shrinkage of differently heated and cooled arm of a welded joint. The weld metal subjected to the highest temperature tends to contract more than all other areas, but this contraction is hindered by the cooler parts of the joint. Thus, the weld metal is subjected to tensile stresses in the longitudinal direction, and it increases with increase in yield strength of the material as a result of decrease in temperature. Due to the shrinkage effect, stresses also arise in a direction perpendicular to the seam. The final stresses remaining after cooling, are called residual stresses, due to the shrinkage process. The magnitude to the transverse stress component at the centerline of the seam ($Y = 0$) is much lower than the longitudinal stress component. Comparing different materials, higher residual stresses brought about by welding should be found in materials with a larger product of Young's modulus and coefficient of linear thermal expansion.

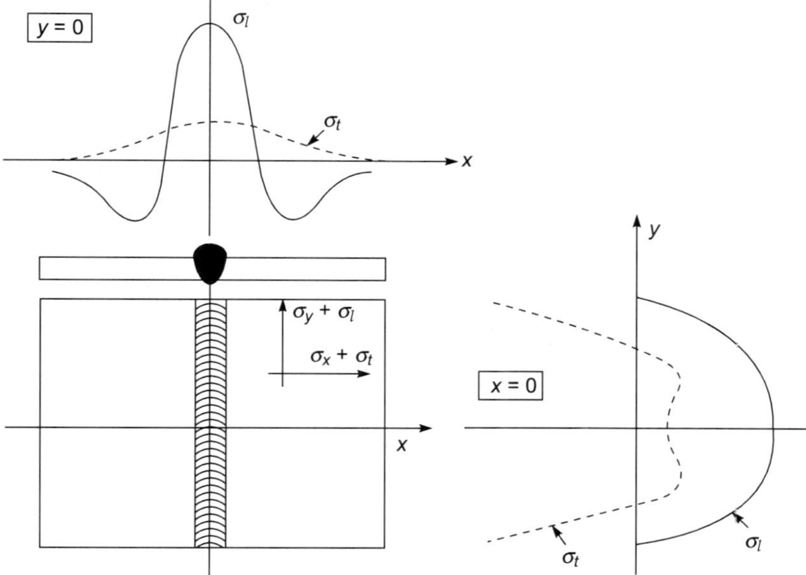

Figure 32.3 Longitudinal and transverse residual stresses showing shrinkage along lines parallel (y-axis) and perpendicular (x-axis) to seam

(b) Residual Stresses Owing to the More Rapid Cooling of the Surface

During the cooling process, the surface layers of the weld and the highly heated areas close to it may cool more rapidly—even with air cooling—than the interior of the weld. Due to this reason, temperature differences are present not only over the width of the joint but also across the thickness. As time progresses, an increasing temperature difference develops between the surface and the interior. Thus, thermal stresses arise over the cross section, which can lead to inhomogeneous plastic deformation and eventually to residual stresses. The more rapid cooling of the surface is called quenching effect, and the residual stresses resulting from this process are quenching residual stresses. Plastic deformation at elevated temperatures is necessary for the formation of quenching residual stress. If the surface quenching effects were to be the only source of residual stresses, compressive residual stresses would be obtained at the surface of the highly heated areas, especially the seam, and would be in equilibrium with the tensile residual stresses in the inner part of the seam.

(c) Residual Stresses Owing to a Phase Transformation

During cooling of welded steel plate phase transformation from austenite to ferrite, bainite or martensite will occur, either at a certain temperature or over a temperature range. Due to transformation, there is an increase in specific volume and so the material in the seam and the HAZ, which is being transformed, tends to expand. But the expansion is hindered by the cooler material not being transformed. Thus, the area being transformed is subject to compression, if the transformation temperature is sufficiently low, so that the material has a marked yield strength after transformation. Therefore, compressive residual stresses parallel to the seam should be expected in this area and for reasons of equilibrium, tensile residual stresses should exist in the region not being transformed. However, the transformation will occur with different heating temperatures at different times. Therefore, only a narrow strip of the weld material being transformed, would be subjected to compressive residual stresses. In all areas being transformed earlier, for instance at the top—tensile residual stresses will increase if this transformation occurs mainly at low temperature after the weldment has attained strength. Transformation residual stresses will appear, especially if bainite or martensite is formed. These transformation stresses diminish as the area being transformed becomes wider.

(d) Superimposition of Residual Stresses Owing to Shrinkage, Quenching and Transformation

In reality, the various stress sources are not independent of one another. At least two or even three different kinds of stress in a cooling welded joint are superimposed, leading to a complicated total state of stress. Therefore, statements concerning the finally resulting residual stress state can be made only on distinct assumptions. The most simple is that a linear superimposition of residual stresses owing to different sources can describe the total residual stress state.

Figure 32.4 schematically illustrates the linear superimposition of residual stresses owing to shrinkage, quenching and transformation. In all instances of the transverse stress components plotted as a function of the distance from the centerline of the weld, maximum tensile residual stresses are not present at the centerline of the seam but at the top. These tensile stress maxima represent the residual stresses owing to shrinkage. The difference between the maximum tensile stress and the stress value at the centerline represents either the surface quenching type of residual stress, the transformation type or both together.

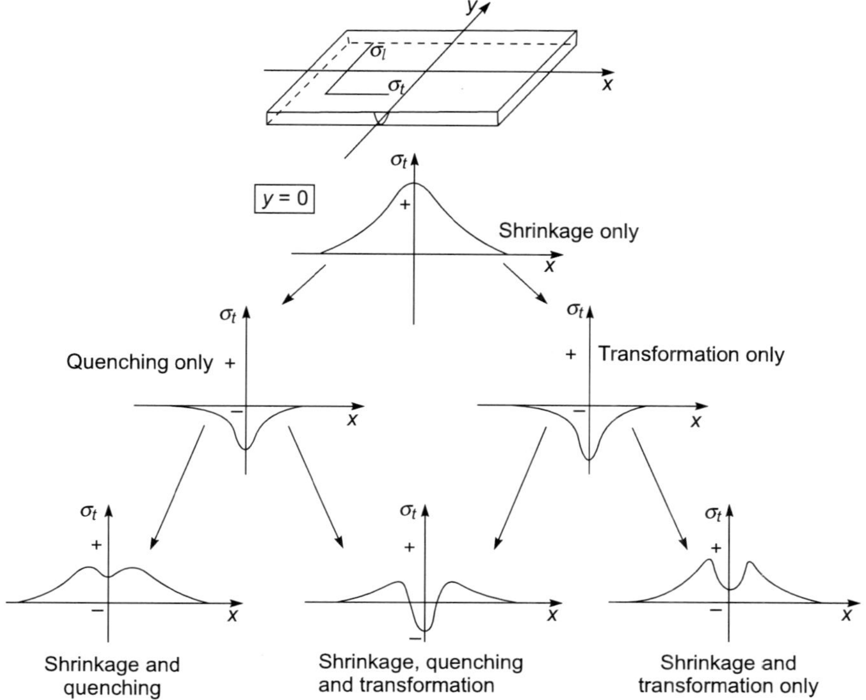

Figure 32.4 Schematic representation of superimposition of transverse residual stresses owing to shrinkage, quenching and transformation

32.4 EFFECTS OF RESIDUAL STRESSES ON PERFORMANCE

Residual stresses decrease the fracture strength of welded structures only when certain conditions exist; however, the loss of strength can be drastic under these conditions. In general, the effects of residual stresses are significant if fractures take place under low applied stress.

32.4.1 Changes of Residual Stresses in Weldments Subjected to Tensile Loading

Changes in residual stresses when a welded butt joint is subjected to tensile loading is shown in Figure 32.5. Curve *A* shows the lateral distribution of longitudinal residual stress in the as-welded condition. When uniform tensile stress σ_1 is applied, the stress distribution will be shown by curve *B*. The stresses in areas near the weld reach yield stress, and most of the stress increase occurs in areas away from the weld. When the tensile applied stress increases to σ_2, the stress distribution will be as shown by curve C. As the level of applied stress increases, the stress distribution across the weld becomes more even (i.e., the effect of welding residual stress on the stress distribution decreases).

When the level of applied stress is further increased, general yielding takes place (i.e., yielding occurs across the entire cross section). The stress distribution at general yielding is shown by curve *D*. Beyond general yielding, the effect of residual stresses on the stress distribution virtually disappears.

The next consideration is the distribution of residual stresses after the tensile loads are released. Curve *E* shows the residual stress which remains after unloading when the tensile stress σ_1 is applied to the weld and then released. Curve *F* shows the residual stress distribution when the tensile stress σ_2 is applied and then released.

Compared to the original residual stress distribution curve *A*, residual stress distributions after loading and unloading, are less severe. As the level of loading increases, the residual stress distribution after unloading becomes more nearly uniform (i.e., the effect of welding residual stress on the overall stress distribution across the welded joint decreases).

Based on this analysis, the effects of residual welding stresses may be summarized as follows:

1. The effect of residual welding stresses on the performance of welded structures is significant only in the phenomena which occur under low applied stress, such as brittle fracture, fatigue, and stress corrosion cracking.
2. As the level of applied stress increases, the effect of residual stresses decreases.
3. The effect of residual stress is negligible on the performance of welded structures under applied stresses greater than the yield strength.
4. The effect of residual stress tends to decrease after repeated loading.

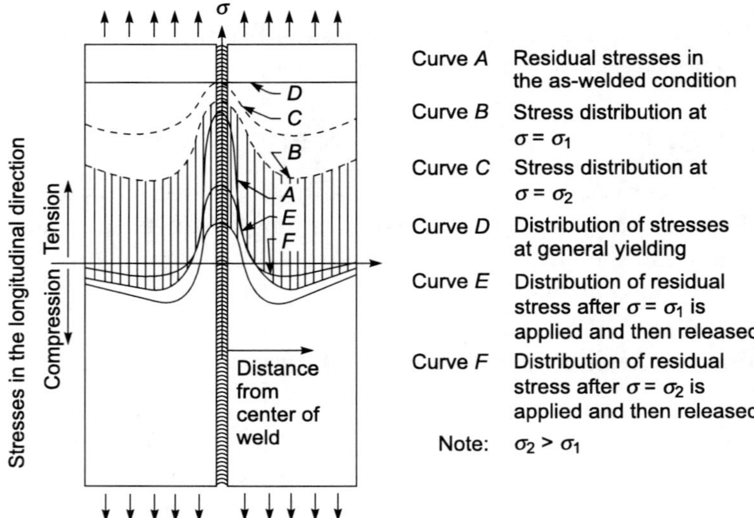

Figure 32.5 Effect of uniform external loads on residual stress distribution of welded butt joints

32.4.2 Brittle Fracture or Unstable Fracture Under Low Applied Stress

Extensive studies have been conducted on the effects of residual stresses on brittle fracture of welded steel structures.

Data obtained from brittle fractures in ships and other structures differ from experimental results obtained with notched specimens. Actual fractures have occurred at stresses far away below the yield strength of the material. However, the nominal fracture stress of a notched specimen is as high as the yield strength, even when the specimen contains very sharp cracks. Under certain test conditions, complete fracture of a specimen has occurred even though the magnitude of applied stress was considerably below the yield stress of the material.

Figure 32.6 shows the general fracture strength tendencies of welded carbon steel specimens at various temperatures, and the effects of a transverse sharp notch and residual stress on fracture strength.

When a specimen does not contain a sharp notch, fracture will occur at the ultimate strength of the material as represented by curve *PQR*. When a specimen contains a sharp notch (but no residual stress), fracture will occur at the stresses represented by curve *PQST*. When the test temperature is higher than the fracture transition temperature, T_f, a high energy (shear type) fracture occurs at high stress. When the test temperature is below T_f, the fracture appearance changes to a low energy (cleavage) type, and the stress at the fracture decreases to near the yield strength. When a notch is located in areas where high residual tensile stresses exist, these various types of fracture can occur:

1. At temperatures higher than T_f, fracture stress is the ultimate strength (curve *PQR*). Residual stress has no effect on fracture stress.
2. At temperatures lower than T_f; but higher than the crack arresting temperature, T_a, 21 may initiate at a low stress, but it will not propagate.
3. At temperatures lower than T, one of two phenomena can occur, depending upon the stress level at fracture initiation:
   (a) If the stress is below a critical stress, *VW*, the crack will stop after running a short distance.
   (b) If the stress is higher than the critical stress, *VW*, complete fracture will occur.

The effect of residual stress on unstable fracture has been analyzed using fracture mechanics concepts. Such an analysis shows that an unstable fracture can develop even from small cracks that would normally be stable, if residual stresses were not present. The sequence of events follows.

When a small subcritical flaw lies in a region free of or containing compressive residual stresses, these stresses do not contribute to the stress intensity at the tip of the discontinuity. If the flaw tip should occur in a region of tensile residual stress, then the residual stress would add to the applied stress and increase the stress intensity around the tip of the discontinuity. The increased stress intensity may cause the flaw to crack, and the crack to extend until the tip is outside the residual stress region. At this point, the crack may arrest or continue to grow, depending on the crack length and the stress intensity. Thus, residual stresses are local stresses and only affect fracture performance within the region of the residual stress field.

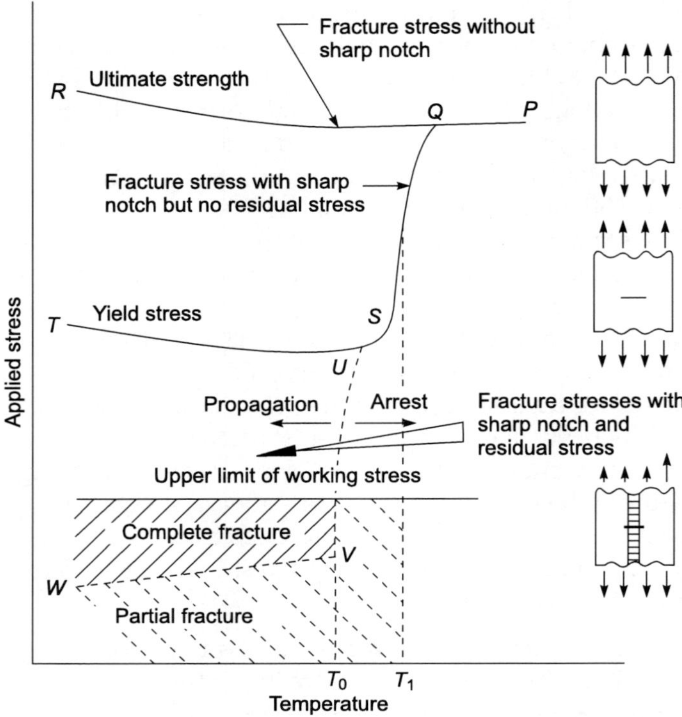

Figure 32.6 Effect of sharp notch and residual stress on fracture strength

32.4.3 Buckling Under Compressive Loading

Failures due to instability or buckling sometimes occur in metal structures, composed of slender bars, beams, or thin plates, when they are subjected to compressive axial loading, bending, or torsional loading. Residual compressive stresses decrease the buckling.

Strength of a metal structure and in addition, initial distortions caused by residual stresses also decrease the buckling strength.

32.4.4 Columns Under Compressive Loading

Residual stresses can significantly reduce the buckling strength of columns, fabricated by welding, particularly when made from universal mill plate. However, the effects are similar to comparable hot rolled columns. Such columns also have very high residual tensile stresses at the intersections of the flanges and webs, and residual compressive stresses at the outer ends of the flanges.

On the other hand, the columns made from oxygen cut plates normally have residual tensile stresses in the outer areas of the flange. Columns fabricated from flame cut plates and stress relieved columns exhibit better buckling performance (Figure 32.7) than columns fabricated from universal mill plates. In this case, the residual stresses that result from preparing plate by oxygen cutting and from subsequent welding into the column section will combine to counterbalance each other from a buckling strength viewpoint. The performance of these columns should be equivalent to hot rolled ones. Almost all welded fabricated columns use flame cut plate rather than universal mill plate.

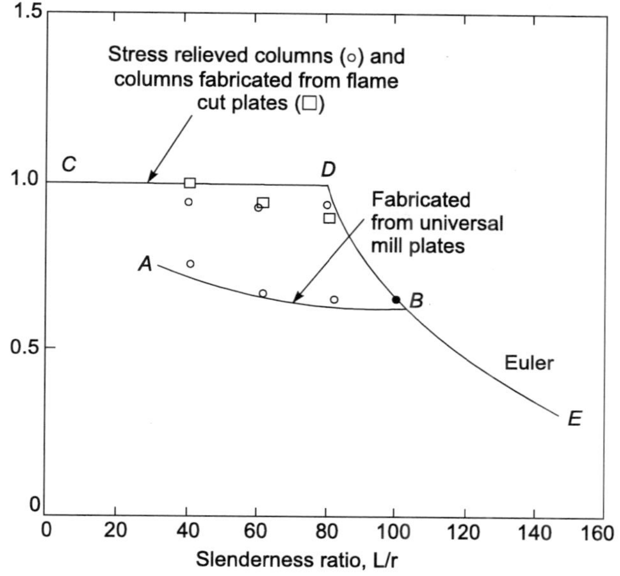

Figure 32.7 Effect of residual stresses on buckling strength on column

32.4.5 Plate and Plate Structures Under Compressive Loading

Residual stresses in the direction parallel to the weld line are compressive in regions away from the weld, while they are tensile in regions near the weld. Consequently, the effects of residual stresses on the buckling strengths of welded structures are complex. Residual stresses may reduce buckling strengths under some conditions and increase them under other conditions.

Both residual stresses and distortion exist in many structures. They create complex effects on the buckling strengths of welded structures. In general, out-of-plane distortion causes considerable reduction in the buckling strength.

32.4.6 Fatigue Strength

Fatigue strength (the number of cycles to fracture under a given load or the endurance limit) increases when a specimen has compressive residual stresses, especially on the specimen surface. For example, experimental studies have shown that local spot heating of certain types of welded specimens increases the fatigue strength.

On the other hand, it is possible that residual stresses are relieved during cyclic loading, and their effects on fatigue strength of weldments are negligible. Surface smoothness is an important characteristic of fatigue fracture because most fatigue cracks originate at the surface. For example, removing weld reinforcement and smoothing surface irregularities, including undercut, are effective in reducing stress concentrations and increasing the fatigue strength of weldments.

32.4.7 Effects of Environment

In the presence of hostile environments, residual stresses can cause cracking in metals without any applied loads. Both ferrous and non-ferrous alloys can be susceptible to stress corrosion cracking. Examples of sensitive combinations of metals and environments are shown in Table 32.1.

TABLE 32.1 Environment producing stress corrosion cracking

Alloy	Sensitive Environment
Low alloy steels	Nitrates, hydroxides, hydrogen sulfide
Chromium stainless steels (greater than 12% Cr)	Halides, hydrogen sulfides, steam
Austenitic stainless steels (13Cr-8Ni)	Chlorides, hydroxides
Aluminum alloys	Sodium chloride, tropical environments
Titanium alloys	Red fuming nitric acid, chlorinated hydrocarbons

The susceptibility of high strength steel to hydrogen induced cracking has been used to indicate the distribution of residual stresses in welds. The crack pattern along a welded butt joint fabricated from oil quenched and tempered SAE 4340 steel plates is shown in Figure 32.8. Longitudinal and transverse cracking are apparent. The longitudinal cracks are in the heat affected zone, and the transverse cracks are in the heat affected zone and the base metal. The tensile residual stresses were transverse to the weld in the heat affected zone, as indicated by the longitudinal cracks. The residual tensile stresses were longitudinal further from the weld, as indicated by the transverse cracks. The transverse cracks are longer and more numerous near the mid length of the weld than near the ends of the weld. This shows that the residual stresses were lower near the ends.

Cracking can occur in weldments, even without external loading, when the material is embrittled by exposure to certain environments and residual stresses are present. Stress corrosion cracking is a brittle type fracture and has been observed in a number of ferrous and non-ferrous alloys exposed to certain sensitive environments containing nitrates, chlorides, hydroxides, hydrogen sulphide, etc.

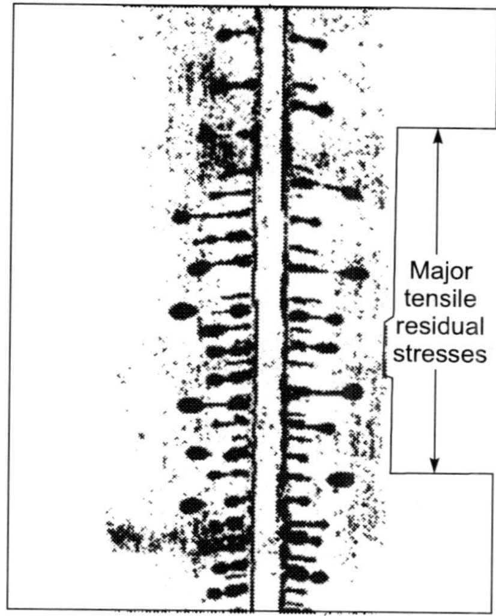

Figure 32.8 Crack pattern in a simple butt joint

32.5 METHODS OF REDUCING RESIDUAL STRESSES

Residual stresses have their origin in complex thermal and/or mechanical interactions. Consequently, their reduction or removal demands the use of mechanical or thermal treatments. The various mechanical and thermal treatments performed on weldments to reduce residual stresses include proof stressing, peening, vibratory conditioning, preheating, postweld thermal treatments and so forth. Thermal and mechanical treatments are necessary to maintain or restore the properties of the base metal affected by the heat of welding.

Among the many properties controlled or improved by suitable treatments are:

1. Distortion during welding.
2. Reduction of stresses that could seriously affect the service performance of a weldment.
3. Weldability (which may be improved considerably by preheat treatment, for example).
4. Improvement of dimensional stability and machinability. Of the mechanical and thermal treatments, vibratory conditioning and postweld thermal treatments are dealt here in great detail because of their wide usage.

Mechanical Methods

Mechanical methods of stress relief can be divided essentially into two basic kinds, monotonic overload and vibratory conditioning. The former consists of continuously increasing load to the maximum level required, prior to unloading; the latter utilizes an oscillating load, which, by hysteresis causes changes to occur that cannot be achieved by monotonic loading.

Monotonic Overload

The two methods of relieving stresses by monotonic overload are proof stressing and peening.

Proof Stressing: Investigations indicate that uniform heavy loading of weldments decrease longitudinal residual stresses by the amount of overload. Cylindrical and spherical pressure vessels can be proof stressed readily by hydrostatic loading. Most critical fabrications are subjected to such an overload treatment in a proof test prior to being put into service. The reduction of residual elastic stress is effected by the conversion of stored elastic strain to plastic strain.

Peening: Peening is a local form of overload. Peening produces a local compressively stressed surface layer so that the magnitude and uniformity of stress and the depth of layer, can be held constant over the whole of the component and between individual components.

Vibratory Conditioning

Vibratory conditioning, commonly known as vibratory stress relief (VSR) uses mechanical energy in the form of low and high frequency vibrations to relieve residual stresses in weldments. VSR consists, in essence, of a vibrating device to induce vibration at the natural resonant frequencies of the structure treated. The induced frequency is varied until it coincides with one of the natural frequencies of the structure when resonance occurs, and the amplitude of vibration increased until it is limited by the natural damping characteristics of the vibrating mass. The frequency is then raised and resonance ceases gradually, until another resonant frequency is found, when the process is repeated. The effect of the high amplitude resonant vibration is to induce an overall elastic distortion into the structure, similar to that which might be achieved by mechanical loading. Because the position of the vibrator must correspond to an antinode, several points of application have to be made to get the maximum number of modes of vibration and applied load in a variety of directions. Due to the gradual application of the load in a cyclic manner, the material will follow its cyclic stress/strain curve, which is lower than its monotonic stress/strain curve.

Advantages of VSR

1. The stress reduction caused by VSR is greater for a given applied strain than that achieved by a single overload.
2. VSR is most effective in enabling shape stability of internally stressed items to be achieved for machining purposes.

3. In circumstances where for good metallurgical reasons, heat treatment is not desirable, a mechanical treatment such as VSR is the only alternative.
4. Also, where it provides a technically viable alternative, VSR tends to be much cheaper, quicker and more convenient than thermal methods.
5. A variety of loading patterns can be obtained in complex structures which would be very difficult or impossible using a direct mechanical loading device.

Disadvantages of VSR

1. VSR is capable of causing fatigue or sudden fracture if it is applied to structures which contain severe welding defects or brittle regions, e.g., HAZ areas, which are not capable of withstanding mechanical strain.
2. VSR by its very nature is most unlikely to give any significant metallurgical benefits, and may in some circumstances, cause mechanical and/or metallurgical damage.

Thermal Methods

Thermal methods of stress relieving, commonly known as 'thermal stress relief' (TSR), are defined as the uniform heating of a structure to a temperature below the critical range, holding it at this temperature for a predetermined period of time, followed by uniform cooling. Stresses are reduced to a level just below the yield point of the material at the temperature of the stress relief treatment. The residual stress remaining in a material after thermal stress relief depends on the rate of cooling. Uneven cooling from stress relief to ambient temperatures may undo much of the value of the heat treatment and result in additional stresses within the weldment. The percentage of relief of internal stresses depends on steel type, composition, or yield strength. The effect of varying time and temperature is shown in Figure 32.9. The temperature reached during the stress relief heat treatment has a far greater effect in relieving stresses than the length of time the specimen is held at that temperature. The closer the temperature to the critical or recrystallization temperature, the more effective it is in the removal of residual stresses, provided proper heating and cooling cycles are employed.

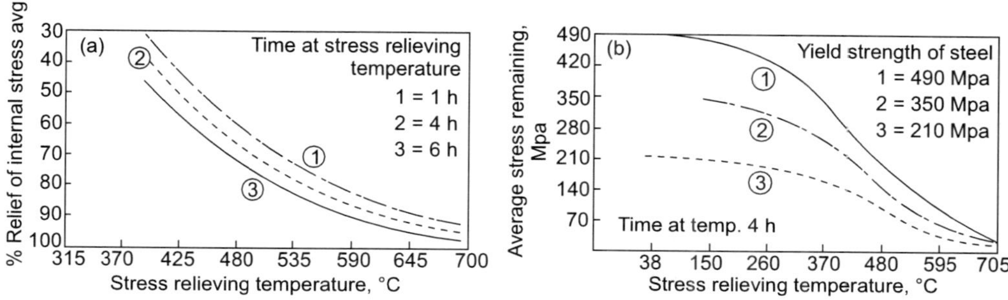

Figure 32.9 Effect of temperature and time on stress relief of steel structures, effect of hold time (left) and (right) effect of material yield strength

Advantages

1. Thermal stress relief improves mechanical properties of HAZ.
2. There is a reduction of 'metallurgical notch' effect resulting from abrupt changes in hardness or other microstructural discontinuities around holes.
3. The resistance of the components to the propagation of cracks improves, especially in the HAZ adjacent to tile weld.
4. The corrosion resistance of the weld is greatly improved.

Disadvantages

1. In thick structures, there is reduction in yield strength if treatments are extended.
2. Uneven heating can lead to distortion and warpage in the component while an uneven cooling rate may result in additional stresses within the weldment.

32.6 METHODS OF RELIEVING WELDING RESIDUAL STRESSES

Design Considerations

Smallest sized weld of smallest length: Larger the amount of weld metal deposited, more the shrinkage and higher the residual stresses.

Fillet weld favorable over butt weld: The force flow diversion in the case of fillet weld reduces the stiffness of the joint with the result that the transverse residual stresses build up to a lesser extent.

Prefer division of structure into subassemblies: As the subassembly can be fabricated in the shop with higher dimensional accuracy and only the final joining may be done at site, which leads to reduced overall stresses.

Material considerations: The selected material should have good weldability under the given fabrication conditions, as material characteristics also have profound effect on welding residual stresses:

- The welding residual stresses increase with the increase in melting point of the material.
- The residual stresses increase with the increase in thermal diffusivity of a material, which characterizes the rate of temperature equalization with transient heat flow condition.
- Welding residual stresses increase with the increase in coefficient of thermal expansion.
- Residual stresses ion weld also increase with the increase in modulus of elasticity.
- Higher the yield limit, higher the possibility of sustaining residual stresses.

Preheating: Welding residual stresses can be reduced by preheating, as:

(a) It reduces cooling rate of the HAZ, which improves the microstructure.
(b) It reduces diffusion of hydrogen, which prevents cold cracking.

(c) Preheating lessens heat input by welding and reduces width of the fusion zone.
(d) Preheating becomes necessary for materials with high thermal diffusivity (Al, Cu) and materials having low ductility at higher temperatures (cast iron).

Welding Procedure (Figure 32.10)

The residual stresses can also be controlled by applying following welding procedures:

(a) Back-Step Welding: In back-step welding, first of all short tack welds are made along the weld line at predetermined intervals. Then, the weld layer is deposited by back-step method from one track to another, opposite to the overall welding direction.

(b) Skip Welding: In this method, a short length of weld metal is deposited in one part of the seam. Then the next length is done some distance away, keeping the sections as far away from each other as possible, thus localizing the heating.

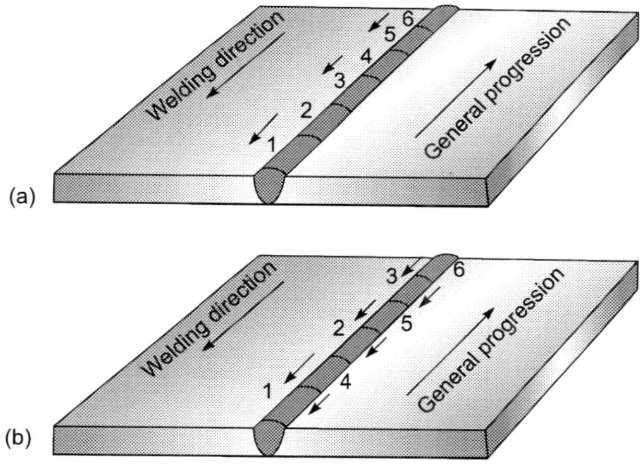

Figure 32.10 Welding procedure

Welding Sequence: Correct welding sequence can reduce the development of residual stresses and avoid fracture. The prime principle is to assure unrestrained transverse shrinkage for butt welds.

Postweld Heat Treatment (PWHT)

(a) The best method to relieve stresses is to anneal the weldments. But this method is limited to small components which can be accommodated in furnaces.
(b) For big structures, localized heat can be applied using electrical heaters, gas/oil burners or flexible insulated pads to perform the stress relieving operation.

Stress Relief by Peening: It consists of lightly hammering the weld and/or the surrounding parent metal in order to relieve stresses present.

32.7 MEASUREMENT OF RESIDUAL STRESSES IN WELDMENTS

Table 32.2 classifies techniques for measuring residual stresses into the following groups:

1. Stress-relaxation
2. X-ray diffraction
3. Cracking

In the stress relaxation techniques (group A-1), residual stresses are determined by measuring elastic strain release. This takes place when residual stresses are relaxed by cutting the specimen into pieces or by removing a piece from the specimen. In most cases, electric or mechanical strain gages are used for measuring the strain release. A variety of techniques exist depending upon the methods of sectioning specimens to determine residual stresses. Some techniques are applicable primarily to plates, while others are applicable to cylinders, tubes, or three-dimensional solids.

TABLE 32.2 Classification of techniques for measuring residual stresses

<table>
<tr><td rowspan="3">A-1 Stress relaxation using electric and mechanical strain gages</td><td>Techniques applicable primarily to plates</td><td>1. Sectioning technique using electric resistance strain gages
2. Gunnert technique
3. Mathar-Soete drilling technique
4. Stablein successive milling</td></tr>
<tr><td>Techniques applicable primarily to solid cylinders and tubes</td><td>5. Heyn-Bauer successive machining technique
6. Mesnager-Sachs boring-out technique</td></tr>
<tr><td>Techniques applicable primarily to three dimensional solids</td><td>7. Gunnert drilling technique
8. Rosenthal-Norton sectioning technique</td></tr>
<tr><td colspan="2">A-2 Stress relaxation using apparatus other than electric and mechanical strain gages</td><td>9. Grid system dividing technique
10. Brittle coating drilling technique
11. Photoelastic coating drilling technique</td></tr>
<tr><td colspan="2">B. X-ray diffraction</td><td>12. X-ray film technique
13. X-ray diffractometer technique</td></tr>
<tr><td colspan="2">C. Cracking</td><td>14. Hydrogen induced cracking technique
15. Stress corrosion cracking technique</td></tr>
</table>

Strain release during stress relaxation (group A-2) can be determined using a grid system, brittle coatings, or photoelastic coatings. An inherent disadvantage of the stress relaxation techniques is that they are destructive; the specimen must be sectioned either entirely or partly. Nevertheless, the stress relaxation techniques are most widely used for measuring residual stresses in weldments primarily because they provide reliable, quantitative data.

Elastic strains in metals can be determined by measuring the lattice parameter using X-ray diffraction techniques. Since the lattice parameter of a metal in the unstressed state is known, elastic strains in the metal can be determined non-destructively without machining or drilling. Two techniques are available at present: the X-ray film technique and the X-ray diffraction technique.

With the X-ray diffraction techniques, surface strains can be determined in a small area, say to a depth of 0.004 inch (0.01 mm) and an area of several square millimeters. These techniques are the only ones suitable for measuring residual stresses as in ball bearings and gear teeth and also surface residual stresses after machining or grinding.

X-ray diffraction techniques have several disadvantages. First, they are rather slow processes. At each measuring point, measurements must be made in two directions for the film technique. Secondly, measurement is not very accurate, especially when applied to heat treated materials in which the atomic structures are distorted.

Techniques are available for studying residual stresses by observing cracks in the specimen. The cracks may be induced by hydrogen or stress corrosion.

The cracking techniques are useful for studying residual stresses in complex structural models which have complicated residual stress distributions. However, they provide qualitative rather than quantitative data.

Measurement by Stress Relaxation Techniques

The stress relaxation techniques are based on the fact that strains taking place during unloading are elastic, even when the material has undergone plastic deformation. Therefore, it is possible to determine residual stresses without knowing the history of the material.

There are five techniques for measuring residual stresses based on stress relaxation techniques that can be used for weldments. The first two techniques apply primarily to thin plates; the next two techniques apply primarily to solids. The first four techniques employ electric or mechanical strain gages, while the fifth one employs a photoelastic coating. The range of application and the advantages and disadvantages of each of the five techniques are summarized in Table 32.3.

Sectioning Technique for a Plate Using Electric Resistance Strain Gages

With this technique, electric-resistance strain gages are mounted on the surface of the test structure or specimen. A small piece of metal containing the gages is then

TABLE 32.3 Stress relaxation techniques for measuring residual stresses

Technique	Applications	Advantages	Disadvantages
1. Sectioning technique using electric resistance strain gages	Relative all-around use, with the measuring surface placed in any position.	Reliable method, simple principle, high measuring accuracy.	Destructive. Gives average stresses over the area of the material removed from the plate; not suitable for measuring locally concentrated stresses. Machining is sometimes expensive and time consuming.
2. Mathar-Soete drilling technique	Laboratory and field work use and on horizontal, vertical, and overhead surfaces.	Simple principle, causes little damage to the test piece; convenient to use on welds and base metal.	Drilling causes plastic strains at the periphery of the hole which may displace the measured results. The method must be used with great care.
3. Gunnert drilling technique	Laboratory and field work use. The surface must be substantially horizontal.	Robust and simple apparatus. Semi-distractive damage to the object tested can be easily repaired.	Relatively large margin of error for the stresses measured in a perpendicular direction. The underside of the weldment must be accessible for the attachment of a fixture. The method entails manual training.
4. The Rosenthal-Norton sectioning technique	Laboratory measurements.	Fairly accurate data can be obtained when measurements are carried out carefully.	Troublesome, time consuming, and completely destructive method.
5. Photoelastic coating-drilling technique	Primarily a laboratory method, but it can also be used for field measurements under certain conditions.	Permits the measurement of local stress peaks. Causes little damage to the material.	Sensitive to plastic strains, which sometimes occur at the edge of the drilled hole.

removed from the structure, as shown in Figure 32.11. The strains that take place during the removal of the piece are determined. The piece removed is small enough so that the remaining residual stress may be neglected and therefore, the measured strains are the elastic strains resulting from the residual stress distribution of the specimen. Stated mathematically as

$$\bar{\epsilon}_x = -\epsilon'_x$$

$$\bar{\epsilon}_y = -\epsilon'_y$$

$$\bar{\gamma}_{xy} = \gamma'_{xy} \tag{32.1}$$

where, ϵ'_x, ϵ'_y and γ'_{xy} are the elastic strain components of the residual stress and $\bar{\epsilon}_x, \bar{\epsilon}_y$ and $\bar{\gamma}_{xy}$ are the measured values of the strain components. The negative sign indicates that the strains are in opposite directions. Then residual stresses can be determined by equation 32.2 for σ_x, σ_y and τ_{xy}. Thus,

$$\sigma_x = \frac{-E}{1-\nu^2}\left[\bar{\epsilon}_x + \nu\bar{\epsilon}_y\right]$$

$$\sigma_y = \frac{-E}{1-\nu^2}\left[\bar{\epsilon}_y + \nu\bar{\epsilon}_x\right]$$

$$\tau_{xy} = -G\gamma_{xy} \tag{32.2}$$

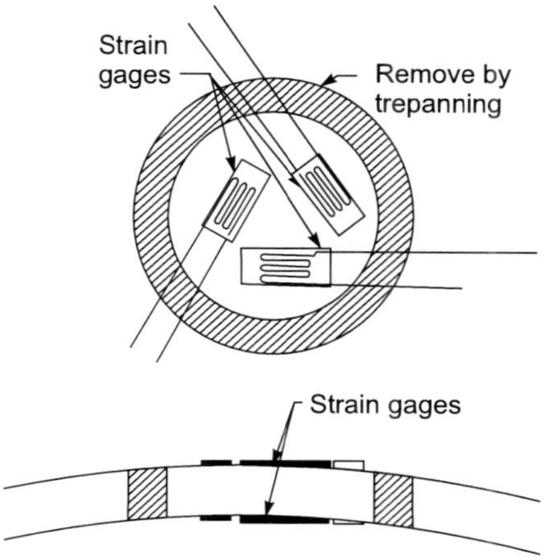

Figure 32.11 Complete stress relaxation technique applied to a plate

It is advisable to make strain measurements on both surfaces of the plate because residual stresses may be caused by bending. The mean value of strains measured on both surfaces represents the normal stress component, while the difference between the strains on both surfaces represents the bending stress component.

Mathar-Soete Drilling Technique

When a small circular hole is drilled in a plate containing residual stresses, those stresses in areas outside the hole are partially relaxed. It is possible to determine residual stresses that existed in the drilled area by measuring stress relaxation in areas outside the drilled holes. The hole method of measuring stress was first proposed and used by Mathar and was further developed by Soete.

A common way to determine residual stresses is to place strain gages at 120 degrees from each other, and drill a hole in the center, as shown in Figure 32.12. The magnitudes and directions of the principle stresses are calculated by measuring strain changes at the three gages.

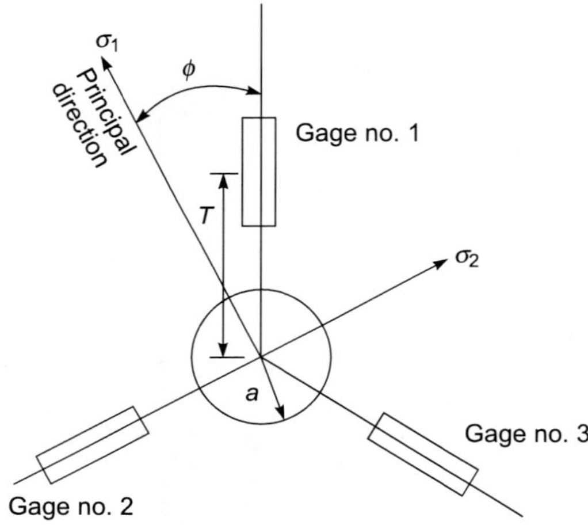

Figure 32.12 Degree star arrangement of strain gages for the Mather-Soete drilling technique

Gunnert Drilling Technique

Four 3 mm holes located on a circle with a 9 mm diameter are drilled through the plate perpendicular to the surface at the measuring point, as shown in Figure 32.13. The diametrical distance between these holes at different levels below the surface of the plate is measured by means of a specially designed mechanical gage. The perpendicular distance between the plate surface and the gage location at different levels below the surface is also measured. A groove of 16 mm from the measuring point is then trepanned around the holes in steps, and the same measurements are made. Formulas are given to calculate residual stresses in the interior of the specimen from the measurement data.

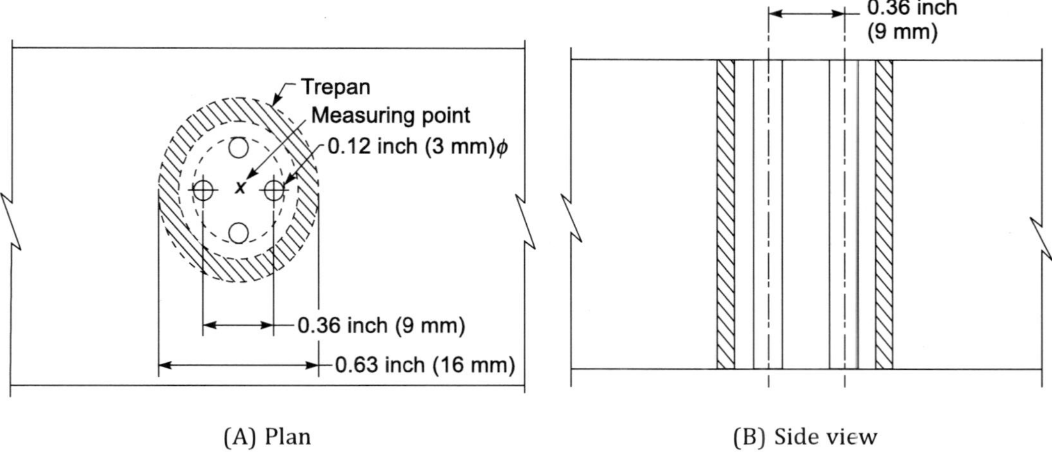

(A) Plan

(B) Side view

Figure 32.13 Gunnert drilling method

Rosenthal-Norton Sectioning Technique

Rosenthal-Norton proposed a technique for determining residual stresses in a thick weldment. Two narrow blocks with full thickness of the plate, one parallel to the weld and the other transverse to the weld, are cut from the weld. Then, any residual stresses still left in the narrow blocks are measured. Formulas are used to estimate residual stresses in the interior of the weldments from (1) strain changes, which take place while cutting the narrow blocks, and (2) residual stresses, which are left in the blocks.

32.8 DISTORTION IN WELDS

Any unwanted physical change or departure from specifications in a fabricated structure for component, as a consequence of welding, is called welding distortion. It is caused by the non-uniform expansion and contraction of the weld metal and adjacent base metal during the heating and cooling cycle of the welding process.

During welding when a metal is heated, expansion takes place. If this expansion is resisted, deformation of the metal takes place. Also, during cooling of the metal after welding, contractions take place, which if resisted, generates stress. If this stress causes movement of the metal, distortion occurs. If the stress causes no movement, it is left as the residual stress. These stresses may result in plastic deformation, if the yield stress is exceeded, which will again lead to distortion.

32.8.1 Factors Affecting Distortion

Material Properties

(a) The higher the value of coefficient of thermal expansion of a material, the greater the distortion is likely to be.
(b) Low thermal conductivity of a material leads to high thermal gradients, and high thermal gradients lead to high distortion because distortion depends upon internal or external restraint.
(c) The higher the yield strength of the parent material, the greater the residual stresses available for causing distortion.
(d) The modulus of elasticity is a measure of stiffness of a material, so that higher the modulus, the more it will resist distortion.

Welding Processes and Procedure

The effects of welding processes and procedures on distortion can be summarized as:

(a) A concentrated heat source gives less distortion than a spread heat source.
(b) High welding speeds (less spread of heat) gives less distortion.
(c) Deep penetration welds give less distortion.
(d) Single-pass welding gives less distortion than multipass welds.
(e) Amount of weld metal should be kept to a minimum to avoid distortion.
(f) Wet bead contours should be flat or only slightly convex.
(g) Maximum welding speed should be used to minimize heat spread and build-up.
(h) Edge preparation should be such that it requires minimum amount of weld material. It should be as uniform as possible for consistent shrinkage.
(i) Close fit-up also reduces the amount of weld metal and hence, weld metal shrinkage.
(j) Mechanized, single-pass, high-speed welds reduce the tendency to distort.
(k) For manual welds, back step technique or skip welding technique should be used.

Preheat and Postweld Heat Treatment

Requirements

- Substances after weld, cool very fast since the surrounding metal is cold. This increases hardness and can cause embrittlement in HAZ.
- Preheat: Metal is heated just before welding. The area nearby is of higher temperature, leaving no place for the weld heat to go, thereby ensuring the temperature doesn't fall quickly.

- Postweld heat treatment: Even after preheating, there are chances that residual stresses occur. Welded component is heat treated to remove residual stresses.

Aims of Preheating

- Reduce heat losses from weld area.
- Prevent martensite formation, reducing cracking.
- Reduce distortion and residual stresses by reducing expansion and contraction rates.
- Increase welding speed by removing grease, scale, etc.
- Allow sufficient time for hydrogen to diffuse out of weld.

Postweld Heat Treatment

The component is heat treated after welding for:

- **Postweld Stress Relieving:** Residual stresses formed on welding have to be relieved to ensure the safety of component.

  Full annealing: Whole workpiece is heated. Done if workpiece is small.
- **Sub-Critical/Low Temperature Annealing:** Local stress relieving around the weld, at low temperatures.
- **Reformation of Structure:** Postweld reformation of structure is necessary for attaining desired microstructure in mild steel and low-alloy steel. It can dissolve the precipitates that may cause defects.
- **Stabilization of Structure:** Rapid cooling with welding can disrupt the structure, and can make a metastable structure, which changes to stabilized structure either naturally (natural ageing) or artificially. Annealing or normalizing or subcritical heat treatments could be done to give a stable structure, avoiding hydrogen embrittlement, which occurs in natural ageing.

Example: Al 6061 heat treatment (Table 32.4)

Aluminum 6061 is an alloy which has very low strength, that makes it versatile for machining. After welding, we have to heat treat it to make it 6061-T6, which is highly stronger as compared to the parent metal.

Process

- Component is machined and welded.
- It is heated to 600°C
- Hold it at that temperature for about 1 hour.

- Quench it in water. This is now known as 6061-T4.
- Reheat the component to 200°C.
- Hold it at that temperature for 1 hour.
- Cool it in air. This is known as 6061-T6.

TABLE 32.4 Comparison of strength of 6061 after heat treatment

Property	Alloy - 6061		
	Temper		
	'O'	T4	T6
UTS (kg/mm^2)	15 (max.)	18 (min.)	26 (min.)
Y.S. (kg/mm^2)/0.2% P.S. (kg/mm^2)	11 (max.)	11 (min.)	24 (min.)
% Elongation	16 (min.)	16 (min.)	8 (min.)
Hardness (BHN)	35 (max.)	35 (max.)	85 (min.)

32.8.2 Fundamental Types of Distortion

1. Longitudinal shrinkage
2. Transverse shrinkage
3. Angular distortion
4. Longitudinal distortion; bowing or bending
5. Rotational distortion
6. Buckling and twisting

The distortion found in fabricated structures is caused by three fundamental dimensional changes that occur during welding: (1) transverse shrinkage that occurs perpendicular to the weld line, (2) longitudinal shrinkage that occurs parallel to the weld line, and (3) an angular change that consists of rotation around the weld line. These dimensional changes are illustrated in Figure 32.14.

Figure 32.14 (a) shows transverse shrinkage in a simple welded butt joint. The distribution of the longitudinal residual stress, as discussed earlier, is shown again in Figure 32.14 (b). This stress causes longitudinal shrinkage. This shrinkage of the weld metal and base metal regions adjacent to the weld is restrained by the surrounding base metal.

Figure 32.14 (c) shows the angular change that occurs in a butt joint. Non-uniformity of transverse shrinkage in the thickness direction causes this rotation. Figure 32.14 (d) shows the angular change that occurs in a fillet weld. Here, the angular change is caused by the unbalance of shrinkage on opposite sides of the flange member.

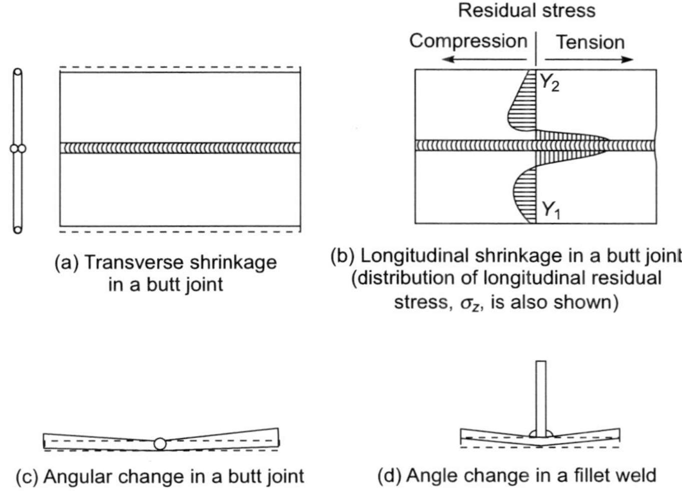

Figure 32.14 Fundamental dimensional changes that occur in weldments

Transverse Shrinkage in Butt Joints

In Figure 32.14 (a), the shrinkage is uniform along the weld. However, transverse shrinkage that occurs in welded butt joints, in actual structures, is usually much more complex. The major factors causing non-uniform transverse shrinkage in welded butt joints are (1) rotational distortion during welding, and (2) joint restraint. The amount of transverse shrinkage of carbon and low-alloy steel welds may be estimated using the following equation:

$$S = (0.2A_w)/t + 0.05d$$

where,

S = Transverse shrinkage, inch
A_w = Cross-sectional area of weld, inch
t = Thickness of plates, inch
d = Root opening, inch

Rotational Distortion of Butt Joints

Rotational distortion of butt joints during welding is affected by both heat input and welding speed. When mild steel plates are joined by shielded metal arc welding, the unwelded portion of the joint tends to close because of the relatively low welding speeds of the process. This is shown in Figure 32.15 (a). When similar steel plates are joined with the submerged arc welding process, the unwelded portion of the joint tends to open, as shown in Figure 32.15 (b). This is a result of thermal expansion of the heated edges ahead of the arc. To maintain desired root opening, tack welds for a submerged arc welded joint must be large enough to withstand stresses caused by the rotational distortion as welding progresses along the joint.

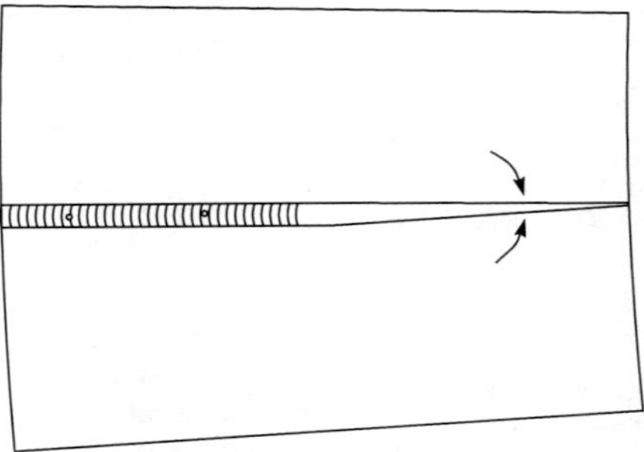

(a) Unwelded portion of the joint closes
(in shielded metal arc welds)

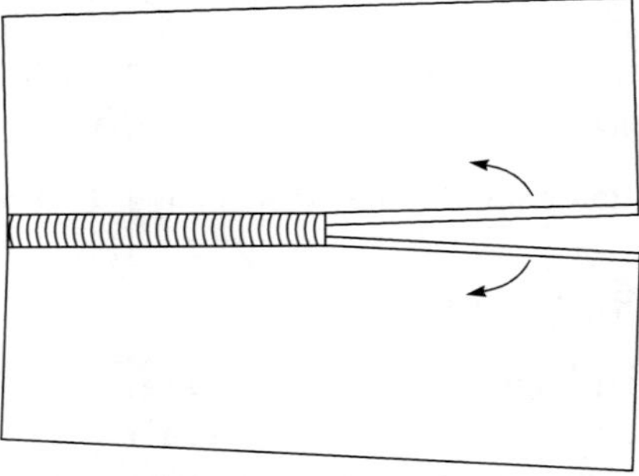

(b) Unwelded portion of the joint opens
(submarged arc welds)

Figure 32.15 Rotational distortion

32.8.3 Rules for Minimizing Distortion during Welding (Figure 32.16)

1. Design and fabricate balanced weld on each side of the neutral axis.
2. Avoid using oversized fillet welds.
3. Use double sided weld rather than single sided and use minimum bevel angles, to reduce the amount of weld metal.
4. Use minimum gap sizes.
5. In no-fatigue sensitive areas, use intermittent fillet welds, wherever possible.
6. Use double fillet welds rather than full penetration butt welds.
7. Use clams, strong back up plates, jigs and fixtures.
8. Deposit few weld runs alternatively on each side of the joint in double V butt welds.
9. Use iron powder MMA electrodes.
10. Use block welding to prevent movement.
11. When block welding thick plate, butter the sides of the preparation and build up the butter progressively towards the center of the joint, so that most of the joint can contract transversely before the joint bridges.

12. First weld the joints that cause interchange most contraction
13. Make use of subassemblies
14. Make frequent dimensional checks during welding, and if distortion is evident, change the welding sequence or the clamping arrangement accordingly.

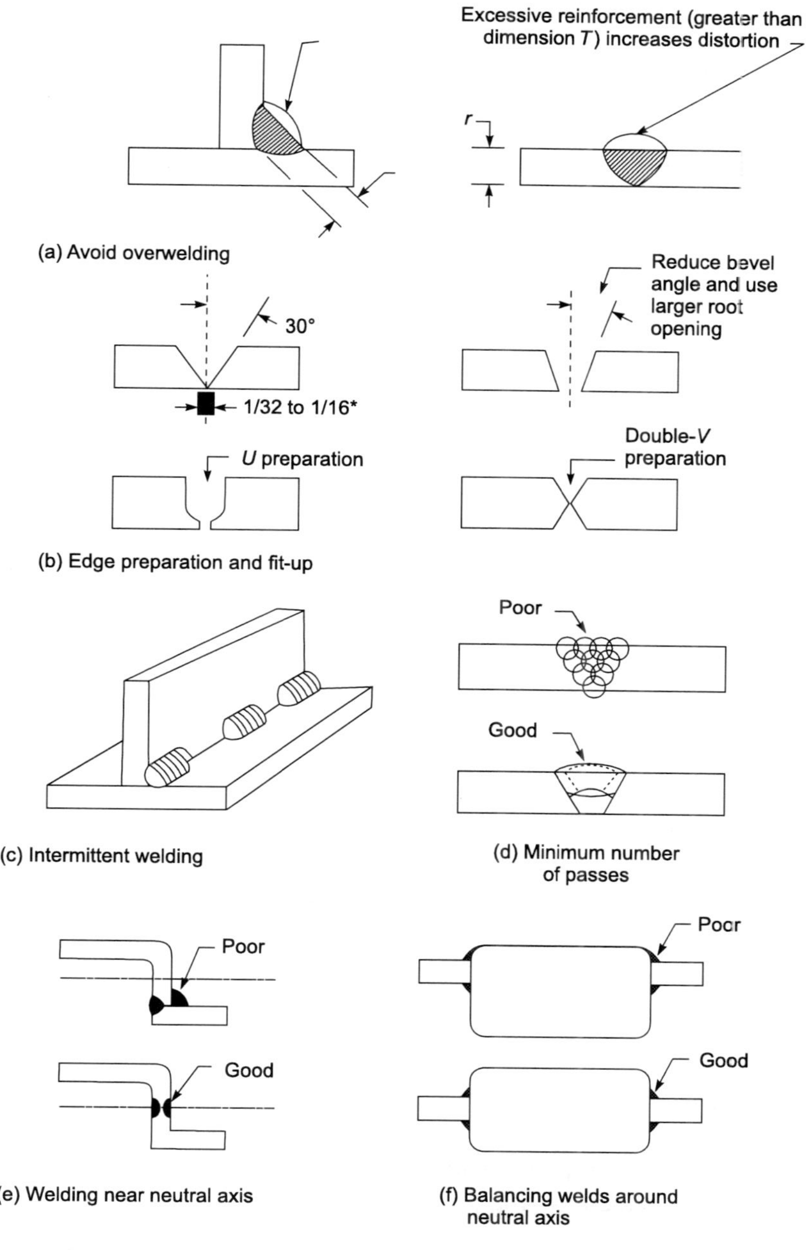

Contd.

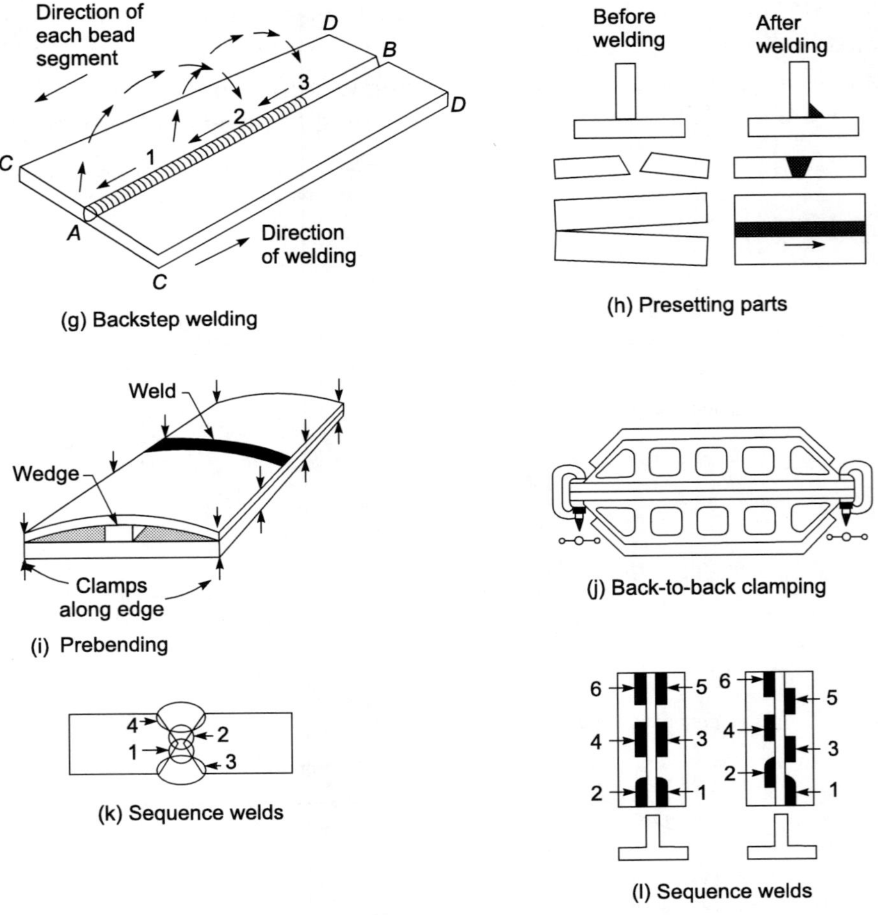

Figure 32.16 Various ways to minimize distortion

Review Questions

1. What is the basic mechanism of residual stresses in weldment?
2. What are the types of residual stresses?
3. Explain the different sources of residual stresses.
4. Explain the methods of reducing residual stresses.
5. Explain vibratory conditioning.
6. State the factors which affect distortion.
7. State fundamental types of distortion.
8. Discuss transverse shrinkage in butt joints.
9. Discuss about rotational distortion of butt joints.
10. State all the rules for minimizing distortion during welding.

33 CHAPTER

Weld Defects

33.0 INTRODUCTION

Performance of welded structures during their service depends on quality of their weld, which in turn, depends on the presence or absence of weld defect. Weld defect refers to any departure from specified requirements. It results in reducing the strength of joints and can even cause failure. A weld defect can be detrimental to the component, if it is situated in a region of high stress. If it is in a region of flow stress, it may be turned into imperfection, flaw or discontinuity. A welding defect is any flaw that compromises the usefulness of a weldment. Performance and longitivity of the weld depend on the presence or absence of defects in weld joints. Improper weld parameters and wrong welding procedures introduce defects in the weld metal and HAZ. Defects impair the strength of weld joints. A defective weldment fails under load. There is a great variety of welding defects.

According to American Society of Mechanical Engineers (ASME), welding defect causes are broken down into the following percentage:

- 41% poor process conditions
- 32% operator error
- 12% wrong technique
- 10% incorrect consumables, and
- 5% bad weld grooves

33.1 CAUSES OF WELD DEFECTS

- Failure to follow the requirements of welding specifications
- Use of substandard materials and equipment
- Poor skills of welder
- Design or drawing error
- Careless material preparation

33.2 CLASSIFICATION OF WELD DEFECTS

International Institute of Welding broadly classifies weld defects into six groups. They are cracks, cavities, solid inclusions, incomplete fusion or penetration, imperfect shape and miscellaneous defects. The miscellaneous defects include excessive spatter, uneven ripples, etc.

General classification of welding defects is shown in Figure 33.1.

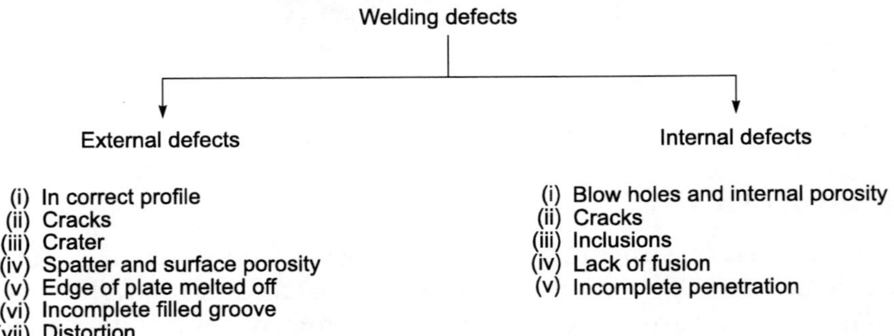

Figure 33.1 General classification of defects

33.2.1 Cracks in Welds

It is a discontinuity produced by fracture, generated mainly from the stresses during cooling. It is the most detrimental type of defect. It can be found on the surface or inside, either in the weld metal or HAZ. During loading, they can easily propagate especially through stress concentration regions and can cause failure easily during impact loading or low temperature loading.

Cracks can broadly be classified into two, based on the type of formation:

1. Cold cracks
2. Hot cracks

Factors Responsible for Cold Cracking

1. Hydrogen content in the weld metal
2. Welding joint restraint and high thermal sensitivity
3. Presence of impurities
4. Embrittlement of the heat affected zone in low-alloy steel
5. High welding speed and low current density
6. Weld of insufficient sectional area

Factor Responsible for Hot Cracking

1. High current density
2. Weld metal composition having wide freezing range

3. Distribution of heat and thermal stresses in the weld metal
4. Impurities such as sulphur and high carbon or nickel content and those which form low freezing point liquid films
5. Joint restraints and high thermal sensitivity
6. Electrode crack sensitivity
7. Dilution of weld metal
8. Welding parameter, which stimulate dendritic growth

Depending on Location and Orientation, Cracks can be Classified into

- Longitudinal cracks
- Transverse cracks
- Crater cracks
- Underbead cracks
- Toe cracks

Solidification Cracking

These cracks occur only in weld metal (Figures 33.2 and 33.3).

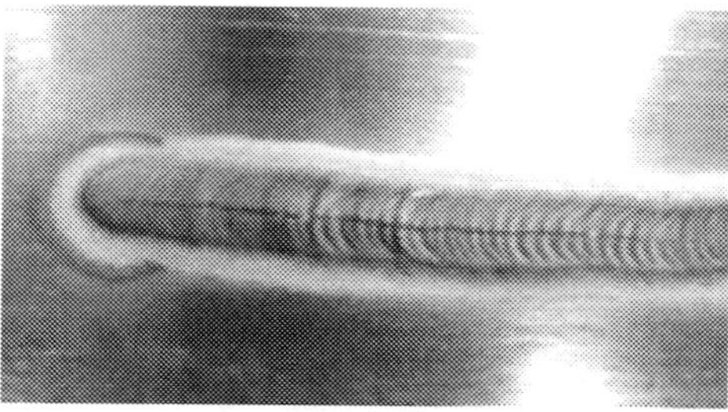

Figure 33.2 Solidification cracking

Identification: They are visible to naked eye. Normally appear as straight line through the center of weld. They can have branches in the end.

Causes

- Weld on cooling contracts, causing the cracks due to compression.
- Accelerated by the presence of impurities, which reduce freezing point of liquid at the center, which on solidification, cause crack.

Remedies

- Increased manganese content in filler.
- Increased amount of filler in weld.
- Less joint gaps.

- Cleaning contaminants.
- Low welding speeds, reducing thermal stresses.
- Maintaining depth to width as 2:1.

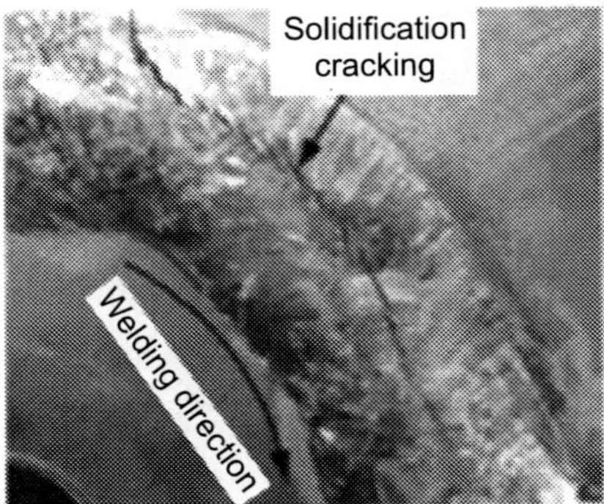

Figure 33.3 Solidification cracking

Liquation Cracks (Figure 33.4)

These cracks found in heat-affected zone. They occur in processes having high heat input, such as electroslag or submerged arc welding. They also occur when we weld steel with a high ratio of sulphur and phosphorus with a low manganese content.

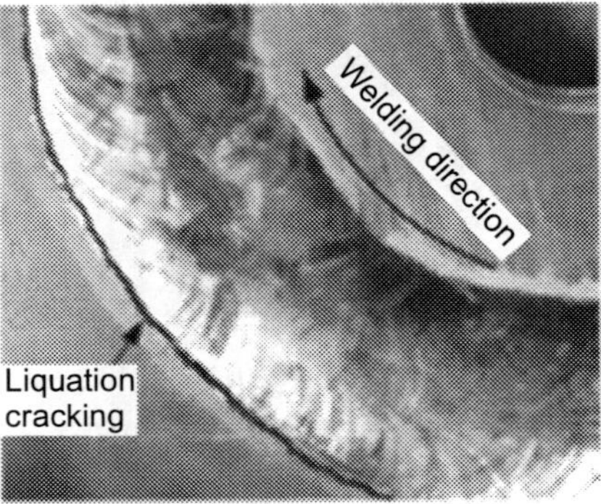

Figure 33.4 Liquation cracking

Causes

- Impurities like sulphur and phosphorus in the grain boundaries become molten and form brittle films.
- These films on contraction break causing microcracks in HAZ.

Remedies

- Avoiding the presence of impurities in parent metal.
- Low heat input welding processes should be used with less dwell time at higher temperature.

Hydrogen Cracks (Figure 33.5)

They are known as cold cracking and occur immediately while welding or after a short time. This defect occurs commonly in ferrite steels.

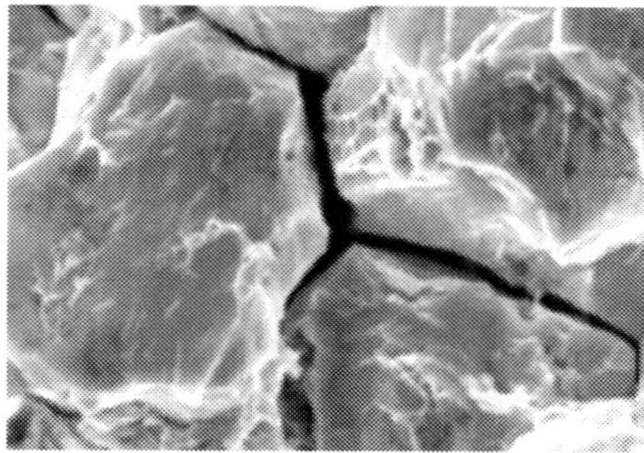

Figure 33.5 Microscopic view of hydrogen embrittlement

Causes

- Hydrogen which is generated by welding process.
- Weld metal hydrogen – moisture in flux, oil, grease, etc.
- Hard brittle structure susceptible to cracking.
- Increased amount of carbon in the parent metal.
- Higher thickness can increase cooling rate and retain more hydrogen.
- Stress concentration.

Remedies

- Preheating and postweld heating.
- Use of austenitic and nickel alloy weld metal, which have higher solubility of hydrogen.
- Reduction of weld metal hydrogen.
- Drying the electrode before welding.
- Reduce stress in stress concentration areas.
- Uniform heating.

Longitudinal Cracks (Figure 33.6)

It may occur in the weld metal and usually runs through the centre of weld, which can be an extension of crater crack at the end of weld or of a subsurface crack often from root pass as shown in the figure. It can occur in base metal also, runs parallel to weld bead. It occurs due to poor weld penetration into sides of butt joints. For fillet welds, start from toe of weld or proceed from root of weld to base metal.

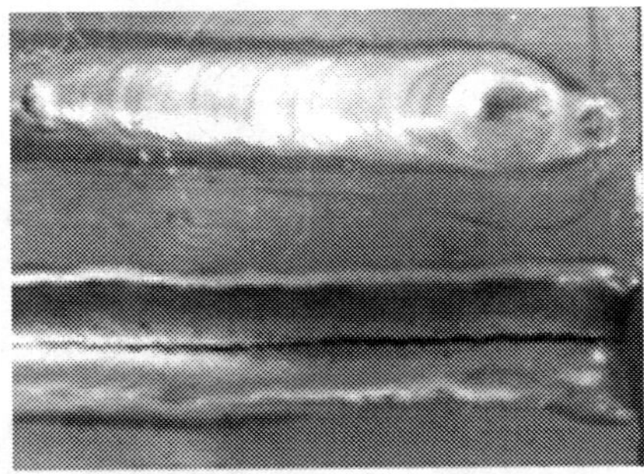

Figure 33.6 Longitudinal crack

Transverse Cracks (Figure 33.7)

- Found in weld metal and base metal.
- In weld metal, runs perpendicular to weld axis
  - Due to tight restriction or due to shrinkage during cooling.
- In base metal, runs perpendicular to weld axis, placed in HAZ
  - Common in welding of steels with high hardenability
  - May not be seen until weld metal is cooled.

Figure 33.7 Transverse crack

Crater Cracks (Figure 33.8)

- Star-shaped cracks, found in crater of the welds.
- Small cracks at the point of arc breaking.
- Propagate easily into the weld bead.
- Progress only to the edge of the weld crater.
- Act as starting point for longitudinal weld cracks.

Close-up wiew of a fine cracks (arrows) in a fillet weld

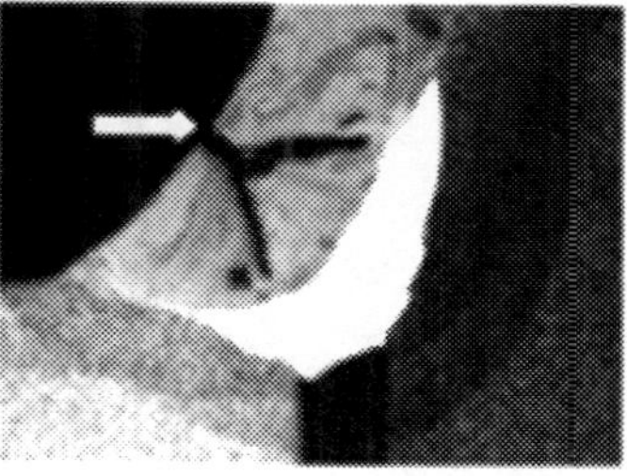

A cross section through the cracked weld reveals a *Y*-shaped crack in the center of the weld deposit

Figure 33.8 Crater crack

Causes

- During interrupted welding due to asymmetric contraction and expansion.
- Incorrect weld ending technique used.
- Weld bead with excessive concave surface can promote this.

Remedies

- By starting each weld properly after interruption.
- By properly filling crater before ending the weld.
- Crater should be shielded from supply gas until it completely solidifies.
- Avoid discontinuous welding cycles in materials with high thermal expansion.

Underbead and Toe Cracks (Figure 33.9)

- Underbead cracks are surface cracks in base metal under or near welds.
- Toe cracks are appear on plate surface adjacent to welds.

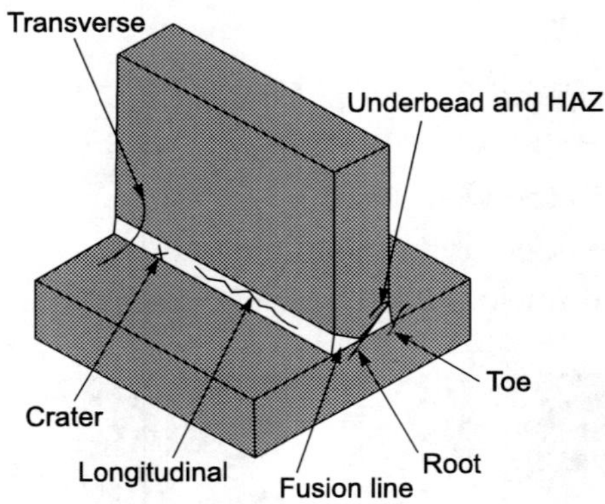

Figure 33.9 Undercut

Causes

- Rapid cooling of high carbon or alloy content steels.
- Hydrogen pick up during welding.

Remedies

- Use of low-hydrogen welding electrodes and fluxes.
- Slower cooling rates, achieved by slow welding speed or by preheating, allow more hydrogen to escape.
- High welding currents and slow welding speeds reduce the rate of cooling.

33.2.2 Visual Surface Weld Defects

The weld defects which can be easily observed by naked eye or low magnification lens are given below:

1. Surface cracks
2. Crater cracks
3. Distortion/warpage
4. Incorrect or bad bead profile
5. Dimensional deviation/incorrect weld size
6. Overlaps or rolls
7. Undercuts
8. Underfill/suck-back
9. Excess penetration/excess reinforcement
10. Start/stop points
11. Interruptions
12. Unfilled craters

13. Burn-through
14. Surface porosity
15. Surface oxidation
16. Uneven and coarse ripples
17. Pock marks/flat spots/dents
18. Excessive spatter
19. Arc strike
20. Bad surface appearance
21. High–low/off-set

1. Surface Cracks (Figure 33.10)

(a) Causes

1. High concentration stresses.
2. Poor fit-up and incorrect welding procedure.
3. Poor edge quality.
4. High welding speed – Al alloys.
5. Deep penetration process – CO_2 welding, SAW, narrow gap.

(b) Remedy: Fill the crater to match the bed shape.

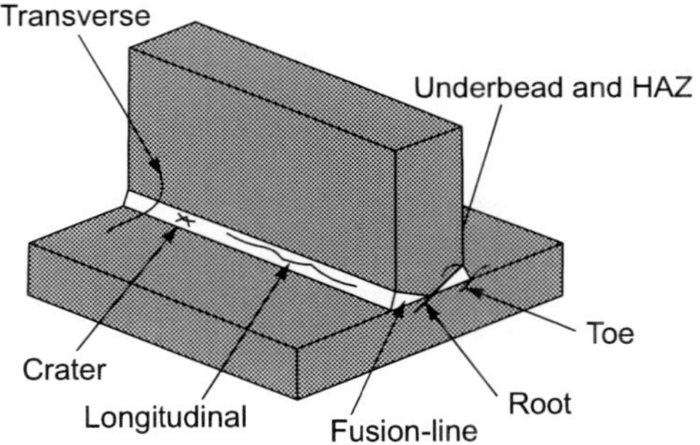

Figure 33.10 Surface and crater cracks

2. Crater Cracks

(a) Causes

1. Sudden interruption in welding operation.
2. High thermal expansion of base metal, e.g., austenitic stainless steel.

(b) Remedy: Do crater filling to convex shape manually or with crater filler in automatic welding.

3. *Distortion/Warpage (Figure 33.11)*

(a) Causes

1. Uneven beating of workpieces.
2. Weld metal shrinkage.
3. Faulty metal procedure.
4. Excessive weld metal.

(b) Remedies

1. Mechanized, single-pass, high speed welding, minimize distortion.
2. Jigging of joints.
3. Proper sequencing in welding.

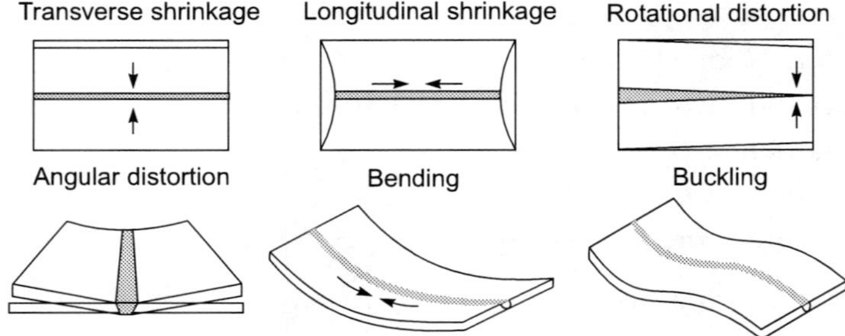

Figure 33.11 Distortion/warping of weld plates

4. *Bad Weld Profile (Figure 33.12)*

(a) Causes

1. Poor welding skill.
2. Excessive/deficient weld metal.

(b) Remedy: Skilled operator, particularly for the root run in multirun welds.

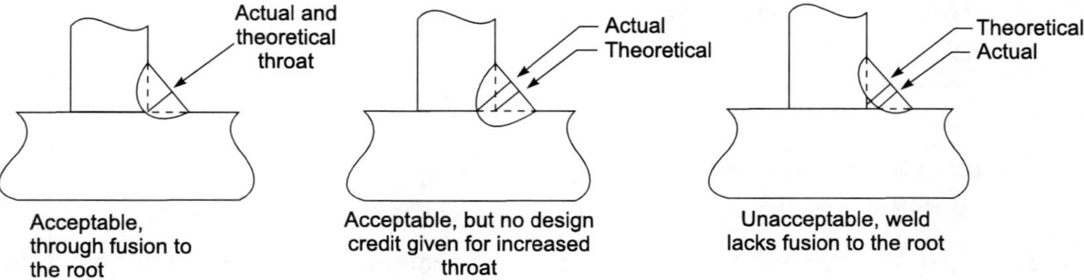

Figure 33.12 Acceptable and unacceptable weld profiles

5. Dimensional Defects (Figure 33.13)

(a) Causes

1. Variation in supply voltage.
2. Wire slippage in feed rolls.
3. Variation in welding speed.
4. Wrong work angle.
5. Poor skill of operator.

(b) Remedies

1. Deficient welds can be rewelded.
2. Excessive weld metal may be chipped off, if it interferes with assembly, etc.

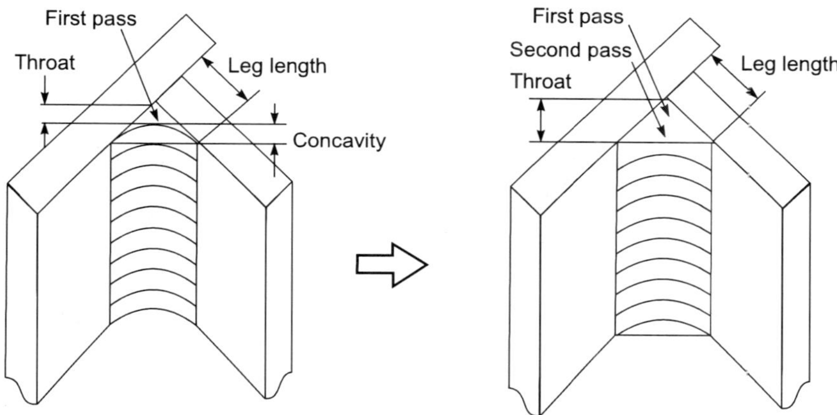

Figure 33.13 Dimensional defect (concavity)

6. Overlaps/Rolls

(a) Causes

1. Excessive welding current.
2. Wrong tilt of electrode in fillet welds.
3. Electrode in offset too much from 12'o' clock position for circumferential welds.

(b) Consequence: Cause lack of fusion, cracks, etc. in the overlapped portion; notch effect, stress concentration, etc.

(c) Remedies

1. To be chipped off.
2. Weld bead ground to proper shape.

7. Undercuts (Figure 33.14)

Undercut basically occurs in the parent metal nearby the weld and the basic cause is due to high current, excessive arc length and when speed of the welding is too high. Improper manipulation of electrode and dampness in electrode and arc blow are also responsible for undercut.

(a) Causes

1. Excessive welding current and arc voltage.
2. Excessive electrode tilt in fillet welds.
3. Wrong positioning of cooling blocks in ESW.
4. Too large electrode diameter.
5. Arc blow in DC weld.

(b) Remedies

1. To weld up.
2. Use AC power.

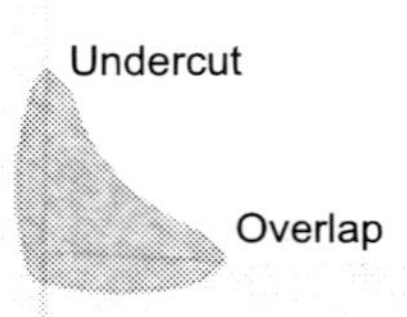

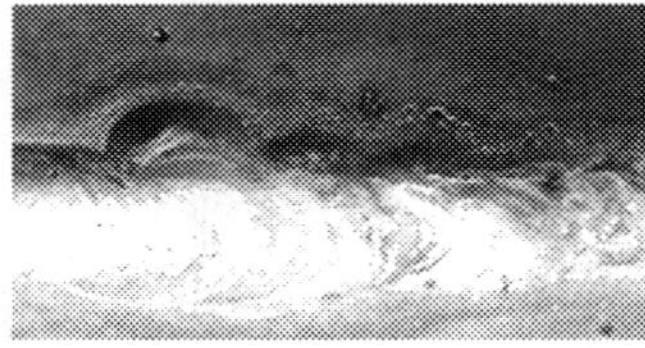
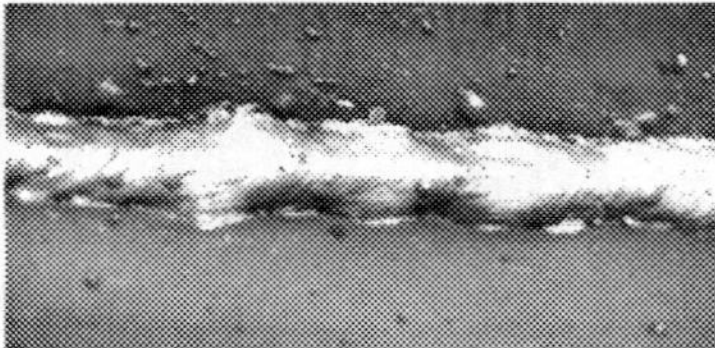

Figure 33.14 Undercut and overlap

8. Underfill/Suck-back (Figure 33.15)

(a) Causes

1. Use of backing gas/flux.
2. Excessive root gap.

(b) Consequence: Not acceptable for dynamic loading and/or subzero service condition.

(c) Remedy: Add additional layer of weld metal to fill up the depression.

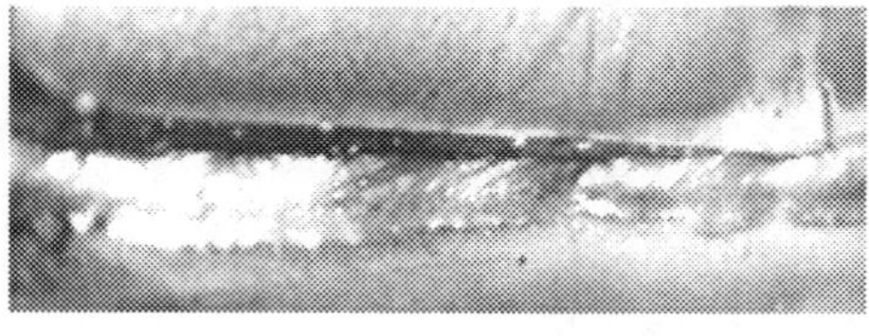

(a)

(b)

Figure 33.15 (a) Underfill, (b) Suck-back

9. Excessive Reinforcement/Penetration (Figure 33.16)

Excessive reinforcement is caused by the low skill of operator or low welding speed in the automatic welding operation. Maximum reinforcement is limited to 3 mm or 20% of the plate thickness.

Excessive penetration can be formed due to the high welding current, very small root face, etc. It is recommended to chip off excess metal and give proper shape to the weld bead to avoid these defects.

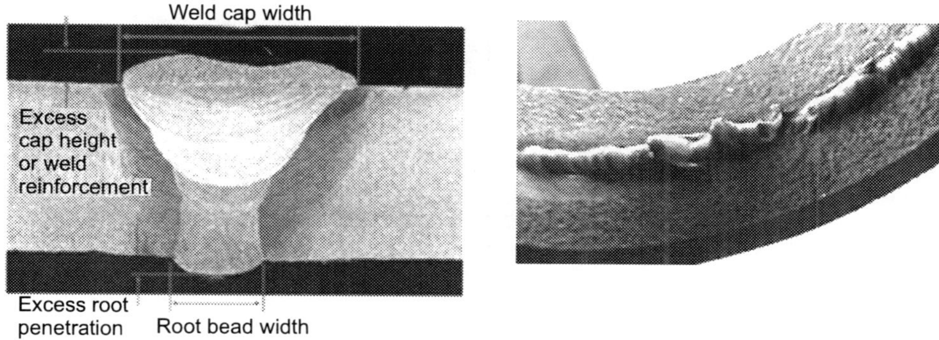

Figure 33.16 Excessive penetration/reinforcement

Causes

1. Incorrect joint preparation.
2. High current used.
3. High arc length.

Remedies

1. Joint should be prepared properly.
2. Correct current setting to be used.
3. Proper control of arc length.

10. Lack of Penetration (Figure 33.17)

This defect occurs when the weld metal does not extend to the required depth into the joint root.

Causes

1. Low amperage.
2. Low preheat.
3. Tight root opening or root gap and bevel angle is too small.
4. Very large diameter of electrode.
5. Fast travel speed.
6. Short arc length.

Remedies

1. Proper joint preparation.
2. Proper selection of electrode diameter.
3. Proper selection of process parameter.

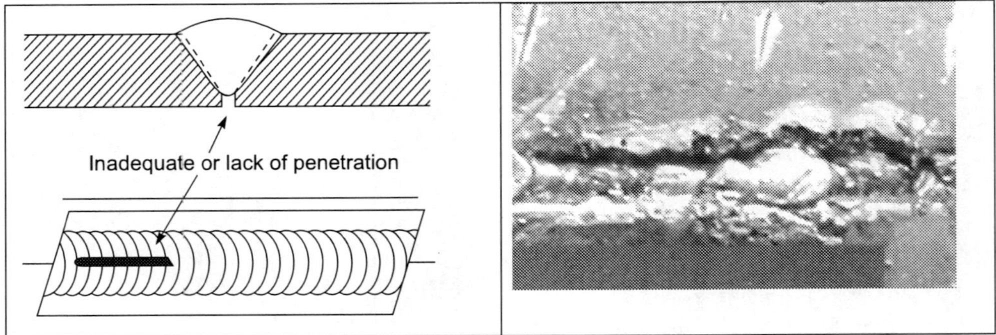

Figure 33.17 Lack of penetration

11. Start/Stop Point (Figure 33.18)

These types of defects are found in SMAW operations. Improper filling of crater in weld metal and abrupt interruption in welding operations are the causes of these defects. Welding is ended by smoothly tapering off the operation to avoid this defect.

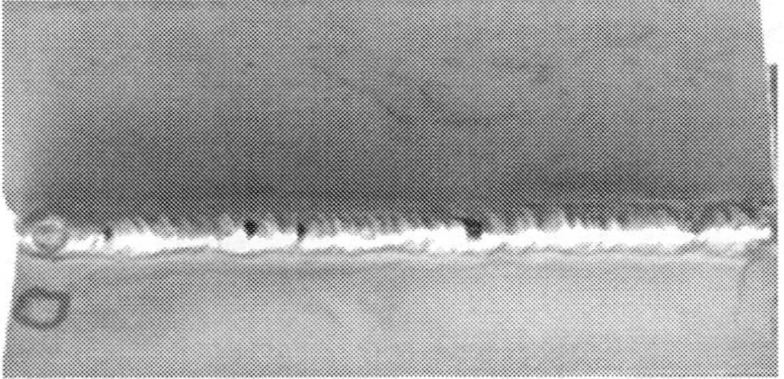

Figure 33.18 Start or stop point

12. Sudden Interruption

Sudden interruption in the welding operation due to power failure or accidental stoppage can cause porosity and slag inclusion in the weld. Lack of fusion and high stress concentration are some other consequences of the sudden stoppage of welding. The weld shape and the dimension may also change.

13. Unfilled Craters (Figures 33.19 and 33.20)

The craters remain unfilled when there is an abrupt interruption of the arc. The defects formed may act as a source of crack formation, lead to an undercut, if terminated at the parent metal.

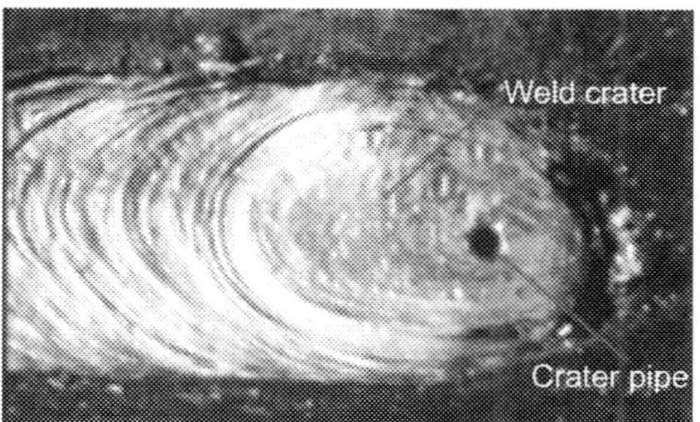

Figure 33.19 Unfilled crater in butt weld

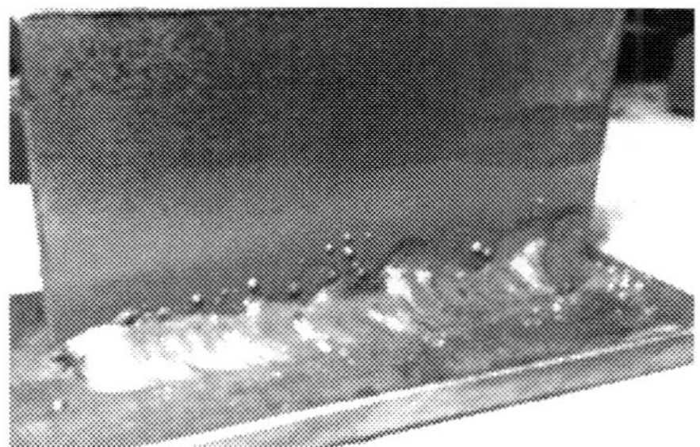

Figure 33.20 Unfilled crater in fillet weld

14. Burn-Through (Figure 33.21)

A burn-through is a hole burned through the parent metal in a single pass weld or a root run in a multirun weld. Excessive root gap, insufficient root face, excessive welding current, low welding speed are the reasons for the burn-through defects. These defects are avoided by the correct edge preparation, using suitable weld parameters and chip-back and weld-up.

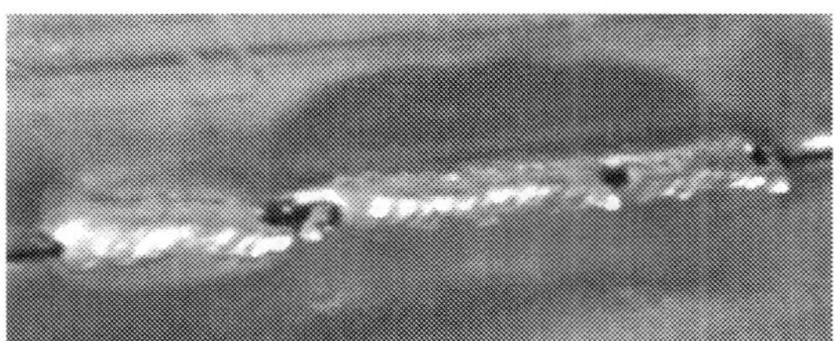

Figure 33.21 Burn-through

15. Surface Porosity (Figure 33.22)

Surface porosities are formed because of the use of undried coated electrodes and expired shelf life electrodes. They are cavity-type discontinuities formed by gas entrapment during solidification or in thermal spray deposit. Surface porosities cause decrease of strength in the joint, stress concentration and many more problems. Fresh electrodes are used after duly drying, to desist the surface porosity.

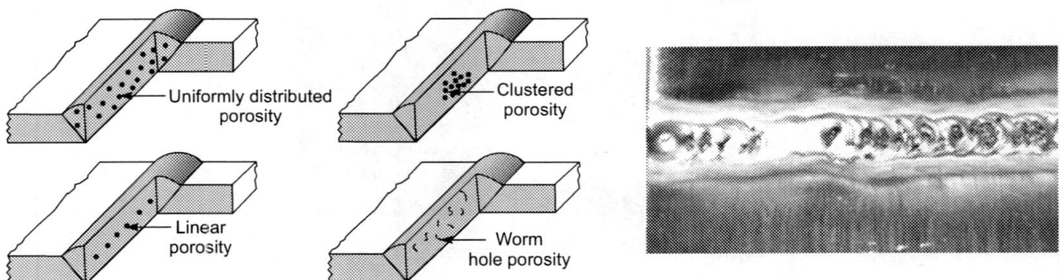

Figure 33.22 Surface porosity

Causes

1. High welding speed.
2. Dirty base metal.
3. Damp and damaged electrode.
4. Improper arc length.
5. More hydrogen, nitrogen and oxygen content in welding atmosphere.

Remedies

1. Reduced welding speed to allow time for gases to escape.
2. Use of properly coated electrode.
3. Use of filler metal with proper deoxidizers.

16. Surface Oxidation (Figure 33.23)

It is the formation of oxide layer around the weld of readily oxidizing metals, like Al, Ti, Mg and their alloys. These are produced due to the improper and inadequate gas shield envelope. By providing proper and adequate shielding of welding zone, oxidation can be eliminated. There will be difference in the color in the heat affected zone due to the interference produced by the oxides.

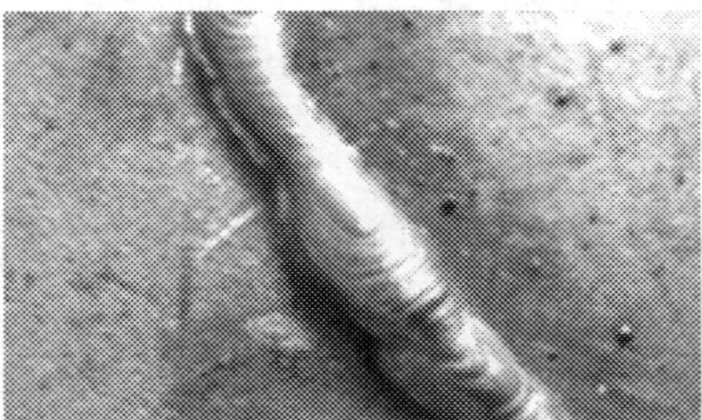

Figure 33.23 Surface oxidation

17. Uneven and Coarse Ripples

These are caused by the inconsistent solidification pattern due to the inconsistent heat input. The defects are reduced by making the welding parameters smooth, regular and consistent.

18. Flat Spots/Pock Marks/Gas Dents (Figure 33.24)

These defects are found in arc welding operations, mainly in SMAW and FCAW. Use of high welding current, high moisture content in the flux, excessive gas pressure and the gas formation by reaction between molten flux and molten weld metal are the main reasons for the defects. They appear as slight depressions and may lead to lack of fusion at these flat spots in multilayer welding operations. This defect can be eliminated by providing proper control of welding current gas flow rate and moisture content of flux.

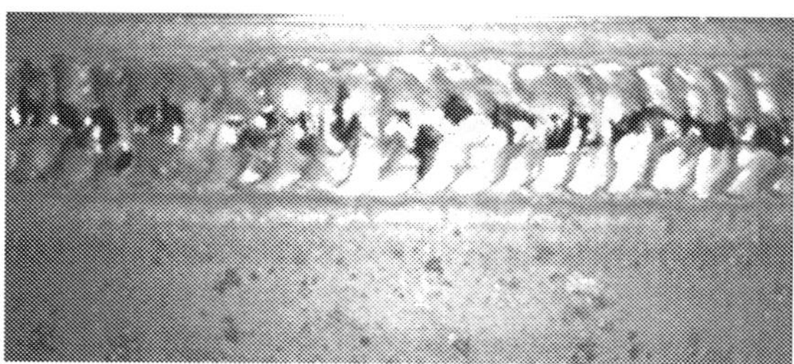

Figure 33.24 Flat spots/pock marks/gas dents

19. Excessive Spatter (Figure 33.25)

Here, the metal particles expelled during fusion welding do not form a part of the weld. Arc blow, wrong selection of electrode, Poor welding technique, and heavy welding current are the main causes for excessive splatter. It impairs the quality of the weld metal such that the metal surface will not take paint well. Anti-spatter spray on the base plate adjacent to the weld is used to reduce excessive spatter.

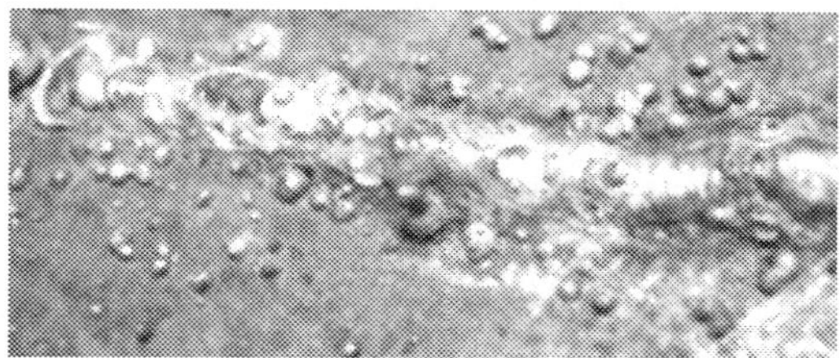

Figure 33.25 Spatter

20. Arc Strikes

Arc strikes are the short duration arc strikes at unwanted places on the work. Faulty ground clamp or inadvertent strikes of the electrode or the electrode holder against the work causes the defects. It can lead to initiation of cracks in bending or cyclic loading and also lead to hard spots. The defects are removed by chipping and preheating, if possible, before rewelding in the case of low-alloy steels.

21. Bad Weld Surface Appearance

These defects are the result of bad workmanship like, the variation in width of the weld, humps in the surface, occasional depressions, uniform ripples, overfilled craters, etc. These cannot be avoided without improving the welder's skill.

22. High-Low Offset (Figure 33.26)

The poor fit-up of the welding metals like the unmatching edges putting together and welding, the waviness of plates are the causes of these defects. To avoid these, more tack welds and welding jigs and clamps/fixtures are used.

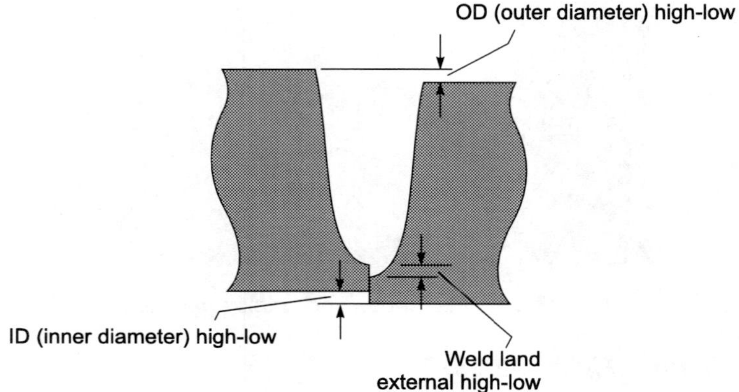

Figure 33.26 High low off-set

Review Questions

1. What is meant by defect in weldment?
2. What are the causes for defects in weldment?
3. Write the classification of weld defects.
4. Write the factors responsible for hot and cold cracks.
5. What is hydrogen cracking?
6. What are the causes for undercut?
7. What are the causes for lack of penetration?
8. What are the remedies for underfill?
9. What are dimensional defects?
10. How do we overcome porosity problems in weldment?

34
CHAPTER

Inspection and Testing of Welds

34.0 INTRODUCTION

Present-day engineering industry relies heavily on the integrity of welds for adequate and reliable performance of components, structures and plants. Weld integrity depends on the base material, specifications and welding processes. Reliability of weld performance is evaluated by measurement and control of weld properties. It is accepted widely that testing, measurement and control of welds should be optimized based on fitness-for-purpose approach, taking into account the welding processes and economical aspects of ensuring the desired levels of reliability. Recent advances in test techniques for ensuring the desired quality have met high technological demands.

34.1 DIFFERENCE BETWEEN TESTING AND INSPECTION

Testing is specially referred to the physical performance of the operations to find out quantitative measure of certain properties, such as mechanical, physical, etc. Testing can be of destructive or non-destructive type. Inspection refers to the examination of welding structure. Inspection may be qualitative and involves only visual observation of correctness of functioning/dimensions. Inspection can be done before welding, during welding and after welding. Inspection is the activity of examining the product or its components to determine if they meet the design standards, testing is a procedure in which the item is observed during operation, in order to determine whether it functions properly for a reasonable period of time.

Inspection Before Welding

1. Check the welding joint drawing, such as dimensions, tolerance, process specification.
2. Choose a suitable welding process by easy method.
3. Set the welding parameters, such as current, voltage, frequency, polarity, etc.
4. Select defect-free material as per specifications.

5. Selection of proper size, proper flux coated electrode as per specification.
6. Select the jig and fixture of proper size as per welding joint complexity.
7. Proper arrangement of welding joint cooling and ventilation as well as smoke.

Inspection in Between the Welding

1. Check whether welding groove and edge preparation details are correct or not.
2. Tack weld should be of adequate size, length and pitch.
3. Method of welding to be used.
4. Welding position. Fill-up gap should be as per the procedure.
5. Slag on the welding joint should be properly removed from each pass in multipass arc welds.

Inspection After Completion of Welding

1. Determine the properties of weld and weld quality.
2. Find out the suitability and functioning through non-destructive or destructive testing.

34.2 WELD IMPERFECTIONS AND IMPORTANCE OF THEIR EVALUATION

Historically, welding has replaced riveted construction in engineering structures. It is now scarcely possible to design an industrial structure without a welded joint. No weld is completely perfect. Welds may be compared to small castings, except that weld metal cools much more rapidly, mainly due to heat sinks provided by the base metals. This results in thermal stresses that may lead to cracking and also due to the entrapment of gases or foreign materials within the weld. These and other defects may cause premature failure of the weld in service.

Defects in welded joints can be of three types:

1. Physical discontinuities
2. Microstructural defects
3. Defects related to residual stress and distortion

The relevance and importance of defects are best understood through fracture mechanics concepts, wherein the following parameters are important:

(a) Defect size
(b) Defect shape
(c) Defect location
(d) Loading, including both externally imposed and arising out of the presence of residual stresses.

The objective of a good and effective testing program is to detect defects as specified by the design based on fitness-for-purpose.

34.3 TENSILE PROPERTIES: STRENGTH AND DUCTILITY

Tension and bend tests are frequently used to evaluate the breaking strength and ductility of a metal and to determine that the metal meets applicable specification requirements. Welded joints contain metallurgical and compositional differences that result from the welding process. It is important to know the effects of these changes on mechanical properties. Therefore, tension and bend tests are frequently made to determine the suitability of the welded joint for service. They are also often used to qualify welding procedures or welders according to specific code requirements.

34.3.1 Tension Tests

Base Metal Tension Test

Longitudinal or transverse test (Figure 34.1): The strength and ductility of metals are generally obtained from a simple uniaxial tensile test, in which a machined specimen is subjected to an increasing load while simultaneous observations of extension are made. Longitudinal specimens are oriented parallel to the direction of rolling, and transverse specimens are oriented perpendicular to the rolling direction. The load can be plotted against elongation, customarily as shown in Figure 34.2. The stress (load divided by original area) is plotted against the strain (elongation divided by the original gage length). Several important engineering properties can be determined from the tensile test. These properties are the yield strength, tensile strength, and ductility, expressed in percentage elongation, and percentage reduction in cross-sectional area.

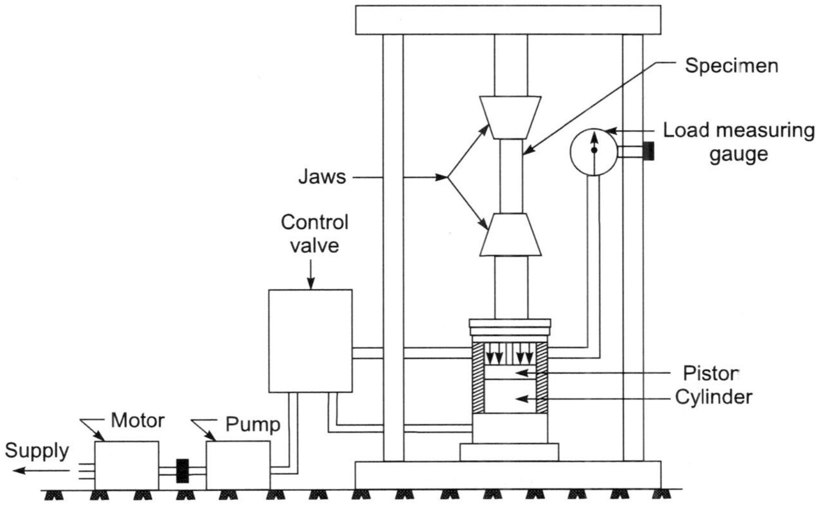

Figure 34.1 Tensile testing machine

The yield strength shown by the engineering stress-strain curve is generally the strength at some arbitrary amount of extension under load or a permanent plastic strain (offset). For certain steels that exhibit such a phenomenon, yielding takes place at the point where plastic extension occurs with no increase in load (upper yield point, Figure 34.2). The ultimate tensile strength is the maximum stress on the stress-strain curve. It can be calculated by dividing the maximum load by the original cross-sectional area. Typical stress-strain curves for several commercial steels are shown in Figure 34.3.

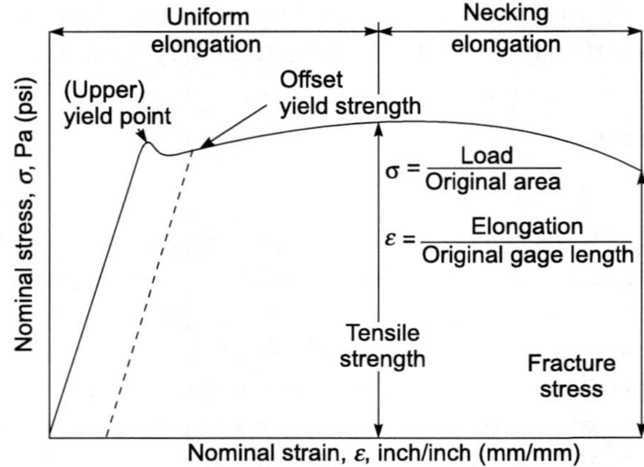

Figure 34.2 Engineering stress-strain curve for low-carbon steel

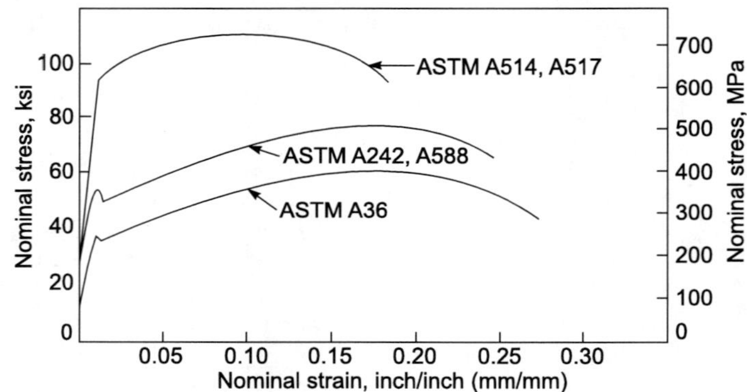

Figure 34.3 Nominal stress-nominal strain curves for various constructional steels

Two standard measurements of ductility are derived from an engineering tension test—elongation and reduction of area expressed in percent. The percent elongation is the ratio of the increase in gage length to its original gage length. Because a measurement of percent elongation includes both uniform strain and local or necking strain, this ductility value is affected by the choice of original gage length and by the geometry of the specimen cross section. Therefore, comparisons of percent elongations should be made only for identical specimens. The percent reduction of area is the ratio of the decrease in cross-sectional area after fracture to the original cross-sectional area.

The details of specimen preparation and test procedures for smooth tension tests for base metals are described in ASTM A370, standard methods and definitions for mechanical testing of steel products.

Weld Tension Test

The tension testing of welds is somewhat more involved than testing base metal because a weld test section is heterogeneous, containing weld metal, heat affected zone, and unaffected base metal. To obtain an accurate assessment of the strength and ductility of welds, several different specimens and orientations may be used, as shown in Figure 34.4. In some cases, the weld reinforcement is left intact on the test specimen.

All-Weld-Metal Test: To determine the tensile properties of weld metal, the test specimen orientation is parallel to the axis of the weld, and the entire specimen is machined from the weld metal. The chemical composition of the weld metal will be affected by the joint penetration.

If the purpose of the test is to qualify a filler metal, then melting of the base metal should be minimized when making the test weld. This procedure is described in AWS A5.1, specification for covered carbon steel arc welding electrodes and other filler metal specifications. If the purpose of the test is to determine properties of the weld metal in a particular weldment, the welding process and procedure used in actual fabrication should be employed to make the test weld.

The mechanical properties measured and reported in all-weld-metal tension test are tensile strength, yield strength, elongation and reduction in area.

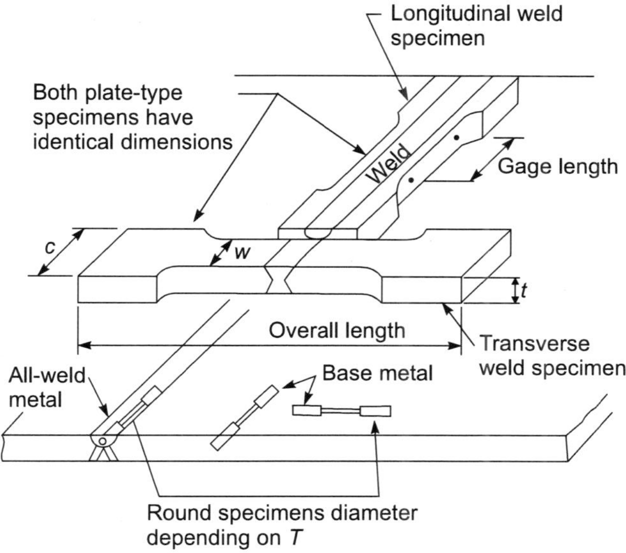

Figure 34.4 Typical test specimens for evaluation of welded joints

Transverse Weld Test: Interpretation of test result for a welded joint as a whole is not possible in the transverse weld specimen. The reduced section of this specimen contains base metal, heat affected zones and weld metal. When all of these materials are simultaneously subjected to the same stress, the one with the lowest strength will elongate and break. For example, if the weld metal strength is higher than that of the unaffected base metal, then failure will occur outside the weld area, and no quantitative information about the weld metal strength is provided by the test. Thus, this test should not be used to make quantitative comparison of weld metals.

The most common use of transverse weld tensile tests is to qualify welding procedures. The purpose of the test is to verify that the welding procedure will produce welds that equal or exceed design strength requirements. Only the tensile strength and the location of the fracture are normally reported for transverse weld tension tests.

Longitudinal Weld Test: In the longitudinal weld test, the direction of loading of the specimens is parallel to the weld axis, but the test differs from an all-weld-metal test specimen in that the reduced cross section contains weld, heat affected zone and base metal. During testing, all of these zones must strain equally and simultaneously. Weld metal, regardless of strength; elongates with the base metal until failure occurs. Low weld or heat affected zone ductility may initiate fracture at strength levels below that of the base metal. Only the tensile strength of longitudinal weld tests is normally reported. Elongation may be measured as an indication of minimum ductility of the joint.

Tension-Shear Test

Fillet Weld Shear Test: Tension-shear tests determine the shear strength of fillet welds. The test specimens are usually intended to represent actual weldments and therefore, the welds are prepared using production procedures. Two specimen types, transverse and longitudinal, are shown in Figure 34.5.

To avoid rotation and bending stresses during testing, transverse shear specimens are tested as double lap joints.

The shear strength of the weld metal, calculated as shown in Figure 34.6 and the location of fracture are the test results normally reported.

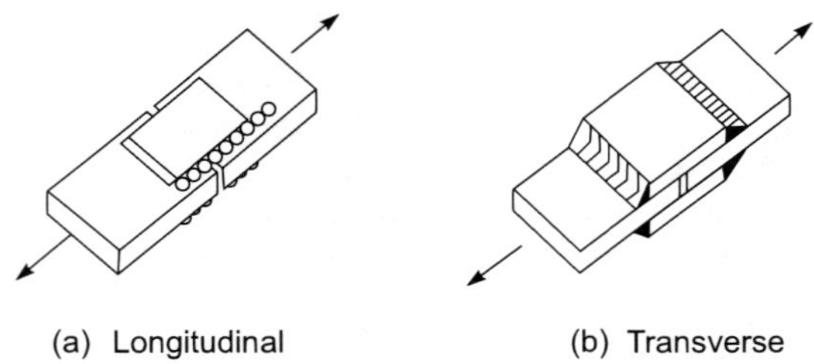

Figure 34.5 Longitudinal and transverse fillet weld shear stress specimens

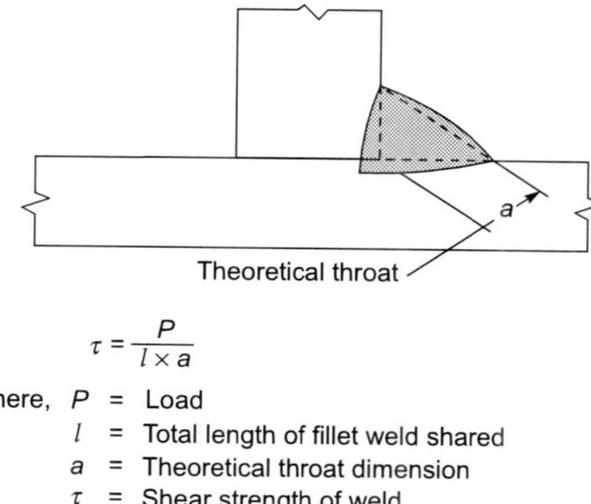

Figure 34.6 Method of determining shear strength of fillet welds

34.3.2 Bend Test (Figure 34.7)

The bend test assists in determining the soundness of the weld metal, the weld junction and the heat affected zone. The test shows the quality of the welded joint. Any cracking of the metal will indicate false fusion or defective penetration.

The stretching of the metal determines, to some extent, its ductility. Fractured surface shows the crystalline structure. Large crystals usually indicate wrong welding procedure or poor heat treatment after welding. A good weld has small crystals. The bend test is an easy and useful method of comparing one welded joint with another of the same type and of revealing abnormalities and defects at or near the surface in tension.

Types of Bend Tests

Bend tests may be categorized as:

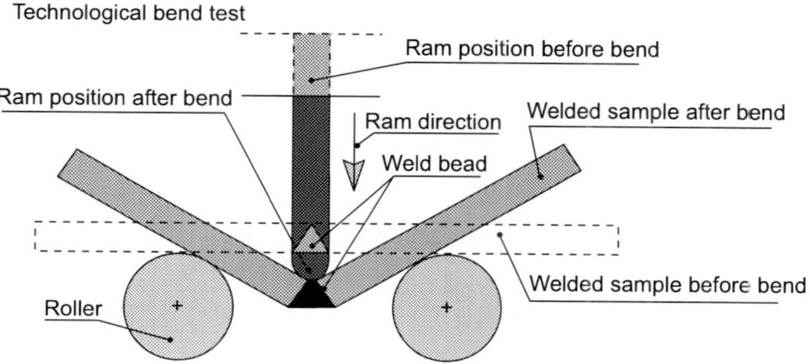

Figure 34.7 Bend test

1. Free bend tests
2. Guided bend tests

Bend tests may be further classified as:

1. Transverse bend tests
   (a) Root bend test
   (b) Face bend test
2. Longitudinal bend tests
3. Side bend tests

Free Bend Tests (Figure 34.8)

Free bend test determines the ductility of weld metal. Free bend test may be conducted on a tensile testing machine or a vice capable of exerting sufficiently large compressive force. For bend tests, the test pieces are cut from the plate to include the weld. The top of the weld is ground or machined so that it becomes flush with the base metal surface. This specimen may be bent initially by using a device. After giving an initial bend onto the specimen, it is placed in another fixture. The bending is continued until a crack or open defect, exceeding 1.5 mm, in any direction appearing on the convex surface of the specimen. If no crack appears, the specimen shall be bent double.

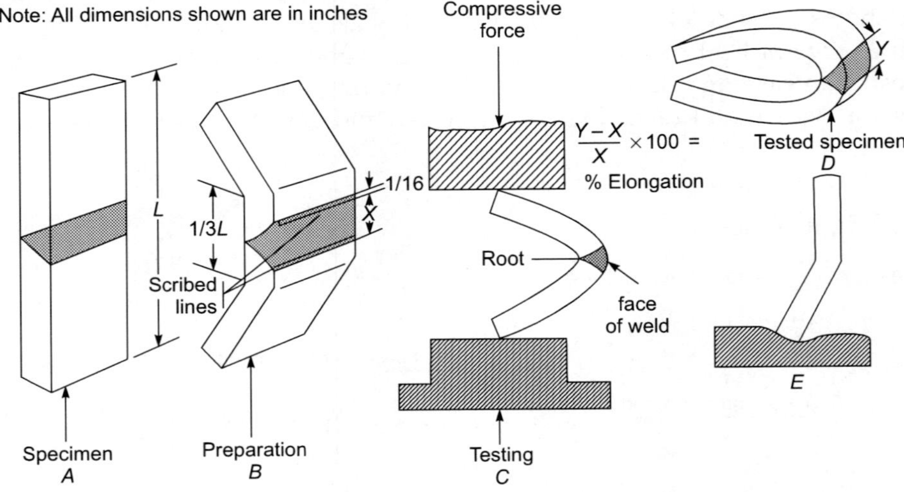

Figure 34.8 Free bend test of welded metal

Guided Bend Tests (Figure 34.9)

A guided bend test shows surface imperfections near and in the weld bead. A guided bend test is performed on the specially designed jig. The specimen to be tested is first ground smooth so that weld reinforcement is removed. The specimen is then placed across the die supports and bent by depressing the plunger until it forms the shape of 'U'.

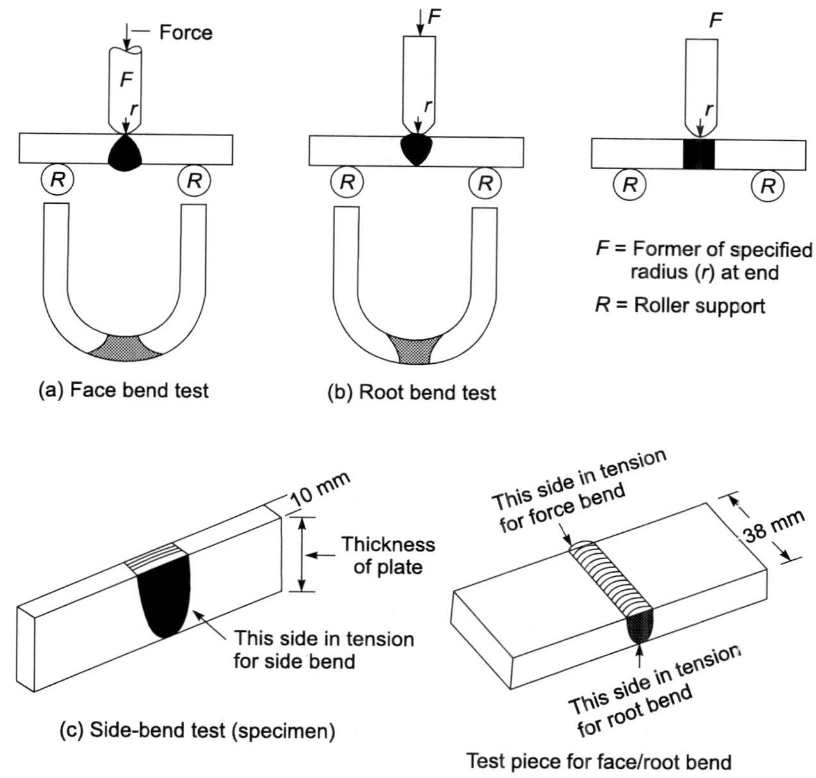

Figure 34.9 Guided bend test

Transverse Bend Test

A transverse bend test is useful in qualifying welders because it quite often reveals the presence of defects that are not detected in tension tests. However, in this test, non-uniform properties along the length of the specimen can cause non-uniform bending.

The specimen for transverse bend test is of the full thickness of the material at the welding joint and the upper and lower surface of the weld are dressed flush with the base metal surface. The specimen is bent by the movement of a plunger or a former. Suitable gage marks, if scribed at the outside surface of the specimen, help estimate the percentage elongation.

In a face bend test, the specimen is placed with its face down. The former is depressed until the piece becomes 'D' shaped in the die. If upon examination, cracks greater than 3 mm appear in any direction, the weld is considered to have failed.

Face bend tests are used to inspect the degree of fusion, the absence or presence of inclusions and the weld porosity, if any.

In a root bend test, the specimen is placed in the jig with the root down or in just the reverse position of the face bend test. The results may show no cracks to be acceptable. Root bend tests are used primarily to determine the degree of weld penetration.

Longitudinal Bend Test

The problems of weld mismatch can be avoided by using longitudinal bend specimens in which the weld runs the full length of the bend specimen; the bend axis being perpendicular to the weld axis.

In longitudinal bend tests, all sorts of welded joints are strained equally and simultaneously. This test is generally used for evaluating of joints in dissimilar metals. Specimens for longitudinal bend tests are prepared in the same manner as for transverse bend tests.

Side Bend Test

Side bend specimens strain the entire weld cross section and are thus, especially useful for exposing defects near mid thickness, that might not contribute to failure in face or root bend tests. A side bend test determines the soundness of the welded joint in cross section. The test is used for relatively thick sections (over 19 mm), since in this test, the entire weld thickness may be included in the test.

The width of the test specimen shall be the full thickness of the material at the welded joint and the upper and lower surfaces of the weld shall normally be dressed flush with the surface of the base metal. The specimen is placed over the supports and it is bent by the downward movement of the former. The specimen is examined after the test and type and location of flaws present, if any, are noted.

34.4 IMPACT TEST

Impact testing becomes essential in order to study the behavior of welded objects under dynamic loading. An impact test determines the behavior of welds when subjected to high rates of loading, usually in bending.

An impact test gives an indication of the relative toughness of the material. Toughness is defined as the resistance of a metal to fracture after plastic deformation begins. The purpose of impact testing is to determine the amount of impact a specimen will absorb before fracturing.

In an impact test, a specimen machined or surface ground and notched is struck and broken by a single blow in a specially designed testing machine. The quantity measured is the energy absorbed in breaking the specimen by a single blow. The ideal impact test would be one in which all the energy of the a blow is transmitted to the test specimen.

Types of Impact Tests

The two basic types of impact tests are:

1. The Charpy test (Figure 34.10 (a))
2. The Izod test (Figure 34.10 (b))

The Charpy specimen is placed in the vice so that it is just a simple beam supported at the ends, whereas the Izod specimen is placed in the vice such that it is in the form of a cantilever.

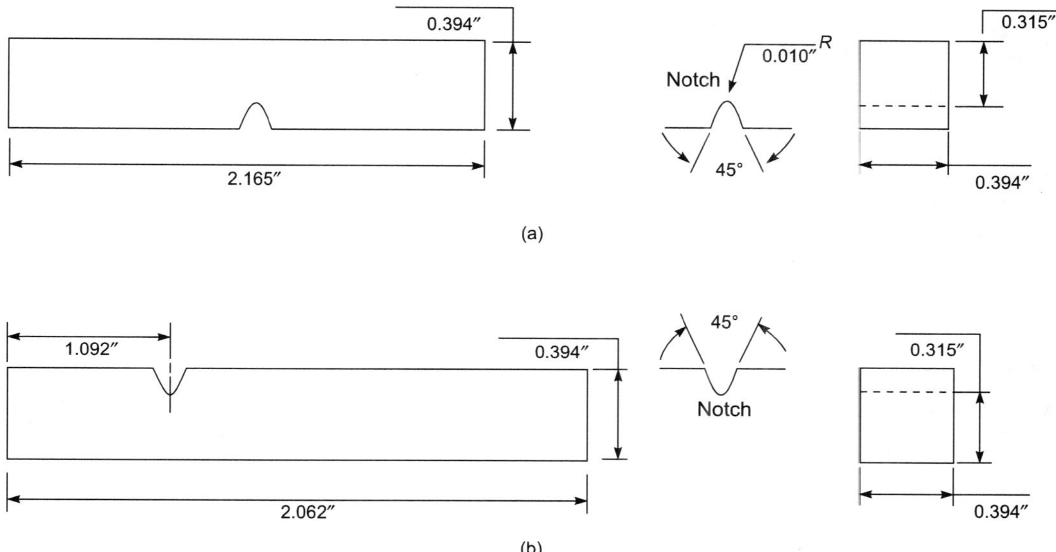

Figure 34.10 Impact tests (a) Charpy test (b) Izod test

Test Procedure (Figure 34.11)

1. The swinging pendulum weight is raised to standard height depending upon the type of specimen to be tested.
2. With reference to the vice holding the specimen, the higher the pendulum, the more potential energy it has got.
3. As the pendulum is released, potential energy is converted into kinetic energy until it strikes the specimen.
4. The Charpy specimen is hit behind the V-notch, while the Izod specimen placed with the V-notch facing the pendulum will be hit above the V-notch.
5. A portion of energy possessed by the pendulum is used to rupture the specimen and the pendulum rises on the other side of the machine to a height lower than its initial height on the opposite side of the impact testing machine.

6. The energy consumed in breaking the specimen is the weight of the pendulum times the difference in height of the pendulum from mean position.
7. This energy in ft-lbs or m-kg is the notched impact strength and can be read from the dial of the impact testing machine.

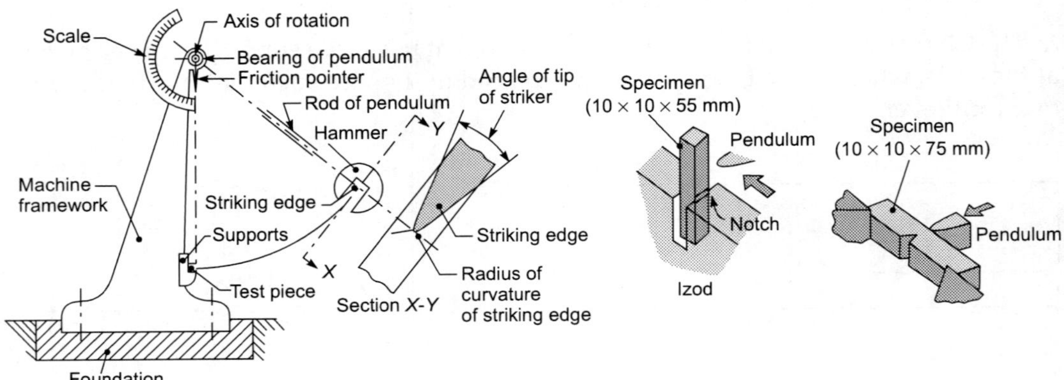

Figure 34.11 Test setup for Izod and Charpy tests

34.5 NICK BREAK TEST

A nick break test involves breaking the weld joints to examine the fractured surface for internal defects, such as:

1. Gas pockets
2. Slag inclusions
3. Porosity

Test also determines weld ductility and degree of fusion.

Procedure

The test specimen is cut transversely to the welded joint and shall have the full thickness of the plate at the joint. The excess weld metal and penetration bead shall be left intact. Slots are sawed at each end of the specimen to be tested.

The specimen is placed upright on two supports and force is applied on the weld, either by a press or by sharp blows on a hammer until a fracture occurs between the two slots. Figure 34.12 shows the nick break test arrangement.

A visual inspection of the fractured surfaces is carried out in order to find defects, if any. If any defect exceeds 1.5 mm in size or the number of gas pockets exceeds 1 cm^2, the piece has failed the test.

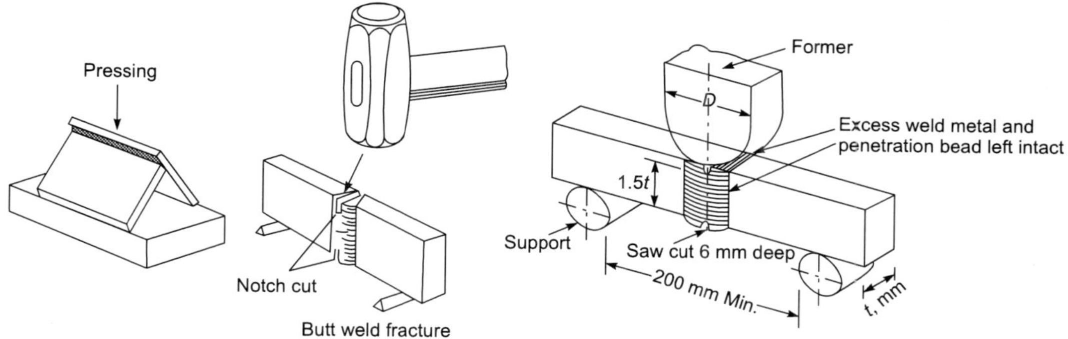

Figure 34.12 Nick break test specimen

34.6 HARDNESS TEST

The hardness test gives an idea of the resistance of wear of the weld metal. This is important with respect to the components, which have been built up and have to withstand abrasive wear. Hardness values can give information about the metallurgical changes caused by welding.

Methods of Hardness Testing

Commonly used test methods for finding hardness are:

1. The Brinell test
2. The Rockwell test
3. The Vickers test

The weld specimen in whose case hardness is to be tested is ground, polished and etched to show clearly the weld metal area. Hardness is determined on specific areas of interest, including the weld centerline, face or root regions of the weld deposit, the heat affected zone and base metal.

Brinell Hardness Test

It consists of pressing a hardened steel ball into a test specimen. According to ASTM specifications, a 10 mm diameter ball is used for this purpose. Lower loads are applied for measuring hardness of soft materials and vice versa.

Procedure of Hardness Testing

Specimen is placed on the anvil; the hand wheel is rotated so that the specimen along with the anvil moves up and contacts the ball. The desired load is applied mechanically or hydraulically and the ball presses into the specimen.

The diameter of indentation made in the specimen by the pressed ball is measured by the use of a micrometer microscope, having a transparent engraved scale in the field of view.

The indentation diameter is measured at two places, right angled to each other, and the average of the two readings is taken. The Brinell hardness number (BHN), which is the pressure per unit surface area of the indentation in kg/sq. m, is calculated as follows:

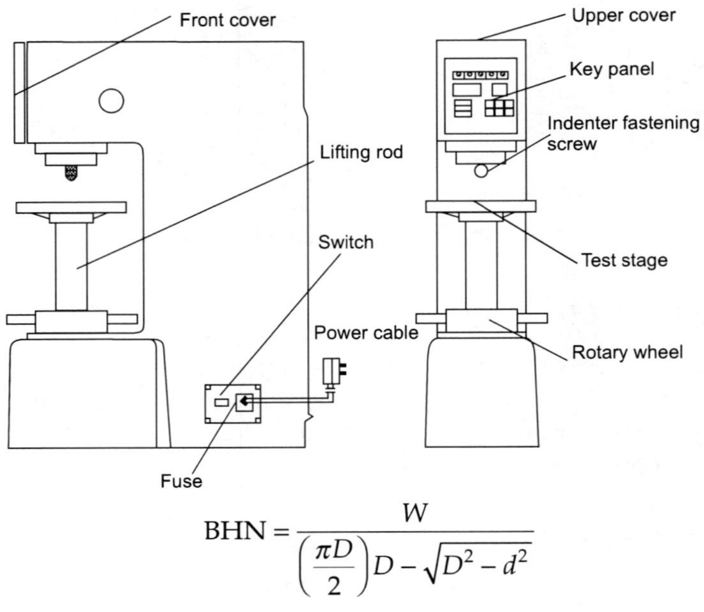

$$\text{BHN} = \frac{W}{\left(\frac{\pi D}{2}\right) D - \sqrt{D^2 - d^2}}$$

where,

W: Load on indenter (kg)
D : Diameter of steel ball (mm)
d : Avg. measured diameter of indentation (mm)

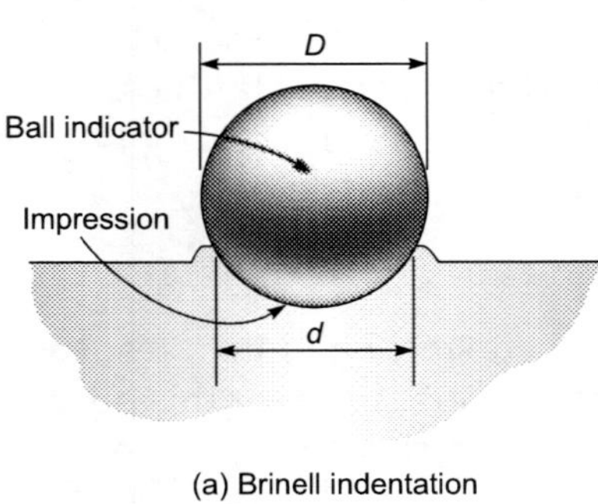

(a) Brinell indentation

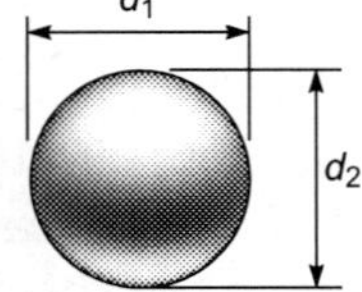

(b) Measurement of impression diameter

Figure 34.13 Brinell hardness test

Rockwell Hardness Test (Figure 34.14)

Rockwell hardness testing differs from Brinell testing in that the indenter and loads are smaller and therefore, the resulting indentation on the specimen is smaller and shallower. Rockwell testing is suitable for materials having hardness beyond the scope of Brinell testing.

Rockwell testing is faster when compared to Brinell testing because diameter of the indentation need not be measured. The Rockwell machine gives arbitrary direct reading. Unlike Brinell testing, Rockwell testing needs no surface preparation of the specimen whose hardness is to be measured.

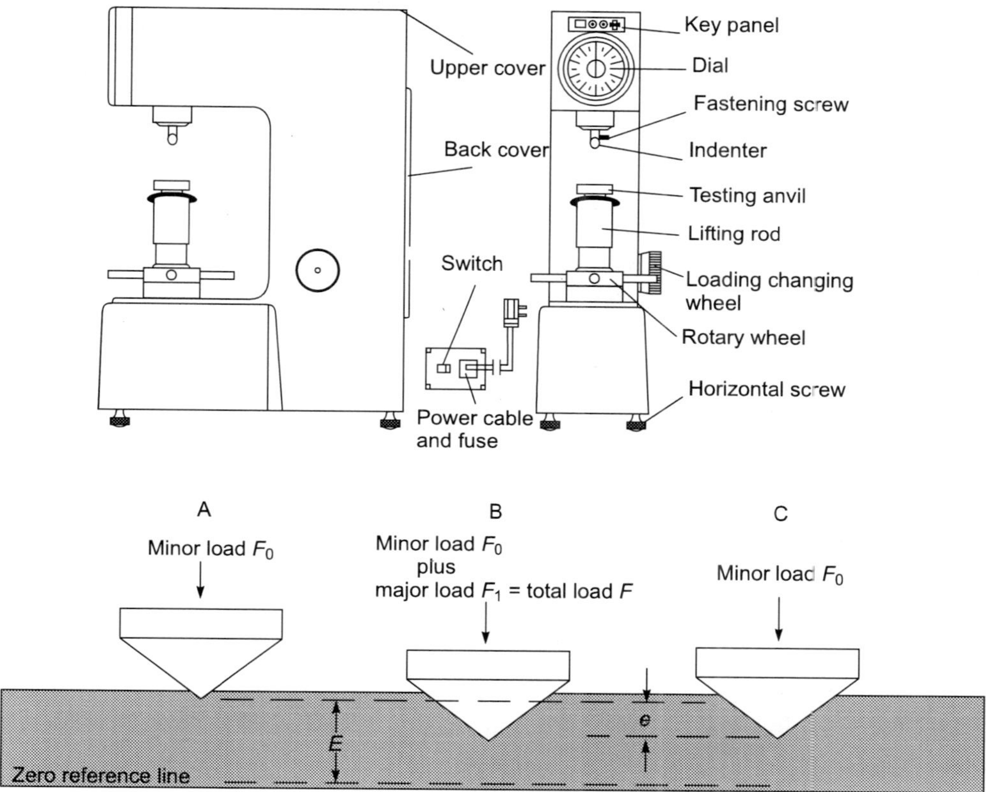

Figure 34.14 Rockwell hardness test

Procedure for Measuring Hardness

1. Test piece is placed on the machine.
2. Handwheel is turned, thereby raising the test piece up against the steel ball indenter till the needle on the dial reads 0. This applies minor load.

3. Major load is applied by pressing the crank provided on the right hand side of the machine.
4. The crank is turned in the reverse direction, thereby withdrawing major load but leaving minor load applied.
5. Hand wheel is rotated and test piece is lowered.

At this stage, the hardness of the test piece material can be directly read from the dial scale. There are two scales on a Rockwell testing machine; B Scale and C Scale.

B Scale uses a steel ball indenter, whereas a diamond cone penetrator is employed for measuring hardness on C Scale. B Scale is for testing materials of medium hardness, such as low and medium carbon steels in annealed condition. The working range of the scale is from 0 to 100. C Scale is used for testing materials harder than B 100.

C Scale is commonly used for testing the hardness of alloy cast irons. In Rockwell hardness testing, the minor load for all cases is 10 kg, whereas major loads for scales C and B are 150 and 100 kg, respectively.

Vickers Hardness Test (Figure 34.15)

In Vickers hardness test, a known load (P) from 1 to 120 kg is applied for a specified time to the surface of the material through a square-based pyramid diamond, having 136 degrees between opposite faces. The two diagonals of the resulting square indentation on the test piece are measured with a micrometer microscope and averaged (D mm). The Vickers hardness number is calculated as follows:

$$VHN = 1.854\frac{P}{D^2}$$

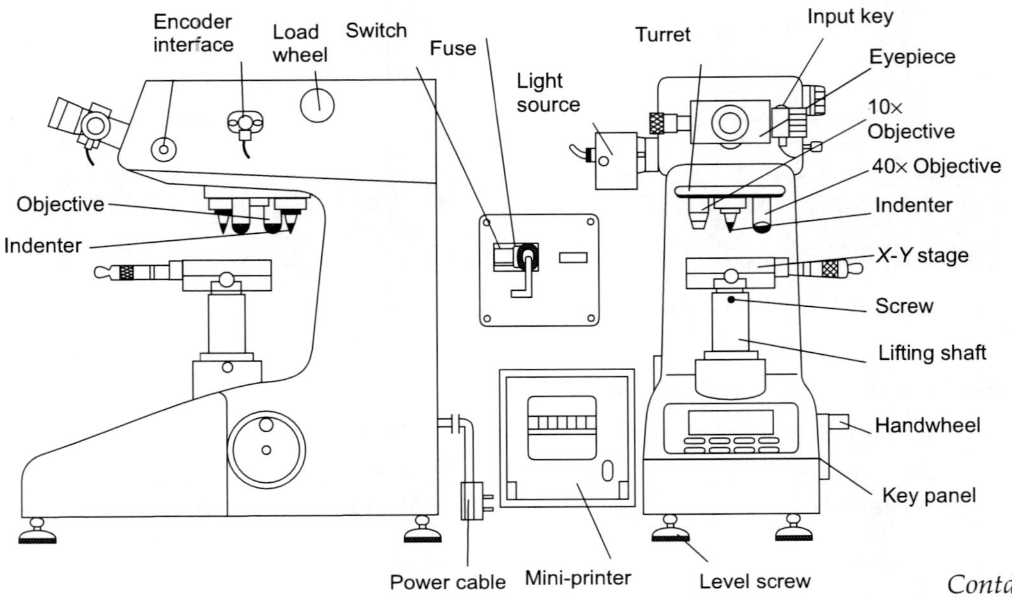

Contd.

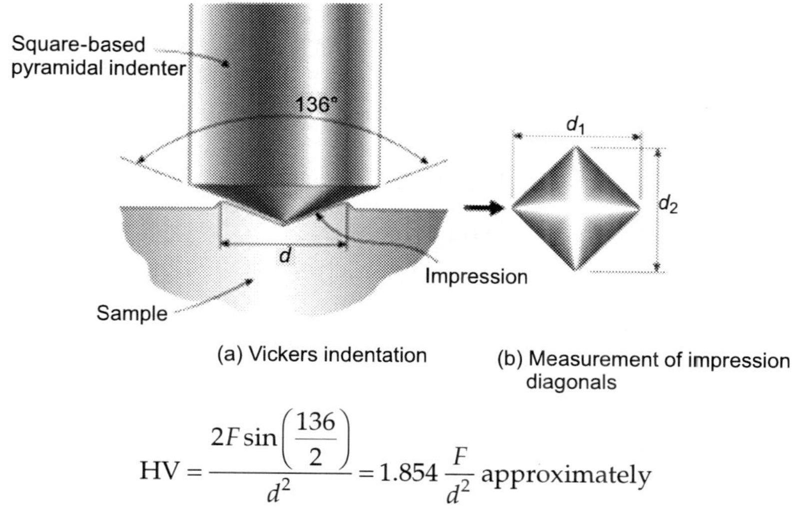

$$HV = \frac{2F\sin\left(\frac{136}{2}\right)}{d^2} = 1.854\frac{F}{d^2} \text{ approximately}$$

Figure 34.15 Vickers hardness test

34.7 ETCH TEST

An etch test involves inspecting the welded test specimen after polishing and etching the same with a chemical reagent, e.g., a diluted acid.

Types of Etch Test

There are two types of etch tests, namely:

1. Macroetch examination
2. Microetch examination

Macroetch Examination: After preparing the specimen by polishing and etching, it is examined either by naked eye or by low power magnification up to ×15. Macroexamination gives a broad picture of the specimen by studying relatively large sectioned areas.

Macroexamination in welded specimen reveals:

1. Cracks
2. Slag inclusions
3. Blowholes
4. Shrinkage porosity
5. Penetration of the weld
6. The boundary between the weld metal and the base metal, etc.

Microetch Examination: After preparing the specimen by polishing and etching, it is examined under a microscope at magnifications from ×20 to ×2000.

Microetch examination involves areas much smaller than those considered in macroetch examination and brings out information that can never be revealed by macroexamination.

Microexamination in a welded specimen determines:

1. Cracks and inclusions of microscopic size.
2. Grain boundaries and solidification structures of weld metal, heat affected zone and the base metal.
3. Distribution of microconstituents in the weld metal.
4. The quality of heat treatment, etc.

Preparation of Test Specimen

1. The specimen shall be the full thickness of the material at the welded joint and the weld reinforcement and penetration bead shall be left intact. The specimen shall contain a length of the joint of at least 10 mm and shall extend on each side of the weld for a distance that includes the heat affected zone and some base metal portion.
2. Specimen after being cut from the plate is filed or ground, to obtain flat surface on the specimen.
3. Intermediate and fine grinding is carried out using emery papers of progressively finer grades, i.e., of grades of 200 grit, 320 grit, 400 grit and 600 grit (from coarse to fine).
4. Rough and fine polishing of the specimen is carried out on a rotating polishing wheel. Fine polishing removes the scratches and very thin distorted layer remains on the specimen from the rough polishing stage.
5. The specimen is then etched in order to make visible the grain boundaries, heat affected zone, the boundary between the weld metal and parent metal, and so on. Etching imparts unlike appearances to the metal constituents and thus, makes structure apparent under the microscope.

   Etching is done either by:

   (a) immersing the polished surface in the etching reagenting, or
   (b) rubbing the polished surface gently with a cotton swab wetted with the etching reagent.

34.8 NON-DESTRUCTIVE TESTING TECHNIQUES

Non-destructive tests are applied to welded components to determine their suitability for the service conditions to which they will be subjected. These tests neither break nor alter the structure or appearance of the welded component. Non-destructive tests have the ability to detect invisible sub-surface defects.

Non-destructive tests (NDT) make components more reliable and safe. Although non-destructive tests do not provide direct measurements of mechanical properties, yet they are extremely useful in revealing defects in components that could impair their performance when put in service.

34.9 CHOICE OF NDT TECHNIQUES TO EVALUATE FITNESS-FOR-PURPOSE

For many highly stressed components, failure occurs in an elastic manner, involving fast fracture. The capability to detect planar defects is the first criterion that decides the relative merits of various NDT techniques, since fracture mechanics concepts indicate the prime importance of detecting and measuring planar defects. Hence, the NDT techniques that are most appropriate are those that are:

- sensitive to planar defects, whatever their orientation and position is
- sensitive to surface breaking defects
- capable of discriminating planar from non-planar defects

Comparison between destructive and non-destructive tests is given in Table 34.1.

TABLE 34.1 Comparison between destructive and non-destructive tests

	Destructive Testing	Non-Destructive Testing
Purpose	It is carried to find properties and behavior of specimen under different loads.	It is used to find properties of material and to find out defects.
Specimen	Specimen is damaged during test.	Specimen is not damaged during test.
Defects	Defects cannot be found using destructive testing.	Defects are found using NDT.
Cost	More costly.	Less costly.
Example	Bending test, tensile testing, compression testing, impact testing, etc.	Ultrasonic testing, liquid dye penetrant method, eddy current testing.

Non-destructive tests applied to welds are:

1. Visual inspection
2. Leak or tightness test
3. X-ray and gamma ray radiography
4. Magnetic particle inspection
5. Liquid dye penetrant test
6. Ultrasonic inspection
7. Eddy current testing

34.9.1 Visual Inspection

Visual inspection is probably the most widely used among the non-destructive tests. It is simple, easy to apply, quickly carried out and usually low in cost. Even though a component is to be inspected using other NDT methods, a good visual inspection should be carried out first. A simple visual test can reveal gross surface defects, thus leading to an immediate rejection of the component and consequently, saving much time and money, which would otherwise be spent on more complicated means of testing. It is often necessary to examine the weld joint for the presence of finer defects. For this purpose, visual methods have been developed to a very high degree of precision. With the advent of CCD cameras, microprocessors and computers, visual examination can be carried out reliably and with minimum cost. Image processing, pattern recognition and automatic accept/reject choice are used when large number of components are to be assessed. Visual inspection (VI) has wide applications for inspection of wrought, cast and welded materials. However, for welds, this is all the more important as at various stages of welding, visual inspection gives useful information. Many characteristics of a weld can be evaluated by visually examining a completed weld but much can be learnt by observing the weld as it is being made. For many non-critical welds, integrity is verified principally by visual inspection. Even when other non-destructive methods are used, visual inspection still constitutes an important part of quality control. Visual inspection should be done before, during and after welding. Visual inspection is useful for assessing the following:

- dimensional accuracy,
- conformity of welds to size, fit-up and control requirements,
- acceptability of weld appearance with regard to surface roughness, weld spatter, undercuts and overlaps, and
- Imperfections and cracks on the observed surfaces.

Visual examination can help detecting the following flaws on the surface of the welded structure:

1. Porosity, blowholes, pipes, exposed inclusions, unfused welds, unfilled craters, etc.
2. Surface cracks in the weld metal, heat affected zone or the parent metal.
3. Undercutting, burning or overheating of the base metal adjoining the weld metal.
4. Improper profile and dimensional inaccuracy of the welds.
5. There may exist excessive convexity or concavity, overlap, excessive reinforcement, unequal leg length, excessive penetration bead, shrinkage grooves, incompletely filled grooves, etc. Door weld appearance, i.e., irregular ripple marks, weaving faults, chipping and peening marks, spatter, surface roughness, etc.

Although visual inspection (VI) is a valuable method, it is unreliable for detecting subsurface flaws. Therefore, judgement of weld quality must be based on information from bulk material, in addition to that from surface indications. Capabilities of VI can be enhanced considerably by using simple gadgets and instruments for viewing, dimensional measurements, etc. In spite of the developments in the field of visual aids, there is no decision making electronic computer that can emulate the human brain. Intrascopes are available to permit entry and inspection of internal surfaces through access openings and the human inspector is equipped to supply a wealth of primary NDT results.

Optical Aids Used for Visual Inspection

The use of optical instruments in visual inspection is beneficial and is recommended to magnify defects that cannot be detected by the unaided eye and to permit visual checks of areas not accessible to the unaided eye. In performing visual/optical checks, it is of utmost importance to know the type of defects that may develop and to recognize the areas where such failures may occur. Magnifying devices and lighting aids should be used wherever appropriate. The general area should be checked for cleanliness, presence of foreign objects, and corrosion damage. In many cases, the area to be inspected should be cleaned before the visual examination.

Microscope

An optical microscope is a combination of lenses used to magnify the image of a small object. Minute defects and details of a fine structure on a surface can be detected more easily with the help of a microscope. The practical upper limit of the magnifying power of a simple microscope is in the region of 10×. Optical microscopes are used to evaluate cracks with respect to their shape and orientation. Low power microscopes are used at a magnification of 2 to 20× to examine external cracks on unprepared surfaces. Observation of microstructural details alongside the crack location usually requires a magnification of 100 to 500× after metallographic preparation.

Borescope

As the name implies, a borescope is an instrument designed to enable an observer to inspect the inside of a narrow tube, bore or chamber. Borescope consists of precision built-in illumination system having a complex arrangement of prisms and plain lenses, through which light is passed to the observer with maximum efficiency. The light source located in front or ahead of the objective lens provides illumination for the part being examined. Borescopes are available in numerous models from 2.5 to 6.19 mm in diameter and a few metres in length. Optical systems are generally designed to provide direct, right-angle, retrospective and oblique vision. The choice of inspection angle is determined by the flaw type and location. In most borescopes, the observed visual area is approximately 25 mm in diameter and 25 mm from the object. The size of the visual field usually vanes with the diameter, for a given magnification system.

Endoscope

The endoscope is much like a borescope except that it has a superior optical system and a high-intensity light source. Various viewing angles, as discussed in the case of borescope, can be used. A unique feature of the endoscope is that objects are constantly in focus from about 4 mm to infinity. Endoscopes are available in diameters down to 1.7 mm and in lengths from 100 to 1500 mm.

Flexible Fibre Optic Borescope (Flexiscope)

Flexible fibre-optic borescopes permit manipulation of the instrument around corners and through passages with several directional changes. Woven stainless steel sheathing protects the image-relaying fibre optic bundle during repeated flexing and maneuvering. These devices are designed to provide sharp and clear images of pans and interior surfaces that are normally impossible to inspect. Remote end-tip deflection allows the viewer to thread the fibroscope through a complex series of bends. High resolution CCD camera based fibroscopes are available for examination of the inner surface of the tubes.

34.9.2 Leak Testing

Leak testing is one of the non-destructive testing methods used for detection and location of leaks in either pressurized or evacuated systems and components as a result of pressure differential between the two regions. The word "leak" refers to the physical hole that exists and does not refer to the quantity of fluid passing through that hole. A leak may be a crack, crevice, fissure, hole or passageway that admits fluid or lets fluid escape. The word "leakage"refers to the flow of fluid through a leak without regard to the physical size of the hole through which flow occurs. "Leakage rate" refers to the rate of fluid flow per unit of time under a given set of conditions, and is expressed in units of mass per unit of time.

Leaks are a special type of flow that can have tremendous importance, where they influence the safety or performance of engineering systems. Leak testing is performed for three basic reasons:

1. To prevent material leakage loss, which interferes with system operation.
2. To prevent environmental contamination hazards caused by accidental leakage.
3. To detect unreliable components and those whose leakage rates exceed acceptance standards.

The end purpose of leak testing is to ensure reliability and serviceability of components and to prevent premature failure of systems containing fluids under pressure or vacuum.

The term "minimum detectable leak" refers to the smallest hole or discrete passage that can be detected and "minimum detectable leakage rate" refers to the smallest detectable fluid flow rate, generally known as sensitivity of the test system. The sensitivity of the instrument is the amount of leakage required for a leak testing instrument to give a minimum detectable signal. The instrument sensitivity is independent of test conditions, but when an instrument is applied to a test, the sensitivity of the test depends on the existing conditions of pressure, temperature and fluid flow.

Leak Testing Units

The leak is measured by how much leakage it will pass under a given set of conditions, as leakage varies with conditions. At a given temperature, the product of pressure and volume of the given quantity of gas is proportional to its mass. Therefore, leakage rate is often expressed as the product of some measure of pressure and volume per unit time. The basic SI units used in leak testing are: pressure in Pascal (1 N/m^2), volume in m^3 and time in seconds.

The fluid quantity can thus be expressed as Pascal cubic meter ($Pa{\cdot}m^3$) and the fluid leakage rate as Pascal cubic meter per second ($Pa{\cdot}m^3/s$) or in terms of std.cc/s (0.1 $Pa{\cdot}m^3/s$). SI unit for vacuum is Pa (132 Pa is 1 mm of Hg = 1 torr).

Types of Leaks

There are two basic types of leaks; real leak and virtual leak. A real leak is an essentially localized leak, i.e. a discrete passage through which fluid may flow. Such a leak may take the form of a tube, a crack, an orifice, etc. The flow can be of permeation type or diffusion type. Virtual leaks are leaks that involve the gradual desorption of gases from surfaces or components within a vacuum system.

Leak Testing Methods

Leak testing methods can be classified according to the pressure and fluid in the system. The commonly used methods of leak testing of systems at pressure are (i) pressure change, (ii) acoustic, (iii) bubble leak testing, (iv) flow detection, (v) specific gas detectors, (vi) quantity loss determination, etc. Leak testing is also performed by creating vacuum inside the component and applying the tracer gas on the other side to indicate the location and quantity of leak. The major factors that determine the choice of the leak testing methods are the physical characteristics of the system, the tracer fluid, the size of the anticipated leak and the reason for the testing. Leak testing can be employed for a variety of components and different ways, ranging from a small hermetically sealed component to very large vessels and piping systems. If the volume under test is filled with fluid at a pressure higher than the atmospheric pressure, then the test involved is called vacuum method. It is convenient under certain circumstances to carry out vacuum tests.

Pressure Change Method

The component or assembly is either pressurized or evacuated to a known value and then sealed. After a known time of hold, the pressure drop or pressure rise, if any, is measured, while maintaining the volume and temperature constant. The leakage rate can be arrived at from these. Atmospheric air and nitrogen are often used as pressurizing fluids in leak testing. Their fluid pressure serves to create a pressure differential, which causes the pressurizing gases to flow, by various mechanisms, through leaks. In general, higher the differential pressure, the greater the rate of leakage.

The leakage rate Q (Pa·m^3/s) can be calculated using the relationship,

$$Q = (P_1 - P_2) \cdot V/\Delta t$$

at constant volume (V in m^3) and temperature (T in K). $(P_1 - P_2)$ is the pressure difference (in absolute) and Δt is the time of hold (sec). Depending upon the leak tightness required and the sensitivity of the differential pressure gage, the time required for the reading of pressure drop varies. It should also be noted that the sensitivity depends on the volume of the component also.

The above is valid under the conditions that volume and the temperature do not change during the test, as it is a short duration test. When the same is conducted under varying temperature conditions, which require measurement of both gage pressure and temperature, and the barometric pressure is assumed to be one standard atmosphere (101.3 kPa), the pressure loss per unit of time is then determined from the initial gage pressure P_1 and temperature T_1 and the final gage pressure P_2 and final temperature T_2, by the equation

$$\frac{\Delta P}{\Delta t} = \frac{\left[P_1 - P_2\left(\frac{T_1}{T_2}\right)\right]}{\Delta t}, \text{ where } P_1, P_2 \text{ and } T_1, T_2 \text{ are absolute.}$$

If they are measured as pressure in kPa, temperature in °C, and time in seconds, then the equation will be:

$$\frac{\Delta P}{\Delta t} = \frac{\left[(P_1 + 101) - (P_2 + 101)\left(\frac{T_1 + 273}{T_2 + 273}\right)\right]}{\Delta t} - \left(\frac{kPa}{s}\right)$$

$$Q = \left(\frac{\Delta P}{\Delta t}\right) \cdot V$$

Evacuated Systems

Any pressure less than standard atmospheric pressure (101 kPa) is some form of vacuum. An enclosure is said to be under vacuum, if its internal pressure is less than that of the surrounding atmosphere. The vessels and components to be tested under vacuum should be designed for vacuum application. While designing such vessels, care has to be taken to provide vacuum compatible weld joints, material of construction to minimize outgassing, etc.

The vacuum hold test or pressure rise test is a pressure change leakage measurement test, performed on a system evacuated below the atmospheric pressure. It is mostly done by evacuating the vessel to an absolute pressure ranging from 10 Pa to 1 mPa (10^{-1} to 10^{-5} mm of Hg or torr). The leakage rate test is performed by isolating the system under test, after it has been evacuated to the required absolute pressure (vacuum). Then the pressure and the surface temperature of the system are observed for a specific time to determine the rate of pressure rise per unit of time for the system. Figure 34.16 shows the schematic of the test setup on a smaller test system. The leakage sensitivity increases with the decrease of volume and the sensitivity diminishes as the size or volume increases. In addition, the location of the leak cannot be detected by this test. Moreover, the sensitivity of the leakage rate test increases directly with elapsed time during the test.

Bubble Emission Technique

When one fluid gets trapped in another fluid, a bubble is formed. Most common examples are air bubbles in water and soap bubbles in air. Even small pressure differences are enough to form bubbles with liquids of low surface tension. This is taken due advantage of in the selection of liquid and to increase the sensitivity.

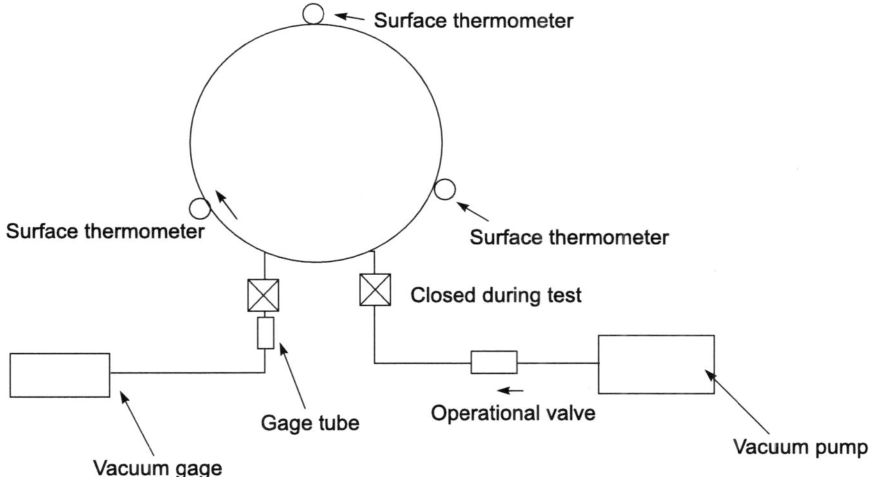

Figure 34.16 Arrangement of equipment for pressure rise leakage rate testing of an evacuated system (vacuum retention test)

In vacuum retention test, a gas pressure differential is established across a pressure boundary to be tested. A test liquid is then placed in contact with the lower pressure side of the pressure boundary. The gas leakage through the pressure boundary can then be detected by observing bubbles formed in the detection liquid at the exit points of leakage through the pressure boundary. This method provides immediate indication of the existence and location of large leaks (10^{-3} to 10^{-5} Pa m^3/s). Somewhat longer inspection time periods may be needed for detection of small leaks (10^{-5} to 10^{-6} Pa m^3/s), where the bubble indication forms slowly. In bubble leak testing, the rate of bubble formation, size of bubble formed, and rate of growth in the size of individual bubbles provide means for estimating the size of leaks. The bubbles are continuously supplied by gas and bubble formation or flow rate is proportional to the square of the pressure.

Bubble detection techniques can be divided into three major classifications:

1. **Liquid Immersion Technique:** The pressurized test object is submerged in the test liquid. Bubbles then form at the exit point of the leakage and tend to rise towards the surface of immersion bath.
2. **Film Application Technique:** A thin layer of test liquid is flowed over the low pressure surface of the test object; best suited for large volume/surface area components. For detection of small leaks, the test liquid should form a thin, continuous, wetted film, covering all areas to be examined.
3. **Foam Application Technique:** Used for detection of large leaks, in which the applied liquid forms thick suds or foam. When large leaks are encountered, the rapid escape of gas "blow a hole" through the foam blanket, revealing the leak location.

Sub-classifications of this technique are:

1. Controlled heating of sealed objects and small objects and small components to cause internal gas expansion. This increases the pressure differential to aid gas flow outwards.
2. Applying partial vacuum above the surface of the test liquid also results in increased pressure differential across the system boundaries. Refer Figure 34.17 (vacuum chamber technique).

Oils are more sensitive medium than water. Because of this, it is common to test electrical components in a bath of hot glycerine. When testing by reducing the pressure above the liquid, several precautions must be observed, as the reduced pressure brings the liquid closer to boiling point. The test vessel must be thoroughly cleaned to increase surface wetting, to prevent bubbles clinging to its surface and to prevent contamination of the fluid. If water is used, it must be of distilled or de-ionized grade. A small amount of wetting agent normally is added to water to reduce surface tension.

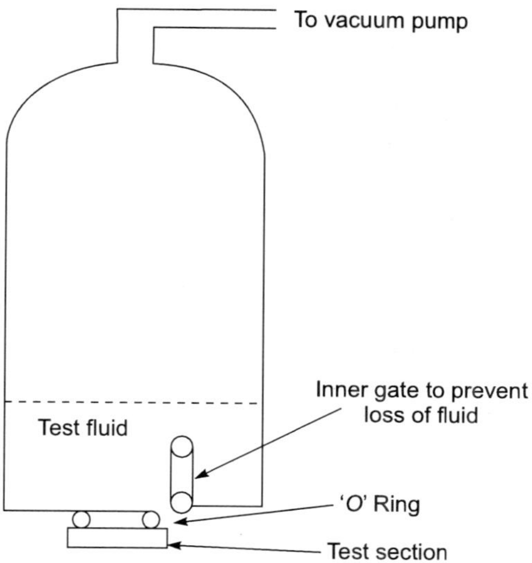

Figure 34.17 Vacuum chamber technique for providing pressure differential across leaks during bubble

Immersion testing may be used on any internally pressurized component that would not be damaged by the test liquid. Bubble testing methods are quite expensive and require little operator skill.

Vacuum box technique is used for detecting of leaks via through-thickness discontinuities in welds and pressure boundaries of systems containing air atmospheric pressure. Different forms of vacuum boxes can be designed and made as per the configurations of the area to be tested to accommodate joints like corner joint, circumferential pipe and tank joints, etc.

Ultrasonic Leak Detection

In this method, use is made of the fact that any gas at high pressure leaking out into the atmosphere produces acoustic disturbances in the air in different frequencies. High wave intensities are observed to be produced in the ultrasonic range of 40 kHz. The detector comprises the probe or a receiving transducer, which picks up the ultrasound and converts it into an electrical signal, a preamplifier to amplify the low signal, a converter circuit and a display system to indicate the amplified output.

The sensitivity of the system depends on the pressure of the gas used, the distance of the leak from the probe and the sensitivity of the circuit. The system must be calibrated with a standard leak. This method can be applied on (1) pipeline carrying chemicals, (2) air conditioning units, (3) pressurized cables, (4) boilers, (5) gas cylinders, etc.

Halogen and Sulphur Hexa-Fluoride (SF 6) Detectors

Halogen-Diode Leak Detection

In the case of closed loop refrigeration systems, there is a need for detecting leakage of Freon and other refrigerant gases. Halogen detectors meet this requirement. In halogen-diode testing, a leak detector is used that responds to most of the gases containing chlorine, fluorine, bromine or iodine. Therefore, one of these halogen-compound gases is used as tracer. The most popular halogen tracer gases are the two refrigerant gases; R-12 (dichlorodifluoromethane, CCl_2F_2) and R-22 (monochlorodifluoromethane, $CHClF_2$). Some common trade names for these refrigerant gases are freon, genetron, isotron, etc. When a vessel is pressurized with such a tracer gas or mixture of a halogen compound and air or nitrogen, the sniffing probe of the leak detector is used to locate the leak. Three types of halogen leak detectors used are: (1) the halide torch, (2) the heated anode halogen detector and (3) the electron capture detector.

The halide torch is used to locate leaks in systems pressurized with air containing halogen tracer gas. The halide torch consists of a burner connected to a tank of halide-free fuel, such as acetylene gas or alcohol (Figure 34.18). Some of the air for combustion is drawn into the flame through a tube near the bottom of the burner, which heats a copper plate. A flexible extension of this air intake is a snifer tube used as a probe to locate leaks. When the open end of this tube passes near a halogen tracer gas leak, some of the gas is drawn into the flame. The flame is pale blue, if only air is pulled into the burner through the suction hose. If small amounts of vapor containing halogen compounds enter the suction tube, the flame turns green. The sensitivity of the halide torch is approximately 10^{-5} Pa m^3/s.

The heated anode halogen leak detector (shown in Figure 34.19) makes use of a red hot platinum and ceramic heater element, which emits positive ions. These ions are collected on a negatively charged cylindrical cathode to provide a leak signal. The presence of small traces of halogen vapors increases the emission of positive ions markedly. It is this increase in positive ion emission that is measured to indicate the presence of a leak.

The sensitivity of halogen detectors operating at atmospheric pressure is about 10^{-9} $Pa{\cdot}m^3/s$, but this will vary depending on the specific gas that is used. All the detectors mentioned above, include a control unit and a probe through which air is drawn. When searching for leaks from an enclosure pressurized with a tracer gas, the probe tip is moved over joints and seams suspected of leaking. When the probe passes over or close to a leak, the tracer gas is drawn into the probe with air and through a sensitive element, where it is detected. The leak signal is either audible or visual.

The major advantage of these detectors is applicability to various types of refrigeration systems with adequate sensitivity. Other advantages are the simple construction, ruggedness, portability and low cost. They can be operated by unskilled personnel.

SF6 Detector

SF6 is a highly electronegative gas with a large electron capture cross section. It is also safe to handle, inert chemically and non-toxic in nature. The actual sensitive element is an electron capture detector employing a disc containing tritium as an anode-cum-beta emitter and a stainless steel mesh at a negative potential with respect to the tritium source. The beta particles (electrons) continuously emanating from the tritium disc produce a large amount of ionization in the region and the standing current is in the order of 10^{-9} A. When the detector cell volume contains some electronegative components like SF6, freon or oxygen, the electrons are captured and there is a reduction in the standing current. The schematic arrangement of the detector is shown in Figure 34.18. This method is simple and the equipment is also inexpensive and portable. The sensitivity is limited by the background current due to atmospheric oxygen.

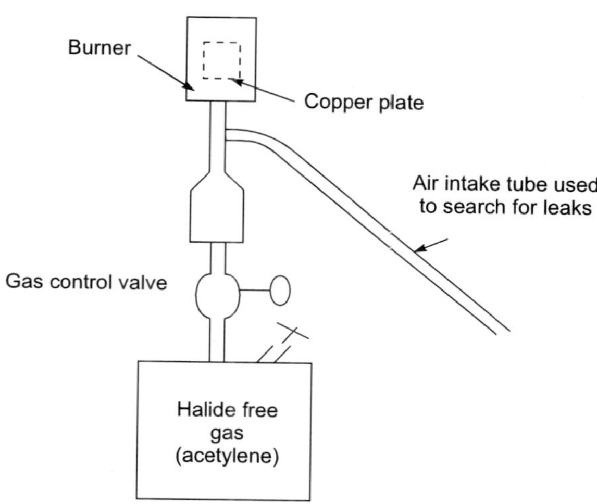

Figure 34.18 Halide torch for leak detection

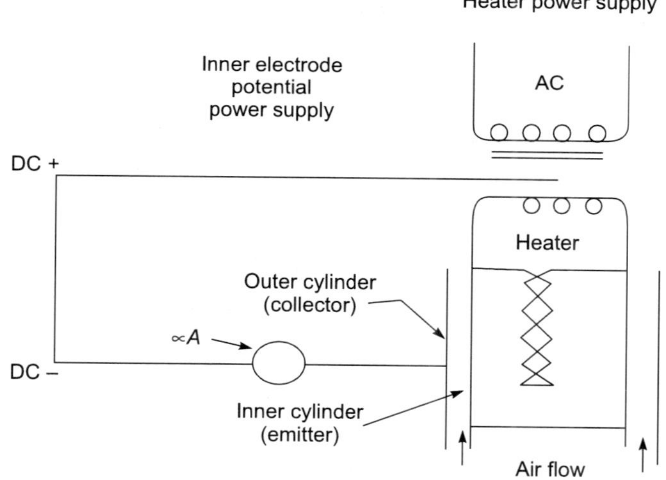

Figure 34.19 Basic circuit of heated anode halogen

34.9.3 Radiography Using X-Ray and Gamma Ray (Figure 34.20)

Radiography is one of the most useful non-destructive tests, which can be applied for assessing the quality of the welded joints. Radiograph has been used for inspecting welds of all types and thickness ranging from minute welds in electronic components to welds up to half meter thick employed in heavy fabrications. Radiography can detect flaws and discontinuities in welds such as:

(i) Cracks
(ii) Porosity and holes
(iii) Slag, flux or oxide inclusions
(iv) Lack of fusion between the parent metal and the weld metal
(v) Incomplete penetration

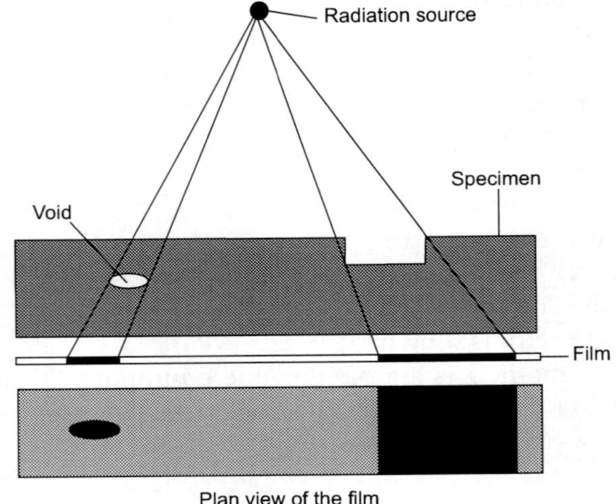

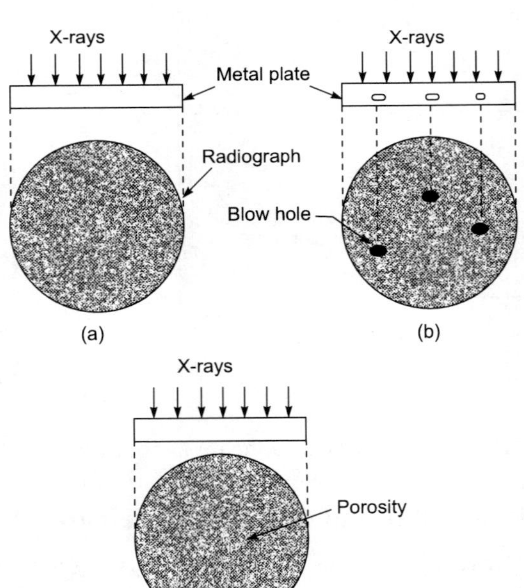

Figure 34.20 Radiographic testing

Principle

Radiography is based upon exposing the components to short wavelength radiations in the form of X-rays (wavelength less than 0.001×10^{-8} cm to about 40×10^{-8} cm) or γ-rays (wavelength about 0.005×10^{-8} cm to 3×10^{-8} cm) from a suitable source, such as an X-ray tube or cobalt-60.

The characteristic feature of X-rays and γ-rays which makes them to work is their power to penetrate matters opaque to light. X-rays operating at 400,000 volts can inspect steel objects having thickness up to 62 mm. γ-rays given off by radium and radioactive isotopes, such as cobalt-60, iridium-192 and caesium-167, can penetrate and thus, inspect joints of bigger thickness than examined by X-rays.

X-Ray Radiographic Procedure (Figure 34.21)

X-rays are produced in an X-ray tube, where a (cathode) filament provides electrons which proceed towards the target (anode); strike and are suddenly stopped; a part of their kinetic energy is converted to energy of radiation or X-rays.

The portion of the weldment where defects are suspected is exposed to X-rays emitted from the X-ray tube. A cassette containing X-ray film is placed behind in contact with the weldment, perpendicular to the rays. During exposure, X-rays penetrate the welded object and thus, affect the X-ray film.

Since most defects possess lesser density than the sound parent metal, they transmit X-rays better than the sound metal does; therefore the film appears to be darker where defects are in line of the X-ray beam. The exposed and developed X-ray film showing light and dark areas is known as radiograph (or precisely known as an exograph).

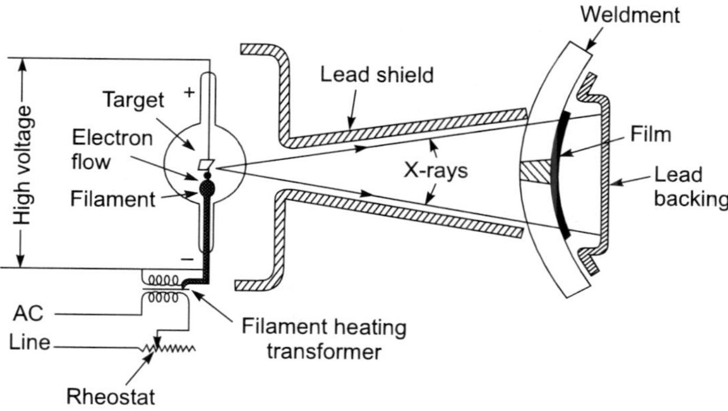

Figure 34.21 Radiography procedure

Gamma (γ-ray) Radiography Procedure (Figure 34.22)

γ-ray radiography differs from X-ray radiography in the following respects:

(i) γ-rays are used for detecting defects in plates thicker than those inspected by X-rays.
(ii) Owing to less scatter, γ-rays are more satisfactory than X-rays for examining objects of varying thickness.
(iii) X-rays are better than γ-rays for detecting small defects in weldment sections less than about 50 mm.
(iv) X-ray method is faster than γ-ray method.
(v) γ-ray can inspect a number of welds at one time unlike X-ray method.
(vi) γ-ray equipment being small possesses better probability and convenience of use for certain field inspections.

Gamma radiations, a product of radioactive decay, are extensively used in testing of welded objects. Radium and its salts decompose at a constant rate, giving out γ-rays, which are of much shorter wavelength and more penetrating.

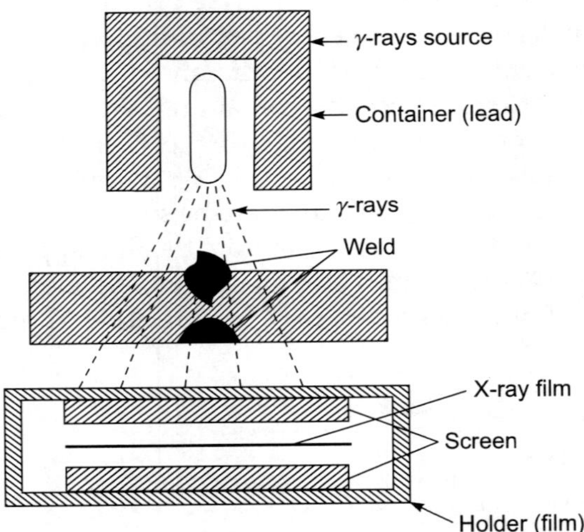

Figure 34.22 Gamma-ray radiography

Radium and radon were originally employed as γ-ray sources but more convenient sources are available in the form of isotopes, like cobalt-60. It is an isotope produced by neutron irradiation and can be used in place of radium; it is much cheaper as well.

Unlike X-rays, γ-rays from its source are emitted in all directions; therefore, a number of separate welded objects having cassette containing film, fastened to the back of each object, are disposed in a circle around the source placed in a central position. This way many welded objects can be radiographed simultaneously and overnight exposure may be taken without continuous supervision.

Operating Parameters

Wavelength of radiation	0.001 – 0.015 nm
	0.01 – 1 nm (1.25 MeV – 80 KeV)
Portability	Good (except for containers)
Access	Good
Exposure time	1 second – 24 hours
Thickness range	Up to 250 mm
Minimum defect size	1% of thickness

Radiographic Examination Techniques

The choice of a particular technique is generally based on the sensitivity requirement. The discerning of minute discontinuities in the test object is possible only when a proper technique is applied. The sensitivity of the radiographic technique is influenced by:

1. Radiation
2. Film used
3. Source-to-film distance
4. Radiation beam alignment

The selection of a particular technique is based on the knowledge about the following factors:

1. Test object – material, thickness and configuration
2. Fabrication
   (a) Welds – edge preparation and welding process
   (b) Casting – foundry techniques, including mould preparation, location of vents, gates, etc.
3. Anticipated location and nature of the discontinuities
4. Critical and vulnerable locations
5. Area of the inspection coverage
6. Sensitivity level required

Butt Welds

Single Wall Technique: In this technique, radiation passes through a single wall of the object and the image of the wall is obtained the radiographic film. The radiation source is located on one side of the object and the radiographic film on the other side. This technique is applicable whenever both top and bottom sides of the weld are accessible. This is followed for plate welds and large diameter vessels, as shown in Figure 34.23.

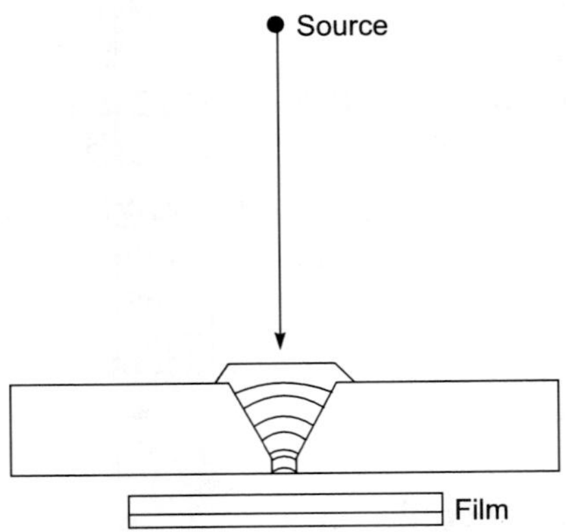

Figure 34.23 Radiographic setup for butt

Double Wall Technique: In this technique, radiation passes through two walls of the object and the image of one or both the walls are obtained in the recording medium, according to the case. The double wall image techniques are used mainly for pipe welds where both side accessibility is not available. In these techniques, both the radiation source and the film will be kept outside. The double wall technique is further divided into two, namely double wall single image (DWSI) and double wall double image (DWDI) techniques, according to the size of the pipe to be radiographed. DWDI is used for the pipes below 89 mm OD (outer diameter) and DWSI is used for above 89 mm OD. In the DWDI technique, the source is kept inclined or offset to avoid the overlap of the top and bottom weld images. A minimum of two exposures are required to cover the weld length. In the case of DWSI, the source is kept near one wall of the pipe and a number of exposures made to cover the full weld length. The examination setup for both the types is given in Figure 34.24.

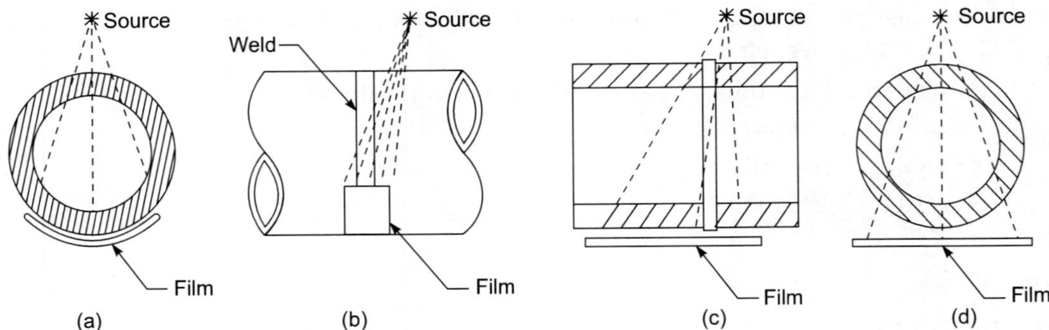

Figure 34.24 Double wall radiographic exposure

Fillet Welds

All the fillet welds, nozzle tee joints and other types can be examined using the exposure techniques as given in Figure 34.25.

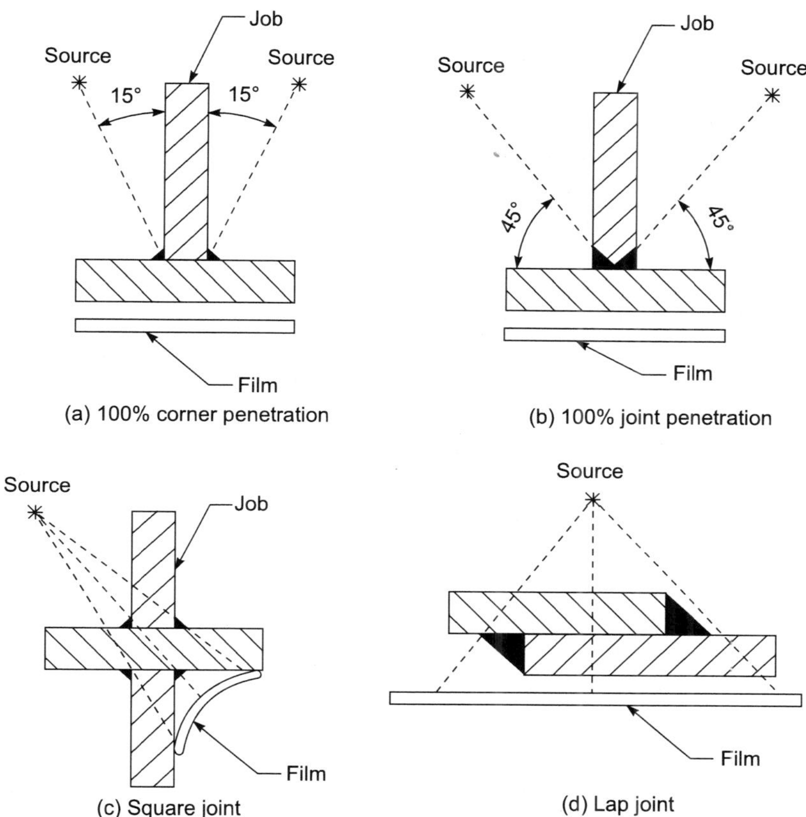

Figure 34.25 Radiographic exposure techniques for fillet welds

Some Special Techniques

Radiographic Examination of Multiwall Configuration

Construction Details of the Components (Piping and Vessels): In fast breeder reactors (FBRs) with liquid sodium as coolant, the major portion of the piping and vessels of the primary and the secondary systems are fabricated with a complex construction of core (pipe/vessel) positioned in a concentric double envelope (guard pipe or vessel) without any spacers between them. At times, a number of core pipes are assembled in a common double envelope. The concentricity of the double envelope with respect to the core, also has to be maintained during the construction.

The double envelope configuration poses many problems during the welding of seams, both at shop and at site conditions. Radiographic examination was found an appropriate technique in addition to ultrasonic examination for achieving stringent NDE (non-destructive examination) requirements. The radiographic quality level stipulated for the core components was more stringent than that specified in the corresponding clauses of international codes, like ASME Sec. III.

Multiwall Image Technique: The site welds of core pipes and the double envelopes of the system constitute a multiwall configuration, being four walls minimum at any cross section. It was observed that the pipes were amenable to radiographic and ultrasonic examinations independently. The quality requirements stipulated for the core components demand the most stringent sensitivity levels. Achieving such a quality level mainly depends on the judicious selection of radiation source adaptable within the restricted site conditions. The core pipe welds were successfully radiographed by employing conventional double wall single or double image techniques according to the pipe size. The double envelope pipe joint examination could be carried out by multiwall exposures involving a minimum of four walls of the pipes using multiwall image technique. The interpretable weld length per radiograph is arrived by evaluating the radiographic density, within the acceptable range, taking density value 2.0 as optimum. It was essential to take more number of exposures to cover the entire weld circumference compared to the usual number of exposures with the conventional double wall techniques specified for the same pipe size. The multiwall exposure techniques are illustrated in Figures 34.26 and 34.27.

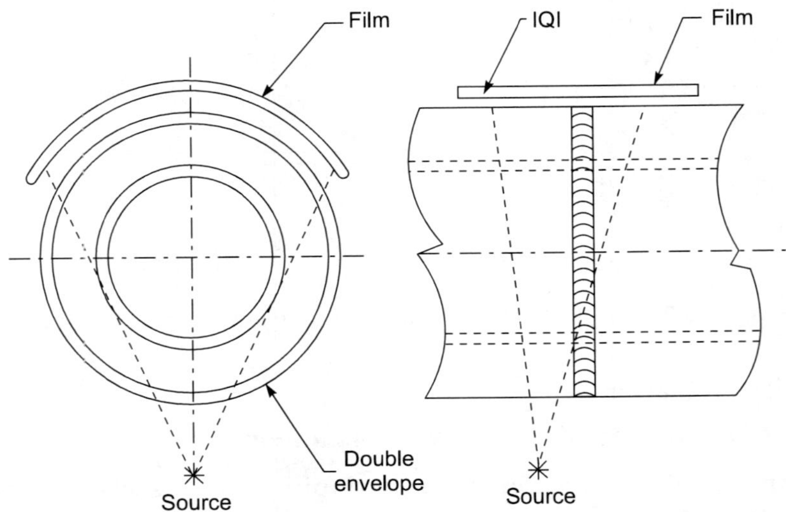

Figure 34.26 Multiwall radiographic exposure setup for defect assessment

Measurement of Annular Gap in Pipelines: A uniform gap between the core and double envelope was a prime requisite for such configurations as explained earlier. Hence, the necessity arises for proper assessment of annular gap after welding. In the absence of access for direct measurement, the only method available was radiography. The assessment of the gap being the main consideration over sensitivity, the gamma source was considered a better alternative, as it provides clear outer and inner wall edges, compared to an X-ray source. The pipes were exposed in two perpendicular directions, using the multiwall image technique, directing the gamma ray beam normal to the expected annular gap, as shown in Figures 34.26 and 34.27. The actual annular gap was arrived at by applying an appropriate correction factor corresponding to the radiographic setup. After studying the radiographs, it was noted that a non-uniform gap existed between the core and the envelope due to welding distortion. It was also observed that at some sectors of the piping system, the double envelope was touching the core pipe. Such sectors were rectified by selective heating of the weld joint and the minimum gap was ascertained by radiographic examination.

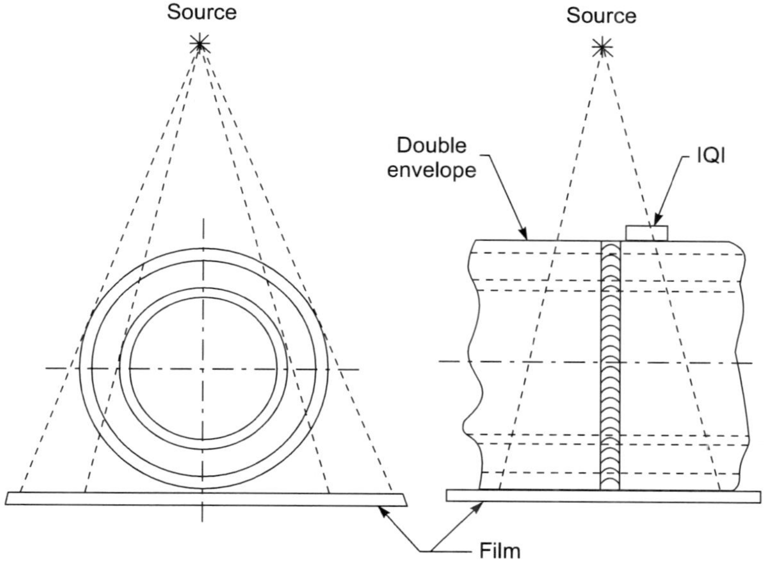

Figure 34.27 Multiwall radiographic setup for assessment of gap

Advantages of X-Ray Radiography

(i) A permanent record of defects in a welded object is obtained.
(ii) Reference standards for defects are available.

Disadvantages of X-Ray Radiography

(i) The equipment is costly.
(ii) Trained operator is required.
(iii) The method involves radiation hazards.

Applications of X-Ray Radiography

(i) Pressure vessels and boilers.
(ii) Penstocks.
(iii) Aircraft and ship structures.

Advantages of γ-Ray Radiography

(i) A permanent record of defects in a welded object is obtained.
(ii) Reference standards for defects are available.
(iii) Low initial cost.
(iv) This is a good method for testing at the site.

Disadvantages of γ-Ray Radiography

(i) Trained operator is required.
(ii) The method involves radiation hazards.
(iii) γ-ray source loses strength continuously.
(iv) γ-ray radiography has lower sensitivity and definition than X-ray radiography.

Applications of γ-Ray Radiography

(i) Pressure vessels and boilers.
(ii) Penstock and pipework.
(iii) Ship building.
(iv) Structural steel work.

34.9.4 Magnetic Particle Inspection

This method is used on magnetic ferrous weldments for detecting invisible surface or slight subsurface defects. Deeper subsurface defects are not satisfactorily detected because the influence of distorted lines of magnetic flux (owing to discontinuity) on the magnetic particles spread over the job surface becomes weaker with the distance, so that sensitivity falls away rapidly with the depth.

The defects commonly revealed by magnetic particle inspection are quenching cracks, thermal cracks, seams, laps, grinding cracks, overlaps, non-metallic inclusions, fatigue cracks, hot tears, etc.

Principle of the Method (Figure 34.28)

When a piece of metal is placed in a magnetic field and lines of magnetic flux get intersected by a discontinuity, such as a crack or slag inclusions in a job, magnetic poles are induced on either side of the discontinuity.

The discontinuity causes an abrupt change in the path of magnetic flux flowing through the job normal to the discontinuity, resulting in a local flux leakage field and interference with the magnetic lines of force. This local flux disturbance can be detected by its effect upon magnetic particles, which are attracted to the region of discontinuity and pile up and bridge over the discontinuity.

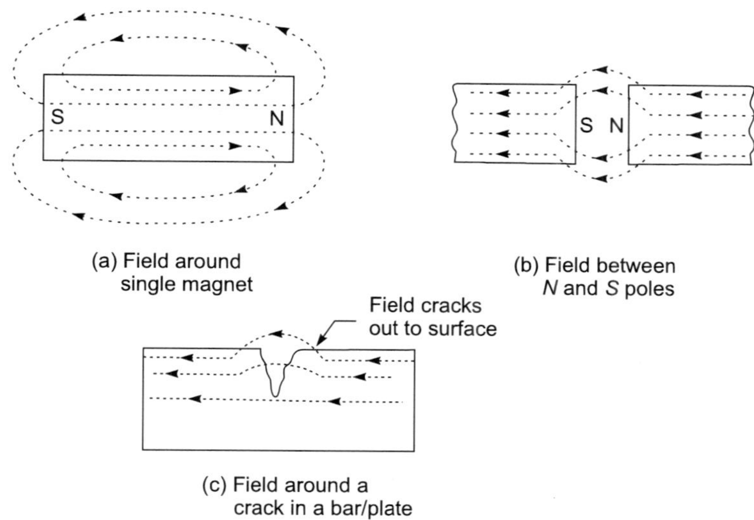

Figure 34.28 Magnetic particle inspection

A surface crack is indicated (under favorable conditions) by a line of fine particles following the crack outline and a subsurface defect by a fuzzy collection of magnetic particles on the surface near the discontinuity.

Maximum sensitivity of indication is obtained when the discontinuity lies in a direction normal to the applied magnetic field and when the strength of magnetic field is just enough to saturate the section being inspected.

Procedural steps involved in magnetic particle inspection are:

- Magnetizing the component part (e.g., the welded part).
- Applying magnetic particles on the component part.
- Locating the defects.

For detecting of surface and subsurface defects in welded components, liquid penetrant testing or magnetic particle testing is widely used. In case of ferromagnetic materials, magnetic particle technique has been preferred, since it also detects subsurface flaws that are not open to the surface. Because of this advantage over liquid penetrant, it has become customary to specify magnetic particle testing for all ferromagnetic materials.

This method is based on the principle that when a ferromagnetic material under test is magnetized, discontinuities that lie in a direction, generally transverse to the field, cause a leakage field around the discontinuity. The procedure adopted for the magnet particle testing is shown in Figure 34.29. When finely divided ferromagnetic powder is sprinkled over the surface, some of these particles will be gathered and held by the leakage field. This magnetically held collection of particles form an outline of the discontinuity and indicate its location, shape and extent. To get the highest sensitivity, fluorescent magnetic particles suspended in oil using full wave DC continuous technique, are employed.

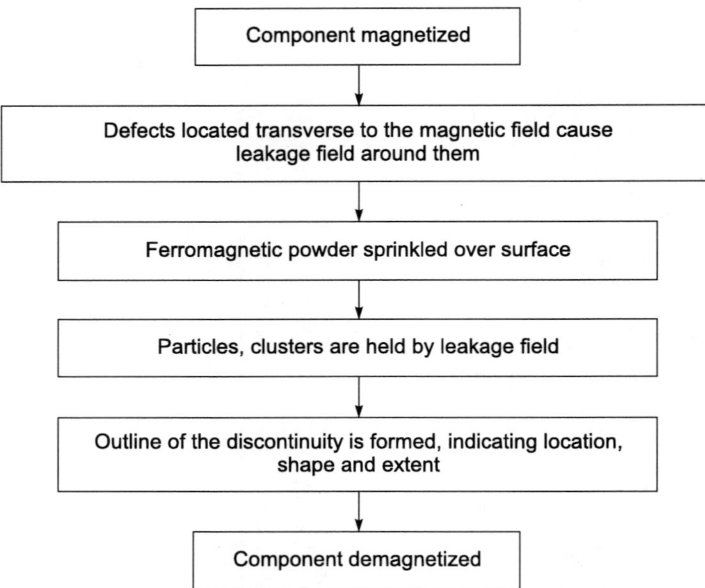

Figure 34.29 Sequence of magnetic particle testing

The test method consists of magnetization of the component applying magnetic powder, examination of powder particles and demagnetization of the component. Magnetic particle testing is a sensitive means of locating small and shallow surface cracks in ferromagnetic materials. Indications may be produced at cracks that are large enough to be seen by the naked eye. Wide cracks do not produce a particle pattern, if the surface opening is too wide for the particles to bridge. If the discontinuity is fine, sharp, and close to the surface, such as a long stringer of non-metallic inclusions, a sharp indication will be produced. If the discontinuity lies deeper, the indication is less distinct. Magnetic particle indications are produced directly on the surface of the part, and constitute magnetic pictures of actual discontinuities. There is little or no limitation on the size or shape of the part being inspected. Ordinarily, no elaborate precleaning is necessary and cracks filled with foreign materials can be detected.

Magnetization Methods

In magnetic particle testing, the magnetic particles may be applied to the part while the magnetizing current is flowing or after the current has ceased to flow, depending largely on the retentivity of the part. The first technique is known as the continuous method and the second one is known as the residual method. The residual method can be used only on materials having sufficient retentivity. Usually, the harder the material, the higher the retentivity. The continuous method is the only method used on low-carbon steels or iron, having little or no retentivity.

Magnetizing Current

Both direct current and alternating current, are suitable for magnetizing parts. The strength, direction and distribution of magnetic fields are greatly affected by the type of current that is used for magnetization. Fields produced by direct current generally penetrate the cross section of the part, whereas the fields produced by alternating current are confined to the metal at or near the surface of the part, which is commonly known as the skin effect. Therefore, alternating current should not be used for detecting subsurface discontinuities. Most satisfactory source of direct current is the rectification of alternating current configurations of magnetic parts omitable for inspection of pipe and rod geometrics, as shown in Figures 34.30 and 34.31, respectively.

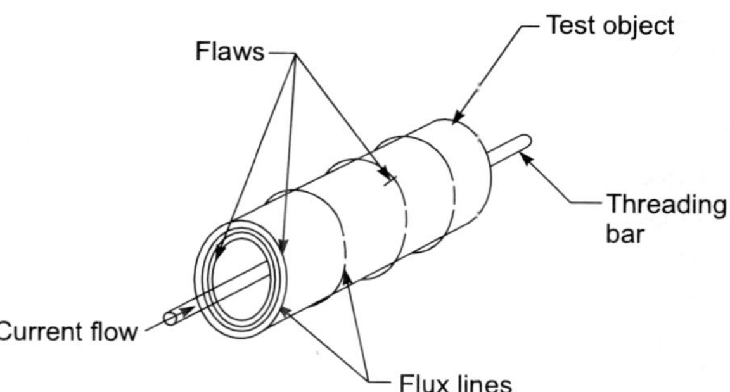

Figure 34.30 Circular magnetization using single turn of coil

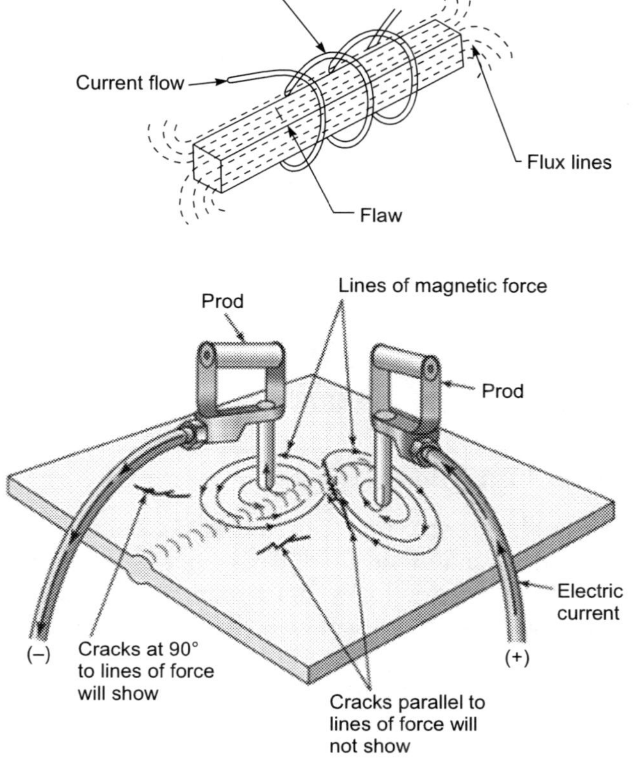

Figure 34.31 Longitudinal magnetization using coil

Inspection of Weldments

Many weld defects are open to the surface and are readily detectable by magnetic particle inspection using prods and yokes (Figure 34.32). For detection of subsurface discontinuities, such as slag inclusions, voids and inadequate joint penetration at the root of the weld, prod magnetization is the best, using either alternating current, direct current or half-wave current.

Positioning of a yoke with respect to the direction of the discontinuity sought, is different from the corresponding positioning of prods. Because the field traverses a path between the poles of the yoke, it must be placed on opposite sides of the weld bead to locate cracks parallel to the weld axis and adjacent to the bead to locate transverse cracks. Prods are spaced adjacent to the weld for parallel cracks and on opposite sides for transverse cracks.

For applications in which holding of prod contacts by hand is difficult or tiring, prods incorporating magnetic clamps or faches, that hold the prods magnetically to the part to be tested, are available. The prods carrying the magnetizing current are held firmly to the part by an electromagnet. Both prods may be attached by the magnets, or one of the prods may be held magnetically and the other by hand.

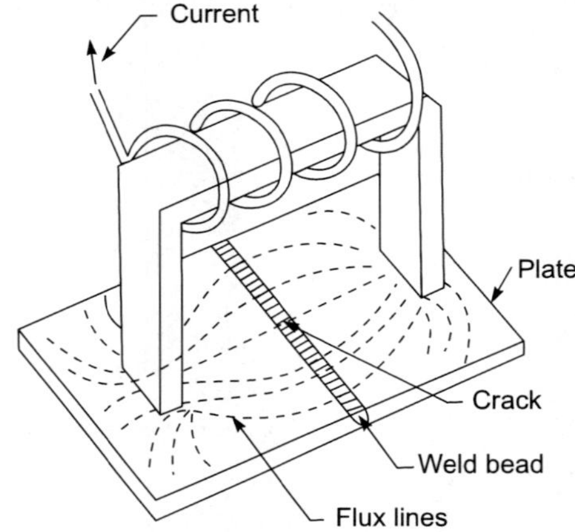

Figure 34.32 Use of yoke for magnetization

Demagnetization after Inspection

All ferromagnetic materials after having been magnetized, retain some magnetic field, which is known as residual magnetic field. This field is negligible in magnetically soft materials. However, in magnetically hard materials, it may be comparable to the intense fields associated with the special alloys used for permanent magnets. Demagnetization, easy or difficult, depends upon the type of material. Metals having high coercive force are difficult to demagnetize. There are many reasons for demagnetizing a part after magnetic particle inspection. For example, during subsequent machining of a part, chips may adhere to the surface being machined and adversely affect the surface finish, dimensions and tool life. During electric arc welding operations, strong residual magnetic fields may deflect the arc away from the point at which it should be applied.

Methods of Demagnetization

1. **Thermal Method:** A ferromagnetic component can be demagnetized if it is raised above its curie temperature (for example, 1023 K for iron) but often, this is not practicable or convenient. This can be employed where postweld heat treatment to a temperature above curie temperature follows magnetic particle inspection.
2. **AC Circular Field Demagnetization:** This is useful for large parts. This is similar to AC coil method in that the field reversal is provided by the cyclic nature of the current. In this method, the desired field is obtained by passing current through the part where the current intensity is gradually reduced to zero.

AC or DC Yoke Method: This is suitable for parts having very high coercive force. Some yokes are similar in operation to the AC coil method whereby, the part is passed between pole faces and then withdrawn. A modified version of this uses a solenoid type electromagnet.

34.9.5 Liquid Penetrant Inspection

Liquid penetrant inspection (LPL) is another NDT method to detect surface and subsurface defects in welded materials. This method is used in root pass and subsequent passes to detect surface defects so that repair work can be undertaken to remove the defects in the weld. In this method, a liquid penetrant is applied to the surface of the product for a certain predetermined time during which die penetrant seeps through the surface opening defects by capillary action. The excess penetrant is removed from the surface. The surface is then wiped with solvent, dried and a developer applied to it. The developer indicates the presence as well as the location, size and nature of discontinuity. The procedural sequence adopted for liquid penetrant inspection is shown in Figure 34.33. Care should be taken so that chemical contents in the liquid penetrant do not affect the material. This method is the best for inspection of all types of surface cracks, porosity, laminations and lack of bond at exposed edges or joined materials and leaks in welded tubes and tanks. It has been used with excellent success on ferrous and non-ferrous materials, ceramics, powder metallurgy products, weldments, glass as well as on some plastics and synthetic materials.

Types of Penetrants

Liquids having good penetrating ability and potent color dyes are required to achieve the desired sensitivity. These are clearly visible under either white light or ultraviolet light. The amount of penetrant that can enter extremely fine surface discontinuities is quite minute. The visibility of the penetrant brought out of the flaw must be extremely high. The contrast between the penetrant and the developer or surface of the pan should be as great as possible. Some penetrants are water-washable and can be removed from the surface by washing with ordinary tap water. Other penetrants are removed with special solvents.

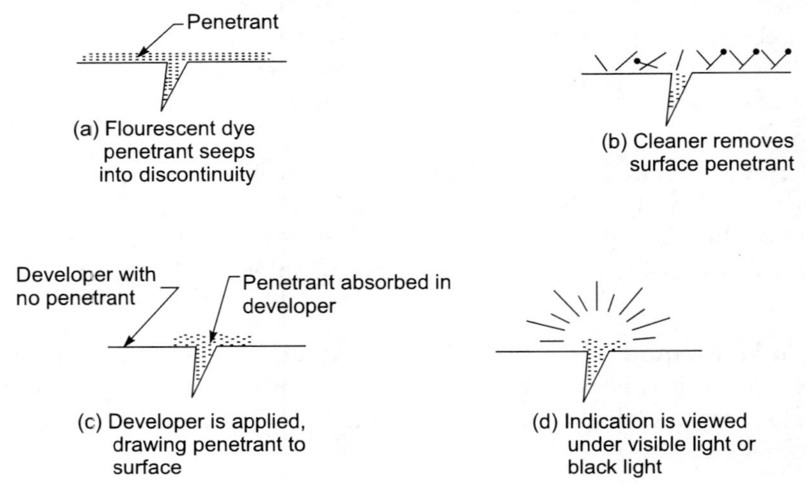

Figure 34.33 Sequence of liquid penetrant testing

Types of Developers

Two types of developers are employed:

Dry developer consists of a dry light colored powdery material. It is applied to the surface of the parts after removal of the excess penetrant and drying of the part.

Wet developer consists of a powdered material, suspended in a suitable liquid, such as water or a volatile solvent.

Penetration Time

The penetration time varies considerably depending upon the type of penetrant used, type of materials to be inspected, the sensitivity desired and the type of defects to be found. The time may vary from as little as one minute to one hour.

Inspection

Inspection is carried out by viewing the part for color contrast between the penetrant drawn out from a defect and the background surface. In the case of fluorescent dyes, viewing is done under ultraviolet or black light. The important thing is to look for very small amounts of penetrant that indicate discontinuities.

Postemulsifiable Fluorescent Penetrant System

This is the most sensitive of all penetrant systems; this procedure locates wide and shallow flaws as well as tight cracks. It requires only a short penetration time and thus, is ideal for high speed production work. On the other hand, emulsification requires an extra operation, which increases the cost. Also, it requires a water supply and facilities for inspection under black light.

Solvent Removable Fluorescent Penetrant System

This system employs a procedure similar to that used for the post-emulsifiable fluorescent system, except that excess penetrant is removed with a solvent. This system is especially recommended for spot inspection or where water cannot be conveniently used. It is more sensitive than the water-washable system but extreme caution is needed and the additional time required for application of the solvent, often precludes its use.

Water Washable Fluorescent System

This system is reliable and reasonably economical. It can be used for both small and large workpieces and is good on rough surfaces. However, it cannot reliably reveal open shallow flaws and in some cases, will not locate the light cracks because it is less sensitive than the two fluorescent systems described above. Other limitations are that the inspection must be carried out whether there is adequate water supply and whether a black light can be used.

Steps Involved in Fluorescent Penetration Test (Figure 34.34)

1. First of all, clean the surface that is to be inspected.
2. Now apply a layer of fluorescent penetrant by brush or by dipping and allow it to penetrate for approximately 30 to 60 minutes.

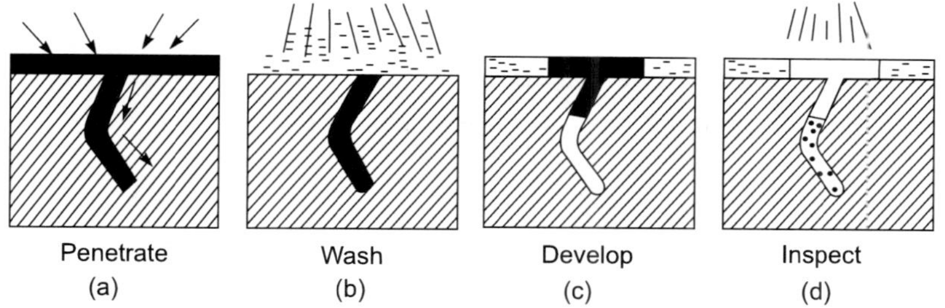

Figure 34.34 Steps involved in fluorescent penetrant inspection

3. Now clean the penetrant surface by water spray to remove the penetrant layer from surface that is to be inspected.
4. Apply developer. A developer acts as a blotter to draw penetrant out of the cracks, etc., and enlarge the size of cross-sectional area of penetrant indication.

5. The surface is viewed under a source of black light, whose wavelength is approximating 3650 A. Black light causes penetrant to flow in the dark.

Operating Parameters

Portability	Excellent (for aerosols)
Access	Good
Minimum defect size	0.025 mm wide
Time	30 minutes approx.

Materials

- Any – Non-porous.

Typical Welding Applications

- Root runs in pipe butt welds.
- Leak paths in containers.

Advantages

- Low cost.
- Direct indication of defect location.
- Initial examination by unskilled labor.
- Relative ease of use.
- Can be used on wide range of material types.
- Large areas or large volumes of parts/material can be inspected rapidly and at low cost.
- Parts with complex geometries can be inspected.
- Indications are produced directly on the surface of the part, providing visual image of the discontinuity.
- Initial equipment investment is low.
- Aerosol spray cans make the equipment easily portable.

Limitations

- Surface defects only detected.
- Defects cannot readily be rewelded due to trapped dye.
- Rough welds produce spurious indications.
- Requires relatively smooth non-porous material.
- Precleaning is critical. Containments can mask defects.
- Requires multiple operations under controlled conditions.
- Chemical handling precautions are necessary (toxicity, fire, waste).

- Metal smearing from machining, grinding and other operations inhibits detection; material may need to be etched prior to inspection.
- Postcleaning is necessary to remove chemicals.

Safety

- Dye and propellant gases have low flash points.

34.9.6 Ultrasonic Inspection (Figure 34.35)

Ultrasonic inspection is employed to detect and locate internal defects such as cracks, porosity, inclusions, lack of fusion and incomplete penetration. Ultrasonic vibration can be used to locate defects in ferrous and non-ferrous metallic objects as well as in plastics and ceramics.

Ultrasonic inspection for flaw detection makes use of acoustic waves with frequencies in the range between 20 kHz and 20 MHz, which can be transmitted through solids (even liquid and air as well) and get reflected by the substance defects. Ultrasonic waves form a basis for detection, location and size estimation of defects.

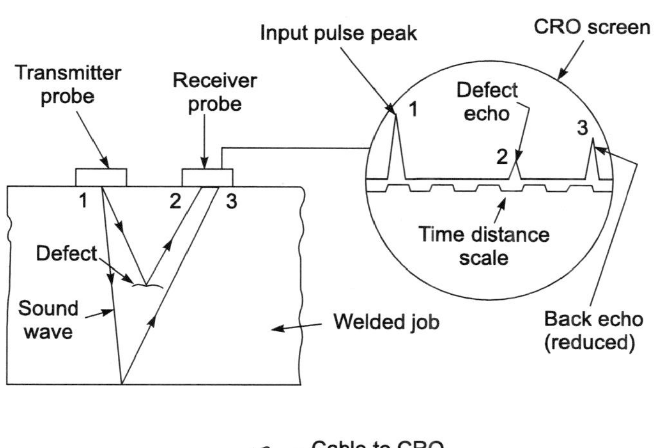

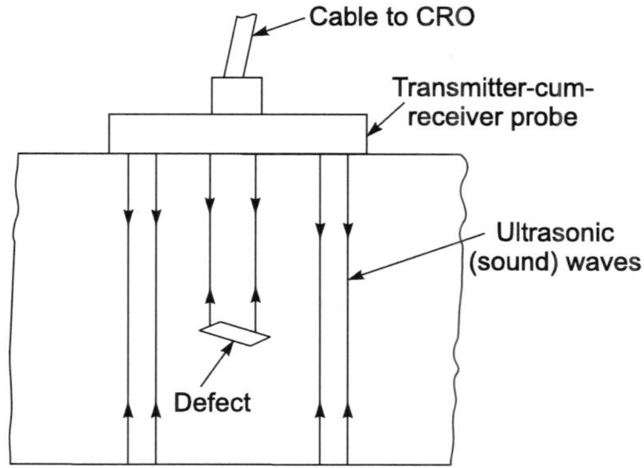

Figure 34.35 Ultrasonic inspection

Principle of Operation

Ultrasonic waves are generated by piezoelectric effect using a quartz. When a high frequency alternating current is impressed across a surface of quartz crystal, the crystal expands during the first half and contracts when the electric field is reversed, thus, producing mechanical vibrations.

The surface of job to be inspected is made smooth, so that ultrasonic waves can be efficiently transmitted from the probe to the job. Before transmission, an oil film is provided in between the probe and the job surface, to ensure that there is proper contact between them and to facilitate better transmission of waves from the probe into the surface of the object to be tested. For operation, ultrasonic wave is introduced into the metal and the time interval between transmission of the outgoing and reception of the outgoing and reception of the incoming signals is measured with a cathode ray oscilloscope (CRO). To start with, as the wave is sent out from the transmitter probe, it strikes the upper surface of the job and makes a sharp peak or echo at the left hand side of the CRO screen.

If the job is sound, this wave will strike the bottom surface of the same, get reflected and indicated by a peak towards the CRO screen. In case a defect exists in between the top and the bottom surfaces, most of the beam striking this defect will get reflected from the defect, reaching the receiver probe and indicating the peak on the CRO screen before the peak given by the waves striking the far end of the job and returning. The distance of the defect from the surface, where a transmitter probe is applied, can be determined with the help of a time distance scale in the form of a square wave constantly shown on the oscilloscope.

Operating Parameters

Probe frequency	1-5 MHz
Portability	Good
Access	Good (can be battery operated)
Thickness range	5-500 mm
Minimum defect size	5 mm wide

Characteristics of Ultrasonic Waves

In many respects, a beam of ultrasound is similar to a beam of light, both are waves obey a general wave equation. Each travels at a characteristic velocity in a given homogeneous medium; a velocity that depends on the properties of the medium and not on the properties of waves. Like a beam of light, ultrasonic beams are reflected from surfaces, refracted when they cross a boundary between two substances that have different characteristic sound velocities, and diffracted at edges or around obstacles.

Types of Ultrasonic Waves

Ultrasonic waves are classified on the basis of the mode of vibration of the particles of the medium with respect to the direction of propagation of the waves, namely, longitudinal, transverse and surface waves.

Longitudinal Waves: These are called compression waves. In this type of ultrasonic wave, alternate compression and rarefaction zones are produced by vibration of particles parallel to the direction of propagation of the wave. Because of its easy generation and detection, this type of ultrasonic wave is most widely used in ultrasonic testing. Almost all the ultrasonic energy used for the testing of materials originates in this mode and then, is converted to other modes for special test applications. This type of wave can propagate in solids, liquids and gases.

Transverse Waves: This type of ultrasonic wave is called a transverse or shear wave because the direction of particle displacement is at right angles or transverse to the direction of propagation. For all practical purposes, transverse waves can only propagate in solids.

Surface or Rayleigh Waves: Surface waves were first described by Rayleigh and that is why, they are called Rayleigh waves. These waves can travel only along the surface bounded on one side by strong elastic forces of the solid and on the other side, by nearly non-existent elastic forces between gas molecules. The waves have a velocity of approximately $90 V_0$ of that of an equivalent shear wave in the same material and these can propagate only in a region no thicker than about one wavelength below the surface of the material.

Types of Transducers, Probes/Sensors

Conventionally, ultrasonic waves are generated by piezoelectric transducers that convert high frequency electrical signals into mechanical vibrations. These mechanical vibrations form a wave front, which is coupled to the component being inspected through the use of a suitable medium (couplant). Several wave modes can be used for inspection depending upon the orientation and location of discontinuities that exist. The three most common piezoelectric materials used are quartz, lithium sulphate and polarized ceramics. The most common ceramics are barium titanate, lead metaniobate and lead zironate titanate. The most important characteristics of ultrasonic probes are sensitivity, resolution, dead zone and near-field effects.

Ultrasonic Flaw Detector

The most common technique employed is the pulse echo technique. The basic equipment comprises an ultrasound pulse generator, a receiver, and its amplification and display system. Depending on the display of information, pulse echo equipment can be subdivided into three groups, namely, A-scan, B-scan and C-scan.

Test Techniques

Techniques of ultrasonic testing are either of the contact type or the immersion type. In the contact type, the probe is placed in direct contact with the test system with a thin liquid film used as couplant for better transmission of ultrasonic waves into the test specimen. In the immersion method, a waterproof probe is used at some distance from the test specimen and the ultrasonic beam is transmitted into the material through a water path or water column. Contact techniques use normal beam techniques, angle beam techniques and surface wave techniques. Immersion testing techniques are mostly used in the laboratory and for large installations doing automatic ultrasonic testing.

Ultrasonic Testing of Weldments

There are two aspects that distinguish ultrasonic testing (UT) of welds from UT of other products, like forgings, castings, pipes, etc. They are: (a) area of interest is well defined and limited (weld and heat affected zone) and (b) knowledge of a specific set of defects whose probable location and orientation is known. Most welds fall into one of the following categories: (1) butt weld (2) Tee-weld and (3) nozzle weld. Various butt weld configurations are shown in Figure 34.36.

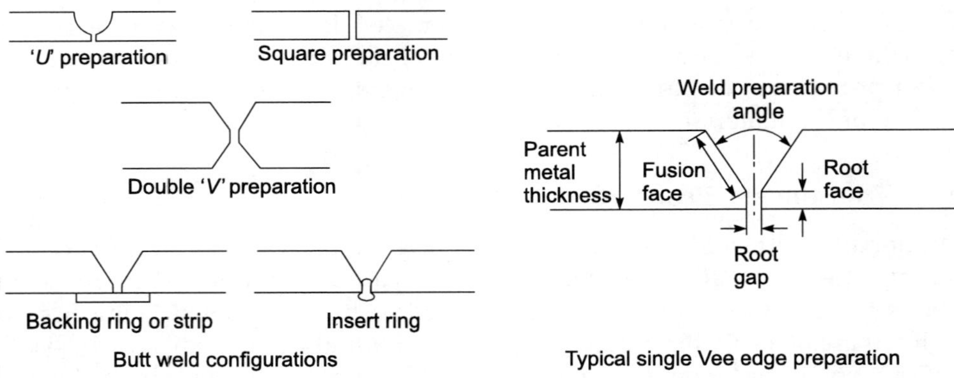

Figure 34.36 Various butt weld configurations

Typical ultrasonic testing procedure for examining weld root and weld body is shown in Figure 34.37. Welds of certain materials, like austenitic stainless steel and nickel based alloys, pose serious problems for ultrasonic testing. These problems are primarily due to the acoustic anisotropy of these materials and the cast structure of the weld. Heavy scattering of the ultrasonic beam, false indications and wrong judgment of position and size of the defect can be encountered in these materials. This results from heavy attenuation of the beam as well as "noise" signals reaching the probe. The combined effect of these, results in loss of sensitivity and low signal to noise ratio.

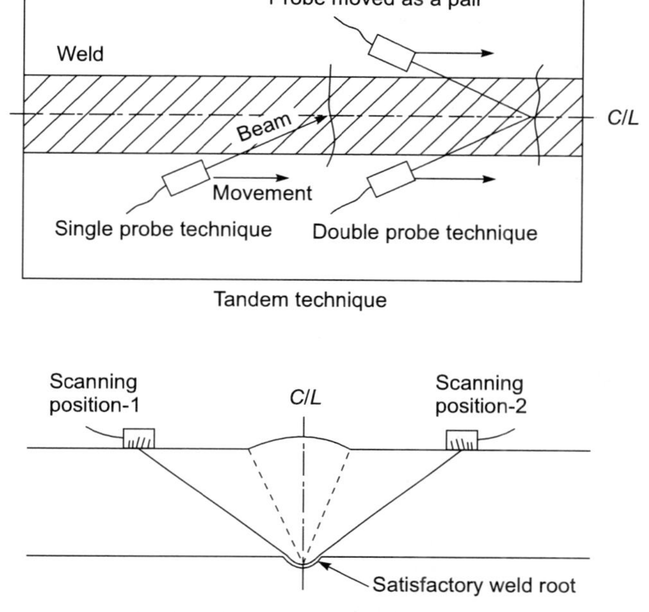

Tandem technique

Detection of weld centre line

Figure 34.37 Ultrasonic test scanning methods for weld inspection

Advantages of Ultrasonic Inspection

(i) It is a fast and reliable method of non-destructive inspection.

(ii) Ultrasonic inspection involves low cost and high speed of operation.

(iii) Big weldments can be systematically scanned for initial detection of major defects.

(iv) The sensitivity of ultrasonic flaw detection is extremely high.

Disadvantages of Ultrasonic Inspection

(i) Surface to be tested must be ground smooth and clean.

(ii) Skilled and trained operator is required.

(iii) It is not suitable for examining weldments of complex shapes or configurations.

Applications of Ultrasonic Inspection

(i) Inspection of large weldments, castings and forging, for internal soundness, before carrying out expensive machining operation.
(ii) Inspection of moving strip or plate as regard to its thickness.
(iii) Routine inspection of locomotive axles and wheel pins for fatigue cracks.
(iv) Inspection of rails for bolt-hole breaks without dismantling rail-end assemblies.

34.9.7 Eddy Current Testing (Figure 34.38)

An AC coil is brought close to the weldment to be tested. The AC coil induces eddy currents in the welded object. These eddy currents produce their own magnetic field that opposes the field of the AC coil. The result is an increase in the impedance of the AC coil. If there is any flaw in the weldment, as soon as the coil passes over the flow coil, there is a change in the coil impedance, which can be wired to give a warning light or sound and thus, the flaw and its location can be determined.

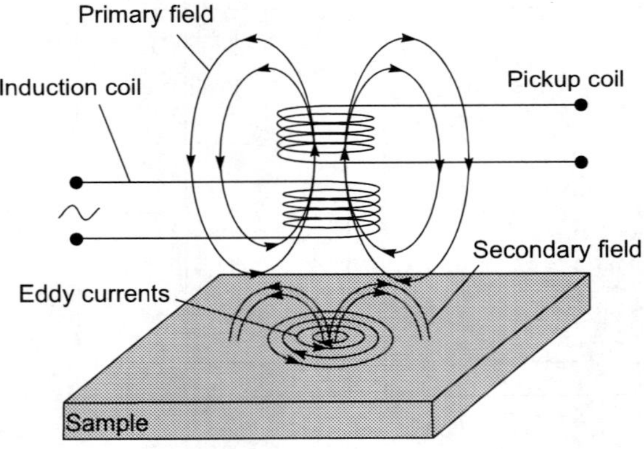

Figure 34.38 Eddy current testing

Procedure

For generating eddy currents, the test piece is brought into the field of a coil carrying alternating current. The coil may encircle the part, may be in the form of the probe, or in the case of tubular shapes, may be wound to fit inside a tube or pipe. The eddy current in the metal test piece also sets up a magnetic field, which opposes the original magnetic field. The impedance of the exciting coil or of a second coil, coupled to the first and in close proximity to the piece, is affected by the presence of the induced eddy current.

A secondary coil is often used for convenience, and is called a sensing pickup coil. In the case of a crack or unwelded seam, the discontinuity must be oriented nearly normal to the eddy current flow to disturb it. The change in the coil impedance caused by the presence of a discontinuity can be measured, and is used to give an indication of the extent of the defects.

The coil may encircle the core, may be in the form of a probe, or in the case of tubular shapes, may be wound to fit inside a tube or pipe. Typical applications are shown in Figure 34.39.

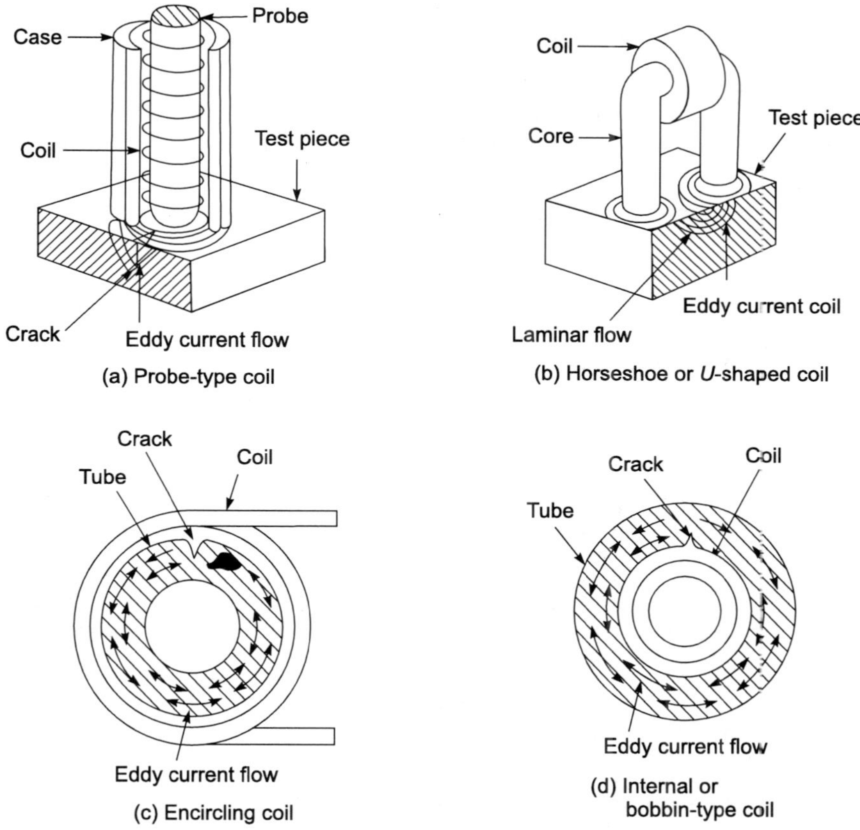

Figure 34.39 Type of coils used in eddy current testing

Operating Parameters

Current	500-10,000 A (AC or DC)
Supply load	1-100 kVA
Portability	Good
Access	Restricted
Minimum defect size	0.025 mm wide at surface
Testing time	10-80 seconds

Advantages of Eddy Current Testing

(i) The coil or probe does not require contact with the surface to be tested.
(ii) The method can be used to test both ferrous and non-ferrous weldments.

Disadvantages of Eddy Current Testing

(i) The method is limited to materials with good electrical conductivity.
(ii) The method is difficult to set and interpret.

Applications of Eddy Current Testing

Eddy current testing is primarily used for continuous inspection of seamless and welded piping and tubing during production. Testing of ferromagnetic steel, austenitic stainless steel, copper alloy and nickel alloy tubular products are covered by ASTM specifications.

All the non-destructive techniques discussed have been summarized in the Table 34.2.

TABLE 34.2 Summary of non-destructive testing techniques

Method	Defects detected	Advantages	Limitations
Visual	Inaccuracies in size and shape. Surface cracks, porosity, etc.	Easy to apply at any stage of fabrication and welding.	Does not provide a permanent record.
Dye penetrate	Surface cracks which may be missed by the naked eye.	Easy to use. No equipment required. Low cost.	Only surface cracks detected with certainty.
Magnetic particle	Surface cracks which may be missed by the naked eye.	Relatively low cost. Portable. Gives clear indications.	Can be used only on ferromagnetic metals.
Radiography	Porosity, slag inclusions, and lack of penetration.	Can be controlled to give reproducible results.	Expensive equipment. Requires high level of skills.
Ultrasonic	All sub-surface defects, laminations.	Very sensitive, can detect defects too small to be discovered by other methods.	Permanent record is difficult to obtain. Requires high level of skills.

34.9.8 Residual Life Assessment of Welded Structures

All fabricated structures are designed for a stipulated service life, and are considered safe for use within this estimated service life. But, the service life has to be checked from time to time as once the structure is put to use, it may also fail before that time due to unforeseen reasons and on the other hand, it may continue to be safe for use even after the stipulated service time is over. Therefore, to ensure safe working without having to deal with unexpected failures, it is important to inspect the structures at regular intervals to assure their soundness. This becomes more critical with welded structures, as welds often have some discontinuities or defects, which may lead to premature failure. The defect can be some type of crack or cavity, which if once detected, becomes important to repair. However, if it is not possible to repair the defect, then measures have to be taken to assess the residual life of the component so that necessary steps can be taken to replace it before the defect reaches the critical stage, leading to catastrophic failure of the entire structure.

All the equipment/components of welded products/structures are, although, fabricated in accordance with an inspection code relevant to the particular equipment/component, for example, pressure vessels, pipes or storage tanks. These construction codes provide the rules for the design and construction of the process equipment and lay down workmanship standards that limit initial flaws. The inspection codes contain warning and notes on the possibility of certain type of environmentally induced damage. But none of these codes provide guidelines to determine the fitness of the equipment or for assessing the remaining life after the equipment has spent a period in service, which may cause its damage or deterioration failure.

Fitness for Service (FFS): A component is deemed to be fit for service, if it can serve satisfactorily under a given set of process conditions for a reasonable period of time to be considered for economic operation. It is based on evaluating some acceptable critical sizes of defects/cracks or material deterioration beyond which the equipment cannot be accepted to be suitable for continued service. These acceptable critical sizes of defects/cracks and material deterioration are based on the laid down scientific principles of sound weld/structure design, detailed stress and failure analysis, etc.

Residual Life Assessment (RLA): Residual life assessment is the process of estimation of the time period for which a structure/component can continue to be used for service or deemed fit for service. Thus, fitness for service (FFS) analysis is important for any residual life assessment exercise.

Thereby, both fitness for service (FFS) and residual life assessment (RLA) techniques are gaining immense popularity in view of findings that relatively large proportion of the process equipment, like vessels in service, either contained many manufacturing imperfections (defects) or were damaged in service or are susceptible to damage under the existing service conditions.

Review Questions

1. What is the difference between testing and inspection?
2. Explain tension test of weldments.
3. Explain bending test of weldments.
4. Compare destructive and non-destructive tests.

5. Explain liquid penetration method used for testing weld.
6. Explain ultrasonic method of testing weldment.
7. What is gamma radiography?
8. Explain the eddy current method of testing weldment.
9. Explain the magnetic particle method of testing weldment.
10. How do we assess residual stresses in weldment?

35 CHAPTER

Weldability

35.0 INTRODUCTION

Welding can be defined as a process of joining metallic parts by heating to a suitable temperature with or without applying of pressure.

According to American Welding Society, *weldability* has been defined as the capacity of a metal to be welded under the fabrication conditions imposed into a specific suitably designed structure and to perform satisfactorily in service.

Weldability is the ease of welding of the given material by a given process for a given purpose under specified conditions and is determined by the quality of the weld joint, effort required and cost of developing the weld joint. The weldability of the metal is not an intrinsic property, as it is influenced by welding procedure, purpose of the weld joints and fabrication conditions.

Steel undergoes certain changes as a result of being exposed to heat input in any form of welding. Some changes can be permanent, like structural transformation during heating and cooling and changes in shape and size due to thermal stresses.

Good Weldability: Steel is said to possess *good weldability* if it can be welded without any dangerous consequences resulting from these changes.

Limited Weldability: Steel is said to possess *limited weldability* if changes due to a normal welding process are in danger of causing failure in a welded component, or if actual defects, such as cracking, occur during welding. In most of the cases, this type of steel can be welded without risk, by following certain precautions and carrying out certain pre- or postwelding treatments.

The term "*unweldable steels*" is unrealistic because any steel can be welded, provided correct metallurgical conditions are chosen, sometimes these conditions may seem impossible to realize in practical production work.

In this chapter, it is intended to discuss the weldability of more often materials and their alloys used in engineering. These include:

- Carbon steels
- Low-steel alloys

- Stainless steels
- High alloy steels
- Cast iron
- Aluminum and its alloys

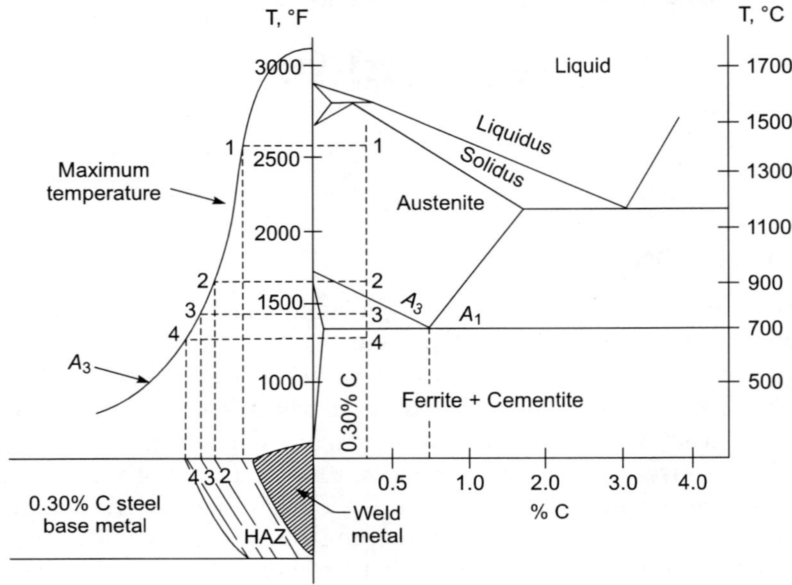

Figure 35.1 Approximate relationships among peak temperature, distance from weld interface, and the iron-carbon phase diagram

35.1 FACTORS AFFECTING WELDABILITY OF STEEL

Weldability is a function of many interrelated factors. These may be put together into three important factors:

1. Process and technique (processes like TIG, MIG, etc.; procedures like preparation, cleaning, process parameters, postweld treatment, etc.).
2. Composition of parent plate.
3. Joint design and size (like service condition, temperature, loading, etc.).

1. Influence of Process

Each process gives a characteristic intensity of power. Processes that offer high power intensity offer advantages in fusion welding because essential melting can be obtained without excessive heat inputs with consequent thermal expansion of parent metal. Usually, successful welding depends on feeding filler wire into weld pool, which carries a deoxidant (or ferrite forming element), hence processes which do not use filler wire are limited in application.

We can define a welding process in terms of following two factors:

(a) Heat input in the process
(b) Extent of preheat required

Heat Input (Figure 35.2)

Heat input is a relative measure of the energy transferred during welding. It is a combination of voltage, amperage and travel speed.

$$\text{Heat input} = \frac{\text{Voltage} \times \text{amperage}}{\text{Travel speed}}$$

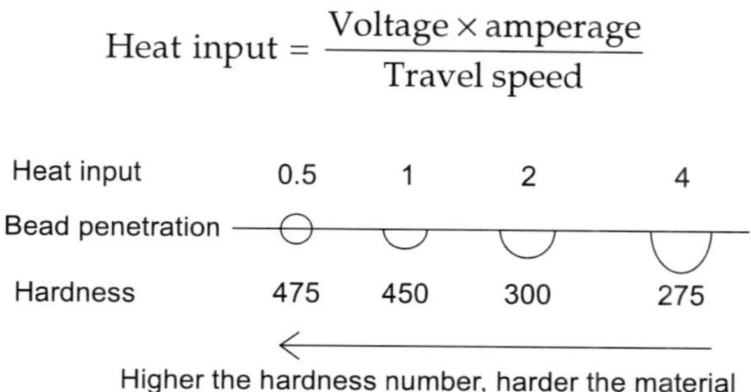

Figure 35.2 Heat input

Toughness – resistance to bending and impact
Hardness – resistance to penetration

Parameters of welding, which get affected when heat input is varied, are:

1. **Cooling Rate:** More the heat input, slower will be the cooling rate for a given thickness of steel.
2. **Weld Size:** More the heat input, more melting of base metal and the filler will result in more surface area of weld bead.

Influence of heat input on mechanical properties is shown in Figure 35.3. Table 35.1 gives material properties versus heat input for SMAW.

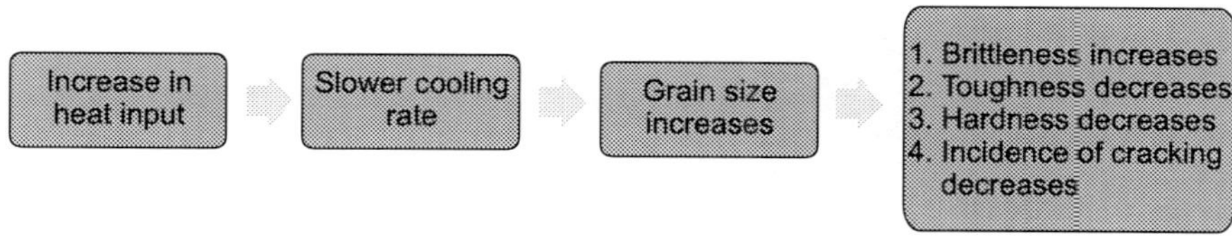

Figure 35.3 Influence of heat input on mechanical properties

TABLE 35.1 Material properties versus heat input for SMAW*

Properties	Change
Yield strength	↓30%
Tensile strength	↓10%
Percentage elongation	↓10%
Notch toughness (CVN)	↓10% (for 15 < H < 50 kJ/in) ↓50% (for 50 < H < 110 kJ/in)
Hardness	↓10%

*SMAW with a heat input range of 15 to 110 kJ/in.

Preheating

Four general factors that must be accessed to determine the level of preheat:

1. Material type
2. Combined thickness
3. Heat input potential (welding process)
4. Joint type

2. Influence of Composition

The composition of the steel and its effect on weldability may be divided into two parts:

1. **Segregation Effects, Particularly that of Sulfur**

   When a steel solidifies, there is a tendency for the iron to solidify fast and for the alloying elements to be accumulated in the centre of the ingot. This configuration is retained even after prolonged and severe rolling and results in high concentrations of sulfur in the central layers of the plate. These layers have little strength and may crack, if stressed.

2. **Tendency of the Steel to Harden**

   The hardening characteristics of steel are mentioned in the heat treatment section. As the carbon or alloy content of the steel increases, ductility decreases and cracking tendency increases.

Weld and HAZ Cracking

Main Causes of Cracking

(a) High levels of carbon and other alloy elements result in brittle zones around the weld.

(b) High cooling rates after welding increases the hardness, which increases the susceptibility to cold cracking.

(c) Joint restraint prevents contraction after welding, which can lead to internal cracking.

(d) Hydrogen in the weld bead can lead to hydrogen induced cold cracking.

(e) Contaminants like sulfur and phosphorus cause segregation or solidification cracking.

(f) Lamellar tearing caused by layering inclusions during rolling resulting in deterioration of the through-thickness properties.

Onset of Cracking (Figure 35.4)

Cracking usually occurs in unmelted areas adjacent to a weld bead (in the heat affected zone). This zone is made up of an area that has been heated to above about 800°C (to become austenite), and an area heated above 723°C (and partially transformed). The fully austenitic region just adjacent to the fusion zone is often the most brittle area of the weld due to grain growth that has occurred while the metal was at this temperature. The grain size gradually decreases, moving away from the fusion zone to the partially transformed region. Grain refinement occurs here, which is considered beneficial, the grain size being smaller. The composition of steel, particularly the carbon level, determines the hardness and brittleness of both, the weld and the HAZ. On cooling, austenite transforms to ferrite. Carbon is soluble in austenite, but less so in ferrite. Iron carbides are precipitated when this transformation takes place. These can be arranged in lamellar form (pearlite), which gives the structure, reasonable strength and ductility.

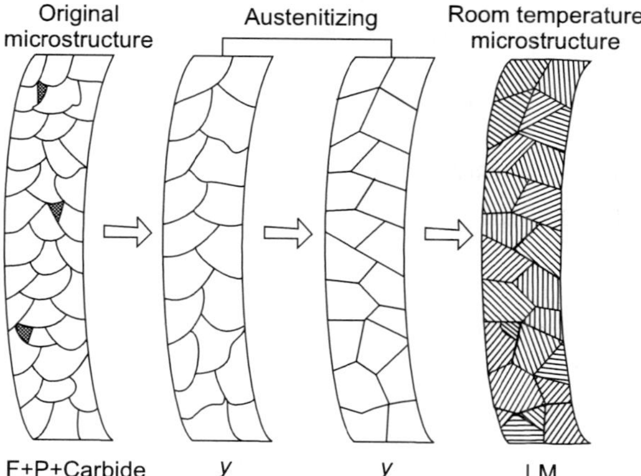

Figure 35.4 Onset of cracking

The HAZ can be the region where the highest cooling rates occur, particularly in thicker section plate. High cooling rates tend to lock the carbides in a solid solution, forming martensite. This leads to high internal stress and hard brittle structure (Figure 35.5).

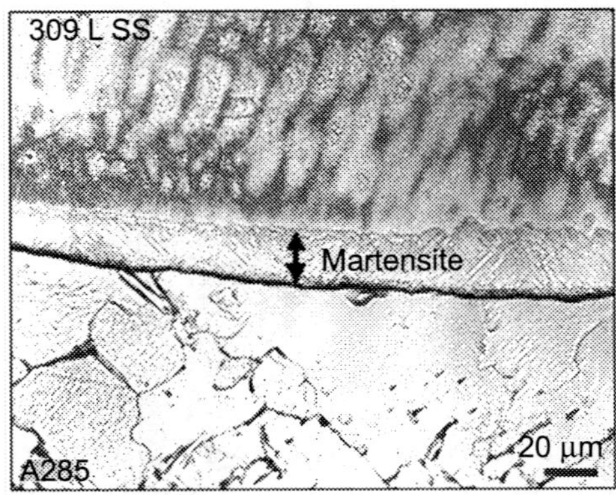

Figure 35.5 Cooling rate effect

3. Influence of Joint Design and Size

Proper design of welding joints according to the required size of assembly is one of the key factors that affect weldability.

(a) Square Edge Butt Joint

- It is easiest to prepare.
- It can be welded with and without any filler material.
- Extreme care in positioning is required while welding light gage material without a filler metal.

(b) Flange Type Butt Joint

- It is used where reinforcement is desired.
- It is suitable only for relatively thin materials (1.5 to 2 mm).

(c) Lap Joint

- It needs no edge preparation.
- Only close contact along the entire length is needed.

(d) T Joint

- It needs additional filler to get necessary built-up for proper welding.
- Thickness of base and welding current should be properly adjusted to have maximum perfection.

Typical Weldability Defects

(a) Solidification Cracking (Figure 35.6)

Also known as hot cracking or centerline cracking. It always occurs in the weld metal.

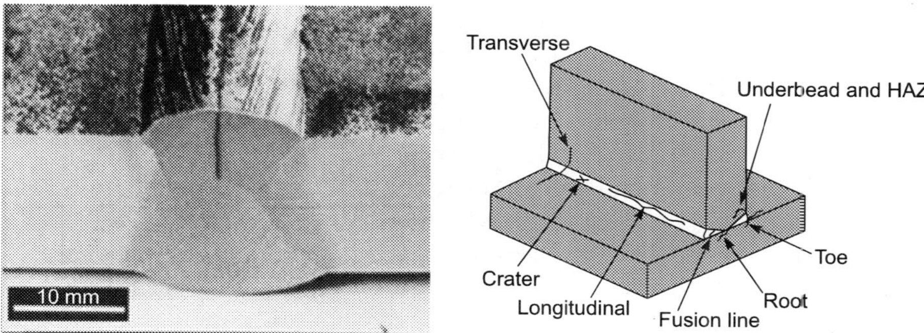

Figure 35.6 Solidification cracking

Causes (for carbon manganese steels)

1. Sulphur in the parental plate
2. Shape of the weld

1. Sulphur in the parental plate

During solidification, the centre of the weld is the last to cool. Sulphur from the parental plate forms iron sulphide. The iron sulphide forms a thin liquid film on the grain boundaries, which has a lower melting point and as the weld cools, it is pushed towards the center of the weld. The iron sulphide has very little tensile strength, so as it cools, the weld is under stress and cracking results.

Causes

1. More sulphur
2. Phosphorus contamination

Avoidance

1. Reduce sulphur
2. Cleaning
3. Add manganese to the parental plate via the fillet wire
4. Reduce dilution by reducing the weld size

2. Shape of the Weld

Deep and narrow weld preparations are susceptible. As weld cools, grains grow from the extremities towards the center of the weld and may interfere each other's growth in a narrow path. While this occurs, grain impingement takes place and causes voids between the grains and results in cracking.

Avoidance: Wider weld preparation (increases the included angle or gap size).

(b) Hydrogen Cracking (Figure 35.7)

Also known as underbead cracking, hard zone cracking and hydrogen-induced cold cracking (HICC). This phenomenon can occur during welding of hardenable steels, and is normally a HAZ phenomenon. Hydrogen is more soluble in austenite than ferrite and hence, it can easily be picked up by the weld metal. When ferrite is formed, the hydrogen solubility decreases and hydrogen diffuses to the HAZ, where it may contribute to crack propagation.

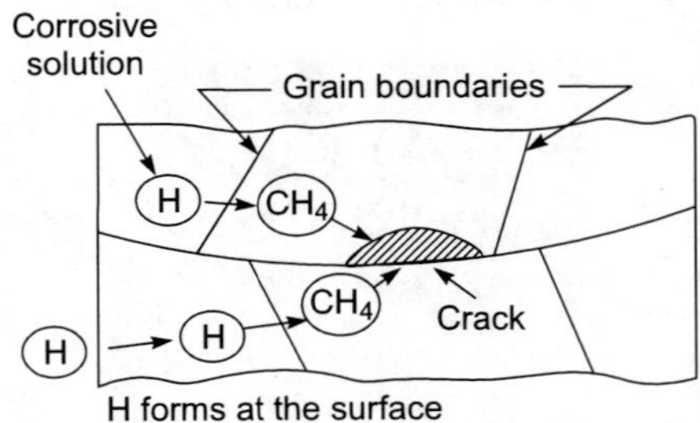

Figure 35.7 Hydrogen cracking

Factors Involving Hydrogen Cracking

1. Hydrogen >15 mL hydrogen per 100 mg of weld metal
2. Temperature <350°C
3. Hardness >400 VPN (Vickers pyramid number)
4. Stress >50% yield

Avoid any one factor and hydrogen cracking will not occur.

Avoidance

1. Hydrogen
   - (a) Parent material dried (free from moisture)
   - (b) Low hydrogen rods (baked to less than 5 mL per 100 mg)
   - (c) Use lower potential hydrogen process (hydrogen potential-electron < TIG < MIG/MAG < SAW, MMA (low H) < MMA (rutile))
   - (d) Heat treatment (pre- and postweld)
2. Temperature
   - (a) Postweld heat treatment (keep at higher temperature as long as possible, to help hydrogen diffuse, NDT as late as possible, 72 hours)
3. Hardness
   - (a) Preheating (to reduce the cooling rate)
   - (b) Keep interpass temperature to a minimum
   - (c) Higher heat input (gives a larger grain size)
   - (d) Postweld heat treatment tempers hard HAZ, removes hydrogen, and reduces stress.

4. Stress
   (a) Joint design
   (b) Fit up
   (c) Distortion control
   (d) Stress relieve treatment (pre- and postweld)

   On high micro-alloy steels, it may occur in the weld metal, otherwise only in the HAZ.

(c) Lamellar Technique

A step like crack, that occurs in thick section tee joints and closed corner joints.

Causes

Poor through thickness ductility in parent plate where high shrinkage strains (through thickness stress) act through the plate thickness in combination with parent metal inclusions. The inclusions can be lamellar inclusions, small micro-inclusions or tramp elements. It only occurs on the leg with stress on the non-rolling direction/through thickness.

Avoidance

1. Joint design (use preformed tees, use open corners)
2. Use buttering layer (possibly more ductile and will take up some shrinkage)
3. Use Z grade steel (tested in the short transverse direction using a STRA test-short transverse reduction area)

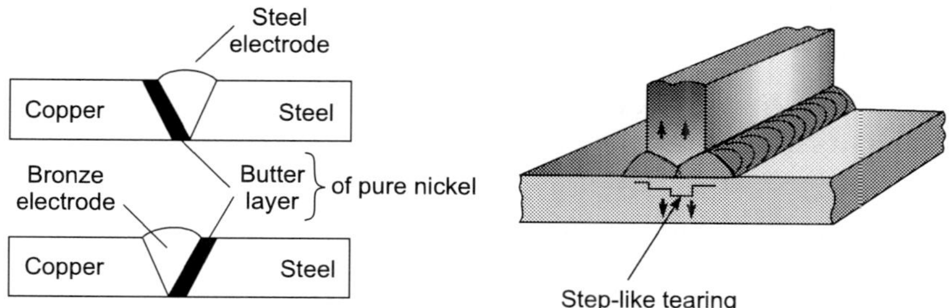

Figure 35.8 Avoidance of cracking

(d) Weld Decay in Austenitic Stainless Steel (Figure 35.9)

It occurs in the HAZ, due to the precipitation of chromium carbides. In the 550 to 850°C range, the chromium leaches out of the grain and reacts with the carbon to form chromium carbide, due to the chromium having a great affinity for carbon. This results in chromium deficiency in the grain. The chromium is in the grains for corrosion resistance, so when the chromium is lost, decay (rusting) occurs quickly.

The longer the zone is in the temperature range of 550 to 850°C, the greater the area is susceptible to decay. The temperature range is usually just off the weld.

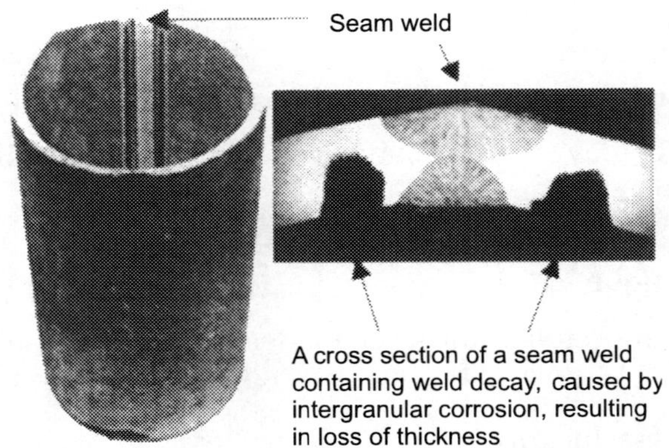

Figure 35.9 Weld decay in austenitic stainless steel

Avoidance

1. Reduce carbon: Use low carbon rods, for example, 316L. But low carbon reduces strength, so a stabilized stainless steel could be used with high carbon content but with titanium and niobium added. (Carbon has greater affinity for titanium and niobium, therefore titanium and niobium carbides are produced instead of chromium carbides).
2. Lower heat input
3. Quench cool: less time in the 550°C to 850°C temperature range
4. Keep interpass temperature low

The keyword is precipitation of chromium. The material is then said to be sensitized. Put in seawater, it rusts.

35.2 WELDABILITY OF CARBON STEELS

Carbon steel is principally a mixture of iron and carbon, with very small amounts of silicon, sulfur, phosphorus, and manganese. The carbon is added to iron in varying amounts to harden or strengthen the steel.

As carbon content increases, hardness and tensile strength increase whereas ductility, plasticity and malleability decrease. Also, as the carbon content increases, weldability decreases.

Table 35.2 groups carbon content, weldability, problems faced during welding, measures to improve weldability and their typical usage.

TABLE 35.2 Summary of weldability of carbon steels

Group	Content %	Weldability	Problems Faced	Measures Taken to Improve Weldability	Typical Usage
Low-carbon steels	0.15 maximum	• Excellent weldability. • No preheat or postheat necessary.	Produces porosity.	Slower welding speeds.	Welding electrodes, rivets and nails, softer easily formed shapes.
Mild carbon plain steels	0.15-0.30	• Readily weldable. • No preheat, interpass, or postheat necessary.	Prone to underbead cracking after welding (fillet welds).	Slightly reduced speeds and currents.	Plate, angle and bar stock for general fabrication. Parts where good plasticity and ductility is required.
Medium carbon steels	0.30-0.50	• Preheat, interpass, temperature controls and postheat required.	High weld hardness, susceptibility to cracking, tendency for brittle failure.	• Low hydrogen electrodes are used. • Filler metal for arc welding used. • Stress relief PWHT recommended.	Used for machine parts, gears and where parts may be hardened by heat treatment.
High carbon steels	0.5-1.0	• Poor. • Preheat, interpass, temperature controls and postheat required.	High weld hardenability and susceptibility to cracking in the HAZ region.	• Preheat and interpass retard the formation of martensite in weld. • Low hydrogen process and filler metal used. • Should be welded in annealed conditions and heat treated.	Springs, dies, railroad track, many tools, band saws and knives (particularly where sharp edge is required).

35.3 WELDABILITY OF LOW ALLOY STEELS (FIGURE 35.10)

Low alloy steels are carbon steels that have additional elements added to produce a classification of steel. Although carbon is the main alloy that affects hardenability and weldability, other elements also play an important role in weldability of steel.

TABLE 35.3 Types of low alloy steels and their properties

Steels	Contents	Properties
High strength low alloy steels (HSLA)	C: 0.15-0.22% Mn: 0.85-1.60% Less than total 1% of Cr, Cu, Cb, Mo, Ni, N, P, Ti, V, Zr and Si.	• Provide better mechanical properties and greater resistance to corrosion. • Mainly ferritic steels and yield strength lies between 410-785 MPa. • Weldability similar to that of mild steels. • Preheat and control of hydrogen required. • Can be welded by all common arc welding processes.
Quenched and tempered steels (Q & T)	C: 0.10-0.25% Small amounts of Mn, Si, Cr, Ni, Mo, V, Ti, B, Cu and Cb.	• High yield strength constructional steels. (560-880 N/mm^2). • Favored the formation of tempered martensite and low bainite structures. • Water quenched and tempered to enhance strength and toughness. • Excellent weldability and do not require heat treatment. • Preheat may reduce the cooling rate of HAZ, which can decrease the strength.
Nickel steels	C: 0.10-0.25% Ni: 5-9%	• High strength and extreme toughness. • Weld in heat treated and PWHT not required. • During SMAW, high Ni-Cr-Fe type electrodes are used. • During SAW, compatible material electrode wire with neutral type welding flux is used. • During GMAW, high Ni-Cr electrodes are used and shielding gas mixture of 90% He + 10% Ar recommended.
Chromium-molybdenum steels	C: less than 0.2% Cr: 0.5-9% Mo: 0.5-1%	• Good weldability and high hardenability. • Cr provides oxidation and corrosion resistance and Mo increases strength at elevated temperatures. • Creep resistant steels. • Requires preheating to prevent hardening and cracking. • Welded by SMAW, GTAW, GMAW, SAW, FCAW, ESW, EBW, laser welding, friction welding, resistance welding and brazing. • PWHT reduces welding stresses and increases the ductility and toughness of the weld metal and HAZ.

Formula applied to classify steel to roughly determine the hardenability and hence, the weldability and need for preheating are as follows:

$$CE = \%C + \%Mn/6 + \%Ni/15 + \%Cr/6 + \%Mo/4 + \%V/5$$

where, CE = Carbon equivalent
C = Carbon
Mn = Manganese
Ni = Nickel
Cr = Chromium
Mo = Molybdenum
V = Vanadium

Figure 35.10 Microstructure of low alloy steel (heat carburized)

Here, we will discuss:

I. High strength low alloy (HSLA) steels
II. Quenched and tempered steels
III. Nickel steels
IV. Chromium-molybdenum steels

Table 35.3 gives types of low-alloy steels and their properties.

35.4 WELDABILITY OF STAINLESS STEELS

Stainless steel is a steel alloy with a minimum of 10.5% chromium content by mass. There are different grades and surface finishes of stainless steel to suit the environment the alloy must endure. Stainless steel is used where both the properties of steel and corrosion resistance are required.

Following are the properties of stainless steels:

- Do not rust and strongly resist attack by a great many liquids, gases and chemicals.
- Good low temperature toughness and ductility.
- Good strength properties and resistance to scaling at high temperatures.

Important Notes

- **Chromium** provides the basic corrosion resistance to stainless steels.
- **Nickel** reduces the thermal conductivity and decreases the electrical conductivity. They are non-magnetic and have austenitic microstructure.

- **Manganese** is added to conserve nickel. These steels have austenitic microstructure and are non-magnetic.
- **Molybdenum** is added to improve the creep resistance of the steel at elevated temperatures and increase resistance to pitting and corrosion in many applications.

Stainless steels are, however, difficult to weld than mild carbon steels due to the following differences:

- Lower melting temperature.
- Lower coefficient of thermal conductivity.
- Higher coefficient of thermal expansion.
- Higher electrical resistance.

35.5 AUSTENITIC STAINLESS STEELS (FIGURE 35.11)

Composition: Total of 24% of Cr, Ni, Mn with
Cr: above 16%, Ni above 7%.

Properties

- Cr provides resistance to corrosion, and oxidation
- Ni and Mn stabilize the austenitic phase sufficiently to retain most of it when steel is rapidly cooled to room temperature.
- Not hardenable by heat treatment and non-magnetic in the annealed condition.
- They have about 45% higher thermal coefficient of expansion, higher electrical resistance and lower thermal conductivity than mild carbon steels.
- Melting point is lower than that of mild carbon steels.

Recommendations to Perform Welding

- High travel speed to reduce heat input and carbide precipitation, and minimize distortion.
- Because of lower thermal conductivity and melting temperature, welding current is lower.
- Tack welds should be twice as often as normal.
- Distortion reducing techniques, such as back-step welding, skip welding, and wandering sequence, should be used.
- It is difficult to completely avoid buckling and distortion.

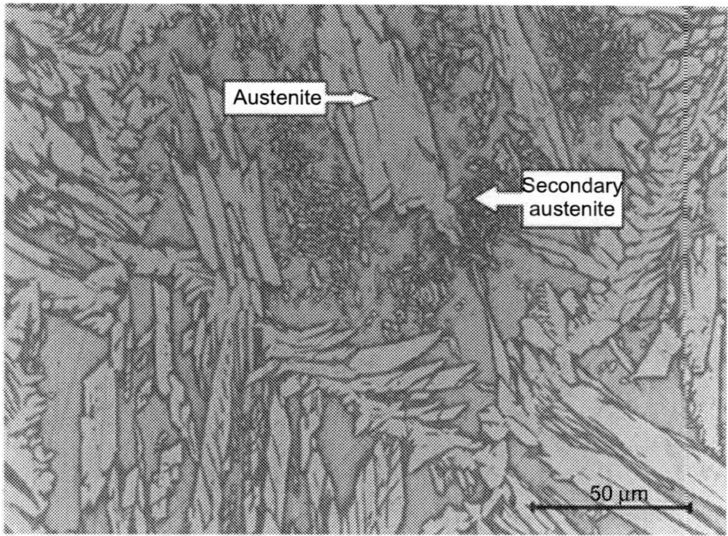

Figure 35.11 Microstructure of austenitic steel after welding

35.6 FERRITIC STAINLESS STEELS (FIGURE 35.12)

Composition: Cr: 14-27% and no Ni.

Properties

- They are not hardenable by heat treatment and are magnetic.
- They do not allow formation of austenitic, so remain ferritic.
- Coefficient of thermal expansion is lower than the austenitic types and is same as mild steel.
- Lower Cr types show tendencies towards hardening, with a resulting martensitic type structure at grain boundaries of the weld area, which lowers the ductility, toughness and corrosion resistance.

Recommendations to Perform Welding

- Welding processes that tend to increase carbon pickup are not recommended, such as oxyfuel gas, carbon arc, and gas metal arc with CO_2 shielding gas processes.
- For heavier sections, preheat of 200°C is beneficial.
- To restore full corrosion resistance and improved ductility after welding, annealing at 760 to 820°C, followed by water or air quench is recommended.
- Toughness can be improved by cold working the weld.
- Sometimes, service demands impact resistance, an austenitic stainless steel filler metal should be used.

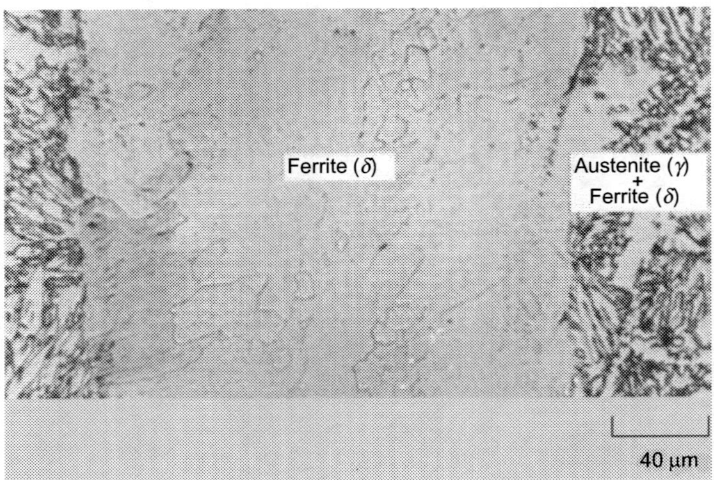

Figure 35.12 Microstructure of ferrite stainless steel

35.7 MARTENSITIC STAINLESS STEELS

Composition: Fe-Cr-C alloys, 12% chromium and 0.12% C.

Properties

- They are hardenable by heat treatment and are magnetic.
- Tempered martensite gives steel good hardness and high toughness; used largely for medical tools (scalpels, razors and internal clamps).
- Untempered martensite is low in toughness and therefore, brittle.

Recommendations to Perform Welding

- The types with over 0.15% carbon, tend to be air hardenable and therefore, preheat and postheat of weldments are required.
- A preheat temperature range of 230 to 290°C is recommended.
- Postheating should perform immediately in the range of 650 to 760°C, followed by slow cooling.
- If preheat and postheat are not possible, then an austenitic stainless steel filler metal should be used.
- Welding processes that tend to increase carbon pickups are not recommended.
- Increased carbon content increases crack sensitivity in the weld area.

Welding Procedures (Figures 35.13 and 35.14)

- Flux cored electrode wires have been developed for welding stainless steels.
- For shielded metal arc welding, there are two types of electrode coatings; lime type and titanium type.

- The **lime-type electrodes** are used only with direct current electrode positive (reverse polarity). **The titanium electrode** can be used with alternating current and with direct current electrode positive.
- However, titanium-type is smoother, has more welder appeal, and operates better in the flat positions.
- The lime-type electrodes are more crack resistant and are slightly better for out-of-position welding.
- The width of weaving should be limited to 2.5 times the diameter of the electrode core wire.
- Covered electrodes for **SMAW** must be stored at normal room temperature in a dry area. They are susceptible to moisture pickup.
- **GTAW** is widely used for thinner sections. 2% tungsten is recommended and the electrode should be ground to a taper. Argon is used for shielding.
- **GMAW** is used for thicker materials, since it is a faster welding process. The spray transfer is used for flat position welding and argon is used as a shielding gas with 2% or 5% oxygen. Oxygen helps producing better wetting action of the weld.
- The short-circuiting transfer can also be used for thinner materials. In this case, CO_2 shielding or the 25% CO_2 plus 75% Ar mixture is used.

Figure 35.13 Lime-type electrode

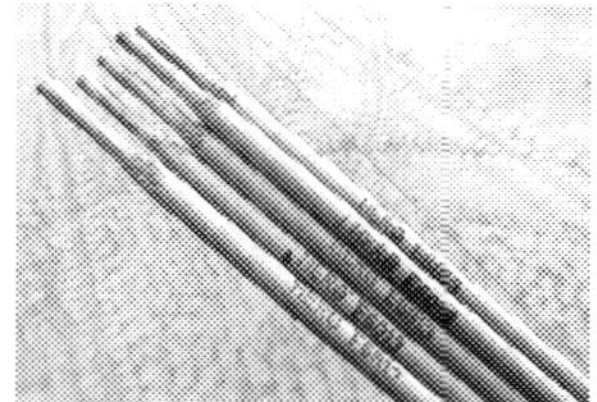

Figure 35.14 Titanium-type electrode

Note:

- For all welding operations, the weld area should be cleaned and free from all foreign material, oil, paint, dirt, etc.
- The welding arc should be as short as possible, when using any of the arc processes.

35.8 WELDABILITY OF ALUMINUM AND ITS ALLOYS (FIGURE 35.15)

Aluminum is a silvery white metal and it has the following characteristics:

- It is a light-weight metal, with density about a third that of steel or brass.
- Aluminum is a good conductor of electricity. On weight basis aluminum is better conductor than copper.
- Aluminum has higher resistance to corrosion owing to the protector conferred by the thin but tenacious film of oxide, which forms on its surface.

- Aluminum is a good conductor of heat.
- Aluminum is very ductile.
- Aluminum is non-magnetic.
- Melting point of pure aluminum is about 659°C.

Aluminum forms high strength alloys in conjunction with other metals, such as Cu, Cr, Ni, Zn, Fe, Zn, Mn, Si and Mg. Some of these aluminum alloys are more than four times as strong as the same weight of mild steel. They are malleable and ductile, exhibit excellent toughness and become stronger at temperatures below the range but they do not work well at temperatures of the order of 300 to 400°C. Aluminum and its alloys can be: (a) cast, (b) forged, (c) welded, (d) extruded, and (e) rolled, etc.

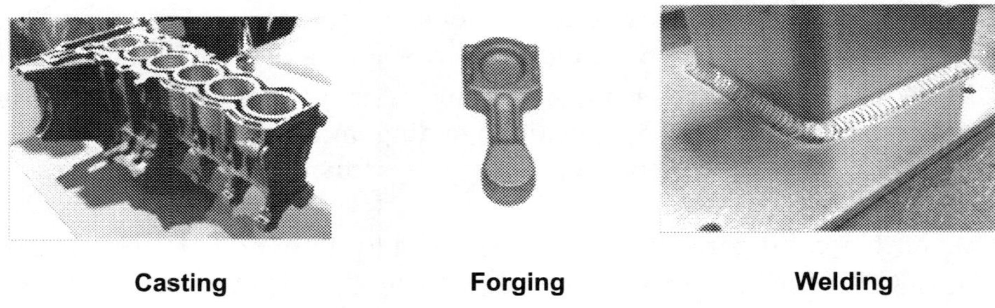

Figure 35.15 Weldability of aluminum and its alloys

Uses of aluminum and its alloys

Aluminum and its alloys are frequently used for:

- Transportation industry structural framework, engine parts, trim and decorative features, hardware, doors, window frames, tanks, furnishings and fittings.
- Trains, trucks, buses, automobile cars and aeroplanes, use many component parts made up of aluminum alloys.
- Overhead conductors and heat exchanger parts.
- In food industry, aluminum alloys find their application as food preparation equipment (pans, etc.), refrigeration, storage containers, bakery equipment, shipping containers, etc.
- In architectural field, aluminum alloys find uses, such as window frames, doors hardware, roofing, coping, sills, railings, fasteners, lighting fixtures, solar shading, grills, gratings, etc.
- Cryogenic applications.
- As heavy duty structures, such as dragline booms, travelling cranes, hoists, conveyor supports, bridges, etc.

- In process industries, parts made up of aluminum and its alloys are used to handle organic chemicals, petrochemicals and drugs. Tanks, drums, pipes, heat exchangers, gratings, smoke stacks, drilling towers, precipitators, centrifuges, valves, fittings, etc., are produced from aluminum alloys.

Welding Characteristics of Aluminum and its Alloys

I. The first and the foremost consideration is the effect of the thin film of oxide, which is always present on the surface of aluminum. This film contains moisture, which may react during fusion welding, with the liquid metal in the weld pool, to form further oxide and to liberate hydrogen, which can cause porosity.

   The oxide film on the surface of the base metal is removed chemically or mechanically (by brushing, scrubbing, etc.) before welding. The oxide film forms rapidly during welding also. This difficulty is overcome by:

   (a) Using a suitable flux in gas welding and brazing.

   (b) Employing a suitable thickly coated electrode in metallic arc welding.

   (c) Use of DCRP (electrode+) has been found effective for MIG welding, while AC is used for TIG welding of aluminum.

   In MIG welding, the oxide layer is removed by the arc cleaning action when the aluminum base metal forms the negative pole. If DCRP is used in TIG welding, the tungsten electrode will be overheated and the base metal may not melt adequately.

   Therefore, AC is used for TIG welding. In AC TIG welding, the cleaning of oxide is carried out in the positive half cycle and in the negative half, the electrode gets cooled down while the base metal gets heated up, aiding adequate fusion in the joint interface.

II. Aluminum being a very good conductor of heat, dissipates heat at a very fast rate from its joint being welded to the adjacent base metal, which compensates the heat loss from the joint being welded:

   (a) Nozzle size slightly bigger than that used for gas welding steel is employed for welding aluminum.

   (b) Currents slightly larger than those used for welding mild steel of the same thickness, are recommended for arc welding aluminum.

   (c) Thicker workpieces are preheated before welding.

III. Aluminum has high coefficient of linear expansion. Unless some precautions are taken, the welding heat may cause distortion and buckling. Proper joint design, edge preparation and preheating will counteract this tendency.

IV. Aluminum is weak when hot and thus, extra care is required when welding thinner sections, which may buckle.

V. In case of aluminum alloys at the solidus, both strength and ductility are low, so that when, welded under restraint the metal is likely to crack. Cracks may develop either in the weld metal, due to its low strength around the solidus of the freezing pool or in the overheated zone of the parent metal, due to fusion of low melting point constituents.

   Thus, it is reasonable to expect the pure aluminum and alloys of low alloy content to be less prone to cracking than the alloys with high additions. Heat treatable alloys of the duralumin type exhibit a greater tendency to crack during welding than the pure aluminum or the non-heat treatable alloys.

VI. Cold worked or peened weld metal is attacked more readily than annealed metal, especially under severe corrosive conditions, but corrosion may be less on a weld that has been hammered smooth than on an unhammered weld because the smoother surface exposes a smaller area.

VII. Aluminum does not show any color change on heating and without experience or control, it may be difficult to judge when the metal begins to melt or when the proper welding temperature has been reached. Two indicators that the welder, however, should watch are:

   (a) The melting of the dry flux.
   (b) Blistering of the metal surface, this shows that a proper welding temperature has been attained.

Processes Used for Welding of Aluminum and its Alloys

The following methods can be employed for welding aluminum and its alloy components:

1. Oxygas welding
2. Metallic arc welding
3. MIG welding
4. TIG welding
5. Resistance welding
6. Solid state welding
7. Carbon arc welding
8. Atomic hydrogen welding
9. Brazing

35.9 WELDABILITY OF CAST IRON (FIGURES 35.16-35.18)

Cast iron contains 2 to 4% carbon, which is above the solubility limit for steel alloys. It contains either inclusions of pure carbon, graphite or hard phases.

Gray iron contains flakes of graphite that cause its fractured surface, when broken, to appear to be gray in color.

Ductile iron, or nodular iron, contains spheroids of graphite that are nodular-shaped and have better ductility than others.

Figure 35.16 Gray cast iron

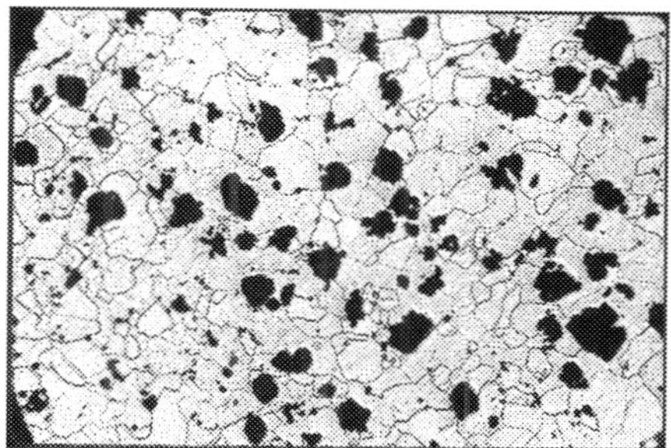

Figure 35.17 Malleable cast iron

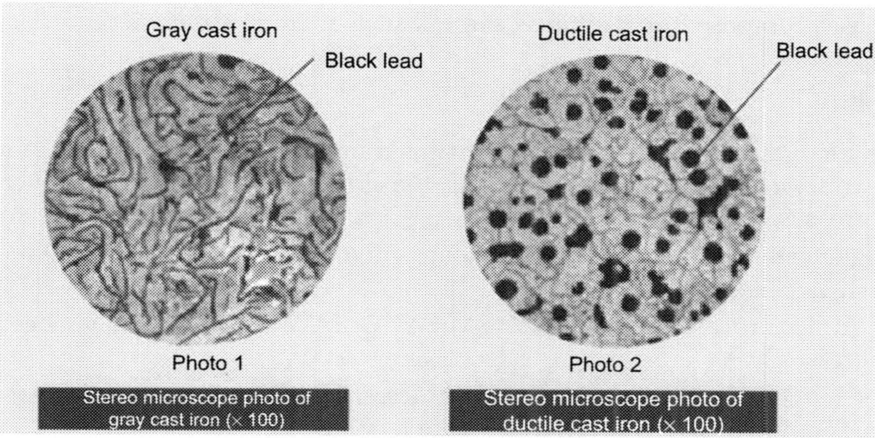

Figure 35.18 Comparison of gray cast iron and ductile cast iron

Primary Nodules

Because ductile iron is treated during melting to cause the graphite to form as nodules in the molten state, these nodules are known as primary nodules.

Temper Carbon or Graphite Rosettes

If a low-carbon, low-silicon cast iron is poured as carbide, or "white iron", and then heat treated to precipitate and agglomerate carbon, the graphite is often referred to as "temper" carbon or graphite "rosettes" and the iron is called malleable iron.

Alloying Elements

- Cast iron usually contains silicon, in addition to iron and carbon.
- Some elements added in very small amounts tend to stabilize various phases within the matrix or microstructure and may impart performance characteristic, such as oxidation resistance or strength, at elevated temperature.
- They also contain undesirable elements, such as phosphorus and sulphur, which may have various effects on the microstructure and weldability of cast iron.

Matrix or Microstructure

- The matrix of each iron is influenced by both, the composition and the cooling rate, from casting, heat treating or welding.
- Under the condition of very slow cooling, equilibrium condition favor the formation of graphite plus ferrite.
- At fast cooling rate, quench conditions are said to prevail, which result in hard phases composed of combined carbon, such as iron carbide and martensite.

Weldability

- As the matrix of each cast iron contains more combined carbon in the form of pearlite, iron carbide and martensite, the ability to be welded without cracking decreases.
- Higher preheat is often required along with other crack-discouraging methods, such as peening.
- Welding processes like OFW, SMAW, GMAW & FCAW, have been used successfully.
- The higher deposition rate welding processes are the most appealing for higher productivity because they have severe heating and cooling rates.
- Cast iron with totally ferritic matrices exhibit best weldability and the ductile iron with spheroidal graphite inclusions enjoy best weldability.

Recent Accomplishment in Cast Iron Development: Ferritic Si/Mo ductile iron. It provides excellent automatic weldability, requiring no preheat.

35.10 WELDABILITY OF PLASTICS

Most commonly used plastics are either thermoplastics or thermosetting plastics. Thermoplastics could be compared to wax. They are capable of remelting and changing shape. Thermosetting plastic could be compared to an egg when boiled; an egg becomes solid and sets, it cannot be brought back into liquid condition and cannot be reshaped. Thermoplastics are weldable while thermosettings are not weldable but can be joined by adhesive bonding.

A number of widely used plastics can be welded as they are thermoplastics. The most common of these are polyvinyl chloride (PVC), polythene, polypropylene, acrylonitrile budaddiene styrene (ABS) and acrylics. Such plastics can be wildly melted by melting the surfaces to be joined and allowing them to solidify while in contact. Plastics containing volatile compound form gas bubbles, which cause the formation of defects in the weld made.

Friction welding machines can be used to produce excellent weld in circular cross-sectional components. The most common method of welding plastics uses hot gases as a source of heat and torches similar to oxyfuel torch.

Welding torches for plastics are designed to let a compressed gas flow through electrically heated coils, which raise the gas temperature between 175 and 315°C. The hot gas passes through an orifice, forming a narrow gas, stream which can be directed to the surfaces to be joined (Figures 35.19 and 35.20).

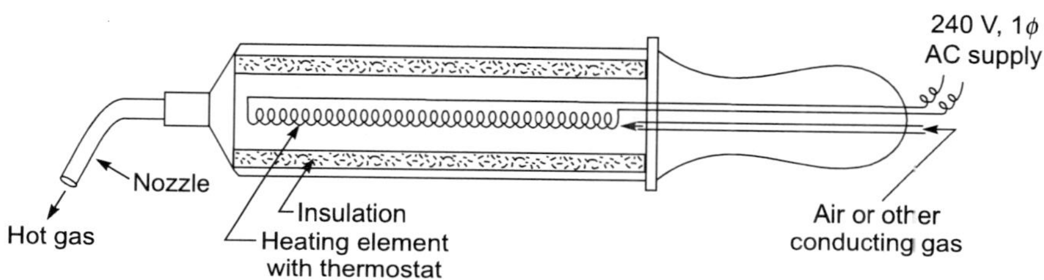

Figure 35.19 Electrically heated plastic welding torch

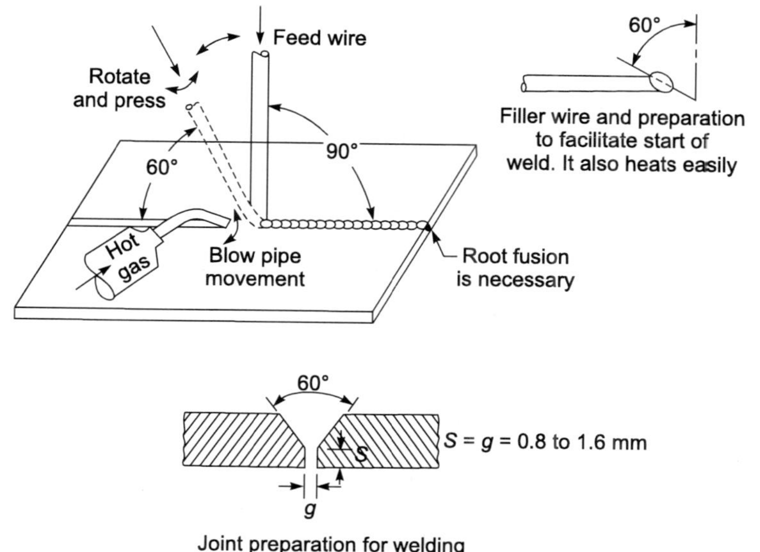

Figure 35.20 Manual hot gas torch welding

TABLE 35.4 Manual welding force on filler rod (intermittent)

Filler Rod Dia. (mm)	Approx. Load (kg)
2.4	1.0
3.2	1.8
4.8	3.0

Daily power requirements exceed 500 W for the heating element. Gas/air flow is of the order 280 L/min which can be supplied by $\frac{1}{4}$ horsepower compressor motors. Some plastics (example, polythene) are easily oxidized for such situation, heated compressed nitrogen gas gives best result. The force on the welding filler rod is important for welding of plastics and given in Table 35.4.

Fortunately, there is a wide margin between the softening (melting) temperature and the burning or charging temperature for thermoplastics. It is still advisable to use thermostat and melting temperatures that give best result, as the filler material does not change shape. With little practice, a welder can deposit excellent beads.

35.10.1 Hot Air Welding of PVC Plastics

Plastics are finding surprisingly new and diversified applications, replacing metals and ceramics. From ordinary toys and utensils to the complicated precision heart valves.

Plastics have a combination of desirable properties; they have high strength-to-weight ratio, corrosion resistance against most of the corrosive media, low cost and ability to take good finish.

Plastic structure can be fabricated by welding. Thermoplastics are the only weldable plastics, as they maintain their molecular structure even after repeated heating. The common thermoplastics are: acrylics, fluorocarbons, shellac, asphalt, nylon, polyethylenes, polyvinyles and protein substances. Among the above, the rigid polyvinyle chloride has sufficient resistance against corrosion, strong acid, alkalies and organic solvents. It is, therefore, the most common thermoplastic in use these days. The term "welding of plastic" is still rarely known among the engineers because of the fact that the use of plastic is still not very common, like carbon tetrachloride and adhesive, like areldite. PVC, however, is almost insoluble in most of the organic solvents. Though, it is slightly soluble in carbon tetrachloride but the action is very slow. There are certain other limitations too, in the way of joining plastics by the method other than welding. With the help of welding, adequate strength at the joining is achieved in minimum time. Within a few minutes, after welding any welded joint can be handled with reasonable care, facilitating rapid and economical fabrication of plastic structures.

35.10.2 Welding Action

Unlike metals, the welding action in plastic takes place due to the adhesive bonding at high temperatures, between the parent material and the filler rod. There is no mixing or puddling action as is common in the metallic weld pools. The melt of plastic is quite viscous and has poor flow properties, while good flow properties are essential for obtaining homogeneous welds. The surface of the parent material and the filler rod are heated and brought near to the melting temperature and by the application of pressure, the filler rod gets adhered to the joining (weld bead) surface to be joined. Thus, a homogeneous weld bead is not obtained but the filler rod gets adhered to the material in its neighbourhood and gives a defect-free non-porous joint.

35.10.3 Equipment

The tool used for hot gas welding, resembles in appearance with the ordinary welding torch (Figure 35.18). Direct flames char the material (PVC) and, therefore, hot gases are used for welding purposes. The torch consists of a main body, which contains a heating element. At one end of the body, there is an inlet hose connector for gas and a handle for gripping the torch, while the other end has a nozzle through which hot gas is available for use. The welding gas (usually air) enters the torch at some pressure and gets heated while passing over the heating element and comes out of the exit nozzle at the desired temperature. The gas temperature is controlled by providing in the heating element circuit, a thermostat valve, which controls the 'on' and 'off' period of the current fed to the element, thus regulating the temperature of the gas to a desired value depending upon the parent plate thickness. The torch may also be heated by using a fuel gas. A sectioned view of the torch used is shown in Figure 35.18.

35.10.4 Welding of PVC Plastic Using Hot Air Technique

For welding of PVC sheets, hot-air technique is most commonly used. Air is easily available and gives good results with PVC. Air flow needed for the process can be obtained by using a small compressor, with automatic tripping device to obtain constant pressure. Supply of air pressure can be measured by a mercury manometer, shown in Figure 35.21.

Rigid PVC sheets in common use are of 3 mm thickness and can be welded by using 3 mm filler rods. Air temperature is controlled by using a simmerstat that controls the amount of current in the heating coil (Figure 35.21).

Welding Transverse Speed: It depends upon air, temperature, nozzle distance from the plate and filler rod. It is manipulated by the experienced welder to obtain quality welds. Rod is fed to the plate at an angle of 90°. A fixture can be made if required to guide the filler rod at 90° and keep the nozzle torch at an angle of 45° with the joining line (Figure 35.20). Milling machine table could be used to obtain uniform transverse speed. A large number of transverse speeds are possible with this arrangement.

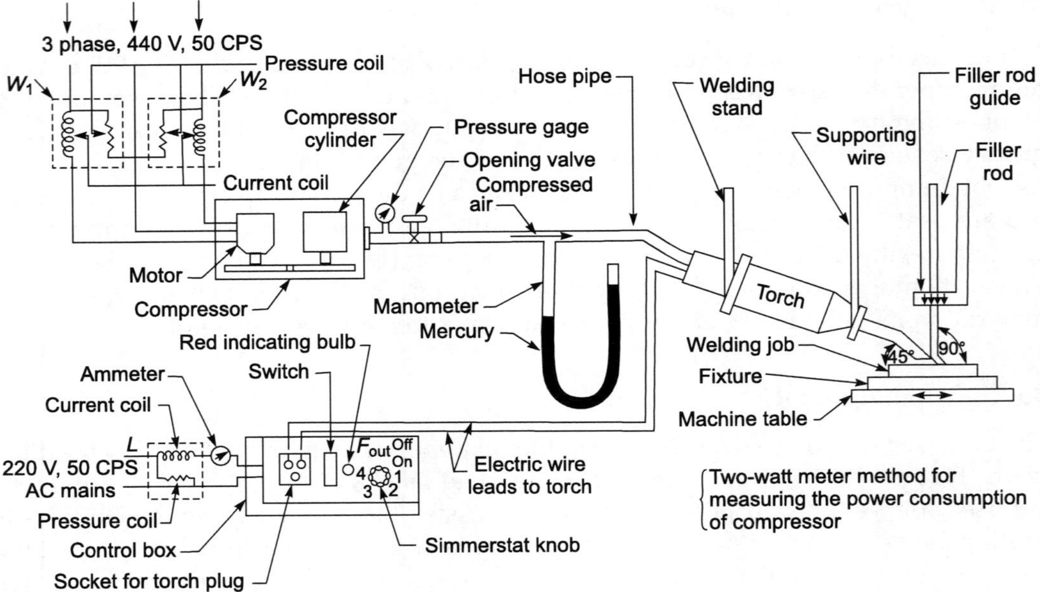

Figure 35.21 Block diagram of welding setup

Gap Distance: There is a slight variation of temperature with change of gap distance. This could be noticed from Figure 35.22 that there is a slight variation of gap distance between the torch nozzle and plate due to hand welding, will not appreciably affect the weld quality.

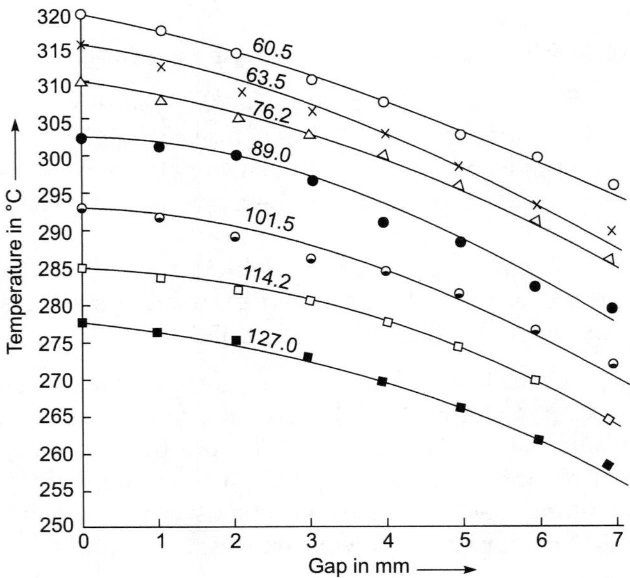

Figure 35.22 Gap distance between torch and the job versus temperature of hot air

Welded Joints. Two types of welded joints in general use are:

(a) Butt joints
(b) Double strap fillet joint

To obtain a butt joint, the plates to be joined are bevelled at 60° angle, cleaned, assembled over a backing plate and climb to the machine table-vice. The compressor is started, torch is switched on, and air pressure is regulated to about 100 mm of mercury. When a constant temperature of the system is achieved, a tank well is made at the starting and simultaneously hitting the base plate and filler rod. Estimating surfaced uses, it is possible to slightly rotate the filler wire in place slight pressure is applied to the filler rod to affect proper adhesion. The table is then moved away from the torch. The pressure on the rod is maintained with slightly rotary motion on the filler wire as shown in Figure 35.19. This is a matter of practice on the part of the welder. After completion of one pass, the table is stopped and the filler rod is cut. The process is repeated for subsequent passes as needed to fill the joining groove completely. Satisfactory belts have been obtained at a transfer speed of 50 mm/min.

Similar procedure is adopted for obtaining double strap fillet joint except that the assembly of the piece to be welded is tilted through an angle of 45°, to facilitate the heating of fillet properly. The fillet in this position serves as a 90° V-groove angle and heat is equally distributed in the plates to be joined.

Testing of Joints

Dumbbell-type test specimens have been proposed in literature for finding out the strength of plastic sheets with no mention about testing of the joint strength in welds. Dumbbell-type specimen shown in Figure 35.23 has been used for testing the strength of the parent plate as well as that of the butt weld specimen by some investigators. And effect can be avoided by removing and discarding a strip of 35 mm wide from both the sides of the weld. These test specimens can be tested on a 20 tonne universal testing machine using flat grips and may render the gripping difficult in the flat jaws. Tight strong grips can be obtained by making creations on both the sides of the specimen near the ends.

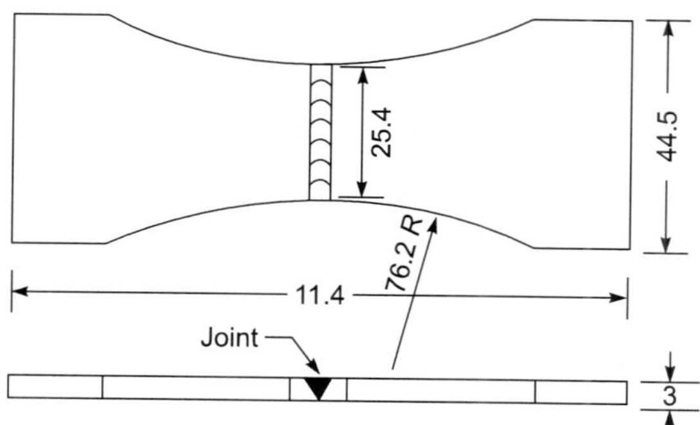

Figure 35.23 Test specimen for butt joint

Straight test piece is used for testing the strength of double strap fillet joint, as shown in Figure 35.24. The testing procedure is same as in the case of butt welded joints.

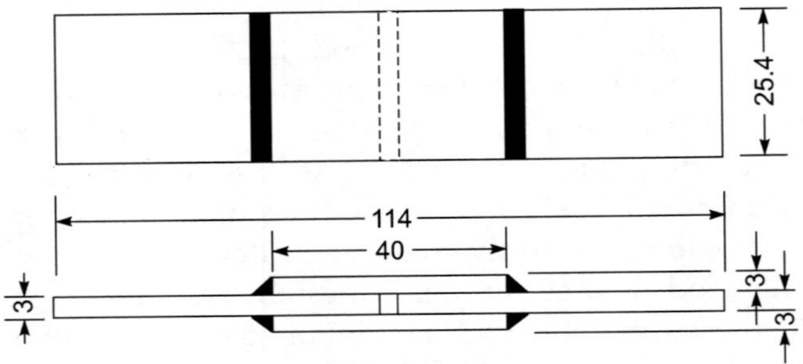

Figure 35.24 Test specimen for double strap fillet joint (all dimensions in mm)

Review Questions

1. Define the term weldability?
2. What is good and limited weldability?
3. What factors affect weldability of steel?
4. Explain weld decay in austenitic stainless steel.
5. Discuss about weldability of carbon steels.
6. Discuss about weldability of stainless steels.
7. Discuss about weldability of ferrite stainless steel.
8. Explain the processes used for welding of aluminum.
9. Discuss weldability issues in cast iron.
10. Discuss weldability of plastics.

36 CHAPTER

Repairs and Maintenance Welding

36.0 INTRODUCTION

Repair and maintenance of parts and components are major activities in any process industry. Repair welding can be carried out as a logical procedure that ensures the part is usable and safe. If repairs are done with proper care or precautions, one can avoid premature failures, large warranty claims, safety of property and personnel. Repair welding and surfacing are both considered in the field of maintenance welding and are covered together since they are both done by the same welders. Often, it is extremely difficult to separate what is considered repair welding from maintenance welding, and surfacing can be included in both situations. The same basic factors apply to both, weld repair and surfacing. Repaired parts may be more serviceable than the original parts, since they can be reinforced and the weaknesses of the original parts corrected. It is often more economical to weld repair since the delay in obtaining the replacement part could be excessive and the cost of the new part would normally exceed the cost of repairing the damaged part. This chapter deals with the repair welding and surfacing. Repair procedure, rebuilding and overlaying, selection of welding process, hardfacing, etc. are discussed below.

36.1 REPAIR PROCEDURE

A written procedure is required for all but simplest jobs. The procedure should be similar to a procedure developed for welding a critical part. It includes different steps as follows:

(a) Identify the material to be welded
(b) Selection of the welding process
(c) Preparation for welding
(d) Repair welding
(e) Postweld operation

The material can be identified in a number of ways, such as referring to the drawing of the part and the specifications, clues from the maintenance shop, by appearance, laboratory analysis of the metal from the piece filed from the metal, etc.

The selection of welding process depends upon:

(i) Type of metal
(ii) Thickness of metal
(iii) Position of welding
(iv) Base material properties (strength, etc.)

Surfacing based on the knowledge and experience, need for rework or resurfacing is the maintenance of surface characteristics. It helps to rebuild the surfaces of the worn out part without reducing the strength of the part and service life is also increased.

36.1.1 Preparation for Welding

The factors considered and decisions made before starting to weld are:

(a) Safety

- Welding location must be surveyed.
- All safety regulations and considerations must be satisfied.
- Removal of combustible materials from area.
- Draining of fuel tanks of construction equipment, aircraft, boats, trucks, etc.
- Elimination of toxic materials such as thick coats of lead paint, plastic covering of metals.

(b) Cleaning (Figures 36.1 and 36.2)

- Immediate work area must be clean from contaminants.
- Removal of dirt, grease, oil, rust, paint, plastic coverings, etc., from the surface of the parts being welded.
- Method of cleaning depends on the material to be removed and location of workpiece.
- Steam cleaning is recommended for construction and production equipment.
- Solvent cleaning is also employed for production equipment.
- Blast cleaning with abrasives is used for removing heavier impurities.
- Solvent dip cleaning is used for small particles.
- Power tool cleaning with brushes, grinding wheels, disc grinding, etc., can be employed.

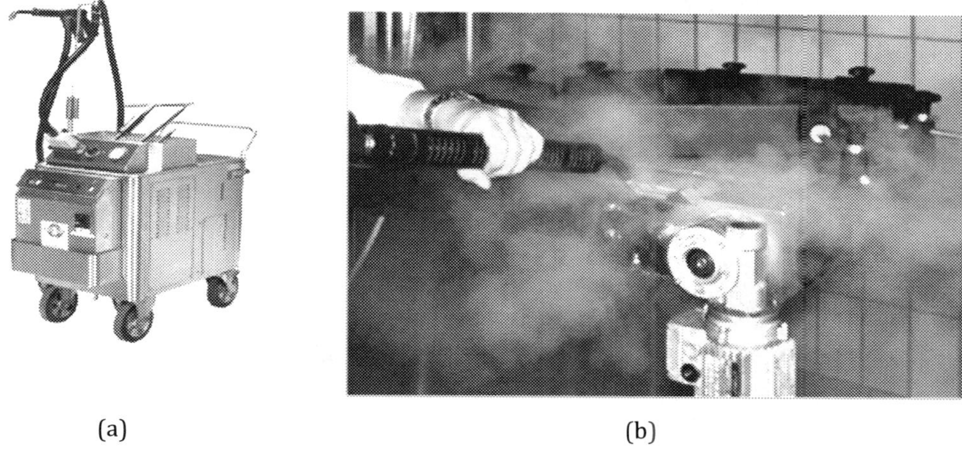

(a) (b)

Figure 36.1 (a) Steamer equipment (b) Steam cleaning

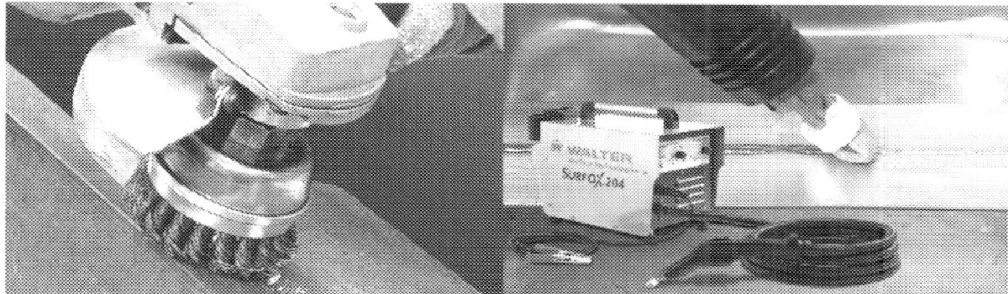

Figure 36.2 Power tool cleaning using brushes and grinding wheels

(c) Disassembly

- It may be required except the simplest repair jobs.
- It may require in case of lubrication lines, instrument tubing, wiring, etc.
- It is necessary to disassemble major components, such as machinery from machinery frames, etc.

(d) Protection of Adjacent Machinery and Machined Surfaces

- The parts not involved in repairing procedure should be protected from weld spatter flame cutting sparks and other foreign materials.
- Asbestos cloth is used to protect machined surfaces.
- Sheet metal guards or baffles protect adjacent machinery.
- Protective material should be properly secured using wires, clamps or temporary bracing.
- Machine surfaces within five feet of welding operation should be protected.

(e) Bracing and Clamping

- Used for complex repair jobs.
- Load on the parts to be welded can be reduced using clamps.
- Bracing is employed when main structural members are to be cut to support load.
- Braces are temporarily welded to structure.

(f) Layout Repair Work

- Remove metal to get full penetration weld.
- Contains the metal to be removed by cutting or gouging.
- Minimum amount of metal should be removed.
- Should be selected in such a way that welding can be balanced.
- Make sure welding should be made from comfortable position.

(g) Preheating

To obtain a crack-free weld metal, the preheating temperature is most important as is the interpass temperature. Table 36.1 gives preheating temperature for different materials.

Preheating reduces:

- the risk of hydrogen cracking,
- the shrinkage stress, and
- the hardness in the heat affected zone (HAZ).

The need for preheating increases with the following factors:

- the carbon content of the base material,
- the alloy content of the base material,
- the size of the workpiece,
- the initial temperature,
- the welding speed, and
- the diameter of the consumable.

One way to find the preheating temperature is to calculate the carbon equivalent, C_{eq}.

$$C_{eq} = \%C + \%Mn/6 + (\%Cr + \%Mo + \%V)/5 + (\%Ni + \%Cu)/15$$

Higher the value of C_{eq}, higher will be the preheating temperature.

TABLE 36.1 Recommended preheating temperatures

Base Material / Filler Material	Plate Thickness, mm	Steel C_{eq} <0.3 <180 HB, °C	Low Alloy C_{eq} 0.3–0.6 200–300 HB, °C	Tool Steel C_{eq} 0.6–0.8 300–400 HB, °C	Chromium Steel 5–12% Cr 300–500 HB, °C	Chromium Steel >12% Cr 200–300 HB, °C	Stainless Steel 18/8 Cr/Ni –200 HB, °C	Manganese Steel 14% Mn 250–500 HB, °C
Low alloy 200–300 HB	≤20	–	100	150	150	100	–	–
	>20 ≤60	–	150	200	250	200	–	–
	>60	100	180	250	300	200	–	–
Tool steel 300–450 HB	≤20	–	100	180	200	100	–	–
	>20 ≤60	–	125	250	250	200	–	○
	>60	125	180	300	350	250	–	○
12% Cr steel 300–500 HB	≤20	–	150	200	200	150	–	×
	>20 ≤60	100	200	275	300	200	150	×
	>60	200	250	350	375	250	200	×
Stainless steel 18/8 25/12 200 HB	≤20	–	–	–	–	–	–	–
	>20 ≤60	–	100	125	150	200	–	–
	>60	–	150	200	250	200	100	–
Mn steel 200 HB	≤20	–	–	–	×	×	–	–
	>20 ≤60	–	–	•100	×	×	–	–
	>60	–	–	•100	×	×	–	–
Co-based type 6 40 HRC	≤20	100	200	250	200	200	100	×
	>20 ≤60	300	400	•450	400	350	400	×
	>60	400	400	•500	•500	400	400	×

(h) Cutting and Gouging (Figure 36.3)

- Oxygen fuel gas cutting torch is most commonly used.
- Gauging tips are selected on the basis of the geometry of the joint.
- Proper power sources and carbons should be selected for the volume of the metal to be removed.
- Carbon deposit over the surface should be avoided.
- Mechanical chipping and grinding are employed in case where torch or carbon cannot be used.

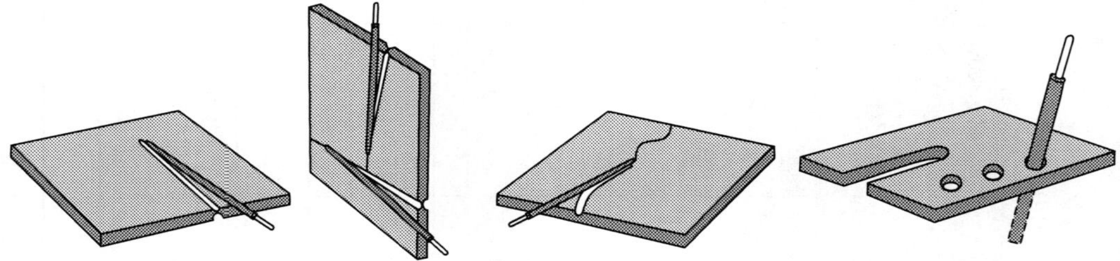

Figure 36.3 Cutting and gouging

(i) Grinding and Cleaning

- Burnt areas and oxides are removed and surfaces are smoothened.
- Grind the surface prior to welding to clean bright metal surface.
- Magnetic particle inspection is done in suspicion of additional cracks.
- Make sure that all defects have been removed.

36.1.2 Repair Welding

Logical sequence to make sure that all factors are considered and adequately provided:

1. **Welding Procedure**
    - Must be available for welders.
    - Include process, specific filler metals, preheat required, and other information.
2. **Welding Equipment**
    - Availability without delays.
    - Standby equipment is also required.
    - Includes sufficient electrode holder, grinders, wire feeders, cables, etc.

3. **Materials**
   - Availability of materials for the entire job.
   - Include insert pieces, reinforcing pieces, etc.
   - Include fuel for preheat and interpass temperature, shielding gases, fuel for engine powered machines.
4. **Alignment Markers**
   - Center punch marks made across the joint at various locations prior to welding.
5. **Welding Sequences**
   - Should be described in procedure.
   - Include block welding, back-step sequence welding, wandering sequence welding and peeling.
   - Sufficient number of welders assigned to the job.
6. **Safety**
   - Ventilation must be provided when fuel gases are used for preheating.
   - Welding suits, glasses, gloves are compulsory.
7. **Weld Quality**
   - It should be checked continuously.
   - Weld should be smooth, no notches, good reinforcement, etc.
   - Grinding should be done to maintain smooth flowing contours.

36.1.3 Postweld Operation

Weld should be allowed to cool and it should not be exposed to winds or drafts, nor the machinery loads, until the temperature of the repaired part reaches the normal ambient temperature.

It involves:

1. **Inspection**
   - Finished weld is inspected for smoothness and quality.
   - Includes non-destructive testing, like magnetic particle, ultrasonic, X-ray.
   - Repaired weld should be of high quality.
2. **Clean-up Operation**
   - Removal of strong backs, bracing, and protective covers, etc.
   - Weld stubs, weld spatter, weld slag, and other residue are also removed.
   - Grinding dust should be removed as it can cause critical damage to the joints and bearings.

3. **Repainting**
   - After cleaning, the area is painted and other areas should be greased for the operation of machinery.
4. **Reassembly**
   - The pieces of machinery taken away are returned and reassembled.

36.2 REBUILDING AND OVERLAYING (FIGURE 36.4)

- Rebuilding and overlaying with weld metal or spray metal are both considered surfacing operations.
- Surfacing is the deposition of filler metal on a base metal to obtain the desired dimensions or properties.
- Overlay is considered to be a weld or spray metal deposit that has specific properties, sometimes unlike the original surface properties.
- Rebuilding is used to bring parts back to their original dimensions and properties, such as the rebuilding of worn shafts, repair of parts that were machined undersize, etc.
- Overlay surfacing is used to return the part to original dimensions but with weld metal having particular properties to reduce wear, erosion, corrosion, etc.
- It is possible to thermal spray certain materials that cannot be deposited with the welding processes and applied to the ceramic sprayed coatings or other non-metallic materials.

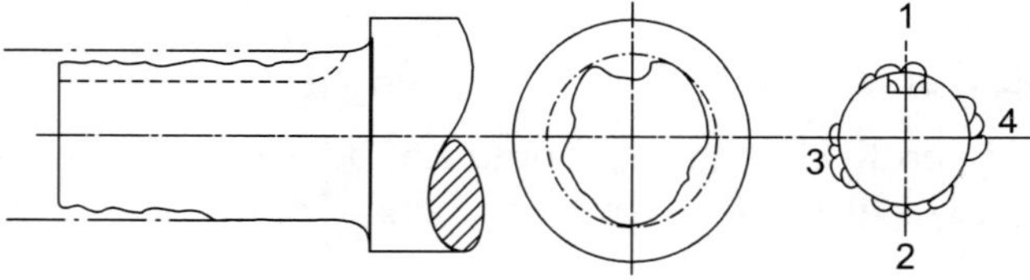

Figure 36.4 (a) Rebuilding a worn shaft

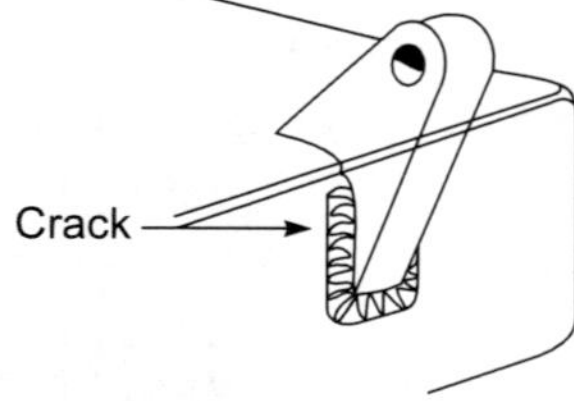

Figure 36.4 (b) Repair of cracked part

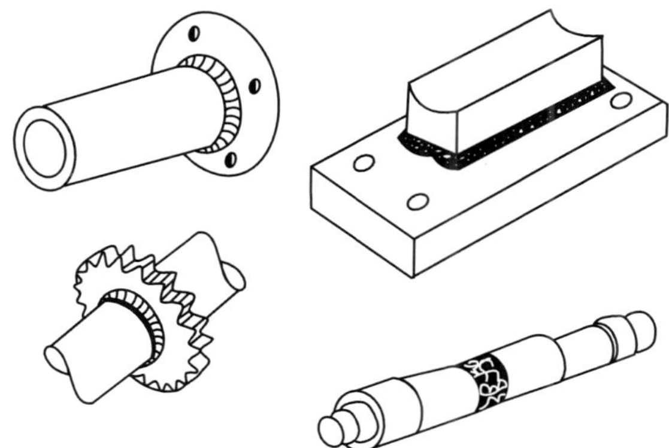

Figure 36.4 (c) Repair of machine parts

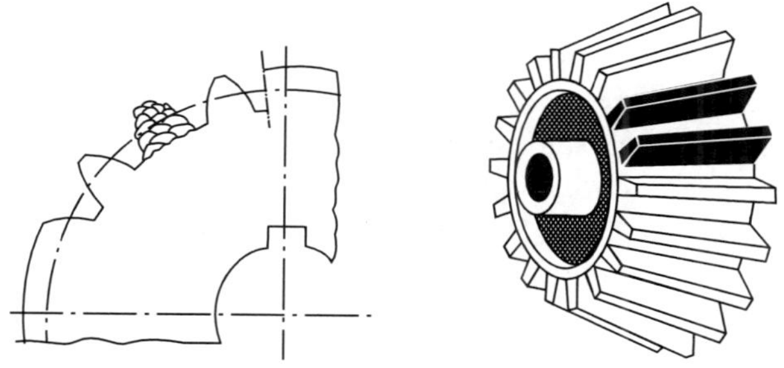

Figure 36.4 (d) Repair of broken teeth in gear

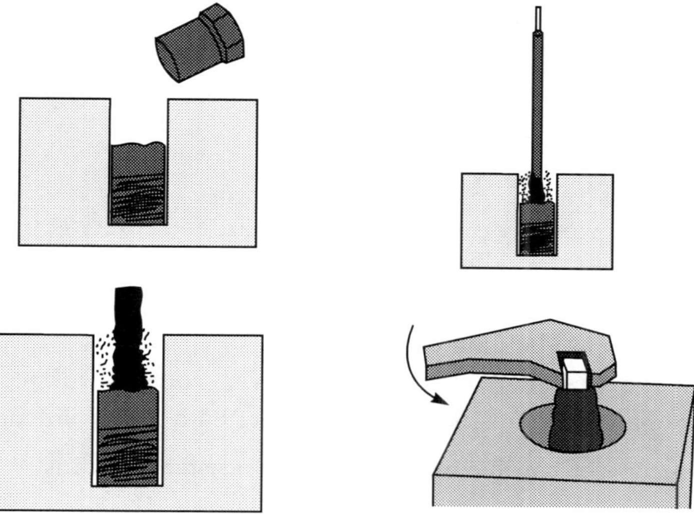

Figure 36.4 (e) Extraction of broken bolt

36.2.1 Selection of Welding Process

- *The selection of the welding process* and the welding procedure and technique are as important as the selection of the deposit alloy. Almost all the arc welding processes and several others can be used for the application of hard facing weld metal.
- *The shielded metal arc welding* process is probably the most commonly used welding processes for hard facing. It can be used in the field and in the shop and can be applied to small and large parts in any position.
- *Submerged arc welding* is also used for many applications but it is restricted to welding in the flat position. Most often, it is used for plant operations and not used in the field. It is often used for repeating applications when the same part is surfaced on a routine basis.
- *Flux-cored arc welding* with and without shielding gas, is a popular semi-automatic welding process. It can be used in the field or in the shop and is not restricted to the flat position.
- *Gas tungsten arc welding* process is used for many smaller applications, usually for shop work in which the part can be brought to the shop and manipulated and moved for ease of welding.
- *Plasma arc welding* is also used much in the same manner as gas tungsten arc. It does have a higher temperature and for this reason, can be used in certain cases where gas tungsten arc welding is not applicable. It is again restricted to the smaller type of jobs.
- *The electroslag welding* process is also used for certain special applications. It has been widely used for rebuilding crusher hammers. These can be rebuilt with special fixturing and done quite rapidly with the electroslag process.
- *Oxyacetylene welding* is widely used for specialized cobalt alloys on relatively thin edges.

36.3 HARDFACING

Hardfacing is a metal working process by which harder or tougher material is added to the base metal through welding, for the purpose of restoring dimensions of worn surfaces of used implements or to increase the life of a new tool or machine.

- It is also known as cladding or overlaying.
- It consists of depositing weld overlay to cover worn out surface.
- It provides hardness, erosion, galling, impact with less maintenance.

The main purpose of hardfacing is to provide resistance against:

- Wear or abrasion
- Impact

- Erosion
- Galling
- Cavitation
- Heat corrosion

36.3.1 Hardfacing: Operation and Parameters

The deterioration of surfaces is the major problem faced in industry. Hardfacing can be a solution to the wear problems due to different reasons. Steps involved in it are:

1. Determine the exact composition of the base metal as it influences the surfacing method.
2. Determine the type and degree of the wear on the base metal-to-metal.
3. Determine welding position of the job.
4. Estimate area that needs to be hardfaced.
5. Select appropriate hardfacing alloy and method.
6. Prepare the surface to be hardfaced.
7. Apply hardfacing alloy to the base metal and impart proper surface finish.

The factors which should be taken care of for the proper hardfacing procedure are:

- In many cases, two separate materials might be required for hardfacing: the buildup alloy, which is used when the part is to be reclaimed or is excessively worn, and the hardfacing alloy. The hardfacing alloy will be much more expensive than buildup alloy.
- Another factor to be considered is dilution. This is the dilution of hardfacing alloy with base metal. Excessive dilution reduces the effectiveness of hardfacing.
- Like all other welding, fabrication, surfacing or repair operations, the base metal part has to be cleaned properly by the required method: like grinding, shot-blasting, etc.
- Another consideration is relative position between finished surface and worn surface. If there is only small wearing, excessive material has to be removed from the base metal to make sure that there is enough space for sufficient number of hardfacing layers.
- The factors that are needed to be considered for welding of the base metal in the original fabrication like preheating, cooling, etc., are to be considered for overlaying processes also.

36.3.2 Surfacing Methods Using Welding

The following welding processes can be used for hard facing:

1. Oxyacetylene
2. Shielded metal arc

3. Submerged arc welding
4. Metal inert gas
5. Tungsten inert gas
6. Plasma welding

1. Oxyacetylene Welding

- Surfaces using oxyacetylene welding produce high quality surfacing deposits. The quality depends upon the choice of gas mixtures, flame characteristics, etc.
- Before hardfacing, the base metal surface is thoroughly cleaned of dirt, grease, etc. For hardfacing of dry steel surfaces, no flux is required. But for cast iron, it is required.
- After cleaning, the surface to be hardfaced is heated to red hot with the torch.
- For hardfacing nickel alloys, a neutral flame is used. The surface of the base metal appears watery when melting begins, this condition is known as sweating.
- As the base metal melting begins, the hardfacing alloy is deposited drop-by-drop and blends immediately with the sweating zone and spreads.
- This process is continued till the area to be hardfaced is fully covered with the alloy. The hardfacing layer should be allowed to cool slowly to produce crack-free and residual stress-free deposit.
- After cooling, the surface may be finished using grinding to remove high spots and to correct the dimensions.
- The depth of the fused hardfacing alloy does not exceed 0.25 mm. Due to this minimal penetration, the dilution of the hardfacing alloy is kept at minimum.
- For hardfacing with cobalt base alloys, a carburizing flame is used.
- For hardfacing of cast iron, the hardfacing alloy is rubbed on the heated surface of the job to break through the surface crust of cast iron. This is done as sweating of cast iron does not take place.

Advantages

1. Very thin layer of hardfacing can be done.
2. Smooth, high quality deposits are obtained.
3. Can be easily employed in field work.
4. Equipment needed is simple and investment cost is less.

Limitations

1. High deposition rates compared to arc welding cannot be obtained.
2. Skilled welder is required.
3. Speed of process is slower than arc welding.

Applications: Surfacing of steam valves, chain saw bars, and wood cutting blades.

2. Shielded Metal Arc Welding

- The equipment needed is same as in manual metal arc welding. Parameters such as welding current, voltage, travel speed, position of electrode relative to the job, etc., are adjusted so that the hardfacing alloy is deposited with the lowest heat, keeping dilution to a minimum.
- The surface to be hardfaced is initially cleaned of all foreign matters. Preheating is not required for low carbon and mild steels, except for large sections.
- Medium and high carbon steels need to be preheated. Cast iron is preheated at a slow and uniform rate to about 260 to 370 degrees.

A medium long arc is maintained so that:

1. Electrode coating does not touch base metal.
2. It preheats the base metal ahead to keep hardfaced material molten, long enough for impurities to float to the surface.

   Usually direct current in reversed polarity is used. Electrode should be held at an angle of 45°.

Advantages

1. Low cost equipment, great versatility and convenient.
2. Hardfacing can be done in almost all locations.
3. Welder can easily cover irregular areas.

Disadvantages

1. Low deposition rates.
2. High labor costs for per kilogram of alloy deposited.

Applications: Slide valves, power shovels, teeth of rooters, etc.

3. Submerged Arc Welding Hardfacing

- The equipment required is similar to that employed in shielded arc welding. The electrode used is in the form of a continuous wire (solid or tubular) or a solid strip. One or more electrodes may be used for high deposition surfacing.
- Metals hardfaced using this process are: carbon and low alloy steels, stainless steel, cast iron, nickel and its alloys. A constant voltage power source is usually preferred.

Advantages

1. Weld metal quality is consistently high. It is used when surfacing requires strength, toughness or abrasion resistance.
2. High deposition rates.
3. Elimination of weld spatter.

Disadvantages

1. SAW is limited surfacing of round or flat parts. Not usually used in field work.
2. Due to high penetration, the dilution of hardfacing metal takes place, along with large heat generation.
3. Parts made of austenitic manganese steel cannot be hardfaced using this method.

Applications: Hardfacing of oil-field drill collars, tractor rollers and idlers, blades for hot shearing.

4. MIG Welding

- The equipment required is same as that of MIG welding, except for an attachment used in case of automatic process.
- Pulsed DC current is used to obtain better control of electrode melting rate and bead rate.
- Carbon dioxide, argon or helium is used either singly or with small amounts of oxygen as shielding gas to prevent weld metal oxidation.
- The process takes place very fast, but not used in a large manner as some hard facing alloys are not present in wire forms.

Applications

1. Artillery shell banding with gliding metal.
2. Surfacing lands between the piston ring groove in heavy duty hydraulic pistons with aluminum bronze.

5. TIG Welding

- The equipment used for hardfacing is same as that used in TIG welding.
- Hardfacing can be done in almost all positions. Most of the ferrous and non-ferrous base metals can be hardfaced. The alloys used in hardfacing are high alloy steels, nickel and its alloys, cobalt and its alloys. The alloy electrodes are used in forms like wrought, cast or tubular.
- Flux is not usually used as argon/helium protects the deposit. For hard-facing, the filler used is in powdered form, whereas wire/rod fillers are used for surfacing.

Applications: Surfacing drill pipe joints, hardfacing of valve gates.

6. Plasma Welding

- It involves high cost of equipment. The filler material is forced through an orifice to attain sonic velocity in a powdered form along with the plasma jet.

- The process generates penetration as low as 0.125 mm and dilution between 5 and 30%.

Advantages of Hard Facing by Welding Methods

1. It is a good method for reclaiming used parts, saves considerable time and money.
2. It reduces downtime of equipment.
3. Cheaper than replacing original parts and should be used wherever possible.
4. Avoids excess cost of transportation to repair facility as hardfacing can be done on the site.
5. Original equipment manufacturers are now adopting hardfacing of wear parts on new equipment, to provide better service life of equipment.

Review Questions

1. Write steps involved in repair procedure of welds.
2. Discuss rebuilding and overlaying.
3. How do we repair a broken shaft?
4. How do we repair a worn out gear?
5. What is the purpose of hardfacing?
6. Discus various surfacing methods used in welding.
7. What are the advantages of oxyacetylene welding surfacing?
8. Discuss about submerged arc welding hardfacing.
9. Discuss about plasma welding hardfacing.
10. What are the advantages of hardfacing by welding methods?

37

CHAPTER

Multiple-Choice Questions

1. The British code for visual inspection requirements is:
   (a) BS4872 (b) BS499
   (c) BS4870 (d) None of the above

   Answer: (d) British code for visual inspection requirement is BS 5289.
2. A code of practice for visual inspection should include the following:
   (a) Before, during and after welding activities
   (b) Before welding activities only
   (c) After welding activities only
   (d) None of the above

   Answer: (a)
3. Incomplete root penetration in a butt joint could be caused by:
   (a) Excessive root face width (b) Excessive root gap size
   (c) Low current setting (d) Both (a) and (c)

   Answer: (d) Excessive root face width will need more energy to melt and since heat generated is proportional to square of current, both low current and excessive root face width will cause incomplete root penetration.
4. Incomplete root fusion would certainly be caused by:
   (a) Linearm misalignment (b) Incorrect tilt angle
   (c) Differing root face widths (d) All of the above

   Answer: (d) In case of linear misaligned portion, the misaligned portion will not fuse, incorrect tilt angle will provide more heat to one side and hence, the other side won't fuse correctly. In case of different root width, thinner section will melt more and the fusion in the thicker section will be incomplete.
5. When visually inspecting a completed single Vee butt weld cap, you would certainly assess:

(a) Cap height
(b) Toe blend
(c) Weld width
(d) All of the above

Answer: (d)

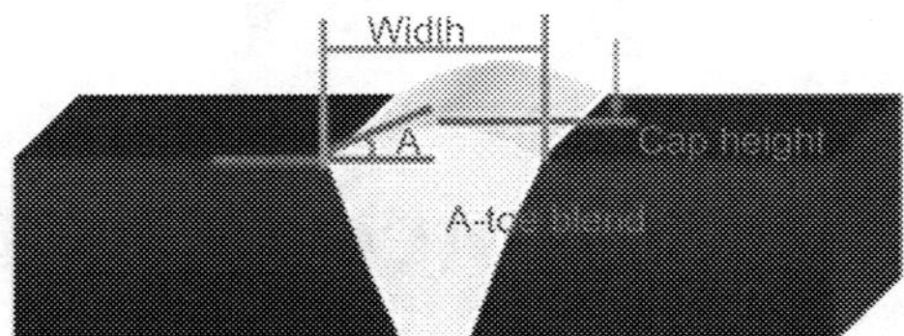

6. You notice a very 'veed' ripple shape. This is most likely caused by:
   (a) Poor consumable choice
   (b) Welding position
   (c) Excessive travel speed
   (d) All of the above

   Answer: (c) Excessive travel speed will prevent adequate fusion with parent metal, so elongated ripples will be formed.
7. Toe blending is important as it may affect:
   (a) Corrosion
   (b) Fatigue life
   (c) Overlap type defects
   (d) All of the above

   Answer: (b) If a very high toe angle is there, then there will be a sharp transition which will cause high local stress concentrations and this will reduce fatigue life.
8. Slag inclusions would occur with:
   (a) Manual metal arc
   (b) Metal inert gas
   (c) Submerged arc welding
   (d) Both (a) and (c)

   Answer: (d) In case of manual metal arc (MMA) welding, flux coated welding electrode is used and in case of SAW, welding zone is submerged in flux, so in these two welding techniques slag inclusions will be there.
9. Undercut is principally caused by:
   (a) Excessive amps
   (b) Excessive volts
   (c) Excessive travel speed
   (d) All of the above

   Answer: (d)
10. Undercut is normally assessed by:
   (a) Its depth
   (b) Its length
   (c) Its blending
   (d) All of the above

   Answer: (d)
11. A welding procedure is useful to:
   (a) Give information to the welder
   (b) Give information to the inspector
   (c) Give confidence to a product
   (d) All of the above

   Answer: (d)

12. An essential variable may:
    (a) Change the properties of a weld
    (b) Influence the visual acceptability
    (c) Require reapproval of a weld procedure
    (d) All of the above

    Answer: (d)

13. A magnifying glass may be used during visual inspection, but BS 5289 states that its magnification should be:
    (a) Up to 5∅
    (b) 2 to 2.5∅
    (c) 5 to 10∅
    (d) None of the above

    Answer: (b)

14. When visually inspecting a fillet weld, it would normally be sized by:
    (a) The leg lengths
    (b) The actual throat thickness
    (c) The design throat thickness
    (d) Both (a) and (c)

    Answer: (d)

15. A planar defect is:
    (a) Incomplete fusion defect
    (b) Slag inclusion
    (c) Incomplete penetration
    (d) Both (a) and (c)

    Answer: (d) Planar defects are defects whose width is small as compared to height. Both (a) and (c) have this characteristic.

16. Penetrant inspection and magnetic particle inspection are mainly used:
    (a) To aid visual inspection
    (b) Because the application says so
    (c) To confirm visual uncertainties
    (d) All of the above

    Answer: (d)

17. Defects outside the limits specified in a standard should always be:
    (a) Repaired
    (b) Reported to 'a senior person'
    (c) Assessed along with other defects
    (d) All of the above

    Answer: (b)

18. MIG welding tends to be susceptible to lack of fusion problems. This is because of:
    (a) Poor maintenance of equipment
    (b) Incorrect settings
    (c) Poor inter-run cleaning
    (d) All of the above

    Answer: (d)

19. Manual metal arc (MMA) electrodes can be grouped into three main types. These are:
    (a) Basic, cellulosic and rutile
    (b) Neutral, cellulosic and rutile
    (c) Basic, cellulosic and neutral
    (d) None of the above

    Answer: (a)

20. The main cause of porosity in welded joints is:
    (a) Poor access (b) Loss of gas shield
    (c) 'Dirty' materials (d) All of the above

    Answer: (d)
21. Cracks in welds may be due to:
    (a) Solidification problems (b) Hydrogen problems
    (c) Excessive stresses (d) All of the above

    Answer: (d) Hydrogen released into the heat affected zone (HAZ) region may cause cracks, cracks will also be caused because weld bead in the final stage of solidification has insufficient strength to withstand the contraction stresses generated as the weld pool solidifies.
22. A weave technique may give rise to:
    (a) Better profiles (b) Improved toe blending
    (c) Improved ripple shape (d) All of the above

    Answer: (c)

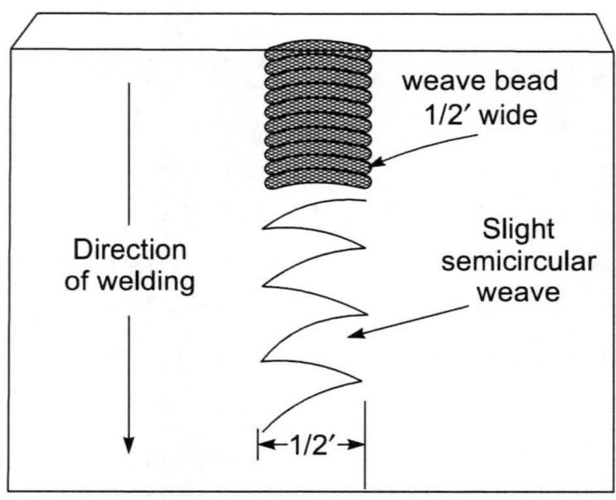

23. With reference to a root penetration bead, you would certainly assess:
    (a) Root fusion and penetration (b) Root concavity
    (c) Burn-through (d) All of the above

    Answer: (d)
24. In a fatigue failure, the appearance of the fracture surface is characteristic. It would be:
    (a) Rough and torn (b) 'Chevron'-like
    (c) Smooth (d) None of the above

    Answer: (c)

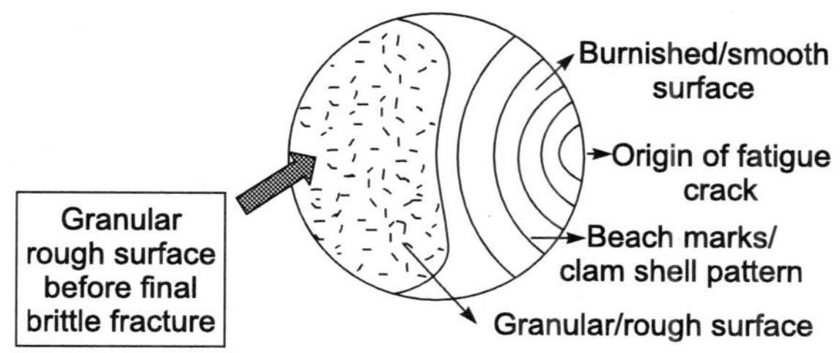

25. Stray arcing may be regarded as a serious defect because:
    (a) It may reduce the thickness of the component
    (b) It may cause liquation cracks
    (c) It may cause hard zones
    (d) All of the above

    Answer: (d)
26. Overlap in welds could be influenced by:
    (a) Poor welding technique (b) Welding process
    (c) Welding position (d) All of the above

    Answer: (d)
27. Flame cut preparations may, during welding, increase the likelihood of:
    (a) Cracking (b) Misalignment problems
    (c) Inclusions (d) All of the above

    Answer: (d)
28. Macroscopic examination requires any specimen to be inspected:
    (a) Once, after etching (b) Twice, before and after etching
    (c) Using microscope (d) None of the above

    Answer: (b)
29. Which of the following may be classed as a more serious defect(s)?
    (a) Slag inclusions (b) Fusion defects (interrun)
    (c) Fusion defects (surface) (d) Porosity

    Answer: (a) Slag will cause discontinuity in weld bead and hence, can cause stress concentrations.
30. A code of practices is:
    (a) A standard for workmanship only
    (b) A set of rules for manufacturing a specific product
    (c) Levels of acceptability of weldment
    (d) None of the above

    Answer: (b)

31. Movement of the arc in manual metal arc (MMA) welding by magnetic forces is called:
   (a) Arc deviation
   (b) Arc misalignment
   (c) Arc blow
   (d) Arc eye

   Answer: (a)
32. A metallurgical problem most associated with submerged arc welding is:
   (a) Hydrogen cracking in the heat affected zone (HAZ)
   (b) Solidification cracking in the weld metal
   (c) Hydrogen cracking in the weld metal
   (d) Lamellar tearing in the weld metal

   Answer: (b)
33. Oxy pressure and nozzle size in flame cutting would influence:
   (a) The temperature required for cut initiation
   (b) The ability to cut stainless steels
   (c) The depth of the cut obtainable
   (d) None of the above

   Answer: (c)
34. The main use of arc cutting/gouging processes is:
   (a) The cutting of single bevel preparations
   (b) The removal of deposited welds
   (c) The cutting of single U-type preparations
   (d) The cutting/gouging of non-ferrous materials only

   Answer: (b)
35. Which of the following process joins metals plastically?
   (a) Friction welding
   (b) Resistance welding
   (c) Plasma welding
   (d) All the above

   Answer: (a) In friction welding, a lateral force is given to plastically displace and fuse the two parts which are rotating.
36. Which electrode classification would be relevant to American Welding Society (AWS) A5.1-81?
   (a) E 6013
   (b) E 5133
   (c) E 7018-G
   (d) Fleet weld 5

   Answer: (a)
37. Which of the following coatings is associated with stove welding?
   (a) Rutile
   (b) Cellulosic
   (c) Basic
   (d) Oxidizing

   Answer: (b)
38. A common gas mixture used in metal inert gas (MIG) welding nickel alloys to combine good levels of penetration with good arc stability, would be:

(a) 100% CO_2
(b) 100% argon
(c) 80% argon and 20% CO_2
(d) 98% argon and 2% oxygen

Answer: (b)

39. Which type of submerged arc welding (SAW) flux is more resistant to moisture absorption?
    (a) Fused
    (b) Agglomerated
    (c) Basic
    (d) All of the above, having same resistance

    Answer: (a)

40. The flame temperature of oxy/acetylene mixture gas is given as:
    (a) 3200°C
    (b) 2300°C
    (c) 5000°C
    (d) None of the above

    Answer: (a)

41. A large grain structure in steels is said to produce:
    (a) Low ductility values
    (b) Low fracture toughness values
    (c) High fracture values
    (d) High tensile strength

    Answer: (b) Relation independently derived by Cottrell and Petch provides the fracture stress to be proportional to the ratio of surface energy by its square root of grain size for a given material.

42. The likelihood of brittle fracture in steels will increase with:
    (a) A large grain formation
    (b) A reduction of in-service temperature to sub-zero temperatures
    (c) Ferritic rather than austenitic steels
    (d) All of the above

    Answer: (d) Large grains will lower the fracture stress value.

43. Repair welding is often more difficult than production welding due to:
    (a) The material being ingrained with in-service contaminants
    (b) Restricted access within the repair area
    (c) The possible position of the weld repair
    (d) All of the above

    Answer: (d)

44. Hydrogen cracking in the weld metal is likely when welding:
    (a) Carbon manganese steels
    (b) Stainless steels
    (c) Micro-alloyed steels HSLA (high-stregth low-alloy)
    (d) Low-carbon steels

Answer: (c) The weld metal high-strength low-alloy (HLSA) transforms from austenite to ferrite at a lower temperature than the parent steel, meaning that any hydrogen in the heat affected zone (HAZ) is rejected into the still austenitic weld metal, which has a high solubility for hydrogen and high chance of hydrogen cracking can be there?

45. EN 288 standard would refer to which of the following?
   (a) Welder approval testing
   (b) Welding equipment
   (c) Welding procedure approval
   (d) Consumables for welding

   Answer: (c)

46. Porosity is caused by:
   (a) Entrapped slag in the solidifying weld
   (b) Entrapped gas in the solidifying weld
   (c) Entrapped metallic inclusions in the solidifying weld
   (d) None of the above

   Answer: (b)

47. In a bend test, the face of the specimen is in tension and the root is in compression. What type of test is being carried out?
   (a) A root bend test
   (b) A side bend test
   (c) A face bend test
   (d) None of the above

   Answer: (c)

48. Ultrasonic testing is more advantageous in detecting which of the following weld imperfections, over other non-destructive testing (NDT) methods?
   (a) Lack of sidewall fusion
   (b) Surface undercut
   (c) Incompletely filled groove
   (d) Overlap

   Answer: (a) Ultrasonic testing can measure internal defects.

49. Tempering is often carried out to regain toughness after which of the following processes?
   (a) Annealing
   (b) Normalizing
   (c) Hardening
   (d) Stress relieving

   Answer: (c)

50. The presence of iron sulphide in the weld metal is most likely to produce which of the following upon contraction of the weld?
   (a) Solidification cracking
   (b) Hydrogen cracking
   (c) Intergranular corrosion
   (d) Stress corrosion cracking

   Answer: (a)

51. Austenitic stainless steel electrodes are generally smaller in length than mild steel electrodes because:
   (a) High amperage is used
   (b) Shelf life will be decreased

(c) Their electrical conductivity is less than that of steel
(d) They are more expensive

Answer: (c) Due to their higher resistance, voltage will be dropped if long ones are used.

52. The voltage necessary to maintain an arc during metal arc welding is termed:
(a) Mains supply voltage
(b) Arc current
(c) Arc voltage
(d) Open circuit voltage

Answer: (c)

53. When manual metal arc (MMA) welding low carbon steel, which electrode will give the greatest deposition rate?
(a) Hydrogen controlled
(b) Cellulosic
(c) Rutile
(d) Iron powder

Answer: (d)

54. Inherent rectification of the electrical output is produced in the arc when tungsten inert gas (TIG) welding using:
(a) AC with a suppressor
(b) AC without a suppressor
(c) DC with reverse polarity
(d) DC with straight polarity

Answer: (b) When welding aluminum with a tungsten electrode on AC polarity, there is a tendency for the current to be changed to DC. This condition (inherent rectification) increases as the oxide film increases. A DC suppressor unit allows the AC current but prevents what is known as the DC component, thus a suppressor is only necessary when welding aluminum, magnesium, and their alloys.

55. Gamma rays and X-rays are part of a family of waves called:
(a) Acoustic waves
(b) Light waves
(c) Electromagnetic waves
(d) Transverse waves

Answer: (c)

56. A measure of the accuracy of a radiograph as an non-destructive testing (NDT) tool is given by its:
(a) Intensity
(b) Density
(c) Sensitivity
(d) Exposure

Answer: (c)

57. A surface breaking crack will be detected during a magnetic particle inspection if it is:
(a) At right angles to the lines of flux
(b) Parallel to the lines of flux
(c) At 25° to the lines of flux
(d) All of the above

Answer: (a)

58. The advantage of ultrasonic non-destructive testing for the examination of weldments is:

(a) It can be used to locate flaws
(b) It can be used to size flaws
(c) It has a high sensitivity to planar flaws
(d) All of the above

Answer: (d)

59. Under normal contract conditions, weld procedure approval tests for pipework are:
(a) Mandatory
(b) Dependent on site and weather conditions
(c) Dependent upon the contractor's confidence in his procedures
(d) Only required when manual metal arc (MMA) welding is used

Answer: (a)

60. Hydrogen controlled electrodes were developed principally for:
(a) The prevention of porosity
(b) The prevention of cracking
(c) The enhancement of arc voltage
(d) Their ease of arc starting

Answer: (b) Hydrogen embrittlement is a series issue. Hydrogen controlled electrode will reduce this.

61. Generally, the most suitable method of detecting lack of sidewall fusion is:
(a) Ultrasonics
(b) MPI (magnetic particle inspection)
(c) Radiography
(d) Penetrant inspection

Answer: (a) Ultrasonic testing can detect internal defects.

62. Hot shortness is a term used to indicate:
(a) Lamellar tearing
(b) Solidification cracking
(c) Hydrogen cracking
(d) None of the above

Answer: (b)

63. Cobalt, as an isotope, would generally be used on:
(a) Thin material
(b) Tee joints
(c) Plate thicknesses greater than 25 mm
(d) All of the above

Answer: (c)

64. In welding procedure terms, a change in essential variable means:
(a) Requalification of the weld procedure
(b) Possible changes in the weld's microstructure
(c) Possible changes in the mechanical properties
(d) All of the above

Answer: (d)

65. Weld symbols placed on a dotted line in accordance with International Standards Organisation (ISO) requirements means:
    (a) Weld on 'arrow' side
    (b) Weld on 'other 'side
    (c) Weld on site
    (d) Full penetration required

    Answer: (b)
66. A welding inspector's main attributes include:
    (a) Knowledge and experience
    (b) Literacy
    (c) Honesty and integrity
    (d) All of the above

    Answer: (d)
67. Technically, a code of practice is:
    (a) A standard
    (b) A 'set of rules' for the manufacture of a product
    (c) Related to welder and weld procedure approval
    (d) All of the above

    Answer: (b)
68. The correct term for 'cap height' is:
    (a) Reinforcement
    (b) Cap profile height
    (c) Excess weld metal
    (d) All of the above

    Answer: (c)
69. A tensile test will assess:
    (a) Impact values
    (b) Stress
    (c) Strain
    (d) Both (b) and (c)

    Answer: (d)
70. The important point of high temperature steels is that:
    (a) They can withstand creep failure
    (b) They may suffer reheat cracking problems
    (c) They may suffer loss of toughness
    (d) All the above

    Answer: (d)
71. An austenitic stainless steel may suffer:
    (a) Weld decay
    (b) Sensitization
    (c) Solidification cracking
    (d) All of the above

    Answer: (d)
72. Carbon equivalent values are useful to determine:
    (a) Weldability aspects
    (b) Crack sensitivity aspects

(c) Typical mechanical properties
(d) All of the above

Answer: (d)

73. A basic electrode would normally:
    (a) Have superior mechanical properties
    (b) Require baking before use
    (c) Not be used on low carbon steels
    (d) Both (a) and (b)

    Answer: (d) When exposed to atmosphere, they absorb moisture rapidly, so baking is needed before use.

74. When referring to tungsten inert gas (TIG) welding, the shielding gas could be:
    (a) Argon and hydrogen
    (b) Argon and helium
    (c) Argon and nitrogen
    (d) All of the above

    Answer: (d)

75. When referring to metal inert gas (MIG) welding, the shielding gas would be:
    (a) Argon
    (b) Argon + 1% oxygen
    (c) Argon + 20% carbon dioxide
    (d) None of the above

    Answer: (a)

76. Submerged arc utilizes:
    (a) Deep penetration characteristic
    (b) High deposition rates on DC+
    (c) Flat welding only
    (d) None of the above

    Answer: (a)

77. Ultrasonics would be preferred over radiography due to:
    (a) Ability to find most defects
    (b) Lower skill requirement
    (c) Ability to detect laminations
    (d) Both (a) and (c)

    Answer: (d)

78. The most serious defect types are:
    (a) Planar
    (b) Cracks
    (c) Lack of fusion
    (d) All of the above

    Answer: (d)

79. Manual metal arc (MMA) welding of low alloy steels is more likely to be performed with:
    (a) Rutile electrodes
    (b) Cellulosic electrodes
    (c) Iron powder electrodes
    (d) Basic hydrogen controlled electrodes

    Answer: (d)

80. Which of the following defects is more common to welds deposited by CO_2 welding than welds deposited by manual metal arc (MMA)?
   (a) Slag inclusions (b) Excess penetration
   (c) Lack of sidewall fusion (d) Tungsten inclusions
   Answer: (c)
81. Which defect would you expect to get in tungsten inert gas (TIG) welds in non-deoxidized steel?
   (a) Undercut (b) Porosity
   (c) Tungsten inclusions (d) Linearm in alignment
   Answer: (b) Because the oxygen in the steel will combine with carbon and CO is formed and in tungsten inert gas (TIG) welding, they won't get time to escape so it will cause porosity.
82. Which of the following can arise from copper inclusions in a ferritic steel weld?
   (a) Weld metal cracks (b) Heat affected zone (HAZ) cracks
   (c) Lamellar tearing (d) Porosity
   Answer: (a) Copper inclusion causes hot cracking because of its lower melting point and low solubility in Fe (iron).
83. Which of the following is likely to give the highest impact strength in ferritic weld metal?
   (a) Cellulosic electrodes
   (b) Submerged arc with acid flux
   (c) Spray transfer CO_2 (Carbon dioxide) welding
   (d) Basic coated MMA electrodes
   Answer: (b)
84. You suspect that ferritic steel plates contain cracks in the prepared edges. What NDT method would you use to check this?
   (a) Radiography (b) Magnetic particle inspection
   (c) Penetrant inspection (d) Ultrasonic flaw detection
   Answer: (b)
85. Which of the following defects would you not expect to find by visual inspection of welds?
   (a) Linear slag inclusions (b) Undercut
   (c) Overlap (d) Linearm in alignment
   Answer: (a) Because it is an internal defect.
86. Stress relieving is not helpful in which of the following cases?
   (a) Improving resistance to stress corrosion cracking
   (b) Improving dimensional stability after machining
   (c) Lowering the peak residual stress
   (d) Softening the steel
   Answer: (b)

87. What is the maximum hardness usually recommended for the heat affected zone of a medium strength ferritic steel weld?

(a) 100 units (b) 350 units

(c) 500 units (d) 750 units

Answer: (b)

88. What effect does mid thickness laminations in steel plate normally have when they are located within a weld heat affected zone (HAZ)?

(a) Cause lamellar tearing

(b) Fuse together to form a bond

(c) Affect the weld metal composition

(d) Cause internal tearing on a microscale

Answer: (a)

89. The permanent backing material for manual metal arc (MMA) welding of low carbon steel should be made from:

(a) Copper (b) Low carbon steel

(c) Quenched and tempered (QT) steel (d) Cast iron

Answer: (b)

90. The overall length of a pipeline can be affected by:

(a) Transverse shrinkage (b) Longitudinal shrinkage

(c) Angular shrinkage (d) Circumferential shrinkage

Answer: (b)

91. The weld dimension used to indicate the minimum strength of a fillet weld is:

(a) Leg length (b) Throat thickness

(c) Width of bead (d) Length of weld element

Answer: (b)

92. An electroslag weld requires what heat treatment to improve the grain structure?

(a) Annealing (b) Stress relieving

(c) Normalizing (d) Quench and tempering

Answer: (c)

93. The most common type of failure associated with sharp fillets, notches and undercut is:

(a) Crystallization (b) Fatigue

(c) Corrosion (d) Brittle fracture

Answer: (b) Because of higher stress concentrations at these defects.

94. Weld decay in stainless steels can be avoided by:

(a) Stress relieving

(b) Slow cooling after welding

(c) Addition of more manganese to the steel

(d) Addition of titanium to the steel

Answer: (d) Weld decay is caused by the formation of chromium carbide and thus, consuming a portion devoid of chromium which is susceptible to failures. Titanium reacts with the carbon to form the corresponding carbides thereby, preventing chromium depletion.

95. An eutectoid mixture in steel is:

(a) A mixture of ferrite and austenite

(b) A mixture comprising a substitutional solid solution

(c) Pearlite

(d) Ledeburite

Answer: (c)

96. Low alloy steels having a high carbon equivalent before welding will require:

(a) A reduction in carbon content
(b) High preheat temperatures
(c) Low preheat temperatures
(d) No preheating

Answer: (b)

97. The electrodes for welding low alloy steels should be:

(a) Used with a low current value

(b) One size larger than for general-purpose electrodes

(c) Used for welding in the flat position only

(d) Heated in a drying oven before use

Answer: (d) For getting devoid of moisture.

98. The purpose of preheating low alloy steel pipes before electric arc welding is to:

(a) Refine grain structure
(b) Relieve internal stress
(c) Retard rapid cooling
(d) Regulate excessive expansion

Answer: (c)

99. Welder qualification tests are designed to:

(a) Test the correctness of the welding procedure

(b) Test the welder's skill

(c) Prove the weldability of the parent material

(d) All of the above

Answer: (b)

100. In positional manual metal arc (MMA) welding on pipe work, welders are having difficulty in obtaining good capping profiles when welding in the overhead position. Would you:

(a) Advise them to increase the current

(b) Advise them to increase the voltage

(c) Ask for a new welding team
(d) Suggest the use of a smaller diameter electrode

Answer: (d)

101. You have a macrosection of a 'T' butt joint that shows a step-like defect lying outside the visible HAZ (heat affected zone). What would this defect possibly signify?
   (a) HAZ cracking
   (b) Toe cracking
   (c) Lamination
   (d) Lamellar tearing

   Answer: (d)

102. Which electrode deposits weld metal with the greatest ductility and resistance to cracking?
   (a) Rutile
   (b) Cellulosic
   (c) Basic
   (d) Oxidizing

   Answer: (c)

103. Which one of the following is not helpful for minimizing angular distortion during welding?
   (a) Use of double 'V' weld prep using balanced welding technique
   (b) Presetting of workpiece
   (c) Applying postweld heat soak
   (d) Changing from a single 'V' prep for thick material

   Answer: (c)

104. Argon purging on the root side is necessary in the TIG (tungsten inert gas) welding of stainless steel to:
   (a) Obtain full penetration
   (b) Obtain full fusion
   (c) Avoid porosity in the root
   (d) Obtain a satisfactory weld surface finish

   Answer: (c)

105. Which of the following can arise from copper inclusions in a mild steel weld?
   (a) Weld metal cracks
   (b) HAZ (heat affected zone) cracks
   (c) Lack of fusion
   (d) Porosity

   Answer: (a) Copper inclusion causes hot cracking because of its lower melting point and low solubility in Fe (iron).

106. Stress relief is not helpful in which of the following cases?
   (a) In improving resistance to stress corrosion?
   (b) In improving dimensional stability after machining?
   (c) In lowering the peak residual stresses?
   (d) In softening the metal?

   Answer: (b)

107. Stray arc strikes are undesirable since they:
    (a) Leave a poor surface finish
    (b) Cause weld metal cracking
    (c) Reduce corrosion resistance
    (d) Cause local hardening and cracking in the parent material

    Answer: (d)

108. Cold cracking is most likely to occur in a weldment if:
    (a) The rate of cooling is too fast
    (b) The rate of cooling is too slow
    (c) It lacks ductility at high temperatures
    (d) Impurities are present at its grain boundaries

    Answer: (a)

109. Chromium, when added to steel as an alloying element, has the effect of making the alloy more:
    (a) Ductile (b) Plastic
    (c) Hardenable (d) Malleable

    Answer: (c)

110. When depositing weld metal, fusion will take place at the sides of the joint resulting in an admixture between weld metal and parent metal. This alloying effect is known as:
    (a) Diffusion (b) Absorption
    (c) Dilution (d) Migration

    Answer: (c)

111. Percentage elongation of a metal undergoing a tensile test is a measure of:
    (a) Elasticity (b) Plasticity
    (c) Ductility (d) Malleability

    Answer: (c)

112. When a longitudinal load is put on a lap joint, the stress set up is normally:
    (a) Shear stress (b) Tensile stress
    (c) Compressive stress (d) Residual stress

    Answer: (a)

113. When a metal is subjected to a fluctuating load, a condition of cyclic stressing can be set up, which eventually can result in structural breakdown known as:
    (a) Tensile failure (b) Fatigue failure
    (c) Yield failure (d) Shear failure

    Answer: (b)

114. What happens to the mechanical properties of steel if the carbon content is increased to 0.5%?
   (a) The material becomes softer
   (b) Malleability is increased
   (c) The tensile strength is increased
   (d) Ductility is increased

   Answer: (c)

115. Columnar growth takes place when a metal is:
   (a) Cold
   (b) Losing heat
   (c) Being heated
   (d) Being rolled

   Answer: (b)

116. If a low carbon steel pipe has to carry a liquid, care must be taken when making the butt welds to ensure penetration is not excessive because it:
   (a) Reduces the flow rate of the liquid
   (b) May increase the rate of corrosion
   (c) Can contaminate the liquid
   (d) May cause excessive pipe wear

   Answer: (a)

117. When a steel suffers hot shortness, it is mostly due to the presence of:
   (a) Sulphur
   (b) Phosphorous
   (c) Silicon
   (d) Manganese

   Answer: (a)

118. When a steel is heated to above its upper critical temperature, the structure produced is:
   (a) Martensite
   (b) Austenite
   (c) Pearlite
   (d) Sorbite

   Answer: (b)

119. The type of crystal normally found in a single run arc weld in the as welded condition is:
   (a) Equiaxed
   (b) Polycrystalline
   (c) Dendritic
   (d) Columnar

   Answer: (d)

120. The first subzone in the heat affected zone of the parent metal nearest the weld deposit will consist of:
   (a) Large crystal grains
   (b) Small crystal grains
   (c) Elongated crystal grains
   (d) Distorted crystal grains

   Answer: (a)

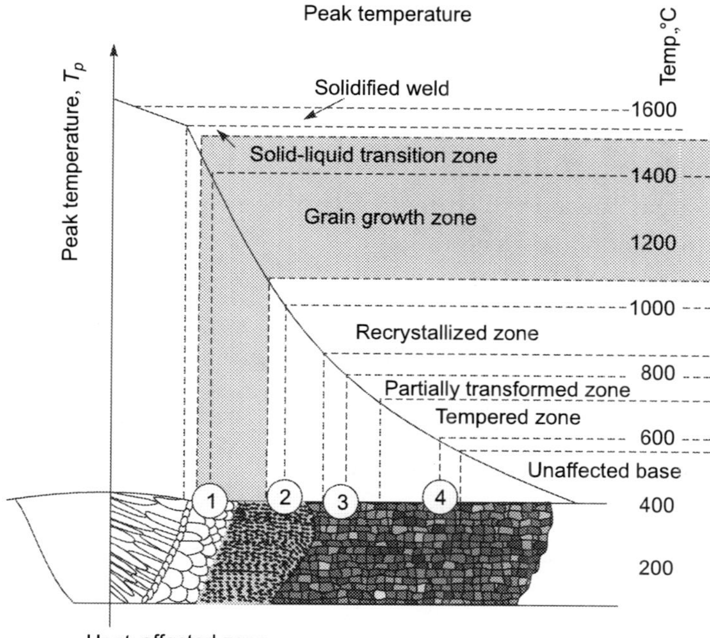

121. Pipe welding codes are set up by:
   (a) Welding operators
   (b) State governments
   (c) Associations, societies, insurance companies, manufacturers and the military
   (d) Construction unions

   Answer: (c)

122. The different grain structure between the weld deposit and the base metal can be determined by:
   (a) A face bend test
   (b) A root bend test
   (c) A hardness test
   (d) An etching test

   Answer: (d)

123. A root bend test is used to test the amount of weld:
   (a) Ductility
   (b) Elongation
   (c) Hardness
   (d) Penetration

   Answer: (a)

124. What would be observed if a fillet weld is sectioned and macro-etched?
   (a) The grain of the other beads is coarser than the final bead
   (b) The penetration and fusion into the root is very deep

(c) Each bead appears to be distinctly separated from the adjoining beads
(d) The grain structure remains the same in all passes

Answer: (c)

125. What is the most common cause of failure in root bend tests?
    (a) Too high a current setting
    (b) Too long a pause in the down cycle of the weave
    (c) Lack of fusion and penetration
    (d) Too high a travel speed

    Answer: (c)

126. The purpose of a neck break specimen is to provide a test for:
    (a) Tensile strength and fracture appearance
    (b) Ductility and fracture appearance
    (c) Elongation and fracture appearance
    (d) Soundness and fracture appearance

    Answer: (d)

127. Which organization publishes the most commonly used code for boiler and pressure vessel welding?
    (a) American Welding Society
    (b) American Society of Mechanical Engineers
    (c) American Petroleum Institute
    (d) American National Standards Institute

    Answer: (b)

128. A low hydrogen electrode, according to BS 639, would contain:
    (a) No hydrogen
    (b) Less than 15 mL of hydrogen per 100 grams of deposited weld metal
    (c) Between 15 mL and 25 mL of hydrogen per 100 grams of deposited weld metal
    (d) Less than 25 mL of hydrogen per 100 grams of deposited weld metal

    Answer: (b)

129. The second run in a three-run butt weld using the stove pipe technique is known as the:
    (a) Filling run (b) Hot pass
    (c) Intermediate run (d) Sealing run

    Answer: (b)

130. You could determine that an electrode is cellulosic by its:
    (a) BS 639 coding (b) Color
    (c) Trade name (d) BS 499 coding

    Answer: (a)

131. Which type of electrode coating gives the most voluminous gas shield?
   (a) Rutile
   (b) Basic
   (c) Oxidizing
   (d) Cellulosic

   Answer: (d)

132. Which of the following steels is likely to be more susceptible to hydrogen cracking?
   (a) Carbon equivalent of less than 0.25%
   (b) Carbon equivalent of 0.35%
   (c) Carbon equivalent of 0.38%
   (d) Carbon equivalent of 0.43%

   Answer: (d)

133. Preheating and interpass heating are used primarily for:
   (a) Aiding fusion
   (b) Reducing hydrogen content of weld preparation prior to welding.
   (c) Ensure a fine grain size
   (d) Slow down the cooling rate after welding

   Answer: (b)

134. Submerged arc welds made with recycled flux are liable to:
   (a) Porosity
   (b) Coarse grain size
   (c) Undercut
   (d) Incomplete penetration

   Answer: (a)

135. Incomplete penetration in a single 'V' butt joint could be caused by:
   (a) Too large a root gap
   (b) Too small a root gap
   (c) Too high a heat input
   (d) Too small a root face

   Answer: (b)

136. In submerged arc welding, which of the following width to depth ratios would be likely to result in solidification cracking?
   (a) 1 : 3
   (b) 3 : 1
   (c) 2 : 1
   (d) 1 : 1

   Answer: (a)

137. You are responsible for controlling welding on site. A large incidence of porosity has been reported in recent welding. Would you investigate?
   (a) The electrode type
   (b) Power source
   (c) Electrode storage
   (d) Day temperature

   Answer: (c) If electrode is not properly stored, there is a high chance of moisture absorption.

138. The main reason why all adhering scales should be removed when the pipe end preparation is made by oxygas cutting is....

(a) Oxidization of the weld metal is minimized
(b) The speed of welding is increased
(c) Pipe bore alignment is made easier
(d) Reduction of the weld deposit is prevented

Answer: (a)

139. When manual metal arc welding low carbon steel, which electrode covering will give the greatest degree of penetration?
(a) Iron powder
(b) Rutile
(c) Cellulosic
(d) Low hydrogen

Answer: (c)

140. When tungsten arc gas shielded welding stainless steel, which one of the following should be used?
(a) Alternator
(b) AC transformer
(c) DC generator
(d) Constant potential rectifier

Answer: (c) It will provide constant melting rate and hence, good quality weld.

141. Which gas should be used when tungsten arc gas shielded welding aluminum alloys?
(a) Nitrogen
(b) Carbon dioxide
(c) Argon-carbon dioxide mixture
(d) Argon

Answer: (d)

142. The most common type of defect found in a structure when it is undergoing service is:
(a) Fatigue cracking
(b) Crystallization
(c) Weld decay
(d) Stress fracture

Answer: (a)

143. In the examination of a welded aluminum joint, macro-etching may reveal:
(a) Lack of inter-run penetration
(b) Carbon pick-up
(c) Weld decay
(d) Micro-cracks

Answer: (a)

144. MMA welds made with damaged electrode coatings are subject to:
(a) Porosity
(b) Undercut
(c) Excessive penetration
(d) Excessive bead height

Answer: (a)

145. Which physical test is more likely to reveal heat affected zone (HAZ) embrittlement?
(a) Transverse tensile
(b) All weld tensile
(c) Root bend
(d) Charpy impact

Answer: (d)

146. Which of the following destructive tests is not normally required for welder approval?
   (a) Bend tests
   (b) Macro examination
   (c) Impact tests
   (d) Fracture tests

   Answer: (c)

147. Too large a diameter of filler rod should not be used to make a welded joint because:
   (a) Excess reinforcement profile will be difficult to obtain
   (b) The included bevel angle will have to be reduced
   (c) Root fusion may be difficult to obtain
   (d) The gap setting will have to be changed

   Answer: (c)

148. If pipe bores are not matched correctly, it can result in:
   (a) Lack of root penetration
   (b) Incorrect gap setting
   (c) Excessive root faces
   (d) Overheating during welding

   Answer: (a)

149. A correctly made tack weld should slope from the middle to the ends in order to:
   (a) Aid better penetration at the join-up
   (b) Prevent porosity at the join-up
   (c) Reduce the electrode size required
   (d) Reduce the overall consumable consumption

   Answer: (a)

150. Two low carbon steel pipes, 150 mm diameter and 6 mm wall thickness, are to be butt welded using the TIG (tungsten inert gas) process. To ensure a full strength joint, which of the following preps is most suitable?
   (a) Open single bevel
   (b) Open single Vee
   (c) Open square preparation
   (d) Closed square preparation

   Answer: (b)

151. The ability of a material to withstand a load pulling it apart is called its ______________.

Answer: Tensile strength

152. The ability of a material to be stretched out without breaking is called ______________.

Answer: Ductility

153. An Izod impact machine is used to give indication of the ______________ of a material.

Answer: Toughness

154. The ability to withstand indentation is called ______________.

Answer: Hardness

155. Lack of ductility is called ______________.

Answer: Brittleness

156. The property of a metal to return to its original shape is called ______________.

Answer: Elasticity

157. Increase in carbon content causes an ______________ in strength and hardness.

Answer: Increase

158. When carbon percentage increases, there is a decrease in ______________.

Answer: Ductility

159. Low carbon steel contains less than ______________ carbon.

Answer: 0.2%

160. Low ductility in a weld metal could result in ______________.

Answer: Cracking

161. Alloying is used to ______________ mechanical and physical properties of a steel.

Answer: Increase

162. Sulphur and phosphorus are not alloying elements; they are ______________.

Answer: Impurities

163. Alloying allows designers to use ______________ sections and still have the same strength.

Answer: Smaller or thinner

164. An alloy that contains a high percentage of chromium and nickel would have resistance to ______________.

Answer: Corrosion

165. Quenching a carbon or low alloy steel will result in an ______________ in hardness and a ______________ in ductility.

Answer: Increase/Decrease

166. The hard constituent that results when steel is quenched is called ________.
Answer: Martensite

167. The tough laminated structure that is formed on slow cooling of ferrite and iron carbide (cementite) is called ____________.
Answer: Pearlite

168. The amount of marten site formed depends on the speed of ____________ and the percentage of ____________.
Answer: Cooling/Carbon

169. After quenching, the structure may be improved by reheating to 200 to 300°C. This is called ____________.
Answer: Tempering

170. Small percentages of chromium will increase the strength and ____________, while a small percentage of nickel will increase ____________.
Answer: Hardness/Roughness

171. When hydrogen control is specified for a manual metal arc welding project, the electrode would normally be:
(a) Cellulosic
(b) Iron oxide
(c) Acid
(d) Basic
Answer: (d)

172. You would with certainty recognize a hydrogen controlled flux covered electrode from its:
(a) Color
(b) Length
(c) Trade name
(d) AWS/BS639 code letter
Answer: (d)

173. When manual metal arc welding is being carried out on an open construction site, which groups of welder are most likely to require continuous monitoring?
(a) Concrete shuttering welding team
(b) Pipe welding teams
(c) Plater welders
(d) Plant maintenance welders
Answer: (b)

174. In welding workshop, you noticed manual metal arc (MMA) welding electrodes, stripe of flux, are being used as filler wire, for tungsten inert gas (TIG) welding. You would object because:
(a) It is too expensive
(b) The wire would be too thick
(c) The weld metal composition may be wrong
(d) The wire is too short
Answer: (c)

175. When open site working, serious porosity in manual metal arc welding is brought to your attention. What would you investigate?
   (a) Electrode type (b) Power plant type
   (c) Electrode storage (d) Day temperature

   Answer: (c)

176. The steel composition in structural contract is changed form 0.15% carbon, 0.6% manganese, to 0.2% carbon, 1.2% manganese. Would this influence the incidence of:
   (a) Porosity (b) Cracking in the weld area
   (c) Under-cut for fillet welds (d) Lack of root fusion defects

   Answer: (b)

177. Which one of the following alloys is non-magnetic?
   (a) 4.0% chromium molybdenum (b) 12.0% chromium
   (c) Austenitic stainless steel (d) 9.0% nickel steel

   Answer: (c)

178. When the tungsten inert gas (TIG) welding austenitic stainless steel pipe, argon gas backing is called for to:
   (a) Prevent oxidation
   (b) Prevent underbead cracking
   (c) Prevent porosity
   (d) Control the penetration bead shape

   Answer: (a)

179. Preheating of carbon steel manual metal arc welding is carried out to minimize the risk of
   (a) Scattered porosity (b) Worm hole porosity
   (c) Parent metal cracking (d) Lack of penetration

   Answer: (c)

180. In United Kingdom (UK) practice, BS 499 Part 2 specifies that the drawing dimension quoted for a fillet weld is the:
   (a) Leg length (b) Actual throat thickness
   (c) Weld width (d) Lack of penetration

   Answer: (a)

181. For open site manual metal arc welding, the following equipment is available. Which would you choose for safe working?
   (a) Single operator transformer
   (b) Multi-operator transformers

(c) AC/DC composite power unit
(d) Diesel engine driven motor generator

Answer: (d)

182. If submerged welding to be used to make butt welds, which would you be most critical of:
(a) The root gap tolerance
(b) The angle of penetration
(c) The root face width
(d) The gas cut finish

Answer: (a)

183. During carbon dioxide (CO_2) welding, the arc length is most likely to be affected by:
(a) The wire diameter
(b) The current return connections
(c) The gas flow rate
(d) The torch to work angle

Answer: (b)

184. Preheating for arc welding applies to:
(a) Assembly welding only
(b) Assembly and tack welding
(c) Joint over 25 mm thick only
(d) Cruciform welds only

Answer: (b)

185. Which of the following statement is correct?
(a) Preheating increases hardness
(b) Preheating increases cooling
(c) Preheating increases dilution
(d) Preheating increases shrinkage stress

Answer: (c)

186. You see a welder using oxyacetylene flame along with feathered inner cone. What would be the effect of this on carbon steel?
(a) The weld could be hard and brittle
(b) The weld metal could be too soft
(c) There will be no effect on the weld
(d) The weld have under cut

Answer: (a)

187. A welder qualification test is to verify:
(a) The skill of welder
(b) The quality of the materials
(c) The non-destructive procedures
(d) The manufacturing methods

Answer: (a)

188. A fabricating procedure calls for fillet welds to be blended in by grinding. This influences:
   (a) Heat affected zone (HAZ) cracking
   (b) Fatigue life
   (c) Residual stresses
   (d) Yield strength

   Answer: (b)

189. Bend test specimens have been taken from a 25 mm thick carbon steel butt weld. Which would show lack of inter-run fusion?
   (a) Side bend
   (b) Root bend
   (c) Face bend
   (d) Guided bend

   Answer: (a)

190. Lamellar tearing has been occurred in steel fabrication. Before welding could it have been found by:
   (a) X-ray examination
   (b) Dye penetrant
   (c) Ultrasonic inspection
   (d) It would not have been found by any inspection method

   Answer: (c)

191. You are to oversee arc welding of some machine fittings and find that they are cadmium plated. Would You:
   (a) Permit it to proceed
   (b) Permit it to proceed with fume extraction
   (c) Stop the operation at once
   (d) Advise the welder to drink milk and proceed

   Answer: (c)

192. One reason for excluding hydrogen from weld metal is to prevent the weld from:
   (a) Cracking
   (b) Cooling slowly
   (c) Cooling quickly
   (d) Expanding

   Answer: (a)

193. When a metal regains its original shape while a stress acting upon it is removed, the metal is said to have:
   (a) Ductility
   (b) Plasticity
   (c) Malleability
   (d) Elasticity

   Answer: (d)

194. Proof stress is used when non-ferrous metals are undergoing tensile tests to determine the equivalent:

(a) Tenacity
(b) Elasticity
(c) Yield strength
(d) Tensile strength

Answer: (c)

195. To test a component for vibrational loading, a suitable mechanical test would be:
(a) Impact
(b) Tensile
(c) Compressive
(d) Fatigue

Answer: (d)

196. The main reason for preheating medium high carbon steels before cutting by oxyfuel gas technique is to:
(a) Improve the quality of cut
(b) Increase the cutting speed
(c) Refine grain structure
(d) Prevent hardening and cracking

Answer: (d)

197. One purpose of microscopic examination of weld is to establish the:
(a) Strength of weld
(b) Number of alloying elements
(c) Grain size
(d) Number of runs used

Answer: (c)

198. The predominant structure of a hypereutectoid steel that has been quenched above its upper critical point will be:
(a) Austenite
(b) Martensite
(c) Troostite
(d) Sorbite

Answer: (b)

199. When weld metal refinement takes place in a multirun deposit, it is known by the term:
(a) Weld annealing
(b) Weld refining
(c) Weld normalizing
(d) Weld recrystallization

Answer: (b)

200. One advantage of metal gas arc shielded welding is:
(a) Can be used in draughty locations without protection
(b) Produces a deposit low in hydrogen content
(c) Any welding position can be welded with spray transfer
(d) Fine spatter at nozzle restricting gas flow

Answer: (b)

201. Which of the following symbols indicates the depth of weld penetration required on the joint as per BS 499 standards?

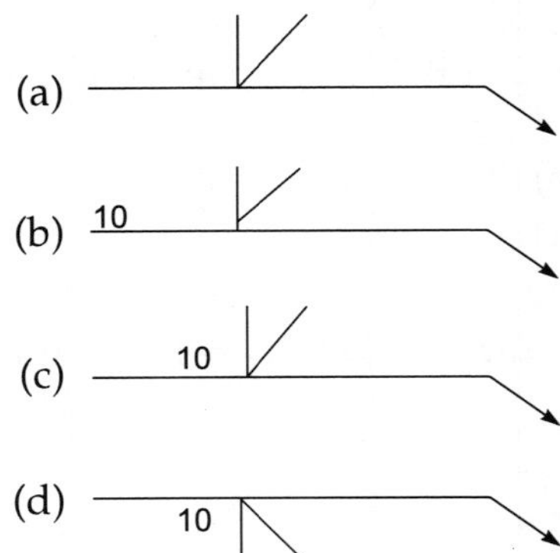

Answer: (c) or (d)

202. Which of the following symbols would indicate that a weld has to be finished on the 'other' side of the weld?

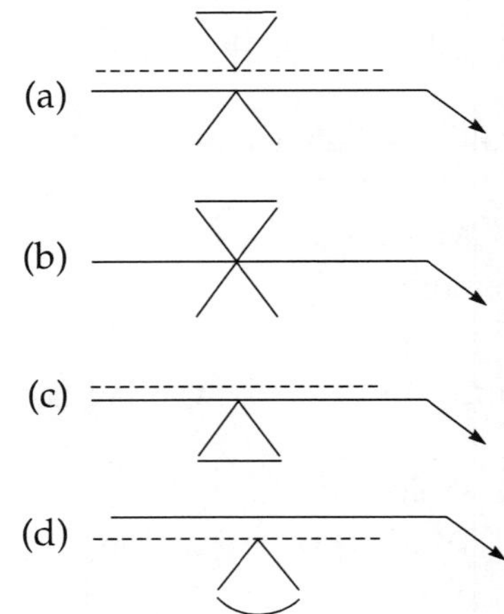

Answer: (b)

203. The use of flux with gas shielded metal arc welding allows:
    (a) Sheet metal to be welded
    (b) A stable arc when using high current densities
    (c) Aluminum to be welded
    (d) Less dilution of the weld by parent metal

    Answer: (b)

204. In manual metal arc (MMA) welding, what parameter is used for the controlled penetration into the base material?
  (a) Voltage (b) Welding speed
  (c) Iron powders in the coating (d) Current
  Answer: (d)

205. In the welding of butt joint from one side, which of the following controls the profile of the root bead?
  (a) Root face (b) Bevel angle
  (c) Root gap (d) One of the above
  Answer: (c)

206. What type of power source characteristic is required for manual welding?
  (a) Constant voltage (b) Flat characteristic
  (c) Drooping characteristic (d) Motor generator
  Answer: (c)

207. Which of the following destructive tests would indicate the toughness of weld metal/parent metal – HAZ (heat affected zone).
  (a) Macro (b) Nick break
  (c) Hardness (d) Charpy Vee-notch
  Answer: (d)

208. Degreasing components are essential for quality welding but some agents may:
  (a) Cause corrosion problems (b) Give off phosgene gas
  (c) Leave residues (d) All of the above
  Answer: (d)

209. Which of the following chemical elements has the greater effect on the hardenability of a steel plate?
  (a) Molybdenum (b) Chromium
  (c) Titanium (d) Carbon
  Answer: (d)

210. In MIG/CO_2 welding, which parameter gives the greatest control of weld appearance during dip transfers or short-circuiting welding?
  (a) Wire sick-out length (b) Amperage
  (c) Wire feed speed (d) Inductance
  Answer: (d)

211. In Manual metal arc (MMA) welding, the slag produced can be varied to suit the welding position; which type of slag would be required for welding in the horizontal vertical (HV) position?
  (a) Fluid (b) Viscous
  (c) Semifluid (d) None of the above
  Answer: (b)

212. The weld metal deposits of manual metal arc electrode achieve their mechanical strength through?
    (a) The core wire
    (b) The flux coating
    (c) Iron powder with the flux coating
    (d) None

    Answer: (b)

213. What constituent is needed in coating of an electrode to prevent formation of porosity in welding of rimming steel?
    (a) Iron powder
    (b) Calcium fluoride
    (c) Silicon
    (d) Calcium carbonate

    Answer: (c)

214. Welds made with high heat inputs show a reduction in one of the following properties?
    (a) Ductility
    (b) Toughness
    (c) Fatigue strength
    (d) Mechanical strength

    Answer: (b)

215. In the welding of austenitic pipe work, the borer is usually purged with Ar to?
    (a) Prevent formation of porosity in the weld
    (b) Prevent burn-through in the root run
    (c) Prevent oxidation of the rood bead
    (d) Eliminate the formation of hydrogen

    Answer: (c)

216. In X-ray, the quality of the radiograph negative is assessed by the?
    (a) Density of the film
    (b) Image quality indicator (IQI)
    (c) Kilovolt-ampere (kVA)
    (d) Standoff distance

    Answer: (b)

217. A steel described as QT (quenched and tempered), will have improved tensile properties, if it has?
    (a) Had control of chemical composition
    (b) Been heat treated
    (c) Been quality tested
    (d) Been vacuum melted

    Answer: (b)

218. Which one of the following steels would give rise to the formation of porosity when autogenously welded with an arc process?
    (a) Fully killed steel
    (b) Semi-killed steel
    (c) Rimming steel
    (d) Fine grained steel

    Answer: (c)

219. In submerged arc welding, the use of excessively high voltage would result in?
    (a) Insufficient flux melting
    (b) Excessive flux melting
    (c) Slag removal difficulties
    (d) Spatter

    Answer: (b)

220. Cellulosic electrodes are often used when welding the root pass of pipes in the field because:
    (a) Hydrogen control is needed
    (b) Iron powder in the electrode
    (c) Higher arc voltage can be obtained
    (d) Shorter arc length can be achieved

    Answer: (b)

221. In the welding of austenitic stainless steels, the electrode and plate material can be purchased with low carbon contents. The reason for this is to prevent:
    (a) Cracking of the HAZ
    (b) The formation of chromium carbides
    (c) Cracking in the weld metal
    (d) Distortion

    Answer: (b)

222. Submerged arc fluxes can be supplied in two forms. These are:
    (a) Sintered and agitated
    (b) Agitated and fused
    (c) Crushed and agglomerated
    (d) Fused and agglomerated

    Answer: (d)

223. In a steel which has improved creep properties at elevated temperature, which one of the following elements helps in this improvement?
    (a) Tungsten
    (b) Manganese
    (c) Molybdenum
    (d) Carbon

    Answer: (c)

224. Welding a steel plate of CE of 0.45 would require preheating to?
    (a) Prevent the formation of sulphides
    (b) Prevent hardening in the HAZ
    (c) Prevent the formation of carbides
    (d) To improve mechanical properties in the weld

    Answer: (b)

225. Which of the following processes uses the "keyholing" system of fusion?
    (a) Friction welding
    (b) Diffusion bonding
    (c) Electron beam welding
    (d) Autogenous TIG welding

    Answer: (c)

226. In friction welding, the metal at the interface is in the?
   (a) Liquid state
   (b) Solid state
   (c) Plastic state
   (d) Elastic state

   Answer: (c)

227. Welding procedures may require welds to be deposited at a controlled rate heat input. High heat inputs would?
   (a) Have poor profile
   (b) Have larger grain size
   (c) Have high hardness in the HAZ
   (d) Have low elongation properties

   Answer: (b)

228. In a tensile test, a brittle material would be indicated if the fractured surface?
   (a) Shows reduction in size
   (b) Is flat and featureless
   (c) Breaks in the weld metal
   (d) Breaks in the parent material

   Answer: (b)

229. What destructive test would be required to ascertain the likelihood of cracking in the heat affected zone of a weld?
   (a) Nick break
   (b) Side bend test
   (c) Charpy input
   (d) Macro test

   Answer: (b)

230. In submerged arc welding, excessive arc voltage may cause?
   (a) Excessive penetration
   (b) Change in weld metal composition
   (c) Narrow weld width
   (d) Excessive bead profile

   Answer: (b)

231. Penetration is increased by: **(GATE 1990, 1991)**
   (a) Increasing welding current and welding speed
   (b) Increasing welding current and decreasing welding speed
   (c) Decreasing welding current and welding speed
   (d) Decreasing welding current and increasing welding current

   Answer: (b)

232. At small variations of arc length at operating conditions, the manual metal arc welding transformer provides nearly: **(GATE 1990)**
   (a) Constant power
   (b) Constant power factor
   (c) Constant voltage
   (d) Constant current

   Answer: (c)

233. For gas welding a particular job using a neutral oxyacetylene flame, the acetylene consumption is 10 L. The oxygen consumption from the cylinder in liters is:

(a) 5
(b) 10
(c) 15
(d) 20

Answer: (b)

234. For resistance spot welding of 1.5 mm thick steel sheets, current required is of the order: **(GATE 1992)**

(a) 10 amp
(b) 100 amp
(c) 1000 amp
(d) 10,000 amp

Answer: (d)

235. In explosive welding process, the ______ (maximum/minimum) velocity of impact is fixed by the velocity of sound in the ______ (flyer/target) plate. **(GATE 1992)**

Answer: Maximum/flyer

236. In DC welding, the straight polarity (electrode negative) results in: **(GATE 1993)**

(a) Lower penetration
(b) Lower deposition rate
(c) Less heating of workpiece
(d) Smaller weld pool

Answer: (b)

237. The electrodes used in arc welding are coated. This is not expected to:

(a) Provide protective atmosphere to weld
(b) Stabilize the arc
(c) Add alloying element
(d) Prevent electrode from contamination

Answer: (d)

238. The ratio of acetylene to oxygen is approximately ______ for neutral flame used in gas welding: **(GATE 1994)**

Answer: 1:1

239. Preheating before welding is done to:

(a) Make the steel softer
(b) Burn away oil, grease from plate surfaces
(c) Prevent cold cracks
(d) Prevent plate distortion

Answer: (b) and (c)

240. Two plates of the same metal having equal thickness are to be butt welded with electric arc. When the plate thickness changes, welding is achieved by:

(a) Adjusting the current
(b) Adjusting the duration of the current
(c) Changing the electrode size
(d) Changing the electrode coating

Answer: (a)

241. The temperature of carburizing flame in gas welding is ______ that of a neutral or an oxidizing flame.

(a) Lower than (b) Higher than
(c) Equal to (d) Unrelated to

Answer: (a)

242. Which of the following welding process does not use consumable elelctrode? **(GATE 2002)**

(a) GMAW (b) GTAW
(c) SAW (d) None

Answer: (b)

243. In oxyacetylene gas welding, temperature at inner core of flame is around: **(GATE 2003)**

(a) 3500°C (b) 3200°C
(c) 2900°C (d) 2550°C

Answer: (b)

244. Strength of a brazed joint: **(GATE 2005)**

(a) Decreases with increase in gap between the two joining surfaces
(b) Increases with increase in gap between the two joining surfaces
(c) Decreases up to certain gap between the two joining surfaces, beyond which it increases
(d) Increases up to certain gap between the two joining surfaces, beyond which it decreases

Answer: (d)

245. Which one of the following is a solid state welding?

(a) GTAW (b) Resistance spot welding
(c) Friction welding (d) SAW

Answer: (c)

246. Which one among the following welding uses non-consumable electrode? **(GATE 2011)**

(a) Gas meal arc welding (b) Submerged arc welding
(c) Gas tungsten arc welding (d) Flux coated arc welding

Answer: (c)

247. The major difficulty during welding of aluminum is due to its: **(GATE 2014)**

(a) High tendency of oxidation (b) High thermal conductivity
(c) Low melting point (d) Low density

Answer: (a)

248. In solid state welding, the contamination layer between the surfaces to be welded is removed by: **(GATE 2014)**

(a) Alcohol (b) Plastic deformation
(c) Water jet (d) Sand blasting

Answer: (b)

249. With the heat affected zone (HAZ) in fusion welding process, the work material undergoes: **(GATE 2014)**
    (a) Microstructural changes but does not melt
    (b) Neither melting nor microstructural changes
    (c) Both melting and microstructural changes after solidification
    (d) Melting and retains the original microstructure after solidification

    Answer: (a)

250. In welding brass with oxyacetylene flame, the type of flame used is ________ **(GATE 1989)**

    Answer: Oxidizing flame

251. The mode of metal transfer in CO_2–MIG welding at low temperature is: **(GATE 1990)**
    (a) Spray transfer
    (b) Short circuiting type of transfer
    (c) Globular transfer
    (d) Droplet transfer

    Answer: (d)

252. With increasing joint thickness, the tensile strength of a brazed joint: **(GATE 1990)**
    (a) Continuously decreases
    (b) First decreases and then increases
    (c) Continuously increases
    (d) First increases and then decrease

    Answer: (d)

253. Match the following: **(GATE 1990)**

Welding process	Heat sources
A. Thermal welding	1. Electric arc
B. Projection welding	2. Mechanical work
C. MIG welding	3. Exothermic chemical reaction
D. Friction welding	4. Ohmic resistance

    (a) A-3, B-4, C-1, D-2
    (b) A-2, B-3, C-1, D-4
    (c) A-3, B-4, C-1, D-2
    (d) A-3, B-1, C-4, D-2

    Answer: (a)

254. In TIG welding a _______ and ________ electrode is used. **(GATE 1990)**
    (a) non-consumable/consumable
    (b) coated/bare

    Answer: non-consumable/bare

255. High alloy steel components are preheated before welding to reduce: **(GATE 1991)**

(a) Heat affected zone
(b) Total energy consumption
(c) Total time of welding
(d) Welding stresses

Answer: (b) and (c)

256. In spot welding of aluminum welding current should be _______ (higher/lower) and weld time should be kept _______ (higher/lower) than for MS sheet for same thickness. **(GATE 1991)**

Answer: Higher/lower

257. Continuous rails for Indian railways are welded by _______ welding process. **(GATE 1991)**

Answer: Thermit welding

258. _______ is the non-fusion method of joining dissimilar metal with metal whose melting temperature is greater than 400°C. **(GATE 1991)**

Answer: Brazing and braze welding

259. The joint configuration best suited for adhesive bonding is: **(GATE 1992)**

(a) Butt
(b) Fillet
(c) Lap
(d) Spot

Answer: (c)

260. The type of coated electrode most widely used for low carbon steel is: **(GATE 1992)**

(a) Cellulose
(b) Acidic
(c) Rutile
(d) Oxide

Answer: (c)

261. Filler material is ________ in resistance welding and heat generated in the process is directly proportional to ________. **(GATE 1992)**

(a) Used/not used
(b) Square of the current/cube of the current

Answer: Not used/square of current

262. The composition of an oxyacetylene flame for cutting of steel contains more oxygen. True or False? **(GATE 1994)**

Answer: True

263. Which of the following powders should be fed for effective oxyfuel cutting of stainless steel? **(GATE 2001)**

(a) Steel
(b) Aluminum
(c) Copper
(d) Ceramic

Answer: (a)

264. High speed electron beam is focused on the weld spot using: **(GATE 2003)**

(a) Vacuum lens
(b) Inert gas lens
(c) Optical lens
(d) Magnetic lens

Answer: (d)

265. In resistance welding, heat is generated due to the resistance between: **(GATE 2003)**
   (a) Electrode and workpiece
   (b) Asperities between touching plates
   (c) Two dissimilar metals being in contact
   (d) Interatomic force

   Answer: (b)

266. The current in amperes used in resistance spot welding of plain carbon steel sheets (1 to 3 mm thick) lies within the range: **(GATE 2004)**
   (a) 10 to 50 (b) 50 to 500
   (c) 500 to 5000 (d) 5000 to 50000

   Answer: (d)

267. The metal powder used in thermit welding of steel is: **(GATE 2004)**
   (a) Al (b) Cu
   (c) Pb (d) W

   Answer: (a)

268. Which of the following welding processes results in small heat-affected zone? **(GATE 2011)**
   (a) Shielded metal arc welding (b) Gas welding
   (c) Laser beam welding (d) Thermit welding

   Answer: (c)

269. In resistance seam welding, the electrode is in the form of: **(GATE 2011)**
   (a) Cylinder (b) Flat plate
   (c) Coil of wire (d) Circular disc

   Answer: (d)

270. Brazing and soldering are: **(GATE 2014)**
   (a) Plastic joining methods (b) Homogeneous joining methods
   (c) Autogenous joining methods (d) Heterogeneous joining method

   Answer: (d)

271. In a linear arc welding process, the heat input per unit length is inversely proportional to:
   (a) Welding current (b) Welding voltage
   (c) Welding speed (d) Duty cycle of the power source

   Answer: (c)

272. Which two of the following joining processes are autogeneous?
   (i) Diffusion welding (ii) Electroslag welding
   (iii) Tungsten inert gas welding (iv) Friction welding
   (a) Diffusion welding and friction welding
   (b) Diffusion welding and TIG
   (c) Electrosag and diffusion welding
   (d) TIG and electrosag

   Answer: (a)

273. Under optimal conditions of process, the temperature experienced by copper workpieces in fusion welding, brazing and soldering are such that: **(GATE 2016)**

(a) $T_{welding} > T_{soldering} > T_{brazing}$
(b) $T_{soldering} > T_{welding} > T_{brazing}$
(c) $T_{welding} > T_{brazing} > T_{soldering}$
(d) $T_{soldering} > T_{brazing} > T_{welding}$

Answer: (c)

274. Match the following: **(GATE 1991)**

	Welding Process		Heat Transfer
A	Thermit welding	1	Chemical reaction
B	Seam welding	2	Exothermic chemical reaction
C	MIG welding	3	Electric spark
D	Friction welding	4	Ohmic resistance
		5	Mechanical work
		6	Electric arc

Answer: A-2, B-4, C-6, D-5

275. Match the following: **(GATE 1996)**

	Application		Type of Welding
A	Welding of aluminum	1	Submerged arc welding
B	Ship building	2	Electron beam welding
C	Joining of HSS drill bit to shank	3	TIG welding
D	Deep penetration precision welds	4	Friction welding
		5	Gas welding
		6	Electric arc

Answer: A-3, B-1, C-4, D-2

276. Resistance spot welding is performed on two plates of 1.5 mm thickness with 6 mm diameter electrode using 15,000 A current for 0.25 sec. Assuming the interface resistance to be 0.0001 ohm, the heat generated is ______

(a) 5625 W (b) 8437 W
(c) 22,500 W (d) 33,750 W

Answer: Heat generated = $I^2Rt = 15000^2 \times 0.0001 \times 0.25 = 5625$ W – S = 5625 joules

277. Match the following: **(GATE 2003)**

	Work Material		Type of Joining
A	Aluminum	1	Submerged arc welding
B	Die steel	2	Soldering
C	Copper wire	3	Thermit welding
D	Titanium sheet	4	Atomic hydrogen welding
		5	Gas tungsten arc welding
		6	Laser beam welding

Answer: A-5, B-4, C-2, D-6

278. Two 1 mm thick steel sheets are to be spot welded at a current of 5000 A. Assuming effective resistance of 200 micro-ohms and current flow time of 0.2 second, heat generated, the process will be ______ **(GATE 2004)**
   (a) 0.2 joules (b) 1 joule
   (c) 5 joules (d) 1000 joules

   Answer: Heat generated $= I^2Rt = 5000^2 \times 200 \times 10^{-6} \times 0.2 = 5625 \text{ W} - \text{S} = 1000$ joules

279. Welding enables joining of:
   (a) Only plastics (b) Only metal alloys
   (c) Most of metals and plastic (d) None of these

   Answer: (c)

280. Common methods of welding involve:
   (a) Only heating (b) Only pressing
   (c) Heating and pressing (d) None of these

   Answer: (c)

281. In gas welding, how many flames are generated:
   (a) One (b) Two
   (c) Three (d) Four

   Answer: (c)

282. Acetylene cylinder is painted:
   (a) Maroon (b) Black
   (c) Yellow (d) Red

   Answer: (a)

283. Oxygen gas cylinder is painted:
   (a) Maroon (b) Black
   (c) Yellow (d) Blue

   Answer: (b)

284. In oxyacetylene welding, heat is obtained by:
    (a) Electric arc
    (b) Gas flame
    (c) Chemical reaction
    (d) None of these
    Answer: (b)
285. Acetylene is contained in the cylinder as:
    (a) A gas
    (b) A solid
    (c) Dissolved acetone
    (d) All of the above
    Answer: (c)
286. Following gas flame is generally used in gas welding:
    (a) Oxygen acetylene
    (b) Oxygen natural
    (c) Oxygen cool gas
    (d) None of these
    Answer: (a)
287. Which flame is commonly used for welding of cast iron?
    (a) Neutral flame
    (b) Oxidizing flame
    (c) Carburizing flame
    (d) None of the above
    Answer: (a)
288. An acetylene generator is used in:
    (a) High pressure welding
    (b) Low pressure welding
    (c) High pressure cutting
    (d) All of these
    Answer: (b)
289. In low pressure welding, acetylene is drawn from generator at:
    (a) 0.023 kg/cm^2
    (b) 0.5 kg/cm^2
    (c) 1.2 kg/cm^2
    (d) Any one of the above
    Answer: (a)
290. In high pressure welding, oxygen is drawn from the cylinder at:
    (a) 0.35 kg/cm^2
    (b) 0.5 kg/cm^2
    (c) 1.0 kg/cm^2
    (d) None of these
    Answer: (a)
291. O_2 cylinder, acetylene cylinder, pressure regulator, blow pipes, etc., are used in:
    (a) Electric arc welding
    (b) Gas welding
    (c) Forge welding
    (d) MIG welding
    Answer: (b)
292. Filler rod is used in:
    (a) Electric arc welding
    (b) Gas welding
    (c) Forge welding
    (d) Seam welding
    Answer: (b)
293. Temperature produced by an oxyacetylene flame is:
    (a) 2000°C
    (b) 2300°C
    (c) 3200°C
    (d) 2600°C
    Answer: (c)

294. A neutral flame is generated when the preparation of oxygen by flame is:
 (a) Equal by acetylene (b) Less than acetylene
 (c) More than acetylene (d) All of the above
 Answer: (a)

295. Which flame is used for welding of brass?
 (a) Neutral flame (b) Oxidizing flame
 (c) Carburizing flame (d) None of these
 Answer: (b)

296. Carburizing flame used for welding of:
 (a) Mild steel (b) Aluminum
 (c) Lead (d) Monal metal
 Answer: (a)

297. Temperature produced by oxy-hydrogen flame is about:
 (a) 1600°C (b) 2400°C
 (c) 3000°C (d) 3700°C
 Answer: (b)

298. An oxyacetylene cutting torch carries:
 (a) One passage for oxygen and one for acetylene
 (b) Single passage for both oxygen and acetylene
 (c) One passages for oxygen and for acetylene
 (d) Two separate passages for oxygen and one for acetylene
 Answer: (d)

299. Electric arc welding is a:
 (a) Chemical process
 (b) Pressure welding process
 (c) Fusion welding process
 (d) Non-conventional welding process
 Answer: (c)

300. Function of flux coating in arc welding:
 (a) Alloying element (b) Shielding effect
 (c) Arc stabilization (d) Help in controlling the bead
 (e) All of these
 Answer: (e)

301. Transformer is used in:
 (a) Gas welding (b) DC welding
 (c) AC welding (d) None of these
 Answer: (c)

302. An arc blow is encountered in:
   (a) DC welding
   (b) AC welding
   (c) By magnetic forces
   (d) Both (a) & (c)

   Answer: (d)

303. An arc blow can be in the form of:
   (a) Side blow
   (b) Forward blow
   (c) Backward blow
   (d) All of these

   Answer: (d)

304. Polarity is fixed in:
   (a) AC welding
   (b) DC welding
   (c) Gas welding
   (d) All of these

   Answer: (b)

305. Consumable electrode is used in:
   (a) TIG welding
   (b) MIG welding
   (c) Spot welding
   (d) Friction welding

   Answer: (b)

306. Non-consumable electrode is used in:
   (a) TIG welding
   (b) MIG welding
   (c) Spot welding
   (d) Friction welding

   Answer: (a)

307. MIG welding is used for welding of:
   (a) Al and its alloys
   (b) Stainless steel
   (c) Cu and its alloys
   (d) Heat-resistant alloy
   (e) All of these

   Answer: (e)

308. In CO_2-MIG welding, the shielding gas used is:
   (a) N_2
   (b) O_2
   (c) CO_2
   (d) H_2O_2

   Answer: (c)

309. MIG welding employs only
   (a) DC with straight polarity
   (b) AC
   (c) DC with reverse polarity
   (d) None of these

   Answer: (c)

310. For butt welding 40 mm thick plate when expected quantity of such job is 5000/month over a period of 10 years, choose the best suitable welding process out of the following available alternative:
   (a) MIG welding
   (b) Electron beam welding
   (c) Submerged arc welding
   (d) Oxyacetylene gas welding

   Answer: (b)

311. In gas welding of mild steel using an oxyacetylene flame, the total amount of acetylene used was 10 litres. The oxygen consumption from cylinder is: **(GATE 1989, 1991)**

(a) 5 litre (b) 10 litre
(c) 15 litre (d) 25 litre

Answer: (b)

(For welding mild steel, neutral flame is used)

312. Purpose of a gas welding flux is to:

(a) Remove the elements from parent metal
(b) Lower the melting point of the metal
(c) Remove the oxides from the surface of the metal
(d) Lower the melting point of oxide

Answer: (c)

313. Temperature of a carburizing flame in gas welding is ______ that of a neutral or an oxidizing flame. **(GATE 2002)**

(a) Equal to (b) Higher than
(c) Lower than (d) None of these

Answer: (d)

314. Flux are used while welding is:

(a) To clean the joints
(b) To increase rate of welding
(c) To prevent oxidation of metal during welding
(d) All of these

Answer: (b)

315. In oxyacetylene gas welding for complete combustion, the volume of oxygen required/unit ton of acetylene is:

(a) 1 (b) 1.5
(c) 2.5 (d) 3.0

Answer: (b) & (d)

316. Braze and bronze are welded by:

(a) Neutral flame (b) Oxidizing flame
(c) Carburizing flame (d) None of these

Answer: (b)

317. Flux coating of an electric arc electrode has a melting point:

(a) Same as the metal being welded (b) Higher than the metallic core
(c) Lower than the metallic core (d) Same as the metallic core

Answer: (c)

318. Two plates of the same metal having equal thickness are to be butt welded with electric arc welding. When the plate thickness changes, welding is achieved by:

(GATE 2001)

(a) Changing the electrode coating
(b) Adjusting the current
(c) Adjusting the direction of current
(d) Changing the electrode size

Answer: (d)

319. In arc welding, electric arc is produced between the job and the electrode due to:

(a) Constant resistance
(b) Current
(c) Voltage
(d) All of these

Answer: (b)

320. Electrode used in arc welding is coated electrode. The coating is not expected to:

(a) Prevent electrode from contamination
(b) Provide protective atmosphere to weld
(c) Stabilize the arc
(d) Add alloying elements

Answer: (a)

321. Hard-zone cracking in low-alloy steel welding is a result of an absorption of

(a) H_2
(b) N_2
(c) O_2
(d) C

Answer: (d)

322. The open circuit voltage (OCV) in arc welding ranges:

(a) 40 to 80 V
(b) 100 to 110 V
(c) 200 to 220 V
(d) 400 to 440 V

Answer: (d)

323. The voltage-current characteristic of a DC generator for AVC welding is a straight line between an open circuit voltage of 80 V and short circuit current of 300 A. A generator setting for maximum power will be:

(a) 40 V and 300 A
(b) 40 V and 150 A
(c) 80 V and 150 A
(d) 80 V and 300 A

Answer: (a)

324. In DC welding, the straight polarity gives:

(a) Less heating of workpiece
(b) Smaller weld pool
(c) Lower penetration
(d) Lower depositing rate

Answer: (b)

325. Correct sequence of the given materials in decreasing order of their weldability is:

(a) Al, Cu, C.I., M.S.
(b) M.S., Cu, C.I., Al
(c) C.I., M.S., Al, Cu
(d) Cu, C.I., M.S., Al

Answer: (b)

326. Electrode beam welding can be carried out in:

(a) Vacuum
(b) Shielding gas environment
(c) Open air
(d) A pressurized inert gas chamber

Answer: (a)

327. Cast iron, titanium alloys, refectory metals and Mn-bronze can be easily welded using:

(a) Carbon arc welding
(b) MIG welding
(c) Metal arc welding
(d) Resistance welding

Answer: (b)

328. In carbon arc welding, _________ is used.

(a) A carbon electrode
(b) A tungsten electrode
(c) A metal electrode
(d) No electrode

Answer: (a)

329. TIG welding is suitable for:

(a) Refractory metal
(b) Zirconium alloy
(c) Aluminum alloys
(d) Titanium alloy
(e) None of these

Answer: (c)

330. In TIG welding:

(a) Only AC can be used
(b) Only DC can be used
(c) AC and DC can be used
(d) None of these

Answer: (c)

331. A granular fusible flux is used for shielding the arc in:

(a) Submerged arc welding
(b) TIG welding
(c) MIG welding
(d) All of these

Answer: (a)

332. Alternative name of submerged arc welding is:

(a) Subarc welding
(b) Submerged melt welding
(c) Hidden arc welding
(d) All of these

Answer: (d)

333. In resistance welding for heating the base metal:
(a) No current is needed
(b) No voltage is required
(c) 5 to 10 volts, voltage is required
(d) None of these
Answer: (c)

334. Pressure applied for completion of weld in electric resistance welding is:
(a) 1 to 5 MPa
(b) 25 to 50 MPa
(c) 5 to 10 MPa
(d) 10 to 25 MPa
Answer: (b)

335. Which of the following arc is an electric resistance welding process:
(a) Butt welding
(b) Seam welding
(c) Flash welding
(d) Percussion welding
Answer: (c)

336. In resistance welding, the turns of primary winding of the transformer can be changed to:
(a) Increase voltage
(b) Vary the input current
(c) Regulate the current
(d) None of these
Answer: (c)

337. A series of continuous spots is produced in:
(a) Spot welding
(b) Projection welding
(c) Seam welding
(d) All of these
Answer: (c)

338. Seam welding is not recommended for welding of:
(a) Coated steel
(b) Ni and its alloys
(c) Cu and its alloys
(d) Al and its alloys
Answer: (c)

339. Electrode used in resistance welding is made out of:
(a) Al base alloys
(b) Cu base alloys
(c) Stainless steel base alloys
(d) Mild steel based
Answer: (b)

340. Sheet thickness for making lap joint through spot welding is:
(a) 2 to 3 mm
(b) 0.025-1.25 mm
(c) >2 mm
(d) >5 mm
Answer: (b)

341. Resistance butt welding is used to weld the metal piece:
(a) Of dissimilar cross section
(b) Dissimilar metal
(c) Of same cross section
(d) All of these
Answer: (c)

342. Compared to resistance butt welding, flash welding is capable of:
   (a) A stronger joint
   (b) Less current consumption
   (c) Faster welding
   (d) All of these

   Answer: (d)

343. Seam welding is best suitable to weld up to the thickness of:
   (a) 0.25 to 3 mm
   (b) 4 to 6 mm
   (c) 6 to 10 mm
   (d) 5 to 12 mm

   Answer: (a)

344. Projection welding is a variation of spot welding in which:
   (a) A continuous series of spot welds is produced
   (b) Multiple spot welds are produced at projections
   (c) Only round surface can be projection welded
   (d) All of these

   Answer: (b)

345. Percussion welding can be used:
   (a) For joining wires of dissimilar metals
   (b) For sheets of all thicknesses
   (c) For wire thickness of 0.05-0.38 mm
   (d) All of these

   Answer: (a) & (c)

346. An electrode sometimes is used in:
   (a) Seam welding
   (b) Percussion welding
   (c) Spot welding
   (d) Projection welding

   Answer: (c)

347. Melting point of a solder is
   (a) >that of parent metal
   (b) =that of parent metal
   (c) <that of parent metal
   (d) All of these

   Answer: (c)

348. In a soldering process, the soft solder is an alloy of
   (a) Sn + Zn
   (b) Sn + Pb
   (c) Cu + Zn
   (d) Cu + Zn

   Answer: (b)

349. Melting point of a brazing filler road
   (a) >that of parent metal
   (b) =that of parent metal
   (c) <that of parent metal
   (d) <than the melting point of parent metal

   Answer: (d)

350. Inclusion in weld occurs due to:
   (a) Metal between the bead and base metal
   (b) Failure to squeeze out the slag
   (c) Poor cleanliness
   (d) All of these

   Answer: (b)

351. Crack, a welding defect, may occur:
   (a) In the weld
   (b) In base metal
   (c) Unequal shrinkage of weld
   (d) All of these

   Answer: (d)

352. Spatters, a welding defect, may:
   (a) Spoil the appearance of the base metal and weld surface
   (b) Increase the weld strength
   (c) Reduce the weld strength
   (d) All of these

   Answer: (a)

353. Main reason for occurrence of an undercut in welding is:
   (a) Wrong inclusion of electrode
   (b) High current
   (c) Long arc
   (d) All of these

   Answer: (d)

354. Two pipes of inner diameter 100 mm and outer diameter 110 mm, each joint by flash butt welding using 30 V power supply at the interface, 1 mm of material melts from each pipe, which has a resistance of 42.4 ohms. If the unit melt energy is 64.4 MJ/m^3, then time required for welding in second is:
   (a) 1
   (b) 5
   (c) 10
   (d) 20

   Answer: (c)

355. Match List-I with List-II and select the correct answer using the code given below the lists:

List-I	List-II
(A) Atomic hydrogen welding	(1) Two pieces are brought together and power is switched on
(B) Plasma arc welding	(2) Nugget is formed at the interface of two plates
(C) Spot welding	(3) Gas is ionized
(D) Flash welding	(4) Inert gas shielded arc welding

Code:

(a) A B C D
    4 3 2 1

(b) A B C D
    1 3 2 4

(c) A B C D
4 2 3 1

(d) A B C D
1 2 3 4

Answer: (a)

356. Match List-I with List-II and select the correct answer using the code given below the lists:

List-I (Basic form of welding)	List-II (Weld Symbols)
(A) Fillet	(1) ◺
(B) Square butt	(2) ⊤⊤
(C) Spot weld	(3) ⚹
(D) Seam weld	(4) XXXX

Code:

(a) A B C D
4 3 2 1

(b) A B C D
4 3 2 1

(c) A B C D
4 2 3 1

(d) A B C D
4 2 3 1

Answer: (a)

357. The ratio between oxygen and acetylene gases for neutral flame in welding is:
(a) 2 : 1
(b) 1 : 1
(c) 1 : 2
(d) 4 : 1

Answer: (b)

358. In oxyacetylene gas welding, the volume of oxygen required/unit volume of acetylene is:
(a) 1
(b) 1.5
(c) 2
(d) 2.5

Answer: (d)

359. In an inert gas welding process, the commonly used gas is:
(a) Hydrogen
(b) Krypton
(c) Helium or argon
(d) Oxygen

Answer: (c)

360. In MIG welding, metal transfers in the form of:
(a) Molecules
(b) Weld pool
(c) Molten drop
(d) Fine spray of metal

Answer: (c)

361. Which one of the following welding processes consist of minimum heat-affected zone?
(a) Shielded metal arc welding
(b) Laser beam welding
(c) Metal inert gas welding
(d) Tungsten inert gas welding

Answer: (b)

362. Gray cast iron is usually welded by:
   (a) Arc welding
   (b) TIG welding
   (c) Gas welding
   (d) MIG welding
   (e) PAW

   Answer: (c)

363. Match List-I with List-II and select the correct answer using the code given below the lists:

List-I	List-II
(A) Laser welding	(1) Uniting large arc sheets
(B) Friction welding	(2) Repairing large parts
(C) Ultrasonic welding	(3) Welding a rod to flat surface
(D) Explosive welding	(4) Fabrication of nuclear reactor components
	(5) Welding very thin materials

   Code:

	A	B	C	D
(a)	5	4	3	2
(b)	1	4	2	5
(c)	1	3	4	2
(d)	5	3	4	1

   Answer: (a)

364. In atomic hydrogen welding, hydrogen acts as:
   (a) One of the gases to generate the flame
   (b) A heating agent
   (c) An effective shielding gas protecting the weld
   (d) A lubricant to increase the flow characteristics of weld metal

   Answer: (b)

365. Preheating before welding is done to: **(GATE 1996)**
   (a) Prevent plate distortion
   (b) Make the plate softer
   (c) Burn away oil, greases, etc., from the plate surface
   (d) Prevent cold cracks

   Answer: (d)

366. Which one of the following is a solid state joining process?
   (a) Gas tungsten arc welding
   (b) Friction welding
   (c) Submerged arc welding
   (d) Resistance spot welding

   Answer: (b)

367. Pinch effect in welding is the result of:
   (a) Surface tension of the molten metal
   (b) Electric force
   (c) Electromagnetic force
   (d) Expansion of gases in the arc

   Answer: (c)

368. In resistance welding, heat is generated due to resistance between:
   (a) Interatomic forces
   (b) Electrode and workpiece
   (c) As parities between touching plates
   (d) Two dissimilar metals being in contact

   Answer: (b)

369. Arc in the arc welding is created by:
   (a) Contact resistance
   (b) Current
   (c) Voltage
   (d) Frequency

   Answer: (a)

370. Which one is not a fusion welding process?
   (a) Gas welding
   (b) Arc welding
   (c) Brazing
   (d) Resistance welding

   Answer: (d)

371. Which of the following welding process involves exothermal chemical reaction for the purpose of developing a high temperature?
   (a) Thermit welding
   (b) Arc welding
   (c) Pressure welding
   (d) Explosion welding

   Answer: (a)

372. Maximum heat in resistance welding is at the:
   (a) Tip of negative electrode
   (b) Tip of positive electrode
   (c) Top surface of the plate at the time of electric contact with electrode
   (d) Interface between the two plates being joined

   Answer: (d)

373. Temperature developed during a thermit welding process is of the order of:
   (a) 1500°C
   (b) 2000°C
   (c) 2500°C
   (d) 3500°C

   Answer: (c)

374. Electrode used in welding is coated. This coating is not expected to: **(GATE 1994)**
   (a) Stabilize the arc
   (b) Prevent electrode from contamination

(c) Provide protective atmosphere to weld
(d) Add alloying elements

Answer: (b)

375. In arc welding, DC reverse ploarity is used to bear greater advantage in:
(a) Overhead welding
(b) Flat welding lap joint
(c) Edge welding
(d) Flat welding of butt joint

Answer: (a)

376. Which of the following coating material of an arc welding electrode contains:
1. Deoxidizing agent
2. Arc stabilizing agent
3. Slag forming agent,

Select the correct answer using the code giving below:
(a) 1, 2 & 3
(b) 1 & 2
(c) 1 & 3
(d) 2 & 3

Answer: (a)

377. Which of the following are the major characteristics of submerged arc welding:
1. High welding speed
2. High deposition rate
3. Low penetration
4. Low clean liners

Select the correct answer using the code given below:
(a) 2 & 3
(b) 1, 2 & 3
(c) 3 & 4
(d) & 2

Answer: (d)

378. In ultrasonic welding, the frequency range is generally:
(a) 100 to 400 Hz
(b) 500 to 5000 Hz
(c) 6000 to 20,000 Hz
(d) 20,000 to 50,000 Hz

Answer: (c)

379. Which of the following welding process consists of minimum heat affected zone?
(a) Metal inert gas welding
(b) Shielded metal arc welding
(c) Laser beam welding
(d) Ultrasonic welding

Answer: (c)

380. Size of heat affected zone will increase with:
1. Increased starting temperature
2. Increased welding speed
3. Increased thermal conductivity of the base metal
4. Increase in base metal thickness.

Which of the statement given below is correct:

(a) 1, 2 & 3 (b) 1 & 3

(c) 1 & 4 (d) 2 & 3

Answer: (b)

381. Which one of the following statement is correct:
   (a) AC can be used for GTAW process
   (b) No flux is used in gas welding of mild steel
   (c) Borax is the commonly used flux coating on welding electrode
   (d) Laser beam welding employs a vacuum chamber and thus, avoids use of a shielding method

   Answer: (d)

382. Aircraft body has:
   (a) Bolted structure (b) Welded structure
   (c) Precast structure (d) Riveted structure

   Answer: (d)

383. Which metal has good weldability compared to rest?
   (a) Stainless steel (b) Cast steel
   (c) Cast iron (d) Carbon steel

   Answer: (d)

384. Which one of the following welding methods provides the highest heat flux (W/mm^2) **(GATE 2019-S1)**
   (a) Axy-acetylene gas welding (b) Laser beam welding
   (c) Plasma arc welding (d) Tungsten inert gas welding

   Answer: (b)

385. The type of weld represented by the shaded region in the figure below is **(GATE 2018-S1)**

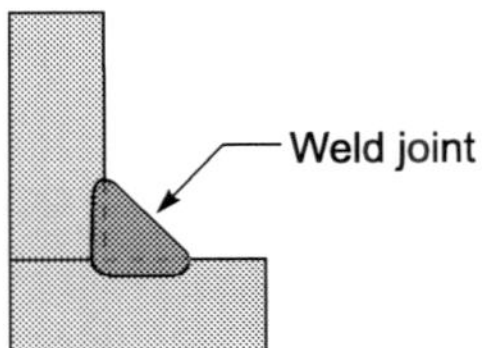

   (a) Spot weld (b) Groove weld (c) Fillet weld (d) Plug weld

   Answer: (c)

386. Two plates each of 6 mm thickness are to be butt welded. Consider the following processes and select the correct sequence in increasing order of size of the heat affected zone. **(GATE 2020-S2)**

1. Arc welding
2. MIG welding
3. Laser beam welding
4. Submerged arc welding

(a) 1-4-2-3 (b) 3-4-2-1 (c) 4-3-2-1 (d) 3-2-4-1

Answer: (d)

387. In a fabrication shop, a T-joint is to be made between two plates using a chin intermittent, double fillet weld of 6 mm leg length. The intermittent welds are 40 mm long and spaced 80 mm apart, edge to edge. The welding symbol of the joint is **(GATE 2005)**

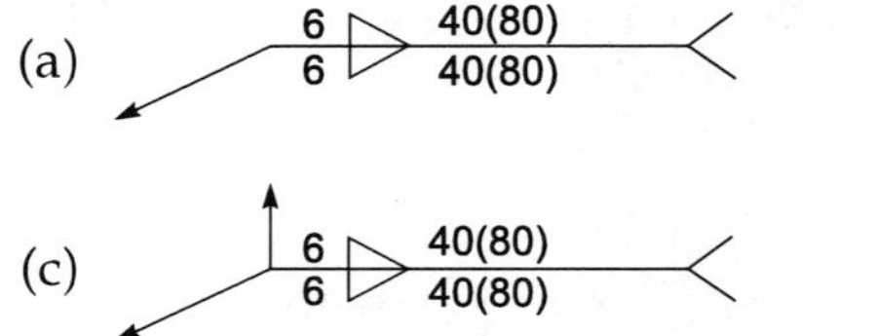

(b) 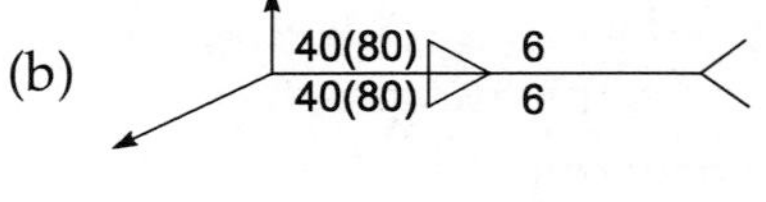

(d) 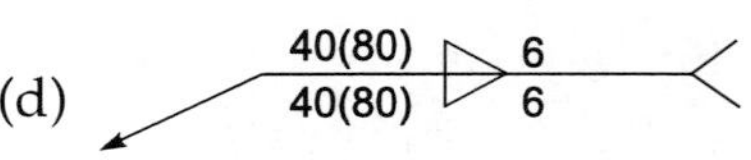

Answer: (d)

388. The cold cracking susceptibility of the heat affected zone in an arc weld is influenced by **(GATE 2003)**

(P) Entrapped hydrogen
(Q) Residual stresses
(R) Martensitic transformation
(S) Slag inclusion

Answer choices are

(a) P and S (b) P Q and S (c) P Q and R (d) R and S

Answer: (c)

389. Which of the following process can be used for welding of aluminum alloys **(GATE 2003)**

(P) Submerged arc welding
(Q) Gas metal arc welding
(R) Electroslag welding
(S) Gas tungsten arc welding

Choices are

(a) P and Q (b) Q and S (c) Q and R (d) R and S

Answer: (b)

In GTAW, GMAW, inert gas is used as shielding gas. Sometimes SAW is used.

390. Match the following: **(GATE 2003)**

	Welding		Heat Source
A	Arc welding	1	Diffusion
B	Friction welding	2	Polarity
C	Solid-state welding	3	Focusing
D	Laser welding	4	Kinetic energy

Answer: A-2, B-4, C-1, D-3

391. Match the following: **(GATE 2003)**

	Welding		Heat Source
A	Spatter	1	Damp electrodes
B	Distortion	2	Arc blow
C	Slag inclusions pass welding	3	Improper cleaning
D	Porosity	4	Poor joint selection

Answer: A-2, B-4, C-3, D-1

392. Match the following: **(GATE 2003)**

	Welding		Heat Source
A	Highly reactive metals	1	Ultrasonic
B	Distortion	2	Arc blow
C	Slag inclusions pass welding	3	Improper cleaning
D	Porosity	4	Poor joint selection

Answer: A-2, B-4, C-3, D-1

393. The welding process which uses a blanket of fusible granular flux is

(a) Tungsten inert gas (TIG) welding (b) Submerged arc welding (SAW)
(c) Electroslag welding (ESW) (d) Thermit welding (TW)

Answer: Submerged arc welding (SAW) uses blanket of fusible granular flux. Hence, the correct option is (b).

394. Match the following: **(GATE 1990)**

	Welding		Heat Source
A	Thermit welding	1	Electric arc
B	Projection welding	2	Mechanical work
C	MIG welding	3	Exothermic chemical reaction
D	Friction welding	4	Ohmic resistance

(a) A-3 (b) B-4 (c) C-1 (d) D-2

Answer: (a) For thermit welding, we get heat required for melting from exothermic reaction, whereas in MIG welding heat is due to electric arc. In friction welding, heat is due to work.

395. In an arc welding process, welding speed is doubled. Assuming all the other parameters to be constant, the cross-sectional area of the weld bead will ———.
(GATE 2017-S2)

(a) Increase by 25% (b) Increase by 50%
(c) Reduce by 25% (d) Reduce by 50%

Solution:

Since all the process parameters are constant: Material deposition rate = Constant material deposition rate = Area of weld × Welding speed

MD rate = $A_1 \times V_1$

MD rate = $A_2 \times V_2 = A_2 \times 2 \times V_1$

$A_1 \times V_1 = A_2 \times 2 \times V_1$

$$A_2 = \frac{A_1}{2}$$

$$\text{Percentage change} = \frac{A_2 - A_1}{A_1} = \frac{\frac{A_1}{2} - A_1}{A_1} = -0.5 \text{ or } -50\%$$

Hence, area reduces by 50% and option (d) is the correct answer.

396. In linear arc welding processes, the heat input per unit length is inversely proportional to **(GATE 2015-S2)**

(a) Welding current (b) Welding voltage
(c) Welding speed (d) Duty cycle of the power source

Answer: (c)

Heat input $Q = CVI$

C = fraction of time during which arc is on = L/v

L = length of weld

v = welding speed

$$\frac{Q}{L} = \frac{VI}{v} \Rightarrow Q \propto \frac{1}{v}$$

So heat input is inversely proportional to welding speed.

397. Which pair among the following solid-state welding processes uses heat from external source **(GATE 2008)**

(P) Diffusion welding (Q) Friction welding
(R) Ultrasonic welding (S) Forge welding

(a) P and R (b) R and S (c) Q and S (d) P and S

Answer: (b) Heat supplied is external in ultrasonic and forge welding.

398. Consider the following statement. **(GATE 2005)**

The size of the heat affected zone (HAZ) will increase with

(a) Increased starting temperature
(b) Increased welding speed
(c) Increased thermal conductivity of the base metal
(d) Increase in base metal thickness

Which statements given above are correct?

(a) 1, 2 and 3 (b) 1 and 3 (c) 1 and 4 (d) 2 and 3

Answer: (b) If we increase the speed, heat transfer decreases and increase in thickness increases the thermal resistance and heat transfer decreases and HAZ reduces.

399. Consider the following statement: **(GATE 2005)**

The magnitude of residual stress in welding depends on

(a) Metal melted/deposited
(b) Design of weldment
(c) Support and clamping of components
(d) Welding processes used

Which statements given above are correct?

(a) 1, 2 and 4 (b) 1, 2 and 3 (c) 1 and 3 (d) 2 and 3

Answer: (b)

400. Directions for the assertion (A) and reason (R) below, choose the correct alternative: **(GATE 2005)**

(a) Both A and R are true and R is the correct explanation of A
(b) Both A and R are true and R is the not correct explanation of A
(c) A is true and R is false
(d) A is false but R is true

Assertion (A): A sound welded joint should not only be strong enough but should also exhibit a good amount of ductility.

Reason (R): Welding process is used for fabricating mild steel components only.

Answer: (c) Weld should exhibit ductility so as to withstand impact load but R is not the correct explanation and also welding process is not limited to MS only.

401. **Assertion:** In electron beam welding process, vacuum is essential process parameter: **(GATE 2005)**

Reason (R): Vacuum provides a highly efficient shield on the welding zone.

Answer: (b) Vacuum is used to avoid electron dispersion and not for providing shield.

402. Match the following: **(GATE 1992)**

	Welding		Heat Source
A	Highly reactive metals	1	Ultrasonic
B	Long straight butt joint in thick sections	2	Thermit
C	Site weld of rails	3	Electron beam
D	MS rods to plate like component	4	Submerged arc
		5	Stud

Answer: A-3, B-4, C-2, D-5

403. Match the following: **(GATE 1989)**

	Welding		Heat Source
A	Cellulose coating	1	Alloying agent
B	Rutile coating	2	Absorb gases
C	Iron powder coating	3	Forming slag
D	Basic coating	4	Shielding gas

Answer: A-4, B-3, C-1, D-2

404. Which kind of resistance is experienced in upset butt welding?

(a) Electric resistance (b) Magnetic resistance
(c) Thermal resistance (d) Air resistance

Answer: (a)

Explanation: In the making of an upset butt welding, there are jaws made of copper, into which the part to be welded is put, and hence a solid contact is made. At this point of contact, while the current flows, it gets transformed into heat because of electric resistance.

405. Which of the following can be easily welded from flash butt welding process?

(a) Tin (b) Lead
(c) Cast irons (d) Carbon steel

Answer: (d)

Explanation: In flash butt welding process, the welding of materials like steels and other iron alloys are easy to weld, except for cast iron. The welding of materials like tin, lead, zinc and antimony cannot be welded using flash butt method.

406. Electrodes used in spot welding are made up of ———.

(a) Only copper (b) Copper and tungsten
(c) Copper and chromium (d) Copper and aluminum

Answer: (d)

Explanation: The electrodes that are employed in the spot welding are required to possess a high amount of electrical and thermal conductivity. Hence, they are primarily made up of copper or copper and tungsten or copper and chromium.

407. How are the metals to be welded connected to each other in spot welding?

(a) Electric contact (b) Magnetic field
(c) Mechanical pressure (d) Direct contact

Answer: (c)

Explanation: In spot welding process, the metal or the alloy parts which are to be welded, are heated in their zones. To bring these parts in contact with one another, mechanical pressure is applied, causing them to connect.

408. Which of the following method is not used for applying pressure in spot welding process?

(a) Hand lever (b) Foot lever
(c) Air pressure (d) Hydraulic cylinder

Answer: (a)

409. What is the maximum power supply needed for the working of spot welding process?

(a) 135 kVA (b) 140 kVA (c) 145 kVA (d) 150 kVA

Answer: (d)

Explanation: Spot welding can be used for all types of ductile metals. It can also be employed for structures of sheet metals, and can be applied for making of boxes and cans. For its applications, the maximum power that can be needed for this process is 150 kVA.

410. What is the minimum power supply needed for the working of spot welding process?

(a) 10 kVA (b) 14 kVA (c) 6 kVA (d) 22 kVA

Answer: (a)

411. Upset butt welding is majorly used in the making of automobile parts.

(a) True (b) False

Answer: (b)

Explanation: Upset butt welding is mainly used for the manufacture of welding bars, rods, wires or tubing. For the manufacture of automobile parts, such as axles, wheels or frames, flash butt welding is extensively used.

412. In flash butt welding, the forced out metal is called flash.

(a) True (b) False

Answer: (a)

Explanation: In flash butt welding process, there is a light contact between the edges. A high voltage is supplied when the flashing action takes place. In this process, the metal that is forced out, is called flash.

413. Which of the following is not included in weldability?

(a) Ability of mechanical soundness
(b) Serviceability of joint
(c) Strain relieving brittleness
(d) Metallurgical compatibility of metal

Answer: (c)

Explanation: For the making of good fabricated composition of a structure, metal with good weldability gets readily welded. Weldability includes the ability of the metal or the alloy to be having soundness in its mechanical properties, metallurgical compatibility of metal and serviceability of the joint.

414. Which effect is not seen in alloys on weldability?
   (a) Control of ductile-malleable transformation temperature
   (b) Provision for oxidation to molten metal
   (c) Reduction in segregation
   (d) Grain refinement

Answer: (b)

Explanation: On weldability, there are some effects seen in the alloying elements. Some of them are control of ductile-malleable transformation temperature, reduction in segregation, grain refinement and provision for deoxidation to the molten metal.

415. On which of the following factor, does weldability not depend?
   (a) Boiling point
   (b) Melting point
   (c) Thermal expansion
   (d) Thermal conductivity

Answer: (a)

Explanation: Weldability depends on five factors of the metal or alloying element. The factors being, melting point, thermal expansion, thermal conductivity, surface condition and change in the microstructure.

416. Which of the following is not a way of correcting the metal which does not have good weldability properties?
   (a) Procedure of welding
   (b) Provision of fluxing material
   (c) Provision of filler material
   (d) Procedure of adding filler materials

Answer: (d)

Explanation: When a particular metal or an alloy does not have desirable welding characteristics, then it can be corrected by proper fluxing material, correct filler material and a perfect welding procedure.

417. What does HAZ stand for?
   (a) Helium aerated zone
   (b) Heated area zone
   (c) Heat affected zone
   (d) Heat allowed zone

Answer: (c)

Explanation: HAZ stands for heat affected zone. This zone is generally the base of the metal or a thermoplastic material. This zone is not melted but the properties are altered by the welding process.

418. With an increase in the heat input of arc welding method, how does it affect the welding speed?
   (a) Increases
   (b) Decreases
   (c) Remains same
   (d) No relation

Answer: (b)

Explanation: In arc welding process, for calculating the efficiency, we use the formula for heat input. According to the formula, heat input is inversely related to the speed of welding, that is, with an increase in heat input, the welding speed decreases.

419. With an increase in the heat input of arc welding method, how does it affect the voltage applied?
(a) Increases (b) Decreases
(c) Remains same (d) No relation
Answer: (a)

420. In fusion welding, welded pieces are kept together under pressure.
(a) True (b) False
Answer: (b)
Explanation: In fusion welding, the welded pieces are not kept under pressure, instead they are welded first and then allowed for some time to cool down. It is in pressure welding, that the welded pieces are kept under pressure for some time.

421. Materials having high thermal conductivity are difficult to melt.
(a) True (b) False
Answer: (a)
Explanation: Materials that have a medium melting point are considered to be welded easily and those materials which have their thermal conductivity very high are the toughest to be melted.

422. Why is carbon used in carbon arc welding?
(a) Generation of more heat at electrode tip
(b) Generation of less heat at electrode tip
(c) Provides coating
(d) Fixed polarity is maintained
Answer: (b)
Explanation: Carbon is used in carbon arc welding, at the negative terminal of the cathode. The reason for employing carbon at the negative terminal is that, a lesser amount of heat is generated at the electron tip than at the workpiece.

423. What is the temperature at the negative terminal in carbon arc welding?
(a) 3000°C (b) 3100°C (c) 3200°C (d) 3300°C
Answer: (c)

424. The process of carbon arc welding is not used for which of the following material?
(a) Copper (b) Bronze (c) Aluminum (d) Iron
Answer: (d)
Explanation: The process of carbon arc melting is widely used for copper and copper alloys, hence bronze and brass are treated by this method. Aluminum can also be treated likewise, but not iron.

425. Which gas is used as a protection in carbon arc welding?
(a) Carbon dioxide (b) Carbon monoxide
(c) Carbon tetrafluoride (d) Methane
Answer: (b)
Explanation: In carbon arc welding, protection has to be provided to the molten weld. This protection is provided in the form of an envelope by using a long arc giving carbon monoxide gas.

426. What is the temperature at the positive terminal in carbon arc welding?
(a) 3800°C (b) 3900°C (c) 4000°C (d) 4100°C
Answer: (b)
Explanation: Carbon is used in carbon arc welding, at the negative terminal of the cathode and the metal which is being worked upon functions as the positive terminal. The temperature of the electrode at the positive terminal is 3900°C.

427. What is the temperature of the negative electrode in metal arc welding?
(a) 2200°C (b) 2300°C (c) 2400°C (d) 2500°C
Answer: (c)
Explanation: In metal arc welding, the temperature of negative electrode is higher than the temperature of positive electrode. The temperature produced in a negative electrode of a metal arc weld is about 2400°C.

428. What is the temperature of the positive electrode in metal arc welding?
(a) 2300°C (b) 2400°C (c) 2500°C (d) 2600°C
Answer: (d)

429. Which bead character is seen due to too high welding current?
(a) Pilling of metal (b) Pilling of weld metal
(c) Excess splatter (d) Small bead formation
Answer: (c)
Explanation: On provision of too low current, there is pilling of metal seen. When the voltage provided is very high, beads formed are too high. When the speed of weld is less, pilling of weld metal takes place and when welding currents are too high, excessive splatter formation occurs.

430. A smooth and regular bead cannot be formed in metal arc welding.
(a) True (b) False
Answer: (b)
Explanation: Before the welding starts, the machine has to be adjusted very correctly. The current to be supplied and the voltage to be applied should not to be too low or high, and the welding speed should be normal. By doing this, regular and smooth beads will be formed.

431. Carbon is intended to be mixed in small amounts in the weld, in carbon arc welding.
(a) True (b) False
Answer: (b)

Explanation: In carbon arc welding, the carbon is present at the negative electrode, where the electrode temperature is at 3200°C. This temperature is less than that at the positive electrode. Carbon electrode is kept at a lower temperature on the negative electrode, to avoid the mixing of carbon in the weld. If the carbon gets mixed with the weld, the weld might end up becoming brittle.

432. Which of the following is not a type of arc welding?

(a) Plasma (b) Electroslag
(c) Submerged (d) Air-acetylene

Answer: (d)

Explanation: Among the following, plasma arc, electroslag and submerged arc are few of the types of arc welding. Air-acetylene is one of the forms of gas welding processes.

433. Which of the following is not a type of resistance welding?

(a) Seam (b) Projection (c) Electroslag (d) Spot

Answer: (c)

Explanation: Among the following, seam, projection and spot are few of the types of resistance welding. Electroslag is one of the forms of gas welding processes.

444. Which of the following materials are not suited for cold welding?

(a) Brass (b) Steel (c) Silver (d) Gold

Answer: (b)

Explanation: Cold welding process is not very suitable for alloys like steel, the reason being, cold welding cannot take place excellently in ferrous materials. Steel contains high amount of iron, hence it cannot be cold welded.

445. What is the cross-sectional shape of a thermoplastic welding rod?

(a) Triangular (b) Hexagonal (c) Square (d) Pentagonal

Answer: (a)

Explanation: A welding rod or a thermoplastic welding rod is used mainly in plastic welding. The rod should not be very porous or it may catch voids. This rod has a cross-sectional shape triangular or circular.

446. Which of the following is not a type of related processes?

(a) Soldering (b) Percussion
(c) Brazing (d) Hard-pressing

Answer: (b)

Explanation: Among the following, soldering, pressing and hard-pressing are few of the types of related processes. Percussion is one of the forms of resistance welding processes.

447. Laser is classified under newer beam.

(a) True (b) False

Answer: (a)

Explanation: There are various types of welding processes, one of them is a newer welding process. Under this, there are two types, one is electron beam and another is the use of laser.

448. Which of the following is not a type of solid-state welding?

(a) Projection (b) Ultrasonic (c) Friction (d) Diffusion

Answer: (a)

Explanation: Among the following, ultrasonic, friction and diffusion are a few related processes. Percussion is one of the forms of resistance welding processes.

449. What is the density of the plastic used in speed tip welding?

(a) 916 kg/m^3 (b) 923 kg/m^3 (c) 946 kg/m^3 (d) 982 kg/m^3

Answer: (c)

Explanation: The plastic which is used in the process of speed tip welding is polypropylene or polypropene. It is classified under thermoplastics materials. The density of polypropylene is 946 kg/m^3 and it starts melting at 130°C.

450. Which of the following could be considered a related process?

(a) Explosive (b) Oxy-hydrogen welding

(c) Oxy-acetylene cutting (d) Electron beam

Answer: (c)

Explanation: Among the following, oxy-acetylene cutting is considered a related process. Explosive is a type of solid-state welding. Electron beam is a type of newer welding and oxy-hydrogen welding is a form of gas welding.

451. A liquid phase has to be used in cold welding.

(a) True (b) False

Answer: (b)

Explanation: It is hot welding that a liquid or a molten state is used for the joining of two pieces of metal. But in cold welding, there is no use of a liquid phase, that is, joining is done without using heat.

452. In which run weld, in ordinary steels, the grain refinement takes place?

(a) First (b) Second (c) Third (d) Fourth

Answer: (b)

Explanation: A multi-run process causes a good amount of grain refinement. With every proceeding run weld, it normalizes its previous run. In ordinary steels, grain refinement takes place in the second run.

453. What is the minimum strength of mild steel arc welds on addition of microelements in the slag?

(a) 43 kgf/mm^2 (b) 39 kgf/mm^2

(c) 45 kgf/mm^2 (d) 48 kgf/mm^2

Answer: (a)

Explanation: When there is an addition of some microelements in the slag along with a few materials being present which are not metals, there is an increase in the strength of the mild steel arc weld. The minimum amount of strength goes up to 43 kgf/mm^2, which is 430 N/mm^2.

454. Which of the following is affected by weld contour?

(a) Fatigue (b) Creep

(c) High pressure property (d) Low pressure property

Answer: (a)

Explanation: The contour present on the welds forms notches. The contours in the form of notches then affect the low temperature properties of the structure and also the fatigue properties.

455. Thermite contains ———.

(a) Zinc (b) Copper

(c) Aluminum (d) Manganese

Answer: (c)

Explanation: The reaction that takes place in between compounds and elements is known as welding with solid reactants. These reactions are exothermic, that is, give out heat as a product. Thermit reaction is a common example for this, which is a mixture of aluminum and rust.

456. Which of the following property is considered for controlling the porosity of the metal?

(a) Solid solubility (b) Liquid solubility

(c) Gas solubility (d) Flame solubility

Answer: (c)

Explanation: The amount of solubility of gas in liquid or solid welded metals is considered to be an essential criteria for the control of porosity of the weld. The gas reactions which occur, then regulate porosity.

457. Slag used in fusion welding contains manganese.

(a) True (b) False

Answer: (a)

Explanation: Slags used in welding are generally a combination of silica materials and some carbonate compounds. In fusion welding, some amount of bead control has to be done, which is done by the slags, for this purpose, to improve its performance, microelements like manganese are added.

458. Which of the following property is responsible for weld metal cracking?

(a) Stress (b) Strain

(c) Temperature conditions (d) Pressure conditions

Answer: (a)

Explanation: The stresses which get developed in welding during shrinkage are of importance. These stresses are considered to be an important factor which are responsible for the cracking that occurs in metal welds.

459. A thermit reaction produces heat up to ———.

(a) 4800°F (b) 4000°F (c) 5600°F (d) 5300°F

Answer: (b)

460. What is the maximum strength of mild steel arc welds on addition of microelements in the slag?

(a) 53 kgf/mm^2 (b) 49 kgf/mm^2

(c) 57 kgf/mm^2 (d) 61 kgf/mm^2

Answer: (c)

Explanation: When there is an addition of some microelements in the slag along with a few non-metallic inclusions, there is an increase in the strength of the mild steel arc weld. The maximum amount of strength goes up to 57 kgf/mm^2, which is 570 N/mm^2.

461. Slag used in fusion welding has nitrogen.

(a) True (b) False

Answer: (a)

Explanation: Slags used in welding are generally a combination of silica materials and some carbonate compounds. In fusion welding, some amount of bead control has to be done, which is done by the slags, for this purpose, to improve its performance, microelements like nitrogen are added.

462. In an iron thermit, how many moles of aluminium react with three moles of iron oxide to give nine moles of iron?

(a) 6 (b) 7 (c) 8 (d) 9

Answer: (c)

Explanation: The thermit process, which is used for welding depends on the chemical reaction between oxide of iron and aluminum metal. According to this reaction, when 8 moles of aluminum react with 3 moles of iron oxide, then, 4 moles of aluminum oxide and 9 moles of iron are formed.

463. In the given figure of thermit welding apparatus, what does the star mark represents?

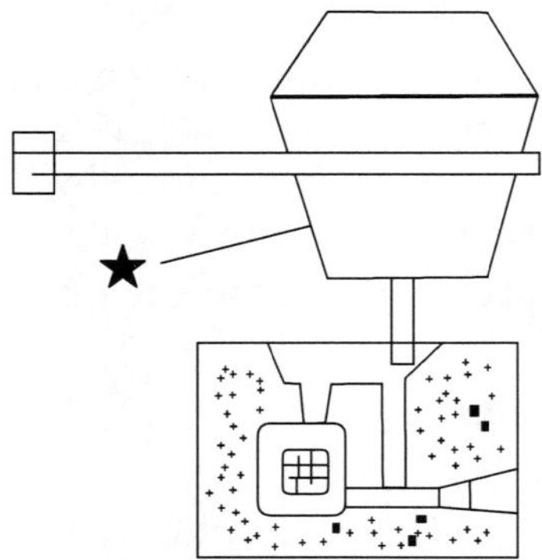

(a) Sand plug (b) Crucible
(c) Wax pattern (d) Slag basin

Answer: (b)

Explanation: Around the parts which are to be united, superheated liquid steel is poured, which then effects in forming the weld. This process, hence, is a fusion welding process. The part represented by the star mark is the crucible.

464. What is the magnitude of temperature produced in a thermit reaction?
(a) 2500°C (b) 3000°C (c) 3500°C (d) 4000°C

Answer: (b)

Explanation: Thermit welding is called fusion welding. A superheated melt operates in this type of melt. The heat that is produced has the temperature approximately twice than that of the melting point of steel, which is about 3000°C.

465. Thermit welding is used for making rail tracks.
(a) True (b) False

Answer: (a)

Explanation: A superheated melt operates in this type of melt. The production of pipes, rail tracks, shafts or big gears can be done using thermit pressure welding. Thermit welding is called fusion welding.

466. Which of the following types of fuel gas is commonly used in gas welding?
(a) Biogas (b) Coal gas (c) Acetylene (d) Methane

Answer: (c)

Explanation: Gas welding is usually referred to as oxy-welding or oxy-fuel welding which is a process of joining metallic materials by application of heat produced by the gas flame. In gas welding, acetylene is commonly used as fuel gas which is mixed with proper proportion of oxygen in a mixing chamber of welding torch. Gas welding can produce a hot flame of temperature about 3000-3500°C.

467. In gas welding, the joint can be made even much stronger than the original.
(a) True (b) False

Answer: (a)

Explanation: In gas welding, various metals like cast iron, steel, aluminum, copper, etc., can be used for joining operation. First, they are brought into a molten state and then a fusion takes place between two pieces of metal in such a way that the point of fusion closely approaches the strength of the metal fused. In this process, the union or joint can be made even much stronger than the original.

468. Acetylene is a hydrocarbon which is a molecule made up of two carbon atoms and four hydrogen atoms.
(a) True (b) False

Answer: (b)

Explanation: In gas welding, acetylene is used as a fuel gas which is a hydrocarbon like propane, methane, etc. However, it generally differs from those hydrocarbons in a manner that its molecule is made up of two carbon atoms and two hydrogen atoms. And the carbon atoms are joined with hydrogen atoms by making a triple bond.

469. Acetylene is a blue color gas which mixed with oxygen to produce a hot flame.

(a) True (b) False

Answer: (b)

Explanation: In gas welding, acetylene is a colorless gas with a very distinctive odor and is highly combustible when it is mixed with oxygen. Acetylene gas is formed by mixing the calcium carbide with water and then it is composed, which forms a compound of carbon and hydrogen having the chemical formula, C_2H_2.

470. Gas welding is mainly suited for welding of thin sheets, small diameter pipes and tubes.

(a) True (b) False

Answer: (a)

Explanation: Gas welding process is mainly suited for the welding of a thin sheet, tubes, and small diameter pipe because it cannot provide very high temperature as compared to that provided by the electric arc welding. This method of joining is also used for repairing work, maintenance and in body shops.

471. Acetylene is a non-toxic gas which can be kept or stored in a cylinder at any pressure.

(a) True (b) False

Answer: (b)

Explanation: In gas welding, acetylene is used as a fuel gas which produces a hot flame when it is mixed with oxygen. Acetylene may be a non-toxic gas, but if it is stored in a cylinder at more than the pressure of 2 bars, then the acetylene becomes self-explosive, that's why the pressure of stored acetylene should be maintained with great care.

472. Single stage gas regulator cannot be used in cold weather for regulating the acetylene or oxygen gas.

(a) True (b) False

Answer: (a)

Explanation: Single stage converter is used to reduce the pressure from the cylinder pressure to the welding pressure, which can take place in a single stage. However, it tends to freeze in cold weather. This is because an abrupt expansion and resulting drop in initial pressure makes rapid cooling of the gas and may cause ice formation in the presence of moisture.

473. In gas welding, the hoses are mainly used for making the connection between gas torch and regulators.

(a) True (b) False

Answer: (a)

Explanation: In gas welding, the hoses are mainly used for making connections between the gas torch and the regulators, which should be made strong, light and flexible enough to make torch movements smooth. In general, welding hose can be of two types that are twin welding hose and double welding hose.

474. A flashback is a condition in gas welding, when the pure methane gas is used as fuel gas instead of acetylene.

(a) True (b) False

Answer: (b)

Explanation: In gas welding, a flashback is a condition in which the flame burns inside the gas torch. Flashback is generally caused by improper mixing of gases, which further increases the rate of flame propagation to high extent that the flame goes back to the mixing chamber. If it is not stopped, the flame may burn the mixture and may travel back into the cylinder.

475. A welding torch is mainly used for mixing and burning the gases in the desired proportions.

(a) True (b) False

Answer: (a)

Explanation: In gas welding, a welding torch is mainly used for mixing oxygen and acetylene in the desired proportions. It is also used for burning the mixture at the end of the tip, and also provides a mean for moving and directing the flame front. Welding torch can be of high-pressure type or low-pressure type.

476. At what temperature, does fusion welding takes place?

(a) 850°C (b) 900°C (c) 950°C (d) 1000°C

Answer: (b)

Explanation: The process of fusion welding can be used for joining metal to a metal or a metal can be joined to a ceramic. For the joining of this process, a temperature of 900°C is provided for the approach.

477. Which of the following is not a type of diffusion welding?

(a) Gas-pressure welding (b) Vacuum fusion welding
(c) Eutectic fusion bonding (d) Eutectoid fusion welding

Answer: (d)

Explanation: Diffusion welding has three basic techniques which are incorporated by it. These three basic techniques are: gas-pressure welding, vacuum fusion welding and eutectic fusion welding.

478. What is the minimum frequency used in ultrasonic welding?

(a) 10,000 Hz (b) 20,000 Hz (c) 30,000 Hz (d) 40,000 Hz

Answer: (b)

Explanation: The process of ultrasonic welding is mainly used for the joining of similar metals or dissimilar metals using vibration energy, by way of high frequency. The minimum frequency needed for this purpose is 20,000 Hz.

479. What is the maximum frequency used in ultrasonic welding?

(a) 30,000 Hz (b) 40,000 Hz (c) 50,000 Hz (d) 60,000 Hz

Answer: (d)

480. In what conditions are the parts kept before welding in ultrasonic welding method?

(a) Low static pressure (b) High static pressure

(c) Low moving pressure (d) High moving pressure

Answer: (a)

Explanation: In ultrasonic welding method, the parts which are to be joined, are kept clamping to one another in between of supporting member and the welding tip. This clamping is done at a low static pressure.

481. What is the maximum thickness that can be welded in ultrasonic welding?

(a) 1.8 mm (b) 2.1 mm (c) 2.3 mm (d) 2.5 mm

Answer: (d)

Explanation: Weldings such as spot welding or continuous seam welding can be done through ultrasonic welding. The maximum thickness that can be achieved varies depending on the metal. The maximum thickness that can be welded is 2.5 mm.

482. Electric currents are passed in ultrasonic welding.

(a) True (b) False

Answer: (b)

Explanation: In ultrasonic welding method, the parts which are to be joined, are kept clamped to one another in between of supporting member and the welding tip. In doing this, no current is required to be passed through the metal, and usually, no heat is needed for this.

483. What is the minimum thickness that can be welded in ultrasonic welding?

(a) 0.58 mm (b) 0.21 mm (c) 0.38 mm (d) 0.92 mm

Answer: (c)

484. Which of the following is not a factor for explosive welding?

(a) High relative velocity (b) Less amount of plastics

(c) Proper orientation (d) High pressure

Answer: (b)

Explanation: While considering for explosive welding, the factor that one needs to understand is that, there has to be high relative velocity. There should be proper orientation and it needs to be kept under high pressure.

485. α is the angle between target plate and flyer plate.

(a) True (b) False

Answer: (a)

Explanation: The workpiece which has its position fixed is called the target plate and the other plate is called the flyer plate. The distance between the two plates is denoted by *d* and the angle between them is denoted by α.

486. Oxy-acetylene process is most suited for which of the following process of joining?

(a) Metal wires (b) Metal sheets
(c) Metal tubes (d) Metal bars

Answer: (b)

Explanation: Oxy-acetylene process is most suited for the joining of metal sheets or for joining metal plates which are having a thickness of 2 mm to 50 mm. While joining, the molten metal flows together to form a continuous joint.

487. For what thickness of a material, in an oxy-acetylene is a welding rod used?

(a) 10 mm (b) 15 mm (c) 20 mm (d) 25 mm

Answer: (b)

Explanation: Oxy-acetylene processes are mainly used for joining metal sheets or metal plates which have thickness of 2 mm to 50 mm. When the thickness of the material is more than 15 mm, an additional filler metal is added like a welding rod.

488. What is the flame temperature of city gas?

(a) 1500°C (b) 1600°C (c) 1700°C (d) 1800°C

Answer: (c)

Explanation: City gas is a type of fuel gas that is used for the welding of thin sheets of steel which have a thickness of up to 3 mm. The flame temperature of city gas is approximately about 1700°C.

489. In a neutral flame, what is the temperature in the inner core?

(a) 3200°C (b) 2100°C (c) 1250°C (d) 2700°C

Answer: (a)

Explanation: When equal amount of oxygen and acetylene is made to react in a torch, a neutral flame gets produced. The inner core has a temperature of 3200°C, whereas the tip of the neutral flame has a temperature of 1250°C.

490. What is the flame temperature of water gas?

(a) 2500°C (b) 2800°C (c) 2300°C (d) 3100°C

Answer: (c)

Explanation: Water gas is a type of fuel gas that is used for the welding of thin sheets of steel which have a thickness of up to 8 mm. The flame temperature of water gas is approximately about 2300°C.

491. Carburising flame has excess of oxygen supply.
(a) True (b) False
Answer: (b)

492. What is the flame temperature of methane?
(a) 2000°C (b) 1200°C (c) 2700°C (d) 3300°C
Answer: (a)

493. Oxidizing flame has equal volumes of oxygen and acetylene supply.
(a) True (b) False
Answer: (b)
Explanation: A neutral flame is the one in which there is an equal amount of supply of oxygen and acetylene. A carburising flame is the one in which the supply of acetylene is much more than that of oxygen and an oxidizing flame is the one where there is ample supply of oxygen compared to acetylene.

494. What is the flame temperature of acetylene?
(a) 1700°C (b) 2200°C (c) 2800°C (d) 3200°C
Answer: (d)
Explanation: Acetylene gas is a type of fuel gas that is used for the welding of thin sheets of steel which have a thickness of up to 50 mm. The flame temperature of acetylene gas is approximately about 3200°C.

495. What is the flame temperature of hydrogen?
(a) 2000°C (b) 1200°C (c) 2700°C (d) 3300°C
Answer: (c)

496. Which gas is used for the removal of oxygen layer formed on molten iron?
(a) Hydrogen (b) Oxygen
(c) Carbon dioxide (d) Chlorine
Answer: (b)
Explanation: In oxygen cutting method, under considerable amount of pressure, when a jet of oxygen gas is made to blow on the particles of molten iron oxide, the oxygen layer present on the molten iron gets removed.

497. Up to what amount of carbon content in steels, can they be treated by oxygen cutting?
(a) 0.3% (b) 0.2% (c) 0.7% (d) 0.4%
Answer: (c)
Explanation: For a steel to be treated by oxygen cutting process, the carbon content that should be available in terms of percentage composition is around 0.7%, that is, a high alloy steel can be treated by this process.

498. Up to what temperatures are high alloy steels heated before treating them by oxygen cutting?
(a) 675°C (b) 725°C (c) 775°C (d) 825°C
Answer: (a)

Explanation: In oxygen cutting process, not all steels can be treated by this method. Only high alloy steels can be treated by this method. These steels need to be preheated before oxygen cutting. The temperature range up to which these steels are preheated is about 650°C to 700°C.

499. How many openings are available in the cutting torch used in oxygen cutting process?

(a) 4 (b) 5 (c) 6 (d) 7

Answer: (d)

Explanation: Just like any other gas welding equipment, the welding torch gets substituted by a cutting torch in the oxygen cutting process. The total number of openings available in the torch is six, with a large opening at the center, making the count to seven.

500. What is the maximum oxygen pressure that can be applied to steel?

(a) 13 atm (b) 14 atm (c) 15 atm (d) 16 atm

Answer: (b)

501. What is the minimum oxygen pressure that can be applied to steel?

(a) 3 atm (b) 4 atm (c) 5 atm (d) 6 atm

Answer: (a)

Explanation: In oxygen cutting process, not all steels can be treated by this method. Only high alloy steels can be treated by this method. The pressure supply of oxygen should not be low, as it may lead to the formation of slag on the bottom of the cut. The minimum oxygen pressure that can be applied to steels is 3 atm or 300 kN/m^2.

502. What is the initial oxygen pressure provided through the lance?

(a) 1 atm to 2 atm (b) 2 atm to 3 atm
(c) 3 atm to 4 atm (d) 4 atm to 5 atm

Answer: (a)

Explanation: In oxygen lance cutting, jet of oxygen is passed through steel pipes, and by doing this, holes are produced. For doing this, the jet of oxygen is passed through the lancer under the pressure of 1 atm to 2 atm.

503. What should the outer diameter of a thick-walled lance be, for having good results?

(a) 13 mm to 15 mm (b) 15 mm to 17 mm
(c) 17 mm to 19 mm (d) 19 mm to 21 mm

Answer: (c)

Explanation: In oxygen lance cutting, when the lance is made up of a thick wall, with its outer diameter of 17 mm to 19 mm, and inner diameter of 8 mm to 6 mm, respectively, then good results can be obtained.

504. Cast irons are best treated using oxygen cutting process.

(a) True (b) False

Answer: (b)

Explanation: Cast irons cannot be treated using oxygen cutting process. The reason being that the melting point of cast irons is a lot lower than its ignition temperature. The same cannot be said for high alloy steels.

505. Oxygen cutting process can be applied to metals having a melting point higher than their oxidization temperature.

(a) True (b) False

Answer: (b)

Explanation: It can be counted among one of the drawbacks of oxygen cutting process, that the metal which is to be treated must have its oxidizing temperature higher than its melting temperature, otherwise the metal would get melted before getting oxidized.

506. What is the order of temperature of heat produced in an electric arc furnace?

(a) 3000°C – 4000°C (b) 4000°C – 5000°C
(c) 5000°C – 6000°C (d) 6000°C – 7000°C

Answer: (d)

Explanation: An electric arc furnace generates a lot of heat when the working metal comes in contact with the electrode. The heat that is generated in this joining process is of the order of 6000°C to 7000°C.

507. The distance from the center of arc to the tip of electrode is called ———.

(a) Arc distance (b) Arc length (c) Arc crater (d) Arc depth

Answer: (b)

Explanation: The distance from the center of the arc to the tip of the electrode, right till the bottom of the crater is defined to be the arc length. The arc length is considered to be one of the most essential variables in welding process.

508. What is the length of the arc generally?

(a) 1 mm – 2 mm (b) 2 mm – 3 mm
(c) 3 mm – 4 mm (d) 4 mm – 5 mm

Answer: (c)

509. In arc welding process, the high quality alternator functions at ——— frequency.

(a) 250 Hz (b) 300 Hz (c) 350 Hz (d) 400 Hz

Answer: (d)

Explanation: In an arc welding process, there are alternators used for producing high current and function at lower frequency, and there are alternators which are of high quality and work at high frequency. The functioning frequency of these alternators is 400 Hz.

510. Which type of electrode is designed for melting quickly?

(a) Fast fill (b) Fast freeze (c) Quick fill (d) Fill freeze

Answer: (a)

Explanation: Electrodes which are classified into three types, fast fill, fast freeze and fill freeze. Among these, the fast fill electrodes are considered to be able to operate at extremely high welding speeds, and hence fast fill electrodes are designed for quick melting.

511. What amount of electricity gets converted into heat energy in an arc welding process?

(a) 0.25 kWh (b) 0.75 kWh (c) 1 kWh (d) 2 kWh

Answer: (c)

Explanation: In an arc welding process, there is a direct conversion of electrical energy into heat energy. An approximate amount of 1 kilowatt-hour of electrical energy gets converted into heat energy for this process.

512. Arc welding machine uses only DC supply.

(a) True (b) False

Answer: (b)

Explanation: Arc welding machines are able of working on both AC as well as DC supply. But mainly arc welding machines work on DC supply only, where the positive terminal is connected to the welding machine.

513. Which type of electrode is designed for melting quickly?

(a) Fast fill (b) Fast freeze (c) Quick fill (d) Fill freeze

Answer: (b)

Explanation: Electrodes which are classified into three types, fast fill, fast freeze and fill freeze. In fast freeze electrodes, a filler metal is supplied. This filler element gets solidifies and provides a variety in positions of the weld.

514. Which material is not used as an iron coating on the electrode used in arc welding?

(a) Cellulose (b) Iron powder

(c) Calcium fluoride (d) Steel

Answer: (d)

Explanation: Materials that are used as coating on the electrode, that are used in arc welding, are cellulose, iron powder, calcium fluoride and rutile. Iron is used for ease of weld, but not steel, though it mainly contains iron.

515. The negative pole of the electrode burns away faster than the positive pole.

(a) True (b) False

Answer: (b)

Explanation: During the operation of the arc welding process, the positive pole of the electrode experiences two-thirds of the heat and the negative pole experiences one-third the heat, hence the positive pole burns away faster than the negative pole.

516. Which of the following methods cannot be used for welding carbon steel?
(a) Arc welding (b) Gas welding
(c) Ultrasonic welding (d) Forge welding

Answer: (c)

Explanation: For the welding of carbon steel material, one may use the arc welding method. To gas weld carbon steel is also a feasible option and carbon steels can be treated using forge welding also, but it cannot be welded using ultrasonic welding.

517. High alloys steels are preheated up to what temperature?
(a) 350°C (b) 400°C (c) 450°C (d) 500°C

Answer: (b)

Explanation: High alloy steels are those steels which contain high amount of carbon in them. These alloys need to be preheated before welding in order to avoid cracking. The temperature up to which these alloys are preheated is 400°C.

518. Up to what percentage of carbon content in steel is not required to preheat it?
(a) 0.25% – 0.3% (b) 0.3% – 0.35%
(c) 0.35% – 0.4% (d) 0.4% – 0.45%

Answer: (a)

Explanation: Low carbon steels are the ones containing low amount of carbon in them, which does not require preheating. The amount of carbon content needed in low carbon steels, is 0.25% to 0.3%, if preheating is to be avoided.

519. Which of the following material is not used in alloy steels?
(a) Molybdenum (b) Nickel (c) Chromium (d) Sodium

Answer: (d)

Explanation: In alloy steels, there is a big composition used up by iron. Along with iron, there vitally has to be carbon in alloy steels. Along with carbon, there is a presence of other elements such as, molybdenum, nickel and chromium in small amounts.

520. Which among the following methods is best suited for stainless steels?
(a) Electric butt welding (b) Seam welding
(c) Flux coated arc welding (d) Oxy-hydrogen welding

Answer: (a)

Explanation: There are a few methods of treating stainless steels for welding, such as, any of the metal arc welding methods or the oxy-acetylene welding method, but the one best suited for this alloy is electric butt welding.

521. Stainless steels are annealed at what temperature?
(a) 650°C – 700°C (b) 700°C – 750°C
(c) 750°C – 800°C (d) 800°C – 850°C

Answer: (c)

Explanation: Stainless steels are hardened by constant heating and cooling and the best method suited for this alloy is electric butt welding. As soon as the stainless steel alloy is electric butt welded, it needs to be annealed at a temperature between 750°C and 800°C.

522. Which of the following is not a material used for making an electrode in metal arc welding, for treating stainless steels?

(a) Niobium (b) Columbium (c) Titanium (d) Potassium

Answer: (d)

Explanation: Stainless steels are best welded using the electric butt welding method and then followed by annealing at around 750°C to 800°C. Stainless steels can also be welded using the metal arc welding methods. For this, the electrode which is needed is made up of niobium, titanium and columbium. These materials avoid the occurrence of the defect called weld decay.

523. What is the temperature needed for preheating in cast irons?

(a) 550°C (b) 600°C (c) 650°C (d) 700°C

Answer: (b)

Explanation: Cast irons, upon the removal of welding heat, get solidified very quickly. This solidification takes place due to the surrounding cold mass and the air around. Because of such quick cooling, the carbon gets retained and hard metal stays in the weld. To overcome this, cast irons have to be preheated to 600°C.

524. Low carbon steels need not be preheated before welding.

(a) True (b) False

Answer: (a)

Explanation: Carbon steels which contain carbon in a very low amount are not compulsorily required to be preheated before undergoing welding if the large runs are made. This is because, in low carbon steels, unlike high carbon steels, there is no cracking seen on welding.

525. Carbon steels can be fusion welded.

(a) True (b) False

Answer: (b)

Explanation: Carbon steels are best treated using forge welding, resistance welding, arc welding or by gas welding. Carbon steels essentially cannot be fusion welded, because of the occurrence of cracks which may cause due to carbon pick up.

526. What is the capacity of an automatic welding machine?

(a) 100 to 200 A (b) 300 to 400 A

(c) 500 to 700 A (d) 800 to 3000 A

Answer: (d)

Explanation: An automatic welding machine has a high amount of capacity. Its capacity ranges between 800 and 300 A. This type of machine is not portable and is suitably used for heavy welding processes.

527. What is the capacity of a light manual welding machine?

(a) 100 to 200 A (b) 300 to 400 A
(c) 500 to 700 A (d) 800 to 3000 A

Answer: (a)

Explanation: A light manual welding machine has one of the lowest capacity among all types of welding machines. Its capacity ranges between 100 and 200 A. This machine is portable and is best suited for light welding processes.

528. What is the current required in an Oxygen's INDARC 400 (S) transformer?

(a) 20 amp to 100 amp (b) 40 amp to 250 amp
(c) 50 amp to 400 amp (d) 60 amp to 500 amp

Answer: (c)

Explanation: In an Oxygen's INDARC 400 (S) transformer, there is a hand wheel provided which is used for varying the output, for altering the air gap in the core and results in regulating the current between 50 amperes and 400 amperes.

529. What is the efficiency of an AC welding transformer?

(a) 0.6 (b) 0.7 (c) 0.8 (d) 0.9

Answer: (c)

Explanation: AC welding is currently gaining considerable ground. The efficiency of an alternating current welding transformer varies from 0.8 to 0.85 and the efficiency of AC outfits is in the range of 0.3 to 0.6.

530. What is the amount of electric energy required per kg in DC welding operation?

(a) 1 – 4 kWh (b) 3 – 7 kWh (c) 4 – 9 kWh (d) 6 – 10 kWh

Answer: (d)

Explanation: The amount of electrical energy consumed required per kg for a DC welding operation, is 6 kWh to 10 kWh. Direct current provides the advantage of distribution of the heat needed inside the arc.

531. Which material is not used for making non-consumable electrodes?

(a) Carbon (b) Graphite (c) Sodium (d) Tungsten

Answer: (c)

Explanation: Non-consumable electrodes, are those electrodes which do not get consumed as the welding operation proceeds. These electrodes are primarily made up of materials like carbon, graphite and tungsten.

532. Which of the following characteristic does not define a coated electrode?

(a) Maintenance of arc
(b) Protection from oxygen
(c) Protection from nitrogen
(d) Preventing other elements into molten metal

Answer: (d)

Explanation: A coated electrode is a further classification of a consumable electrode. The purpose of this electrode is to maintain the arc, to shield the molten metal from oxygen and nitrogen present in the air and to provide other alloying elements into the molten metal.

533. What is the amount of carbon contained by a soft steel wire?
   (a) 0.1% to 0.18%
   (b) 0.15% to 0.22%
   (c) 0.18% to 0.26%
   (d) 0.2% to 0.27%

Answer: (a)

Explanation: A soft steel wire contains about 0.1% to 0.18% of carbon. Soft steel wire also has other contents in it like phosphorus and sulfur in it. The amount of phosphorus is soft steel is around 0.0025%.

534. What is the percentage of carbon present in low alloy steel?
   (a) 0.15%
   (b) 0.2%
   (c) 0.25%
   (d) 0.3%

Answer: (c)

Explanation: The amount of carbon that is present in a soft steel wire is around 0.1% to 0.18% and the amount of carbon in terms of percentage composition in low alloy steels is 0.25%.

535. The presence of arc blower makes good quality welds.
   (a) True
   (b) False

Answer: (b)

Explanation: A flow of large amount of current causes the generation of arc blower. The space around the arc and the metal always experience magnetic fields, causing the currents to get squared. This results in an unstable behavior by the arc blower, which is not desirable for good welds.

536. Cross-wire welding is ———.
   (a) Multi-spot welding process
   (b) Continuous spot welding process
   (c) Used to form mesh
   (d) Used where additional strength is desired
   (e) None of the above

Answer: (c)

537. In gas tungsten arc welding, the following polarity is used
   (a) Direct current straight polarity (DCSP)
   (b) Direct current reverse polarity (DCRP)
   (c) Alternating current high frequency (ACHF)
   (d) All of the above

Answer: (d)

538. The voltage needed in resistance welding does not depend upon ———
(a) Composition
(b) Area
(c) Thickness of weld
(d) Length of weld
Answer: (d)

539. Which of the following statement(s) is/are true for resistance welding?
(i) The time for which current flows is very important
(ii) After switching off the current, the pressure is maintained until the weld cools
(iii) Water is circulated through hollow electrodes to cool the electrodes
(a) (i) and (ii)
(b) (i) and (iii)
(c) (ii) and (iii)
(d) (i), (ii) and (iii)
Answer: (d)

540. Which of the following is not resistance welding?
(a) Spot welding
(b) Butt welding
(c) Pressure welding
(d) Percussion welding
Answer: (c)

541. In spot welding, the tip diameter of electrode is about
(a) $\sqrt{t}$
(b) $\sqrt{2t}$
(c) $2\sqrt{t}$
(d) $\sqrt{3t}$
Answer: (a)

542. Which of the following gas mixtures is not used in gas tungsten arc welding?
(a) Argon-Helium
(b) Argon-Nitrogen
(c) Argon-Hydrogen
(d) Argon-Carbon dioxide
Answer: (b)

543. Which welding process is used to join two thick plates in one single pass?
(a) Oxy-acetylene welding
(b) Gas tungsten arc welding
(c) Gas metal arc welding
(d) Electroslag welding
Answer: (d)

544. Which flame is suitable for cutting operations?
(a) Oxidizing flame
(b) Carburising flame
(c) Neutral flame
(d) None of the above
Answer: (a)

References

1. Amstead, B.H., Ostwald, P.F. and Begeman, M.L. Manufacturing Process, John Wiley, N.Y., 1976.
2. Banerjee, A. Welding Science Metallurgy and Technology, NCBA Publications, 2007.
3. Anon. A Better Way to Weld Underwater, Ocean Industry Manual, 1917.
4. Anon. Underwater Welding and Cutting, US Navy Technical Manual, 1953.
5. Allum C.J. Overcoming Problems in Hyperbaric Welding, Welding and Metal Fabrication, 1983.
6. Amson, J.C. An Analysis of the Gas Shielded Consumables Metal Arc Systems, Physics of the Welding Arc, A Symposium, London, 1966.
7. Baker, H.W. Modern Workshop Technology, BI Publication, 1956.
8. Raj, B., Shankar, V. and Bhaduri, A.K. Welding Technology for Engineers, Narosa, 2009.
9. Biley, A.F. The Effect of Gas Metal Arc Parameter in Seawater on Welding Current, Arc Voltage, Bead Geometry and Soundness, NCEL Preliminary, 1971.
10. Blazynski, T.Z. Explosive Welding Forming and Compaction, Applied Sc. Pub., 1983.
11. Brown, A.J. Fundamental Research on Underwater Welding, MIT Report, 1974.
12. Stinchcomb, C. Welding Technology Today, Prentice Hall, 1989.
13. Cooksey, C.J. Metal Transfer in Gas Shielded Consumable Metal Arc Welding, Physics of the Welding Arc, A Symposium, London, 1966.
14. DeGarmo, P.E. Materials and Processes in Manufacturing, MC Millan, 1983.
15. Delaune, P.T. Offshore structural repair using specification for underwaters welding, Welding Journal, Vol. 66, 2, pp. 32-43, 1987.
16. Dennary, F. Electric Field Distribution in Welding Arc, Physics of Welding Arc, A Symposium, London, 1966.
17. Department of Army, USA, Welding Theory and Application, TC-9-237, May 1993.

18. Dipl.-Ing. Schultz, H. Electron Beam Welding, Abington Publications, 2004.
19. Erohin, A.A. Heat Balance of Electrode Droplet Melting Process, A Symposium, London, 1966.
20. Foster, V.R. Guide to Underwater Welding, Vol. 27,135, pp. 88-92, 1958.
21. Garg, G.D. A Textbook of Welding Technology, Katson Publications, 2015.
22. Galerne, A. Underwater Welding and Painting, Symposium on Underwater Welding Cutting and Hand Tools, Battelle, 1967.
23. Ghosh, A. and Mallik, A.K. Manufacturing Science, East-West Press, 1985.
24. Giachino, J.W. Welding Skills and Practices, American Technical Society, 1967.
25. Gourd, L.M. Principles of Welding Technology, Edward Arnold, London, 1980.
26. Groover, M.P. Fundamentals of Modern Manufacturing, Wiley, 2010.
27. Haim, G. Welding of Plastics, Crosby Lockwood and Son Limited, 1947.
28. Hamasaki, M. On the studies of underwater fire cracker welding, J.of Japan Welding Soc. Vol. 42, 1, pp. 666-673, 1973.
29. Zhang, H. Resistance Welding, CRC Press, 2011.
30. Hobart Institute of Welding Technology, Shielded Metal Arc Welding, USA, 2012.
31. Houldcroft, P.T. Submerged Arc Welding, Abington Publications, 1977.
32. Ishizaki, K. A Method of Evaluating Metal Transfer Characteristics of Welding Electrodes, Physics of Welding Arc, A Symposium, London, 1966.
33. Jagadeesha, T. Welding Technology, Lecture Notes, AIT Chickmagalur, 2005.
34. Jagadeesha T. DST-SERB Workshop on Welding, NIT Calicut, 2019.
35. Cornu, J. Advanced Welding Systems, Springer, 1986.
36. Goldak, J.A. and Akhlaghi, M. Computational Welding Mechanics, Springer, 2005.
37. John, K.C. Metal Casting and Joining, PHI Publications, 2015.
38. Norrish, J. Advanced Welding Processes, IOP Publications, 1992.
39. Kalpakjean, S. Manufacturing Engineering and Technology, Pearson, 2005.
40. Kinney, G.F. Engineering Properties and Applications of Plastics, John Wiley, 1957.
41. Khan, M.I. A Study of Hardfacing Undermagnetic Field, Proc., ISME Conference, Delhi, 1979.
42. Khan, M.I. Welding Science and Technology, New Age Publications, 2007.
43. Khanna, O.P. A Textbook of Welding Technology, Dhanpat Rai Publications, 2015.
44. Weman, K. Welding Process Handbook, CRC Press, 2003.
45. Weman, K. MIG Welding Guide, Woodhead Publications, 2006.
46. Lancaster, J.F. The Physics of Welding, Pergamon Press, UK, 1984.
47. Lancaster, J.F. The Metallurgy of Welding, George Allen and Unwin Ltd., 1980.

48. Lancaster, J.F. Influence of Heat Flow on Metal Transfer in MIG Welding, Physics of Welding Arc, ASymposium, 1962.
49. Little, R.L. Welding and Welding Technology, Tata McGraw Hill, 1976.
50. Liming L. Welding and Joining of Magnesium Alloys, Woodhead Publications, 2010.
51. Mandal, N.R. Welding and Distortion Control, Narosa, 2004.
52. Mandal, N.R. Ship Construction and Welding, Springer, 2017.
53. Mandal, N.R. and Adak, M. Fusion zone and HAZ prediction through 3D simulation of welding thermal cycle, J. of Mech. Behaviour of Materials, Vol. 12, 6, pp. 401-414, 2001.
54. Kazem, M.D. Advances in Friction Stir Welding &Processing, Woodhead Publications, 2004.
55. Milner, D.R. Introduction to Welding and Brazing, Pergamon Press, 1968.
56. Miller Report (Welding Company), Guidelines for TIG Welding, 2008.
57. Ahmed, N. New Developmentsin Advanced Welding, Woodhead Publications, 2005.
58. Needham, J.C. Control of Transfer in Aluminium Consumable Electrode Welding, Physics of Welding Arc, ASymposium, London, 1966.
59. Neumon, J.A. Welding of Plastics, Reinhold Chapman and Hall, 1959.
60. Nikolaev, G. Advanced Welding Processes, MIR Publications, Russia, 1977.
61. Kumar, N. FSW of Dissimilar Alloys and Materials, BH Publications, 2015.
62. Blodgett, O.W. Design of Welded Structures, Lincoln Arc Welding Foundations, 1976.
63. Parmar, R.S. Welding Engineering and Technology, Khanna Publications, 2015.
64. Potapevaski, A.G. Melting and Transfer of Metal During Underwater Welding with Fine Electrode, Welding, Vol. 2, 12, pp. 110-113, 1967.
65. Mishra, R.S. Friction Stir Welding and Processing, ASM International, 2007.
66. Hilton, R. Welding Design and Process, W. J. Mackay, London, 1950.
67. Lofting, R. Welding, CrowoodPress, Marlborough, Wiltshire, 2013.
68. Klevin, R.J. Welding, Nova Publications, New York, 2011.
69. Messler, R.W. Principles of Welding, Wiley VCH Publications, 1999.
70. Rykalin,N.N. Laser Machining and Welding, MIR Publications, Russia, 1978.
71. Resistance Welders Manufacturing Alliance (RWMA), Resistance Welding Manual, 2003.
72. Serdjuk, G.B. Magnetic Forces in Arc Welding Metal Transfer, Physics of Arc, ASymposium, London, 1966.
73. Som A. Advanced Welding Technology, Acme Learning, 2016.
74. Shinanda, K. Welding and Cutting FGR, Welding and Fabrication, July 1982.

75. Srinivasan, N.K. Welding Technology, Khanna Publications, 2016.
76. Stalker, A.W. A Preliminary Study of Manual Metal Arc Welding, Report 3412/58, 1974.
77. Farnsworth, S.R. Welding for Dummies, Wiley Publications, 2010.
78. Tewari S.P. Advanced Welding Technology, Katson Publications, 2011.
79. Tweeddale, J.G. Welding Fabrication, Vol. 1-3, Elsevier Publication, 1994.
80. Udin, H. Welding for Engineers, John Wiley and Sons, 1954.
81. Van der Willingen, P.C. Contact arc welding, Welding, Vol. 25, 5, pp. 313-320, 1946.
82. Verhaeghe, G. Moving contact arc welding,International Conference Exploiting Advances in Arc Welding Technology, pp. 98-104, March 1998.
83. Waugh, R.A. Underwater metallic arc welding, Welding, Vol. 33, 10, pp. 531-534, 1946.
84. Welding Handbooks, American Welding Society, Volume 1, 2, 3 and 4, 1976.

Index

F

G

N

O

P

R

S

X

Z